MW00837941

THE PHYSICS AND CHEMISTRY OF MATERIALS

THE PHYSICS AND CHEMISTRY OF MATERIALS

Joel I. Gersten
Frederick W. Smith

The City College of the City University of New York

A WILEY-INTERSCIENCE PUBLICATION

JOHN WILEY & SONS, INC.

New York • Chichester • Weinheim • Brisbane • Singapore • Toronto

Copyright © 2001 by John Wiley & Sons, Inc. All rights reserved.

Published simultaneously in Canada.

For ordering and customer service, call 1-800-CALL-WILEY.

Library of Congress Cataloging-in-Publication Data:

Gersten, Joel I. (Joel Irwin)
    The physics and chemistry of materials / Joel I. Gersten, Frederick W. Smith.
      p. cm.
    ISBN 0-471-05794-0 (cloth : alk. paper)
    1. Solid state chemistry. 2. Solid-state physics. 3. Materials. I. Smith, Frederick W.
    (Frederick William), 1942– II. Title.
  QD478 .G47 2001
  541′.0421 — dc21
                                            2001026009

Printed in the United States of America.

10 9 8 7 6 5 4 3 2 1

For Harriet and Françoise

CONTENTS

SECTION II PHYSICAL PROPERTIES OF MATERIALS

SECTION III CLASSES OF MATERIALS

SECTION IV SURFACES, THIN FILMS, INTERFACES, AND MULTILAYERS

As science has become more interdisciplinary and impinges ever more heavily on technology, we have been led to the conclusion that there is a great need now for a textbook that emphasizes the physical and chemical origins of the properties of solids while at the same time focusing on the technologically important materials that are being developed and used by scientists and engineers. A panel of physicists, chemists, and materials scientists who participated in the NSF Undergraduate Curriculum Workshop in Materials in 1989, which addressed educational needs and opportunities in the area of materials research and technology, issued a report that indicated clearly the need for advanced textbooks in materials beyond the introductory level. Our textbook is meant to address this need.

This textbook is designed to serve courses that provide engineering and science majors with their first in-depth introduction to the properties and applications of a wide range of materials. This ordinarily occurs at the advanced undergraduate level but can also occur at the graduate level. The philosophy of our approach has been to define consistently the structure and properties of solids on the basis of the local chemical bonding and atomic order (or disorder!) present in the material. Our goal has been to bring the science of materials closer to technology than is done in most traditional textbooks on solid-state physics. We have stressed properties and their interpretation and have avoided the development of formalism for its own sake. We feel that the specialized mathematical techniques that can be applied to predict the properties of solids are better left for more advanced, graduate-level courses.

This textbook will be appropriate for use in the advanced materials courses given in engineering departments. Such courses are widely taught at the junior/senior level with such titles as "Principles of Materials Science & Engineering," "Physical Electronics," "Electronics of Materials," and "Engineering Materials." This textbook is also designed to be appropriate for use by physics and chemistry majors. We note that a course in materials chemistry is a relatively new one in most chemistry undergraduate curricula but that an introductory course in solid-state physics has long been standard in physics undergraduate curricula.

To gain the most benefit from courses based on this textbook, students should have had at least one year each of introductory physics, chemistry, and calculus, along with a course in modern physics or physical chemistry. For optimal use of the textbook it would be helpful if the students have had courses in thermodynamics, electricity and magnetism, and an introduction to quantum mechanics.

As the title indicates, the range of topics covered in this textbook is quite broad. The 21 chapters are divided into five sections. The range of topics covered is comprehensive, but not exhaustive. For example, topics not covered in detail due to lack of space include biomaterials, a field with a bright future, and composites, examples of which are discussed only within specific classes of materials. Much more material is presented than can be covered in a one-semester course. Actual usage of the text in courses will be discussed after the proposed subject matter has been outlined.

Following an introduction, which emphasizes the importance of materials in modern science and technology, Section I, on the "Structure of Materials," consists of four chapters on the structure of crystals, bonding in solids, diffraction and the reciprocal lattice, and order and disorder in solids.

Section II, on the "Physical Properties of Materials," consists of six chapters on phonons; thermally activated processes, phase diagrams, and phase transitions; electrons in solids: electrical and thermal properties; optical properties; magnetic properties; and mechanical properties.

Section III, titled "Classes of Materials," consists of eight chapters on semiconductors; metals and alloys; ceramics; polymers; dielectric and ferroelectric materials; superconductors; magnetic materials; and optical materials. In each chapter the distinctive properties of each class of materials are discussed using technologically-important examples from each class. In addition, the structure and key properties of selected materials are highlighted. In this way an indication of the wide spectrum of materials in each class is presented.

Section IV, titled "Surfaces, Thin Films, Interfaces, and Multilayers," consists of two chapters covering these important topics. Here the effects of spatial discontinuities in the physical and chemical structure on the properties of materials are presented, both from the point of view of creating materials with new properties and also of minimizing the potential materials problems associated with surfaces and interfaces.

Section V, titled "Synthesis and Processing of Materials," consists of a single chapter. Representative examples of how the structure and properties of materials are determined by the techniques used to synthesize them are presented. "Atomic engineering" is stressed. The tuning of structure and properties using postsynthesis processing is also illustrated.

Problem sets are presented at the end of each chapter and are used to emphasize the most important concepts introduced, as well as to present further examples of important materials. Illustrations are employed for the purpose of presenting crystal structures and key properties of materials. Tables are used to summarize and contrast the properties of related groups of materials. The units used throughout this textbook are SI units, except in cases where the use of electron volt, cm^{-1}, poise, etc., were felt to be too standard to ignore.

We have created a home page at www.wiley.com that provides a valuable supplement to the textbook by describing additional properties of materials, along with additional examples of current materials and their applications. Chapter W22 on our home page emphasizes the structural and chemical characterization of materials, as well as the characterization of their optical, electrical, and magnetic properties. As new materials and applications are developed, the home page will be regularly updated.

Since this text will likely be used most often in a one-semester course, we recommend that Chapters 1–4 on structure be covered in as much detail as needed, given the backgrounds of the students. A selection of chapters on the properties of materials (5–10) and on the classes of materials (11–18) of particular interest can then be covered. According to the tastes of the instructor and the needs of the students, some of the remaining chapters (surfaces; thin films, interfaces, and multilayers; synthesis and processing of materials) can be covered. For example, a course on engineering materials could consist of the following: Chapters 1–4 on structure; Chapter 6 on thermally activated processes, etc.; Chapter 10 on mechanical properties; Chapter 12 on metals and alloys; Chapter 13 on ceramics; Chapter 14 on polymers; and Chapter 21 on synthesis and processing.

Physics majors usually take an introductory course in solid-state physics in their senior year. Therefore in such a course it will be necessary to start at "the beginning," i.e., Chapter 1 on the structure of crystals. Students in MS&E or engineering departments who have already taken an introductory course on materials can quickly review (or skip) much of the basic material and focus on more advanced topics, beginning with Chapter 5 on phonons, if desired, or Chapter 7 on electrons in solids.

We owe a debt of gratitude to our colleagues at The City College and City University who, over the years, have shared with us their enthusiasm for and interest in the broad and fascinating subject of materials. They include R. R. Alfano, J. L. Birman, T. Boyer, F. Cadieu, H. Z. Cummins, H. Falk, A. Genack, M. E. Green, L. L. Isaacs, M. Lax, D. M. Lindsay (deceased), V. Petricevic, F. H. Pollak, S. R. Radel, M. P. Sarachik, D. Schmeltzer, S. Schwarz, J. Steiner, M. Tamargo, M. Tomkiewicz, and N. Tzoar (deceased). Colleagues outside CUNY who have shared their knowledge with us include Z. L. Akkerman, R. Dessau, H. Efstathiadis, B. Gersten, Y. Goldstein, P. Jacoby, L. Ley, K. G. Lynn, D. Rahoi, and Z. Yin. Our thanks also go to our students and postdocs who have challenged us, both in our research and teaching, to refine our thinking about materials and their behavior.

Special thanks are due to Gregory Franklin who served as our editor at John Wiley & Sons for the bulk of the preparation of this textbook. His unflagging support of this effort and his patience are deeply appreciated. Thanks are also due to our current editor, George Telecki, who has helped us with sound advice to bring this project to a successful conclusion. We acknowledge with gratitude the skill of Angioline Loredo who supervised the production of both the textbook and supplementary Web-based material. We have appreciated the useful comments of all the anonymous reviewers of our textbook and also wish to thank all the authors who granted permission for us to use their artwork.

Finally, we gratefully acknowledge the constant support, encouragement, and patience of our wives, Harriet and Françoise, and our families during the years in which this textbook was prepared. Little did we (or they) know how long it would take to accomplish our goals.

JOEL I. GERSTEN
FREDERICK W. SMITH

New York City

▰▰▰ INTRODUCTION

The study of materials and their properties and applications is an important part of modern science and technology. As may be expected for such a wide-ranging subject, the study of materials is a multidisciplinary effort, encompassing segments of physics, chemistry, and essentially all branches of engineering, including aerospace, chemical, civil, electrical, and mechanical. In addition, the relatively new discipline of materials science and engineering focuses directly on the study of the properties and applications of materials.

Materials can be classified as being either natural or artificial, the latter corresponding to materials, not found in nature, that are prepared by humans. Important natural materials have included organic materials such as wood, ivory, bone, fiber, and rubber, along with inorganic materials such as minerals and ceramics (stone, flint, mica, quartz, clay, and diamond) and metals such as copper and gold. Different eras of civilization have been given names corresponding to the materials from which tools were made: for example, the Stone Age, the Chalcolithic (Copper–Stone) Age, the Bronze Age, and the Iron Age. Recently, the dominant technological materials have been manufactured, such as steels as structural materials and the semiconductor Si for electronics.

Although the use of solid materials extends to prehistory, the systematic study and development of materials have begun much more recently, within the last 100 years. Development of the periodic table of the elements in the nineteenth century and the resulting grouping of elements with similar properties played a crucial role in setting the stage for the development of materials with desired properties. The discovery that x-rays could be used to probe the internal structure of solids early in the twentieth century also played a key role in accelerating the study of materials.

The study of materials as presented in this book begins with in-depth discussions of the structure of materials in Chapters 1 to 4 and of the fundamental principles determining the physical properties of materials in Chapters 5 to 10. Following these discussions of structure and properties, which apply to all materials, eight essentially distinct classes of materials are discussed in Chapters 11 to 18, with emphasis placed on their special properties and applications. The surfaces of materials, interfaces between materials, and materials in the form of thin films and multilayers are then discussed in Chapters 19 and 20. A discussion of the synthesis and processing (S&P) of materials follows in Chapter 21, with emphasis both on general issues and also on the S&P of specific materials.

In addition to the text material, supplementary material for all the chapters can be found at the Wiley Web site (www.wiley.com). This material includes a wide range of additional discussions of the properties and applications of materials. Also, experimental techniques used for the characterization of a wide range of materials properties are discussed. The following topics are reviewed briefly in the appendices appearing at the Web site: thermodynamics, statistical mechanics, and quantum mechanics.

The supplementary material can be downloaded through a standard Web browser at the following address: ftp://ftp.wiley.com/public/sci_tech_med/materials/. You may also find a link to the site and other resources from the Wiley Electrical Engineering software supplements Web page at: http://www.wiley.com/products/subject/engineering/electrical/software_supplem_elec_eng.html.

The eight classes of materials discussed in this book include semiconductors, metals and alloys, ceramics, polymers, dielectrics and ferroelectrics, superconductors, magnetic materials, and optical materials. Our discussions of these materials are meant to provide an introduction and solid grounding in the specific properties and applications of each class. Although each class of materials is often considered to be a separate specialty and the basis for a distinct area of technology, there are, in fact, many areas of overlap between the classes, such as magneto-optical materials, ceramic superconductors, metallic and ceramic permanent magnet materials, semiconductor lasers, dilute magnetic semiconductors, polymeric conductors, and so on.

There have been many materials success stories over the years, including the high-T_c superconductors, a-Si:H in photovoltaic solar cells, Teflon and other polymers, optical fibers, laser crystals, magnetic disk materials, superalloys, composite materials, and superlattices consisting of alternating layers of materials such as semiconductors or metals. These materials, most of which have found successful applications, are described throughout.

Our understanding of the structure of materials at the atomic level is well developed and, as a result, our understanding of the influence of atomic-level microstructure on the macroscopic properties of materials continues to improve. Between the microscopic and macroscopic levels, however, there exists an important additional level of structure at an intermediate length scale, often determined by defects such as grain boundaries, dislocations, inclusions, voids, and precipitates. Many of the critical properties of materials are determined by phenomena such as diffusion and interactions between defects that occur on this intermediate structural level, sometimes referred to as the mesoscopic level. Our understanding of phenomena occurring on this level in the heterogeneous (e.g., polycrystalline, amorphous, and composite) materials that are used in modern technology remains incomplete. Many of the properties of materials that are critical for their applications (e.g., mechanical properties) are determined by phenomena occurring on this level of microstructure.

Useful materials are becoming more complex. Examples include the high-T_c copper oxide–based ceramic superconductors, rare earth–based permanent magnets, bundles of carbon nanotubes, and even semiconductors such as Si–Ge alloys employed in strained layers and superlattices. Recent and continuing advances in the design and manipulation of materials atom by atom to create artificial structures are revolutionary steps in the development of materials for specific applications. This area of nanotechnology is an important focus of this book.

As we enter the twenty-first century and the world population and the depletion of resources both continue to increase, it is clear that the availability of optimum materials will play an important role in maintaining our quality of life. It is hoped that textbooks such as this one will serve to focus the attention of new students, as well as existing researchers, scientists, and engineers, toward the goals of developing and perfecting new materials and new applications for existing materials.

STRUCTURE OF MATERIALS

CHAPTER 1

Structure of Crystals

1.1 Introduction

The physical properties of solid-state materials are determined by three principal factors:

1. The properties of constituent atoms (masses, atomic numbers, electron configurations, ionization energies, etc.)
2. The local interactions of atoms with each other in the solid state (i.e., the nature of the bonding and the resulting nearest-neighbor configurations of atoms)
3. The arrangement of atoms in space to form a three-dimensional solid

Many important properties of solids, including their response to electric and magnetic fields, also involve the correlated motions of electrons and their spins. These properties include superconductivity and magnetism, among others, and require additional information for their analysis.

The nearest-neighbor (NN) configurations of atoms mentioned above are referred to here as *local atomic bonding units*. Their structure determines the *short-range order* (SRO) of the solid and the arrangement of the atoms in space constitutes the *long-range order* (LRO). If the LRO in the solid is perfect (i.e., if the arrangement in space of the atoms is perfectly *periodic*), the solid is said to be a *perfect crystal* or, simply, a *crystal*. Most solid-state materials are actually far from being structurally perfect and possess deviations from both SRO and LRO. The lack of structural order and its effect on the properties of the solid are important themes throughout this book.

To illustrate the significance of the three factors listed above, some examples of solid-state materials whose properties differ due to one or more of these factors are given next. First, consider the two crystalline forms of pure carbon: diamond, an insulator, and graphite, a semimetal. The distinctive physical properties and characteristic LRO of diamond and graphite differ significantly, due to the differences in SRO present in the two crystals: tetrahedral (fourfold) local coordination of C atoms for diamond and trigonal (threefold) local coordination of C atoms for graphite. In contrast, crystals of metallic Al and insulating Ar share the same types of SRO and LRO but nevertheless, have very different physical properties, due to the different outer-electron configurations of Al and Ar atoms. Solids comprised of the same atoms and having different LROs but essentially the same SRO are the crystalline and amorphous forms of Si and SiO_2.

As distinguished from the properties of more disordered forms of solids such as glasses and other amorphous materials, the properties of crystals depend on both the local atomic bonding and the fact that in crystals, the local configurations of atoms are arranged periodically in space. These two fundamental aspects of the structure of crystals — short-range local chemical bonding and long-range periodic array of atoms — are introduced and discussed in this chapter.

The three-dimensional lattice and its unit cell are introduced and defined in a formal way. Together, they serve as the geometric frame of reference for the discussion of crystals. The concept of *crystal structure* and its relationships to the periodic lattice and to local atomic bonding units are then presented and illustrated using well-known crystals as examples. The various forms of chemical bonding that occur between the atoms (or ions) within local atomic bonding units are discussed in Chapter 2 and examples are given of crystals possessing each type of bonding. A summary of some important properties and parameters of atoms which influence their behavior in solid-state materials are also presented in Chapter 2. Additional interesting and important examples of bonding in solids and the resulting crystal structures are presented in later chapters, where specific classes of solid-state materials are discussed in more detail.

INTRODUCTION TO LATTICES

A *lattice* is simply a periodic set of points in space, in principle infinite in extent but which for our purposes can often be limited to the volume enclosed by the crystal. The focus here is on three-dimensional lattices. Two-dimensional lattices are important in discussions of crystal surfaces and are introduced in Chapter 19. A convenient way to define a lattice is in terms of the set of *translation vectors* **R** which can be used to generate the lattice points.

1.2   Translation Vectors

As a concrete example of a three-dimensional lattice, consider one of the simplest possible cases, the *simple cubic* (SC) lattice shown in Fig. 1.1. Starting with an arbitrary lattice point as the origin (0,0,0) of the coordinate system, all other possible points of the SC lattice are generated by using the translation vectors **R**, which originate at (0,0,0) and terminate at the lattice points $(x, y, z) = (n_1 a, n_2 a, n_3 a)$. Here a is the

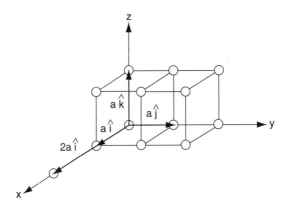

Figure 1.1. Simple cubic lattice and two unit cells. Note that the vector $2a\hat{\mathbf{i}}$ is not a primitive translation vector of this lattice.

lattice constant, the distance between NN adjacent points of the SC lattice. The values of n_1, n_2, and n_3 allowed are the set of all integers. In terms of the unit vectors $\hat{\mathbf{i}}$, $\hat{\mathbf{j}}$, and $\hat{\mathbf{k}}$ along the x, y, and z Cartesian axes, the translation vectors $\mathbf{R}$ for the SC lattice can be written as

$$\mathbf{R} = n_1 a\hat{\mathbf{i}} + n_2 a\hat{\mathbf{j}} + n_3 a\hat{\mathbf{k}}. \tag{1.1}$$

For the simple cubic lattice the three vectors $a\hat{\mathbf{i}}$, $a\hat{\mathbf{j}}$, and $a\hat{\mathbf{k}}$ shown in Fig. 1.1 are the *fundamental* or *primitive translation vectors* of the lattice since all points of the lattice can be generated using the vectors $\mathbf{R}$ defined in Eq. (1.1). These primitive translation vectors $a\hat{\mathbf{i}}$, $a\hat{\mathbf{j}}$, and $a\hat{\mathbf{k}}$ serve to define the lattice. Note that the vector $2a\hat{\mathbf{i}}$ shown in Fig. 1.1 is not a primitive translation vector for the SC lattice since only every other lattice point along the x-axis can be reached using the vector $2a\hat{\mathbf{i}}$.

The definition of a lattice is now extended to the most general case, defined by the translation vectors

$$\begin{aligned}\mathbf{R} &= n_1 a\hat{\mathbf{u}}_1 + n_2 b\hat{\mathbf{u}}_2 + n_3 c\hat{\mathbf{u}}_3 \\ &= n_1 \mathbf{u}_1 + n_2 \mathbf{u}_2 + n_3 \mathbf{u}_3,\end{aligned} \tag{1.2}$$

where, in general, $a \neq b \neq c$.[†] The three nonorthogonal unit vectors $\hat{\mathbf{u}}_1$, $\hat{\mathbf{u}}_2$, and $\hat{\mathbf{u}}_3$ are defined as shown in Fig. 1.2, with $\alpha \neq \beta \neq \gamma \neq 90°$.[‡] This most general three-dimensional lattice with primitive translation vectors $\mathbf{u}_1 = a\hat{\mathbf{u}}_1$, $\mathbf{u}_2 = b\hat{\mathbf{u}}_2$, and $\mathbf{u}_3 = c\hat{\mathbf{u}}_3$, is known as the *triclinic lattice*.

The property of the lattice that follows from its definition in terms of the translation vectors $\mathbf{R}$ is known as *translational symmetry*. This is the symmetry that all lattices possess. As a result of translational symmetry, the lattice appears identical when viewed from any lattice point. More generally, any two points in a lattice, defined by the vectors $\mathbf{r}$ and $\mathbf{r}'$, are identical (i.e., have identical surroundings), if they can be connected by a translation vector $\mathbf{R}$ (e.g., if $\mathbf{r}' = \mathbf{r} + \mathbf{R}$). As a result, in a perfect crystal, all the properties of the crystal are identical at points $\mathbf{r}$ and $\mathbf{r}'$ (except, of course, for those

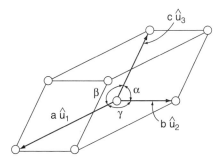

Figure 1.2. Triclinic lattice and unit cell. This is the general lattice in three dimensions, with $a \neq b \neq c$ and $\alpha \neq \beta \neq \gamma \neq 90°$.

[†] The notation $a \neq b \neq c$ means that these three lengths are all different from each other (i.e., that $a \neq b$, $a \neq c$, and $b \neq c$).
[‡] The notation $\alpha \neq \beta \neq \gamma \neq 90°$ means that these three angles are all different from each other (and from $90°$).

points in a finite perfect crystal that lie near its surface). If the lattice also appears the same from every lattice point after being rotated about a certain axis through an angle of $360/n$ degrees, the lattice also has an axis of n-fold *rotational symmetry* in addition to its translational symmetry.

The full symmetry of a given crystal structure has importance for the physical properties of a crystal since it is expected that any macroscopic physical property has, at the least, the symmetry of the point group of the space group that the crystal possesses. The 32 possible crystallographic point groups are described in Chapter 15. For a more complete discussion of these symmetries and of the 230 possible space groups, see the books by Ashcroft and Mermin (1976), Burns (1985), and Burns and Glazer (1990), or books on crystallography.

1.3 Unit Cells

The cube of edge length a and volume a^3 is the conventional *unit cell* of the SC lattice and can be used to generate the entire lattice. This is done by moving (i.e., by translating) the unit cell through space using the translation vectors **R** so that the entire volume occupied by the SC lattice is covered just once.

Lattice points lie at the eight corners of the cubic unit cell, and each lattice point is shared by the eight unit cells, which touch at their single common corner. When lattice points lie only at the corners of the unit cell, as for the SC unit cell shown, the unit cell is then a *primitive unit cell*. This primitive unit cell, which need not in general be cubic, is the smallest volume that can be used to generate the lattice. This definition of a primitive unit cell is not unique since other primitive unit cells can be defined, including one with a lattice point at its center. This is the *Wigner–Seitz cell* described in Chapter 3. Often, a larger unit cell which is not primitive is chosen as the conventional unit cell for a particular lattice. Examples of and reasons for this use are given later.

The volume V of a unit cell defined by the three primitive translation vectors $\mathbf{u}_1$, $\mathbf{u}_2$, and $\mathbf{u}_3$ is given by $V = |\mathbf{u}_1 \cdot (\mathbf{u}_2 \times \mathbf{u}_3)| = |a\hat{\mathbf{u}}_1 \cdot (b\hat{\mathbf{u}}_2 \times c\hat{\mathbf{u}}_3)|$.

1.4 Bravais Lattices

Bravais in 1845 showed on the basis of symmetry arguments that there exist only 14 distinct lattice types in three dimensions, two of which have already been mentioned: the simplest, SC, and the most general, triclinic. The 14 *Bravais lattices*, together with their conventional unit cells shown in Fig. 1.3, are usually divided into seven systems. These seven systems are characterized by the special relationships given in Table 1.1, which involve the lengths of the sides a, b, and c and the angles α, β, and γ. These seven lattice systems are discussed next.

Cubic. There are three cubic Bravais lattices: *simple cubic* (SC), *body-centered cubic* (BCC), and *face-centered cubic* (FCC). The BCC lattice is obtained by placing an additional lattice point at the center of the cubic unit cell; the FCC lattice is obtained by placing additional lattice points at the centers of the six faces of the cubic unit cell. The conventional cubic unit cells chosen for these three lattices are shown in Fig. 1.3 and illustrate their essential symmetries, as listed in Table 1.1. The BCC and FCC lattices, however, have smaller, noncubic primitive unit cells which can also be used

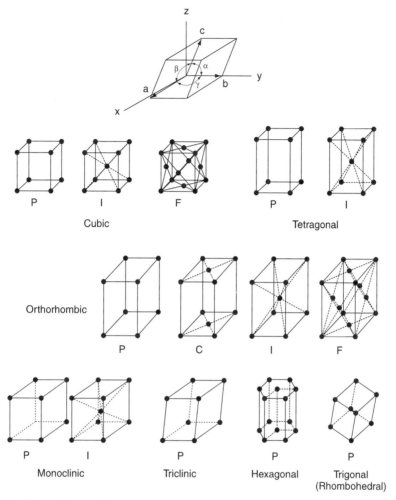

Figure 1.3. The 14 Bravais lattices in three dimensions. See Table 1.1 for the relationships between the sides a, b, and c and between the angles α, β, and γ. P, primitive; I, body-centered; F, face-centered; C, base-centered.

to generate these lattices. The primitive units cells for BCC and FCC are shown in Figs. 1.4 and 1.5, respectively. The corresponding primitive translation vectors are also given. Note that these primitive unit cells for BCC and FCC have lattice points only at their corners. It can readily be shown that the conventional BCC and FCC cubic unit cells are two and four times greater in volume, respectively, than the corresponding primitive unit cells.

The BCC and FCC lattices are quite important since the majority of solid-state materials have crystal structures based on these two lattices. Also, many elements crystallize with atoms or ions located at the lattice points of the BCC or FCC lattices.

Tetragonal. There are only two tetragonal Bravais lattices, *simple tetragonal* (ST) and *body-centered tetragonal* (BCT). These correspond to SC and BCC lattices that

TABLE 1.1   Bravais Lattices: 14 Distinct Lattice Types in Three Dimensions

System (Essential Symmetry)	Number of Members	Lattices	Special Relationships
Cubic (four three-fold axes)	3	Simple, body-centered, face-centered	$a = b = c, \alpha = \beta = \gamma = 90°$
Tetragonal (one four-fold axis)	2	Simple, body-centered	$a = b \neq c, \alpha = \beta = \gamma = 90°$
Orthorhombic (three orthogonal two-fold axes)	4	Simple, body-centered, face-centered, base-centered	$a \neq b \neq c, \alpha = \beta = \gamma = 90°$
Monoclinic (one two-fold axis)	2	Simple, body-centered	$a \neq b \neq c, \alpha = \beta = 90° \neq \gamma$
Triclinic (none)	1	Simple	$a \neq b \neq c, \alpha \neq \beta \neq \gamma \neq 90°$
Trigonal or rhombohedral (one three-fold axis)	1	Simple	$a = b = c, \alpha = \beta = \gamma < 120°$ and $\neq 90°$
Hexagonal (one six-fold axis)	1	Simple	$a = b \neq c, \alpha = \beta = 90°, \gamma = 120°$

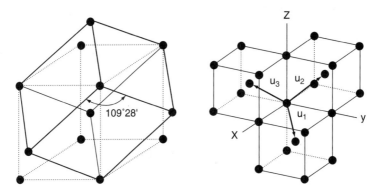

Figure 1.4. Trigonal primitive unit cell and a set of primitive translation vectors $\mathbf{u}_1$, $\mathbf{u}_2$, and $\mathbf{u}_3$ for the body-centered cubic (BCC) lattice. Also shown is the conventional cubic unit cell. Note that $\mathbf{u}_1 = a(\hat{\mathbf{i}} + \hat{\mathbf{j}} - \hat{\mathbf{k}})/2 = \sqrt{3}\,a\hat{\mathbf{u}}_1/2$, $\mathbf{u}_2 = a(-\hat{\mathbf{i}} + \hat{\mathbf{j}} + \hat{\mathbf{k}})/2 = \sqrt{3}\,a\hat{\mathbf{u}}_2/2$, $\hat{\mathbf{u}}_3 = a(\hat{\mathbf{i}} - \hat{\mathbf{j}} + \hat{\mathbf{k}})/2 = \sqrt{3}\,a\hat{\mathbf{u}}_3/2$. (After C. Kittel, *Introduction to Solid State Physics*, 7th ed., copyright 1996 by John Wiley & Sons, Inc. Reprinted by permission of John Wiley & Sons, Inc.)

have been either elongated or compressed along one axis, so that, for example, $a = b \neq c$ with $\alpha = \beta = \gamma = 90°$. This special axis is often referred to as the c-axis of the tetragonal lattice. The face-centered tetragonal lattice can be shown to be equivalent to the BCT lattice and therefore is not a distinct Bravais lattice.

Orthorhombic. There are four *orthorhombic* Bravais lattices with $a \neq b \neq c$ and $\alpha = \beta = \gamma = 90°$: *simple, body-centered, face-centered*, and *base-centered*.

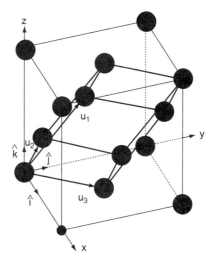

Figure 1.5. Trigonal primitive unit cell and primitive translation vectors $\mathbf{u}_1$, $\mathbf{u}_2$, and $\mathbf{u}_3$ for the face-centered cubic (FCC) lattice. Also shown is the conventional cubic unit cell. Note that $\mathbf{u}_1 = a(\hat{\mathbf{j}} + \hat{\mathbf{k}})/2 = \sqrt{2}\,a\hat{\mathbf{u}}_1/2$, $\mathbf{u}_2 = a(\hat{\mathbf{i}} + \hat{\mathbf{k}})/2 = \sqrt{2}\,a\hat{\mathbf{u}}_2/2$, $\mathbf{u}_3 = a(\hat{\mathbf{i}} + \hat{\mathbf{j}})/2 = \sqrt{2}\,a\hat{\mathbf{u}}_3/2$.

Monoclinic. There are two monoclinic Bravais lattices: *simple monoclinic* and *body-centered monoclinic*. These are ordinarily formed using three primitive translation vectors of unequal length ($a \neq b \neq c$), one of which is perpendicular to the plane of the other two, with $\alpha = \beta = 90° \neq \gamma$.

Triclinic. There is a single *triclinic* Bravais lattice, with $a \neq b \neq c$ and $\alpha \neq \beta \neq \gamma \neq 90°$. This is the Bravais lattice with minimum symmetry. As mentioned earlier, the triclinic lattice can be considered to be the most general Bravais lattice.

Trigonal (rhombohedral). There is also a single *trigonal* Bravais lattice, generated from the SC lattice by stretching one of the body diagonals of the cubic unit cell. The trigonal lattice therefore has $a = b = c$ and $\alpha = \beta = \gamma < 120°$, $\neq 90°$. For the special case with $\alpha = \beta = \gamma = 60°$, the trigonal lattice actually has the same symmetry as the FCC lattice. As seen in Fig. 1.5, the FCC primitive unit cell is, in fact, a trigonal cell with angles of 60°.

Hexagonal. The single hexagonal Bravais lattice is *simple hexagonal* and has a conventional hexagonal unit cell (Fig. 1.6), which contains three primitive hexagonal unit cells. The primitive unit cell has $a = b \neq c$, $\alpha = \beta = 90°$, and $\gamma = 120°$. The hexagonal lattice has a six-fold symmetry axis, whereas the trigonal lattice has only a three-fold symmetry axis.

Some additional properties of lattices are discussed next. Still others will become apparent when specific crystal structures based on these lattices are described later in this chapter.

1.5 Lattice Axes, Planes, and Directions

To be able to discuss effectively the properties of lattices and crystal structures, it is necessary to provide ways to specify some important geometrical properties of lattices:

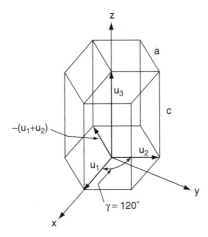

Figure 1.6. Primitive unit cell and primitive translation vectors $\mathbf{u}_1$, $\mathbf{u}_2$, and $\mathbf{u}_3$ for the hexagonal lattice. Also shown is the conventional hexagonal unit cell. $\mathbf{u}_1 = a\hat{\mathbf{u}}_1 = a\hat{\mathbf{i}}$, $\mathbf{u}_2 = a\hat{\mathbf{u}}_2 = a(\cos 120°\hat{\mathbf{i}} + \sin 120°\hat{\mathbf{j}}) = a(-\hat{\mathbf{i}} + \sqrt{3}\hat{\mathbf{j}})/2$, $\mathbf{u}_3 = c\hat{\mathbf{u}}_3 = c\hat{\mathbf{k}}$.

the axes of the lattice coordinate system, planes containing lattice points, and directions in the lattice.

Axes of the Lattice. In the same way that different unit cells can be chosen for a given lattice, it is also true that different lattice coordinate systems and sets of lattice axes can be chosen. In general, the axes of a lattice are taken to lie along the edges of its unit cell. For example, when a conventional cubic unit cell is chosen, the axes are chosen to be the x, y, and z axes. The understandable convenience of the use of orthogonal axes makes the choice of the conventional cubic unit cells for the BCC and FCC lattices quite natural.

Lattice Planes. It is important to have a simple way of specifying or labeling planes in a lattice since the corresponding planes of atoms in a crystal are important, for example, in a discussion of diffraction effects. The labeling procedure universally used for this purpose employs the *Miller indices*. The specification of the Miller indices for a set of parallel lattice planes is essentially a two-step process:

1. The three points or intercepts where one of the set of lattice planes in question intersects the lattice axes are located. This is illustrated in Fig. 1.7 for the case of a cubic lattice. The plane shown intersects the three orthogonal axes at $x = 3a$, $y = 2a$, and $z = 3a$. Only the three numerical factors (i.e., 3, 2, 3) are kept.
2. The reciprocals of these three numbers are taken:

$$3, 2, 3 \rightarrow \tfrac{1}{3}, \tfrac{1}{2}, \tfrac{1}{3}.$$

The reciprocals are transformed to the three corresponding smallest integers:

$$\tfrac{1}{3}, \tfrac{1}{2}, \tfrac{1}{3}, \rightarrow \tfrac{2}{6}, \tfrac{3}{6}, \tfrac{2}{6}, \rightarrow 2, 3, 2.$$

The resulting set of integers (2,3,2), or simply (232), are the Miller indices (hkl) of the lattice plane shown in Fig. 1.7.

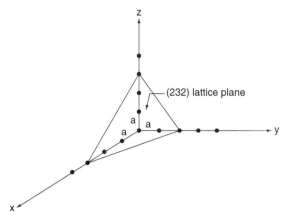

Figure 1.7. Miller indices of lattice planes. The intercepts of the lattice plane shown are at $x = 3a$, $y = 2a$, and $z = 3a$. The Miller indices (hkl) of this plane are (232), obtained as described in the text.

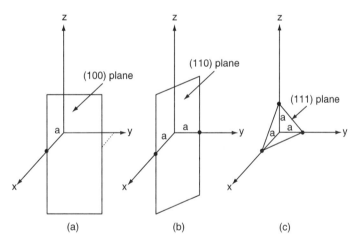

Figure 1.8. Important planes of a cubic lattice. The intercepts for these planes are at (a) $(a, \infty, \infty) \rightarrow (hkl) = (100)$, (b) $(a, a, \infty) \rightarrow (hkl) = (110)$, and (c) $(a, a, a) \rightarrow (hkl) = (111)$.

The following important points concerning Miller indices are worth noting:

1. The Miller indices of an arbitrary lattice plane are specified by three integers, taken in the general case to be h, k, and l. Actually, (hkl) refers not to a single plane but to the entire set of identical planes in the lattice that are parallel to the given (hkl) plane. For cubic lattices the perpendicular distance between adjacent parallel lattice planes is given by $d = a/\sqrt{h^2 + k^2 + l^2}$.

2. When a lattice plane is parallel to one of the axes of the lattice, the corresponding intercept is taken to be at infinity. The corresponding Miller index is therefore zero. Some examples are given in Fig. 1.8.

3. A lattice plane with a negative intercept has a corresponding negative Miller index. This is indicated with a bar over the index in question [e.g., $(\bar{h}\,\bar{k}\,\bar{l})$].

4. The notation {hkl} is used when referring to a set of related lattice planes. For example, the {100} set of planes refers to the six faces of a cube, given by (100), (010), (001), ($\bar{1}$00), (0$\bar{1}$0), and (00$\bar{1}$). Similarly, the eight faces of a regular octahedron correspond to the {111} set of planes (see Fig. 1.13).

5. In general, the important lattice planes in a crystal are those with low Miller indices [e.g., (100), (110), and (111)], since these are the planes that will have the highest concentrations of atoms (per unit area).

6. Specification of the Miller indices for lattice planes in hexagonal lattices involves the intercepts of the plane with four axes instead of three. Three of these are the nonorthogonal axes that lie in the xy-plane, separated from each other by 120°. These three axes are in the $\mathbf{u}_1$, $\mathbf{u}_2$, and $-(\mathbf{u}_1 + \mathbf{u}_2)$ directions shown in Fig. 1.6. The fourth axis is the c axis (i.e., the usual z-axis), which is orthogonal to the other three. Once the four intercepts are determined, the procedure for finding the indices ($hkil$) proceeds as outlined earlier. The six faces of the conventional hexagonal unit cell are therefore the (10$\bar{1}$0), (01$\bar{1}$0), (1$\bar{1}$00), ($\bar{1}$010), (0$\bar{1}$10), and ($\bar{1}$$\bar{1}$00), planes (i.e., the {1100} set of planes). The top and bottom planes of the unit cell are (0001) planes. Note that the index i corresponds to the $-(\mathbf{u}_1 + \mathbf{u}_2)$ direction and is equal to $-(h + k)$.

Directions in the Lattice. The direction from the origin of the lattice to the point reached by the translation vector $\mathbf{R} = n_1 a\hat{\mathbf{u}}_1 + n_2 b\hat{\mathbf{u}}_2 + n_3 c\hat{\mathbf{u}}_3$ is referred to as the $[n_1, n_2, n_3]$ direction or simply $[n_1 n_2 n_3]$. The smallest set of integers n_1, n_2, and n_3 corresponding to a given direction are the direction indices for that direction. In cubic crystals the direction perpendicular to the set of (hkl) planes is the $[hkl]$ direction. For example, the [100] direction (i.e., the $+x$ axis), is perpendicular to the (100) plane, as illustrated in Fig. W1.1 at our Web site.[†] A set of lattice directions related by symmetry is indicated by $\langle n_1 n_2 n_3 \rangle$. For example, the lattice axes correspond to the $\langle 100 \rangle$ set of directions. Note that directions in hexagonal lattices are given by $[hkil]$, where, again, $i = -(h + k)$.

LOCAL ATOMIC BONDING UNITS AND CRYSTAL STRUCTURES

When the same atom or group of atoms is associated with each point of one of the lattices described earlier, a specific *crystal structure* results. In this section some important examples of the wide variety of existing crystal structures are introduced and discussed. Additional examples of crystal structures are given in Chapters 11 to 18, where the following classes of solid-state materials are described: semiconductors, metals and alloys, ceramics, polymers, dielectrics and ferroelectrics, superconductors, magnetic materials, and optical materials.

The first questions to ask about any crystal structure deal with what its lattice is and how the atoms are arranged relative to each lattice point. One approach to answering such questions is based on the equation

$$\text{crystal structure} = \text{lattice} + \text{basis}, \tag{1.3}$$

[†] Supplementary material for this textbook is included on the Web at the resource site (ftp://ftp.wiley.com/ public/sci_tech_med/materials). Cross-references to elements of the Web material are prefixed by "W."

which is often used as the operational definition of a crystal structure. Here the term *basis* refers to the set of atoms that lie within or are associated with the unit cell chosen for the structure. Although Eq. (1.3) is technically correct and in fact provides all the information that is ordinarily needed for most purposes (e.g., for the calculation of the scattering of waves from a crystal presented in Chapter 3), this approach often is not particularly helpful in addressing the important question of why the atoms form a particular crystal structure in the first place. Missing from this approach is any information about the local bonding between the atoms which determines the structure and hence the physical properties of the crystal.

To emphasize the role that local bonding plays in determining crystal structure, the following expression is used to supplement the information provided by Eq. (1.3):

$$\text{crystal structure} = \text{local atomic bonding units} + \text{lattice.} \tag{1.4}$$

Here the term *local atomic bonding unit* refers to one of the smallest groupings or configurations of atoms, which serves to demonstrate some important aspects of the bonding in the crystal.

1.6 Local Atomic Bonding Units

Local atomic bonding units consist in general of a central atom (or molecule) and its NN atoms (or molecules). The central atom can either be neutral, as in the rare gas solid Ar, or can have a net charge, either positive or negative, as in the ionic solid NaCl (Na^+Cl^-). The important bonding units in most solid-state materials contain from as many as 12 NNs to as few as 2 NNs to the central atom. Bonding units with central atoms that have 12, 8, 6, 4, 3, and 2 identical NNs are described in this section.

The notations $A–A_n$ and $A–B_n$ are used to identify local bonding units, with A indicating the central atom and A_n or B_n the n NN atoms. Here n is the *coordination number* ($n = CN$). The bonding units described here are idealized in the sense that the NN A or B atoms are all assumed to be the same distance from and bonded with equal strength to the central A atom. In many crystals, however, the local atomic bonding units are distorted so that some of the NN atoms are closer than others. Examples of structures with distorted bonding units are given later. The concept of local atomic bonding units will be useful even in amorphous or disordered materials where no lattice or LRO exists. In such materials a type of SRO can still exist if the local atomic bonding units retain their identity (e.g., the same number of NNs as in the crystal, even if the bond lengths and bond angles are distorted from their crystalline values).

Important information for the local atomic bonding units described here is summarized in Table 1.2, specifically the number and identity of the NNs, examples of specific bonding units in real solids, and the *coordination polyhedra* or regular geometrical figures which are often used to represent the bonding units. Note that these coordination polyhedra illustrate the symmetry of the bonding units but are not in general the unit cells for the structures.

These bonding units are each described below and the distinguishing features of each are stressed. In general, the central A atom and the NN A and B atoms are considered to be rigid spheres in contact with each other. Atoms (and ions) are, in fact, somewhat compressible, so that this model of atoms as rigid spheres will need to be modified in order to understand many crystal structures. The types of bonding occurring between the atoms or ions in these units are described in Chapter 2.

TABLE 1.2 Local Atomic Bonding Units in Solid-State Materials

Coordination Number n	Bonding Unit	Examples	Coordination Polyhedra
12	$A-A_{12}$(cub)	Al, Ar, C_{60}	Cubo-octahedron
	$A-A_{12}$(hex)	Mg	Twinned cubo-octahedron
	$A-A_{12}$(icos)		Icosahedron
	$A-B_{12}$(icos)	$Sn-Nb_{12}(Nb_3Sn)$	Icosahedron
8	$A-A_8$	Na, Cr, W	Cube
	$A-B_8$	$Cs^+-Cl_8^-$ (CsCl)	Cube
6	$A-A_6$	Po	Octahedron
	$A-B_6$	$Na^+-Cl_6^-$ (NaCl)	Octahedron
	$A-B_6$	$C-Fe_6$ (Fe_3C)	Triangular prism
4	$A-A_4$	C (diamond), Si	Tetrahedron
	$A-B_4$	$Ga-As_4$ (GaAs)	Tetrahedron
		$Si-O_4(SiO_2)$	Tetrahedron
		$Cu-O_4$ (CuO)	Square
3	$A-A_3$	C (graphite)	Triangle or
	$A-B_3$	$B-N_3$ (BN)	pyramid
		$N-Si_3$ (Si_3N_4)	
2	$A-A_2$	S, Se	Link or
		CH_2 (polyethylene)	bridge
	$A-B_2$	$O-Si_2$ (SiO_2)	

$A-A_{12}$(cub), $A-A_{12}$(hex), $A-A_{12}$(icos), and $A-B_{12}$(icos). An atom can be bonded to the 12 identical NN atoms in an $A-A_{12}$ bonding unit in at least three distinct ways, of which the cubic $A-A_{12}$(cub) and hexagonal $A-A_{12}$(hex) cases are referred to as close-packed units. In close-packed units the *packing fraction*, defined as the fraction of space occupied by hard-sphere atoms, has its maximum possible value for identical atoms of 0.74. The structures of these bonding units are shown in Fig. 1.9, with the central A atom in both cases first placed in contact with six other A atoms, all lying in the same plane. Note that pairs of adjacent NN A atoms in the plane are also in contact with each other.

Six additional A atoms then are placed in contact with the central A atom, with three in a plane above and three more in a plane below the original plane. This can be done in two distinct ways, as shown in Fig. 1.9b and c. In Fig. 1.9b the three atoms in the upper plane lie in depressions adjacent to the central A atom and directly above the three A atoms in the lower plane. The structure of the resulting $A-A_{12}$ unit is consistent with a hexagonal lattice so the notation $A-A_{12}$(hex) is used. The coordination polyhedron for the $A-A_{12}$(hex) unit is a twinned *cubo-octahedron* with 14 sides, 8 being equilateral triangles and 6 being squares. In Fig. 1.9c the three atoms in the upper plane are displaced from those in the lower plane by placing them into the remaining three depressions adjacent to the central A atom. The notation $A-A_{12}$(cub) is used for this unit since all the atoms lie on an FCC lattice. In addition, these planes of atoms coincide with (111) planes of the FCC lattice. The cubo-octahedron shown in Fig. 1.10 is the coordination polyhedron for the $A-A_{12}$(cub) unit and corresponds to a cube with the eight corners cut off. The cubic symmetry of this bonding unit is

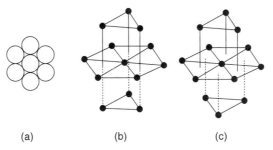

(a)   (b)   (c)

Figure 1.9. Structures of $A-A_{12}$(hex) and $A-A_{12}$(cub) bonding units. (*a*) A central A atom is in contact with six NN A atoms, all in the same plane (viewed from above). (*b*) An $A-A_{12}$(hex) bonding unit is obtained when the three upper A atoms lie directly above the three lower A atoms, as shown. For clarity, the atoms are not shown in contact with each other. (*c*) An $A-A_{12}$(cub) bonding unit obtained when the three upper A atoms are displaced laterally from the three lower A atoms, as shown.

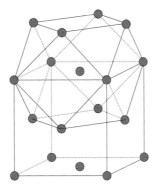

Figure 1.10. The cubo-octahedron shown is the coordination polyhedron for the $A-A_{12}$(cub) bonding unit. The FCC cubic unit cell is also shown. (After W. B. Pearson, *The Crystal Chemistry and Physics of Metals and Alloys*, copyright 1972 by John Wiley & Sons, Inc. Reprinted by permission of John Wiley & Sons, Inc.)

evident from this figure, where the atoms are reduced from their normal size and so are not in contact with each other.

The coordination polyhedron for the $A-A_{12}$(icos) and $A-B_{12}$(icos) icosahedral units is the regular *icosahedron* containing a central A atom (Fig. 1.11). Here A (or B) atoms are located at the 12 vertices of the regular icosahedron, which has 20 triangular faces. An icosahedron therefore consists of 20 tetrahedra sharing a common vertex and has a total of six five-fold symmetry axes. For the $A-A_{12}$(icos) unit shown, the 12 NN A atoms will not be in contact with each other unless there is some compression at their points of contact with the central A atom. Bonding units related to these icosahedral units and which also have coordination polyhedra with triangular faces include those with a central atom surrounded by 14, 15, or 16 NN atoms (i.e., $A-B_{14}$, $A-B_{15}$, and $A-B_{16}$). These are known as CN 14, CN 15, and CN 16 polyhedra, It should be noted that the set of B atoms in these units can also include some A atoms. These units, together with $A-A_{12}$(icos) and $A-B_{12}$(icos), are found in topologically close packed

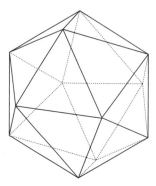

Figure 1.11. The regular icosahedron is the coordination polyhedron for the $A-A_{12}$(icos) and $A-B_{12}$(icos) bonding units. (After W. B. Pearson, *The Crystal Chemistry and Physics of Metals and Alloys*, copyright 1972 by John Wiley & Sons, Inc. Reprinted by permission of John Wiley & Sons, Inc.)

phases (e.g., *Frank–Kasper* and *Laves* phases). For a discussion of these phases, see Pearson (1972).

A–A₈ and A–B₈. An A atom can be bonded to eight other A atoms or to eight B atoms in a symmetric way when the central A atom is placed at the body-centered position of a cube and the eight NN A or B atoms are placed at the corners of the cube (Fig. 1.12). The *cube* is therefore the coordination polyhedron for the $A-A_8$ and $A-B_8$ bonding units. When the B atoms are larger than the A atom, they will come into contact with each other when the radius of a B atom, r_B, is equal to 1.366 times r_A, the radius of an A atom. This establishes the requirement that the *radius ratio $r_B/r_A \leq$* 1.366 for the central A atom to remain in contact with all eight of its NN B atoms in the $A-B_8$ unit. This requirement can also be expressed as $r_A/r_B \geq 0.732$ $(= \sqrt{3} - 1)$. For $r_A/r_B < 0.732$, the $A-B_8$ bonding unit may become unstable, resulting in a possible transformation to an $A-B_6$ unit. The reasons for this instability in ionic crystals are presented in Chapter 2, where ionic bonding is discussed.[†] The radius ratios for several crystal structures are discussed further in Chapter 2.

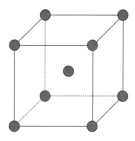

Figure 1.12. The $A-A_8$ bonding unit is shown with the central A atom at the body-centered position and with the eight NN A atoms at the corners of a cube. The cube is the coordination polyhedron for these bonding units.

[†] See Table 2.3 for a summary of the radius ratios for various bonding units and crystal structures.

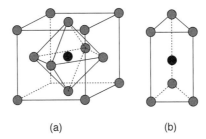

(a)   (b)

Figure 1.13. A–A$_6$ and A–B$_6$ bonding units. (*a*) The central A atom is at the body-centered position and the six NN A or B atoms are at the face-centered positions of a cube. The regular octahedron shown is the coordination polyhedron for these bonding units. (*b*) An A–B$_6$ unit in the form of a body-centered triangular or trigonal prism is shown. The central A atom is in contact with three B atoms in a lower plane and also with three B atoms in an upper plane.

A–A$_6$ and A–B$_6$. An A atom can be bonded to six other A atoms or to six B atoms in a symmetric way when the central A atom is placed at the body-centered position of a cube and the six NN A or B atoms are placed at the six face-centered positions of the same cube (Fig. 1.13a). When the positions of these six NN atoms are connected as shown, the resulting regular *octahedron* with eight identical triangular faces is seen to be the coordination polyhedron. Using an argument similar to that presented above for the A–B$_8$ unit, it is required that $r_B/r_A \leq 2.414$ (or that $r_A/r_B \geq 0.414 = \sqrt{2} - 1$) for the A atom to remain in contact with all six NN B atoms in the A–B$_6$ unit.

   Another example of an A–B$_6$ bonding unit is the body-centered triangular or trigonal *prism* (Fig. 1.13*b*). This unit consists of a central A atom in contact with three B atoms in a lower plane and three B atoms in an upper plane.

A–A$_4$ and A–B$_4$. The symmetric bonding of an A atom to four other A atoms or to four B atoms can be visualized by placing the central A atom at the body-centered position of a cube and the four NN A or B atoms at four of the eight corners of the cube in such a way that all four NN atoms are separated from each other by the diagonals of the cube faces (Fig. 1.14). The coordination polyhedron formed when the positions of the four NN atoms are connected is the regular *tetrahedron* with vertex angles of 109.47°. For the central A atom to remain in contact with its four NN B atoms in an A–B$_4$ unit, it is required that $r_B/r_A \leq 4.45$ (or $r_A/r_B \geq 0.225 = \sqrt{3/2} - 1$). This

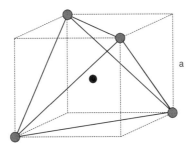

Figure 1.14. The A–A$_4$ and A–B$_4$ bonding units are shown with the central A atom at the body-centered position and with the four NN A or B atoms at four of the eight corners of a cube. The regular tetrahedron shown is the coordination polyhedron for these bonding units.

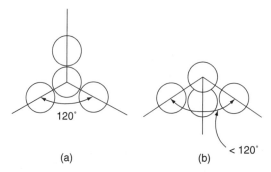

Figure 1.15. Structures of A–A$_3$ and A–B$_3$ bonding units. (*a*) planar triangular structure with central A atom and three NN A or B atoms lying in the same plane; (*b*) pyramidal structure with central A atom lying above the triangular base of three NN A or B atoms.

Figure 1.16. Structures of A–A$_2$ and A–B$_2$ bonding units: (*a*) linear structure with central A atom and two NN A or B atoms lying in a straight line; (*b*) nonlinear structure.

stability criterion may have some validity for crystals in which the approximation of spherical atoms or ions is reasonable. It is of less use in covalent crystals with the diamond or zincblende crystal structures, where the bonding tends to be directional and the resulting charge distributions are non spherical. Another type of A–B$_4$ bonding unit has square-planar symmetry, with the A atom surrounded by four NN B atoms at the corners of a square.

A–A$_3$ and A–B$_3$. The A–A$_3$ and A–B$_3$ bonding units can be formed with all the atoms lying in the same plane (Fig. 1.15*a*), or with the central A atom removed from the plane of the other three atoms (Fig. 1.15*b*). In the first case, the central A atom lies at the center of an equilateral triangle, while in the second case it lies above the triangular base at the vertex of a pyramidal bonding unit. The condition $r_B/r_A \leq 6.46$ (or $r_A/r_B \geq 0.155 = \sqrt{4/3} - 1$) can be shown to apply to the first case.

A–A$_2$ and A–B$_2$. A central A atom can be bonded to just two NN A or B atoms in a linear unit (Fig. 1.16*a*), or in a unit with a bond angle of less than 180° (Fig. 1.16*b*). These bonding units are called *linking* or *bridging units* since they can correspond either to a single link in a long chain of similar units or to a bridging unit connecting two larger bonding units.

1.7 Crystal Structures

Important crystal structures based on a single type of bonding unit are described next. Examples of solid-state materials with each crystal structure are also given. The types of bonding occurring within local atomic bonding units and in crystals are discussed in Chapter 2. Other important crystal structures based on more than one type of bonding unit and some based on the $A-B_{12}$ and $A-A_2$ bonding units are presented in later chapters.

As indicated in Eq. (1.4), crystal structures will be defined by specifying how the local atomic bonding units are placed on a given lattice. The viewpoint taken here is that the lattice of a given crystal structure simply allows the bonding units to fill space efficiently, with low strain.

Crystal Structures Based on A–A₁₂(hex). The central A atom and the 12 NN A atoms of the $A-A_{12}$(hex) bonding unit all lie on adjacent points of an hexagonal Bravais lattice, with the planes of atoms parallel to the (0001) lattice planes (see Fig. 1.6). The resulting crystal structure (Fig. 1.17*a*) is known as *hexagonal close-packed* (HCP). In the HCP crystal structure it can be seen that every atom is at the center of an $A-A_{12}$(hex) bonding unit. The hexagonal primitive unit cell for HCP (Fig. 1.17*b*) contains a basis of two identical A atoms at the positions (0,0,0) and $(\frac{2}{3}, \frac{1}{3}, \frac{1}{2})$. The coordinates refer to the primitive translation vectors $\mathbf{u}_1$, $\mathbf{u}_2$, and $\mathbf{u}_3$.

The close-packed planes of atoms in the HCP crystal structure are the (0002) planes of the lattice and are stacked along the c direction in the sequence **ABABAB**.... This stacking sequence is consistent with the structure of the $A-A_{12}$(hex) unit, where the central A atom and six of its NN atoms lie in an **A** plane. The three A atoms above the **A** plane and the remaining three A atoms below the **A** plane lie in **B** planes.

The ratio c/a of the lattice constants for the hexagonal unit cell has the value $\sqrt{\frac{8}{3}} =$ 1.633 for the ideal HCP crystal structure when the atoms are taken to be hard spheres in contact with each other. The actual c/a ratio for crystals with HCP crystal structures can deviate from this ideal value when the atoms are not perfectly spherical, as when

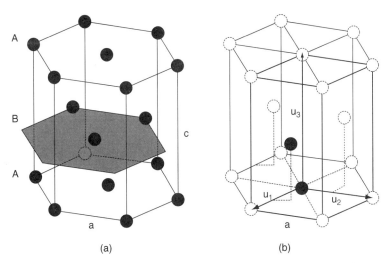

(a)   (b)

Figure 1.17. (*a*) Hexagonal close-packed (HCP) crystal structure based on the $A-A_{12}$(hex) bonding unit; (*b*) primitive unit cell for HCP with a basis of two A atoms at the positions (0,0,0) and $(\frac{2}{3}, \frac{1}{3}, \frac{1}{2})$, where the coordinates refer to the u_1, u_2, and u_3 axes, as shown. (After C. Kittel, *Introduction to Solid State Physics*, 7th ed., copyright 1995 by John Wiley & Sons, Inc. Reprinted by permission of John Wiley & Sons, Inc.)

TABLE 1.3   Elemental Crystals with the HCP Crystal Structure

Element	a (nm)[a]	c/a	Element	a (nm)[a]	c/a
Be	0.229	1.568	Mg	0.321	1.624
Ti	0.295	1.587	Co	0.251	1.623
Zn	0.267	1.856	Zr	0.323	1.593
Tl	0.346	1.598	Gd	0.363	1.591
Ho	0.358	1.570			

[a]Lattice constants are values at room temperature.

bonding between atoms in the solid state involves nonspherical electron orbitals. Some examples of elemental crystals with the HCP crystal structure are given in Table 1.3, along with the lattice constants a and c/a ratios observed at room temperature (as is the case for all the lattice constants given in tables in this chapter). Information on the structures and lattice constants of crystals of the elements is given in Table 1.4. Metallic bonding tends to dominate in elemental crystals that have the HCP crystal structure, with the result that the A atoms are actually positively charged ions (cations). An exception is the inert-gas solid He, where van der Waals bonding occurs between neutral He atoms.

As mentioned earlier, the packing fraction for the $A-A_{12}$(hex) bonding unit and therefore for the ideal HCP crystal structure has the value 0.74. An introduction to the calculation of packing fractions for important crystal structures is presented later.

Several metallic rare earth elements (e.g., La, Pr, Nd, Pm, and Am) can have crystal structures in which the stacking of the close-packed planes of atoms along the c axis is **ABACABAC**.... These crystal structures are known as *double HCP* (DHCP). In DHCP half the planes of atoms have the cubic stacking sequence (**ABCABC**...) while the other half have hexagonal stacking (**ABAB**...). Thus elements with the DHCP crystal structure contain both the $A-A_{12}$(cub) and $A-A_{12}$(hex) local bonding units. The DHCP crystal structure is analogous to the hexagonal 4H–SiC crystal structure in which the stacking of planes of atoms along the c axis is also **ABACABAC**..., where, in this case, $A = A(Si)A(C)$, and so on. The local bonding units in 4H–SiC, however, are tetrahedral $Si-C_4$ and $C-Si_4$ units.

Crystal Structures Based on $A-A_{12}$***(cub).*** The crystal structure based on the $A-A_{12}$(cub) bonding unit is the close-packed *FCC structure*, also known as *cubic close-packed* (ccp). In this crystal structure the central A atom and its 12 NN A atoms all lie on adjacent points of an FCC lattice. In the FCC crystal structure every atom is the center of an $A-A_{12}$(cub) unit. The planes of atoms in the bonding unit are parallel to the (111) planes of the lattice (Fig. 1.18). Choosing an FCC lattice for this crystal structure corresponds to a basis of a single A atom at (0,0,0), while choosing a SC lattice corresponds to a basis of four identical A atoms at (0,0,0), $(0\frac{1}{2},\frac{1}{2})$, $(\frac{1}{2},0,\frac{1}{2})$, and $(\frac{1}{2},\frac{1}{2},0)$. The latter choice confirms that there are four A atoms per conventional cubic unit cell.

The stacking of the (111) planes of atoms in the FCC crystal structure is **ABCABC**... since the atoms in the upper and lower planes within the $A-A_{12}$(cub) bonding unit are displaced from each other. Deviations from this ideal stacking sequence can occur and are known as *stacking faults* (see Chapter 4).

TABLE 1.4   Crystal Structures and Lattice Constants (nm) of the Elemental Solids[a]

1	2	3	4	5	6	7	8	9	10	11	12	13	14	15	16	17	18
1 14K H₂ HCP 0.378 0.616																	2 4K He HCP 0.356 0.580
3 Li BCC 0.351	4 Be HCP 0.229 0.358			5 B trig. 1.017 65.1°	6 C graphite 0.246 0.671	7 36K N₂ cubic 0.566	8 55K O₂ cubic 0.683	9 54K F₂ cubic 0.667	10 30K Ne FCC 0.446								
11 Na BCC 0.429	12 Mg HCP 0.321 0.521			13 Al FCC 0.405	14 Si diamond 0.543	15 P ortho. 0.331 1.048 0.438	16 S ortho. 1.046 1.287 2.449	17 113K Cl₂ ortho. 0.624 0.448 0.826	18 34K Ar FCC 0.532								
19 K BCC 0.532	20 Ca FCC 0.559	21 Sc HCP 0.331 0.527	22 Ti HCP 0.295 0.468	23 V BCC 0.302	24 Cr BCC 0.288	25 Mn cubic 0.891	26 Fe BCC 0.287	27 Co HCP 0.251 0.407	28 Ni FCC 0.352	29 Cu FCC 0.361	30 Zn HCP 0.267 0.495	31 Ga ortho. 0.452 0.766 0.453	32 Ge diamond 0.566	33 As trig. 0.413 54.1°	34 Se hex. 0.437 0.495	35 123K Br₂ ortho. 0.668 0.449 0.874	36 116K Kr FCC 0.581
37 Rb BCC 0.571	38 Sr FCC 0.608	39 Y HCP 0.365 0.573	40 Zr HCP 0.323 0.515	41 Nb BCC 0.330	42 Mo BCC 0.315	43 Tc HCP 0.274 0.439	44 Ru HCP 0.271 0.428	45 Rh FCC 0.380	46 Pd FCC 0.389	47 Ag FCC 0.409	48 Cd HCP 0.298 0.562	49 In tetr. 0.325 0.595	50 Sn tetr. 0.583 0.318	51 Sb trig. 0.451 57.1°	52 Te hex. 0.446 0.593	53 I₂ ortho. 0.727 0.479 0.979	54 162K Xe FCC 0.635
55 Cs BCC 0.614	56 Ba BCC 0.502	57 La DHCP 0.377 1.217	72 Hf HCP 0.319 0.505	73 Ta BCC 0.330	74 W BCC 0.317	75 Re HCP 0.276 0.446	76 Os HCP 0.273 0.439	77 Ir FCC 0.384	78 Pt FCC 0.392	79 Au FCC 0.408	80 227K Hg trig. 0.301 70.5°	81 Tl HCP 0.346 0.552	82 Pb FCC 0.495	83 Bi trig. 0.475 57.2°	84 Po SC 0.337	85 At —	86 Rd —
87 Fr —	88 Ra BCC 0.571	89 Ac FCC 0.531															

(6 C diamond 0.357)

58	59	60	61	62	63	64	65	66	67	68	69	70	71
Ce DHCP 0.368 FCC 0.508 1.186	Pr DHCP 0.367 1.183	Nd DHCP 0.366 1.180	Pm DHCP 0.365 1.165	Sm trig. 0.900 23.2°	Eu BCC 0.458	Gd HCP 0.363 0.578	Tb HCP 0.361 0.570	Dy HCP 0.359 0.565	Ho HCP 0.358 0.562	Er HCP 0.356 0.559	Tm HCP 0.354 0.555	Yb FCC 0.548	Lu HCP 0.351 0.555

90	91	92	93	94	95	96	97	98	99	100	101	102	103
Th FCC 0.508	Pa tetr. 0.392 0.324	U ortho. 0.285 0.587 0.495	Np ortho. 0.666 0.472 0.487	Pu mono. 0.618 0.482 1.096	Am DHCP 0.347 1.124	Cm DHCP 0.350 1.133	Bk DHCP 0.342 1.107	Cf DHCP 0.339 1.102	Es DHCP ?	Fm —	Md —	No —	Lr —

Source: Data from D. R. Lide, ed.-in-chief, *CRC Handbook of Chemistry and Physics*, 79th ed. (CRCnetBASE 1999), CRC Press, Boca Raton, Fla., 1999, pp. 12–19 to 12–21; W. B. Pearson, *The Crystal Chemistry and Physics of Metals and Alloys*, Wiley-Interscience, New York, 1972; and R. W. G. Wyckoff, *Crystal Structures*, Vol. II, 2nd ed., Interscience, New York, 1963.

[a]Values correspond to room temperature, unless stated otherwise.

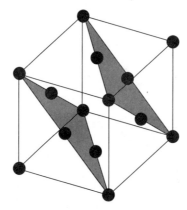

Figure 1.18. Cubic unit cell of the face-centered cubic (FCC) crystal structure based on the $A-A_{12}(cub)$ bonding unit. The planes of atoms in the bonding unit are the (111) planes of the FCC structure.

TABLE 1.5 Elemental Crystals with the FCC Crystal Structure

Element	a (nm)[a]	Element	a (nm)[a]
Al	0.405	Ca	0.559
Ni	0.352	Cu	0.361
Pd	0.389	Ag	0.409
Ir	0.384	Au	0.408
Pb	0.495	Yb	0.548

[a]Lattice constants are values at room temperature.

Some elemental crystals with the FCC crystal structure are listed in Table 1.5. Metallic bonding dominates in these crystals and the A atoms are positively charged ions.

An important property of both hexagonal and cubic close-packed arrays of atoms is the presence of unoccupied *interstitial sites* in the HCP and FCC crystal structures. Even though close-packed arrays of atoms have the maximum possible packing fraction of 0.74 for spheres of a given radius, the fact remains that 26% of the volume of the crystal is in principle still available to be filled, for example, by smaller atoms. Examples of this filling of interstitial sites are given later. These interstitial sites are of two types, tetrahedral and octahedral, according to the symmetry of the NN close-packed atoms surrounding the site.

The tetrahedral and octahedral interstitial sites can easily be identified within the cubic unit cell of the FCC crystal structure (Fig. 1.19). Here the tetrahedron and octahedron of close-packed atoms surrounding the corresponding interstitial sites are shown. A tetrahedral interstitial site is located at the $(\frac{1}{4},\frac{1}{4},\frac{1}{4})$ position within the unit cell. Eight tetrahedral interstitial sites surround each atom in the FCC crystal structure, with the sites oriented along the set of eight $\langle 111 \rangle$ directions.

As shown in Fig. 1.19, an octahedral interstitial site is located at the body-centered position of the unit cell. Identical interstitial sites are also located at the centers of

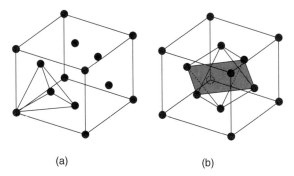

(a)   (b)

Figure 1.19. Unoccupied interstitial sites in the FCC crystal structure: (*a*) example of an interstitial site with tetrahedral symmetry located at the $(\frac{1}{4},\frac{1}{4},\frac{1}{4})$ position within the cubic unit cell; (*b*) interstitial site with octahedral symmetry located at the body-centered position of the cubic unit cell. Identical interstitial sites are also located at the centers of the cube edges.

the cube edges. Each atom in the FCC crystal structure is surrounded by six of these octahedral interstitial sites, which are oriented along the six $\langle 100 \rangle$ directions. As a result, there are two tetrahedral and one octahedral interstitial site per FCC atom. The same is true for the HCP crystal structure, although the interstitial sites are arranged differently around each atom. The occupation of these interstitial sites by cations occurs in minerals where the close-packed array consists of O^{2-} anions. Crystal structures based on A–A$_{12}$(icos) and A–B$_{12}$(icos) bonding units are described in Chapter W1, at our Web site.

Crystal Structures Based on A–A$_8$. The crystal structure based on the A–A$_8$ bonding unit is obtained by placing atoms on every point of the BCC lattice (Fig. 1.20). This results in a BCC crystal structure in which every atom is at the center of an A–A$_8$ bonding unit. When a BCC lattice is chosen for this structure, the basis is simply an A atom at (0,0,0). There are, however, two atoms per conventional cubic unit cell

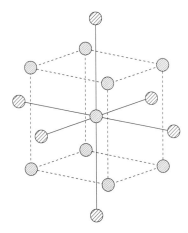

Figure 1.20. Body-centered cubic (BCC) crystal structure based on the A–A$_8$ bonding unit. The eight NN A atoms and the six second-NN A atoms to the central A atom are shown.

TABLE 1.6   Elemental Crystals with the BCC Crystal Structure

Element	a (nm)[a]	Element	a (nm)[a]
Li	0.351	Na	0.429
K	0.532	V	0.302
Cr	0.288	Fe	0.287
Rb	0.571	Nb	0.330
Mo	0.315	Cs	0.614
Ba	0.502	Ta	0.330
W	0.317	Eu	0.458

[a]Lattice constants are values at room temperature.

of this structure, located at (0,0,0) and $(\frac{1}{2},\frac{1}{2},\frac{1}{2})$. The packing fraction for the BCC crystal structure is 0.68, only about 10% less than the value of 0.74 for HCP and FCC. As shown in Fig. 1.20, there are six second-NN A atoms arranged octahedrally with respect to the central A atom.

Some elemental crystals with the BCC crystal structure are listed in Table 1.6. Metallic bonding dominates in these crystals and the A atoms are positively charged ions.

Vacant interstitial sites are also present in BCC crystals (Fig. 1.21). There are six distorted octahedral interstitial sites per cubic unit cell, in the middle of each of the six faces (shared by two cells) and also at the midpoint of each of the 12 edges (shared by four cells). In addition, there are six distorted tetrahedral interstitial sites per cubic unit cell, two in each of the six faces. These octahedral and tetrahedral interstitial sites are distorted because the distances from the site to the surrounding atoms are not all the same. The maximum radii of hard-sphere atoms that can occupy the interstitial sites in BCC crystal structures are smaller than the corresponding radii in FCC crystal

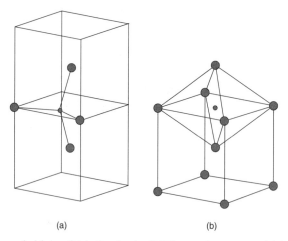

(a)   (b)

Figure 1.21. Unoccupied interstitial sites in the BCC crystal structure: (*a*) interstitial site with distorted tetrahedral symmetry; (*b*) interstitial site with distorted octahedral symmetry located at the face-centered position of the cubic unit cell. Other distorted interstitial sites are located at the centers of the cube edges.

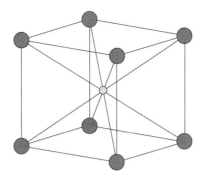

Figure 1.22. Cesium chloride (CsCl) crystal structure based on the $A-B_8$ bonding unit.

structures despite the fact that FCC has a higher packing fraction than BCC. Interstitial carbon atoms in BCC (and FCC) Fe play a crucial role in the properties of cast iron and steels (see Chapter 12).

Crystal Structures Based on A–B$_8$. Crystals that are based on the $A-B_8$ bonding unit are typically binary compounds with a SC lattice and a basis of two atoms in the cubic unit cell: an A atom at (0,0,0) and a B atom at $(\frac{1}{2},\frac{1}{2},\frac{1}{2})$, or vice versa; (Fig. 1.22). This *cesium chloride* (CsCl) crystal structure can be viewed as two interpenetrating SC lattices, with A atoms on one SC lattice and B atoms on the other lattice, displaced from the first by one-half of a body diagonal. As a result, each B atom is at the center of a $B-A_8$ bonding unit that has the same symmetry as that of the $A-B_8$ unit.

The bonding in crystals with the CsCl crystal structure has a strong ionic component. It is therefore important to recognize that, for example, if the A atom is a positively charged ion (cation) such as Cs^+, the B atom will be a negatively charged ion (anion) such as Cl^-. Examples of crystals based on the $A-B_8$ bonding unit and having the CsCl crystal structure are listed in Table 1.7.

Crystal Structures Based on A-A$_6$. When the central A atom of an $A-A_6$ unit is placed at a lattice point, the six NN A atoms will lie on points of a SC lattice. The resulting crystal structure is SC, with every A atom at the center of an $A-A_6$ unit. The SC crystal structure has a packing fraction of only 0.52 and is therefore a rather open structure with a vacant interstitial site with cubic symmetry located at the body-centered position. Since this site could be occupied by other A atoms, as in BCC crystals, or by B atoms, as in crystals with the CsCl structure, it is not surprising that at most one element, polonium (Po), has the SC crystal structure.

TABLE 1.7   Crystals with the CsCl Crystal Structure

Compound	a (nm)[a]	Compound	a (nm)[a]
CsCl	0.411	NH_4Cl	0.387
CuZn (β-brass)	0.294	AlNi	0.288
TlBr	0.397	BeCu	0.270

[a]Lattice constants are values at room temperature.

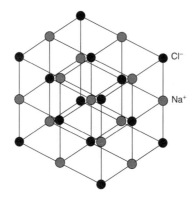

Figure 1.23. Sodium chloride (NaCl) crystal structure based on the A–B_6 bonding unit.

Crystal Structures Based on A–B_6. When the central A atoms of A–B_6 bonding units are placed on the lattice points of an FCC lattice, with the six NN B atoms oriented along the six ⟨100⟩ directions, the resulting crystal structure (Fig. 1.23) is *NaCl*, or *rocksalt*. This crystal structure, in fact, consists of two interpenetrating FCC lattices with A atoms on one lattice and the B atoms on the other lattice. The A and B lattices are displaced from each other by, for example, one half of the lattice constant *a* along the [100] direction or by one half of the body diagonal in the [111] direction. The basis for this structure consists of an A atom at (0,0,0) and a B atom at either $(\frac{1}{2},0,0)$ or $(\frac{1}{2},\frac{1}{2},\frac{1}{2})$. As a result, B atoms are also at the centers of B–A_6 units having the same symmetry as the A–B_6 units.

Another useful way to view the NaCl crystal structure is to recognize that the larger Cl^- anions lie on a close-packed FCC lattice containing two tetrahedral and one octahedral interstitial site per Cl^- ion, as described earlier for the FCC crystal structure. When the smaller Na^+ cations are placed in all the octahedral interstitial sites, the resulting crystal structure is electrically neutral (as required), contains both Na^+–Cl_6^- and Cl^-–Na_6^+ bonding units, and is in fact the NaCl crystal structure. It is apparent that this arrangement of two oppositely charged ions in the NaCl crystal structure makes use of the efficient close-packing solution to filling space.

Table 1.8 presents examples of important crystals with the NaCl crystal structure. Since ionic bonding dominates in these crystals, the A atoms may be considered to be the cations and the B atoms, the anions.

Crystal Structures Based on A–A_4. A–A_4 bonding units can be placed at the lattice points of an FCC lattice so that every A atom is at the center of a tetrahedral

TABLE 1.8   Crystals with the NaCl Crystal Structure

Compound	a (nm)[a]	Compound	a (nm)[a]
NaCl	0.563	LiH	0.408
AgBr	0.577	MgO	0.420
PbS	0.592	MnO	0.443
KCl	0.629	KBr	0.659

[a]Lattice constants are values at room temperature.

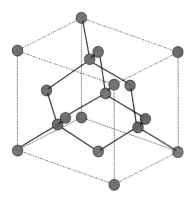

Figure 1.24. Diamond crystal structure based on the $A-A_4$ bonding unit.

$A-A_4$ unit. The resulting crystal structure (Fig. 1.24), known as the *diamond crystal structure*, consists of two interpenetrating FCC lattices of A atoms separated from each other by one-fourth of a body diagonal (i.e., along the [111] direction). With the choice of an FCC lattice the basis consists of two A atoms at (0,0,0) and $(\frac{1}{4},\frac{1}{4},\frac{1}{4})$ within the cubic unit cell. There are eight A atoms per conventional cubic unit cell and the structure is quite open, with a packing fraction of only 0.34. There are a total of eight interstitial sites with tetrahedral symmetry in the cubic unit cell, four on the body diagonal [e.g., at $(\frac{3}{4},\frac{3}{4},\frac{3}{4})$], three on the cube edges, and one in the body-centered position. The stacking of the (111) planes of A atoms in the diamond crystal structure is **AABBCCAABBCC...**, which is a doubling of the FCC sequence of **ABCABC...**. Another view of the diamond crystal structure corresponds to an FCC lattice of A atoms with additional A atoms placed at four of the eight otherwise vacant tetrahedral interstitial sites.

The four elements having the diamond crystal structure are listed in Table 1.9. The bonding in these crystals is covalent. It should be noted that the stable form of Sn at room temperature is white Sn (β-Sn), which has a tetragonal structure and is metallic. Gray Sn (α-Sn) is semimetallic.

A hexagonal crystal structure based on the $A-A_4$ unit which differs from the cubic diamond crystal structure in the distribution of second-NN atoms is also possible, and for the case of carbon is known as *lonsdaleite*. In this "hexagonal diamond" the stacking sequence of planes of atoms is **AABBAABB...**, which is a doubling of the **ABAB...** sequence for the HCP crystal structure.

Crystal Structures Based on $A-B_4$. $A-B_4$ bonding units can be placed at the lattice points of an FCC lattice in the same way as was done for the $A-A_4$ units

TABLE 1.9 Elemental Crystals with the Diamond Crystal Structure

Element	a (nm)^a	Element	a (nm)^a
C	0.3567	Si	0.543
Ge	0.5657	Sn (gray)	0.649

^{a}Lattice constants are values at room temperature.

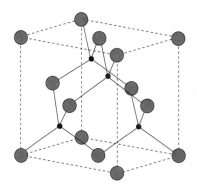

Figure 1.25. Zincblende (sphalerite or cubic ZnS) crystal structure based on the A−B$_4$ bonding unit.

of the diamond crystal structure. The resulting crystal structure, known as *zincblende (sphalerite)* or cubic ZnS, consists of a FCC lattice of A atoms and a second FCC lattice of B atoms displaced from each other by one-fourth of a body diagonal, along the [111] direction (Fig. 1.25). This crystal structure possesses both A−B$_4$ and B−A$_4$ tetrahedral units. With the choice of the FCC lattice the basis consists of an A atom at $(0,0,0)$ and a B atom at $(\frac{1}{4},\frac{1}{4},\frac{1}{4})$, with four A and four B atoms per conventional cubic unit cell. The stacking sequence for the (111) planes of the cubic ZnS structure is

$$\mathbf{A}(Zn)\mathbf{A}(S)\mathbf{B}(Zn)\mathbf{B}(S)\mathbf{C}(Zn)\mathbf{C}(S)\mathbf{A}(Zn)\mathbf{A}(S)\mathbf{B}(Zn)\mathbf{B}(S)\mathbf{C}(Zn)\mathbf{C}(S)\dots,$$

where alternating (111) planes consist of either Zn or S atoms. Cubic ZnS can also be thought of as consisting of an FCC lattice of the larger S "anions" with four of the eight otherwise vacant tetrahedral interstitial sites occupied by the smaller Zn "cations".

Important examples of compounds that have the zincblende or cubic ZnS crystal structure are presented in Table 1.10. Both semiconductors and insulators are represented here, with bonding that has both ionic and covalent components. Additional examples are given in Table 11.9.

A hexagonal crystal structure based on the A−B$_4$ bonding unit also exists and is analogous to the hexagonal crystal structure discussed earlier for the A−A$_4$ unit. This structure, known as *wurtzite* or hexagonal ZnS (Fig. 1.26), can be seen also to consist of tetrahedral A−B$_4$ and B−A$_4$ bonding units. In the wurtzite structure the S "anions" lie on an HCP lattice in which the Zn "cations" occupy four of the eight otherwise vacant

TABLE 1.10   Crystals with the Zincblende (Cubic ZnS) Crystal Structure

Compound	a (nm)^a	Compound	a (nm)^a
ZnS	0.541	ZnSe	0.567
β-SiC	0.435	GaAs	0.565
CdS	0.583	AlAs	0.566
InSb	0.648	AlP	0.545
BN	0.362	GaP	0.545
BP	0.454		

^{a}Lattice constants are values at room temperature.

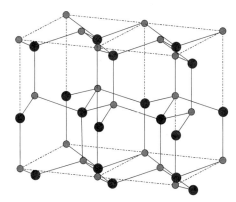

Figure 1.26. Wurtzite (hexagonal ZnS) crystal structure based on the $A-B_4$ bonding unit. (After R. J. Borg and G. J. Dienes, *The Physical Chemistry of Solids*, copyright 1992. Reprinted by permission of Academic Press, Inc.)

tetrahedral interstitial sites. With a hexagonal lattice, the basis of atoms for wurtzite within the hexagonal unit cell (see Fig. 1.17) can be chosen to be A atoms at $(0,0,0)$ and $(\frac{2}{3},\frac{1}{3},\frac{1}{2})$, as for the HCP structure, but also with B atoms at $(0,0,u)$ and $(\frac{2}{3},\frac{1}{3},\frac{1}{2}+u)$ where $u = \frac{3}{8}$ when $c/a = \sqrt{\frac{8}{3}}$. The stacking of planes of atoms perpendicular to the c axis for hexagonal ZnS occurs in the sequence

$$A(Zn)A(S)B(Zn)B(S)A(Zn)A(S)B(Zn)B(S)\ldots.$$

Examples of binary compounds with the wurtzite crystal structure are also presented in Table 11.9.

Crystal Structures Based on $A-A_3$. When the central A atom of a planar $A-A_3$ unit is placed at every point of a HCP lattice, the *graphite* crystal structure results (Fig. 1.27). In this crystal structure every A atom is at the center of a planar $A-A_3$

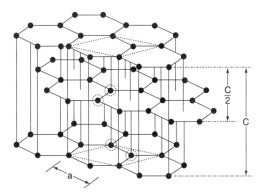

Figure 1.27. Graphite crystal structure based on the $A-A_3$ bonding unit. The hexagonal unit cell containing four identical atoms at $(0,0,0)$, $(\frac{1}{3},\frac{2}{3},0)$, $(0,0,\frac{1}{2})$, and $(\frac{2}{3},\frac{1}{3},\frac{1}{2})$ is indicated. (After R. J. Borg and G. J. Dienes, *The Physical Chemistry of Solids*, copyright 1992. Reprinted by permission of Academic Press, Inc.)

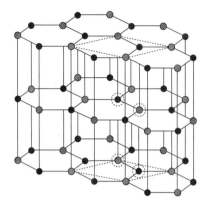

Figure 1.28. Crystal structure of hexagonal BN based on the A−B$_3$ bonding unit. The hexagonal unit cell containing two identical A atoms at (0,0,0) and ($\frac{1}{3}$,$\frac{2}{3}$,$\frac{1}{2}$) and two identical B atoms at ($\frac{1}{3}$,$\frac{2}{3}$,0) and (0,0,$\frac{1}{2}$) is indicated.

unit and the stacking sequence of planes of atoms along the c axis is **ABABAB...** . The lattice constants for graphite are $a = 0.246$ nm and $c = 0.671$ nm. The resulting c/a ratio of 2.73 is quite high and results from the positioning of one-half of the atoms in a given layer directly above and below atoms in the adjacent layers. Choosing a hexagonal lattice and unit cell, the basis of atoms for this structure consists of four A atoms at (0,0,0), ($\frac{1}{3}$,$\frac{2}{3}$,0), (0,0,$\frac{1}{2}$), and ($\frac{2}{3}$,$\frac{1}{3}$,$\frac{1}{2}$). No other element has the graphite crystal structure.

Crystal Structures Based on A−B$_3$. A crystal structure similar to graphite can be formed with every A atom at the center of an planar A−B$_3$ unit and every B atom at the center of a planar B−A$_3$ unit (Fig. 1.28). The stacking sequence of planes of atoms along the c axis is again **ABABAB...** . The basis of atoms corresponding to a hexagonal lattice and unit cell consists of A atoms at (0,0,0) and ($\frac{1}{3}$,$\frac{2}{3}$,$\frac{1}{2}$), as for HCP, and B atoms at (0,0,$\frac{1}{2}$) and ($\frac{1}{3}$,$\frac{2}{3}$,0). The stoichiometry of crystals with this structure is therefore AB. Note that unlike atoms lie above and below each other in this structure. A crystal with this structure is hexagonal boron nitride (α-BN), which is isoelectronic with graphite. With lattice constants $a = 0.2504$ nm and $c = 0.6661$ nm, the c/a ratio for hexagonal BN is 2.66, quite close to that of graphite. Although the crystal structure is similar in appearance to that of graphite, the relative arrangement of atoms in adjacent planes is very different.

It is found that solids typically take on crystal structures that have the highest packing fraction and therefore the highest density consistent with the bonding present in the crystal. These two properties of crystals are defined and illustrated next.

1.8   Packing Fractions and Densities

Packing Fractions. The *packing fraction* of a crystal structure corresponds to the fraction of space filled by its atoms. The assumptions typically made are that the atoms

Each entry lists: atomic number (and temperature where noted), symbol, density (10^3 kg/m^3), atomic concentration (10^{28} atoms/m^3).

Z	Symbol	Density	Atomic Conc.
1 (14K)	H	0.088	5.26
2 (4K)	He	0.209	3.15
3	Li	0.533	4.63
4	Be	1.84	12.33
5	B	2.28	12.70
6	C (graphite)	2.27	11.37
6	C (diamond)	3.52	17.63
7 (30K)	N	1.03	4.41
8 (55K)	O	1.33	5.02
9 (54K)	F	1.70	5.39
10 (30K)	Ne	1.51	4.50
11	Na	0.967	2.53
12	Mg	1.74	4.30
13	Al	2.70	6.02
14	Si	2.33	5.00
15	P	2.71	5.27
16	S	2.07	3.88
17 (113K)	Cl	2.04	3.64
18 (34K)	Ar	1.77	2.66
19	K	0.862	1.33
20	Ca	1.53	2.29
21	Sc	2.99	4.00
22	Ti	4.51	5.66
23	V	6.12	7.23
24	Cr	7.19	8.33
25	Mn	7.47	8.19
26	Fe	7.87	8.49
27	Co	8.84	9.03
28	Ni	8.91	9.14
29	Cu	8.94	8.47
30	Zn	7.14	6.57
31	Ga	5.91	5.11
32	Ge	5.33	4.42
33	As	5.78	4.65
34	Se	4.81	3.67
35 (123K)	Br	4.05	3.05
36 (116K)	Kr	2.84	2.04
37	Rb	1.53	1.08
38	Sr	2.58	1.78
39	Y	4.47	3.03
40	Zr	6.51	4.30
41	Nb	8.58	5.56
42	Mo	10.22	6.42
43	Tc	11.4	7.01
44	Ru	12.36	7.37
45	Rh	12.43	7.27
46	Pd	12.00	6.79
47	Ag	10.51	5.86
48	Cd	8.64	4.63
49	In	7.28	3.82
50	Sn	7.29	3.70
51	Sb	6.69	3.31
52	Te	6.24	2.94
53	I	4.94	2.35
54 (162K)	Xe	3.41	1.56
55	Cs	1.91	0.864
56	Ba	3.60	1.58
57	La	6.15	2.66
58	Ce	6.69	2.87
59	Pr	6.77	2.89
60	Nd	7.01	2.93
61	Pm	7.17	2.98
62	Sm	7.54	3.02
63	Eu	5.24	2.08
64	Gd	7.90	3.03
65	Tb	8.23	3.12
66	Dy	8.55	3.17
67	Ho	8.80	3.21
68	Er	9.05	3.26
69	Tm	9.32	3.32
70	Yb	6.97	2.42
71	Lu	9.84	3.39
72	Hf	13.28	4.48
73	Ta	16.68	5.55
74	W	19.25	6.31
75	Re	21.01	6.80
76	Os	22.22	7.03
77	Ir	22.56	7.07
78	Pt	21.45	6.62
79	Au	19.29	5.90
80	Hg	14.26	4.28
81	Tl	11.87	3.50
82	Pb	11.35	3.30
83	Bi	9.81	2.83
84	Po	9.10	2.62
85	At	—	—
86	Rd	—	—
87	Fr	—	—
88	Ra	5.50	1.47
89	Ac	10.07	2.67
90	Th	11.73	3.04
91	Pa	15.43	4.02
92	U	19.05	4.82
93	Np	20.48	5.20
94	Pu	20.3	5.00
95	Am	13.8	3.42
96	Cm	13.7	3.35
97	Bk	14.7	3.58
98	Cf	15.2	3.65
99	Es	—	—
100	Fm	—	—
101	Md	—	—
102	No	—	—
103	Lr	—	—

^{a}Values correspond to room temperature, unless stated otherwise, and have been calculated using the structural data presented in Table 1.4.

can be represented by hard spheres that are in contact with each other. The definition used here is

$$\text{packing fraction (PF)} = \frac{\text{(no. atoms per unit cell)(volume per atom)}}{\text{unit cell volume}}$$

$$= \frac{N(\text{atom})V(\text{atom}).}{V(\text{unit cell})} \tag{1.5}$$

This equation can easily be generalized for crystal structures containing more than one type of atom, as can be seen in Chapter W1. The volume $V(\text{atom}) = 4\pi r^3/3$ for spherical atoms of radius r.

As an example of the calculation of packing fractions, consider the simple cubic (SC) crystal structure. With one atom per cubic unit cell, it follows that $N(\text{atom}) = 1$ and $V(\text{unit cell}) = a^3$. To find the relationship between the atom radius r and the lattice constant a, note that two atoms are in contact along the edge of the cube in the [001] direction, as shown in Fig. W1.2a at our Web site. It follows that $a = 2r$, so that $V(\text{atom}) = \pi a^3/6$. Therefore,

$$\text{PF(SC)} = \frac{(1)(\pi a^3/6)}{a^3} = \frac{\pi}{6} = 0.52. \tag{1.6}$$

This relatively low PF for the SC crystal structure is due primarily to the presence of a vacant cubic interstitial site at the body-centered position.

Densities. For a crystal with a single type of atom, the number density or concentration of atoms $n(\text{atom})$ and the mass density ρ are defined as

$$n(\text{atom}) = \frac{N(\text{atom})}{V(\text{unit cell})}, \tag{1.7}$$

$$\rho(\text{kg/m}^3) = \frac{N(\text{atom})m(\text{atom})}{V(\text{unit cell})} = n(\text{atom})m(\text{atom}). \tag{1.8}$$

The generalizations to crystals with more than one type of atom is obvious.

The mass densities and atomic concentrations of crystals of the elements are given in Table 1.11. The crystalline element with the highest concentration of atoms is C in the form of diamond with $n = 1.76 \times 10^{29}$ atoms/m^3, a somewhat surprising result given the low packing fraction of 0.34 for this crystal structure. Crystals of Cs have the lowest atom concentrations, less than 1×10^{28} atoms/m^3, due to the large size of the atoms. Mass densities are found to be highest for crystals of the close-packed 5d transition metals Os and Ir, where ρ is over 22,000 kg/m^3. The elemental crystal with the lowest mass density is Li, with $\rho = 533$ kg/m^3.

REFERENCES

Ashcroft, N. W., and N. D. Mermin, *Solid State Physics*, Saunders College, Philadelphia, 1976.

Borg, R. J., and G. J. Dienes, *The Physical Chemistry of Solids*, Academic Press, San Diego, Calif., 1992.

Burns, G., *Solid State Physics*, Academic Press, San Diego, Calif., 1985.

Burns, G., and A. M. Glazer, *Space Groups for Solid State Scientists*, 2nd ed., Academic Press, San Diego, Calif., 1990.

Christman, J. R., *Fundamentals of Solid State Physics*, Wiley, New York, 1988.

Jaffe, H. W., *Crystal Chemistry and Refractivity*, Dover, Mineola, N.Y., 1996.

Kittel, C., *Introduction to Solid State Physics*, 7th ed., Wiley, New York, 1996.

McKie, D., and C. McKie, *Crystalline Solids*, Wiley, New York, 1974.

Pearson, W. B., *The Crystal Chemistry and Physics of Metals and Alloys*, Wiley, New York, 1972.

Phillips, J. C., *Bonds and Bands in Semiconductors*, Academic Press, San Diego, Calif., 1973.

Wyckoff, R. W. G., *Crystal Structures*, I–III., 2nd ed., Interscience, New York, 1963.

PROBLEMS

1.1 Using the primitive translation vectors given in Figs. 1.4 and 1.5, calculate the volumes of the BCC and FCC primitive unit cells.

1.2 Show that the face-centered tetragonal lattice is equivalent to the body-centered tetragonal lattice.

1.3 Show that the spacing $d(hkl)$ between adjacent lattice planes with Miller indices (hkl) is equal to $a/\sqrt{h^2 + k^2 + l^2}$ for cubic Bravais lattices and to $1/\sqrt{h^2/a^2 + k^2/b^2 + l^2/c^2}$ for orthorhombic Bravais lattices.

1.4 Calculate the densities of lattice points in the (100), (110), (111), and (hkl) lattice planes of a simple cubic lattice.

1.5 Write a computer program that will determine the distance $d(n)$ from a given atom to the nth nearest neighbor (NN) in a Bravais lattice. Also compute $N(n)$, the number of nth NNs. Carry out the calculation for the SC, BCC, and FCC lattices.

1.6 Calculate the packing fractions for the following crystal structures: FCC, HCP, and diamond.

1.7 Show that the B atoms in an $A-B_8$ bonding unit come into contact with each other when $r_B = 1.366 r_A$ [i.e., when $r_A = (\sqrt{3} - 1)r_B$]. Here r_A and r_B are the radii of the hard-sphere A and B atoms, respectively. Find the analogous conditions on the radii for the $A-B_6$ and $A-B_4$ bonding units.

1.8 Prove for hard-sphere atoms in the HCP crystal structure that $c/a = \sqrt{8/3} = 1.633$.

1.9 Assuming that the atoms in the FCC crystal structure are hard spheres of radius R in contact with each other, calculate the maximum radii r of the smaller hard-sphere atoms that could occupy the octahedral and tetrahedral interstitial sites in the FCC crystal structure.

Note: An additional problem is given in Chapter W1.

CHAPTER 2

Bonding in Solids

2.1 Introduction

The bonding between atoms in solid-state materials is dominated by *valence electrons*. The electrons in filled shells are more tightly bound to the nuclei and so interact only weakly with nearest-neighbor (NN) atoms. For example, in the case of the Na atom with 11 electrons and a ground-state electron configuration given by $1s^2 2s^2 2p^6 3s^1$, the $1s$, $2s$, and $2p$ electrons are in filled shells. Only the single valence electron in the $3s$ level is ordinarily available for bonding.

In this chapter the nature of the bonding between atoms in the solid state that contributes to the cohesive energies of solids is explored. The *cohesive energy* of a crystal or, more generally, of a solid is defined to be the energy required to overcome the bonding and separate the constituent atoms or ions from each other completely. One of the goals of this chapter is to attempt to explain on the basis of bonding why the local atomic bonding units and crystal structures discussed in Chapter 1 are formed. The goals of this chapter are similar to those of the subject of crystal chemistry (i.e., the study of how atoms become ordered into various crystal structures and the role that the valence electrons play in binding the atoms together in the solid state).[†]

A very general way of illustrating the interactions leading to bonding between pairs of atoms is through the interatomic potential energy $U(r)$, where r is the internuclear distance. The potential $U(r)$ includes the attractive Coulomb interactions between the electrons and the nuclei as well as the repulsive interactions between the electrons and between the nuclei. The net result of these interactions for the H_2 molecule is shown in Fig. 2.1, where the experimental and predicted $U(r)$ curves for two H atoms are presented. The two lower curves correspond to the *bonding* state and the upper curve to the *antibonding* state.

Several features of the potential energy $U(r)$ shown in Fig. 2.1 should be emphasized:

1. For two isolated, noninteracting H atoms, $U(r) \to 0$ as $r \to \infty$, as expected.

2. For the lower, bonding state, the spins of the two electrons associated with the H atoms are antiparallel. This corresponds to the *singlet* spin state of the system of two H atoms. In the upper, antibonding state, the spins of the two electrons are parallel, corresponding to the *triplet* state, and $U(r)$ is predicted to decrease

---

[†] The subject of crystal chemistry is presented more completely in the books by Jaffe (1996) and Pearson (1972).

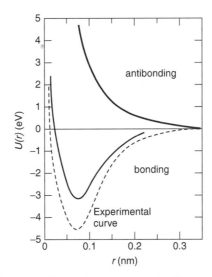

Figure 2.1. Experimental and predicted interatomic potential energy $U(r)$ curves for two hydrogen atoms are shown. The two lower curves correspond to the bonding state and the upper curve to the antibonding state. (After R. J. Borg and G. J. Dienes, *The Physical Chemistry of Solids*, copyright 1992. Reprinted by permission of Academic Press, Inc.)

monotonically with increasing r (except for large r, where there is a weak van der Waals attractive interaction). In this case the system of two interacting H atoms would not be bound.

3. The minimum in $U(r)$ indicates that the H_2 molecule is, in fact, stable and determines the equilibrium separation of the H atoms.

4. The equilibrium separation of the two atoms occurs at $R(H–H) = 0.070$ nm. The depth of the potential well at $r = R(H–H)$ is the bonding or equilibrium *dissociation energy* of the H_2 molecule, $E(H–H) = 4.52$ eV (aside from a zero-point energy correction due to vibration).

5. The rapid rise in $U(r)$ for $r < R(H–H)$ results from the increasing electrostatic repulsion of the two positively charged hydrogen nuclei (i.e., the two protons). In atoms with $Z > 1$, the electrostatic repulsion resulting from the overlap of electrons in closed shells also contributes to the rapid rise in $U(r)$. When the wavefunctions of electrons in closed shells on two different atoms start to overlap, the energies of the participating electrons are modified. The new energy levels for the electrons are often found to lie higher in energy than the original levels. As a result, work must be done to give electrons the additional energy needed in the new levels. The electrostatic repulsion and the need to move electrons to higher energies are therefore closely related. The repulsion occurring between atoms as a result of the spatial overlap of electrons in closed shells is often attributed to the *Pauli exclusion principle*. Since this principle states that multiple occupancy of states is not allowed, it again follows that electrons must be promoted to higher energy levels.

Although the potential energy $U(r)$ shown in Fig. 2.1 is valid in detail only for two H atoms, the existence of a potential well and of a repulsive interaction at smaller separations are also important aspects of the interatomic potential energy found in the bulk and at the surfaces of solids. For example, the repulsive interaction leads to the low bulk compressibility found in solids.

The interactions in the solid state that contribute to the existence of a potential well in the interatomic potential energy $U(r)$ are ultimately due to the attractive electrostatic forces between the negatively charged electrons and the positively charged nuclei. In principle, the spatial distributions and energies of the bonding electrons can be calculated using quantum mechanics and the many-electron Schrödinger equation. One replaces the many-electron problem by an approximate one-electron problem, in which the electron interacts with a self-consistent potential proportional to $U(r)$. In practice this approach does not yet give generally satisfactory results, due to the computational difficulties related to the large number of electrons and nuclei in a solid. The usual approach is to develop instead approximate theories which can predict some of the chemical and physical trends in classes of crystals (such as the alkali halides or FCC metals) in which similar bonding mechanisms dominate.

The five distinct mechanisms for bonding in solids discussed in this chapter are *covalent, metallic, van der Waals, ionic*, and *hydrogen bonding*. The depth of the potential well (i.e., the strength of the bonding) and the equilibrium separation of the atoms clearly depend on the details of each type of bonding. The general features of each bonding mechanism are presented, with more detail to be given in subsequent chapters, where specific classes of materials are discussed. For example, the relevance of covalent bonding to semiconductors, ceramics, and polymers is discussed in Chapters 11, 13, and 14, respectively, while metallic bonding is discussed further in Chapter 12.

To understand the bonding in solid-state materials it is important to be familiar with some aspects of the properties of electrons in *atomic orbitals*, in *hybrid orbitals*, and in *molecular orbitals*. These topics are presented and discussed in Chapter W2, at our Web site.[†] If the reader is unfamiliar with or needs to review these topics, it is recommended that our home page be consulted now.

The bonding between identical atoms in elemental solids is discussed first, followed by a discussion of the bonding between unlike atoms in multielement solids.

BONDING IN ELEMENTAL SOLIDS

Bonding in elemental solids is generally attributed to three distinct mechanisms, which result from three distinctly different types of behavior of the valence electrons as isolated atoms interact to form the solid. In both covalent and metallic bonding the wavefunctions of the valence electrons are delocalized and spread throughout the solid, but with some important differences. In solids in which covalent bonding dominates, the valence electron density is highest between pairs of NN atoms, corresponding to the *covalent bonds*. In solids in which metallic bonding dominates, the valence electron density is spread more uniformly throughout the solid. When each atom in the crystal is neutral and has a closed-shell electron configuration, the bonding results primarily from the van der Waals interaction.

[†] Supplementary material for this textbook is included on the Web at the resource site (ftp://ftp.wiley.com/ public/sci_tech_med/materials). Cross-references to elements of the Web material are prefixed by "W."

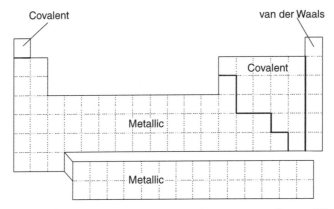

Figure 2.2. Format for the periodic table showing regions corresponding to different dominant bonding mechanisms in the elemental solids.

Covalent, metallic, and van der Waals bonding are discussed in turn. Examples are given of elemental solids in which each type of bonding dominates and also of elemental solids in which more than one type of bonding plays a significant role. The dominant bonding mechanisms are indicated for the elemental solids in Fig. 2.2 using the format of the periodic table. It can be seen that there exist distinct regions in which each bonding mechanism is dominant. This is a clear indication that the valence electron configuration of an atom plays a dominant role in the bonding of the atoms in the solid. This point will become clear as the discussion proceeds.

2.2 Covalent Bonding

When the wavefunctions of the valence electrons occupying unfilled shells on neighboring atoms start to overlap, there can arise an attractive interaction between the atoms in addition to the repulsive interactions between the electrons and between the nuclei. When the attractive interaction is strong enough, a strong covalent bond can be formed which corresponds to the sharing of valence electrons between pairs of atoms. The valence electron charge density calculated for a crystal of Ge is shown in Fig. 2.3a, where it can be seen to have a maximum value midway between the two atoms. Shown here for comparison and later discussion are the charge densities for crystals of GaAs and ZnSe. The attractive interaction giving rise to covalent bonding can be thought of as arising from the attraction of each nucleus toward the enhanced electron charge density associated with the covalent bond and centered midway between the two nuclei. The covalent bond, which is similar to a bonding molecular orbital, is clearly directional, corresponding to a nonspherical distribution of electron charge.

Covalent bonding between atoms can dominate in an elemental solid when the number of valence electrons per atom is equal to (or greater than) the number of available bonding orbitals, given by the number of NNs that each atom has in the solid.[†] This criterion is met near the center of the periodic table, specifically for the

---

[†] The topic of valence is covered in Chapter W2. The existence of more than one possible valence for an atom is discussed there.

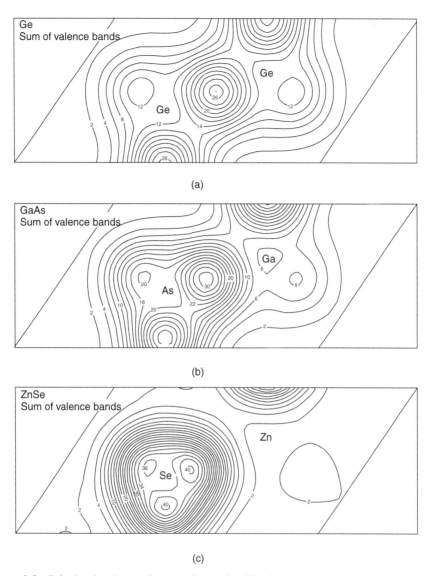

Figure 2.3. Calculated valence electron charge densities between pairs of atoms (or ions) in the following isoelectronic crystals: (*a*) Ge; (*b*) GaAs; (*c*) ZnSe. [From J. P. Walter et al., *Phys. Rev. B* **4**, 1877 (1971). Copyright 1971 by the American Physical Society.]

covalently bonded crystals of C (diamond), Si, Ge, and gray Sn (α-Sn) in group IV and for the elements in groups V and VI (see Fig. 2.2). The four elements in group IV can all have the diamond crystal structure described in Chapter 1. In diamond, for example, the two $2s$ and two $2p$ electrons of each C atom first hybridize, forming four sp^3 hybrid orbitals. In this oversimplified picture each C atom then contributes one of its four valence electrons in an sp^3 orbital to each of the covalent bonds formed with its four NNs in the diamond crystal structure. The absence of covalent bonding and the existence instead of metallic bonding in white Sn (β-Sn) and Pb are discussed in Chapter W2.

When elements crystallize in structures in which the number of valence electrons per atom is greater than the number of available orbitals or NN atoms, the valence electrons not participating in bonding enter into either delocalized bonds (as in the case of the π bonds in graphite) or *nonbonding* [i.e., *lone-pair* (lp)] *orbitals*, as in the case of elements such as P and S in groups V and VI of the periodic table, to be discussed later.

Each covalent bond in an elemental solid contains two electrons with oppositely directed spins, with one electron provided by each atom. These covalent bonds are similar to the σ-bonding molecular orbitals (BMOs) of the methane molecule, CH_4. In the covalent, group IV elements such as diamond and Si, however, the σ BMOs that form the covalent bonds overlap with each other both energetically and spatially. As a result, their energy levels broaden into a continuous *energy band*. The electrons therefore lose their identity as σ BMOs and instead have wavefunctions that extend throughout the solid.

Covalent bonds can result from the overlap of atomic s or p and hybrid sp, sp^2, or sp^3 orbitals on NN atoms. Covalent bonding based on sp^3 hybridization leads quite naturally to formation of the tetrahedral $A-A_4$ local atomic bonding units discussed in Chapter 1. The elemental solids from group IV of the periodic table can all have the diamond crystal structure based on these covalently bonded $A-A_4$ units.

Elemental solids in which the atoms are covalently bonded via sp^2 or sp^2-like hybridization have structures based on $A-A_3$ local atomic bonding units. Examples include C in the form of graphite from group IV and P from group V of the periodic table. For graphite the valence electrons on neighboring C atoms that are not involved in the formation of σ-like covalent bonds instead form π-like covalent bonds. The electrons in these π-like bonds are delocalized over the planes of C atoms in the graphite structure.

The bonding in elemental solids of P, As, Sb, and Bi from group V is also primarily covalent. These atoms have $s^2 p^3$ valence electron configurations. Whereas the $A-A_3$ bonding unit for graphite is planar, the elements from group V have $A-A_3$ units that are pyramidal [i.e., the central A atom lies above (or below) the plane of the other three A atoms]. As a result, atoms in solid As, Sb, and Bi form hexagonal rings, as in graphite, but with the atoms not being coplanar. Instead, three atoms per ring lie above the plane while the remaining three lie below (Fig. 2.4). With only three NN atoms, the two remaining valence electrons per atom for P, As, Sb, and Bi enter into nonbonding lp orbitals.

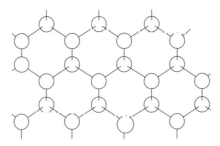

Figure 2.4. Atomic structure of As, Sb, and Bi showing some of the hexagonal rings based on nonplanar $A-A_3$ bonding units.

The bonding via sp^2 hybridization in crystals is essentially effective in only two dimensions (e.g., within the planes of sp^2-bonded C atoms in the case of graphite). The bonding in the third dimension (i.e., between the planes of C atoms), occurs primarily by weak electrostatic interactions and the van der Waals mechanism, discussed later.

Crystals of S, Se, and Te from group VI of the periodic table are covalently bonded and have crystal structures based on $A-A_2$ bonding units. These atoms, which have $s^2 p^4$ valence electron configurations, form covalent bonds with their two NNs, with the remaining four valence electrons not involved in the bonding directly. The two s electrons per atom can remain as core electrons, as for the case of Se, or may occupy nonbonding lp orbitals. The two remaining p electrons per atom will occupy lp orbitals. The $S-S_2$ and $Se-Se_2$ units link together to form S_8 and Se_8 rings, which in the case of Se, are aligned in long, spiral chains. The rings and chains are then bonded to each other primarily via the van der Waals interaction, resulting in rather complicated crystal structures.[†] Solid S can be considered to be a molecular crystal composed of S_8 moleculelike units.

Crystals of B from group III of the periodic table are formed from icosahedral B_{12} units, but without a B atom at the center of the icosahedron. Complex crystal structures are formed, including cubic, rhombohedral, and tetragonal structures. The bonding in crystals of B cannot be completely covalent since there are only $12 \times 3 = 36$ $2s$ and $2p$ electrons available to occupy the 30 bonds between NN B atoms in the B_{12} unit.

It is clear from these discussions that covalent bonding plays a critical role in stabilizing the structures of most elements in groups IV, V, and VI of the periodic table. These observations are summarized in Table 2.1, where the coordination numbers, local atomic bonding units, and resulting structures are given for the covalently bonded elements.

With the exception of C in the form of graphite, it can be seen that the coordination numbers CN of the covalently bonded elements listed in Table 2.1 are equal to $8 - N$. Here N is the number of valence electrons per atom of the element and corresponds to its group number in the periodic table. Also included in Table 2.1 are the group VII elements (i.e., the halogens), which form covalently bonded diatomic molecules. These

TABLE 2.1 Structures of the Covalently Bonded Elemental Solids

Elements	Group Number[a]	Coordination Number CN	Local Atomic Bonding Unit	Resulting Local Structures
B	III	5	$B-B_5$	Icosahedra
C (diamond), Si, Ge, gray Sn	IV	4	$A-A_4$	Tetrahedra (three-dimensional)
C (graphite)	IV	3	$A-A_3$	Planar hexagons
P (black), As, Sb, Bi	V	3	$A-A_3$	Nonplanar hexagons
S, Se, Te	VI	2	$A-A_2$	A_8 rings, spiral chains
F, Cl, Br, I	VII	1	$A-A$	A_2, diatomic molecules

[a]The group number in the periodic table is also the number of valence electrons per atom.

[†] See Wyckoff (1963) and Pearson (1972) for further details of these complicated structures.

molecules retain their identity in crystals of these elements and are bonded to each other via the van der Waals interaction. Crystals in which molecules retain their identity are known as *molecular crystals*.

2.3 Metallic Bonding

It can be seen from Fig. 2.2 that the elemental solids in which metallic bonding dominates lie toward the left and bottom of the periodic table. For the elemental solids in groups IA, IIA, and IIIA, the valence electrons originate from the s and p atomic orbitals. Examples include the *alkali metals* $(Li - 2s^1, Na - 3s^1, \ldots)$ in group IA, the *alkaline earth metals* $(Mg - 2s^2, Ca - 3s^2, \ldots)$ in group IIA, and the *trivalent metals* $(Al - 3s^2 3p, Ga - 4s^2 4p, \ldots)$ in group IIIA. These metals, formed from elements with s and p valence electrons only (and no d electrons), are known as *simple metals*. The *transition metals* near the middle of the periodic table correspond to elements with valence electrons in partially filled $3d$, $4d$, and $5d$ shells. The *rare earth metals* correspond to those with valence electrons in partially filled $4f$ and $5f$ shells.

If covalent bonding were to dominate in crystals of the simple metals, each atom would need a number of NNs equal to its valence (i.e., 1, 2, or 3). Instead, the atoms achieve a state of lower energy by solidifying in structures based on the $A-A_{12}$(cub), $A-A_{12}$(hex), and $A-A_8$ bonding units, with 12, 12, and 8 NNs, respectively. Under these circumstances, where the number of valence electrons per atom is much less than the number of NNs, the valence electrons do not enter well-defined bonding orbitals but rather, are shared among all the atoms (i.e., their wavefunctions extend throughout the crystal). As a result, metallic bonding is nondirectional, at least in the simple metals where d electrons are not present.

Two distinct approaches can be used to gain insight into metallic bonding and the behavior of the valence electrons in crystals of the simple metals. The first is a chemical approach and begins by considering the electron orbitals involved in the bonding of clusters of two or more metal atoms. The second is the approach more usually taken in solid-state physics texts, which begins with delocalization of the electrons throughout the crystal and then attempts to justify the stability of the metal. Metallic bonding in the transition metals is more complicated since the contributions of the d valence electrons must also be considered. Further discussions of the bonding and cohesive energies of the sp metals, transition metals, and rare earth metals are given in Chapter 12.

Starting with the chemical approach, consider first the simple sp metal Na, beginning with the Na atom. Clusters of Na atoms [i.e., Na_n $(n = 2, 3, \ldots)$], will then be considered, and finally, Na metal, $n \to \infty$. Na atoms with atomic number $Z = 11$ have a single $3s$ valence electron in the outermost shell. It will be instructive in the following to consider the fractional average s character $f(s)$ of the valence electrons for Na_n clusters as n increases from 2 to infinity. The quantity $f(s)$ is equal to the fraction of the valence electron wavefunction that has the spherical symmetry of the s state $(l = 0)$ atomic orbital. The $3s$ valence electron of a Na atom is clearly unhybridized and so has $f(s) = 1$.

The Na_2 dimer is the smallest Na cluster and is covalently bonded with the two available $3s$ electrons entering a σ molecular orbital (MO). Na_2 is weakly bound, with a dissociation or *atomization energy* E_a corresponding to $Na_2 \to 2Na$ of only 0.78 eV. Therefore, $E_a = 0.39$ eV per Na atom. The quantity $f(s)$ for Na_2 has been determined to be 0.93, corresponding to a 7% admixture of p ($l = 1$) character in the valence electron wavefunctions.[†]

The bonding electrons in the triangular Na_3 and three-dimensional Na_7 clusters have a greater degree of hybridization, with $f(s) = 0.87$ and 0.63, respectively. Note that the equivalent of $f(s)$ for an electron at the Fermi surface of Na metal is 0.66;[†] see Chapter 7 for a discussion of the free-electron model as applied to electrons in metals. It is apparent that there is a decreasing trend in s character with increasing cluster size as the covalent σ bond for Na_2 is replaced by electron orbitals, which are increasingly delocalized over the larger clusters. For Na_7, the hybridization approaches that of the bulk metal. With only three valence electrons available for bonding, the triangular Na_3 cluster does not possess the total of six valence electrons that would be required if there were σ MO covalent bonds between the three pairs of Na atoms.

As the cluster size increases, the atomization energy E_a increases from 0.39 eV per atom for Na_2 up to 1.13 eV per atom for Na metal.[‡] In the BCC structure of Na metal, where each Na atom has eight NNs, it is apparent that four Na–Na pairs can be associated with each Na atom. Therefore, the bonding energy per Na–Na pair in Na metal is $1.13/4 \approx 0.28$ eV, which is less than the value 0.37 eV for Na_2. The bonding advantage of Na metal is therefore not in having stronger Na–Na bonds than in Na_n clusters but rather in allowing more Na–Na "bonds" to exist within the BCC structure. The valence electrons, on the other hand, are less strongly bound in Na metal than in the Na_n clusters. For example, the first ionization energy IE(1) of Na_2 is 4.93 eV, while the average binding energy $\langle E_B \rangle$ of electrons in a simple metal like Na is $\langle E_B \rangle = W + 2E_F/5$. Here W is the work function and E_F is the Fermi energy (see Chapter 7). With $W = 2.75$ eV and $E_F = 3.24$ eV for Na metal, it follows that $\langle E_B \rangle \approx 4.0$ eV.

To summarize, the enhanced stability of metals such as Na in the solid state as compared to the stability of clusters can be attributed to the stronger bonding of the Na^+ ions by the distribution of delocalized valence electrons rather than to stronger bonding of the electrons in the metallic state.

An alternative way of viewing metallic bonding is to note that the delocalization of an electron over the entire solid lowers the kinetic energy associated with its confinement in a certain region of space. The screening of the electric field of the ion by other valence electrons weakens the attraction to the point where delocalization becomes energetically favorable. Other aspects of the localization of electrons in solids versus their delocalization are discussed further in Chapter W7.

2.4 van der Waals Bonding

An additional contribution to bonding in elemental solids involving neither the sharing of valence electrons, as in covalent bonding, nor their delocalization, as in metallic bonding, arises from the *van der Waals interaction*. The van der Waals potential results from the interaction of the electric-dipole moments of atoms that are produced by quantum-mechanical fluctuations. Although it exists for all atoms, even those without

---

[†] D. A. Garland and D. M. Lindsay, *J. Chem. Phys.*, **80**, 4761 (1984).

[‡] The atomization energy E_a of an element such as Na is equivalent to its cohesive energy (or enthalpy) ΔH_c per atom.

static electric-dipole moments, its effect become dominant at large separations for atoms with filled shells. As a result, the van der Waals interaction dominates the bonding of *inert-gas solids* (Ar, Ne, etc.) composed of atoms having filled shells (and high ionization energies). This interaction also contributes to the bonding between planes of C atoms in graphite and between rings and chains of S, Se, and Te atoms in these elemental solids. Although van der Waals bonding is in principle present in all solids, due to its relative weakness it makes only a minor contribution to the bonding in elemental covalent and metallic solids.

The question naturally arises how attractive interactions can exist, for example, between inert-gas atoms, which are neutral and do not under ordinary circumstances share their electrons. The answer lies in the fact that the distribution of the electrons within an atom is not static or rigid but rather, undergoes quantum-mechanical fluctuations. The average electric-dipole moment μ of the atom is zero since the centers of the average distributions of positive and negative charge coincide. Nevertheless, the instantaneous dipole moment of the atom can be nonzero due to fluctuations that give rise to a nonsymmetric distribution of electrons within the atom. The specific form of the van der Waals interaction resulting from the forces between the fluctuating, induced electric-dipole moments on neighboring atoms is well-known.

The van der Waals interaction results from the second-order perturbation-theoretic expression for the energy shift of two atoms interacting by a dipole–dipole interaction. The interaction is given by

$$V(\mathbf{r}) = \frac{\boldsymbol{\mu}_1 \cdot \boldsymbol{\mu}_2 - 3(\boldsymbol{\mu}_1 \cdot \hat{\mathbf{r}})(\boldsymbol{\mu}_2 \cdot \hat{\mathbf{r}})}{4\pi\epsilon_0 r^3}, \tag{2.1}$$

where μ_i is the electric-dipole moment of atom i, $\mathbf{r}$ the interatomic separation, and $\varepsilon_0 = 8.85 \times 10^{-12}$ C^2/N·m^2 is the permittivity of free space. The resulting interaction energy is

$$U(r) = -\sum_{nn'}' \frac{|\langle nn'|V(\mathbf{r})|00\rangle|^2}{E_n^{(1)} + E_{n'}^{(2)} - E_0^{(1)} - E_0^{(2)}} = -\frac{C}{r^6}, \tag{2.2}$$

where the $E_n^{(i)}$ are the atomic energy levels for atom i. Here C is a constant and the prime indicates that the term $n = n' = 0$ is excluded from the sum.

Several aspects of the van der Waals interaction and the resulting bonding should be emphasized:

1. The interaction represented by $U(r)$ is attractive and leads to the formation of a potential well when the repulsive Coulomb interactions between the atoms are also included.

2. The van der Waals interaction is quite weak and is also very short range, due to the $1/r^6$ dependence of $U(r)$.[†]

[†] In the case of He, the lightest of the inert-gas atoms, the strength of the attractive van der Waals interaction cannot overcome the zero-point motion of the He atoms, which results from the Heisenberg uncertainty principle. As a result, He remains a liquid up to a pressure of about 28 atm at $T = 0$ K.

TABLE 2.2   Classification of Solids According to the Dominant Type of Bonding

Dominant Bonding	Examples	Cohesive Energy (eV/atom)	Characteristic Properties
Covalent	C (diamond), Si, Ge	$\approx 4-8$	Hard, low electrical conductivity
Metallic	Na, Fe, La	$\approx 1-9$	High electrical conductivity
van der Waals	He, Ar, Cl_2	$\approx 0.02-0.3$	Transparent insulators, low melting points
Ionic	NaCl, MgO	≈ 5 (per ion pair)	Low electrical conductivity
Hydrogen	H_2O (ice), HF	≈ 0.1 (per molecule)	Molecular crystals, low electrical conductivity

3. When an empirical repulsive potential of the form $+B/r^{12}$ is added to the attractive van der Waals term, the result is known as the *Lennard-Jones potential*,

$$U(r) = +\frac{B}{r^{12}} - \frac{C}{r^6} = 4\varepsilon \left[\left(\frac{\sigma}{r}\right)^{12} - \left(\frac{\sigma}{r}\right)^6 \right]. \tag{2.3}$$

Here ε and σ are parameters characterizing the strength and range of the interaction, respectively. This potential is often used in calculations of the cohesive energies of inert-gas crystals.

4. The van der Waals interaction as derived above is essentially isotropic (i.e., nondirectional), so that there are no preferred directions in space. The bonding in crystals in which this interaction is dominant (e.g., inert-gas crystals) is maximized when each atom is surrounded by as many NNs as possible. As a result, the inert-gas crystals have the close-packed FCC crystal structure in which the local atomic bonding is based on A–A_{12}(cub) units. In molecular crystals based on molecules such as I_2 or C_6H_6, the bonding between molecules is through the van der Waals interaction. The molecules in these crystals will therefore also be packed together as tightly as their shapes will allow.

Table 2.2 summarizes and illustrates the dominant types of bonding occurring in solids, including the three types of bonding already discussed for the elemental solids (covalent, metallic, and van der Waals) as well as two additional types of bonding found in multielement crystals (ionic and hydrogen bonding). Included in the table are crystals that serve as examples of each type of bonding, the range of cohesive energies typically found in these crystals, and some characteristic properties of the crystals. It should be stressed again that more than one type of bonding contributes to the stability of most of the crystals listed as examples in this table.

BONDING IN MULTIELEMENT CRYSTALS

In addition to the three bonding mechanisms discussed above for elemental solids, two additional types of bonding — ionic bonding and hydrogen bonding — play important roles in the stability of solids containing two or more elements. The former results when valence electrons are transferred between atoms, while the latter occurs when the nucleus of a H atom (i.e., a proton) is attracted to the electrons in a nonbonding [i.e., lone pair (lp)] orbital on neighboring atoms such as oxygen or nitrogen.

2.5 Ionic Bonding

The attraction between ions with charges of opposite sign via the Coulomb interaction is an important source of bonding in multielement solids in which valence electrons have been transferred completely (or at least partially) from one atom to another. In these ionic solids the potential energy of interaction between a pair of oppositely charged ions is given by

$$U_{\text{Coul}}(r) = \frac{(+z_c e)(-z_a e)}{4\pi\epsilon_0 r} = \frac{-z_c z_a e^2}{4\pi\epsilon_0 r}, \tag{2.4}$$

where $+z_c e$ and $-z_a e$ are the net electrical charges of the positively charged cation and the negatively charged anion, respectively, and r is the internuclear separation. Since this Coulomb potential is spherically symmetric, ionic bonding is nondirectional. The classic example of an ionically bonded crystal is NaCl, which has a structure based on the $A-B_6$ bonding unit. Most of the alkali halide crystals (cations: Li^+, Na^+, K^+, Rb^+, Cs^+; anions: F^-, Cl^-, Br^-, I^-) as well as many other ionic crystals have the NaCl crystal structure. The CsCl crystal structure based on the $A-B_8$ bonding unit is also shared by several ionic crystals.

To understand the energetics of ionic bonding, some processes which at least in principle can lead to the formation of an ionic crystal are examined. These processes are illustrated schematically in Fig. 2.5. To ionize a Na atom in the gas phase, an energy equal to the first ionization energy IE(1) = 5.15 eV is required:

$$Na(g) + 5.15 \text{ eV} \rightarrow Na^+(g) + e^-. \tag{2.5}$$

An energy equal to the electron affinity EA = 3.62 eV of the isolated Cl atom is recovered when the electron produced above is captured by a Cl(g) atom:

$$Cl(g) + e^- \rightarrow Cl^-(g) + 3.62 \text{ eV}. \tag{2.6}$$

Thus a net energy of $5.15 - 3.62 = 1.53$ eV is required to produce simultaneously a pair of isolated $Na^+(g)$ and $Cl^-(g)$ ions from Na(g) and Cl(g).

When this pair of $Na^+(g)$ and $Cl^-(g)$ ions interact with each other to form a stable NaCl(g) molecule, a bonding energy of 5.74 eV is recovered as a result of their electrostatic interaction:

$$Na^+(g) + Cl^-(g) \rightarrow NaCl(g) + 5.74 \text{ eV}. \tag{2.7}$$

Note that 5.74 eV is therefore the dissociation energy of the NaCl(g) molecule into ions. If pairs of $Na^+(g)$ and $Cl^-(g)$ ions instead condense to form a crystal of solid NaCl (based on the $Na^+-(Cl^-)_6$ and $Cl^--(Na^+)_6$ bonding units), an even larger cohesive or bonding energy of 8.13 eV per pair of ions is recovered as a result of their mutual electrostatic interaction within the solid:

$$Na^+(g) + Cl^-(g) \rightarrow NaCl(s) + 8.13 \text{ eV}. \tag{2.8}$$

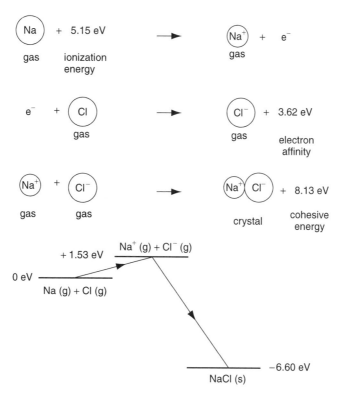

Figure 2.5. Processes involved in the formation of a NaCl crystal from isolated Na and Cl atoms.

More than enough energy is therefore recovered from the electrostatic interaction between the Na^+ cations and the Cl^- anions in the NaCl crystal to pay the cost in energy of creating the cations and anions from isolated neutral atoms in the first place. With respect to $Na(g)$ and $Cl(g)$ atoms, the cohesive energy ΔH_c of NaCl is $8.13 - 1.53 = 6.60$ eV per pair of neutral atoms, while with respect to the Na^+ and Cl^- ions, the cohesive energy is the full 8.13 eV.

In the processes shown in Fig. 2.5, both the Na^+ cations and the Cl^- anions in principle attain stable, filled outer-electron shells: Na gives up its single $3s$ electron and becomes Na^+, which has the $1s^2 2s^2 2p^6$ Ne-like electron configuration, while Cl becomes Cl^-, which has the $1s^2 2s^2 2p^6 3s^2 3p^6$ Ar-like electron configuration. Thus, to a first approximation, all the valence electrons in an ionic crystal occupy atomic orbitals associated with the ions. The actual situation is not as simple, since complete transfer of an integral number of electrons from cations to anions never occurs in practice. Instead, some sharing of the valence electrons between the ions always takes place, resulting in bonding of a mixed ionic–covalent type, discussed later.

A general expression for the electrostatic energy (i.e., the *Madelung energy*) of an ionic crystal is obtained by adding together all the Coulomb interaction energies of the ions. Calculation of the Madelung energy and the corresponding *Madelung constant* are discussed in Chapter W2.

Since ionic bonding is capable of leading to stable crystal structures, it is evident that the structures which ionic crystals assume will be those which enhance the attractive forces between cations and anions while minimizing, to the extent possible, the repulsive cation–cation and anion–anion forces. This is achieved by bringing cations and anions close together and by maximizing the coordination number CN (i.e., the number of NNs of the opposite type) and by keeping like ions apart. Also, in a particular crystal structure, the equilibrium distance d between cations and anions will be determined by the balance between the attractive Coulomb interaction and the repulsive interactions resulting from the overlap of the electrons in the filled shells, as discussed earlier.

It is clear that geometrical considerations, in particular the relative sizes of the cations and anions, can play an important role in determining the crystal structures of ionic solids. This viewpoint is the basis of the *ionic model* for the stability of ionic compounds, which is expressed by Pauling's five bonding rules, enumerated in Chapter 13. To be able to predict which crystal structure a particular ionic compound will take, it will be very useful to have reliable values for the *ionic radii* r_c and r_a (i.e., the radii of the cations and anions, respectively) in the compound in question. It is often convenient as a first approximation to assume that the ions are hard spheres and that the anions are in contact with the central cation. These assumptions are usually made even though the electron clouds of the ions can be deformed due to the presence of their NNs in the structure and also due to the likelihood of some sharing of electrons via covalent bonding. It should also be pointed out that the radius of an ion is not a well-defined quantity since the valence electron orbitals do not have sharp boundaries.

The importance of the relative sizes of the atoms or ions in influencing the stabilities of local atomic bonding units has been discussed in Section 1.6. In ionic solids the repulsion resulting when anions come into contact with each other is the reason for the instability of the bonding unit in question and also of the corresponding crystal structure. Table 2.3 summarizes the ranges of values of the cation/anion radius ratio r_c/r_a, for which the anions in a particular bonding unit or structure remain in contact with the cations and, as a result, do not come into contact with each other. The derivation of these ranges is described in Chapter W13. The assumption made in this ionic model is that ionic crystals will assume structures consistent with the greatest coordination number as long as the larger anions are not in contact with each other. The ionic model is not expected to work well for elements near the center of the periodic table [e.g., B and N in boron nitride (BN)], for which covalent bonding is expected to play an important role.

A great deal of effort has gone into obtaining self-consistent sets of ionic radii. Based on the hard-sphere model, the assumption usually made for ionic crystals is that the cation–anion distances d are equal to the sum of the ionic radii, $d = r_c + r_a$, for the ions in question (i.e., the cation and anion are assumed to be in contact with

TABLE 2.3   Radius Ratios and Crystal Structures

Local Atomic Bonding Unit	Range of Allowed Radius Ratios r_c/r_a (Cation/Anion)	Examples (r_c/r_a)
A–B$_8$ (cube)	1–0.732	CsCl (0.92)
A–B$_6$ (octahedron)	0.732–0.414	NaCl (0.54)
A–B$_4$ (tetrahedron)	0.414–0.225	ZnS (0.40)
A–B$_3$ (triangle or pyramid)	0.225–0.155	B$_2$O$_3$ (0.17), BN (0.13)
A–B$_2$ (link or bridge)	<0.155	

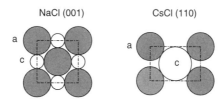

Figure 2.6. Proposed arrangements of cations (c) and anions (a) in contact with each other in the NaCl and CsCl crystal structures.

each other as shown in Fig. 2.6). Selected values of proposed ionic radii from various sources are presented in Table 2.4.[†] A more complete list of proposed ionic radii is presented in Table 2.12. Note that the negatively charged anions are typically much larger than the positively charged cations. To illustrate this, consider Na^+ and F^-, where $r_c(Na^+) \approx 0.10$ nm and $r_a(F^-) \approx 0.135$ nm. Both ions have Ne-like $1s^2 2s^2 2p^6$ electron configurations, but the Na^+ cation with a nuclear charge of $+11e$ pulls its 10 electrons in closer than does the F^- anion with a nuclear charge of $+9e$.

As a test of the validity of the ionic model and the predictions summarized in Table 2.3, consider the case of the five alkali chlorides. The radius ratios for these five compounds, obtained using the ionic radii listed in Table 2.4, are presented in Table 2.5 along with the crystal structures predicted and observed.

It can be seen that as the ionic radius ratios increase from 0.38 for LiCl to 0.92 for CsCl, the coordination numbers of the crystal structures predicted increase from 4 for ZnS (cubic or hexagonal) to 8 for the CsCl crystal structure. This is quite reasonable since as the radius of the cation increases from Li^+ to Cs^+, more anions can be bonded to the central cation without coming into contact with each other. Although the NaCl crystal structure is predicted for only two compounds, NaCl and KCl, it is, in fact, the crystal structure observed for four of the five alkali chlorides. For LiCl, the Li^+ ion with $r_c \approx 0.07$ nm is so much smaller than the Cl^- ion with $r_a \approx 0.18$ nm that LiCl is predicted to have either the cubic or the hexagonal ZnS crystal structure. LiCl is actually observed to have the NaCl crystal structure, which means that the Cl^- ions are likely to be in contact with each other.

The failures of the ionic model in predicting the crystal structures observed for many of the alkali halides may be related either to uncertainties in the ionic radii or to the fact that the electrostatic energies of the NaCl and CsCl crystal structures are so close that other effects related to contributions from van der Waals bonding, covalent bonding, or from the effects of entropy at finite temperatures can reverse the prediction. Ionic radii can take on different values in different crystal structures and also may vary with the coordination number of the ion (see Table 2.4). Nevertheless, the basic model of ionic bonding presented above is a useful one for many crystals.

Pure ionic and covalent bonding correspond to the extreme situations in which valence electrons are either transferred completely from one atom to another or shared equally between two atoms. When the transfer of valence electrons is not complete, both types of bonding can play a role in the determining the structure of the crystal.

[†] Various sets of ionic radii derived by different methods and based on different assumptions have been proposed. These different sets of radii can differ by up to 20% and have been summarized by Jaffe (1996, Chap. 4).

TABLE 2.4 Selected Values of Proposed Ionic Radii

Cation	CN^a	r_{ion} (nm)	Source[b]	Anion	CN^a	r_{ion} (nm)	Source[b]
Li^+	VI	0.068	A	F^-	VI	0.136	A
	VI	0.074	S-P		VI	0.133	S-P
Na^+	VI	0.097	A	Cl^-	VI	0.181	P, A
	VI	0.102	S-P				
K^+	VI	0.133	P, A	Br^-	VI	0.195	P, A
	VI	0.138	S-P				
Rb^+	VI	0.147	A	I^-	VI	0.216	P, A
	VI	0.149	S-P				
Cs^+	VI	0.167	A	O^{2-}	II	0.135	S-P
	VI	0.170	S-P		III	0.136	S-P
Be^{2+}	IV	0.035	A		IV	0.138	S-P
	IV	0.027	S-P		VI	0.140	A, S-P
B^{3+}	III	0.023	A	S^{2-}	VI	0.184	A, P
Si^{4+}	IV	0.026	S-P	Se^{2-}	VI	0.198	P
	VI	0.042	A	Te^{2-}	VI	0.221	P
	VI	0.040	S-P				
Mg^{2+}	VI	0.066	A				
	VI	0.072	S-P				
Al^{3+}	VI	0.051	A				
	VI	0.053	S-P				
Fe^{2+}	IV	0.063	S-P				
	VI	0.077	S-P				
Fe^{3+}	IV	0.049	S-P				
	VI	0.0645	S-P				
Zn^{2+}	VI	0.074	P, A				

[a]CN = coordination number.
[b]Data from H. W. Jaffe, *Crystal Chemistry and Refractivity*, Dover, Mineola, N.Y., 1996, Chap. 4. Original sources: P, L. Pauling, *The Nature of the Chemical Bond*, 3rd ed., Cornell University Press, Ithaca, N. Y., 1960; A, L. H. Ahrens, *Geochim. Cosmochim. Acta*, **2**, 158 (1952); S-P, R. D. Shannon and C. T. Prewitt, *Acta Crystallogr. Sect. B*, **25**, 925 (1969).

TABLE 2.5 Radius Ratios and Crystal Structures of the Alkali Chlorides

Alkali Chloride	$r_c/r_a{}^a$	Crystal Structure	
		Predicted[b]	Observed
LiCl	0.38	Cubic ZnS or hexagonal ZnS (Λ B_4)	NaCl ($A-B_6$)
NaCl	0.54	NaCl ($A-B_6$)	NaCl ($A-B_6$)
KCl	0.73	NaCl ($A-B_6$)	NaCl ($A-B_6$)
RbCl	0.81	CsCl ($A-B_8$)	NaCl ($A-B_6$)
CsCl	0.92	CsCl ($A-B_8$)	CsCl ($A-B_8$)

[a]Determined using radii presented in Table 2.4; data from L. H. Ahrens, *Geochim. Cosmochim. Acta*, **2**, 158 (1952).
[b]Predictions based on criteria presented in Table 2.3.

The coexistence of ionic and covalent bonding in multielement crystals is examined next.

2.6   Mixed Ionic–Covalent Bonding and Ionicity

The two extreme cases of bonding between pairs of atoms in a molecule or in a solid correspond to ionic bonding (i.e., the complete transfer of one or more valence electrons from one atom to the other) and to covalent bonding (i.e., the equal sharing of valence electrons between the two atoms). Although ideal covalent bonds are found in the H_2 molecule and in crystals of C, Si, and Ge, and nearly ideal ionic bonds are found in crystals of some of the alkali halides, such as CsF, the more common situation corresponds to bonds between pairs of unlike atoms in which the participating valence electrons are partially shared and partially transferred. This type of bonding is referred to here as *mixed ionic–covalent bonding*.

Direct illustration of the partial sharing/partial transfer of valence electrons between pairs of atoms is shown in Fig. 2.3. Here the electron densities calculated in the (110) plane are shown for pairs of NN atoms in the three isoelectronic crystals Ge, GaAs, and ZnSe. Each of these crystals is composed of atoms from the fourth row of the periodic table, with pairs of Ge(4)–Ge(4), Ga(3)–As(5), and Zn(2)–Se(6) atoms each having a total of eight valence electrons available for bonding. All three crystals have either the diamond (Ge) or cubic ZnS (GaAs and ZnSe) crystal structures, which are based on tetrahedral coordination.

The contour plots of electron density shown in Fig. 2.3 illustrate the expected completely covalent bonding found in Ge, where the density of the bonding electrons has a maximum midway between the two atoms. Also shown is the increasing transfer of charge from atoms with lower first ionization energies IE(1) and electron affinities EA to atoms with correspondingly higher values: from Ga with IE(1) = 6.00 eV and EA $\approx$ 0.37 eV to As with 9.81 eV and 0.80 eV and from Zn (9.39 eV and $\approx$ 0 eV) to Se (9.75 eV and 2.22 eV).

Another measure of the degree of ionic bonding in a crystal is the *ionicity* f_i, which is the fraction of ionic character in a bond. The ionicity parameter f_i is defined to be zero in crystals such as diamond and Si, in which the bonding is completely covalent, and to be unity in hypothetical crystals with completely ionic bonding. Table 2.6 presents values of f_i for some common crystals of elements and of binary compounds, as proposed by Phillips (1973). It can be seen that tetrahedral coordination based on the A–A_4 and A–B_4 bonding units is favored for the group IV elements and also the group IV–IV, III–V, and II–VI binary compounds with low values of f_i. Note that a transition from tetrahedral (A–B_4) to octahedral (A–B_6) coordination occurs in these crystals at the critical ionicity value $f_i \approx 0.8$. For $f_i < 0.8$ covalent bonding dominates in the A–B_4 bonding unit, while for $f_i > 0.8$ ionic bonding dominates in the A–B_6 bonding unit. A more complete discussion of the Phillips definition of ionicity and of the nature of bonding in semiconductors is presented in Chapter W11.

2.7   Hydrogen Bonding

Hydrogen can be bonded in solids in a variety of ways, including in molecules such as H_2O in ice, in polymeric chains such as polyethylene, $(CH_2)_n$, and in three-dimensional networks such as amorphous SiO_2 as an "impurity." In these cases the $1s$ electron of the H atom forms a relatively strong σ bond of mixed ionic–covalent character with O, C, or Si.

TABLE 2.6   Ionicities of Some Crystals

Crystal	Group Number	Local Atomic Bonding Unit	f_i^a
C, Si, Ge	IV	$A-A_4$	0.00
SiC	IV–IV	$A-B_4$	0.18
BN	III–V	$A-B_4$	0.26
GaAs	III–V	$A-B_4$	0.31
InAs	III–V	$A-B_4$	0.36
AlN	III–V	$A-B_4$	0.45
MgTe	II–VI	$A-B_4$	0.55
ZnS	II–VI	$A-B_4$	0.62
CdTe	II–VI	$A-B_4$	0.72
HgS	II–VI	$A-B_4$	0.79
AgCl	I–VII	$A-B_6$	0.86
NaCl	I–VII	$A-B_6$	0.94

[a] Ionicities proposed by J. C. Phillips, *Bonds and Bands in Semiconductors*, Academic Press, San Diego, Calif., 1973).

A different type of bonding can arise when the H atom in an O–H, C–H, or Si–H bond in the solid state is located near other atoms, such as N, O, or F, which have electrons in nonbonding (i.e., lone pair) orbitals. The type of bonding that results from the electrostatic attraction between the positively charged proton H^+ and the negatively charged electrons in the *lp* orbital on the neighboring atom is known as *hydrogen bonding*. If strong enough, this attraction can lead to the formation of a hydrogen bond between the two atoms. While the hydrogen bond is directional, it can be considered to be essentially ionic, although having some covalent character.

The local atomic bonding units containing H atoms and involving hydrogen bonding can be written as A–H $\cdots$ B where the symbol $\cdots$ indicates a hydrogen bond between hydrogen and atom B. Hydrogen bonding is stronger when both the A and B atoms in the bonding unit have high electron affinities, as is the case for N, O, and F atoms. Under these circumstances the H atom is essentially a bare proton that can be strongly attracted to the nearby *lp* orbitals. Note that the three atoms in this bonding unit need not lie in a straight line.

An example of a crystal in which hydrogen bonding plays an essential role is solid H_2O or ice, where the hydrogen bonding unit can be written as O–H $\cdots$ O. Each oxygen atom in ice is bonded by strong O–H σ bonds with the two H atoms in the H_2O molecule and by weaker H $\cdots$ O hydrogen bonds to two H atoms in neighboring H_2O molecules. For further discussion of the structure and bonding of ice, see Chapter W2.

Hydrogen bonding plays important roles in the stability of a wide variety of molecules and crystals, including DNA, where the helical chains are held together by hydrogen bonds connecting O and N atoms (e.g., N–H $\cdots$ O), and also in minerals such as $Zn(OH)_2$ and diaspore, AlO(OH).

COHESIVE ENERGIES

As defined earlier, the *cohesive energy* (or *enthalpy*) ΔH_c of a solid is the energy required to overcome the bonding and to separate the constituent atoms (or molecules) from each other completely, usually at $T = 0$ K. From this definition it is clear that ΔH_c is equal to the *standard enthalpy of formation* $\Delta_f H°$ (0 K) of the constituent atoms from the solid phase. The energy released when the atoms react to form the solid is also equal to ΔH_c.

The cohesive energies for the elemental solids are presented in Table 2.7. Some interesting trends are evident in these data:

1. For the alkali metals in group I, ΔH_c decreases as Z increases, due to increased shielding of the conduction electrons from the nuclei.

2. In group IV, ΔH_c decreases as Z increases, due to the increasing delocalization of the covalent bonding electrons.

3. Aside from the inert-gas crystals, the lowest ΔH_c is found for Hg (64.5 kJ/mol). The highest ΔH_c is found for W (848 kJ/mol).

4. The cohesive energies listed for the elements H, N, O, F, Cl, Br, and I correspond to the energies required to dissociate a solid composed of diatomic molecules into atoms.

5. The high cohesive energies of Cu, Ag, and Au relative to their neighbors Zn, Cd, and Hg suggest that their d electrons contribute to the bonding even though the d orbitals are often considered to be fully occupied and hence chemically "inert" for the noble metals.

6. For the inert-gas crystals in group VIII, ΔH_c increases as Z increases, due to the stronger van der Waals interactions.

Some representative values of ΔH_c for crystalline compounds are given in Table 2.8 along with the dominant type of bonding found in each material. For ionic compounds such as NaCl the cohesive energy listed is the energy required to separate the compound into ions (e.g., Na^+ and Cl^-). The standard enthalpies of formation of crystals and calculation of the cohesive energy of α-SiO_2 are further discussed in Chapter W2.

There is a close connection between the cohesive energy of a crystal and its *vaporization or sublimation energy* (i.e., the activation energy for the vaporization processes that are discussed in Section 6.3). Examples are given in Chapter W6 of the vaporization of crystals of Fe and Si. The cohesive energy can be obtained from such data as follows. Vapor pressure measurements[†] for Si yield an activation or atomization energy, when expressed in atomic units, of $E_a = 4.55$ eV/atom. This activation energy is the *enthalpy of vaporization* ΔH_{vap}. Since these measurements were carried out from $T = 1200$ to $1350°$C, taking E_a as the cohesive energy neglects the temperature dependence of the standard enthalpy of formation of Si atoms, $\Delta_f H°[Si(g)]$. When the appropriate correction is made using the JANAF tables,[‡] $\Delta H_c[Si(s)] = 4.62 \pm 0.08$ eV/atom $= 445.7 \pm 8$ kJ/mol is obtained[§] at $T = 0$ K.

For crystals in which bonds between NN atoms are the primary source of the cohesive energy, it is reasonable to equate ΔH_c with the total energy of the bonds. Values for the bond energies between pairs of atoms can thus be obtained from ΔH_c.

[†] R. L. Batdorf and F. M. Smits, *J. Appl. Phys.*, **30**, 259 (1959).

[‡] M. W. Chase, Jr., et al. eds., *JANAF Thermodynamic Tables*, American Chemical Society, Washington, D.C., 1985.

[§] B. Farid and R. W. Godby, *Phys. Rev. B*, **43**, 14248 (1991).

TABLE 2.7 Cohesive Energies ΔH_c (kJ/mol and eV/atom) of the Elemental Solids at $T = 0$ K and Atmospheric Pressure

H 216.0 2.24																	He
Li 157.7 1.64	Be 319.8 3.31											B 559.9 5.80	C 711.2 7.37	N 470.8 4.88	O 246.8 2.56	F 76.9 0.80	Ne 1.92 0.020
Na 107.8 1.12	Mg 145.9 1.51											Al 327.6 3.40	Si 451.3 4.68	P 315.7 3.27	S 274.9 2.85	Cl 120.0 1.24	Ar 7.74 0.080
K 89.9 0.93	Ca 177.3 1.84	Sc 376.0 3.90	Ti 470.3 4.87	V 511.0 5.30	Cr 394.5 4.09	Mn 279.4 2.90	Fe 414.0 4.29	Co 423.1 4.39	Ni 427.7 4.43	Cu 336.2 3.49	Zn 129.9 1.35	Ga 276.0 2.86	Ge 369.2 3.83	As 301.4 3.12	Se 226.4 2.35	Br 117.9 1.22	Kr 11.2 0.116
Rb 82.2 0.85	Sr 164.0 1.7	Y 420.5 4.36	Zr 607.5 6.30	Nb 722.8 7.49	Mo 656.6 6.81	Tc 680.0 7.0	Ru 641.0 6.64	Rh 555.6 5.76	Pd 377.4 3.91	Ag 285.4 2.96	Cd 111.9 1.16	In 243.7 2.53	Sn 301.3 3.12	Sb 262.0 2.72	Te 197.0 2.04	I 107.2 1.11	Xe 15.9 0.16
Cs 78.0 0.81	Ba 180.7 1.87	La 431.3 4.47	Hf 618.9 6.42	Ta 781.4 8.10	W 848.1 8.79	Re 769.0 7.97	Os 790.0 8.2	Ir 664.3 6.89	Pt 564.4 5.85	Au 365.9 3.79	Hg 64.5 0.67	Tl 182.8 1.90	Pb 195.9 2.03	Bi 207.4 2.15	Po 145.0 1.5	At	Rn 19.5 0.202
Fr	Ra 159.0 1.6	Ac 406.0 4.2															

Ce 423.4 4.39	Pr 356.7 3.70	Nd 328.6 3.41	Pm	Sm 206.1 2.14	Eu 177.1 1.84	Gd 398.9 4.14	Tb 390.6 4.05	Dy 293.1 3.04	Ho 302.6 3.14	Er 318.3 3.30	Tm 233.4 2.42	Yb 152.8 1.58	Lu 427.8 4.43
Th 602.2 6.24	Pa 606.0 6.3	U 523.9 5.52	Np	Pu 384.0 4.0	Am	Cm	Bk	Cf	Es	Fm	Md	No	Lr

Source: Data from J. D. Cox et al., *CODATA Key Values for Thermodynamics,* Hemisphere, New York, 1989; NBS Tables of Thermodynamic Properties, *J. Phys. Chem. Ref. Data,* **11,** Suppl. 2 (1985); for Po and Pu: K. A. Gschneider, Jr., in F. Seitz and D. Turnbull, eds., *Solid State Physics,* Vol. 16, Academic Press, San Diego, Calif., 1964, p. 275; for inert-gas crystals: C. Kittel, *Introduction to Solid State Physics,* 7th ed., Wiley, New York, 1996.

TABLE 2.8 Cohesive Energies of Some Crystals

Crystal	ΔH_c (0 K) (kJ/mol)[a]	
Covalent Bonding		
β-SiC	1227	
GaAs	651	(298.15 K)
InAs	604	(298.15 K)
Ionic Bonding[b]		
NaCl	786	
CsCl	668	
NaF	924	
Mixed Bonding		
α-Al$_2$O$_3$	3052	
MgO	991	
α-SiO$_2$	1851	

[a]The cohesive energy ΔH_c is calculated using standard enthalpies of formation $\Delta_f H^\circ$ taken from the NBS Tables of Chemical Thermodynamic Properties, *J. Phys. Chem. Ref. Data*, **11**, Suppl. 2 (1982).
[b]The cohesive energy listed here is the energy required to separate the crystal into its constituent ions.

This procedure will now be illustrated for a crystal of Si. The cohesive energy of crystalline Si is given by

$$\Delta H_c[\text{Si}(s)] = \Delta_f H^\circ[\text{Si}(g)] - \Delta_f H^\circ[\text{Si}(s)] \equiv \Delta_f H^\circ[\text{Si}(g)]$$

$$= 451.29 \text{ kJ/mol}, \tag{2.9}$$

where $\Delta_f H^\circ[\text{Si}(g)]$ is the standard enthalpy of formation per mole of Si atoms from Si(s) at $T = 0$ K. One mole of Si has an Avogadro number $N_A = 6.022 \times 10^{23}$ Si atoms and $2N_A$ Si–Si bonds of energy E (Si–Si). Each Si atom has four bonds, each bond being shared by two Si atoms. It follows that

$$\Delta H_c[\text{Si}(s)] = 2N_A E(\text{Si–Si}). \tag{2.10}$$

The Si–Si bond energy is thus found to be

$$E(Si\text{–}Si) = 2.34 \text{ eV}. \tag{2.11}$$

Note that this result is quite close to one-half of the value of $\Delta H_{\text{vap}} = 4.66$ eV/atom for Si, as expected.[†] Further discussion and a table of bond energies are presented in Chapter W2.

SUMMARY OF SOME ATOMIC PROPERTIES AND PARAMETERS

It is clear from the discussion in this chapter that valence electrons play a critical role in the bonding of atoms in solids. Certain properties and parameters pertaining to atoms

[†] P. D. Desai, *J. Phys. Chem. Ref. Data*, **15**, 967 (1986).

(or ions) have already been mentioned including ionization energy, electron affinity, and atomic or ionic radius. Of these important quantities, only the ionization energies and electron affinities are obtained directly from experiment. The other parameters (i.e., valence, electronegativity, and atomic radii) can only be inferred from the measured properties of atoms.

In this section, definitions of these quantities and of the parameter known as electronegativity are presented, along with tables of values for the elements. This information can be useful in understanding and predicting the bonding and resulting crystal structures of a wide variety of solid-state materials.

2.8   Ionization Energy and Electron Affinity

The *first ionization energy* IE(1) of an atom is the energy required to remove an electron from the neutral atom. IE(1) is also known as the *ionization potential*. The energy required to remove a second electron from the resulting cation is the second ionization energy IE(2). Conversely, the *electron affinity* EA of an atom is the energy released when an additional electron is bound to a neutral atom, leading to the formation of a negative ion with charge $-e$. The quantity IE(1) is thus a measure of the ease with which atoms give up electrons (i.e., of their ability to become cations), while EA is the corresponding quantity for the formation of anions.

Values of IE(1) and IE(2) for the elements are presented in Table 2.9, with IE(1) also shown graphically in Fig. 2.7a as a function of atomic number Z. The electron affinities EA for the elements up to $Z = 87$ are presented in Table 2.10 and also in Fig. 2.7b. Trends in these values of IE(1) and EA are discussed in Chapter W2.

2.9   Electronegativity

The relative ability of an atom to attract and bind additional electrons is known as its *electronegativity*. There are several electronegativity scales, some defined in terms of heats of formation and others defined in terms of first ionization energies IE(1) and electron affinities EA. The usefulness of these electronegativity scales lies in their ability to aid our qualitative understanding of the nature of the bonding and structure of crystals, specifically in ionic and mixed ionic–covalent crystals, where the more electronegative atoms such as O and F are anions and less electronegative atoms such as Li and Na are cations.

Pauling derived a widely used electronegativity scale by assigning a number X for the electronegativity of each atom based on bond energy values derived from thermochemical data. Specifically, Pauling's method defines the electronegativity difference $X_A - X_B$ for atoms A and B by the relationship

$$E(A-B) = \frac{E(A-A) + E(B-B)}{2} + k(X_A - X_B)^2. \tag{2.12}$$

Here $E(A-A)$ and $E(B-B)$ are single-bond energies determined from the cohesive energies ΔH_c of the elements, as described earlier, and $E(A-B)$ is the single-bond energy determined from the cohesive energy of the relevant compound. The proportionality constant k in Eq. (2.12) has the value unity when the bond energies are expressed in electron volts and is equal to 23.05 when they are expressed in kcal/mol. The values of electronegativity presented in Table 2.11 have been obtained from Eq. (2.12).

TABLE 2.9 First and Second Ionization Energies of the Atoms (eV)[a]

1	2	3	4	5	6	7	8	9	10	11	12	13	14	15	16	17	18
1 H 13.60 —																	2 He 24.59 54.42
3 Li 5.39 75.64	4 Be 9.32 18.21											5 B 8.30 25.15	6 C 11.26 24.38	7 N 14.53 29.60	8 O 13.62 35.12	9 F 17.42 34.97	10 Ne 21.56 40.96
11 Na 5.14 47.29	12 Mg 7.65 15.04											13 Al 5.99 18.83	14 Si 8.15 16.35	15 P 10.49 19.77	16 S 10.36 23.34	17 Cl 12.97 23.81	18 Ar 15.76 27.63
19 K 4.34 31.63	20 Ca 6.11 11.87	21 Sc 6.56 12.80	22 Ti 6.83 13.58	23 V 6.75 14.66	24 Cr 6.77 16.49	25 Mn 7.43 15.64	26 Fe 7.90 16.19	27 Co 7.88 17.08	28 Ni 7.64 18.17	29 Cu 7.73 20.29	30 Zn 9.39 17.96	31 Ga 6.00 20.51	32 Ge 7.90 15.93	33 As 9.82 18.63	34 Se 9.75 21.19	35 Br 11.81 21.8	36 Kr 14.00 24.36
37 Rb 4.18 27.29	38 Sr 5.69 11.03	39 Y 6.22 12.24	40 Zr 6.63 13.13	41 Nb 6.76 14.32	42 Mo 7.09 16.16	43 Tc 7.28 15.26	44 Ru 7.36 16.76	45 Rh 7.46 18.08	46 Pd 8.34 19.43	47 Ag 7.58 21.49	48 Cd 8.99 16.91	49 In 5.79 18.87	50 Sn 7.34 14.63	51 Sb 8.64 16.53	52 Te 9.01 18.6	53 I 10.45 19.13	54 Xe 12.13 21.21
55 Cs 3.89 23.16	56 Ba 5.21 10.00	57 La 5.58 11.06	72 Hf 6.83 14.9	73 Ta 7.89	74 W 7.98	75 Re 7.88	76 Os 8.7	77 Ir 9.1	78 Pt 9.0 18.56	79 Au 9.23 20.5	80 Hg 10.48 18.76	81 Tl 6.11 20.42	82 Pb 7.42 15.03	83 Bi 7.29 16.69	84 Po 8.42 —	85 At —	86 Rn 10.75 —
87 Fr	88 Ra 5.28 10.15	89 Ac 5.17 12.1															

58	59	60	61	62	63	64	65	66	67	68	69	70	71
Ce 5.54 10.85	Pr 5.46 10.55	Nd 5.53 10.73	Pm 5.55 10.90	Sm 5.64 11.07	Eu 5.67 11.24	Gd 6.15 12.09	Tb 5.86 11.52	Dy 5.94 11.67	Ho 6.02 11.80	Er 6.11 11.93	Tm 6.18 12.05	Yb 6.25 12.18	Lu 5.43 13.9

90	91	92	93	94	95	96	97	98	99	100	101	102	103
Th 6.08 11.5	Pa 5.89 —	U 6.19 —	Np 6.27 —	Pu 6.06 —	Am 5.99 —	Cm 6.02 —	Bk 6.23 —	Cf 6.30 —	Es 6.42 —	Fm 6.50 —	Md 6.58 —	No 6.65 —	Lr — —

Source: Data from D. R. Lide ed., *CRC Handbook of Chemistry and Physics,* 75th ed., CRC Press, Boca Raton, Fla., 1994, pp. 10–205, 10–206.

[a]Top, IE(1); bottom, IE(2).

59

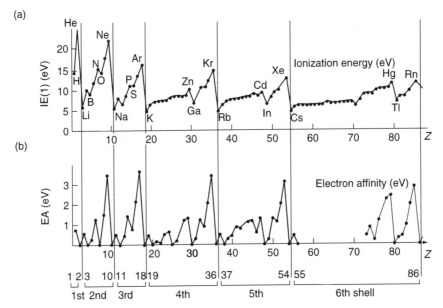

Figure 2.7. First ionization energy IE(1) and electron affinity EA of neutral atoms of the elements as functions of atomic number Z up to $Z = 86$: (a) IE(1) versus Z; (b) EA versus Z. (Data on electron affinities reprinted with permission of W. C. Lineberger.)

The two terms on the right-hand side of Eq. (2.12) are the covalent and ionic contributions to the bond energy $E(A–B)$, respectively. The covalent part is assumed to be given by the average of the bond energies in the elemental solids A and B. In another approach the geometric mean of the bond energies, $[E(A–A)E(B–B)]^{1/2}$, is used instead of the simple average. The ionic part is taken to be proportional to the square of the electronegativity difference $(X_A - X_B)$. Thus the number of electrons transferred between atoms A and B is proportional to $(X_A - X_B)$. It follows that bonds become increasingly ionic as the difference in electronegativity of the constituent atoms increases.

Electronegativities, including the definitions proposed by Mulliken,[†] Sanderson (1976), and Phillips (1973), are discussed further in Chapter W2.

2.10   Atomic Radii: Ionic, Covalent, Metallic, and van der Waals

The sizes or radii of atoms and ions can play critical roles in the structures of multi-element crystals. The effect of the value of the ionic radius on the structure of crystals such as NaCl and CsCl in which the bonding is primarily ionic has already been discussed, both in Chapter 1 and earlier in this chapter. Atomic size effects can also be expected to be important in metallic alloys and compounds, where one of the primary factors determining crystal structure is the filling of space by the constituent atoms.

The size and radius of an atom are determined by the spatial extent of the electrons in its outermost orbitals. Since atoms are not hard spheres but, instead, have electron

---

[†] R. S. Mulliken, *J. Chem. Phys.* **2**, 782 (1934); **3**, 573 (1935).

TABLE 2.10 Electron Affinities EA of the Atoms up to Z = 87 (eV)

1	2		3	4	5	6	7	8	9	10	11	12	13	14	15	16	17	18
1 H 0.754																		2 He <0
3 Li 0.618	4 Be <0												5 B 0.280	6 C 1.26	7 N <0	8 O 1.46	9 F 3.40	10 Ne <0
11 Na 0.548	12 Mg <0												13 Al 0.433	14 Si 1.39	15 P 0.747	16 S 2.08	17 Cl 3.61	18 Ar <0
19 K 0.501	20 Ca 0.025		21 Sc 0.188	22 Ti 0.084	23 V 0.525	24 Cr 0.676	25 Mn <0	26 Fe 0.151	27 Co 0.663	28 Ni 1.16	29 Cu 1.24	30 Zn <0	31 Ga 0.41	32 Ge 1.23	33 As 0.81	34 Se 2.02	35 Br 3.36	36 Kr <0
37 Rb 0.486	38 Sr 0.052		39 Y 0.307	40 Zr 0.426	41 Nb 0.893	42 Mo 0.747	43 Tc 0.55	44 Ru 1.05	45 Rh 1.14	46 Pd 0.562	47 Ag 1.30	48 Cd <0	49 In 0.404	50 Sn 1.11	51 Sb 1.05	52 Te 1.97	53 I 3.06	54 Xe <0
55 Cs 0.472	56 Ba 0.145	57 La 0.47	72 Hf ≈ 0	73 Ta 0.322	74 W 0.815	75 Re 0.15	76 Os 1.08	77 Ir 1.56	78 Pt 2.13	79 Au 2.31	80 Hg <0	81 Tl 0.377	82 Pb 0.364	83 Bi 0.942	84 Po 1.9	85 At 2.8	86 Rd <0
87 Fr 0.46	88 Ra —	89 Ac —																

Source: Data from T. Andersen et al., *J. Phys. Chem. Ref. Data,* **28**, 1511 (1999).

TABLE 2.11   Electronegativities X of the Atoms (Dimensionless)

1	2	3	4	5	6	7	8	9	10	11	12	13	14	15	16	17	18
H 2.1 (2.2)																	He
Li 1.0 (0.98, 0.97)	Be 1.5 (1.57, 1.47)											B 2.0 (2.04, 2.01)	C 2.5 (2.55)	N 3.0 (3.04, 3.07)	O 3.5 (3.44)	F 4.0 (3.98, 4.10)	Ne
Na 0.9 (0.93, 1.01)	Mg 1.2 (1.31, 1.23)											Al 1.5 (1.61, 1.47)	Si 1.8 (1.90, 1.74)	P 2.1 (2.19, 2.06)	S 2.5 (2.58, 2.44)	Cl 3.0 (3.16, 2.83)	Ar
K 0.8 (0.82, 0.91)	Ca 1.0 (1.04)	Sc 1.3 (1.36, 1.20)	Ti 1.5 (1.54, 1.32)	V 1.6 (1.63, 1.45)	Cr 1.6 (1.66, 1.56)	Mn 1.5 (1.55, 1.60)	Fe 1.8 (1.83, 1.64)	Co 1.8 (1.88, 1.70)	Ni 1.8 (1.91, 1.75)	Cu 1.9 (1.90, 1.75)	Zn 1.6 (1.65, 1.66)	Ga 1.6 (1.81, 1.82)	Ge 1.8 (2.01, 2.02)	As 2.0 (2.18, 2.20)	Se 2.4 (2.55, 2.48)	Br 2.8 (2.96, 2.74)	Kr
Rb 0.8 (0.82, 0.89)	Sr 1.0 (0.95, 0.99)	Y 1.2 (1.22, 1.11)	Zr 1.4 (1.33, 1.22)	Nb 1.6 (1.23, 1.22)	Mo 1.8 (2.16, 1.30)	Tc 1.9 (1.36)	Ru 2.2 (1.42)	Rh 2.2 (2.28, 1.45)	Pd 2.2 (1.35)	Ag 1.9 (1.93, 1.42)	Cd 1.7 (1.69, 1.46)	In 1.7 (1.78, 1.49)	Sn 1.8 (1.96, 1.72)	Sb 1.9 (2.05, 1.82)	Te 2.1 (2.01)	I 2.5 (2.66, 2.21)	Xe
Cs 0.7 (0.79, 0.86)	Ba 0.9 (0.89, 0.97)	La 1.1 (1.08)	Hf 1.3 (1.23)	Ta 1.5 (1.33)	W 1.7 (2.36, 1.40)	Re 1.9 (1.46)	Os 2.2 (1.52)	Ir 2.2 (1.55)	Pt 2.2 (2.28, 1.44)	Au 2.4 (2.54, 1.42)	Hg 1.9 (2.00, 1.44)	Tl 1.8 (2.04, 1.44)	Pb 1.8 (2.33, 1.55)	Bi 1.9 (2.02, 1.67)	Po 2.0 (1.76)	At 2.2 (1.96)	Rn
Fr 0.7 (0.86)	Ra 0.9 (0.89, 0.97)	Ac 1.1 (1.0)															

Lanthanides:

Ce	Pr	Nd	Pm	Sm	Eu	Gd	Tb	Dy	Ho	Er	Tm	Yb	Lu
1.1 −1.2 (1.06)	1.1 −1.2 (1.07)	1.1 −1.2 (1.07)	1.1 −1.2 (1.07)	1.1 −1.2 (1.07)	1.1 −1.2 (1.01)	1.1 −1.2 —	1.1 −1.2 —	1.1 −1.2 (1.22)	1.1 −1.2 (1.23)	1.1 −1.2 (1.24)	1.1 −1.2 (1.25)	1.1 −1.2 (1.06)	1.1 −1.2 (1.27)

Actinides:

Th	Pa	U	Np	Pu	Am	Cm	Bk	Cf	Es	Fm	Md	No	Lr
1.3 −1.2 (1.11)	1.5 (1.14)	1.7 (1.38, 1.22)	1.3 (1.36, 1.22)	1.3 (1.28, 1.22)	1.3 —	1.3 —	1.3 —	1.3 —	1.3 —	1.3 —	1.3 —	1.3 —	1.3 —

Source: L. Pauling, *The Nature of the Chemical Bond*, 3rd ed., Cornell University Press, Ithaca, N.Y., 1960. Values in parentheses represent revised values of the Pauling electronegativities and also values from A. L. Allred and E. G. Rochow, *J. Inorg. Nucl. Chem.*, **5**, 264 (1958), and E. J. Little, Jr. and M. M. Jones, *J. Chem. Ed.*, **37**, 231 (1960).

TABLE 2.12 Atomic Radii $(pm = 10^{-12}\ m)^a$

Legend (each box):
r_{ion}	
r_{cov}	
r_{met}	

Z	Element	r_{ion}	r_{cov}	r_{met}
1	H	—	—	—
2	He	(178)	—	—
3	Li	76	—	152
4	Be	45	106	115
5	B	—	88	98
6	C(d)	16	77	92
7	N	13	70	—
8	O	140	66	—
9	F	133	64	—
10	Ne	(158)	—	—
11	Na	102	—	186
12	Mg	72	140	160
13	Al	54	126	143
14	Si	40	118	132
15	P	38	110	—
16	S	184	104	—
17	Cl	181	99	—
18	Ar	(188)	—	—
19	K	138	—	230
20	Ca	100	—	198
21	Sc	75	—	165
22	Ti	86	—	147
23	V	79	—	131
24	Cr	73	—	125
25	Mn	83	—	136
26	Fe	55	—	124
27	Co	65	—	125
28	Ni	69	—	124
29	Cu	77	135	128
30	Zn	74	131	133
31	Ga	62	126	141
32	Ge	53	123	137
33	As	58	118	—
34	Se	198	114	—
35	Br	196	111	—
36	Kr	(205)	—	—
37	Rb	152	—	247
38	Sr	118	—	215
39	Y	90	—	183
40	Zr	72	—	162
41	Nb	72	—	143
42	Mo	69	—	136
43	Tc	65	—	137
44	Ru	68	—	135
45	Rh	67	—	134
46	Pd	86	—	138
47	Ag	115	152	145
48	Cd	95	148	149
49	In	80	144	166
50	Sn	69	140	162
51	Sb	76	136	—
52	Te	221	132	—
53	I	220	128	—
54	Xe	(225)	—	—
55	Cs	167	—	266
56	Ba	135	—	217
57	La	103	—	188
58	Ce	101	—	184
59	Pr	99	—	183
60	Nd	98	—	183
61	Pm	97	—	182
62	Sm	96	—	183
63	Eu	95	—	198
64	Gd	94	—	181
65	Tb	92	—	180
66	Dy	107	—	179
67	Ho	—	—	179
68	Er	89	—	178
69	Tm	88	—	177
70	Yb	99	—	194
71	Lu	86	—	175
72	Hf	71	—	159
73	Ta	72	—	143
74	W	60	—	137
75	Re	63	—	138
76	Os	63	—	136
77	Ir	68	—	136
78	Pt	80	—	139
79	Au	137	—	144
80	Hg	119	148	150
81	Tl	89	147	173
82	Pb	119	146	175
83	Bi	103	145	170
84	Po	97	—	168
85	At	—	—	—
86	Rn	—	—	—
87	Fr	—	—	280
88	Ra	—	—	247
89	Ac	112	—	188
90	Th	94	—	180
91	Pa	104	—	163
92	U	103	—	156
93	Np	101	—	156
94	Pu	100	—	164
95	Am	98	—	173
96	Cm	97	—	175
97	Bk	96	—	171
98	Cf	95	—	170
99	Es	—	—	—
100	Fm	—	—	—
101	Md	—	—	—
102	No	—	—	—
103	Lr	—	—	—

^{a}Top: ionic radius r_{ion} (octahedral coordination with six NNs). Values in parentheses are closed-shell radii obtained from the solid state. [Data from D. R. Lide, ed.-in-chief, *CRC Handbook of Chemistry and Physics*, 79th ed. (CRCnetBASE 1999), CRC Press, Boca Raton, Fla., 1999, pp. 12–14 to 12–16.] Middle: covalent radius r_{cov} (tetrahedral coordination with four NNs and with covalent or mixed ionic–covalent bonding). (Data from L. Pauling, *The Nature of the Chemical Bond*, 3rd ed., Cornell University Press, Ithaca, N.Y., 1960.) Bottom: metallic radius r_{met} (with 8 or 12 NNs, according to crystal structure given in Table 1.4). (From crystal structure data given in Table 1.4 and from W. B. Pearson, *The Crystal Chemistry and Physics of Metals and Alloys*, Wiley-Interscience, New York, 1972, p. 151.)

orbitals that are spread out in space, the radius of an atom is not directly measurable and is not a well-defined quantity. Also, when the charge distribution of an atom or ion is nonspherical, the dimensions of the atom can be different for different directions in the crystal. Despite this, empirical sets of self-consistent atomic radii do convey potentially useful information.

The situation for the radii of atoms bonded in solids is complicated by the fact that interatomic distances and hence atomic sizes can be different for the same atom in crystals in which different types of bonding dominate. For this reason, *ionic, covalent, metallic*, and *van der Waals radii* are often specified for the same atom. It is clear that bonding and crystal structure can influence the radius of an atom just as, in turn, the size of an atom can help to determine the bonding and structure of multielement crystals in which it is a component. This is particularly true in metals, where the "size" of an atom is determined by the number of NNs with which it is in contact. A more useful parameter describing the size of atoms is the atomic volume, which, for a given atom, appears to be more nearly constant in different crystal structures.

Self-consistent sets of radii, r_{ion} for ions with filled outer-shell configurations, r_{cov} for atoms participating in (usually, tetrahedral) covalent bonds, and r_{met} for metallic ions in configurations with 8 or 12 NNs, are presented in Table 2.12. These sets of radii have been obtained self-consistently from the measured lattice constants of a wide range of solids and in some cases from internuclear distances in molecules. Specifically,

TABLE 2.13   van der Waals Radii r_{vdW} for Selected Atoms

Atom	r_{vdW} (nm)
H	0.12
N	0.15
O	0.14
F	0.135
Ne	0.158
P	0.19
S	0.185
Cl	0.18
Ar	0.188
As	0.20
Se	0.20
Br	0.195
Kr	0.205
Sb	0.22
Te	0.22
I	0.215
Xe	0.225

Source: Data from L. Pauling, *The Nature of the Chemical Bond*, 3rd ed., Cornell University Press, Ithaca, N.Y., 1960, p. 260; values for the inert-gas atoms are calculated from the lattice constants of the corresponding inert-gas crystals presented in Table 1.4.

values for r_{ion} have been obtained from highly ionic crystals with the NaCl structure, values for r_{cov} from tetrahedrally coordinated covalent crystals and covalently bonded molecules, and values for r_{met} from crystals of the metallic elements. Trends observed in these radii are discussed in Chapter W2.

The van der Waals atomic radii r_{vdW} are appropriate for neutral atoms with filled outer shells which are effectively in contact with other atoms in solids but which are not bonded to them. Selected values of r_{vdW} are presented in Table 2.13.

REFERENCES

Borg, R. J., and G. J. Dienes, *The Physical Chemistry of Solids*, Academic Press, San Diego, Calif., 1992.

Brown, G. I., *A New Guide to Modern Valency Theory*, SI Edition, Longman Group, London, 1972.

Burdett, J. K., *Chemical Bonding in Solids*, Oxford University Press, New York, 1995.

Burns, G., *Solid State Physics*, Academic Press, San Diego, Calif., 1985.

Christman, J. R., *Fundamentals of Solid State Physics*, Wiley, New York, 1988.

Companion, A. L., *Chemical Bonding*, McGraw-Hill, New York, 1979.

Jaffe, H. W., *Crystal Chemistry and Refractivity*, Dover, Mineola, N.Y., 1996.

Kittel, C., *Introduction to Solid State Physics*, 7th ed., Wiley, New York, 1996.

McKie, D., and C. McKie, *Crystalline Solids*, Wiley, New York, 1974.

Pauling, L., *The Nature of the Chemical Bond*, 3rd ed., Cornell University Press, Ithaca, N.Y., 1960.

Pearson, W. B., *The Crystal Chemistry and Physics of Metals and Alloys*, Wiley, New York, 1972.

Phillips, J. C., *Bonds and Bands in Semiconductors*, Academic Press, San Diego, Calif., 1973.

Sanderson, R. T., *Chemical Bonds and Bond Energy*, 2nd ed., Academic Press, San Diego, Calif., 1976.

Taylor, J. R., and C. D. Zafiratos, *Modern Physics for Scientists and Engineers*, Prentice Hall, Upper Saddle River, N.J., 1991.

Wyckoff, R. W. G., *Crystal Structures*, Vols. I–III, 2nd ed., Interscience, New York, 1963.

PROBLEMS

2.1 Compute the cohesive energies for monatomic crystals of atoms bound together by the Lennard-Jones potential $U(r)$ given in Eq. (2.3). Express the cohesive energy, $\Delta H_c(0\text{ K}) = n(\text{atom})(\text{CN}/2)|U(r_0)|$, in terms of the parameter ε and the equilibrium interatomic distance r_0 in terms of the parameter σ. Here $n(\text{atom})$ is the concentration of atoms and CN is the coordination number. Carry out the calculations for the SC, BCC, and FCC crystal structures.

2.2 Given the following lattice constants for crystals with the NaCl crystal structure, $a(\text{NaCl}) = 0.563$ nm, $a(\text{KCl}) = 0.629$ nm, $a(\text{NaF}) = 0.462$ nm, and $a(\text{KF}) = 0.535$ nm, show that these data are not sufficient to obtain a self-consistent set of ionic radii for the Na^+, K^+, Cl^-, and F^- ions. Why is it not possible to determine a completely self-consistent set of radii from the data given?

2.3 Use the cohesive energy ΔH_c (see Table 2.8) of cubic β-SiC with the zincblende crystal structure to determine the bond energy $E(\text{Si–C})$.

2.4 Calculate the potential energy U of an anion–cation (Na^+–Cl^-) pair resulting from their mutual Coulomb attraction and then compare the result with the cohesive energy ΔH_c of NaCl listed in Table 2.8. The lattice constant of NaCl is $a = 0.563$ nm. (*Hint*: Take into account the fact that each Na^+ ion in NaCl interacts with six NN Cl^- ions, and vice versa.)

2.5 In the structural phase transition from BCC α-Fe to FCC γ-Fe at $T = 912°$C the lattice constant changes from $a(\text{BCC}) = 0.290$ nm to $a(\text{FCC}) = 0.364$ nm. Assuming that the Fe atoms act as hard spheres, which is more nearly constant in α-Fe and γ-Fe — the radius r_{met} or the atomic volume V_{met}?

Note: Additional problems are given in Chapter W2.

CHAPTER 3

Diffraction and the Reciprocal Lattice

DIFFRACTION

In this chapter, *diffraction*, the scattering of a coherent wave by a crystal, is considered. Given the wave–particle duality of nature, the elastic scattering of a beam of particles from a crystal is also discussed. The projectile particles may be x-rays, neutrons, or high-energy electrons. These particles are able to penetrate the bulk of the material and provide information concerning its three-dimensional geometric and chemical structure. Other scattering probes, such as low-energy electrons or atomic beams, are not as penetrating and are more useful for studying the surface region of a solid. Diffraction is sensitive to the translational periodicity presented by the crystal. For a periodic solid, scattering occurs only into a discrete set of directions in space. The strength of the scattering is determined by the nature of both the probe and the target. In the case of x-ray scattering it is the electron density that determines the scattering amplitude. For neutrons it is the nuclear force that is responsible. In both cases the lattice potential determining the dynamics has translational periodicity.

For diffraction patterns to be observable, the characteristic wavelengths of the probe beam must be on the order of or smaller than (twice the) interatomic spacings in the solid. For example, a 10-keV x-ray has a wavelength $\lambda = 0.124$ nm, a thermal neutron (with kinetic energy 1/40 eV) has a de Broglie wavelength $\lambda = 0.181$ nm, and a 100-keV electron has $\lambda = 0.0037$ nm. The wavelength of the x-ray in terms of its energy E is given by $\lambda = hc/E$. For the neutron it is given by $\lambda = h/\sqrt{2m_n T}$, where T is its kinetic energy. For the electron it is given by the relativistic formula $\lambda = h/\sqrt{T(2m_e + T/c^2)}$. Here h is Planck's constant, c is the speed of light, and m_n and m_e are the rest masses of the neutron and electron, respectively.

X-rays were discovered in 1895 by W. Roentgen when he analyzed the emanations produced by the impact of high-energy cathode rays (electrons) on metallic anodes. Evidence for the electromagnetic nature of this radiation was provided in 1912 by W. Friedrich, P. Knipping, and M. von Laue when they passed a beam of x-rays through a crystal and observed a pattern of diffraction spots. These were analogous to the diffraction pattern observed when visible light passed through a pair of crossed diffraction gratings. A systematic study of diffraction spots produced during x-ray reflection allowed W. H. Bragg and W. L. Bragg in 1913 to extract detailed structural and symmetry information for a broad range of crystals. The use of x-ray diffraction techniques has continued to this day. Synchrotron radiation sources provide scientists and engineers with a tunable source of x-ray wavelengths. Large biomolecules may be crystallized and their structures may be determined using x-ray diffraction methods. X-ray diffraction has become the tool of choice for studying the structure of all types

of materials. Although most effective for crystalline materials, where a full characterization is often possible, diffraction also provides useful information about disordered
materials. X-rays are discussed further at our Web site in Chapter W22.[†]

This chapter begins with a discussion of Fourier analysis in one and three dimensions
and introduces the reciprocal lattice. We then proceed to discuss elastic scattering from
both ordered and disordered materials. The latter include polycrystalline and amorphous
solids. An analysis of how the internal structure of the unit cell may be determined is
given at our Web site.

3.1   Fourier Analysis in One and Three Dimensions

The incident beam is taken to be monochromatic (or monoenergetic) and well colli-
mated (unidirectional). It is represented by a plane wave $\psi = \exp[i(\mathbf{k} \cdot \mathbf{r} - \omega t)]$ with
wave vector $\mathbf{k}$ and angular frequency ω. In describing diffraction one must know how
this wave scatters into another wave ψ' with wave vector $\mathbf{k}'$ but with the same frequency
ω. The scattering amplitude is determined by an integral of the form $\int d\mathbf{r} \psi'^* V \psi$, where
V describes some physical property of the crystal which depends on the nature of the
probe. Whatever this quantity is, be it electron density for x-ray scattering, the nuclear
potential for neutron scattering, or the lattice potential for electron scattering, the quantity V will often, but not always, be a periodic function of space. In this section it will
be seen how Fourier analysis allows one to express mathematically what is happening
in x-ray diffraction.

A periodic function in one dimension such that

$$V(x) = V(x \pm a) \tag{3.1}$$

may be represented as a Fourier series:

$$V(x) = \sum_{n=-\infty}^{\infty} V_n e^{i(2\pi n x/a)}, \tag{3.2}$$

where the Fourier coefficients V_n may be found by multiplying through by
$\exp[-i(2\pi m x/a)]$ and integrating over the spatial period. Use is made of the integral

$$\int_0^a e^{i[2\pi(n-m)x/a]} dx = a\delta_{m,n'} \tag{3.3}$$

where $\delta_{m,n}$, called *Kronecker's delta*, is 1 if $m = n$ and 0 otherwise. The formula for
the Fourier coefficients is

$$V_n = \frac{1}{a} \int_0^a V(x) e^{-i(2\pi n x/a)} dx. \tag{3.4}$$

In a similar way a three-dimensional periodic function $V(\mathbf{r})$ may be invariant under
translation by a Bravais lattice vector $\mathbf{R}$:

$$V(\mathbf{r} + \mathbf{R}) = V(\mathbf{r}). \tag{3.5}$$

[†] Supplementary material for this textbook is included on the Web at the resource site (ftp://ftp.wiley.com/
public/sci_tech_med/materials). Cross-references to elements of the Web material are prefixed by "W."

The problem is to generalize the foregoing Fourier analysis technique to three dimensions. To this end, $\mathbf{R}$ is expressed as a linear combination of primitive lattice translation vectors,

$$\mathbf{R} = n_1\mathbf{u}_1 + n_2\mathbf{u}_2 + n_3\mathbf{u}_3, \tag{3.6}$$

where $\{n_1, n_2, n_3\}$ are a set of integers. Introduce a set of primitive reciprocal lattice vectors $\{\mathbf{g}_1, \mathbf{g}_2, \mathbf{g}_3\}$ such that

$$\mathbf{g}_i \cdot \mathbf{u}_j = 2\pi\delta_{i,j}, \qquad i = 1, 2, 3, \quad j = 1, 2, 3. \tag{3.7}$$

Explicit formulas for the $\mathbf{g}_i$ are given by the expressions

$$\mathbf{g}_1 = 2\pi \frac{\mathbf{u}_2 \times \mathbf{u}_3}{\mathbf{u}_1 \cdot \mathbf{u}_2 \times \mathbf{u}_3}, \qquad \mathbf{g}_2 = 2\pi \frac{\mathbf{u}_3 \times \mathbf{u}_1}{\mathbf{u}_1 \cdot \mathbf{u}_2 \times \mathbf{u}_3}, \qquad \mathbf{g}_3 = 2\pi \frac{\mathbf{u}_1 \times \mathbf{u}_2}{\mathbf{u}_1 \cdot \mathbf{u}_2 \times \mathbf{u}_3}. \tag{3.8}$$

A general reciprocal lattice vector is expressed as a linear combination of the primitive reciprocal lattice vectors with integer coefficients

$$\mathbf{G} = j_1\mathbf{g}_1 + j_2\mathbf{g}_2 + j_3\mathbf{g}_3 \tag{3.9}$$

and is thus described by a set of three integers $\{j_1, j_2, j_3\}$. The concept of the reciprocal lattice will play an important role in the understanding of the electron band structure of crystals, as described in Chapter 7.

The Fourier expansion of the crystal potential is

$$V(\mathbf{r}) = \sum_{\mathbf{G}} V_{\mathbf{G}} e^{i\mathbf{G}\cdot\mathbf{r}}, \tag{3.10}$$

where $\{V_{\mathbf{G}}\}$ are a set of Fourier coefficients. Periodicity is obvious since

$$V(\mathbf{r} + \mathbf{R}) = \sum_{\mathbf{G}} V_{\mathbf{G}} e^{i\mathbf{G}\cdot(\mathbf{r}+\mathbf{R})} = \sum_{\mathbf{G}} V_{\mathbf{G}} e^{i\mathbf{G}\cdot\mathbf{r}} e^{i2\pi(j_1n_1 + j_2n_2 + j_3n_3)}$$

$$= \sum_{\mathbf{G}} V_{\mathbf{G}} e^{i\mathbf{G}\cdot\mathbf{r}} = V(\mathbf{r}). \tag{3.11}$$

In the one-dimensional case the basic unit cell is the interval $0 < x < a$. The analogous volume unit in the three-dimensional case is called the *Wigner–Seitz cell*, defined as follows. Select any point in the crystal as a reference point and label it O. Replicate that point by translating it through all possible lattice translation vectors $\mathbf{R}$ and label the resulting set $\{O_{\mathbf{R}}\}$. The Wigner–Seitz (WS) cell is the set of all points $\mathbf{r}$ in the crystal that are closer to O than to any other $O_{\mathbf{R}}$. The boundary of the WS cell has a polyhedral shape. Every point on a face of the polyhedron has a corresponding point on an opposite face that may be reached by a primitive lattice translation vector.

An alternative way of constructing the WS cell is to single out a particular atom, S, and to draw lines to all other atoms occupying the same basis site in other lattice cells. Introduce the perpendicular bisecting planes of these lines. Atom S will then be surrounded by a polyhedral shape. This is also a WS cell.

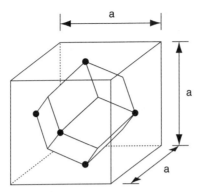

Figure 3.1. Wigner–Seitz cell for the FCC lattice. The heavy dots are at the centers of the faces of the cube.

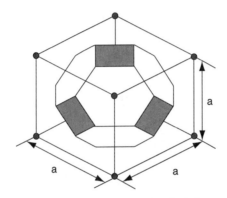

Figure 3.2. Wigner–Seitz cell for the BCC lattice. The shaded squares are on the surface of the cube.

Figure 3.1 illustrates the WS cell for the FCC lattice. It is a regular rhombic dodecahedron whose bounding surface consists of 12 squares perpendicular to the $\langle 110 \rangle$ directions. The WS cell for the BCC lattice is the truncated octahedron given in Fig. 3.2. Its bounding surface consists of six squares oriented perpendicular to the $\langle 100 \rangle$ directions and eight regular hexagons perpendicular to the $\langle 111 \rangle$ directions.

The following orthogonality identity holds:

$$\int_{WS} e^{i(\mathbf{G}-\mathbf{G}')\cdot\mathbf{r}} d\mathbf{r} = V_{WS}\delta_{\mathbf{G},\mathbf{G}'}, \tag{3.12}$$

where the integral $d\mathbf{r}$ is over the volume of the Wigner–Seitz cell. This gives a formula for the Fourier coefficients:

$$V_{\mathbf{G}} = \frac{1}{V_{WS}} \int_{WS} V(\mathbf{r}) e^{-i\mathbf{G}\cdot\mathbf{r}} d\mathbf{r}. \tag{3.13}$$

An explicit formula for the volume of the Wigner–Seitz cell is

$$V_{WS} = |\mathbf{u}_1 \cdot \mathbf{u}_2 \times \mathbf{u}_3|. \tag{3.14}$$

Note that the volume of the WS cell is the same as that of the primitive unit cell defined in Chapter 1.

In the expression for **G** in Eq. (3.9), three integers were introduced, $\{j_1, j_2, j_3\}$. If these integers have a common integer divisor, **G** will be called *reducible*. If there is no common integer divisor, **G** is *irreducible*. Thus the set $\{2,4,6\}$ is reducible, whereas $\{1,2,3\}$ is irreducible.

The reciprocal lattice vectors may be given a simple geometric interpretation. Select an origin O in real space and look at the set of points described by vectors **r** originating at O and satisfying the equation

$$\mathbf{G} \cdot \mathbf{r} = 2\pi \rightarrow \hat{G} \cdot \mathbf{r} = \frac{2\pi}{G}, \qquad (3.15)$$

where $\hat{G}$ is a unit vector and **G** is irreducible. A linear equation of the form $G_x X + G_y Y + G_z Z = 2\pi$ defines a plane in the direct lattice. This plane lies a distance $d = 2\pi/G$ from O and its normal is oriented along **G**. There are, in fact, an infinite set of parallel planes described by the equations

$$\mathbf{G} \cdot \mathbf{r} = 2\pi N \rightarrow \hat{G} \cdot \mathbf{r} = \frac{2\pi N}{G}, \qquad (3.16)$$

where N is any integer and **G** is reducible. The spacing between successive planes is $d = 2\pi/G$, where **G** is irreducible. The significance of these lattice planes will be seen shortly.

Suppose that **G** is irreducible and look at the intersection of the plane defined by $\mathbf{G} \cdot \mathbf{r} = 2\pi$ with the axes defined by the vectors $\{\mathbf{u}_1, \mathbf{u}_2, \mathbf{u}_3\}$. For $\mathbf{r} = \mathbf{u}_1/h$, one obtains $\mathbf{G} \cdot \mathbf{r} = 2\pi j_1/h = 2\pi$, so $j_1 = h$. Similarly, if one writes $\mathbf{r} = \mathbf{u}_2/k$, one obtains $\mathbf{G} \cdot \mathbf{r} = 2\pi j_2/k = 2\pi$, so $j_2 = k$. Finally, for $\mathbf{r} = \mathbf{u}_3/l$, one finds $j_3 = l$. Thus the components of the **G** vector expressed in terms of the primitive reciprocal lattice vectors can be given a simple geometrical interpretation. The intersection of the lattice planes with the primitive axes occurs at coordinates that are inversely proportional to the coefficients j_1, j_2, and j_3. The coefficients (hkl) are referred to as the *Miller indices* of the plane, as discussed in Section 1.5.

3.2 Examples of Reciprocal Lattices

Reciprocal lattices corresponding to the various Bravais lattices are simple to generate. Starting with the class of cubic crystals, the primitive translation vectors for the simple cubic (SC) lattice may be chosen as

$$\mathbf{u}_1 = a\hat{i}, \qquad \mathbf{u}_2 = a\hat{j}, \qquad \mathbf{u}_3 = a\hat{k}. \qquad (3.17)$$

The volume of the unit cell is a^3 and the primitive reciprocal lattice vectors, determined from Eq. (3.8), are

$$\mathbf{g}_1 = \frac{2\pi}{a}\hat{i}, \qquad \mathbf{g}_2 = \frac{2\pi}{a}\hat{j}, \qquad \mathbf{g}_3 = \frac{2\pi}{a}\hat{k}. \qquad (3.18)$$

The orthogonality relations of Eq. (3.7) are obvious.

For the FCC lattice the primitive translation vectors may be chosen in the symmetric form

$$\mathbf{u}_1 = \frac{a}{2}(\hat{j} + \hat{k}), \qquad \mathbf{u}_2 = \frac{a}{2}(\hat{k} + \hat{i}), \qquad \mathbf{u}_3 = \frac{a}{2}(\hat{i} + \hat{j}), \qquad (3.19)$$

as shown in Fig. 1.5. The volume of the unit cell is $a^3/4$. The primitive reciprocal lattice vectors are

$$\mathbf{g}_1 = \frac{2\pi}{a}(-\hat{i} + \hat{j} + \hat{k}), \qquad \mathbf{g}_2 = \frac{2\pi}{a}(\hat{i} - \hat{j} + \hat{k}), \qquad \mathbf{g}_3 = \frac{2\pi}{a}(\hat{i} + \hat{j} - \hat{k}). \quad (3.20)$$

For the BCC lattice the corresponding vectors are

$$\mathbf{u}_1 = \frac{a}{2}(-\hat{i} + \hat{j} + \hat{k}), \qquad \mathbf{u}_2 = \frac{a}{2}(\hat{i} - \hat{j} + \hat{k}), \qquad \mathbf{u}_3 = \frac{a}{2}(\hat{i} + \hat{j} - \hat{k}), \qquad (3.21)$$

as shown in Fig. 1.4. The volume of the unit cell is $a^3/2$ and

$$\mathbf{g}_1 = \frac{2\pi}{a}(\hat{j} + \hat{k}), \qquad \mathbf{g}_2 = \frac{2\pi}{a}(\hat{i} + \hat{k}), \qquad \mathbf{g}_3 = \frac{2\pi}{a}(\hat{i} + \hat{j}). \qquad (3.22)$$

Note that the BCC lattice of lattice constant a in real space generates an FCC lattice of lattice constant $4\pi/a$ in reciprocal space, whereas the FCC lattice of lattice constant a in real space produces a BCC lattice of lattice constant $4\pi/a$ in reciprocal space. For the SC lattice the reciprocal is also SC, but with lattice constant $2\pi/a$. The relation

$$e^{i\mathbf{G}\cdot\mathbf{R}} = 1 \qquad (3.23)$$

is the formula connecting the two lattices, so it is clear that if the $\{\mathbf{R}\}$ vectors form a Bravais lattice, so should the $\{\mathbf{G}\}$ vectors. From the symmetry between $\mathbf{G}$ and $\mathbf{R}$ in this formula, it follows that the reciprocal of the reciprocal lattice is again the direct lattice in real space.

Also note that just as the primitive translation vectors are not unique, neither are the primitive reciprocal lattice vectors. For example, return to the SC lattice and choose

$$\mathbf{u}_1 = a\hat{i}, \qquad \mathbf{u}_2 = a\hat{j}, \qquad \mathbf{u}_3 = a(\hat{i} + \hat{j} + \hat{k}). \qquad (3.24)$$

The volume of the unit cell is still a^3 and the reciprocal lattice vectors are

$$\mathbf{g}_1 = \frac{2\pi}{a}(\hat{i} - \hat{k}), \qquad \mathbf{g}_2 = \frac{2\pi}{a}\hat{j}, \qquad \mathbf{g}_3 = \frac{2\pi}{a}\hat{k}. \qquad (3.25)$$

This is an acceptable set of lattice vectors but is not desirable because it does not conform to the underlying symmetry of the crystal.

The structure in reciprocal space which is analogous to the Wigner–Seitz cell in real space is the first Brillouin zone (FBZ). One imagines a lattice of reciprocal lattice points (defined by $\{\mathbf{G}\}$) in wave vector space ($\mathbf{k}$ space) with the origin chosen as $\mathbf{O}$. All points in $\mathbf{k}$ space closer to $\mathbf{O}$ than to any other $\mathbf{G}$ are in the FBZ. Like the WS cell, the FBZ, has a polyhedral shape. The volume of the FBZ Ω is

$$\Omega = |\mathbf{g}_1 \cdot \mathbf{g}_2 \times \mathbf{g}_3| = \frac{(2\pi)^3}{|\mathbf{u}_1 \cdot \mathbf{u}_2 \times \mathbf{u}_3|^3}|(\mathbf{u}_2 \times \mathbf{u}_3) \cdot (\mathbf{u}_3 \times \mathbf{u}_1) \times (\mathbf{u}_1 \times \mathbf{u}_2)|$$

$$= \frac{(2\pi)^3}{|\mathbf{u}_1 \cdot \mathbf{u}_2 \times \mathbf{u}_3|} = \frac{(2\pi)^3}{V_{WS}}. \qquad (3.26)$$

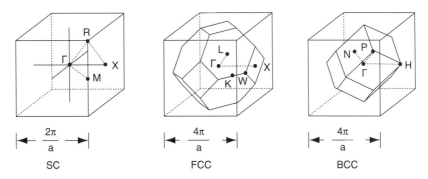

Figure 3.3. First Brillouin zones for the simple cubic, face-centered cubic, and body-centered cubic lattices.

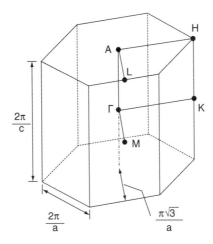

Figure 3.4. First Brillouin zone for the hexagonal lattice.

In Fig. 3.3 the FBZs for the SC, FCC, and BCC lattices are presented. In Fig. 3.4 the FBZ for the hexagonal lattice is sketched. The center of the zone is always called the Γ-*point* and corresponds to $\mathbf{k} = 0$. Other special points in the Brillouin zone are denoted by various alphabetical characters, as shown.

ELASTIC SCATTERING FROM ORDERED AND DISORDERED MATERIALS

In a scattering process there is momentum transfer and energy transfer between the projectile and the solid. An incoming particle with momentum $\mathbf{p}$ is scattered to a state with momentum $\mathbf{p}'$ with the momentum transfer

$$\mathbf{p}' - \mathbf{p} = \hbar\mathbf{q}. \tag{3.27}$$

Similarly, the projectile's energy E is changed to E' with the energy transfer

$$E' - E = \hbar\omega. \tag{3.28}$$

In this chapter attention is restricted to the case of elastic scattering, so that $E' = E$.

For x-rays the energy and momentum are related by $E = pc$. Energy conservation then implies that the magnitude of the momentum is conserved; that is,

$$p' = p. \tag{3.29}$$

For electrons and neutrons the nonrelativistic formula relating energy and momentum is $E = p^2/2m$, with m being the appropriate mass. Energy conservation again implies that $p' = p$. (This remains true even relativistically.) If ϕ denotes the angle between vectors $\mathbf{p}'$ and $\mathbf{p}$, it is readily shown that

$$q = 2\frac{p}{\hbar}\sin\frac{\phi}{2}. \tag{3.30}$$

The scattering geometry is illustrated in Fig. 3.5.

In many situations the amplitude for the wave scattered from a collection of atoms may be expressed as the superposition of scatterings from the individual atoms. Later in the discussion of x-ray scattering (see Chapter W22) it will be seen that the scattering is from electrons, and these are usually associated with given atomic sites. In the case of neutron scattering it is the nuclei that act as scattering centers. For high-energy electrons the scattering is due largely to ion–electron interaction. In these three cases the scattering amplitude may be expressed as a superposition of contributions from the atomic sites {A}:

$$F(\mathbf{q}) = \sum_{\mathbf{A}} f_{\mathbf{A}}(\mathbf{q}) \exp(i\mathbf{q} \cdot \mathbf{A}), \tag{3.31}$$

where $f_A(\mathbf{q})$ is the atomic form factor for a given atom and $\mathbf{q}$ is the wave vector transfer for the projectile. The sum extends over all the atoms in the crystal. The scattering intensity is proportional to the absolute square of the amplitude:

$$I(q) \propto I_0 |F(\mathbf{q})|^2, \tag{3.32}$$

where I_0 is the incident flux. The quantity $I(q)$ is what is measured in a diffraction or scattering experiment. Note that the phase information contained in F is lost in taking this absolute square.

In the analysis above the role of spin or polarization of the probe particles has been neglected. This is an approximation, although usually a good one. Notable exceptions occur, for example, when one uses polarized x-rays or scatters neutrons from

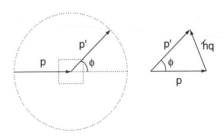

Figure 3.5. Scattering geometry.

magnetic materials. In such cases the spin degrees of freedom must be accounted for, and additional amplitudes need to be introduced. Detailed information concerning the alignment of the nuclear spins in the solid may be obtained from such experiments. Characterization of materials using x-rays is discussed in Chapter W22.

3.3 Crystalline Solids

Consider a crystal defined by a Bravais lattice and a set of basis atoms. The positions of the atoms are

$$\mathbf{A}_j(\mathbf{R}) = \mathbf{R} + \mathbf{s}_j, \tag{3.33}$$

where $\{\mathbf{R}\}$ is a set of Bravais lattice vectors, each pointing to a particular unit cell, and $\{\mathbf{s}_j\}$ is a set of basis vectors locating the atoms within the unit cell. In place of the single subscript $\mathbf{A}$ used before, there are now dual indices, $(\mathbf{R}, \mathbf{s}_j)$. The scattering amplitude is expressed as

$$F(\mathbf{q}) = \sum_{\mathbf{R}} \sum_j f_j(\mathbf{q}) e^{i\mathbf{q}\cdot(\mathbf{R}+\mathbf{s}_j)} = \Phi(\mathbf{q})L, \tag{3.34}$$

where $\Phi(\mathbf{q})$ is the *structure factor* of the basis, defined by

$$\Phi(\mathbf{q}) = \sum_j f_j(\mathbf{q}) e^{i\mathbf{q}\cdot\mathbf{s}_j}, \tag{3.35}$$

and L is the lattice sum,

$$L = \sum_{\mathbf{R}} e^{i\mathbf{q}\cdot\mathbf{R}}. \tag{3.36}$$

In the expression for L, all lattice sites are included. If $\mathbf{R}$ were to be replaced by $\mathbf{R} + \mathbf{R}'$, all lattice sites would still be included and the value of L should not change. Therefore,

$$\sum_{\mathbf{R}} e^{i\mathbf{q}\cdot\mathbf{R}} = \sum_{\mathbf{R}} e^{i\mathbf{q}\cdot(\mathbf{R}+\mathbf{R}')}, \tag{3.37}$$

or

$$\left(1 - e^{i\mathbf{q}\cdot\mathbf{R}'}\right) L = 0. \tag{3.38}$$

Either $\exp(i\mathbf{q}\cdot\mathbf{R}') = 1$, in which case $\mathbf{q}$ is any reciprocal lattice vector or else L vanishes. Thus the scattering amplitude reduces to

$$F(\mathbf{q}) = N\Phi(\mathbf{q}) \sum_{\mathbf{G}} \delta_{\mathbf{q},\mathbf{G}}, \tag{3.39}$$

where N is the number of unit cells in the crystal. The scattering intensity becomes

$$I \sim I_0 N^2 \sum_{\mathbf{G}} \delta_{\mathbf{q},G} |\Phi(\mathbf{q})|^2 = I_0 N^2 \sum_{\mathbf{G}} \delta_{\mathbf{q},G} |\Phi(\mathbf{G})|^2. \tag{3.40}$$

The spectrum consists of a set of discrete points in $\mathbf{q}$ space. The directions of the $\mathbf{G}$ vectors determine the allowed momentum transfers and hence the directions of the diffracted peaks in real space.

The atomic form factors for x-ray scattering are proportional to the Fourier transforms of the electron density $n(\mathbf{r})$:

$$f(\mathbf{q}) \propto \int n(\mathbf{r})e^{-i\mathbf{q}\cdot\mathbf{r}}\,d\mathbf{r}. \tag{3.41}$$

When only a single atom is present, the integral extends over all space. For forward scattering, this is just proportional to the number of electrons, Z:

$$f(\mathbf{0}) \propto \int n(\mathbf{r})\,d\mathbf{r} = Z. \tag{3.42}$$

Thus materials with large atomic numbers will scatter x-rays more efficiently than materials with low Z. With increasing q the form factor tends to fall off. For example, suppose that there is an exponential model for the falloff of n with distance from an isolated atom:

$$n(\mathbf{r}) = \frac{Z}{\pi a^3}e^{-2r/a}, \tag{3.43}$$

where a is on the order of the size of the atom. The corresponding atomic form factor is

$$f(\mathbf{q}) \sim \frac{Z}{1 + (qa/2)^2}, \tag{3.44}$$

which shows the rapid decline of $f(\mathbf{q})$ with q when $qa \gg 1$.

Note that for a monatomic crystal

$$\Phi(\mathbf{G}) = f(\mathbf{G})S(\mathbf{G}), \tag{3.45}$$

where

$$S(\mathbf{G}) = \sum_j e^{i\mathbf{G}\cdot\mathbf{s}_j} \tag{3.46}$$

is called the *geometric structure factor*. If there is only one atom per unit cell, such as in a primitive BCC or FCC lattice, $S(\mathbf{G})$ will simply equal 1. Sometimes, however, it is convenient to choose the conventional cell as being simple cubic and regarding the BCC and FCC lattices as SC lattices with two or four atom bases, respectively. In that case some of the SC diffraction spots will be missing. Thus, for the BCC lattice, let $\mathbf{s}_1 = \mathbf{0}$ and $\mathbf{s}_2 = (a/2)(\hat{i} + \hat{j} + \hat{k})$. For $\mathbf{G} = (2\pi/a)(n_1\hat{i} + n_2\hat{j} + n_3\hat{k})$ it follows that

$$S(\mathbf{G}) = 1 + e^{i\pi(n_1+n_2+n_3)}, \tag{3.47}$$

which is either 2 or 0, depending on whether $n_1 + n_2 + n_3$ is even or odd, respectively. Similarly, for the FCC lattice with $\mathbf{s}_1 = \mathbf{0}, \mathbf{s}_2 = (a/2)(\hat{j} + \hat{k}), \mathbf{s}_3 = (a/2)(\hat{k} + \hat{i})$, and $\mathbf{s}_4 = (a/2)(\hat{i} + \hat{j})$,

$$S(\mathbf{q}) = 1 + e^{i\pi(n_2+n_3)} + e^{i\pi(n_1+n_3)} + e^{i\pi(n_1+n_2)}, \tag{3.48}$$

which is either 4 (if the n_i are all even or all odd) or 0.

As another example of the calculation of the geometric structure factor, consider Si. The crystal is of the diamond crystal structure: an FCC lattice with a two-atom basis. The Si atoms in the basis are at $\mathbf{s}_1 = \mathbf{0}$ and at $\mathbf{s}_2 = a(\hat{i} + \hat{j} + \hat{k})/4$. The geometric structure factor is

$$S(\mathbf{q}) = \sum_j e^{i\mathbf{q}\cdot\mathbf{s}_j} = 1 + e^{i(a/4)(q_x+q_y+q_z)}. \tag{3.49}$$

Setting $\mathbf{q} = \mathbf{G} = j_1\mathbf{g}_1 + j_2\mathbf{g}_2 + j_3\mathbf{g}_3$ and using the primitive reciprocal lattice vectors for the FCC lattice gives

$$q_x = \frac{2\pi}{a}(-j_1 + j_2 + j_3), \qquad q_y = \frac{2\pi}{a}(j_1 - j_2 + j_3), \qquad q_z = \frac{2\pi}{a}(j_1 + j_2 - j_3). \tag{3.50}$$

Thus

$$[S(\mathbf{q})]_{\text{diamond}} = [S(\mathbf{q})]_{\text{FCC}}\left[1 + e^{i(\pi/2)(j_1+j_2+j_3)}\right]$$

$$= \begin{cases} 2 & \text{if } j_1 + j_2 + j_3 \mod 4 = 0 \\ 1+i & \text{if } j_1 + j_2 + j_3 \mod 4 = 1 \\ 0 & \text{if } j_1 + j_2 + j_3 \mod 4 = 2 \\ 1-i & \text{if } j_1 + j_2 + j_3 \mod 4 = 3. \end{cases} \tag{3.51}$$

The notation "$J \mod n$" means "take J modulus n". For positive J one continues to subtract n from J until there is a positive remainder smaller than n. For negative J one adds as many n's as are necessary to make the result positive.

It is worthwhile comparing the results for the monatomic FCC crystal structure with the result for the diamond crystal structure. For FCC crystal structures all the nonvanishing spots have the same intensity. For diamond the spots have intensities 4, 2, 0, and 2.

3.4 Bragg and von Laue Descriptions of Diffraction

There are two equivalent ways of describing how a diffraction pattern is produced, due to Bragg and von Laue, respectively. The Bragg description is based on a wave description of the probe beam and involves constructive interference of the waves. The von Laue picture is based on a particle description of the beam and is closely related to the conservation laws for energy and momentum.

Figure 3.6 illustrates both viewpoints. Beam 1 with wave vector $\mathbf{k}$ is reflected into beam $1'$ with wave vector $\mathbf{k}'$ from lattice plane p_1. Similarly, beam 2, which is parallel to 1, diffracts from lattice plane p_2 into beam $2'$, which is parallel to $1'$. The lattice planes are a distance d apart. The angles between the beams and the plane are θ. Note that the full scattering angle is given by $\phi = 2\theta$. Also note that $k' = k$ for elastic scattering. The path difference between the waves scattered from the two adjacent lattice planes is seen to be $2d\sin\theta$. The Bragg condition for constructive interference is that the path difference is an integer number of wavelengths

$$2d\sin\theta = n\lambda = n\frac{2\pi}{k}, \tag{3.52}$$

where n is an integer ($n = 1$ is the first-order diffraction, $n = 2$ the second order, etc.). This implies that

$$2k\sin\theta = nG_1 = G, \tag{3.53}$$

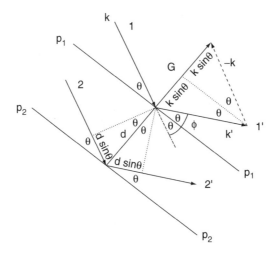

Figure 3.6. Diffraction from two lattice planes.

where $G_1 = 2\pi/d$ is an irreducible reciprocal lattice vector and G is a general reciprocal lattice vector.

It is important to note that the Bragg condition is related only to the lattice and provides no information about the basis atoms or the intensities of the diffraction maxima. Such information is provided by the structure factor of the basis and depends on the atomic form factors.

Von Laue argued that when particles scatter from a periodic structure, momentum will be conserved. Momentum may be transferred to or from the solid in units of $\hbar\mathbf{G}$ and the lattice recoils. However, since the lattice is so massive, the recoil carries with it very little energy. Thus

$$\mathbf{q} = \mathbf{k}' - \mathbf{k} = \mathbf{G}, \tag{3.54}$$

consistent with the Kronecker delta appearing in Eq. (3.40). This is illustrated graphically by the upper right-hand pair of triangles in Fig. 3.6. It is seen explicitly that $G = 2k\sin\theta$, in agreement with the Bragg result.

3.5   Polycrystalline Solids or Powders

Next consider a homogeneous material consisting of a set of randomly oriented and randomly spaced crystallites of finite size, such as is shown in Fig. 3.7 or Fig. 4.1*b* and *c*. The notation used to locate a given atom now needs to be enlarged. In analogy with Eq. (3.33), one has

$$\mathbf{A}_j(\mathbf{R}, \mathbf{U}) = \mathbf{U} + \mathbf{R}_\mathbf{U} + \mathbf{S}_{\mathbf{U}_j}. \tag{3.55}$$

Here $\mathbf{U}$ locates a particular crystallite. Since the orientations of the lattices of the crystallites differ from each other, the lattice vectors $\mathbf{R}_\mathbf{U}$ are labeled by the vector $\mathbf{U}$. Similarly, vectors $\mathbf{S}_{\mathbf{U}_j}$ locating particular atoms within the unit cell now depend on the orientation of the crystallite and are also labeled by $\mathbf{U}$.

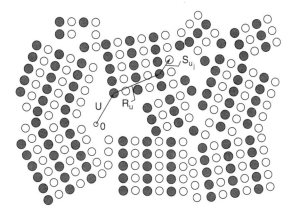

Figure 3.7. Polycrystalline solid.

The scattering intensity is given by

$$I(\mathbf{q}) \sim \left| \sum_{\mathbf{U}} \sum_{\mathbf{R}_U} \sum_{\mathbf{S}_{U_j}} f_j(\mathbf{q}) \exp[i\mathbf{q} \cdot (\mathbf{U} + \mathbf{R}_U + \mathbf{S}_{U_j})] \right|^2. \tag{3.56}$$

Note that the atomic form factor depends only on the type of atom at position $\mathbf{S}_j$.

The $\mathbf{R}_U$ sum may be carried out first. There is a complication arising from the fact that each crystallite has a finite size. To analyze the situation, start with a one-dimensional "crystallite" with $2J + 1$ cells, where the lattice vectors are of the form $R_n = na$, and $n = -J, -J + 1, \dots, +J$. For such an object,

$$\sum_{n=-J}^{J} \exp(iqna) = \exp(-iqJa) + \cdots + \exp(iqJa)$$

$$= \frac{\sin[(J + \frac{1}{2})qa]}{\sin(qa/2)}. \tag{3.57}$$

The square of this function is shown in Fig. 3.8 for several values of J. It has a strong peak at $q = 0$ of size $2J + 1$ and then becomes smaller as qa increases. There are an infinite number of maxima and minima. The first time the function hits zero is at $qa = \pi/(J + 1/2)$. Thus a measure of the width Δq of the central peak gives the crystallite size, through the relation $Ja = 2\pi/\Delta q$. In the limit of large J this function narrows to a delta function, in agreement with the previously obtained result for the single crystal.

For a three-dimensional crystallite in the shape of a parallelipiped, this result may be generalized. Writing $\mathbf{R} = n_1\mathbf{u}_1 + n_2\mathbf{u}_2 + n_3\mathbf{u}_3$ and assuming that the indices range from $-J_i$ to J_i ($i = 1,2,3$), respectively, one obtains

$$\sum_{\mathbf{R}} \exp(i\mathbf{q} \cdot \mathbf{R}) = \prod_{i=1}^{3} \frac{\sin[(J_i + \frac{1}{2})\mathbf{q} \cdot \mathbf{u}_i]}{\sin(\mathbf{q} \cdot \mathbf{u}_i/2)}. \tag{3.58}$$

Thus each diffraction peak is spread out, with the smaller crystallites producing the larger angular spread.

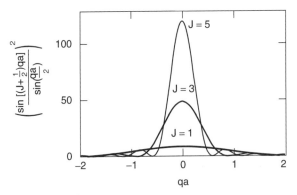

Figure 3.8. Square of the sum in Eq. (3.57) as a function of qa for three values of J: $J = 1$, 3, and 5.

Suppose that crystallites are sufficiently large that the sums may be represented as Kronecker deltas. Then

$$\sum_{\mathbf{R_U}} \exp(i\mathbf{q} \cdot \mathbf{R_U}) = N_c \sum_{\mathbf{G_U}} \delta_{\mathbf{q},\mathbf{G_U}}, \tag{3.59}$$

where N_c is the number of atoms in the crystallite $\mathbf{U}$. Note that the reciprocal lattice vectors $\mathbf{G_U}$ are referred to the orientation of the crystallite at $\mathbf{U}$. Thus one has (using a slightly abbreviated notation)

$$I = N_c^2 \sum_{\mathbf{U}} \sum_{\mathbf{U}'} \sum_{s} \sum_{s'} f_{\mathbf{s}}(\mathbf{q}) f_{\mathbf{s}'}^*(\mathbf{q}) \exp[i\mathbf{q} \cdot (\mathbf{U} - \mathbf{U}' + \mathbf{s} - \mathbf{s}')] \sum_{\mathbf{G_U}} \delta_{\mathbf{G_{U'}q}}. \tag{3.60}$$

For simplicity's sake, the crystallites are all assumed to have the same size. Otherwise, N_c should be interpreted as an average number of atoms per crystallite. The locations of the crystallites are to a large extent random, although they do not overlap with each other. The sum over U' may be decomposed into two sums, one with $\mathbf{U}' = \mathbf{U}$ and the other with $\mathbf{U}' \neq \mathbf{U}$. It will be assumed that there is sufficient phase cancellation in the latter sum to replace it by zero. Then, using Eq. (3.35),

$$I = N_c^2 \sum_{\mathbf{U}} \sum_{s} \sum_{s'} f_{\mathbf{s}}(q) f_{\mathbf{s}'}^*(\mathbf{q}) \exp[i\mathbf{q} \cdot (\mathbf{s} - \mathbf{s}')] \sum_{\mathbf{G_U}} \delta_{\mathbf{G_{U'}q}}$$

$$= N_c^2 \sum_{\mathbf{U}} |\Phi(\mathbf{q})|^2 \sum_{\mathbf{G_U}} \delta_{\mathbf{G_{U'}q}}. \tag{3.61}$$

This shows that the total diffraction intensity is given by an incoherent sum of contributions stemming from the individual crystallites. These crystallites can often be expected to have random orientations. Correspondingly, the reciprocal lattices will also have random orientations in space.

The situation may be represented diagramatically using the Ewald sphere. Energy conservation implies that $k = k'$, so both the initial and final photon wave vectors may be drawn as radii of a common sphere. This is illustrated in Fig. 3.9. In this figure ϕ denotes the scattering angle. Note that it is convenient to draw the vector $\mathbf{k}$ twice. The

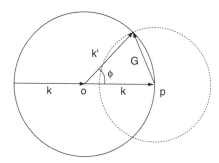

Figure 3.9. Ewald sphere depiction of the diffraction from a polycrystalline solid. The left sphere denotes energy conservation. The right sphere results from averaging over all crystallite orientations.

intersection of the two spheres satisfies the von Laue condition and produces a circle. The scattered x-rays lie on the surfaces of a set of cones whose vertex half-angles are given by

$$G = 2k \sin \frac{\phi}{2} = 2k \sin \theta. \tag{3.62}$$

The result is that the diffraction pattern consists of a set of rings, each produced by a different value of G. The sharpness of these rings is determined by the crystallite size, N_c. In the limit of large N_c the rings will be sharp.

If the temperature of the crystal is raised and the lattice vibrates more vigorously, two things happen. First, the rings still remain sharp, but their intensity diminishes. Second, a thermal diffuse background appears with x-rays being scattered to all angles. A discussion of this effect, the Debye–Waller effect, is given in Chapter W22.

3.6 Elastic Scattering from an Amorphous Solid

In an amorphous solid there is no long-range order, although there may be varying amounts of short-range order, as described in Chapter 4. It differs from the poly-crystalline material, where there is "long"-range order within each crystallite. The factorization of the problem into lattice contributions and basis contributions is no longer possible, and this creates complications.

Suppose that there are N_A atoms located at sites $\mathbf{r}_n$. The scattering intensity for momentum transfer $\hbar\mathbf{q}$ is

$$I(\mathbf{q}) \sim \left| \sum_{n=1}^{N_A} f_n(\mathbf{q}) e^{i\mathbf{q} \cdot \mathbf{r}_n} \right|^2, \tag{3.63}$$

where $f_n(\mathbf{q})$ is the atomic form factor for the nth atom. Writing this as a double sum and separating the diagonal terms yields

$$I(\mathbf{q}) \sim \sum_n |f_n(\mathbf{q})|^2 + \sum_{n,m}{}' f_n(\mathbf{q}) f_m^*(\mathbf{q}) e^{i\mathbf{q} \cdot (\mathbf{r}_n - \mathbf{r}_m)}, \tag{3.64}$$

where the prime implies that $m \neq n$. First consider a monatomic solid for which there is only one atomic form factor $f(\mathbf{q})$ for all n. Introduce the pair or radial distribution function $g(\mathbf{r})$ defined by

$$\rho(\mathbf{r}) = \rho_0(g(\mathbf{r}) - 1) \equiv \frac{1}{N_A} \sum_{n,m} {}' \delta(\mathbf{r} - (\mathbf{r}_n - \mathbf{r}_m)). \tag{3.65}$$

The quantity $\rho(\mathbf{r})$ is the concentration of atoms at a distance r from a central atom, and ρ_0 is the average atomic density. It is assumed that a macroscopic piece of amorphous material has no preferred direction, so $g(\mathbf{r}) = g(r)$ will be isotropic. The expression for I may be rewritten as

$$I(\mathbf{q}) = |f(\mathbf{q})|^2 \left[N_A + \sum_{n,m} {}' e^{i\mathbf{q}\cdot(\mathbf{r}_n - \mathbf{r}_m)} \right]$$

$$= N_A |f(\mathbf{q})|^2 \left[1 + \int d\mathbf{r} \frac{1}{N_A} \sum_{n,m} {}' \delta(\mathbf{r} - (\mathbf{r}_n - \mathbf{r}_m)) e^{i\mathbf{q}\cdot\mathbf{r}} \right]$$

$$= N_A |f(\mathbf{q})|^2 \left[1 + \rho_0 \int d\mathbf{r}(g(r) - 1) e^{i\mathbf{q}\cdot\mathbf{r}} \right]. \tag{3.66}$$

The inverse Fourier transform gives the pair distribution function in terms of the intensity:

$$g(r) = 1 + \frac{1}{\rho_0} \int \frac{d\mathbf{q}}{(2\pi)^3} \left[\frac{I(\mathbf{q})}{N_A |f(\mathbf{q})|^2} - 1 \right] e^{-i\mathbf{q}\cdot\mathbf{r}}$$

$$= 1 + \frac{1}{2\pi^2 r \rho_0} \int_0^\infty \left[\frac{I(q)}{N_A |f(q)|^2} - 1 \right] \sin(qr)\, dq. \tag{3.67}$$

The pair distribution function $g(r)$ gives the relative probability of finding an atom at a distance r from a given atom and can be calculated from the measured structure factor $I/N|f|^2$. The probability for finding the atom closer than an atomic radius is extremely small because of the interatomic repulsion. For large r, in the absence of long-range order, $g(r)$ will approach a constant. Peaks in $g(r)$ versus r indicate popular interatomic distances. The quantity $G(r) = 4\pi\rho_0 r[g(r) - 1]$ is called the *reduced radial distribution function*. The case $g(r) = 1$ corresponds to a continuum rather than a periodic array of discrete atoms. An example of $G(r)$ for the metallic glass $Ni_{0.76}P_{0.24}$ is shown in Fig. 4.11.

A brief discussion of the generalization of WS cells for amorphous solids is given in Section W3.1.

REFERENCES

Azaroff, L. V., *Elements of X-ray Crystallography*, McGraw-Hill, New York, 1968.

Azaroff, L. V., and M. J. Buerger, *The Powder Method in X-ray Crystallography*, McGraw-Hill, New York, 1968.

Cullity, B. D., *Elements of X-ray Diffraction*, 2nd ed., Addison-Wesley, Reading, Mass., 1978.

Henry, N. F. M., and K. Lonsdale, eds., *International Tables for X-ray Crystallography*, International Union of Crystallography, Birmingham, England, 1969.

Klein, C., and C. X. Hurlbut, Jr., *Manual of Mineralogy*, Wiley, New York, 1985.

Klug, H. P., and L. E. Alexander, *X-ray Diffraction Procedures for Polycrystalline and Amorphous Materials*, Wiley, New York, 1954.

Schwartz, L. H., and J. B. Cohen, *Diffraction from Materials, 2nd ed.*, Springer Verlag, New York, 1987.

PROBLEMS

3.1 Prove that $\sum_{\mathbf{R}} \exp(i\mathbf{q} \cdot \mathbf{R}) = 0$, where $\{\mathbf{R}\}$ is a set of Bravais lattice vectors and $\mathbf{q}(\neq \mathbf{0})$ lies within the primitive unit cell of the reciprocal lattice. Also prove the orthogonality identity appearing in Eq. (3.12):

$$\int_{\text{WS}} e^{i(\mathbf{G}-\mathbf{G})'\cdot\mathbf{r}} d\mathbf{r} = V_{\text{WS}} \delta_{\mathbf{G},\mathbf{G}'}.$$

3.2 Find the Fourier coefficients V_n for the periodic functions $V(x) = A\sin(2\pi x/a)$ and $V(x) = B\cos(2\pi x/a)$.

3.3 Prove that the plane defined by the equation $\mathbf{G} \cdot \mathbf{r} = A$ lies a distance $d = A/G$ from the origin and that the normal to the plane is parallel to $\hat{G}$.

3.4 Use the results of Problem 3.3 to generate formulas for the bounding planes of the first Brillouin zones for the FCC, BCC, and HCP crystal structures.

3.5 Determine the structure factor for the basis, $\Phi(\mathbf{q})$, defined in Eq. (3.35), for the cubic ZnS, CsCl, and NaCl crystal structures.

3.6 Draw the x-ray ring patterns produced by diffractions from powders for the SC, FCC, BCC, and diamond crystal structures.

3.7 Given an amorphous solid in which each atom has an electron density described by $n(r) = A\exp(-2r/a)$ and the pair distribution function is the unit step function $g(r) = \Theta(r - b)$, find the expected scattering intensity.

3.8 Sketch the Wigner–Seitz cell for the HCP crystal structure.

3.9 Find the distances from the center of the Wigner–Seitz cells for the BCC, FCC, and HCP crystal structures to the centers of the faces of the cells.

3.10 The primitive translation vectors of the hexagonal lattice can be written as

$$\mathbf{u}_1 = \hat{i}\frac{a\sqrt{3}}{2} + \hat{j}\frac{a}{2}, \qquad \mathbf{u}_2 = -\hat{i}\frac{a\sqrt{3}}{2} + \hat{j}\frac{a}{2}, \qquad \mathbf{u}_3 = c\hat{k}.$$

(a) Show that the fundamental translation vectors of the reciprocal lattice are given by

$$\mathbf{g}_1 = \hat{i}\frac{2\pi}{a\sqrt{3}} + \hat{j}\frac{2\pi}{a}, \qquad \mathbf{g}_2 = -\hat{i}\frac{2\pi}{a\sqrt{3}} + \hat{j}\frac{2\pi}{a}, \qquad \mathbf{g}_3 = \frac{2\pi}{c}\hat{k}.$$

(b) Describe and sketch the first Brillouin zone of the hexagonal lattice.

(c) Prove that the perpendicular distance $d(hkl)$ between adjacent parallel planes in the hexagonal lattice is

$$d(hkl) = \frac{1}{\sqrt{\dfrac{4(h^2 + hk + k^2)}{3a^2} + \dfrac{l^2}{c^2}}}.$$

[*Hint:* Use $d(hkl) = 2\pi/|G(hkl)|$.]

3.11 Find the shortest $\mathbf{G}(hkl)$ for **(a)** the BCC crystal structure, and **(b)** the FCC crystal structure.

Note: An additional problem is given in Chapter W3.

CHAPTER 4

Order and Disorder in Solids

4.1 Introduction

The ideal crystalline solids discussed in Chapter 1 do not and, in fact, cannot actually exist in nature as single crystals in which the positions of the atoms are perfectly ordered in space. *Perfect order* in a solid refers to the identical arrangement of atoms within each local atomic bonding unit and also relative to each lattice point. In real solids at thermodynamic equilibrium, deviations from perfect order in the form of finite concentrations of *defects* such as vacancies and interstitial atoms will always be present. Even though such defects represent a state of locally higher internal energy, they can exist at thermal equilibrium due to the resulting increase of entropy or *disorder*, which leads to a net decrease of the free energy of the solid.

Although only an idealization, the concept of a perfect crystal is nevertheless an extremely useful one. The perfect crystal serves as the natural starting point for discussions of real solids in which the order is imperfect (i.e., in which *structural, chemical,* and *thermal disorder* are all present). Due to the presence of disorder, the physical properties of the solid will no longer be periodic in space, due to the absence of periodicity in the equilibrium positions of the atoms.

In this chapter some important examples of the various types and degrees of structural and chemical disorder present in solids are described, and important examples of ordered and disordered solids are presented and discussed. The wide variety of localized (point) and extended (line and planar) defects that can exist in solids are discussed. The role of thermodynamics in determining equilibrium defect concentrations in solids is discussed. The thermal disorder present in solids and corresponding to the thermally induced oscillations of atoms about their equilibrium positions is the subject of Chapter 5. It should be noted that a solid in which only thermal disorder is present can still be considered to be a periodic solid (i.e., a perfect crystal).

The importance of introducing the topic of disorder at this point follows from observations which indicate that many aspects of the physical behavior of solids are in fact controlled by the nature and concentration of defects present in the solid. The effect of defects on the properties of solids is discussed only briefly in this chapter, with details and specific examples given in other chapters, where the properties of materials and important classes of materials are described. A few important examples include the role that defects play in determining the mechanical properties of solids, such as the tensile strength, in Chapter 10, the control of the electrical properties of semiconductors by dopant impurity atoms in Chapter 11, and the effects of impurities such as carbon and of dislocations and other extended defects on the mechanical properties of metals and alloys in Chapter 12. The influence of stoichiometry on the superconductivity of the perovskite family of high-T_c superconductors and the role of defects in determining flux pinning in the superconducting mixed state are described in Chapter 16. In addition, the role of defects in determining the magnetic hardness of magnetic materials is described in Chapter 17, and defects associated with the surfaces of solids are discussed in Chapter 19.

Defects can be created in solids in a variety of ways. The defects present under equilibrium conditions are generated by the thermal energy present in the solid. Defects can also be created by external means, such as by radiation damage, mechanical treatment, chemical effects due to the environment, or phase transitions. These external means are often used to introduce certain types and concentrations of defects to obtain a desired set of physical properties in the solid.

ORDER AND DISORDER

The crystalline order corresponding to the periodicity described in Chapter 1 can readily be described in terms of the lattice and the local atomic bonding units. Disorder in the solid state is a much more complex subject, due to the wide variety of deviations from periodicity that may be present. *Structural disorder* corresponds to the displacement of atoms from their equilibrium positions in a perfectly ordered crystal while *chemical disorder* corresponds to the occupation by atoms of sites in the structure which are normally occupied by other atoms. Specific examples of structural and chemical defects such as vacancies and impurity atoms are discussed in more detail later in this chapter.

The degree of structural order present in a solid is often specified in terms of the spatial extent of the order (i.e., by the size of the ordered regions). These ordered regions, often referred to as *crystallites*, can be thought of as the remnants of the original perfect order. The term *microstructure* is often used to describe the structure of a crystal on the scale on which deviations from perfect order become evident.

The types of structural order known as polycrystalline, microcrystalline, and nanocrystalline, as well as the structural disorder known as amorphous or noncrystalline, are defined in Table 4.1 in terms of an average crystallite size. Single crystals typically have dimensions on the order of millimeters to centimeters but can be much larger. Solids said to be *polycrystalline* can have crystallite sizes ranging from centimeters down to about 1 mm; crystallite sizes for *microcrystalline solids* extend from about 1000 μm down to about 1 μm. *Nanocrystalline solids*, also known as *nanophase materials*, have crystallite sizes that can extend down to interatomic distances, at which point crystalline order no longer exists and solids are said to be *noncrystalline* or *amorphous*. The term *polycrystalline* is often used to describe any crystalline solid which is not a single crystal. The positions of the boundaries indicated in Table 4.1 between various types of structural order are necessarily arbitrary and serve only as a rough guide.

It is clear from this table that the amorphous state is the limit of maximum structural disorder in a solid, corresponding to the size of an ordered region or crystallite approaching that of an individual atom. Although the amorphous limit is reasonably well defined, there is in practice no unique structural definition of an amorphous solid. Amorphous metals, for example, have structural properties which are quite different from those of amorphous semiconductors or other amorphous solids in which covalent (or nearly covalent) bonding dominates. The types of structural disorder observed in amorphous solids are discussed in more detail in Section 4.3.

TABLE 4.1   Crystallite Sizes and Types of Structural Order in Solids

Crystallite Size	Type of Structural Order
$\approx 1-100$ mm	Polycrystalline (or single crystal)
$\approx 1-1000$ μm	Microcrystalline
$\approx 1-1000$ nm	Nanocrystalline (nanophase)
$\approx 0.1-1$ nm	Amorphous (noncrystalline)

The types of structural order listed in Table 4.1 are illustrated in Fig. 4.1, where a single crystal of Si, polycrystalline Pb, microcrystalline $YBa_2Cu_3O_{7-x}$, and nanocrystalline diamond are shown. Amorphous SiO_2 (a-SiO_2) at its interface with crystalline Si is shown for comparison. The microstructure of these solids is evident and consists of the relative arrangement and orientation of the crystallites. The microstructure of a-SiO_2 corresponds to the local arrangements of the atoms.

It is appropriate at this point to introduce and define some additional terms that are often used to describe the spatial extent of order in solids: short-range order (SRO), intermediate-range order (IRO), and long-range order (LRO).

1. *Short-range order* exists in a solid when the arrangement of NNs for each atom is the same as that found in the crystalline state. SRO is often retained in the amorphous state, where its presence is a strong indication that the type of bonding between atoms found in the crystal exists even in the presence of extreme structural disorder. The existence of SRO does not, however, necessarily imply that the bond lengths and angles found in the disordered solid are the same as those found in the crystal. For example, even though SRO is preserved to a large degree in amorphous covalent solids such as amorphous Si, since practically every Si atom remains at the center of an Si–Si_4 bonding unit, the bond angles within the Si–Si_4 units are distributed continuously over a few degrees around the value of $109.47°$ found in crystalline Si.

2. Although difficult to define precisely, *intermediate-range order* exists in a solid when the order extends beyond a given local atomic bonding unit to neighboring units. An example of IRO can be found in amorphous carbon, a-C, where in addition to trigonal sp^2 coordination corresponding to the C–C_3 bonding units and tetrahedral sp^3 coordination corresponding to C–C_4 bonding units, there can exist sixfold hexagonal rings of carbon atoms (Fig. 4.2). These hexagonal rings in a-C are distorted versions of the regular planar rings found in crystalline graphite.

3. Perfect *long-range order* can exist only in perfectly ordered crystalline solids. Nevertheless, a type of LRO can also be said to exist within ordered regions of a disordered solid when crystallite sizes exceed 1 to 10 nm. With this definition, even the nanocrystalline solids as defined in Table 4.1 may have LRO within small nanocrystallites. In some solids, LRO can exist in two dimensions and can be modified or absent in the third. Examples include superlattice structures consisting of alternating layers of materials, such as the dielectrics $PbTiO_3$ and $SrTiO_3$ (Fig. 4.3). Other examples include liquid crystals, described in Chapter 14, and the atomic structure of surfaces, described in Chapter 19.

(a)

Figure 4.1. Examples of crystalline and amorphous solids. (*a*) Single crystal of Si (about 150 mm in diameter and 125 cm long). (From *Materials and Technology: The Role of Physics in Materials Research.* Copyright by the American Physical Society.) (*b*) Polycrystalline Pb (dimensions: 10 cm wide at the top). (From *ASM Handbook*, 9th ed., Vol. 9, *Metallography and Microstructures*, ASM International, Materials Park, Ohio, 1985, p. 418, Fig. 2.) (*c*) Microcrystalline $YBa_2Cu_3O_{7-x}$. [From B. N. Lucas et al., *J. Mater. Res.*, **6**, 2519 (1991).] (*d*) Nanocrystalline diamond (SEM micrograph). (From R. F. Davis, ed., *Diamond Films and Coatings*, 1993. Copyright 1993 (by William Andrew Publishing, LLC/Noyes Publications.) (*e*) Amorphous SiO_2, shown at its interface with crystalline Si (HRTEM cross-sectional view). (From B. Agius, in J. Kanicki et al., eds., *Amorphous Insulating Thin Films*, Proc. 284, (Materials Research Society, Pittsburgh, Pa., 1993.)

(b)

(c)

Interface _____ 2μm

(d)

Figure 4.1. (*continued*)

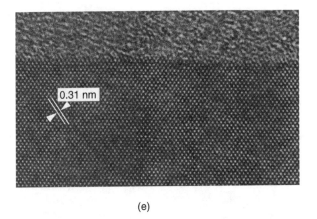

(e)

Figure 4.1. (*continued*)

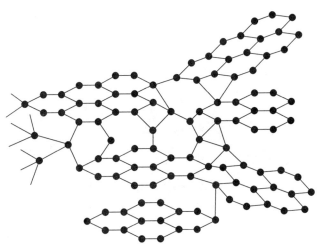

Figure 4.2. Short- and intermediate-range order in amorphous carbon (a-C). SRO, trigonal $C-C_3$ and tetrahedral $C-C_4$ bonding units, IRO, sixfold hexagonal rings of carbon atoms.

A classification of ordered and disordered solids in terms of the range of structural order present has just been given. In addition, solids can be classified as being periodic or nonperiodic and also as being homogeneous or inhomogeneous. Ordered solids are usually assumed to be periodic and homogeneous, while disordered solids are considered to be nonperiodic and often inhomogeneous. These generalizations, although useful, are not always valid. Even though every disordered solid is nonperiodic on a large enough scale, many disordered solids, such as colloids or composites, often contain local, ordered regions. In addition, some disordered solids, such as random alloys and amorphous solids or glasses, can appear to be homogeneous on a long enough length scale, whereas ordered superlattices are certainly inhomogeneous on a macroscopic scale.

To be clear about what is meant by a homogeneous solid, not only must the meaning of the term *homogeneous* be specified but also the length scale on which

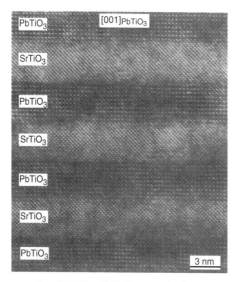

Figure 4.3. HRTEM image of a $PbTiO_3-SrTiO_3$ superlattice taken with the electron beam parallel to the [100] axis of $SrTiO_3$. [Reprinted with permission from J. C. Jiang et al., *Appl. Phys. Lett.*, **74**, 2851 (1999). Copyright 1999 by the American Institute of Physics.]

the homogeneity exists must be given. For example, if a solid is homogeneous on a given length scale, the properties of the solid will be constant in space when probed on that scale. The properties measured will, of course, be an appropriate average of the properties of the components of the solid. For example, a single crystal will be homogeneous on length scales down to about 1 nm, at which point the existence of individual atoms will start to become evident. Also, a nanocrystalline solid such as the diamond film shown in Fig. 4.1*d* with crystallite sizes $\approx$ 10 to 100 nm will appear optically homogeneous in visible light, with wave lengths in the range 400 to 700 nm but will appear inhomogeneous when probed with x-rays having $\lambda \approx 0.1$ nm.

It is thus evident that the homogeneity of a solid depends on the length scale on which its properties are probed. This scale typically corresponds to the wavelength of the probe used (e.g., visible light, x-rays, electrons, neutrons, ultrasound, etc.). Only when probed on a sufficiently small scale will the disorder play a critical role in determining the properties of the solid.

Some additional examples of ordered and disordered solids are presented next, followed by a discussion of current models for the structure of amorphous solids.

4.2 Examples of Ordered and Disordered Solids

Several examples of ordered and disordered solids are presented in Fig. 4.1, where structural order can be seen to extend from the single crystal to polycrystalline, microcrystalline, nanocrystalline solids, and finally, to the limit of maximum disorder found in amorphous solids.

A crystalline alloy composed of two or more elements and possessing long-range structural order can still be disordered in a chemical sense, depending on the spatial arrangement of the atoms. To illustrate this, a stoichiometric binary AB alloy is shown schematically in two dimensions in Fig. 4.4. Examples of such alloys (in three

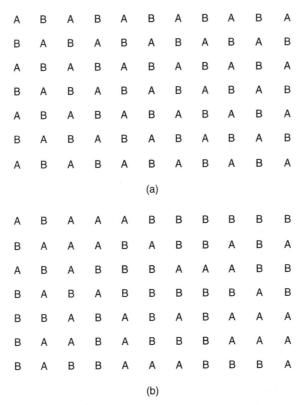

Figure 4.4. Two-dimensional arrangement of atoms in crystalline equiatomic binary alloys such as CuZn. (*a*) The alloy is periodic and therefore both chemically and structurally perfectly ordered. (*b*) The alloy is chemically disordered and hence non-periodic. (Note that the placements of the atoms here were chosen by flipping a coin.)

dimensions) include SiGe and CuZn. In Fig. 4.4*a* the A and B atoms are arranged periodically in space on a lattice so that the alloy is ordered both structurally and chemically. In Fig. 4.4*b* the A and B atoms still occupy lattice sites but are spatially disordered relative to each other. The alloy is structurally ordered but is chemically disordered. Whether a binary AB alloy will be chemically ordered or disordered at a given temperature depends on the strengths of the bonds between pairs of atoms. This topic is discussed further in Chapter 6, where the order–disorder phase transition is described.

Composite solids are multicomponent (or multiphase) mixtures. The composite ceramic solid shown in Fig. 4.5 consists of graphite fibers in a Ag–Cu alloy matrix. Here the graphite fibers are used to control the mechanical properties of the matrix. A second example, shown in Fig. 4.6, consists of nanocrystalline clusters of Au atoms embedded in an amorphous matrix. In addition to being disordered, these composite solids are also *inhomogeneous* on the scale of their microstructure. Composite solids are becoming increasing important technologically because they can be fabricated from a wide variety of components to achieve a desired set of properties. Other examples of multiphase solids discussed elsewhere include the Pb–Sn eutectic alloy shown in Fig. 6.11, steels composed of several phases, such as cementite, pearlite, bainite,

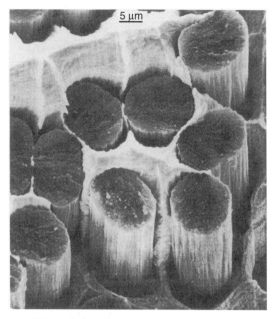

Figure 4.5. Composite solid consisting of graphite fibers in an Ag–Cu alloy matrix. (From *ASM Handbook*, 9th ed., Vol. 9, *Metallography and Microstructures*, ASM International, Materials Park, Ohio, 1985, p. 597, Fig. 30.)

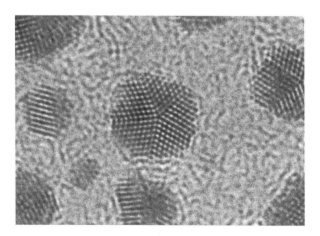

Figure 4.6. Composite solid consisting of nanocrystalline clusters of Au atoms embedded in an amorphous matrix (high-resolution electron micrograph). Several of the Au clusters can be seen to contain dislocations. (Reprinted with permission from M. Flueli and P. Buffat, EPFL-CIME, Lausanne, Switzerland.)

martensite, and acicular ferrite, as well as multiphase ceramics, polymers, superconductors, magnetic materials, and so on.

Colloidal solids consist of aggregates of solid particles with sizes typically ranging from a few nanometers to several micrometers. The shapes of the individual particles depend on their internal structural order, which in turn is usually determined by the conditions of growth. When the particles are amorphous or nanocrystalline, their shapes are generally spherical as a result of growth conditions which tend to be isotropic. When crystalline, the external shapes of the particles can reflect the symmetry of the crystal. Various morphologies observed for colloidal α-Fe$_2$O$_3$ or hematite particles are illustrated in Fig. 4.7, where spherical, spindlelike, cubic, and hexagonal platelike particles are evident. Colloidal solids are clearly inhomogeneous on the scale of the component particle sizes.

Clusters of atoms with sizes in the range 1 to 100 nm and containing from a few tens up to $\approx 10^7$ atoms, sometimes called *nanoclusters* or *nanoparticles*, are a form

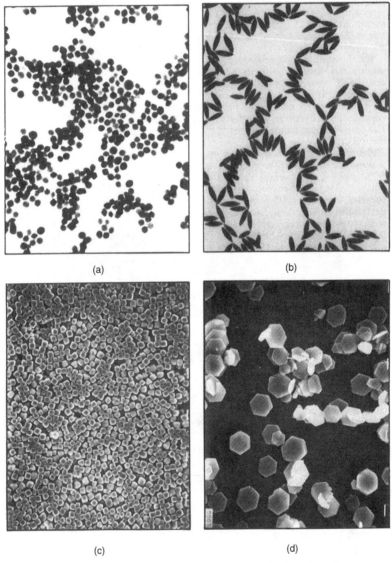

(a)　　　　　　　　　　(b)

(c)　　　　　　　　　　(d)

Figure 4.7. Colloidal α-Fe$_2$O$_3$ particles with various morphologies: (*a*) spheres; (*b*) spindles; (*c*) cubes; (*d*) hexagonal disks. Each bar indicates 2 μm. [From M. Ozaki, *Mater. Res. Soc. Bull.* (12), 35 (1989).]

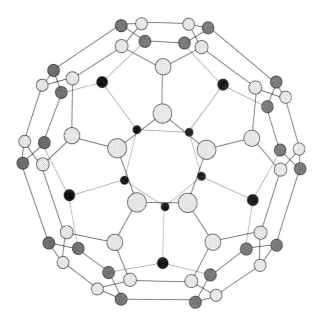

Figure 4.8. C_{60}, a cluster of 60 carbon atoms known as a fullerene.

of solid for which the surface can play a particularly significant role. For example, in a cluster with diameter of 1 nm, about 50% of the atoms are at the surface. Due to its unique structure, shown in Fig. 4.8, the 60 atoms in the fullerene molecule C_{60} are all at the surface of the cluster. Clusters can also be incorporated into a matrix of another material, as illustrated for Au clusters in Fig. 4.6. The dependencies of structure as well as the physical and chemical properties of clusters on their size is currently an area of active research. In addition, the coalescence of these nanoparticles via sintering leads to the formation of nanophase materials, which can have a wide range of interesting properties, including higher mechanical strength than that of their crystalline counterparts.

Multilayered materials or superlattices such as the $PbTiO_3 - SrTiO_3$ superlattice shown in Fig. 4.3 are a type of composite solid in which ordered, essentially two-dimensional layers or thin films are stacked along a third dimension. The interfaces between the individual layers can play a key role in determining the properties of the superlattice. Multilayers, superlattices, and their interfaces are discussed in more detail in Chapter 20. Superlattices are inhomogeneous except on length scales much greater than the thickness of the individual layers.

A *quasicrystalline solid* in the form of a dodecahedral (12-sided) crystallite of an alloy of Al, Cu, and Fe is shown in Fig. 4.9. A quasicrystal is a unique, nonperiodic type of solid possessing unusual order and symmetries not found in crystalline solids. These can include fivefold axes of symmetry, as seen in the figure. Some quasicrystals exhibiting such fivefold symmetry axes are apparently based in part on the icosahedral local atomic bonding unit shown in Fig. 1.11. While icosahedral crystallites cannot fill space when packed together, space filling is possible when they are combined with other geometric shapes. The quasicrystal does not have long-range translational symmetry but is still ordered.

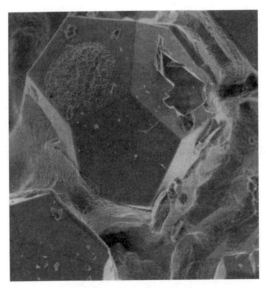

Figure 4.9. Quasicrystalline solid in the form of a dodecahedral (12-sided) crystallite, about 300 μm in diameter. The solid shown is an alloy of Al, Cu, and Fe. [From P. W. Stephens et al., *Sci. Am.*, **264**, 44 (1991). Reprinted by permission of A. P. Tsai et al.]

4.3 Amorphous Solids

Since crystalline solids have already been described in Chapter 1, it is appropriate at this point to indicate what is known about the structures of amorphous or noncrystalline solids. Since an amorphous solid is nonperiodic and lacks long-range order, it cannot be said to have a crystal structure at all. As a result, the properties of amorphous solids tend to be isotropic rather than anisotropic as is often the case for crystalline solids (e.g., graphite). The isotropic nature of amorphous solids is a natural consequence of their lack of structural symmetry.

One view of an amorphous solid is that it represents the limit in which the crystallite size of a nanocrystalline solid has been effectively reduced to the size of a single atom (i.e., the disordered material present at the grain boundaries between the crystallites has become the entire solid). An alternative point of view that amorphous solids are actually nanocrystalline is now believed to be in general incorrect, on the basis of careful structural studies. Specific types of amorphous solids such as amorphous semiconductors and metallic glasses are discussed in more detail in later chapters and at our Web site.[†]

Despite the noncrystallinity of amorphous solids, useful structural models exist which emphasize the short- and intermediate-range order which these materials can possess. Two such models are the *random close-packing* (RCP) model for simple metallic glasses, also sometimes called the *dense random-packing* (DRP) model, and the *continuous random network* (CRN) model for amorphous covalent (or nearly covalent) solids such as amorphous semiconductors and ceramics. The RCP model has

---

[†] Supplementary material for this textbook is included on the Web at the resource site (ftp://ftp.wiley.com/public/sci_tech_med/materials). Cross-references to elements of the Web material are prefixed by "W."

been found to be appropriate for amorphous metals whose crystalline counterparts have structures based on $A-A_{12}(cub)$, $A-A_{12}(hex)$, and $A-A_8$ local bonding units. The CRN model is appropriate for amorphous solids whose crystalline counterparts are based on $A-A_4$, $A-A_3$, and $A-A_2$ as well as $A-B_4$, $A-B_3$, and $A-B_2$ units.

Random Close-Packing Model. To help visualize the random arrangement of atoms that exists in an amorphous metal or metallic glass, consider the experiment in which hard spheres are placed into a container in such a way that they cannot relax locally into one of the most closely packed crystal structures such as FCC or HCP but instead take on a random close-packing structure. A model of a 100-atom portion of a random close-packed solid is shown in Fig. 4.10. In this computer-generated RCP structure the hard-sphere atoms are shown at only 30% of their normal size for clarity. The inability of the system of hard spheres to relax to one of its states of lowest overall energy such as FCC or HCP is a consequence of steric hindrance. The lack of relaxation can result from the atoms having either insufficient time or too low a mobility to move into local $A-A_{12}(cub)$ or $A-A_{12}(hex)$ bonding configurations. In practice these conditions are achieved during the synthesis of real metallic glasses either by using an extremely rapid quenching rate or by using very low temperatures. Metallic glasses are described at our Web site in Chapter W12, and specific processes such as rapid solidification which are used in their synthesis are described in Chapter W21.

When the above-mentioned experiment is actually carried out, for example, by placing identical ball bearings in a rubber balloon, it is found that a packing fraction $PF \approx 0.64$ is obtained quite reproducibly for this system. This packing fraction is 14% less than $PF = 0.74$ for FCC and HCP and is also about 6% below $PF = 0.68$ for BCC. It is, however, considerably higher than $PF = 0.52$ for the SC crystal structure. On the basis of these comparisons one might surmise that the hard spheres in an RCP structure such as that shown in Fig. 4.10 are on the average in contact with between 7 and 8 NNs. Analysis of radial distribution functions obtained from diffraction studies on metallic glasses, however, indicate that the average number of NNs is actually close to 12, as found in the $A-A_{12}$ cubic, hexagonal, and icosahedral bonding units. Apparently, in RCP structures, maximum short-range density is achieved by sacrificing the maximum long-range density found in the FCC and HCP crystal structures.

Diffraction patterns can be thought of as maps of the solid in reciprocal or momentum space. What is really needed is a map of the positions of the atoms in real space. Although such information is unattainable for amorphous solids where no periodicity

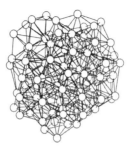

Figure 4.10. Model of a 100-atom portion of a random close-packed solid. [Reprinted with permission from J. A. Barker et al., *Nature*, **257**, 120 (1975). Copyright 1975 by Macmillan Magazines.]

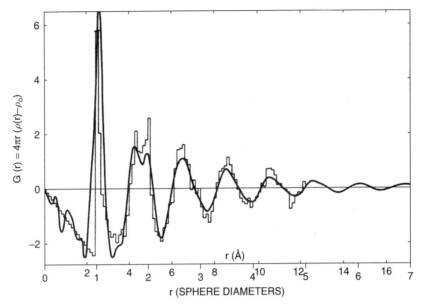

Figure 4.11. Comparisons of the reduced radial distribution functions $G(r) = 4\pi r \rho_0[g(r) - 1]$ derived from x-ray diffraction studies for the metallic glass $Ni_{0.76}P_{0.24}$ and from a corresponding random-close-packing structural model. (From G. S. Cargill III, *Solid State Phys.*, **30**, copyright 1975. Reprinted by permission of Academic Press Inc.)

or lattice exists, the *radial* or *pair distribution function* $g(r)$ defined in Chapter 3 does provide information about the average number of atoms at a distance r from the central atom. In the absence of a crystalline diffraction pattern, $g(r)$ is a convenient way of presenting information about the average arrangement of atoms in an isotropic, amorphous material.

A comparison is made in Fig. 4.11 between the reduced radial distribution function $G(r) = 4\pi r[\rho(r) - \rho_0] = 4\pi r \rho_0[g(r) - 1]$ derived from an x-ray diffraction study of the metallic glass $Ni_{0.76}P_{0.24}$ with that derived from a corresponding RCP structural model[†]. Here ρ_0 is the average atomic density of the amorphous solid. The first peak observed in Fig. 4.11 at $r_1 = D$, where D is the hard-sphere or atomic diameter, corresponds to two NN atoms in contact with each other. The second peak is actually composed of two overlapping contributions which can be attributed to two distinct types of next-NN atomic configurations. The peak at a separation r_2 close to $\sqrt{3}\,D$ corresponds to configurations in which two next-NN atoms are separated by a pair of atoms in contact with each other, while the peak close to $r_3 = 2D$ corresponds to a pair of atoms separated by a single atom, the three lying in a nearly linear configuration. These atomic configurations for the case of a planar, hexagonal array of atoms are shown schematically in Fig. W4.2.

Continuous Random Network Model. Since a type of close packing of atoms observed in the crystalline state is preserved at least locally in amorphous metals, it

---

[†] G. S. Cargill III, Structure of metallic alloy glasses, *Solid State Phys.*, **30**, 227 (1975).

should not be surprising that some features of the bonding and structure of crystalline covalent solids are also preserved in the amorphous state. This expectation follows naturally from the observation that the existence of crystallinelike short-range order is an important aspect of the structures of amorphous solids.

The original idea of representing nonmetallic, vitreous glasses such as SiO_2 and B_2O_3 by a noncrystalline, three-dimensional network of bonded atoms is attributed to Zachariasen[†]. The model proposed by Zachariasen has become known as the *continuous random network* (CRN) model. His specific proposal was that oxide glasses with the generic formula A_nO_m could exist with an energy comparable to (i.e., not much greater than) the corresponding crystals if the following rules were obeyed:

1. An oxygen atom is not bonded to more than two A atoms.
2. The number of oxygen atoms surrounding each A atom must be small (e.g., at most three O atoms in a triangle or four O atoms in a tetrahedron).
3. The oxygen polyhedra share corners with each other but not edges or faces.
4. At least three corners of each oxygen polyhedron must be shared with other oxygen polyhedra, thus ensuring that the glass network is three-dimensional rather that two- or one-dimensional in character.

Rule 1 is consistent with an additional assumption made by Zachariasen that all the normal valences of the atoms must be satisfied in the network. In this case the coordination number of an atom will be equal to $8 - N$, where N is the corresponding column number in the periodic table (e.g., $N = 6$ for oxygen). The lowest-energy version of the CRN model will be the version with the least overall strain in the network and hence with the minimum possible spread of bond angles deviating from the crystalline values. Such a CRN model can also be said to represent a type of ideal amorphous solid in which chemical order (i.e., SRO) is present even though structural order is absent.

Zachariasen's classic schematic model of a CRN of a two-dimensional glass of the A_2O_3 type is shown in Fig. 4.12*a*, with the corresponding "crystal" shown in Fig. 4.12*b*. Examples of the A_2O_3-type compound include group III_2–VI_3 solids such as B_2O_3 and In_2Te_3. The short-range order is preserved in the amorphous version since each A atom still has three O atoms as NNs in an $A-O_3$ bonding unit, each O atom has two A atom NNs in an $O-A_2$ bonding unit, the NN bond lengths are nearly unchanged, and the O$-$A$-$O bond angles are close to $120°$. The essential differences between the amorphous and crystalline versions are that the A$-$O$-$A bond angles, which were all $180°$ in the crystal, now have a spread of values around $180°$ and that the hexagonal, sixfold rings in the crystal are replaced by a distribution of four-, five-, six-, seven-, and eightfold rings in the CRN. The latter result concerning ring statistics demonstrates that intermediate-range order is not preserved in the version of the CRN presented by Zachariasen.

Other examples of amorphous solids for which the CRN is a useful model include the group IV (C, Si, Ge), group V (P, As, Sb), and group VI (Se) elemental amorphous covalent solids, and amorphous alloys of Si with H, C, N, O, and Ge. Hydrogenated amorphous Si, a-Si:H, an amorphous semiconductor with important applications in

---

[†] W. H. Zachariasen, *J. Am. Chem. Soc.*, **54**, 3841 (1932).

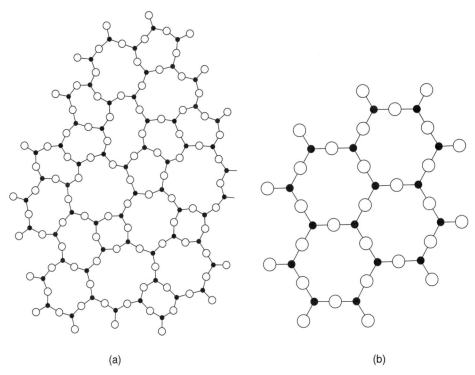

(a)   (b)

Figure 4.12. (*a*) Zachariasen's classic picture of the continuous random network of a two-dimensional glass of the A_2O_3 type; (*b*) corresponding crystalline version. [Reprinted with permission from W. H. Zachariasen, *J. Am. Chem. Soc.*, **54**, 3841 (1932). Copyright 1932 by the American Chemical Society.]

solar cells and flat panel displays, is discussed in Chapter W11. A wide variety of ceramic glasses whose structures are based on CRNs are described in Chapter 13. The synthesis of amorphous covalent solids such as those mentioned earlier is described in Chapter W21.

Native or intrinsic defects can also be present in "ideal" CRNs. Such defects, known as *topological defects*, include vacancies or voids, giving rise to unsatisfied valences or dangling bonds, and *valence alternation pairs* (VAPs). The latter are proposed to be present in a-Se and consist of pairs of Se^+ and Se^- ions. The normal bonding in a-Se consists of $Se-Se_2$ bonding units. It appears to be energetically favorable to have pairs of Se^+-Se_3 and Se^--Se bonding units (i.e., VAPs) rather than dangling bonds in a-Se.

DEFECTS IN SOLIDS

The list of defects that can exist in crystalline and amorphous solids is practically endless. The goal here is to identify, classify, and describe some of the important solid-state defects. Specific examples of defects, defect-related processes, and the influence of defects on the properties of solids are presented in more detail in later chapters where the various properties and classes of materials are discussed. The effects of defects on the properties of solids can be illustrated briefly as follows: *Point defects* such as vacancies and interstitials enhance the diffusion of atoms, *line defects* such as dislocations allow the plastic deformation of solids to occur, and *planar defects* such as grain boundaries promote the hardening of solids.

Before proceeding, it will be useful to distinguish between structural and chemical defects, localized and extended defects, equilibrium and nonequilibrium defects, and static and dynamic defects. To aid in making these distinctions, the structures of several common types of defects are shown schematically in Figs. 4.13 and 4.14. Figure 4.13 represents a two-dimensional crystalline binary AB alloy and demonstrates the differences between structural and chemical defects. Illustrated here are three simple structural defects:

1. A single B atom *vacancy* (monovacancy) known as V_B
2. A single *interstitial* B atom known as I_B
3. An A atom vacancy–interstitial pair known as $V_A - I_A$

Defects 1 and 3 are known as *Schottky* and *Frenkel defects*, respectively. Vacancies and interstitial atoms are point defects and are effectively zero-dimensional. Also shown are three simple chemical defects:

4. A C atom *substitutional impurity* on a site normally occupied by a B atom, known as C_B
5. An *antisite defect* corresponding to a B atom on an A site and an A atom on a neighboring B site, known as $B_A - A_B$
6. An interstitial D atom known as I_D

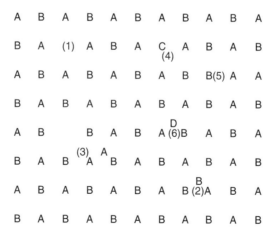

Figure 4.13. Two-dimensional, crystalline binary AB alloy. Three simple structural defects are illustrated: (1) a single B atom vacancy (monovacancy), (2) a single interstitial B atom, and (3) an A atom vacancy–interstitial pair. Also shown are two chemical defects: (4) a C-atom substitutional impurity and (5) an antisite defect corresponding to a B atom on an A site and an A atom on a B site. The D-atom interstitial impurity shown as (6) can be considered to be both a structural and a chemical defect.

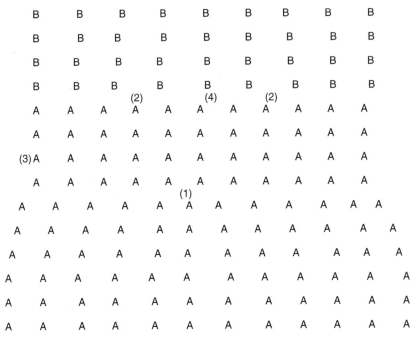

Figure 4.14. Pair of two-dimensional, crystalline solids composed of A and B atoms, respectively, and their interface. The extended defects shown are the edge dislocation (1) and misfit dislocations (2), the external surface (3) of crystal A, and the interface (4) between crystals A and B.

These types of defects are clearly also present in three-dimensional solids. It can be seen that structural defects correspond to missing atoms or to atoms occupying positions in the structure which are normally unoccupied. Chemical defects, on the other hand, can correspond to the occupation of sites in the structure which are normally unoccupied or occupied by atoms of a different element. In addition, one may also have isotopic defects (e.g., A atoms with different isotopic masses).

Figure 4.14 shows a pair of two-dimensional crystalline solids composed of A and B atoms, respectively, and their interface. It illustrates the differences between localized and extended defects. The following extended defects are shown:

1. An *edge dislocation*
2. *Misfit dislocations*

Defects 1 and 2 are linear and hence one-dimensional, while defects 3 and 4 are planar and hence two-dimensional:

3. The external *surface* of crystal A
4. The *interface* between crystals A and B

Examples of equilibrium and nonequilibrium defects are also illustrated in Figs. 4.13 and 4.14. *Equilibrium* or *intrinsic defects* are those defects that are present in solids under conditions of thermodynamic equilibrium. These defects are also known as *native defects. Nonequilibrium defects*, on the other hand, will not ordinarily be present when the solid is in thermodynamic equilibrium. As illustrated in Section 4.6, vacancies will be present in all solids under equilibrium conditions. Other examples of equilibrium defects include interstitials, antisite defects, and even impurity atoms in every real solid. The existence of these defects under equilibrium conditions can be attributed to the lowering of the free energy of the solid, due to the increase in its entropy when defects are present. It follows, therefore, that defect concentrations will increase, usually exponentially, as the temperature of the solid increases. For a variety of reasons, vacancies, interstitials, and other defects can also exist in solids at concentration levels even higher than the equilibrium values at a given temperature. Under these circumstances, these defects can also be referred to as nonequilibrium defects.

Typical nonequilibrium or *extrinsic defects* which are not ordinarily present under equilibrium conditions in solids because their energies grow too fast with the size of the system include dislocations, grain boundaries, stacking faults, twins, and others. These defects can be introduced into solids by external means (e.g., ion implantation, cold working, irradiation, quenching, etc.). These defects are metastable. Nonequilibrium defects can often be eliminated when the solid is annealed at high temperatures for long periods of time. Equilibrium defects cannot in general be removed by annealing.

Many defects in solids are dynamic, especially at high temperatures, and can move through the solid either with or without external influences. For example, vacancies can diffuse in solids when neighboring atoms hop into vacant sites while interstitial atoms can diffuse by hopping to adjacent unoccupied interstitial sites. In addition, vacancies are constantly being created (by an atom hopping from its normal to an interstitial site) and destroyed (by an interstitial atom hopping onto a vacant site). External stress, either mechanical or thermal, can cause dislocations to move in solids. Electric fields can cause charged defects to move. Planar defects such as grain boundaries, surfaces, and interfaces are generally considered to be static, at least at sufficiently low temperatures.

Adding to the complexity is the fact that defects can interact with each other, leading to a wide variety of defect complexes. Such defect–defect interactions can be beneficial, as in the gettering or absorption of point defects (impurities, etc.) by dislocation damage in crystalline Si or the hardening of metals due to dislocation–dislocation interactions. Some specific examples of localized and extended defects are presented next.

4.4 Localized Defects

Localized defects are those defects that do not extend through the solid for appreciable distances and whose sizes range from a single atom or vacancy to perhaps hundreds of atoms or vacancies. Such defects are effectively zero-dimensional and include point defects, the smallest possible defects in solids, as well as some multiatom and multisite defects.

Point Defects. Point defects such as vacancies, interstitials, and impurities, shown schematically in Fig. 4.13, are found in all real solids under equilibrium conditions. Defect concentrations generally increase exponentially with increasing temperature, proportional to $\exp(-E_a/k_BT)$, where E_a is the activation energy for defect generation. The equilibrium defect concentrations in solids will depend on their energies (or enthalpies) of formation and also on their vibrational and configurational entropies, as shown in Section 4.6. It is also important to specify the charge state of a point defect. The charge state (neutral, positive, or negative) can be critical in determining the concentration of the defect since different charge states can have different energies of formation. Although defects can be charged, the solid as a whole must be neutral. The three types of point defects mentioned above are discussed next.

Vacancies. Single vacancies or Schottky defects are neutral in metals such as Al or Fe and in elemental covalent solids such as Si. They can have either positive or negative effective charges in ionic solids such as NaCl or $Fe_{1-x}O$. The presence of the vacancy V_{Al} in a close-packed metal such as Al has little effect on the 12 NN atoms in the $Al–Al_{12}$ bonding unit. The vacancy V_{Si} in Si, on the other hand, results in the breaking of the four covalent bonds between the missing Si atom and its four NNs in the $Si–Si_4$ bonding unit. As a result, some spatial relaxation of the positions of the four NN Si atoms will occur.

In NaCl the positive charge deficit due to a specific concentration of Na cation vacancies V_{Na}^+ will be balanced by appropriate numbers of Na^+ cation interstitials I_{Na}^+ and Cl^- anion vacancies V_{Cl}^- in order to maintain the overall charge neutrality of the solid. Pairs of cation–anion vacancies (e.g., $V_{Na}^+–V_{Cl}^-$) in ionic solids are referred to as *Schottky defects*, and pairs of cation vacancies and cation interstitials are known as *Frenkel defects*. Defects specifically associated with anion vacancies in ionic crystals such as NaCl are *color centers*. These are described in Chapter 8.

Other examples of vacancies in ionic crystals include the Fe^{2+} cation vacancies, V_{Fe}^{2+}, which must exist in $Fe_{1-x}O$. To maintain the overall electrical neutrality of this nonstoichiometric solid, two Fe^{3+} cations are formed for every Fe^{2+} vacancy present. Finally, Na vacancies are created in NaCl when divalent Ca atoms are introduced into the crystal. For every Ca^{2+} ion occupying a Na^+ site in NaCl, a Na^+ vacancy, V_{Na}^+, must be formed. The topic of nonstoichiometry is discussed further in Chapter W4.

Interstitials. An interstitial atom in a solid can be either a *self-interstitial*, if the atom is already a component of the solid, or an *interstitial impurity*, if it is not. Interstitial sites in crystalline solids generally have well-defined sizes and shapes which limit the atoms that can enter them without causing excessive strain in the surrounding structure. In metals with FCC, HCP, and BCC crystal structures, the tetrahedral and octahedral interstitial sites described in Chapter 1 are smaller than the size of one of the host metal atoms. Insertion of a host atom into the interstitial site would generate a large strain energy. As a result, the formation of self-interstitial atoms occurs with a much smaller probability than the formation of vacancies in these close-packed metals. Small impurity atoms such as H, C, and N can, however, enter these interstitial sites rather easily.

As mentioned earlier, when an atom jumps to an interstitial site, leaving a vacant site behind, the resulting vacancy–interstitial pair $V–I$ together are known as a *Frenkel defect*. Defects known as *Frenkel pairs* can be created in metals by ion bombardment. In this case two atoms, the bombarding ion and the atom originally occupying the site, share the site of the original atom. This interstitial configuration is known as a *dumbbell*, with the axis of the dumbbell lying along the $\langle 100 \rangle$ directions in FCC metals and along the $\langle 110 \rangle$ directions in BCC metals.

Impurities. A solid in equilibrium with an external phase such as a gas, liquid, or even another solid can always lower its free energy by accepting some impurity atoms from the external phase. This lowering of the free energy is related to the increase in entropy of the solid as impurities enter the structure, resulting in an increase in disorder.

Depending on their size, impurity atoms can enter a solid either substitutionally or interstitially, as illustrated in Fig. 4.13. Examples of substitutional impurities include B and P as electrically active dopant impurity atoms in Si, Zn in Cu leading to the formation of brass at high enough Zn concentrations, Al for Mg in MgO, and F for Cl in NaCl. Substitutional B and P impurities in Si are usually charged (i.e., B^- and P^+), for reasons that are discussed in Chapter 11, while the substitutional F atom appears as F^- in NaCl, for obvious reasons. Substitutional impurities in metals such as Zn in Cu tend to have the same charge that they would have in their own metallic state (e.g., Zn^{2+}). The simultaneous introduction of two Al^{3+} substitutional impurities in MgO leads to the creation of a single Mg^{2+} vacancy, V_{Mg}^{2+}, in order to preserve overall charge neutrality.

The pair of antisite defects shown in Fig. 4.13 corresponds, for example, to a pair of adjacent Ga and As atoms exchanging sites in crystalline GaAs (i.e., $Ga_{As}-As_{Ga}$). Under these circumstances the Ga and As atoms can also be viewed as a type of substitutional impurity. Important interstitial impurities include hydrogen in Ta and C in Fe. In both cases the impurity is much smaller than the host atom [i.e., $r_{cov}(H) = 0.037$ nm and $r_{met}(Ta) = 0.147$ nm, while $r_{cov}(C) = 0.077$ nm and $r_{met}(Fe) = 0.127$ nm]. It is not clear which radius to use for an atom in an interstitial site since it is not at all obvious, in general, how the interstitial atom is bonded to the surrounding atoms or ions.

All the examples listed above correspond to the usual type of chemical impurities (i.e., impurities corresponding to a different element). Another possible type of impurity is an *isotopic impurity* (e.g., ^{13}C in solid carbon). The ^{13}C isotope normally occurs in carbon at the level of only 1.10 at %, with the balance being ^{12}C. Although ^{13}C and ^{12}C atoms are chemically identical, the higher mass of the ^{13}C nucleus in the diamond structure results in the loss of translational periodicity. The presence of ^{13}C atoms will therefore result in the scattering of phonons in diamond. This additional scattering can be significant since it has been demonstrated recently[†] that isotopically-pure ^{12}C diamond has a thermal conductivity at $T = 300$ K which is about 50% higher than that of normal diamond containing 1.10 at % ^{13}C. Isotopically pure ^{12}C diamond, in fact, has the highest observed thermal conductivity of any solid at $T = 300$ K, over 3000 W/m·K.

Multiatom and Multisite Defects. Under suitable processing conditions such as quenching and annealing, point defects in solids can cluster, with vacancies forming *voids* and *precipitates* forming impurity clusters or inclusions. Interstitials can condense or precipitate to form an extra plane of atoms, thus forming a two-dimensional defect. These clusters and precipitates are also known as *defect complexes*. Examples of Au clusters in an amorphous matrix have already been shown in Fig. 4.6. Oxygen precipitates in crystalline Si are often formed during growth via the Czochralski method (see Chapter W21). Clustering of vacancies can lead first to divacancies and trivacancies and finally, to voids in the structure. The Frenkel defect is also technically a multisite defect, consisting of a vacancy–interstitial pair.

[†] L. Wei et al., *Phys. Rev. Lett.*, **70**, 3764 (1993).

4.5 Extended Defects

Extended defects are not localized but instead, extend through the solid for appreciable distances, typically at least 100 nm. They can be essentially one-dimensional, as in the case of dislocations, or two-dimensional, as in the case of grain boundaries, stacking faults, twins, and surfaces. Some examples of extended defects have already been illustrated (e.g., grain boundaries in polycrystalline Pb in Fig. 4.1*b*, twins in the high-temperature ceramic superconductor $YBa_2Cu_3O_{7-x}$ in Fig. 4.1*c*, and dislocations in Au clusters in Fig. 4.6). Other examples of extended defects and their effects on the properties of materials are described in later chapters. The geometrical structures of these extended defects are discussed next.

Line Defects. *Dislocations* are linear defects corresponding to a disturbance in the structure resulting from systematic deviations of the positions of atoms from their normal sites. The disturbances associated with dislocations typically extend through the structure along a line so that dislocations are essentially one-dimensional. There are two common types of dislocations in crystals: *edge* and *screw*. Other types of dislocation include *disclinations*, which are associated with rotations of the lattice, and *misfit dislocations*, which can exist at the interface between solids having different lattice constants or crystal structures.

Edge dislocations can be thought of as resulting either from the presence of an extra plane of atoms or of a missing plane of atoms in the structure, as illustrated in Fig. 4.15 and again in Fig. 4.16. In either case, the extra (or missing) plane of atoms does not extend through the entire crystal but instead, terminates within the crystal, as shown. The edge dislocation is considered to be localized in the region around the line at the end of the extra plane of atoms where the deviation of atoms from their normal positions is greatest. This line is the *dislocation line*, and the region surrounding it is the *dislocation core*.

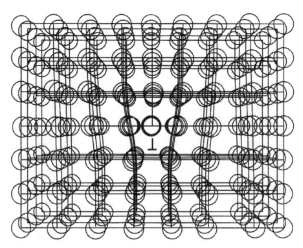

Figure 4.15. How atoms deviate from their normal positions at an edge dislocation in a simple cubic crystal. The dislocation line is perpendicular to the page. (From A. G. Guy, *Introduction to Materials Science*, 1972, copyright 1972 by McGraw-Hill. Reprinted by permission of the McGraw-Hill Companies.)

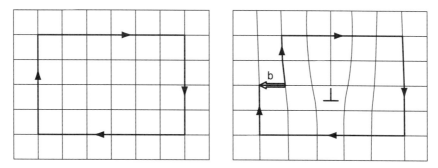

Figure 4.16. Schematic diagram used to define the Burgers vector **b** for an edge dislocation.

Dislocations are usually characterized by the following two properties:

1. The path of the dislocation line in the crystal, the direction of which is defined by a vector **l** tangent to the line
2. The *Burgers vector* **b**, which is a measure of the magnitude of the dislocation

To see how the Burgers vector is determined for an edge dislocation, consider the sketch shown in Fig. 4.16. In this figure a closed, rectangular path in the perfect crystal is compared with a related path in the defective crystal, which encloses the dislocation line and consists of the same number and sequence of steps as the path in the perfect crystal. The path in the defective crystal is open since the same number and sequence of steps does not now lead back to the starting point. The Burgers vector **b** is the lattice vector which closes the path in the defective crystal, having its tail at the starting point of the path and its head at the end, as shown[†]. For an edge dislocation the vectors **l** and **b** are perpendicular to each other.

The other common type of dislocation, the *screw dislocation*, can be thought of as being formed by a shear stress which causes, for example, one side of the crystal to be shifted by a lattice constant relative to the other side. This is shown in Fig. 4.17, where the dislocation line lies along the direction of the relative shift. The Burgers vector of a screw dislocation is therefore parallel to the direction of the dislocation line. The term screw dislocation refers to the fact that the originally parallel planes in the crystal are now joined by a spiral or helical path. Dislocations can also form closed paths within the crystal, in which case they are called *dislocation loops*.

The most general dislocation in a crystal has an angle between **b** and **l** which is between 90°, as for an edge dislocation, and 0°, as for a screw dislocation. The general dislocation can, in fact, be thought of as a combination of an edge and a screw dislocation.

The energy required for the formation of a dislocation is the energy associated with the resulting strain in the crystal. The *line tension* T_L of a dislocation is the energy of the dislocation per unit length and can be shown to be proportional to Gb^2. The quantity G is the shear modulus, a measure of the mechanical strength of the

[†] An alternative but equivalent definition of **b** considers a closed path in the defective material (see Bollmann, 1970, pp. 42–44).

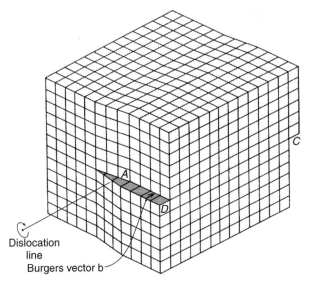

Figure 4.17. Screw dislocation in a crystal. The Burgers vector **b** is also shown and is parallel to the dislocation line. (From W. D. Callister, Jr., *Materials Science and Engineering*, 2nd ed., copyright 1991 by John Wiley & Sons, Inc. Reprinted by permission of John Wiley & Sons, Inc.)

crystal. As a result, dislocations with the smallest possible Burgers vector **b** will be formed preferentially. The smallest lattice vectors and hence the most common types of dislocations are of the $\frac{1}{2}[110]$ type in the FCC, NaCl, and diamond crystal structures, the $[11\bar{2}0]$ type in the HCP crystal structure, and the $\frac{1}{2}[111]$ type in the BCC and CsCl crystal structures.

Misfit dislocations at the interface between two crystals are shown schematically in Fig. 4.14. In this example the lattice constant parallel to the B–A interface in crystal B is larger than the corresponding lattice constant in crystal A. As a result, the planes of B atoms perpendicular to the interface are not aligned with the corresponding planes of A atoms. Instead, the planes of B atoms are distorted at the interface and are no longer equally spaced. The resulting misfit dislocations are spaced uniformly along the interface and correspond to the regions where the lattice distortions are greatest. Misfit dislocations are described in more detail in Chapter W20, where the critical layer thickness for the generation of misfit dislocations at the interface between a thin film and a substrate is derived.

Point defects and dislocations can interact with each other in a variety of ways. Excess vacancies or interstitials present under nonequilibrium conditions can, if sufficiently mobile, coalesce into dislocation loops in metals such as Al, while existing dislocations can act as sources and sinks of vacancies and other point defects in many crystals. Enhanced diffusion of impurities can readily occur along the distorted regions associated with dislocations.

Interactions between dislocations are quite common and can lead to a wide variety of phenomena in which their motions are affected by their interactions with each other. An important example is the plastic deformation of materials occurring under applied stress, discussed in more detail in Chapter 10.

Planar Defects. Planar defects also correspond to disturbances in the crystalline structure resulting from systematic deviations of the positions of atoms from their normal positions. Examples include grain boundaries, stacking faults, twins, and external surfaces. Since the disturbances associated with these defects typically extend through the structure along a plane, they are essentially two-dimensional.

Grain boundaries are the disordered regions in polycrystalline, microcrystalline, and nanocrystalline solids which exist at the interfaces between crystallites having different orientations. Examples of grain boundaries are shown in Fig. 4.1*b–d* and in Fig. 4.14. Grain boundaries are also discussed in Chapter 20.

One can distinguish between low- and high-angle grain boundaries, the angle in question being the angle between a given set of lattice planes in adjoining crystallites. *Low-angle grain boundaries*, where the orientations of the two crystallites differ from each other by only a few degrees, will ideally be composed of a network or array of individual dislocations. An example is shown in Fig. 4.18, where an array of edge dislocations exists to minimize the strain at a low-angle grain boundary, also called a *tilt boundary*. *High-angle grain boundaries* are not likely to be composed of a network of individual dislocations since the cores of adjacent dislocations would overlap.

The structure of grain boundaries on an atomic level is shown schematically in Fig. 4.19, where it can be seen that the disorder extends only over a few interatomic distances and is greater for high-angle than for low-angle grain boundaries. Grain boundaries in crystals can act as sinks for impurities, which often can find a site of lower energy in the disordered region.

Another type of planar defect is the *stacking fault*, which corresponds to a fault or mistake in the stacking sequence of planes of atoms in certain directions in a crystal. Mistakes in the stacking of {111} planes of atoms in crystal structures with FCC lattices such as diamond and cubic ZnS are not uncommon since the NN bonding of atoms in the crystal is unaffected by the stacking fault. As a result, the energy per unit area associated with stacking faults is low compared to other planar defects, such as grain boundaries in these crystals, since the presence of a stacking fault is evident only in changes in the arrangement of next-NN atoms.

As described in Section 1.7, the FCC crystal structure can be generated by the uniform stacking of lattice planes along the ⟨111⟩ directions, the normal sequence of (111) planes being **ABCABCABC.**... Possible stacking faults include a missing

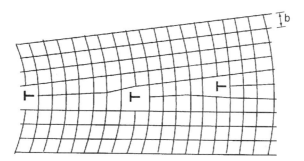

Figure 4.18. Array of edge dislocations at a low-angle grain boundary. The dislocation lines are perpendicular to the plane of the page and lie in the plane of the grain boundary. (From W. D. Callister, Jr., *Materials Science and Engineering*, 2nd ed., copyright 1991 by John Wiley & Sons, Inc. Reprinted by permission of John Wiley & Sons, Inc.)

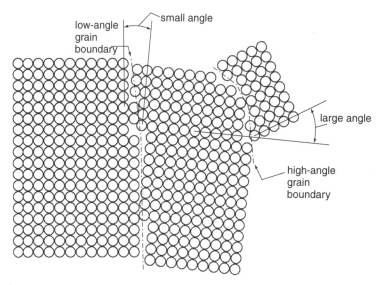

Figure 4.19. Atomic structure at low- and high-angle grain boundaries. The disorder associated with grain boundaries extends only over a few interatomic distances. (From W. D. Callister, Jr., *Materials Science and Engineering*, 2nd ed., copyright 1991 by John Wiley & Sons, Inc. Reprinted by permission of John Wiley & Sons, Inc.)

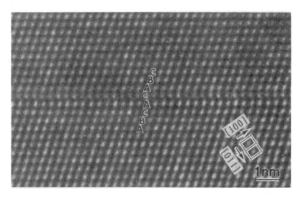

Figure 4.20. Example of a stacking fault of the type **ABCACABC**... (111) is presented in a high-resolution electron micrograph of a crystal of $CoSi_2$. The plane of the stacking fault is perpendicular to the page. (Reprinted from K. Suzuki et al., *Intermetallics*, Vol. 1, p. 21, copyright 1993, with permission from Elsevier Science.)

plane (e.g., a missing **B** plane as in **ABCACABC**...) or an extra plane (i.e., an extra **A** plane as in **ABCABACABC**...). An example of a missing plane is presented in the high-resolution electron micrograph of a crystal of $CoSi_2$ shown in Fig. 4.20. This is known as a *deformation stacking fault* since it can be produced by plastic deformation (i.e., by mechanically shifting one part of the crystal relative to the other). Such a stacking fault can also be formed via the condensation of vacancies, while an extra plane of atoms can result from the condensation of interstitials. Similar stacking faults can also occur in the HCP crystal structure along ⟨001⟩ directions. Stacking faults in

Figure 4.21. Examples of twins are presented in a high-resolution electron micrograph of a mechanically deformed crystal of $CoSi_2$. The planes of the twins are perpendicular to the page. (Reprinted from K. Suzuki et al., *Intermetallics*, Vol. 1, p. 21, copyright 1993, by permission from Elsevier Science.)

which identical planes of atoms are adjacent to each other, such as **ABCAABCABC** . . . or **ABABBABA**. . . , are known as *antiphase boundaries* (APBs). These are generally of high energy and therefore quite unstable.

Another planar defect also related to changes in the normal stacking of lattice planes is the *twin boundary* or *twin*. The characteristic of a twin boundary separating two regions of a crystal is that the atomic arrangements on opposite sides of the twin plane are mirror images of each other. In crystals with an FCC lattice a (111) twin boundary can be equivalent to a special type of stacking fault (e.g., as in the palindrome **ABCABCBCBACBA**. . .). Here the extra **B** plane in the middle of the sequence is the twin boundary, separating the **ABC**. . . sequence on the left from the **CBA**. . . sequence on the right. This type of twin is also known as a *growth stacking fault* since it can occur most readily during crystal growth. Examples of twins are presented in the high-resolution electron micrograph of a mechanically deformed crystal of $CoSi_2$, shown in Fig. 4.21.

The external surface of a crystal also represents a disturbance in the regular crystal structure. The *surface energy* is the energy required to separate the crystal into two parts, thereby disrupting the bonding between the atoms adjacent to the surface. As described in Chapter 19, the atoms at or near to the surface can move from their normal lattice positions to new sites of lower energy, thereby also lowering the surface energy.

4.6 Thermodynamics of Defect Formation: Entropy

Due to the vibrations of atoms about their lattice sites, there exists at any temperature $T > 0$ K a finite probability that a given atom can pass over the energy barrier, which ordinarily binds it to its lattice site. According to kinetic arguments, the temperature-dependent probability for such a jump is proportional to the *Boltzmann factor* $\exp(-E_a/k_B T)$, where E_a is the activation energy for the jump. When the atom passes over this energy barrier, it can either occupy an interstitial site in the lattice or can diffuse to the surface, where it can occupy a surface lattice site. Although interstitial and surface sites are not the lowest-energy sites available to atoms, they do represent local minima in the lattice potential energy.

The intrinsic point defects (vacancies, interstitials, and vacancy–interstitial pairs) have finite concentrations in crystals at equilibrium which are determined by and can be calculated from the laws of thermodynamics. These equilibrium concentrations will be determined by minimizing the free energy of the crystal. This procedure will be shown to yield results identical to those obtained directly from the law of mass action.

The existence of intrinsic defects in crystals follows from the natural tendency for crystals to lower their *Gibbs free energy G* by increasing their *entropy S*. This is evident from the definition of G in terms of the *enthalpy H* and S (see Appendix WA),

$$G = H - TS = U + PV - TS. \qquad (4.1)$$

Here T is the absolute temperature, U is the *internal energy*, and P and V are the pressure and volume, respectively, of the crystal. The PV term is usually neglected or is assumed to remain constant as defects are generated. In an ideal covalent or ionic crystal the enthalpy H is associated with the cohesive energy residing in the covalent bonds or in the electrostatic energy, respectively. The entropy S is a measure of the disorder present in the crystal and in a hypothetical ideal, defect-free crystal represents the disorder present due to the thermal vibrations of the atoms.

At a finite temperature T, the enthalpy H and entropy S will both change when intrinsic defects are generated in the crystal. The Gibbs free energy of the crystal can therefore be written as

$$
\begin{aligned}
G(T, N_d) &= G(T, 0) + \Delta H - T\Delta S \\
&= G(T, 0) + N_d \Delta H_d - T N_d \Delta S_{\text{vib}} - T \Delta S_{\text{con}}(N_d), \qquad (4.2)
\end{aligned}
$$

where $G(T,0)$ is the free energy of the defect-free crystal at temperature T, ΔH the enthalpy needed to form the defects, and ΔS the resulting change in the entropy. Also, N_d is the number of defects, ΔH_d the enthalpy change per defect, ΔS_{vib} the change in *vibrational entropy* per defect, and $\Delta S_{\text{con}}(N_d)$ the change in the *configurational entropy* of the crystal, which is itself a function of N_d. In the case of a vacancy, ΔS_{vib} corresponds to the increase in vibrational entropy resulting from the larger amplitude of vibration of atoms surrounding the vacancy.

The dependencies of these individual contributions to G and of the resulting variation of G with the number of defects N_d are shown schematically in Fig. 4.22. The decrease in G due to the increasing entropy (i.e., the $-T\Delta S_{\text{con}}$ term) is eventually overcome by the increasing enthalpy, so that a minimum in G occurs at a finite N_d. The resulting equilibrium value of N_d will be a function of temperature and for a particular crystal will also depend on ΔH_d and ΔS_{vib}, which can be different for different types of intrinsic defects. These general ideas will now be examined in detail for a specific intrinsic defect, the Frenkel defect in an elemental crystal.

The generation of Frenkel defects (i.e., vacancy–interstitial pairs) in an elemental crystal is a thermally activated process. Let N_L be the number of lattice sites and N_I

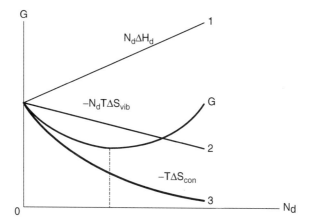

Figure 4.22. Schematic variations of the Gibbs free energy G, enthalpy H, and entropy S of a crystal as functions of defect concentration N_d. (From G. G. Libowitz, in N. B. Hannay, ed., *Treatise on Solid State Chemistry*, Vol. 1, Plenum Press, New York, 1974, Chap. 6.)

the number of interstitial sites in an ideal crystal of element A. At $T = 0$ K all lattice sites will be occupied and all interstitial sites will be empty. As T increases, some A atoms will be thermally excited from their lattice sites to NN interstitial sites. The defect reaction for this process can be written as

$$A + V_I \longleftrightarrow V_A + I_A, \tag{4.3}$$

where A refers to a lattice site occupied by an A atom, V_I to a vacant interstitial site, and V_A and I_A to a vacant lattice site and an interstitial site occupied by an A atom, respectively. Note that N_L and N_I will not change as vacancy–interstitial pairs are created and can be written as

$$N_L = N_L(A) + N_L(V),$$
$$N_I = N_I(A) + N_I(V). \tag{4.4}$$

Here $N_L(A)$ and $N_L(V)$ are the numbers of occupied and vacant lattice sites and $N_I(A)$ and $N_I(V)$ are the numbers of occupied and vacant interstitial sites, respectively. For Frenkel defects an occupied interstitial site is created whenever an A atom leaves a lattice site. It therefore follows that $N_L(V) = N_I(A) = N_d$ (i.e., the number of vacancies is equal to the number of interstitial A atoms).

It will be assumed here that vacancy creation and interstitial creation can be treated as independent processes, that the resulting vacancy and interstitial are well separated, and that there is little chance of their recombination. In the limit of low defect concentrations the Gibbs free energy G can be expressed as a function of $N_L(V)$ and $N_I(A)$ as follows:

$$G = G(0) + N_L(V)G_L(V) + N_I(A)G_I(A) - k_B T \ln \left[\frac{N_L!}{N_L(A)! N_L(V)!} \frac{N_I!}{N_I(V)! N_I(A)!} \right]. \tag{4.5}$$

Here $G_L(V) = \Delta H_d[L(V)] - T\Delta S_{vib}[L(V)]$ and $G_I(A) = \Delta H_d[I(A)] - T\Delta S_{vib}[I(A)]$ are the Gibbs free energies of formation per defect for vacancies and interstitials, respectively, and correspond to the sum of the ΔH_d and $-T\Delta S_{vib}$ terms in Eq. (4.2). The last term in Eq. (4.5) is the configurational entropy, which has the general form $\Delta S_{con} = k_B \ln W$, where W is the total number of possible configurations of the system. As written, ΔS_{con} represents the entropy due to the possible configurations of $N_L(V)$ vacancies distributed among N_L lattice sites and also of $N_I(A)$ interstitial A atoms distributed among N_I interstitial sites. Using Stirling's approximation, $\ln N! \approx N \ln N - N$ (valid for $N \gg 1$), the free energy can be written as

$$
\begin{aligned}
G \approx G(0) &+ N_L(V)G_L(V) + N_I(A)G_I(A) \\
&- k_B T\{N_L \ln N_L - [N_L - N_L(V)]\ln[N_L - N_L(V)] - N_L(V)\ln N_L(V) \\
&+ N_I \ln N_I - [N_I - N_I(A)]\ln[N_I - N_I(A)] - N_I(A)\ln N_I(A)\}.
\end{aligned} \qquad (4.6)
$$

The equilibrium value of $N_L(V)$ is determined by first setting $N_I(A) = N_L(V)$ in Eq. (4.6) and then by minimizing the resulting G with respect to $N_L(V)$ [i.e., by setting $\partial G / \partial N_L(V) = 0$]. The result is

$$
\begin{aligned}
N_L(V) &= [N_L(A)N_I(V)]^{1/2} \exp\left[-\frac{G_L(V) + G_I(A)}{2k_B T}\right] \\
&\approx (N_L N_I)^{1/2} \exp\left[-\frac{G_L(V) + G_I(A)}{2k_B T}\right].
\end{aligned} \qquad (4.7)
$$

The replacements of $N_L(A)$ by N_L and $N_I(V)$ by N_I in Eq. (4.7) are generally quite good approximations since the number of Frenkel defects $N_L(V)$ is always much less than N_L or N_I (except possibly just below the melting point of the crystal). This follows from the fact that the exponential term in Eq. (4.7) is always much less than 1. Experimental results for the number or concentration of intrinsic defects are often expressed in the general form

$$
N_d(T) = N_d(\infty) \exp\left(-\frac{E_a}{k_B T}\right) \qquad (4.8)
$$

which is just the form predicted theoretically for Frenkel defects by Eq. (4.7). Here $N_d(\infty)$ is the defect concentration in the limit of very high temperatures.

Law of Mass Action. A procedure often used to predict equilibrium concentrations of intrinsic defects employs the thermodynamic *law of mass action*, a general rule that expresses the activities or the equilibrium concentrations of species resulting from a chemical or physical reaction in terms of the thermodynamic *equilibrium constant* $K(T)$ for the reaction. *Activities* are dimensionless quantities which are the ratios of the reactivity of a given species (e.g., defect, gas, liquid, or solid) to its reactivity in its standard state. The reactivity of a defect is usually proportional to its concentration.

For the defect reaction described in Eq. (4.3) (i.e., the thermal generation of Frenkel defects), the law of mass action takes the form

$$
\frac{a_L(V)a_I(A)}{a_L(A)a_I(V)} = K_F(T) = \exp\left(-\frac{\Delta G_r}{k_B T}\right), \qquad (4.9)
$$

where the a_L's and a_I's are the activities of the species involved in the reaction and $K_F(T)$ is the equilibrium constant for the defect reaction. The quantity K_F can be expressed in terms of the change in Gibbs free energy ΔG_r per defect for this reaction, as shown. For low numbers or concentrations of defects [i.e., for $N_L(V) \ll N_L(A)$ and $N_I(A) \ll N_I(V)$], the activities of the reacting species can be set equal to their concentrations. It follows that Eq. (4.9) can be written as

$$\frac{N_L(V)N_I(A)}{N_L(A)N_I(V)} = K_F(T). \tag{4.10}$$

An equivalent result can be obtained by noting that $a_L(A)$ for A atoms on normal lattice sites and $a_I(V)$ for vacant interstitial sites can both be set equal to unity and that $a_L(V)$ and $a_I(A)$ can be set equal to $N_L(V)/N_L(A)$ and $N_I(A)/N_I(V)$, respectively.

The change in the Gibbs free energy for the reaction, ΔG_r, is defined in general by

$$\Delta G_r = \sum G(\text{products}) - \sum G(\text{reactants}), \tag{4.11}$$

where the configurational entropy is omitted. For Frenkel defects, $\Delta G_r = G_L(V) + G_I(A)$ from Eq. (4.5). It follows from Eq. (4.10) that

$$\frac{N_L(V)N_I(A)}{N_L(A)N_I(V)} = \frac{N_L^2(V)}{N_L(A)N_I(V)} = \exp\left[-\frac{G_L(V) + G_I(A)}{k_B T}\right], \tag{4.12}$$

using $N_I(A) = N_L(V)$. This prediction can be seen to be equivalent to the result obtained in Eq. (4.7).

Applications of the law of mass action to the creation of Schottky defects (i.e., vacancies) and interstitials are described in Chapter W4.

4.7 Examples of Defect Studies

Vacancies in FCC metals can be studied using a wide variety of experimental techniques. The equilibrium fractional concentration of vacancies $n_v(T) = N_L(V)/N_L(A)$ can be determined in a very direct manner using thermal expansion measurements whenever the contribution of interstitials can be neglected. The concentration of interstitials in close-packed FCC or HCP metals can often be neglected compared to that of vacancies due to the higher enthalpy of formation $\Delta H_f(I)$ of interstitials. The high value of $\Delta H_f(I)$, about 4 eV or more, is due to the elastic strain energy that would be present given the small size of interstitial sites in close-packed metals. Interstitials can be expected to be more important in crystals with diamond crystal structures having correspondingly larger interstitial sites. By measuring both the *macroscopic* thermal expansion $\Delta l/l_0$ and the *microscopic* thermal expansion $\Delta a/a_0$ (by measuring the change Δa in the lattice constant of the unit cell using x-ray diffraction), $n_v(T)$ can be calculated from the expression

$$n_v(T) = 3\left(\frac{\Delta l}{l_0} - \frac{\Delta a}{a_0}\right), \tag{4.13}$$

which is valid as long as $\Delta a \ll a_0$. This expression follows from the fact that while vacancies cause the crystal to expand, they do not contribute to the expansion of the average unit cell. Note that if only Frenkel defects are present in the crystal at equilibrium, the macroscopic and microscopic thermal expansions will be equal.

Experimental results for $n_v(T)$ for the FCC metal Cu are presented in Fig. 4.23 over a wide range of temperatures up to the melting point $T_m = 1084°C$. These data have been fit to a single exponential,

$$n_v(T) = \exp(2.5 \pm 0.2) \exp\left[-\frac{(1.28 \pm 0.03)\ eV}{k_B T}\right]. \qquad (4.14)$$

This indicates that a single intrinsic defect, the simple vacancy, is dominant in Cu over this temperature range. The Gibbs free energy of formation per Schottky defect is given by

$$\Delta G_r = \Delta H_f - T\Delta S_f \qquad (4.15)$$

The enthalpy of formation ΔH_f for a Cu vacancy is thus found to be 1.28 eV. The entropy of formation ΔS_f is $2.5k_B$ and corresponds to an increase in vibrational entropy resulting from the larger amplitude of vibration of the atoms surrounding the vacancy.

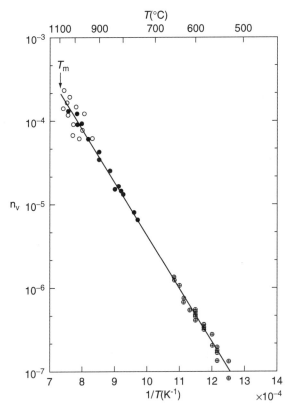

Figure 4.23. Experimental results for the equilibrium fractional concentration of vacancies $n_v(T)$ for the FCC metal Cu over a wide range of temperatures up to the melting point $T_m = 1084°C$. [From A. S. Berger et al., *J. Phys. F: Met. Phys.*, **9**, 1023 (1979). Reprinted by permission of the Institute of Physics.]

TABLE 4.2 Properties of Vacancies in FCC Metals

Metal	ΔH_f (eV)	ΔS_f (k_B)	$n_v(T_m)$ (10^{-4})
Cu	1.28	2.5	2.0
Ag	1.09	1.5	1.7
Au	0.94	1.0	7.2
Al	0.75	2.2	9.0
Pb	≥ 0.53	1.5	≤ 1.5

Source: Cu: A. S. Berger et al., *J. Phys. F: Met. Phys.*, **9**, 1023 (1979); Ag–Pb: R. O. Simmons et al., *Phys. Rev.*, **125**, 862 (1962).

Note that just below T_m, the fractional concentration of vacancies in Cu is about 2×10^{-4}.

The enthalpies and entropies of vacancy formation as well as values of $n_v(T_m)$ for several FCC metals are presented in Table 4.2.

REFERENCES

Bollmann, W., *Crystal Defects and Crystalline Interfaces*, Springer-Verlag, New York, 1970.

Callister, W. D., Jr., *Materials Science and Engineering: An Introduction*, 2nd ed., Wiley, New York, 1991.

Glassy Metals: Magnetic, Chemical, and Structural Properties, Hasegawa, R. ed., CRC Press, Boca Raton, Fla., 1983.

Kittel, C., *Introduction to Solid State Physics*, 7th ed., Wiley, New York, 1996.

Libowitz, G. G., *Defect equilibria in solids*, in N. B. Hannay, ed., *Treatise on Solid State Chemistry*, Vol. 1, Plenum Press, New York, 1976.

Mott, N. F., and E. A. Davis, *Electronic Processes in Non-crystalline Materials*, 2nd ed., Clarendon Press, Oxford, 1979.

Read, W. T., Jr., *Dislocations in Crystals*, McGraw-Hill, New York, 1953.

Zallen, R., *The Physics of Amorphous Solids*, Wiley, New York, 1983.

PROBLEMS

4.1 Take a small box or cylindrical container and measure its volume. Pour marbles or ball bearings into the box until it is full. Determine the volume occupied by the spheres. Compute the packing fraction. Repeat the experiment several times and average the results. Compare your result with the packing fractions for FCC and HCP, BCC, SC, and the value 0.64 obtained for the random packing of hard spheres.

4.2 Draw sketches of the two-atom Frenkel pair interstitial configurations known as "dumbbells" in both FCC and BCC metals.

4.3 Consider the equilibrium concentration of Frenkel defects in a solid.

   (a) Derive the value of $N_L(V)$ given in Eq. (4.7) by first setting $N_I(A) = N_L(V)$ in Eq. (4.6) and then minimizing the resulting Gibbs free energy with respect to $N_L(V)$.

(b) Repeat the derivation using Lagrange multipliers to enforce the constraints of Eq. (4.4).

4.4 Consider a monatomic solid consisting of N atoms. Determine the number of ways, W, that n of the atoms may be removed to form n vacancies. Compute the entropy, given by $S = k_B \ln W$. For the SC, BCC, and FCC crystal structures, compute the entropy for forming NN vacancies.

4.5 Consider a one-dimensional monatomic solid with N atoms and $N_L(V)$ vacancies at temperature $T > 0$ K. Show that the fractional vacancy concentration $n_v(T) = N_L(V)/N$ is given approximately by $n_v \approx \Delta l / l_0 - \Delta a / a_0$. Here $l_0 = N a_0$ is the length of the solid at $T = 0$ K, Δl is the change in length, a_0 is the lattice constant at $T = 0$ K, and Δa is the change in the lattice constant. (Hint: Write the change of length as $\Delta l = l - l_0 = \Delta l_{\text{thermal}} + \Delta l_{\text{vacancies}}$.)

PHYSICAL PROPERTIES OF MATERIALS

CHAPTER 5

Phonons

EXCITATIONS OF THE LATTICE: PHONONS

In the preceding chapters static solids, those essentially at absolute zero in temperature, were considered. As the temperature is increased, the excitations of the solid start playing a more important role. In this chapter, effects of ionic motions on macroscopic properties are studied. One knows that temperature plays a crucial role in determining the physical properties of most materials. As a solid is warmed, structural changes may occur and the solid ultimately melts or vaporizes. The solid may undergo thermal expansion or contraction. There may be one or more phase transitions accompanied by symmetry changes. Intrinsic defects will increase in number and attain a higher degree of mobility. The mechanical properties will change as a function of temperature. The solid has the ability to store internal energy and will be characterized by a specific heat. Heat may be conducted through the material when subjected to a temperature gradient, and just how well this is done is determined by the thermal conductivity. Lattice vibrations will affect the electrical conductivity of the material by scattering the charge carriers. Traditional superconductivity is found to involve the interaction of electrons with lattice motions.

At first sight it looks as if the problem is far more complex for a solid than for a single molecule. After all, a single molecule consisting of s atoms has only $3s$ degrees of freedom, whereas a solid with N such molecules has $3Ns$ degrees of freedom. For a periodic lattice, however, the solid turns out to be no more complex than the single molecule. Translational periodicity permits a factorization of the problem to be made into N separate identical problems, each with $3s$ degrees of freedom.

The chapter begins by introducing the basic lattice excitations, called phonons. We then proceed to employ phonons to study the ionic contributions to the specific heat, thermal expansion, and thermal conductivity of solids. As will be seen phonons provide an appropriate link between the microscopic world, of atoms and their interactions, and the macroscopic world of real materials and their thermal and transport properties.

Phonons also play important roles in melting (Chapter 6), in the electrical resistivity of solids (Chapter 7), in the elastic properties of solids (Chapter 10), in the mobilities of charge carriers, and in the thermoelectric effect (Chapter 11). Neutron scattering, which is the method of choice for studying phonons in the laboratory, is considered later together with other characterization techniques (see Chapter W22 at our Web site[†]).

[†] Supplementary material for this textbook is included on the Web at the resource site (ftp://ftp.wiley.com/public/sci_tech_med/materials). Cross-references to elements of the Web material are prefixed by "W."

5.1   One-Dimensional Monatomic Lattice

Many aspects of lattice excitations are displayed by a one-dimensional solid, so it is useful to begin by studying this system. Consider a chain of N identical atoms of mass M separated from each other by distance a, with successive atoms linked by springs of stiffness constant K. The constants are related to the curvature of the potential energy of interaction between the atoms. So as not to give any atom a special position in the lattice, we impose periodic boundary conditions, (i.e., imagine the Nth and first atoms also linked by such a spring). The atoms may be regarded as if they were lined up on the perimeter of a circle, although for large N any segment of the circle would resemble a straight line. Periodic boundary conditions can also be employed in two dimensions, where the atoms will lie on the surface of a torus, or in three dimensions, where a physical visualization is not possible. The bulk properties of the lattice will not be affected by such boundary conditions.

In Fig. 5.1, atoms are drawn, (a) in their equilibrium positions, and (b) in their instantaneous positions in a lattice with a longitudinally excited wave at an arbitrary time. The amplitude of the displacement of the nth atom is denoted by $u_n(t)$, and the x coordinate of the atom is

$$x_n(t) = na + u_n(t), \qquad n = 1, 2, \ldots, N. \tag{5.1}$$

The nth atom experiences pulls (or pushes) to the right and to the left caused by the stretching (or compressing) of successive springs. Consider the case of Hooke's law and NN interactions only. The equations of motion are

$$M\ddot{u}_n = K(u_{n+1} - u_n) - K(u_n - u_{n-1}), \qquad n = 1, 2, \ldots, \tag{5.2}$$

with the subscript convention that $N + 1 \rightarrow 1$ and $0 \rightarrow N$. Equation (5.2) is really a set of N coupled second-order differential equations whose solution describes the displacements as a function of time. When the displacements are all taken to vary with time as $\exp(-i\omega t)$, the equations reduce to a set of finite-difference equations (instead of the spatial differential equations one would get in continuum mechanics).

In searching for the lowest-frequency excitation, begin by making a wrong guess and assume that the excitation is highly localized with a frequency comparable to that of a diatomic molecule (two masses connected by a spring). This would predict

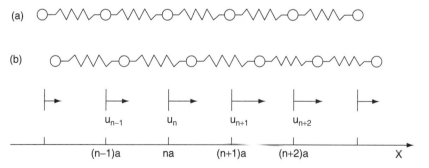

Figure 5.1. Monatomic one-dimensional lattice: (a) equilibrium lattice; (b) instantaneous view of the excited lattice.

a frequency near $\omega = \sqrt{2K/M}$. As will be seen, the frequency may be made much lower by choosing an appropriate delocalized mode of excitation.

Let

$$u_n(t) = A_j \exp[i(k_j na - \omega_j t)]. \tag{5.3}$$

This construct will be referred to as a *lattice wave*. It represents an excitation in which successive oscillators bear a definite phase relationship to the preceding oscillator. The periodic boundary condition $u_n(t) = u_{n+N}(t)$ can be satisfied if the relative phase is an integer multiple of 2π; that is, the wave vector is given by

$$k_j = \frac{2\pi j}{Na}, \tag{5.4}$$

where $j = 1, 2, \ldots, N$. Inserting the expression above for $u_n(t)$ into the equation of motion yields

$$-M\omega_j^2 A_j = KA_j(e^{ik_j a} + e^{-ik_j a} - 2), \tag{5.5}$$

from which it follows that

$$\omega_j = \sqrt{\frac{4K}{M}} \left| \sin \frac{k_j a}{2} \right|. \tag{5.6}$$

For large enough values of N, k_1 and hence ω_1 can be made arbitrarily small. The maximum excitation frequency occurs for $k_j = \pm\pi/a$ (i.e., for $j = N/2$) and is $2\sqrt{K/M}$. Typical values of ω_{max} for solids are $\approx 10^{14}$ rad/s. Thus the lattice frequencies are broadened into a band extending from 0 to ω_{max}. It is customary to regard j as varying between $-N/2$ and $N/2$ instead of between 1 and N. The allowed range of k, between $-\pi/a$ and π/a, is then referred to as the *first Brillouin zone*, as defined in Chapter 3, and the resulting picture is termed the *reduced zone scheme*. This has the advantage that waves traveling in either direction can be treated equivalently. The *extended zone scheme*, in which k is allowed to take on all values (and is therefore highly redundant), is also sometimes used. In Fig. 5.2a one Brillouin zone in the extended zone scheme is shown. Figure 5.2b displays the same information in the reduced zone. In the limit of large N one may drop the subscript j and obtain the dispersion relation

$$\omega(k) = 2\sqrt{\frac{K}{M}} \left| \sin \frac{ka}{2} \right|. \tag{5.7}$$

Some features are worthy of note. First, the number of possible values of k equals the number of cells in the lattice, N. Second, for low k there is a linear dispersion relation

$$\omega(k) = |k|a\sqrt{\frac{K}{M}} = |k|c_s, \tag{5.8}$$

where c_s is to be interpreted as the speed of sound. As expected, this speed increases as the stiffness of the spring increases and decreases as the inertia (M) increases. Typical values of c_s in solids are in the range 10^3 to 10^4 m/s. Third, if a pulse of excitation

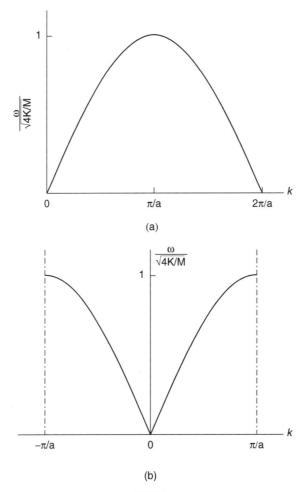

Figure 5.2. (*a*) Dispersion relation $\omega(k)/\sqrt{4K/M}$ for the excitations of the one-dimensional monatomic lattice. One Brillouin zone in the extended-zone scheme is shown. (*b*) First Brillouin zone in the reduced-zone scheme.

were to propagate along the chain, it would travel with a velocity called the *group velocity*, given by

$$c_g = \frac{\partial \omega}{\partial k} = c_s \cos \frac{ka}{2} \text{sgn } (k) \tag{5.9}$$

[sgn (k) is $+1$ or -1 for $k > 0$ or $k < 0$, respectively]. The slope of the dispersion curve vanishes at the zone boundaries ($k = -\pi/a$ and π/a), and the group velocity is zero there.

It should be noted that there are as many modes as there are atoms (i.e., N in this one-dimensional monatomic case). Therefore, phonons are defined only within the first Brillouin zone. If one tried to introduce a scheme in which there were more modes than atoms, one would have a situation where the wavelength of a lattice wave could be shorter than the interatomic spacing, which would be meaningless.

From Eq. (5.9) one notes that $c_g \to 0$ at the zone boundary. This corresponds to Bragg scattering of the lattice wave. At $|k| = \pi/a = 2\pi/\lambda$, the wavelength is $\lambda = 2a$, in agreement with Bragg's law. From Eq. (5.3) it is seen that adjacent atoms are 180° out of phase with each other for $|k| = \pi/a$. It is at the zone boundary where the effect of the lattice is felt most strongly. On the other hand, when $ka \to 0$, so that λ is much larger than a, the lattice may be treated as a continuum.

The model defined by Eq. (5.2) may readily be extended to disordered lattices (see Section W5.1).

Quantization. It is one of those historical curiosities that the quantum theory of fields was discovered before quantum mechanics. In his attempt to interpret the blackbody radiation spectrum, Planck quantized electromagnetic waves, and the resulting particles were called *photons*. Similarly, in their attempt to understand the thermal energy of solids, Einstein and Debye quantized the lattice waves and termed the resulting quanta *phonons*. The procedures are similar. Each mode of a classical field is characterized by a frequency ω and a wave vector $\mathbf{k}$ (and, perhaps, by other labels, such as polarization). The modes are replaced by quanta or particles. The particles carry an amount of energy proportional to the frequency of the mode, $E = \hbar\omega$, and a momentum proportional to the wave vector, $\mathbf{p} = \hbar\mathbf{k}$. Unlike the classical wave, which could have any amount of energy or momentum associated with a given mode, the quantum picture demands that these are quantized (i.e., come in integer multiples, n, of the underlying quantum). Thus a given mode can have $n = 0, 1, 2, \ldots$ quanta, but not any noninteger value. Technical details of the quantization procedure are presented in Appendix W5A.

An important distinction between classical and quantum physics occurs when one asks what the average energy in a mode is for a system at temperature T. In the classical picture each mode is regarded as a one-dimensional harmonic oscillator. According to statistical mechanical arguments, the mean kinetic energy is $k_B T/2$, k_B being the Boltzmann constant. With the potential energy of an oscillator being, on average, equal to the kinetic energy, the average total energy is $\langle E \rangle = k_B T$. In quantum physics, one computes the average energy using the Boltzmann probability factor. The relative probability of attaining an energy E is $\exp(-\beta E)$, with $\beta = 1/k_B T$. The expression for the energy, including the zero-point energy [see Eq. (W5A.20)], is $E(T) = [n(T) + \frac{1}{2}]\hbar\omega$. The average energy becomes

$$
\begin{aligned}
\langle E(T) \rangle &= \frac{\sum_{n=0}^{\infty} (n + \frac{1}{2})\hbar\omega e^{-\beta(n+1/2)\hbar\omega}}{\sum_{n=0}^{\infty} e^{-\beta(n+1/2)\hbar\omega}} \\
&= \frac{\hbar\omega}{2} + \frac{\sum_{n=0}^{\infty} n\hbar\omega e^{-\beta n\hbar\omega}}{\sum_{n=0}^{\infty} e^{-\beta n\hbar\omega}} = \frac{\hbar\omega}{2} - \frac{\partial}{\partial\beta} \ln\left(\sum_{n=0}^{\infty} e^{-\beta n\hbar\omega}\right) \\
&= (\langle n \rangle + \tfrac{1}{2})\hbar\omega.
\end{aligned}
\tag{5.10}
$$

In obtaining this formula the geometric series is summed so that $\sum_0 \exp(-nx) = 1/[1 - \exp(-x)]$. The average occupancy is given by

$$
\langle n(\omega, T) \rangle = \frac{1}{e^{\beta\hbar\omega} - 1}.
\tag{5.11}
$$

The quantity $\langle n(T) \rangle$ is called the *Bose–Einstein distribution function*.

In the high-temperature limit, where $\beta\hbar\omega \ll 1$, this reduces to the classical formula, $\langle E \rangle \approx k_B T$. In this limit $\langle n \rangle \gg 1$. In the low-temperature limit, however, $\langle E \rangle \rightarrow \hbar\omega/2$, which is just the zero-point energy. Effectively, modes whose frequencies are higher than $k_B T/\hbar$ are frozen out of the picture and are not thermally excited.

5.2   One-Dimensional Diatomic Lattice

Having started with the simplest case of a monatomic lattice in one dimension, the next level of complexity is a case in which there is a lattice with a basis — the diatomic lattice in one dimension. Since there are now two atoms per unit cell, one may expect a doubling of the number of modes. The two atoms will be labeled A and B, with masses M_A and M_B, respectively. It will be assumed that there are two spring constants, K and K', and that $M_A > M_B$. The definitions of the spring constants are given in Fig. 5.3.

The equilibrium positions of the A atoms are at locations $x_n = na$ and of the B atoms at $x_n = na + b$, where $n = 1, 2, \ldots, N$. Periodic boundary conditions are again imposed. The displacement of the nth A atom is denoted by u_n and of the nth B atom is denoted by v_n. The geometry is depicted in Fig. 5.4. Again, only longitudinal waves are described.

The equations of motion for the A atoms are

$$M_A \ddot{u}_n = K(v_n - u_n) - K'(u_n - v_{n-1}), \tag{5.12a}$$

and for the B atoms are

$$M_B \ddot{v}_n = -K(v_n - u_n) + K'(u_{n+1} - v_n). \tag{5.12b}$$

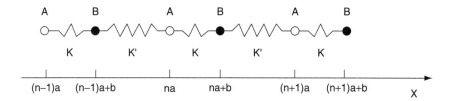

Figure 5.3. Diatomic lattice in one dimension with NN springs.

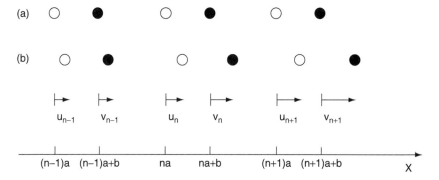

Figure 5.4. Lattice geometry for the diatomic lattice.

Making the substitutions

$$u_n = U \exp[i(nka - \omega t)] \qquad (5.13a)$$

$$v_n = V \exp[i(nka - \omega t)] \qquad (5.13b)$$

permits the reduction of the system to a 2×2 matrix equation

$$\Delta = \begin{bmatrix} U \\ V \end{bmatrix} = \omega^2 \mathbf{M} \begin{bmatrix} U \\ V \end{bmatrix}, \qquad (5.14)$$

where Δ is the force constant matrix,

$$\Delta = \begin{bmatrix} K + K' & -K - K'\eta^* \\ -K - K'\eta & K + K' \end{bmatrix}, \qquad (5.15)$$

and $\mathbf{M}$ is the mass matrix,

$$\mathbf{M} = \begin{bmatrix} M_A & 0 \\ 0 & M_B \end{bmatrix}, \qquad (5.16)$$

where $\eta = \exp(ika)$. Note that $\mathbf{M}$ is a diagonal matrix.

These equations may be simplified by multiplying through by $\mathbf{M}^{-1/2}$ ($\mathbf{M}^{1/2}$ is itself a diagonal matrix with entries $M_A^{1/2}$ and $M_B^{1/2}$), to get

$$\mathbf{M}^{-1/2} \Delta \mathbf{M}^{-1/2} \mathbf{M}^{1/2} \begin{bmatrix} U \\ V \end{bmatrix} = \omega^2 \mathbf{M}^{1/2} \begin{bmatrix} U \\ V \end{bmatrix} \equiv \mathbf{M}' \begin{bmatrix} U' \\ V' \end{bmatrix} = \omega^2 \begin{bmatrix} U' \\ V' \end{bmatrix}, \qquad (5.17)$$

where the first three matrices are combined to form $\mathbf{M}'$. This last equation may be written in the form

$$\mathbf{M}' \begin{bmatrix} U' \\ V' \end{bmatrix} = \begin{bmatrix} A + B & C - iD \\ C + iD & A - B \end{bmatrix} \begin{bmatrix} U' \\ V' \end{bmatrix} = \omega^2 \begin{bmatrix} U' \\ V' \end{bmatrix}. \qquad (5.18)$$

The squares of the mode frequencies are the eigenvalues of the $\mathbf{M}'$ matrix. Thus

$$\omega_\pm^2 = A \pm \sqrt{B^2 + C^2 + D^2}, \qquad (5.19)$$

where

$$A = \frac{1}{2}(K + K')\left(\frac{1}{M_A} + \frac{1}{M_B}\right), \qquad (5.20a)$$

$$B = \frac{1}{2}(K + K')\left(\frac{1}{M_A} - \frac{1}{M_B}\right), \qquad (5.20b)$$

$$C = -\frac{1}{\sqrt{M_A M_B}}(K + K'\cos ka), \qquad (5.20c)$$

$$D = -\frac{K'}{\sqrt{M_A M_B}}\sin ka. \qquad (5.20d)$$

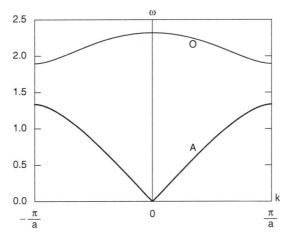

Figure 5.5. Dispersion curves for the acoustic (A) and optic (O) modes in the diatomic lattice. Here $K = K' = 1$, $M_A = 1$, and $M_B = 0.5$, in arbitrary units.

The high-frequency mode, ω_+, is called the *optic mode*; the low-frequency mode, ω_-, is called the *acoustic mode*. The term optic mode stems from the fact that if ions A and B are oppositely charged, this mode could be excited by interacting with infrared radiation. A typical set of dispersion curves is illustrated in Fig. 5.5. The modes do not cross each other and, in fact, there is a gap of forbidden frequencies separating the them. In the gap region there are no propagating modes. The optic mode is fairly flat, whereas the acoustic mode exhibits the linear variation of frequency with wave vector at small values of ka.

The atomic motions in the acoustic and optic modes are fundamentally different. The relative motions of the atoms can be obtained by studying the ratio V'/U':

$$\frac{V'}{U'} = \frac{C + iD}{\omega^2 - A + B}. \tag{5.21}$$

For example, at zone center ($k = 0$), a little algebra shows that

$$\left.\frac{V}{U}\right|_+ = -\frac{M_A}{M_B}, \qquad \left.\frac{V}{U}\right|_- = 1. \tag{5.22}$$

Thus the acoustic mode has the atoms vibrating in phase with each other, whereas in the optic mode the vibrations are 180° out of phase. This is illustrated in Fig. 5.6, where the motions of the respective modes are sketched at a value of ka close to zone center. Of course, in both cases each atom vibrates harmonically in time about its own equilibrium position. Only the relative phasings of neighboring atoms are different. Also note that for the optic mode the heavier atom has the smaller amplitude. For the acoustic mode the amplitudes are the same.

5.3 Phonons: General Case

In the general crystal one is confronted with a three-dimensional lattice and a basis. Suppose that there are s atoms per primitive cell and N cells. There will then be

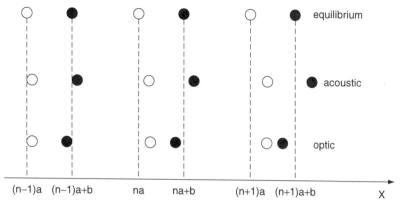

Figure 5.6. Atomic positions in equilibrium and for acoustic and optic modes for $ka \ll 1$ in the diatomic lattice.

$3Ns$ degrees of freedom and hence $3s$ phonon modes. Three are acoustic modes and the remaining $3s - 3$ are optic modes. In analogy to the previous cases, the crystal wave vector of the phonon, **k**, resides in the first Brillouin zone. Each mode has its own set of polarization vectors determining the direction in which the atoms move. In many instances these vectors may be parallel to **k**, in which case they are called *longitudinal modes*. In some instances they are perpendicular to **k**, in which case they are called *transverse modes*. Thus one may talk about LA (longitudinal-acoustic), TA (transverse-acoustic), LO (longitudinal-optic), and TO (transverse-optic) phonons. Of these, the LA mode is the analog of ordinary sound in gases, the TA modes are analogous to transverse waves in a string, and the LO and TO modes are waves of molecular vibrational excitation that propagate through a medium with the degree of excitation varying in a periodic fashion along the wave. In general, lattice waves are pure transverse or longitudinal only in certain high-symmetry directions (e.g., [100], [110], and [111] in cubic crystals). For other directions (e.g., along the [120] direction), the waves will be a mixture of transverse and longitudinal modes. Also, in a given direction the longitudinal and transverse waves in general usually have different speeds (see Chapter 10).

For a free molecule consisting of s atoms, given the atomic masses and force constants, determination of the vibrational frequencies would be straightforward. Normal-mode coordinates would be introduced and the expression for the elastic energy would be rewritten in terms of them. The model would then consist of a set of independent oscillators for $3s - 6$ degrees of freedom (for a nonlinear molecule with $s > 2$). For such a free molecule there are 3 degrees of freedom associated with center-of-mass motion and also 3 degrees of freedom describing the rotation of the molecule in space. (For the case $s = 2$ or for a linear molecule, there are only two rotational degrees of freedom.) In the solid the center-of-mass motion modes become the acoustic phonons. The rotational modes become strongly hindered and are sometimes identified as libration or motions involving rocking back and forth of atoms or groups of atoms.

In Appendix W5B, a general expression [Eq. (W5B.9)] for the phonon dispersion curves is derived. They are dependent on a set of force constants and ionic masses. The force constants could either be obtained from first principles or be determined by fits to the phonon dispersion curves.

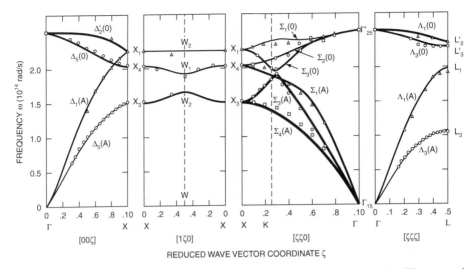

Figure 5.7. Phonon dispersion curves for diamond at $T = 296$ K. [From J. L. Warren et al, *Phys. Rev.*, **158**, 805 (1967). Copyright 1967 by the American Physical Society.]

An example of the phonon dispersion curves for diamond is given in Fig. 5.7. The points were measured using inelastic neutron scattering. The lines are the results of a theoretical calculation using the shell model, in which NN and second-NN interactions are included. Diamond has an FCC lattice with $s = 2$, so there are $3s = 6$ phonon modes. The spectrum is plotted along the ΓX, XW, WX, XK, $K\Gamma$, and ΓL directions of the Brillouin zone (see Fig. 3.3). Note that the acoustic modes grow linearly with k from the Γ point, with slopes equal to the respective speeds of sound. In the XK and $K\Gamma$ directions all six modes are nondegenerate. In other directions shown here there is some degree of degeneracy. For example, along the XW direction, each of the three modes is doubly degenerate. Along the ΓX direction the TA and TO modes are doubly degenerate and the LA and LO modes are nondegenerate. The degeneracies are consistent with the symmetry of the respective modes. Along the ΓX direction (e.g., [001]) there is a four-fold rotational symmetry ($k_x \leftrightarrow k_y$). Therefore, the transverse modes are degenerate. Along the ΓK direction, however, there is only twofold rotational symmetry, so the degeneracy is lifted.

In Fig. 5.8 the phonon dispersion curves are given for graphite. Note the presence of low-frequency phonon modes along the [0001]-ΓA direction (see Fig. 3.4) arising from the weak-binding potential perpendicular to the graphite planar sheets. Due to the anisotropy of the crystal, the degeneracy along the [10$\bar{1}$0]-ΓK direction is broken. Since graphite has $s = 4$, there are 12 modes. Some of the modes are degenerate.

5.4 Phonon Density of States

In describing the excited vibrational states of an atom or a molecule, one generally tabulates the energy levels and degeneracies of the low-lying states. In the case of a macroscopic solid this is impractical since the number of such states becomes too large as the number of atoms in the solid grows to macroscopic proportions. It is therefore

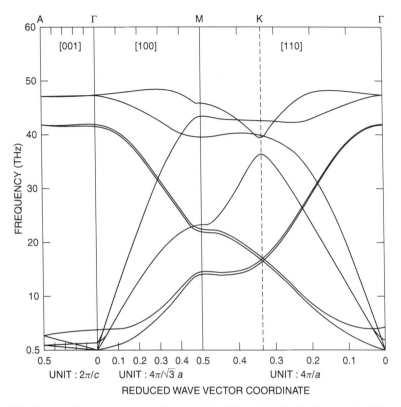

Figure 5.8. Phonon dispersion curves for graphite. [From R. M. Nicklow et al, *Phys. Rev. B*, **5**, 4951 (1972). Copyright 1972 by the American Physical Society.]

useful to introduce the density of states, which is a function giving the number of states per unit energy interval. This is an extensive quantity, proportional to the volume V. Thus one defines

$$\rho(\omega) = \sum_{\mu} \sum_{\mathbf{k}} \delta(\omega - \omega_{\mu}(\mathbf{k}))$$

$$= \sum_{\mu} \int \frac{V \, d\mathbf{k}}{(2\pi)^3} \delta(\omega - \omega_{\mu}(\mathbf{k})), \tag{5.23}$$

with the integral extending over the first Brillouin zone and the μ sum including all modes. The total number of Bravais lattice cells in the crystal is given by

$$N = \int \frac{V \, d\mathbf{k}}{(2\pi)^3}, \tag{5.24}$$

and the total number of modes is

$$3s = \sum_{\mu} 1. \tag{5.25}$$

It is possible to give a geometric interpretation to the density of states. Let $\mathbf{k}_0$ be a point for which $\omega_\mu(\mathbf{k}_0) = \omega$. Then

$$
\begin{aligned}
\omega_\mu(\mathbf{k}) &= \omega_\mu(\mathbf{k}_0) + (\mathbf{k} - \mathbf{k}_0) \cdot \frac{\partial \omega_\mu}{\partial \mathbf{k}} + \cdots \\
&= \omega + (\mathbf{k} - \mathbf{k}_0) \cdot \mathbf{v}_\mu(\mathbf{k}_0) + \cdots,
\end{aligned}
\tag{5.26}
$$

where $\mathbf{v}_\mu$ is called the *group velocity*. Take $\mathbf{k}_0$ as the origin and introduce a rotated set of coordinates $\{\mathbf{k}'\}$ in the Brillouin zone so that the $\hat{k}_3'$ direction is parallel to $\mathbf{v}_\mu$. One may then integrate over k_3' and obtain

$$
\rho(\omega) = \sum_\mu \frac{V}{(2\pi)^3} \int d^2 k' \frac{1}{|\mathbf{v}_\mu(\mathbf{k}_0)|} \Theta(\omega - \omega_\mu^{\min}) \Theta(\omega_j^{\max} - \omega).
\tag{5.27}
$$

The integral extends over the constant energy surface described by $\omega_\mu(\mathbf{k}) = \omega$. The unit step Θ functions ensure that only those modes that contain frequency ω contribute. The inverse of the speed is proportional to the time a particle would spend in a particular section of momentum space as it moved about its orbit.

From this formula it is seen that the integrand becomes infinite whenever the group velocity is zero. These critical points generally occur at the zone boundaries or at zone center for the optic modes. The singularities are integrable but cause features to appear in the density of states called *van Hove singularities*. A description of van Hove singularities is presented in Appendix W5C.

The van Hove singularities are observable as threshold features in the frequency-dependent infrared reflectivity curves taken from solids. The frequencies at which they occur indicate the extremal points in the phonon spectrum and thus permit one to deduce information concerning the elastic constants of solids.

Figure 5.9 illustrates the phonon density of states for diamond. Note that the van Hove singularities appear at the extremal points of the phonon dispersion curves in the Brillouin zone, as shown in Fig. 5.7.

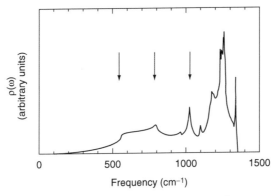

Figure 5.9. Phonon density of states for diamond. The arrows indicate some van Hove singularities. [From R. W. Windl et al, *Phys. Rev. B*, **48**, 3164 (1993). Copyright 1993 by the American Physical Society.]

LATTICE SPECIFIC HEAT OF SOLIDS

When a small amount of heat, dQ, is added to a solid, its temperature rises by dT. The ratio of the incremental heat transferred to the incremental temperature rise dT is called its *heat capacity*. Heat capacity can be measured at constant volume, at constant pressure, or under other thermodynamic constraints. For solids, one generally measures the heat capacity at constant pressure, since it is hard to keep the volume constant. However, the volume expansion involved when a solid is heated is often so small that the work done on the environment is negligible. The various heat capacities are approximately the same. Often, one defines various "specific heats" by dividing the heat capacity by the mass, volume, number of moles, or number of molecules.

Dulong and Petit made a theoretical prediction for the molar specific heat of a monatomic solid based on classical statistical mechanics. The solid was regarded as an assembly of N harmonic oscillators, with as many oscillators as there are degrees of freedom (i.e., $3N$). As mentioned in Section 5.1, the average energy assigned to each oscillator is $k_B T$. The thermal energy of a solid is predicted to be $U = 3Nk_B T = 3nRT$, where $n = N/N_A$ is the number of moles, N_A is Avogadro's number, and $R = N_A k_B$ is the gas constant. Equating (approximately) the heat input to the rise in thermal energy yields the Dulong and Petit prediction for the molar specific heat:

$$c = \frac{1}{n}\frac{dQ}{dT} \approx \frac{1}{n}\frac{dU}{dT} = 3R. \tag{5.28}$$

This law predicts that the molar specific heat has the value $3R$ independent of the nature of the solid and its temperature.

Table 5.1 shows some representative data for the molar specific heats of elements and compounds at room temperature. As may be seen, the values are of the same order of magnitude as $3R = 24.94$ J/mol·K, but the numerical agreement is not particularly good. In particular, the specific heat of carbon is off by a factor of 3 or 4, while that of Si is somewhat smaller than theory predicts. Different crystal forms of the same material can have different values, as exemplified by carbon.

TABLE 5.1 Molar Specific Heat at $T = 298$ K

Material	c_p (J/mol·K)	Material	c_p (J/mol·K)
Ag	25.4	Ni	26.1
Al	24.2	Pb	26.7
Au	25.4	Pt	25.9
C (diamond)	6.23	Si	19.8
C (graphite)	8.52	Sn (α)	27.1
Cu	24.4	Ti	25.1
Fe	25.1	W	24.3
Ge	23.2	Zn	25.4
Hg	28.0	NaCl	50.2
K	29.6	CsCl	52.5
Mg	24.9	Al$_2$O$_3$	79.5
Na	28.2		

Source: Data from D. R. Lide, ed., *CRC Handbook of Chemistry and Physics, 78th ed.*, CRC Press, Boca Raton, Fla., 1997.

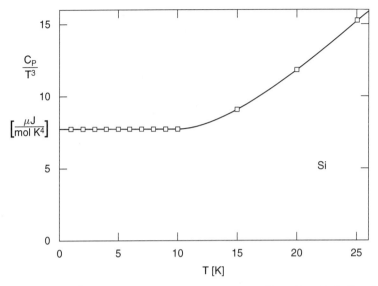

Figure 5.10. Plot of c/T^3 versus T for Si at low temperatures. [Data from P. D. Desai, *J. Phys. Chem. Ref. Data*, **15**, 967 (1986).]

There are three possible causes for the discrepancy. One is due to the use of the harmonic approximation. In reality the use of harmonic oscillators is only an approximation, and anharmonic effects are significant. The second is due to the neglect of the electron contribution to the specific heat, which is described in Chapter 7. This effect is important in metals, especially at low temperatures. The third and most important is due to the need to use a quantum-mechanical description of the excitations of the lattice.

At low temperatures the classical theory breaks down completely. It is found that the specific heat goes to zero as $T \to 0$. For metals, c grows linearly with T, whereas for insulators it varies as T^3. An example of this behavior is given in Fig. 5.10, where the ratio c/T^3 for pure Si is plotted as a function of T at low temperatures. The fact that c/T^3 approaches a constant for small T is indicative of the above-noted trend.

In the following discussion, attention centers on the lattice contribution to the specific heat. The electron contribution is deferred to Chapter 7.

5.5 Specific Heat of Solids

The analysis begins by working in the harmonic approximation but treating the problem using quantum statistical mechanics rather than classical statistical mechanics. Assume that a thermally excited lattice is effectively a "gas" of phonons confined to the interior of the solid.

The phonons are bosons, so any number of phonons may be present in a particular mode. The average number of phonons to be expected in mode μ with crystal wave vector $\mathbf{k}$ is given by Bose–Einstein occupancy:

$$\langle n_\mu(\mathbf{k}, T)\rangle = \frac{\sum_{n=0}^{\infty} n\, e^{-\beta(n+1/2)\hbar\omega_\mu(\mathbf{k})}}{\sum_{n=0}^{\infty} e^{-\beta(n+1/2)\hbar\omega_\mu(\mathbf{k})}} = \frac{1}{\exp[\beta\hbar\omega_\mu(\mathbf{k})] - 1}, \qquad (5.29)$$

where $\beta = 1/k_B T$. The thermal energy of the lattice is equal to the internal energy of the phonon gas (including zero-point motion):

$$U(T) = \sum_{\mathbf{k},\mu} \left[n_\mu(\mathbf{k}, T) + \frac{1}{2} \right] \hbar\omega_\mu(\mathbf{k}). \tag{5.30}$$

Before proceeding with the theory, note at this point where the Dulong and Petit result comes from. At high temperatures $\beta \ll 1$, so

$$n_\mu(\mathbf{k}, T) \longrightarrow \frac{k_B T}{\hbar\omega_\mu(\mathbf{k})} \gg 1. \tag{5.31}$$

If there is only one atom per unit cell and three modes (the acoustic phonons), it follows that

$$U \longrightarrow \sum_{\mu,\mathbf{k}} k_B T = 3N k_B T, \tag{5.32}$$

the same as the Dulong–Petit result.

The general formula for the specific heat results from expressing the sum over states as an integral extending over the first Brillouin zone:

$$U = \sum_\mu \int \frac{V \, d\mathbf{k}}{(2\pi)^3} \frac{\hbar\omega_\mu(\mathbf{k})}{\exp[\beta\hbar\omega_\mu(\mathbf{k})] - 1} + U_0, \tag{5.33}$$

where V is the volume of the crystal. Here the zero-point energy is

$$U_0 = \sum_\mu \int \frac{V \, d\mathbf{k}}{(2\pi)^3} \frac{\hbar\omega_\mu(\mathbf{k})}{2}, \tag{5.34}$$

and the sum over the crystal wave vectors gives the total number N of lattice cells [see Eq. (5.24)]. Note that U_0 is independent of T and therefore will not influence the calculation of the specific heat. It is often convenient to express Eq. (5.33) in terms of the density of states:

$$u(T) = \frac{U - U_0}{V} = \int_0^{\omega_{\max}} d\omega \rho(\omega) \left[n(\omega, T) + \frac{1}{2} \right] \hbar\omega. \tag{5.35}$$

5.6 Debye Theory of Specific Heat

In this section attention is restricted to the case of a monatomic lattice. The Debye calculation of the specific heat utilizes a simplified model for the phonon dispersion relation and assumes that the speed of sound is constant. This effectively treats the solid as if it were a continuum. The dispersion formula is then simply $\omega = kc_s$, c_s being the speed of sound. The first Brillouin zone is reshaped into a sphere of radius k_D of such size as to contain the $3N$ wave vectors in volume V:

$$3N = 3 \int \frac{V \, d\mathbf{k}}{(2\pi)^3} \Theta(k_D - k) = \frac{4\pi k_D^3 V}{(2\pi)^3}. \tag{5.36}$$

The Debye wave vector is therefore given by

$$k_D = \left(\frac{6\pi^2 N}{V}\right)^{1/3}. \tag{5.37}$$

The thermal energy of the lattice, omitting the zero-point energy, is

$$
\begin{aligned}
U &= 3 \int \frac{V\, d\mathbf{k}}{(2\pi)^3} \Theta(k_D - k) \frac{\hbar\omega}{\exp(\beta\hbar\omega) - 1} \\
&= 3 \int_0^{k_D} \frac{V \cdot 4\pi k^2\, dk}{(2\pi)^3} \frac{\hbar\omega}{\exp(\beta\hbar\omega) - 1} \\
&= \frac{3V}{2\pi^2 \beta^4 \hbar^3 c_s^3} \int_0^{\Theta_D/T} dx \frac{x^3}{\exp(x) - 1},
\end{aligned}
\tag{5.38}
$$

where the substitutions $k = \omega/c_s$ and $x = \beta\hbar\omega$ are made, and a parameter called the *Debye temperature* is introduced:

$$\Theta_D = \frac{hc_s k_D}{k_B} = \frac{\hbar\omega_D}{k_B}, \tag{5.39}$$

where ω_D is the Debye frequency.

For high temperatures ($T \gg \Theta_D$) the integral is evaluated trivially to give

$$\int_0^{\Theta_D/T} dx \frac{x^3}{\exp(x) - 1} \approx \int_0^{\Theta_D/T} dx\, x^2 = \frac{1}{3} \left(\frac{\Theta_D}{T}\right)^3. \tag{5.40}$$

In this limit one again obtains the Dulong–Petit result:

$$U \approx 3Nk_B T. \tag{5.41}$$

For low temperatures ($T \ll \Theta_D$), however,

$$\int_0^{\Theta_D/T} dx \frac{x^3}{\exp(x) - 1} \approx \int_0^{\infty} dx \frac{x^3}{\exp(x) - 1} = \frac{\pi^4}{15}. \tag{5.42}$$

Thus

$$U \approx \frac{\pi^2}{10} \frac{Vk_B^4 T^4}{\hbar^3 C_s^3} = \frac{3\pi^4}{5} Nk_B T \left(\frac{T}{\Theta_D}\right)^3 \equiv \frac{A}{4} VT^4, \tag{5.43}$$

so

$$c_v = \frac{1}{V}\frac{dU}{dT} = \frac{12\pi^4}{5}\frac{N}{V} k_B \left(\frac{T}{\Theta_D}\right)^3 \equiv AT^3. \tag{5.44}$$

This cubic variation with temperature is in agreement with experiment at low temperatures for insulators (but not metals). It is not surprising that the Debye model based on a continuum approximation works well at low temperatures, since there only low-energy acoustic phonons with $\omega = kc_s$ are thermally excited.

By comparing Eq. (5.38) with Eq. (5.35), it can be seen that $\rho(\omega) = 3V\omega^2/2\pi^2 c_s^3$. This is in agreement with the phonon density of states for diamond shown in Fig. 5.9. In three dimensions, where the density of states $\rho(\omega) \propto \omega^2$ for small ω, the heat capacity varies as $c_v \propto T^3$ at low T. In two dimensions, $\rho(\omega) \propto \omega$, so $c_v \propto T^2$, and in one dimension where $\rho(\omega) \propto constant$, $c_v \propto T$.

The ratio of the specific heat at temperature T to that at very high ("infinite") temperatures is

$$\frac{c(T)}{c(\infty)} = 12 \left(\frac{T}{\Theta_D}\right)^3 \int_0^{\Theta_D/T} dx \frac{x^3}{\exp(x)-1} - 3\frac{\Theta_D}{T}\frac{1}{\exp(\Theta_D/T)-1}. \tag{5.45}$$

A graph of this function is presented in Fig. 5.11. It is compared with experimental data for Si.

By fitting this curve to experimental specific heat data, values for the Debye temperature may be assigned for various materials. The Debye temperature is the typical temperature at which the phonon modes become unfrozen. At low temperatures not enough thermal energy is available to excite the high-frequency modes, so they remain with approximately zero occupation number. As the temperature is increased and more thermal energy becomes available, these modes may become occupied with one, two, or more phonons. Debye temperatures for various materials are given in Table 5.2. In obtaining Θ_D for metals the electronic contribution to the specific heat is often neglected, since it is relatively small compared to the phonon contribution except at low T.

From Eqs. (5.8), (5.37), and (5.39) it is possible to obtain an approximate formula for the Debye temperature. Consider the case of cubic crystals and denote by $s = Na^3/V$

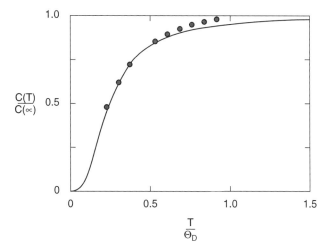

Figure 5.11. Ratio of specific heat at temperature T to the high-temperature limit as a function of T/Θ_D, as given in Eq. (5.45). The dots are data for Si from P. D. Desai, *J. Phys. Chem. Ref. Data*, **15**, 967 (1986). A value of $\Theta_D = 645$ K is used here.

TABLE 5.2 Debye Temperatures of Selected Materials

Material	$\Theta_D(K)$	Material	$\Theta_D(K)$
Ag	227	Ni	477
Al	433	Pb	105
Au	162	Pt	240
C (diamond)	2250	Si	645
C (graphite)	413	Ti	420
Cu	347	W	383
Fe	477	Zn	329
Ge	373	NaCl	275
Hg	72	GaAs	345
K	91	H_2O (solid)	192
Mg	403	BN	1900
Na	156		

Source: Data from I. S. Grigoriev and E. Z. Meilikhov, eds., *CRC Handbook of Physical Quantities*, CRC Press, Boca Raton, Fla., 1997.

the number of atoms in the basis per conventional cubic unit cell. Then

$$\Theta_D = \frac{\hbar}{k_B}(6\pi^2 s)^{1/3} a \sqrt{\frac{K}{M}}. \tag{5.46}$$

This shows that for crystals with high stiffness constants and small ion masses, the Debye temperature will be high. This is the case for diamond with its strong covalent bond, for which $\Theta_D = 2250$ K. On the other hand, $\Theta_D = 105$ K for Pb, with its high ionic mass and weak metallic bond, which is relatively low. Note that for diamond, $s = 8$, whereas for Pb, $s = 4$. Typical values of ω_D are about 10^{14} rad/s.

5.7 Einstein Theory of Specific Heat

The Einstein theory for the specific heat is based on a different assumption about the nature of the phonon spectrum. It assumes that the solid is composed of admixtures of oscillators with definite discrete frequencies. The internal energy of such a system is given by

$$U = \sum_{\mu} N_{\mu} \frac{\hbar\omega_{\mu}}{\exp(\beta\hbar\omega_{\mu}) - 1}, \tag{5.47}$$

where N_{μ} is the number of oscillators with frequency ω_{μ}. At high temperatures this reduces to

$$U \approx \sum_{\mu} N_{\mu} k_B T = 3Ns k_B T, \tag{5.48}$$

where $3Ns$ is the number of degrees of freedom possessed by a solid containing s atoms per unit cell. At low temperatures, $\beta\hbar\omega \gg 1$, the lattice thermal energy becomes exponentially small, that is,

$$U \approx \sum_{\mu} N_{\mu} \hbar\omega_{\mu} \exp(-\beta\hbar\omega_{\mu}), \tag{5.49}$$

in disagreement with the experimental observation of a T^4 behavior. Thus the Einstein model did not enjoy the success of the Debye model at low temperatures, because it failed to treat the acoustic phonons properly.

However, the phonon spectrum often does contain fairly flat optic branches, which may be approximated by oscillators of fixed frequencies. For these modes the Einstein model should provide a good representation. If such a mode were nondegenerate, $N_\mu = N$ for that mode. If the mode is degenerate, $N_\mu = g_\mu N$, g_μ being the degeneracy of the mode.

In practice, a good approximation to the lattice internal energy may be had by treating the acoustic phonons via the Debye approach and the optic phonons according to Einstein. Quantities such as Θ_D and $\{\omega_\mu\}$ are obtained by making a least mean-squares fit to the experimental data. A more exact calculation entails use of the general formula for U and use of the full phonon dispersion curves. Integration over the Brillouin zone and summation over all phonon modes is required.

5.8 Debye–Waller Factor

Thermal motion of the atoms has implications for the diffraction of x-rays and particles, such as electrons or neutrons, from crystals. Until now attention has been restricted to the case where the temperature of the crystal is sufficiently low that the atoms could be regarded as constituting a perfect periodic crystal. Equation (3.40) showed the scattering to occur only when the von Laue diffraction condition is satisfied. As the temperature is increased, however, the atoms undergo thermal fluctuations, and it is no longer obvious that there will be a well-defined diffraction pattern. Even at $T = 0$ K, there will be quantum-mechanical zero-point fluctuations that can upset the symmetry. An instantaneous picture of the atoms will show there to be no translational symmetry. The average positions of the atoms, however, still constitute an ordered lattice.

Debye and Waller have shown that thermal (and quantum-mechanical) fluctuations do not destroy the coherent diffraction but rather, reduce its intensity by an exponential correction factor, $\exp[-2W(T)]$. The diffraction peaks still remain sharp. The missing intensity reappears in what is called *thermal-diffuse scattering*. This is incoherent scattering over a continuous distribution of angles. One way of thinking about it is that the thermally disordered crystal is behaving as if there were two materials present, one perfectly ordered and the other completely disordered. Mathematical details are presented in Section W5.2.

ANHARMONIC EFFECTS

Although the harmonic approximation is useful for describing many lattice phenomena, some require anharmonic effects for their explanation. These include thermal expansion, thermal conductivity, and phonon lifetimes. In addition, anharmonic effects produce significant corrections to the high-temperature specific heat of solids. They are also needed to explain the temperature dependence of the elastic moduli and phonon frequencies and such phenomena as structural phase transitions and melting.

The harmonic potential for a crystal is given in Eq. (W5B.2). It is expressed as a quadratic form in the atomic displacements. Let the αth component of the displacement of the σth atom of unit cell $\mathbf{R}$ be denoted by $u_\alpha^\sigma(\mathbf{R})$. The harmonic potential is

$$U_{\text{harmonic}} = \frac{1}{2} \sum_{\sigma,\sigma'} \sum_{\alpha,\alpha'} \sum_{\mathbf{R},\mathbf{R}'} u_\alpha^\sigma(\mathbf{R}) L_{\alpha,\alpha'}^{\sigma,\sigma'}(\mathbf{R} - \mathbf{R}') u_{\alpha'}^{\sigma'}(\mathbf{R}'), \tag{5.50}$$

where the set of elastic coefficients $\{L\}$ is defined in terms of the second derivatives of the potential evaluated at the equilibrium positions

$$L_{\alpha,\alpha'}^{\sigma,\sigma'}(\mathbf{R} - \mathbf{R}') = \frac{\partial^2 U}{\partial u_\alpha^\sigma(\mathbf{R}) \partial u_{\alpha'}^{\sigma'}(\mathbf{R}')}. \tag{5.51}$$

The indices σ and σ' range over $\{1, 2, \ldots, s\}$ and the indices α and α' over $\{1, 2, 3\}$.

Beyond this there are third-, fourth-, and higher-order terms:

$$U_{\text{anharmonic}} = \frac{1}{3!} \sum_{\sigma,\sigma',\sigma''} \sum_{\alpha,\alpha',\alpha''} \sum_{\mathbf{R},\mathbf{R}',\mathbf{R}''} L_{\alpha,\alpha',\alpha''}^{\sigma,\sigma',\sigma''}(\mathbf{R}, \mathbf{R}', \mathbf{R}'') u_\alpha^\sigma(\mathbf{R}) u_{\alpha'}^{\sigma'}(\mathbf{R}') u_{\alpha''}^{\sigma''}(\mathbf{R}'')$$

$$+ \frac{1}{4!} \sum_{\sigma,\sigma',\sigma'',\sigma'''} \sum_{\alpha,\alpha',\alpha'',\alpha'''} \sum_{\mathbf{R},\mathbf{R}',\mathbf{R}'',\mathbf{R}'''} L_{\alpha,\alpha',\alpha'',\alpha'''}^{\sigma,\sigma',\sigma'',\sigma'''}(\mathbf{R}, \mathbf{R}', \mathbf{R}'', \mathbf{R}''')$$

$$\times u_\alpha^\sigma(\mathbf{R}) u_{\alpha'}^{\sigma'}(\mathbf{R}') u_{\alpha''}^{\sigma''}(\mathbf{R}'') u_{\alpha'''}^{\sigma'''}(\mathbf{R}''') + \cdots. \tag{5.52}$$

For small-amplitude displacements, successive terms in the series become smaller. Generally, one adopts a perturbation theory approach and includes only the lowest-order term necessary to explain a given effect. Note that the fourth-order term is needed to describe phonon–phonon scattering, whereas the cubic term describes processes in which the number of phonons increase or decrease by one.

5.9   Thermal Expansion

Thermal expansion is important in materials science because different materials are often bonded together or are combined to form composite materials. For example, a coating is deposited on a substrate. As the temperature is changed, the dimensions of each material change by an amount determined by its coefficient of thermal expansion. However, the bonded materials constrain each other's expansion. The net result is that a stress is built up at the interface that may induce dislocations or even cracks. The purpose of this section is to see what physical parameters determine thermal expansion.

The thermal expansion of a solid will be analyzed initially using classical physics. Consider an open-ended one-dimensional chain. The instantaneous positions of the atoms are given by $x_n = na + u_n$, where $n = 1, 2, \ldots, N$. The instantaneous length of the chain is

$$L = x_N - x_1 = (N - 1)a + (u_2 - u_1) + (u_3 - u_2) + \cdots + (u_N - u_{N-1})$$

$$= (N - 1)a + z_1 + z_2 + \cdots + z_{N-1}, \tag{5.53}$$

where $z_n = u_{n+1} - u_n$. The lattice potential energy is assumed to be given by a sum of nearest-neighbor (NN) interactions:

$$U = \sum_{n=1}^{N-1} \phi(z_n). \tag{5.54}$$

The average length of the chain is given by

$$
\langle L \rangle = \frac{\int du_1 \cdots \int du_N L \exp(-\beta U)}{\int du_1 \cdots \int du_N \exp(-\beta U)}
$$

$$
= (N-1)a + (N-1)\frac{\int dz\, z \exp[-\beta \phi(z)]}{\int dz \exp[-\beta \phi(z)]}. \tag{5.55}
$$

In the harmonic limit $\phi(z) = \phi_0 + Kz^2/2$. Since $\phi(z)$ grows rapidly with increasing $|z|$, one may take the limits of integration to extend from $-\infty$ to ∞. One then obtains $\langle L \rangle = (N-1)a$, irrespective of the temperature. In the harmonic approximation there is no thermal expansion. In a quantum-mechanical harmonic description thermal energy is carried by the phonons and the phonon frequencies are independent of the volume of the solid. Therefore, heating the solid again does not result in thermal expansion.

Thermal expansion is the result of anharmonic interactions. Rather than attempting a quantum-mechanical perturbation-theoretic solution using the entire potential given by Eqs. (W5B.2) and (5.52), one makes the simple assumption that the phonon frequencies depend on the volume of the crystal. Gruneisen suggested a formula of the form $\Delta\omega/\omega = -\gamma \Delta V/V$, which relates the fractional shift of frequency of a mode to the fractional volume change of the crystal due to expansion or contraction. The parameter γ is called the *Gruneisen constant*. In general, an independent value of γ may be assigned to each mode, but here it is assumed that an average value may be used for all modes. Using this, a formula relating the coefficient of linear thermal expansion, α, to the specific heat, c_v, and the bulk modulus, B, may be found. The parameter γ is a measure of the importance of anharmonic terms and is zero for a harmonic crystal.

The internal energy of the lattice is the sum of its energy at $T = 0$ K and the thermal energy plus the zero-point energy as given by Eq. (5.30):

$$
U(T) = U_{\text{eq}} + \sum_{\mathbf{k},\mu} \left[n_\mu(\mathbf{k}) + \frac{1}{2} \right] \hbar\omega_\mu(\mathbf{k}). \tag{5.56}
$$

The Helmholtz free energy of a solid is obtained from the partition function Z (see Appendix WB):

$$
Z(T) = e^{-\beta F} = \sum_{n_1=0}^{\infty} \sum_{n_2=0}^{\infty} \cdots e^{-\beta[U_{\text{eq}} + (n_1+1/2)\hbar\omega_1 + (n_2+1/2)\hbar\omega_2 + \cdots]}
$$

$$
= e^{-\beta U_{\text{eq}}} \prod_m \frac{e^{-\beta\hbar\omega_m/2}}{1 - e^{-\beta\hbar\omega_m}}, \tag{5.57}
$$

where an abbreviated notation is used for labeling the modes. One finds that

$$
F(T) = U_{\text{eq}} + \sum_{\mathbf{k},\mu} \left[\frac{\hbar\omega_\mu(\mathbf{k})}{2} + \frac{1}{\beta} \ln(1 - e^{-\beta\hbar\omega_\mu(\mathbf{k})}) \right]. \tag{5.58}
$$

Using the thermodynamic formulas

$$\left(\frac{\partial V}{\partial T}\right)_P \left(\frac{\partial P}{\partial V}\right)_T \left(\frac{\partial T}{\partial P}\right)_V = -1, \tag{5.59a}$$

$$\left(\frac{\partial P}{\partial T}\right)_V \left(\frac{\partial T}{\partial P}\right)_V = 1, \tag{5.59b}$$

$$B = -V\left(\frac{\partial P}{\partial V}\right)_T, \tag{5.59c}$$

$$\alpha = \frac{1}{3V}\left(\frac{\partial V}{\partial T}\right)_P, \tag{5.59d}$$

and expressions for c_v and P,

$$C_v = \frac{1}{V}\left(\frac{\partial U}{\partial T}\right)_V = \frac{1}{V}\sum_{\mathbf{k},\mu}\frac{\partial n_\mu(\mathbf{k}, T)}{\partial T}\hbar\omega_\mu(\mathbf{k}), \tag{5.60}$$

$$P = -\left(\frac{\partial F}{\partial V}\right)_T = -\frac{\partial}{\partial V}\left[U_{\mathrm{eq}} + \sum_{\mathbf{k},\mu}\frac{\hbar\omega_\mu(\mathbf{k})}{2}\right] + \sum_{\mathbf{k},\mu}n_j(\mathbf{k})\frac{\partial\hbar\omega_\mu(\mathbf{k})}{\partial V}, \tag{5.61}$$

one finds that

$$\frac{\alpha}{c_v} = -\frac{V}{3B}\frac{\sum_{\mathbf{k},\mu}[\partial n_\mu(\mathbf{k})/\partial T][\partial\hbar\omega_\mu(\mathbf{k})/\partial V]}{\sum_{\mathbf{k},\mu}[\partial n_\mu(\mathbf{k})/\partial T]\hbar\omega_\mu(\mathbf{k})}. \tag{5.62}$$

In the Gruneisen approximation this gives

$$\frac{\alpha}{c_v} = \frac{\gamma}{3B}, \tag{5.63}$$

where γ is the Gruneisen constant. Typically, γ is on the order of unity. For some materials the Gruneisen model is inadequate and γ is temperature dependent.

For low temperatures ($T \ll \Theta_D$) the harmonic approximation is valid and α is very small (because c_v becomes small, in accordance with the Debye model). Equation (5.63) predicts a T^3 behavior for α, which is in fair agreement with experiment. For large T, Eq. (5.63) predicts that α should be independent of T due to the DuLong–Petit law, as observed experimentally.

Table 5.3 presents some values for the thermal expansion coefficient α for representative materials. The values of some of the Gruneisen constants are also given. The general trend is that $\alpha_{\mathrm{polymers}} > \alpha_{\mathrm{metals}} \approx \alpha_{\mathrm{ionic}} > \alpha_{\mathrm{covalent}}$. One notes that for metals γ is in the range 1 to 3. For covalently bonded materials (graphite, Ge, Si) it is an order of magnitude smaller (diamond and SiC being exceptions). This suggests that the metallic bond has a higher degree of anharmonicity than the covalent bond. This is not unreasonable given the extended nature of the metallic bond and the short-ranged strong covalent bond. The value of γ for NaCl is comparable to that of metals, again indicating large anharmonicity for ionically bonded crystals. This, too, is consistent with the long-range nature of the Coulomb interaction. For the group III–V semiconductor InSb, one notes a value for γ intermediate between that of an ionic crystal and a

TABLE 5.3   Thermal   Expansion   Coefficients   at
$T = 298$ **K and Gruneisen Constants for Representative Materials**

Material	$\alpha(10^{-6}\ K^{-1})$	γ
Ag	18.9	2.4
Al	23.1	2.1
Au	14.2	2.8
Cu	16.5	2.0
Fe	11.8	1.8
Mg	24.8	1.5
Ni	13.4	1.6
Pb	28.9	2.4
Pt	8.8	2.3
Ti	8.6	1.2
Steel	11.7	1.8
Diamond	1.18	0.96
Graphite		
$\perp c$ axis	26.7	0.13
$\parallel c$ axis	−1.22	0.13
Ge	5.82	0.14
Si	4.68	0.28
GaAs	5.4	—
AlAs	3.5	—
InSb	4.7	0.80
Si_3N_4	2.7	—
NaCl	39.6	1.7
Vitreous glass	0.41	0.028
Glass		
$(Na_2O)_{0.2}(SiO_2)_{0.8}$	9.0	—
$(Na_2O)_{0.1}(SiO_2)_{0.9}$	5.0	—
Pyrex	3.2	—
α-Alumina		
$\parallel c$ axis	6.7	2.1
$\perp c$ axis	5.0	1.9
α-SiC	4.7	1.6
TiO_2		
$\parallel c$ axis	9.8	—
$\perp c$ axis	7.4	—
α-SiO_2		
$\parallel c$ axis	13.6	0.2
$\perp c$ axis	7.4	0.37
Yttria	8.3	—
PMMA	70	0.51
Polystyrene	70	—
Polypropylene	68	—
Polyisoprene rubber	223	0.75

Source: Data from D. R. Lide, ed., *CRC Handbook of Chemistry and Physics*, 18th ed., CRC Press, Boca Raton, Fla., 1997; and I. S. Grigoriev and E. Z. Meilikhov, eds., *CRC Handbook of Physical Quantities*, CRC Press, Boca Raton, Fla., 1997.

covalent crystal, indicative of the mixed nature of the bond in such crystals. The value of γ for the polymers PMMA and polyisoprene rubber indicate a moderate amount of anharmonicity.

5.10   Thermal Conductivity

If one end of a solid rod is maintained at a higher temperature than the other, heat is conducted in the direction of decreasing temperature. Phonons are generated at the hot end and diffuse toward the cold end, where they are absorbed. The thermal flux (i.e., the energy flow per unit area per unit time) is given by Fourier's law:

$$\mathbf{J_Q} = -\kappa \nabla T, \tag{5.64}$$

where κ is the thermal conductivity. A good thermal conductor has a high κ value, whereas a thermal insulator has a low value. Values κ of at $T = 300$ K are tabulated in Table 5.4 for a variety of materials. The thermal diffusivity, which is defined as $\kappa/C_p\rho$, where C_p is the specific heat per unit mass at constant pressure and ρ is the mass density, is also used to characterize heat flow in solids. The units of κ are W/m·K and the units of the diffusivity are m²/s. In this section the lattice contribution to thermal conductivity is studied, deferring discussion of the electronic contribution in metals to Chapter 7.

It is possible to analyze some aspects of thermal conductivity using classical physics, but at low temperatures quantum-mechanical effects become important, so a semiclassical description is used. One views the thermally activated solid as a phonon gas

TABLE 5.4   Thermal Conductivities of Selected Materials at $T = 300$ K

Material	κ (W/m·K)	Material	κ (W/m·K)
Ag	429	AlN	82
Al	237	Ge	64
Au	317	Si	124
Cu	401	GaAs	56
Fe	80	Fused silica glass	2.0
Mg	156	Pyrex	1.1
Mo	138	α-Alumina	36
Ni	91	Silica	1.4
Pb	35	BeO	210
Pt	72	MgO	36
W	174	β-SiC (at 400 K)	121
Steel	45–65	TiO₂	
Quasicrystal (Al–Cu–Fe)	1.8	∥c axis at 273 K	13
NaCl	6.4	⊥c axis at 273 K	9
Diamond	1000	α-SiO₂	
Graphite		∥c axis	12
⊥c axis	2000	⊥c axis	6.8
∥c axis	5.7		

Source: Data largely from, D. R. Lide, ed., *CRC Handbook of Chemistry and Physics*, 78th ed., CRC Press, Boca Raton, Fla. 1997.

in an inhomogeneous temperature field. Local thermodynamic equilibrium is assumed initially. The internal energy density of this gas, u, depends on $T(\mathbf{r})$ and thus is also a function of $\mathbf{r}$. The phonons propagate with speed c_s in all directions. The phonons come from regions with different temperatures, so they tend to destroy local thermal equilibrium. However, the phonons undergo collisions with other phonons, defects, and electrons, with a mean collision time τ, and this restores the local thermal equilibrium. The phonons then propagate in random directions and the process repeats itself. The net result is that the phonons take a random walk through the lattice.

In addition to phonon scattering processes, one must also take into account the fact that phonons have a finite lifetime. They can decay into two or more phonons at other frequencies. This effect is discussed later in this section.

To simplify the statistical treatment, think of the restoration of thermal equilibrium as occurring at time τ and look at the solid just before this restoration occurs. Consider an area dA at position $\mathbf{r}$ and a volume element $d\mathbf{r}'$ at $\mathbf{r}'$. The geometry is illustrated in Fig. 5.12. The energy coming from $d\mathbf{r}'$ and passing through dA is

$$dU = u(\mathbf{r}')\,d\mathbf{r}'\frac{d\Omega}{4\pi} = u(\mathbf{r}')\,d\mathbf{r}'\frac{dA\cos\theta''}{4\pi|\mathbf{r}-\mathbf{r}'|^2}, \tag{5.65}$$

where $d\Omega$ is the solid angle subtended by dA at $\mathbf{r}'$. For the volume element choose a section of a cone of length $c_s dt$ with apex at $\mathbf{r}$ subtending an angle $d\Omega''$ a distance $|\mathbf{r}''| = |\mathbf{r}-\mathbf{r}'| = c_s\tau$ away: $d\mathbf{r}' = c_s dt|\mathbf{r}-\mathbf{r}'|^2 d\Omega''$. The energy flux is

$$J_Q = -\frac{dU}{dA\,dt} = -\int u(\mathbf{r}+c_s\tau\hat{r}'')\frac{\cos\theta''}{4\pi}c_s\,d\Omega''. \tag{5.66}$$

If the phonon mean free path, $\lambda = c_s\tau$, is much smaller than the typical distance over which T varies, one may expand u and obtain

$$u(\mathbf{r}+c_s\tau\hat{r}'') = u(\mathbf{r}) + c_s\tau\frac{du}{dT}\hat{r}''\cdot\nabla T + \cdots = u(\mathbf{r}) + c_s\tau\frac{du}{dT}\cos\theta''\nabla T + \cdots \tag{5.67}$$

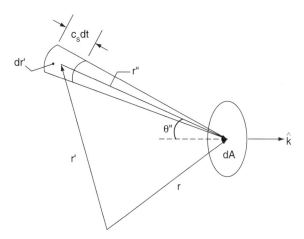

Figure 5.12. Geometry relating to phonon propagation.

Integrating over all solid angles yields

$$\mathbf{J_Q} = -\frac{1}{3}c_s^2\tau\frac{du}{dT}\nabla T. \tag{5.68}$$

Thus the lattice thermal conductivity depends on the product of the mean free path, the speed of sound, and the lattice specific heat per unit volume, $C = du/dT$:

$$\kappa = \tfrac{1}{3}\lambda c_s C. \tag{5.69}$$

The temperature dependence of $\kappa(T)$ is determined by $\lambda(T)$ and $C(T)$. The thermal conductivity data presented in Table 5.4 may be reviewed in light of Eq. (5.69). Note that for the elemental metals (Ag, ..., W) κ tends to be high. However, since there is a significant electronic contribution to κ, one may not directly employ Eq. (5.69), which refers only to the lattice contribution.

Diamond has an exceptionally high value for κ, especially when isotopically pure. This is due in part to the high speed of sound, which may be attributed to the fact that the spring constant, K, associated with the strong sp^3 covalent bond is large. Also, from Eq. (5.8), one sees that c_s tends to be large if M is small. In addition, the high crystalline quality of diamond assures a long mean free path λ. The depth of the covalent bond potential near its minimum implies that the harmonic approximation is likely to be a good one and that anharmonic corrections will be small. As will soon be seen, this works in favor of long phonon lifetimes. Diamond films are often used as heat sinks. AlN, with its high value of κ, is also often used for this purpose.

The importance of bonding is illustrated by the case of graphite. The in-plane value of κ is determined by the sp^2 covalent bond and one sees that the value is comparable to that of diamond. However, the value of κ along the c axis is two orders of magnitude smaller, since it is associated with a weak van der Waals bond. The effective value of κ is much smaller since the overall energy scale for the bonding potential is so much smaller.

The semiconductors Ge, Si, and GaAs have very few electrons that can transport heat, so their κ values are more representative of the lattice contribution. The values are not as high as those of diamond, due to the smaller c_s values. In addition, the NN distances are larger than in C.

Note that κ for steel is considerably less than that for Fe. The presence of interstitial C provides scattering centers that shorten λ. One may estimate the mean free path by the formula $\lambda = 1/n_i\sigma$, where n_i is the concentration of impurity ions and σ is the phonon scattering cross section from these ions. This predicts a degradation of κ with increasing impurity concentration. Inhomogeneous materials will have a lower thermal conductivity than that of their most conducting component material.

Even though the ionic bond in a crystal such as NaCl is moderately strong and the crystal quality can be high, the value for κ is small. The long-range nature of the Coulomb interaction provides a potential with a weak spring constant and with large anharmonicities.

The glasses have small values of κ because of their lack of crystalline order. The phonon mean free path is very small.

Ceramic materials have mixed ionic and covalent bonding, so their κ values assume a range of intermediate values. Note that substantial anisotropies exist in κ for anisotropic materials such as TiO_2 or quartz. This suggests that κ could be more generally described by a tensor than by a scalar.

Typically, metallic quasicrystals possess very small κ values despite the fact that they are highly ordered. The ordering, however, does not include translational periodicity. The mechanism for heat transfer is believed to be different for quasicrystals than for regular crystals. Many quasicrystals consist of icosahedral clusters of atoms with identical orientation but at pseudorandom positions in the lattice. A cluster has a discrete vibrational spectrum rather than a continuum of phonons. Thermal localization occurs; that is, once a cluster is thermally excited it retains that excitation unless the energy can be resonantly transferred to a similar cluster some distance away. Heat transfer proceeds by a series of random jumps. The excitation takes a random walk through the quasicrystal. In place of Eq. (5.69) one has $\kappa = C\Lambda^2/3\tau$, where Λ is the typical jump distance and $1/\tau$ is an attempt frequency. This follows by replacing the phonon speed in Eq. (5.69) by the average velocity Λ/τ and replacing the mean free path by Λ.

Values of κ tend to be small for most polymers, due to the flexible nature of the polymer chains and therefore low speeds of sound. Crystalline polymers would have longer mean free paths than those of glassy polymers. The weak hydrogen bonding in ice gives it a low thermal conductivity value.

The temperature dependence of κ is determined by a variety of factors. As noted in Section 5.6, C varies as T^3 for small T and approaches a constant for large T. The speed of sound, c_s, is roughly independent of T. The problem is to find the behavior of $\lambda(T)$. This turns out to be complicated. Several mechanisms are active, including phonon–phonon collisions, phonon–defect collisions, and phonon–boundary interactions.

Anharmonic effects result in phonon–phonon interactions and are therefore influential in limiting the phonon lifetime and hence the mean free path. These are of three general types: fission, fusion, and scattering, and are illustrated in Fig. 5.13 for third- and fourth-order anharmonicities. Process A is a three-phonon process in which a phonon with crystal wave vector (CWV) $\mathbf{k}_1$ fissions into two phonons with CWVs $\mathbf{k}_2$ and $\mathbf{k}_3$, respectively. Process B denotes another three-phonon process in which phonons with CWVs $\mathbf{k}_2$ and $\mathbf{k}_3$ fuse to form a phonon with CWV $\mathbf{k}_1$. Both A and B result from the cubic terms in the interaction.

Processes C, D, and E are four-phonon processes resulting from the quartic terms in the interaction potential. In process C, phonon $\mathbf{k}_1$ fissions into phonons $\mathbf{k}_2$, $\mathbf{k}_3$, and $\mathbf{k}_4$.

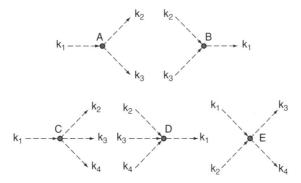

Figure 5.13. Phonon–phonon interactions: A and B are three-phonon processes; C, D, and E are four-phonon processes.

Diagram D represents the corresponding fusion process, with $\mathbf{k}_2$, $\mathbf{k}_3$, and $\mathbf{k}_4$ combining to form $\mathbf{k}_1$. Finally, diagram E denotes the scattering of phonons $\mathbf{k}_1$ and $\mathbf{k}_2$ into $\mathbf{k}_3$ and $\mathbf{k}_4$. Higher-order processes exist, but they are less important.

Associated with these processes are conservation rules for momentum and energy. Overall crystal wave vector must be conserved, up to a reciprocal lattice vector, $\mathbf{G}$. Thus the momentum conservation laws imply the following rules:

$$\mathbf{k}_1 = \mathbf{k}_2 + \mathbf{k}_3 + \mathbf{G} \qquad \text{for } A, B, \qquad (5.70a)$$

$$\mathbf{k}_1 = \mathbf{k}_2 + \mathbf{k}_3 + \mathbf{k}_4 + \mathbf{G} \qquad \text{for } C, D, \qquad (5.70b)$$

$$\mathbf{k}_1 + \mathbf{k}_2 = \mathbf{k}_3 + \mathbf{k}_4 + \mathbf{G} \qquad \text{for } E. \qquad (5.70c)$$

Cases in which $\mathbf{G} = 0$ are called *normal* (N) *processes*, while those in which $\mathbf{G} \neq 0$ are called *umklapp* (U) *processes*. Umklapp processes are required when the final phonon wave vector would lie outside the first Brillouin zone. The vector $\mathbf{G}$ serves to bring the wave vector back inside that zone. The U process corresponding to Eq. (5.70a) is sketched in Fig. 5.14. The dashed square represents the boundary of the first Brillouin zone.

The energy conservation conditions are

$$\omega_1 = \omega_2 + \omega_3 \qquad \text{for } A, B, \qquad (5.71a)$$

$$\omega_1 = \omega_2 + \omega_3 + \omega_4 \qquad \text{for } C, D, \qquad (5.71b)$$

$$\omega_1 + \omega_2 = \omega_3 + \omega_4 \qquad \text{for } E. \qquad (5.71c)$$

These conservation rules prevent a state of thermal equilibrium from being achieved by means of N processes. In such a state the mean value of the phonons' momentum should be constant:

$$\Delta \left\langle \sum_{\vec{k}} \hbar \mathbf{k} \right\rangle = 0. \qquad (5.72)$$

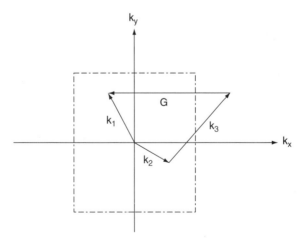

Figure 5.14. Umklapp process in which the reciprocal lattice vector G restores the net crystal wave vector to the first Brillouin zone.

The conservation laws imply preservation of the total crystal momentum, so if the system starts with a net momentum, it retains that momentum. The U-processes, however, do permit the total momentum to change. The relaxation can occur only by adding discrete reciprocal lattice vectors to the center-of-mass motion of the solid, but that is sufficient to destroy the original phonon current.

Consider the collision of a given phonon with other phonons. First, assume that the temperature is high (i.e., $T \gg \Theta_D$). The N and U processes are both about as likely to occur. Process B has a collision rate $1/\tau = n c_s \sigma$, where $n(T)$ is the density of phonons present. The phonon occupancy is

$$\frac{1}{\exp(\beta\hbar\omega) - 1} \approx \frac{k_B T}{\hbar\omega}. \tag{5.73}$$

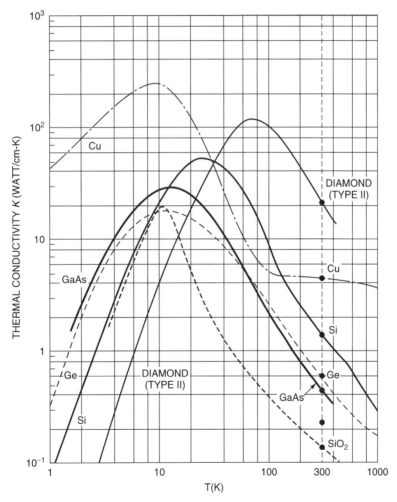

Figure 5.15. Thermal conductivities of various solids as a function of T. (From S. M. Sze, *Physics of Semiconductor Devices*, 2nd ed., copyright 1981 by John Wiley & Sons, Inc. Reprinted by permission of John Wiley and Sons, Inc.)

The cross section is denoted by σ. Process A may also be stimulated by the presence of a preexisting phonon field, in much the same way as an excited atom undergoes stimulated emission in a laser. The stimulated rate is also proportional to n. Processes C, D, and E also have rates going roughly as n^2, since there are two extra phonons available to stimulate the processes. The net result is that the collision lifetime can be expected to vary (roughly) as $\tau \propto 1/T^P$, with p in the range 1 to 2, at high temperatures. The same will be true of the phonon mean free path, $\lambda = c_s\tau$.

At low temperatures it is mainly the low-energy acoustic phonons that are present. Most processes are N processes since there is insufficient energy available to create a phonon sufficiently large that it requires a U process to bring it back into the first Brillouin zone. But due to conservation laws, the N processes are ineffective in relaxing the phonons because, as in process A, the overall direction of the phonon flux is not changed. The rare U processes are responsible for the relaxation. To have a U process, there must first be a thermally activated phonon of high energy, and the probability for this to occur goes as $\exp(-\Theta_D/T)$. Thus the collision lifetime varies as $\tau \propto \exp(\Theta_D/T)$.

At sufficiently low T the mean free path may grow to be larger than some relevant physical length, L, in the solid, such as the distance between impurities, defects, or physical boundaries. In such instances the collision time is determined by these distances (i.e., $\tau \approx L/c_s$).

In Fig. 5.15 the thermal conductivities of Si, diamond, GaAs, Cu, Ge, and SiO_2 are plotted as a function of T. One sees the T^3 behavior at low T and the diminishing of κ at high T. The low-T behavior follows from Eq. (5.69), where $C(T) \propto T^3$ and $\lambda(T) = c_s\tau(T) \approx$ constant, so $\kappa(T) \propto T^3$. For $T > \Theta_D$, $C(T) \propto$ constant, $\lambda(T) \propto 1//T^p$, so $\kappa(T) \propto T^{-p}$. Thus $\kappa(T)$ has a maximum between these two limits, as observed in Fig. 5.15. A high value of Θ_D implies that a high κ can be achieved before U processes start to limit the phonon mean free path.

For the mean free path to be long, the crystal must be defect-free. It must also be isotopically pure. Atoms with a different isotopic mass than the most abundant isotope can act as scattering centers. For example, the thermal conductivity of diamond with the natural abundance ratios of ^{12}C and ^{13}C is significantly less than that of pure ^{12}C.

REFERENCES

Bilz, H., and W. Kress, *Phonon Dispersion Relations in Insulators*, Springer-Verlag, Berlin, 1979.

Born, M., and K. Huang, *Dynamical Theory of Crystal Lattices*, Clarendon Press, Oxford, 1985.

Brillouin, L., *Wave Propagation in Periodic Structures*, Dover, Mineola, N.Y., 1953.

Bruesch, P., *Phonons, Theory and Experiments*, Springer Verlag, Berlin, 1982.

Dove, M. T., *Introduction to Lattice Dynamics*, Cambridge University Press, Cambridge, 1993.

Reissland, J. A., *The Physics of Phonons*, Wiley, New York, 1973.

Srivastava, G. P., *The Physics of Phonons*, Adam Hilger, New York, 1990.

Willis, B. T. M., and A. W. Pryor, *Thermal Vibrations in Crystallography*, Cambridge University Press, Cambridge, 1975.

Ziman, J. D., *Electrons and Phonons*, Clarendon Press, Oxford, 1960.

PROBLEMS

5.1 Calculate the phonon spectrum for a two-dimensional square array of atoms connected by harmonic "springs" to the nearest neighbors. Assume that each spring can only be stretched.

5.2 Use the Debye model to calculate the specific heat of a two-dimensional crystal.

5.3 Consider a one-dimensional harmonic chain with spring constant K and atomic mass M. Assume that one atom is replaced by an isotope with a different mass M'. Compare the resulting spectrum with that of the isotopically pure chain.

5.4 The Toda potential is given by $V(x) = Ae^{-\alpha x} + Bx$. Calculate the thermal expansion coefficient at high temperatures for a one-dimensional chain of atoms interacting by nearest-neighbor interactions through this potential.

5.5 Calculate the specific heat of a system of N identical harmonic oscillators according to the Einstein theory from Eq. (5.47). Assume that the common frequency of all N oscillators is ω_0.

5.6 For a solid with a simple cubic crystal structure and lattice constant $a = 0.3$ nm, calculate the value of the Debye wave vector k_D and compare it with the wave vector at the Brillouin zone edge, $k_{BZ} = \pi/a$. If this crystal is a cube with side 1 mm, calculate the density of allowed phonon modes in wave vector space, assuming two transverse and one longitudinal mode per value of $\mathbf{k}$.

5.7 Use the assumptions of the Debye model to derive the following formula for the thermal conductivity:

$$\kappa(T) = \frac{\hbar^2}{2\pi c_s k_B T^2} \int_0^{\omega_D} \frac{\omega^4 \tau(\omega) \exp(\beta\hbar\omega)}{[\exp(\beta\hbar\omega) - 1]^2} d\omega$$

where $\tau(\omega)$ is an assumed frequency-dependent phonon relaxation time. Evaluate for $T \ll \Theta_D$, assuming that $\tau(\omega)$ is a constant, τ_0.

5.8 Calculate the thermal average of the product of two displacements of a one-dimensional monatomic chain $\langle u_n u_{n'} \rangle$. Repeat for the general crystal and compute $\langle u_\alpha^\sigma(\mathbf{R}) u_{\alpha'}^{\sigma'}(\mathbf{R'}) \rangle$. Show that both of these products grow linearly with temperature.

5.9 Consider the one-dimensional anharmonic lattice with a nearest-neighbor atom–atom interaction given by

$$V(x) = \frac{K}{2}x^2 + \frac{\alpha}{24}x^4,$$

where α is a small parameter. Make a mean-field approximation and replace the anharmonic interaction by an effective harmonic interaction

$$\langle u_n u_{n'} u_{n''} u_{n'''} \rangle = \langle u_n u_{n'} \rangle u_{n''} u_{n'''} + \langle u_n u_{n''} \rangle u_{n'} u_{n'''} + \langle u_n u_{n'''} \rangle u_{n'} u_{n''}.$$

Use the results of Problem 5.8 to obtain the T dependence of the phonon frequencies.

CHAPTER 6

Thermally Activated Processes, Phase Diagrams, and Phase Transitions

6.1 Introduction

This chapter provides an introductory survey of some of the important thermodynamic and kinetic properties of solids. These include the thermally activated processes of diffusion and vaporization, equilibrium phase diagrams, and structural phase transitions. An introduction to these properties is critical for achieving an understanding of a wide range of phenomena observed in solid-state materials. In addition, they are important for the synthesis and processing of materials, as discussed in Chapter 21. Specific examples of phase diagrams and phase transitions are also provided in the chapters on different classes of materials. Additional aspects of processes involved in the synthesis and processing of materials, including reactions and reaction rates, equilibrium constants, and thermodynamic activities, are described in Chapter 21.

THERMALLY ACTIVATED PROCESSES

Thermally activated processes in solids have rates $R(T)$, which usually increase exponentially with increasing temperature. These rates are in general described by *Arrhenius law* expressions, for example,

$$R(T) = A \exp\left(-\frac{E_a}{k_B T}\right), \tag{6.1}$$

where A is a constant (which may be weakly temperature dependent), E_a is the activation energy for the process, and T is the absolute temperature. The thermally activated generation of point defects such as vacancies and interstitials in solids is described in Section 4.6. The processes of diffusion and vaporization, which are related to the motions of atoms and defects in the bulk and on the surface of a solid, respectively, are discussed next. Surface diffusion is also described in Chapter W19 at our Web site.[†]

6.2 Diffusion

When the concentrations of specific atoms or point defects vary spatially in a solid, the system may not be in thermodynamic equilibrium. The thermally activated motion

---

[†] Supplementary material for this textbook is included on the Web at the resource site (ftp://ftp.wiley.com/ public/sci_tech_med/materials). Cross-references to elements of the Web material are prefixed by "W."

of atoms and defects through solids often tends to eliminate concentration gradients via the mass transport processes known as *diffusion*. The thermodynamic driving force for diffusion, at least in chemically homogeneous crystals containing a single chemical species, is the increase in the configurational or mixing entropy resulting from increased randomness on the macroscopic scale when the concentration gradients are eliminated. This type of diffusion (e.g., of Si in bulk Si), is known as *self-diffusion*.

In chemically inhomogeneous solids such as binary alloys, there can exist an additional driving force for diffusion: namely, the tendency for atoms to move preferentially in directions in which they lower their energy (i.e., toward sites where they are more tightly bound). In these situations the driving force is the gradient of the chemical potential of the diffusing species. Chemical diffusion (e.g., of C in bulk Fe), can actually lead to greater concentration gradients and even to physical and chemical segregation of different chemical species. Diffusion that occurs in a direction opposite to that of a concentration gradient is known as *uphill diffusion*. Examples of such processes include phase separation and spinodal decomposition, discussed in Section 6.7.

Diffusion plays a key role in many solid-state phenomena, such as crystal growth, structural phase transitions, zone refining, annealing, sintering, nucleation, precipitation, and solid-state chemical processes such as corrosion and oxidation. In addition, ionic conductivity, order–disorder transitions, internal friction, and the photographic process also involve diffusion processes directly. Most of these processes and phenomena are discussed here or in later chapters. In particular, the topic of surface diffusion is discussed in Chapter W19. Crystal growth, zone refining, annealing, sintering, nucleation, and precipitation are discussed both in Chapter 21 and Chapter W21.

The presence and mobility of defects such as vacancies and interstitials often play critical roles in diffusion processes, as follows:

1. The diffusion of atoms is greatly enhanced when they exchange places with neighboring vacancies (Fig. 6.1 at a). Known as the *vacancy mechanism*, this is equivalent to the migration of vacancies.

2. An interstitial atom can diffuse via the *interstitialcy mechanism* in which it moves onto a neighboring lattice site as the atom occupying that site moves to an adjacent interstitial site. Interstitialcy diffusion via this indirect mechanism therefore corresponds to a sequence of jumps of an atom, usually a self-interstitial, between alternate interstitial and lattice sites (Fig. 6.1 at b).

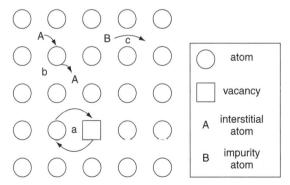

Figure 6.1. Defects such as vacancies and interstitials in solids often participate in diffusion processes, as shown.

3. Interstitial impurity atoms can diffuse by jumping directly from one interstitial site to a neighboring one. This is known as the *direct interstitial mechanism* and is illustrated in Fig. 6.1 at c. Note that this mechanism is distinct from the interstitialcy mechanism described above.

In addition to bulk or volume diffusion through lattice sites, diffusion can also occur along dislocations and grain boundaries and across the surfaces of solids. At lower temperatures the latter diffusion mechanisms typically take place at much higher rates than does bulk lattice diffusion.

Fick's Laws. The basic mechanism for self-diffusion in a chemically homogeneous system can be readily illustrated by the following simple one-dimensional picture. Two adjacent planes of atoms separated by a distance a are shown in Fig. 6.2. The plane on the left contains N_1 atoms per unit area, while that on the right contains $N_2 < N_1$ atoms per unit area. At a finite temperature an atom jumps from one lattice or interstitial site to another at a rate $R(T)$. There is in general a potential barrier that must be overcome, as shown in Fig. 6.3. Here E_a is the activation energy for the jump. The rate $R(T)$ is proportional to the product of an atomic vibrational frequency f_{vib} and the appropriate

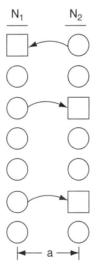

Figure 6.2. Two adjacent planes of atoms separated by a distance a.

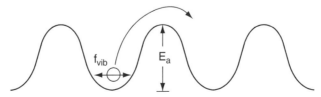

Figure 6.3. A potential barrier must be overcome for an atom to jump from one site to another. The activation energy is E_a for the jump.

Boltzmann factor, that is,

$$R(T) \approx f_{\text{vib}} \exp \left(-\frac{E_a}{k_B T} \right). \tag{6.2}$$

If $R/2$ of these jumps are to the right and $R/2$ are to the left, the net flux of atoms moving between the two adjacent planes in Fig. 6.2 is given by

$$J = \frac{N_1 R}{2} - \frac{N_2 R}{2} = -\frac{aR}{2} \frac{dN}{dx} = -\frac{a^2 R}{2} \frac{dC}{dx} = -D \frac{dC}{dx}, \tag{6.3}$$

where $N_2 = N_1 + a(dN/dx)$ and $C(x) = N(x)/a$ is the volume concentration of atoms. The flux is therefore proportional to the concentration gradient, with a coefficient of proportionality D which is the *self-diffusion coefficient* or *diffusivity*, with units of m^2/s. The quantity D is given by

$$D(T) = \frac{a^2 R(T)}{2} \approx \frac{a^2 f_{\text{vib}}}{2} \exp \left(-\frac{E_a}{k_B T} \right). \tag{6.4}$$

Equation (6.3) is *Fick's first law of diffusion* and, as stated earlier, applies to solids that are chemically homogeneous. The vector generalization of this equation is $\mathbf{J} = -D \nabla C$. Omitted from this simple discussion is the necessity for the existence of an adjacent vacant site into which an atom can jump. This additional factor is discussed later.

A more general form of Fick's first law, which recognizes that spatial variations of the *chemical potential* μ_i determine the diffusion of a given species, is

$$\mathbf{J}_i = -L_i \nabla \mu_i. \tag{6.5}$$

This expression corresponds to a special, approximate case of the Onsager equations, which play a central role in the thermodynamics of irreversible processes. The Onsager relations for the transport of charge and heat in solid materials are presented in Section 7.5 and in more detail in Chapter W22. In the more general theory, diffusion due to temperature gradients, known as the *Soret effect*, can also be treated. Thus at equilibrium when net diffusion ceases, the chemical potentials μ_i of all species will be uniform in space, whereas concentration gradients may still be present, as in a two-phase alloy. Since concentration gradients are much easier to measure than gradients of chemical potential, the version of Fick's first law presented in Eq. (6.3) is preferred to that given in Eq. (6.5), as is the description of diffusion processes in terms of diffusion coefficients D. The flux of diffusing species can also be written as

$$\mathbf{J}_i = C_i \mathbf{v}_i = -C_i \mu_{\text{mob}}(i) \nabla \mu_i \tag{6.6}$$

where C_i is the concentration of the diffusing species, v_i its drift or mean velocity, and $\mu_{\text{mob}}(i)$ its mobility, with units of s/kg. Thus L_i in Eq. (6.5) is given by $C_i \mu_{\text{mob}}(i)$.

Fick's second law of diffusion concerns situations in which the concentration gradient dC/dx is not constant in space. In this case the flux J varies with position. It can be shown, by requiring conservation of mass, that the time rate of change of $C(x, t)$ due to diffusion is given by the *continuity equation*[†]

$$\frac{\partial C}{\partial t} = -\nabla \cdot \mathbf{J}. \tag{6.7}$$

In one dimension this becomes the diffusion equation and is Fick's second law,

$$\frac{\partial C}{\partial t} = -\frac{\partial J}{\partial x} = \frac{\partial}{\partial x}\left(D\frac{\partial C}{\partial x}\right). \tag{6.8}$$

The diffusion coefficient D is in general a tensor quantity and can be anisotropic in noncubic crystals.

In chemically inhomogeneous systems such as binary alloys, chemical diffusion can occur as a result of the chemical inhomogeneities that are present. The concentration changes occurring in such systems can be described by a net chemical diffusion coefficient D_c. The relationship between D_c and the self-diffusion coefficients D_i of the individual diffusing species depends on the details of the system (e.g., a substitutional solid solution, ionic solid, etc.). More details can be found in the references listed at the end of this chapter.

Specific examples of the calculations of concentration profiles in solids due to diffusion in from the surface or boundary are described in Chapter W6.

Atomic Nature of Diffusion. Additional physical insight into the macroscopic diffusion coefficient D can be gained by considering the atomic nature of the diffusion process. The diffusion of atoms or point defects results from the random movement of atoms from one lattice or interstitial site to another. This motion is taken to be a *random walk* process, as illustrated in Fig. 6.4, where each jump is assumed to be independent of all others.[‡] This analysis leads to the following expression for the net

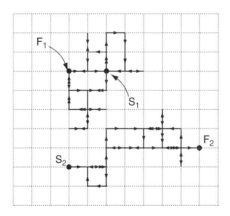

Figure 6.4. The random walk of atoms corresponding to their random movement from one lattice or interstitial site to another is illustrated for two atoms on an otherwise empty square lattice. S, start, F, finish. The direction of the jumps has been determined by tossing a die.

[†] The continuity equation for surface diffusion is derived in Chapter W19.

[‡] The case of a self-avoiding random walk and its relationship to the structure of polymers is described in Chapter W14.

flux of atoms in one dimension:

$$J(x, t) = C\frac{\langle X \rangle}{t} - \frac{\langle X^2 \rangle}{2t}\frac{\partial C}{\partial x},\tag{6.9}$$

where higher-order terms in $\langle X^n \rangle$ and higher-order derivatives of the concentration gradient are not considered. Here X is the net displacement of an atom and $\langle X \rangle$ and $\langle X^2 \rangle$ are averages over all possible paths. Note that $\langle v \rangle = \langle X \rangle / t$ is the mean drift velocity of an atom.

For the case of self-diffusion in a chemically homogeneous solid, $\langle X \rangle = 0$ and Eq. (6.9) reduces to Fick's first law, with the self-diffusion coefficient D given by

$$D = \frac{\langle X^2 \rangle}{2t}.\tag{6.10}$$

This is the *Einstein diffusion equation*, which provides a basis for the atomic theory of diffusion. Note in three dimensions and in an isotropic or cubosymmetric material that $\langle r^2 \rangle = \langle X^2 \rangle + \langle Y^2 \rangle + \langle Z^2 \rangle = 3\langle X^2 \rangle$. The average distance an atom diffuses in time t is given approximately by $L = \sqrt{\langle X^2 \rangle} \approx \sqrt{Dt}$.

When ions diffuse in the presence of an electric field E, it is no longer true that $\langle X \rangle = 0$ since there is now a preferred direction for their motion. The ions will therefore have a net drift velocity $\langle v \rangle = \mu_{mob}E$, where μ_{mob} is the ionic mobility. The mobility and the diffusion coefficient are related by

$$D = \frac{k_B T}{q}\mu_{mob},\tag{6.11}$$

where q is the charge of the diffusing ion. This expression, obtained by setting the total current equal to zero at equilibrium, is known as the *Einstein relation*.

If there are on the average N jumps of distance r_0 in time t, the jump rate $R = N/t$ and $\langle r^2 \rangle = Nr_0^2$. It follows that

$$D = \frac{fRr_0^2}{6},\tag{6.12}$$

where f is the *correlation factor* ($0 \leq f \leq 1$). The factor f accounts for the fact that in certain situations the directions of successive jumps can be correlated. For example, an atom that jumps into a neighboring vacant lattice site is, of course, more likely on the next jump to jump back into the vacant site that it had left. In the simple picture used earlier, which led to Eq. (6.4) for D, the possibility of correlated motions was ignored and f was set equal to 1.

As an example of the usefulness of Eq. (6.10), consider the case of self-diffusion via the vacancy mechanism in an FCC crystal such as Al. In the FCC crystal structure the number of NN sites is 12 at a distance of $\sqrt{2}a/2$, where a is the lattice constant. The jump rate R can be written as the product of the jump frequency f_{jump} and the probability $p(V)$ that one of the 12 NN sites is vacant. If $n_v(T)$ is the fractional vacancy concentration (see Section 4.7), $p(V) = 12n_v(T)$ and $R = 12n_v(T)f_{jump}$. Using Eq. (6.12) and $r_0^2 = a^2/2$ for an FCC lattice, the following result is obtained:

$$D(T) = \frac{12fn_v(T)f_{jump}(a^2/2)}{6} = fn_v(T)a^2 f_{jump}.\tag{6.13}$$

It should be noted that the temperature dependence of D comes from $n_v(T)$, f_{jump}, and possibly the correlation factor f. Differences of several orders of magnitude in the $D(T)$ observed for self-diffusion in some materials can be attributed to defect concentrations, such as $n_v(T)$. These can also differ by orders of magnitude in a given material due to extrinsic effects such as mechanical damage or intentional doping, as in Si.

Experimental results for D are often expressed in the following general forms:

$$D(T) = D_0 \exp\left(-\frac{E_a}{k_B T}\right), \tag{6.14}$$

or

$$D(T) = \sum_i D_{0i} \exp\left(-\frac{E_{ai}}{k_B T}\right) = \sum_i D_i(T) n_{di}(T). \tag{6.15}$$

Here E_a is the experimentally determined activation energy, D_i the diffusivity of a given diffusing species, and n_{di} the corresponding concentration of defects. Equation (6.15) is appropriate when diffusion occurs by two or more parallel mechanisms. The parameter E_a and the prefactor D_0 can each be controlled by several factors, some of which have been discussed earlier. These factors are summarized next.

The jump frequency f_{jump} is the product of a vibrational frequency, usually taken to be the Debye frequency $\omega_D/2\pi$, and the Boltzmann factor $\exp(-G_m/_B T)$, where $G_m = H_m - TS_m$ is the Gibbs free energy of migration. The quantities H_m and S_m are the enthalpy and entropy of migration, respectively, for the process under consideration. Note that a Gibbs free energy is used here because the process occurs under conditions of constant temperature and pressure. Thus

$$f_{jump} = \frac{\omega_D}{2\pi} \exp\left(\frac{S_m}{k_B}\right) \exp\left(-\frac{H_m}{k_B T}\right). \tag{6.16}$$

It is shown in Section 4.6 that the vacancy concentration $N_L(V)$ increases exponentially with increasing T, so that

$$n_v(T) = \exp\left(-\frac{\Delta G_f}{k_B T}\right) = \exp\left(\frac{\Delta S_f}{k_B}\right) \exp\left(-\frac{\Delta H_f}{k_B T}\right), \tag{6.17}$$

where $\Delta G_f = \Delta H_f - T\Delta S_f$ is the Gibbs free energy of formation of the defect, in this case a vacancy.

Ignoring any temperature dependence of the correlation factor f, the following expressions, valid for vacancy diffusion in FCC crystals, are obtained from Eqs. (6.13) to (6.17),

$$D_0 = f a^2 \frac{\omega_D}{2\pi} \exp\left(\frac{\Delta S_f + S_m}{k_B}\right) \tag{6.18}$$

$$E_a = \Delta H_f + H_m. \tag{6.19}$$

Note that through the factor ω_D, $D_0 \propto M^{-1/2}$, where M is the mass of the diffusing atom or ion. The final expression for $D(T)$ is obtained by substituting these expressions for D_0 and E_a into Eq. (6.14). It should be noted that $f = 0.7815$ for self-diffusion in FCC lattices. Values of f for other crystal structures and diffusion mechanisms are given by Le Claire (1976).

Note also that ΔH_f is absent from the activation energy (i.e., $E_a = H_m$), when defects are not involved in the diffusion process, as in the case of direct interstitial diffusion (since interstitial sites are not defects), or when the defect involved has a concentration that is independent of T.

An interesting effect can occur in a chemically inhomogeneous system such as the binary alloy $A_{1-x}B_x$ when A atoms diffuse preferentially in one direction and B atoms in the opposite direction. If the A and B atoms have different diffusion coefficients D_A and D_B, there must exist a net flux of vacancies opposite to the direction of the faster-diffusing atoms in order to conserve the overall number of atoms per unit volume. Since all atoms will have a greater tendency to jump opposite to the direction of this so-called *vacancy "wind"*, the vacancy wind will increase the diffusion rate of the faster-diffusing component and decrease the diffusion rate of the slower-diffusing component.

Examples of diffusion (i.e., self-diffusion in Cu and self- and impurity diffusion in Si), are presented in Chapter W6 to provide illustrations of diffusion in elemental metals and semiconductors.

6.3 Vaporization

Atoms on the surface of a solid undergo a variety of thermally activated processes, including diffusion both across the surface and into the bulk. At any finite temperature a surface atom can be thermally excited out of its surface potential well into the vapor phase. The rate of *vaporization* (i.e., *sublimation*), is observed to increase exponentially with temperature. Vaporization is a particularly simple example of a vapor–solid reaction and is therefore of fundamental interest. Discussion of the surface processes that control vaporization rates is deferred to Chapter W19.

A system consisting of a solid and its vapor is in equilibrium when the flux of atoms or molecules desorbing from the surface is equal to the flux returning from the vapor and condensing onto the surface, that is, when

$$J_{\text{vap}} = J_{\text{con}}. \tag{6.20}$$

The reaction describing the equilibrium of a solid composed of A atoms with its vapor is

$$A(\text{solid}) \longleftrightarrow A(\text{vapor}). \tag{6.21}$$

The corresponding law of mass action can be written as

$$\frac{a(\text{vapor})}{a(\text{solid})} = K(T) = \frac{P_{\text{eq}}(T)}{1\,\text{atm}} = \exp\left(-\frac{\Delta_r G^o}{k_B T}\right), \tag{6.22}$$

where $\Delta_r G^o = G(\text{vapor}) - G(\text{solid})$ is the change in Gibbs free energy and $K(T)$ is the equilibrium constant for the vaporization reaction (6.21). The activity $a(\text{vapor})$ is equal to the equilibrium vapor pressure $P_{\text{eq}}(T)$ of A(vapor), normalized to atmospheric pressure (1 atm $= 1.013 \times 10^5$ Pa), while $a(\text{solid})$ is equal to unity, by convention, since the solid is in its standard state. Using the incident flux $J_{con} = P_{eq}(T)/\sqrt{2\pi mk_BT}$ (particles per square meter per second) from kinetic theory (see Chapter W19), the following result is obtained using Eqs. (6.20) and (6.22):

$$J_{vap}(T) = \frac{P_{eq}(T)}{\sqrt{2\pi mk_BT}} = \frac{1\,\text{atm}}{\sqrt{2\pi mk_BT}} \exp\left(-\frac{\Delta_r G^o}{k_BT}\right). \qquad (6.23)$$

In this derivation it has been assumed that there are no kinetic limitations on the vaporization process, such as sticking or vaporization coefficients that are less than unity. Under these conditions, Eq. (6.23) will also give the flux of atoms evaporating from the surface under nonequilibrium conditions, such as vaporization into a vacuum. Vaporization studies of the elements provide the standard Gibbs free energy of formation $\Delta_f G^o$ of the vapor species, since in this case $\Delta_r G^o = \Delta_f G^o$.

A kinetic approach to vaporization must consider the thermal activation of surface atoms out of their surface potential wells. The rate for this desorption process is given by the product of an atomic vibrational frequency f_{vib} and the appropriate Boltzmann factor $\exp(-\Delta G_{des}/k_BT)$. The flux of atoms leaving the surface is therefore

$$J_{vap}(T) = n_s(T) f_{vib} \exp\left(-\frac{\Delta G_{des}}{k_BT}\right), \qquad (6.24)$$

where $n_s(T)$ is the equilibrium surface concentration of A atoms and ΔG_{des} is the Gibbs free energy of desorption. Equations (6.23) and (6.24) provide alternative ways of describing the vaporization process. The vaporizations of the elemental solids Fe and Si are described in Chapter W6.

EQUILIBRIUM PHASE DIAGRAMS

The equilibrium phase diagrams of materials provide the "road maps" that are necessary for successful design, synthesis, and processing of materials. In addition, they provide important insights into the properties of materials. Equilibrium phase diagrams provide a very useful way of presenting information concerning which phases of a single- or multicomponent system are stable under equilibrium conditions as a function of temperature, pressure, composition, and possibly other external variables, such as magnetic or electric field. This is essential since many important materials, such as steels, ceramics, and a wide range of composite solids, are heterogeneous (i.e., they have two or more phases that coexist with each other). Phase diagrams should be used with caution since many materials are actually prepared in a metastable state. For example, the compound Fe_3C, cementite, is present in Fe–C alloys, in which graphite would be the stable form of carbon. Glasses and other amorphous materials are also clearly metastable. For such materials phase diagrams that illustrate the metastable phases should be consulted.

In this section equilibrium phase diagrams in the temperature–pressure (T–P) plane for pure, single-component substances and in the temperature–composition (T–x) plane for two-component, binary systems are examined. The phases to be considered include solid, liquid, and vapor, with the emphasis here on the solid phase or phases that may be present. Ternary phase diagrams such as the Si–Al–O phase diagram are introduced and described in Chapter W13.

6.4 Pure Substances

The equilibrium temperature–pressure phase diagram for pure Fe is shown in Fig. 6.5. Three stable, homogeneous solid phases, α-Fe (BCC), γ-Fe (FCC), and δ-Fe (BCC), are present in addition to the liquid and vapor phases. The following observations can be made:

1. There exist regions in the T–P plane where only single phases are present, lines on which two phases coexist, and points (the triple points) at which three phases coexist. The coexistence of phases at equilibrium for fixed T and P occurs when their Gibbs free energies G are equal.
2. Phase transitions occur when, by varying the temperature T and/or pressure P, a line in the phase diagram is crossed. For example, the melting transition from δ-Fe to liquid Fe occurs at $T_m = 1538°C$ at atmospheric pressure.
3. The equilibrium vapor pressure $P_{eq}(T)$ of the solid is the pressure corresponding to the vaporization line (i.e., the line separating pure vapor from pure solid).

A useful relationship for expressing the slope of the temperature–pressure lines separating two phases in the equilibrium phase diagram is known as the *Clausius–Clapeyron equation*. The Gibbs free energies of two phases are equal on the line in the phase diagram separating them. The Clausius–Clapeyron equation for the slope of this line can be derived using the standard thermodynamic relation $dG = V\,dP - S\,dT = 0$:

$$\frac{dP}{dT} = \frac{\Delta S}{\Delta V}. \tag{6.25}$$

Here ΔS and ΔV are the differences in entropy and volume, respectively, between the two phases at the point where the slope is evaluated.

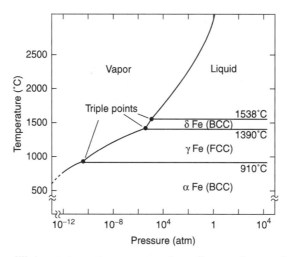

Figure 6.5. The equilibrium temperature–pressure phase diagram for pure Fe is shown. Three stable solid phases, α-Fe, γ-Fe, and δ-Fe, are present in addition to the liquid and vapor phases. (From W. G. Moffat, G. W. Pearsall, and J. Wulff, *Structure and Properties of Materials*, Vol. 1, copyright 1964 by John Wiley & Sons Inc. Reprinted by permission of John Wiley & Sons, Inc.)

As an example of the use of this equation, consider the melting transition, in which case $\Delta S_m = S_l - S_s$ and $\Delta V_m = V_l - V_s$. For most melting transitions both the entropy and volume are higher in the liquid (l) than in the solid (s) phase. It follows, then, that $dP/dT > 0$, as shown in Fig. 6.5. Important exceptions to this rule are Si and H_2O, where $V_l < V_s$, so that ΔV_m is negative and $dP/dT < 0$.

As discussed in Section 6.6, the changes in enthalpy ΔH_m and entropy ΔS_m that occur at the first-order melting-phase transition are related by $\Delta H_m = T_m \Delta S_m$, where ΔH_m is the latent heat of fusion for the transition. Using this result, the Clausius–Clapeyron equation can also be written as

$$\frac{dP}{dT} = \frac{\Delta H_m}{T_m \, \Delta V_m}. \tag{6.26}$$

6.5 Binary Systems

True *binary systems* are solids containing just two elements (e.g., Cu–Ni, Si–Ge, and Pb–Sn). Other systems that contain more than two elements or phases, including mixtures or alloys of InAs and GaAs, of Fe and Fe_3C, and of SiO_2 and Al_2O_3, are considered to be *pseudobinary*. Discussions of these particular pseudobinary systems are deferred to Chapters 11, 12, and 13, respectively.

For a given composition x in a binary system such as $Cu_{1-x}Ni_x$, a $T–P$ phase diagram can be constructed that contains the same type of information presented in Fig. 6.5 for the Fe $T–P$ phase diagram. In addition, temperature-composition ($T–x$) phase diagrams and even three-dimensional $T–P–x$ diagrams can be presented if sufficient thermodynamic data are available. The $T–x$ phase diagrams for a binary solid-solution alloy and also for a eutectic alloy exhibiting limited mutual solubility are presented and discussed next. Phase diagrams for alloys exhibiting intermediate phases or compounds which are discussed in other chapters include Fe–Fe_3C in Chapters 12 and 21, La–Cu in Chapter W12, SiO_2–Al_2O_3, SiO_2–Na_2O, and SiO_2–CaO–Na_2O in Chapter 13, and Al–Cu in Chapter W21.

Binary Solid-Solution Alloys. In *binary solid-solution alloys*, also known as *isomorphous alloys*, the two elements are completely soluble in each other over the entire range of compositions and atoms of the two elements readily substitute for each other on a given lattice site. The alloys therefore exhibit the same crystal structure for all compositions. As expected, these solid solutions are formed between elements not only having the same crystal structures but also having lattice constants that are not too different from each other. These conditions and the two additional conditions that the elements should not form compounds with each other and should have the same valence are known as the *Hume–Rothery rules* after the English metallurgist who first proposed them. These rules are discussed in Chapter 12.

Standard examples of binary solid solutions include Cu–Ni and Si–Ge alloys. Cu and Ni share the FCC crystal structure with lattice constants $a(Cu) = 0.361$ nm and $a(Ni) = 0.352$ nm, a 2.5% difference. Thus there is not much strain and hence strain energy associated with substituting one element for the other. Cu and Ni have outer electron configurations $3d^{10}4s$ and $3d^84s^2$, respectively. The chemical valences of Cu are 1 or 2, and those of Ni are 2 or 3. The first ionization energies of Cu and Ni are 7.72 and 7.63 eV, respectively, indicating that both atoms have equal difficulty in supplying electrons to form the metallic bond. Chemically, Cu and Ni are quite similar. Si and Ge both have the diamond crystal structure with $a(Si) = 0.543$ nm and $a(Ge) = 0.565$ nm, a 4% difference.

The equilibrium $T-x$ phase diagram for $Cu_{1-x}Ni_x$ alloys at atmospheric pressure is shown in Fig. 6.6. Three regions are present, two of which are homogeneous single-phase regions, l = liquid and α = solid, as well as an intermediate heterogeneous two-phase region, $\alpha + l$, in which liquid and particles of solid coexist. The two-phase region is bounded from above by the *liquidus line* and from below by the *solidus line*. The liquidus and solidus lines merge at $x = 0$ and $x = 1$ at the melting points of the two component elements. The compositions of the alloy in the single-phase liquid l and solid α regions are given by the overall alloy composition x. In the two-phase region at a temperature T, however, the liquid and solid have compositions x_l and x_s which are given by the liquidus and solidus lines, respectively. These compositions differ from each other and also from the overall alloy composition x. The weight fractions of the two elements in a given phase can be determined from the atomic fractions x and $1 - x$ and the atomic weights of the elements.

The thermodynamics of phase diagrams is an important subject since the details of the phase diagram (e.g., the locations of the liquidus and solidus lines), are in principle determined by the Gibbs free energies $G(T,P,x)$ of the liquid, the solid, and the "mixture" of liquid and solid in the two-phase region. Predicting phase diagrams is in practice quite difficult, due to lack of experimental data for the relevant thermodynamic quantities. Nevertheless, some important general statements can be made; for example, absolutely pure substances are thermodynamically impossible since, under equilibrium conditions, any small addition of solute to a pure substance always lowers the Gibbs free energy of the system. It is thus incorrect to say that any substance is completely insoluble in another.

The phase boundaries in a binary system may be mapped out by preparing a set of liquid alloys with varying compositions and allowing them to cool slowly. As long as

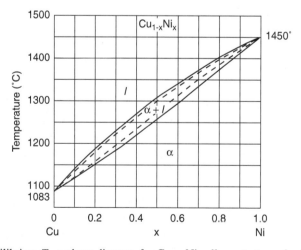

Figure 6.6. Equilibrium $T-x$ phase diagram for $Cu_{1-x}Ni_x$ alloys at atmospheric pressure. Also shown as dashed curves are the liquidus and solidus lines predicted from thermodynamics using melting data for pure Cu and Ni. [Reprinted with permission from H. Seltz, *J. Am. Chem. Soc.*, **56**, 307 (1934). Copyright 1934 by the American Chemical Society.]

the material consists of a single phase the temperature will fall off exponentially in time. When a phase boundary is crossed, however, the liquid is gradually converted to solid as the latent heat of fusion is conducted away. The time dependence is no longer simply exponential during this transition. Solid particles nucleate and grow until the liquid is completely solidified, at which point the temperature again falls exponentially, approaching that of the surrounding medium. The phase boundaries may be located from the temperatures where exponential behavior terminates and begins again.

In any two-phase region such as that shown in Fig. 6.6, the relative amounts of the two coexisting phases can readily be determined from the following statement of the conservation of the number of atoms or moles of B atoms in an $A_{1-x}B_x$ alloy:

$$n_s x_s + n_l x_l = (n_s + n_l)x = nx. \qquad (6.27)$$

Here n_s and n_l are the number of moles of solid and liquid present, respectively. In this equation $n_s x_s$ and $n_l x_l$ are the number of moles of B present in the solid and liquid phases, respectively, and nx is the total number of moles of B present. Equation (6.27) can be solved for the fraction of moles present in the solid phase, yielding

$$\frac{n_s}{n} = \frac{x - x_l}{x_s - x_l}. \qquad (6.28)$$

It follows that

$$\frac{n_l}{n} = \frac{x_s - x}{x_s - x_l}. \qquad (6.29)$$

The geometrical interpretation of these equations can be seen in Fig. 6.7, in which a portion of the two-phase region has been enlarged. The compositions x, x_s, and x_l are shown, as is the horizontal isothermal tie line connecting x_l and x_s. Thus the number n_s of moles of solid is proportional to $x - x_l$, the length of the line connecting x and x_l. For this reason, Eqs. (6.28) and (6.29) are known as the *lever laws*.

In practice, the single-phase solid α region in the $Cu_{1-x}Ni_x$ phase diagram is rarely homogeneous (i.e., the Cu and Ni atoms are not completely mixed on an atomic or

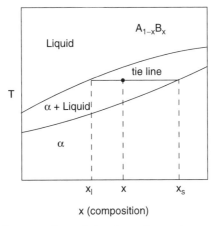

Figure 6.7. Geometrical interpretation of the lever law using an enlarged portion of the two-phase region of the phase diagram of a binary solid-solution alloy.

even on a microscopic level). This is due to the fact that even if carried out very slowly, solidification tends to be a nonequilibrium process, with the result that the resulting solid is inhomogeneous and metastable. For example, as the temperature is lowered and the $\alpha + l$ region is entered, solidification of the solid α phase begins (according to the lever law) with a composition different from that of the original, homogeneous liquid. The portions of the liquid alloy to solidify first will be those which, due to statistics or to local clustering effects, contain more of the higher-melting-point material, in this case Ni. As T is lowered further, the solid α phase formed will include an increasingly smaller fraction of Ni atoms and so will have a different composition from the solid α phase formed initially. Significant concentration gradients will be present by the time the alloy is completely solidified unless the temperature is sufficiently high for the normally slow solid-state diffusion to result in a homogeneous solid α phase with uniform concentration. Another result of nonequilibrium cooling is that the temperature at which all of the liquid is finally converted to solid lies below the solidus curve of the equilibrium phase diagram.

The microstructure of the solidified alloy will in general be polycrystalline, consisting of crystallites centered at the nuclei of the solid phase. The scale of the microstructure (i.e., the average size of the crystallites), will be determined by the cooling rate, with smaller crystallites corresponding to higher cooling rates. In the limit of extremely rapid cooling, the solid can be amorphous, corresponding to a crystallite size of one or just a few atoms. The technique of rapid solidification is discussed in Chapter W21.

The fact that $x_s > x$ in the two-phase region of $Cu_{1-x}Ni_x$ alloys indicates that the solid α phase is always enriched in Ni atoms relative to the overall alloy composition x. This behavior is consistent with the fact that Ni has a higher melting temperature than Cu, indicating that Ni atoms are more stable than Cu atoms in the solid state. It follows that the liquid phase will be enriched in Cu atoms since $x_l < x$. This enrichment of the solid phase as a result of solidification is the basis of the float-zone purification techniques used to purify Si and other materials, as described in Chapter W21.

Referring once more to Fig. 6.6, note that there is a qualitative difference between the regions labeled α or l and the region $\alpha + l$. In the single-phase α or l regions, alloys can be prepared with any value of x and at any temperature T. There are thus two degrees of freedom available in the system. In the $\alpha + l$ region, however, once T is fixed, the possible values of x are also fixed at x_s and x_l. There is thus only one degree of freedom available (e.g., T). Gibbs put forth a rule, derived in Chapter W6, for determining the number of degrees of freedom F that a thermodynamic system can have. The number F is determined by the number of components, C, and the number of phases, P, by the following relationships:

$$F = \begin{cases} C - P + 1 & \text{if pressure is held fixed} \\ C - P + 2 & \text{if pressure can be varied.} \end{cases} \tag{6.30}$$

Thus in Fig. 6.6 in the single-phase α or l regions, $C = 2$ (Cu and Ni), $P = 1$ (α or l), pressure is held fixed, so $F = 2 - 1 + 1 = 2$. Thus both T and x can be varied independently in these single-phase regions. In the $\alpha + l$ region, however, $C = 2$ (Cu and Ni), $P = 2$ (α and l), the pressure is again held fixed, so $F = 2 - 2 + 1 = 1$. Thus only T or x can be varied independently; that is, if T is fixed in the $\alpha + l$ region, the compositions of the α and l phases are also fixed at x_s and x_l, respectively.

The binary solid-solution alloys $A_{1-x}B_x$ discussed here are called *ideal* when the two components mix in a completely random way in both the solid and liquid phases. This occurs when there is no preference for the bonding of A atoms, for example, either with other A atoms or with B atoms. The same statement can also be made for B atoms. In terms of NN chemical bonds, in an ideal solid solution the energy of an A–B bond is the average of the energies of the A–A and B–B bonds:

$$E(\text{A–B}) = \frac{E(\text{A–A}) + E(\text{B–B})}{2} = \langle E \rangle. \tag{6.31}$$

Under these circumstances the bond reaction

$$\text{A–A} + \text{B–B} \leftrightarrow 2\text{A–B} \tag{6.32}$$

will be neither exothermic nor endothermic and the enthalpy of mixing ΔH_{mix} (i.e., the change in enthalpy when mixing occurs), will be zero when the two components A and B are mixed. The driving force for mixing will then be due to the entropy of mixing through the term $T\Delta S_{\text{mix}}$ in the Gibbs free energy of mixing $\Delta G_{\text{mix}} = \Delta H_{\text{mix}} - T\Delta S_{\text{mix}}$ (assuming that the lattice enthalpy and entropy do not change upon mixing).

Note that for an ideal solid solution the Gibbs free energy is

$$G^{\text{id}}(x) = G^{M}(x) - T\Delta S_{\text{mix}}(x). \tag{6.33}$$

Here $G^{M}(x)$ is the Gibbs free energy of a mechanical mixture of the two components [i.e., $G^{M}(x) = (1-x)G_A + xG_B$, where G_A and G_B are the Gibbs free energies of the pure components]. Since the A and B atoms are bonded completely randomly in an ideal solution, it follows for $N = N(\text{A}) + N(\text{B})$ that (see Section 6.7 for the details of the calculation of the entropy of mixing ΔS_{mix})

$$\Delta S_{\text{mix}}(x) = k_B \ln W = -Nk_B[(1-x)\ln(1-x) + x\ln x]$$
$$= -k_B[N(A)\ln a_A + N(B)\ln a_B]. \tag{6.34}$$

Here $N(\text{A}) = xN$ and $N(\text{B}) = (1-x)N$ are numbers of atoms and $a_A = 1 - x$ and $a_B = x$ are the activities of the two components. Since the quantity in brackets in Eq. (6.34) is always negative, $G^{\text{id}}(x) < G^{M}(x)$ for all x and the alloy will thus be more stable as a single solution than as a phase-separated mixture.

For an ideal binary solid-solution alloy, the phase diagram (i.e., the locus of points of the liquidus and solidus curves), can be predicted from thermodynamics, using as input the melting points T_m and the enthalpies or latent heats of fusion ΔH_m of the pure components at T_m. The curves predicted for the Cu–Ni alloy system are shown in Fig. 6.6 along with the experimental curves. The agreement is reasonable but indicates that Cu–Ni alloys are in fact not completely ideal.

In practice, no binary solid-solution alloy can ever be completely ideal since the A and B atoms cannot be identical in every respect chemically and physically and still be different atoms. In the same sense, no gas, no matter how dilute, is ever completely ideal. In real, nonideal solutions, $E(\text{A–B})$ is no longer equal to $\langle E \rangle$ and the bonding in the solid will therefore no longer be perfectly random, especially at low temperatures, where the randomizing effects of the $T \Delta S_{mix}$ term are weaker. Two situations can then occur: When $E(A-B) > \langle E \rangle$, there will be a tendency for compound formation to occur and when $E(A-B) < \langle E \rangle$, there will be a tendency for phase separation or clustering of like atoms to occur. In both cases the effects of entropy will still be to randomize the bonding. These two situations are discussed in more detail in Section 6.7, where order–disorder transitions in alloys are described.

Regular solutions are a type of nonideal solution in which the excess Gibbs free energy of solution ΔG^{xs}, defined by $\Delta G^{xs} = G^S - G^{id}$, can be approximated by the excess enthalpy ΔH^{xs} (i.e., the excess entropy ΔS^{xs} can be neglected). Here G^S is the Gibbs free energy of the regular solution, and G^{id} for an ideal solution has been defined in Eq. (6.33). It follows that

$$G^S = G^M + \Delta H^{xs} - T \Delta S_{mix}. \tag{6.35}$$

In a regular solution the activities a_A and a_B of the two components are in general no longer equal to the corresponding fractional concentrations as was the case in Eq. (6.34). In this case activity coefficients γ_A and γ_B are instead defined by $a_i = \gamma_i x_i$, where γ_i can be greater than or less than 1. The reader is referred to the text by Gaskell (1995) for further discussions of activity coefficients in nonideal or regular solutions. The assumptions used to define regular solutions can be expected to be valid in close-packed solids such as metals.

A method that can be used to calculate the excess enthalpy ΔH^{xs} of regular solutions is the *quasichemical* approach, in which an interaction energy or "bond" energy $E(X-Y)$ is assigned to every pair of NN atoms X and Y in the solid. Thus

$$\Delta H^{xs} = \Sigma_i N_i E_i(X-Y), \tag{6.36}$$

where N_i is the number of atom pairs with energy E_i. Improvements on this pair approximation are made by considering larger clusters of atoms (e.g., tetrahedra). The properties of the alloy are then expressed as cluster expansions.[†] The application of the quasichemical approach to order–disorder phase transitions in alloys is presented in Section 6.7.

Binary Eutectic Alloys. When the two elements in a binary alloy are sufficiently different either chemically or physically, they will not form homogeneous solid solutions for all compositions. This occurs when the Gibbs free energy versus composition curve has two minima, in which case the coexistence of two distinct phases will be more stable than a single, homogeneous phase. In this situation a simple *binary eutectic phase diagram* can result. Eutectic phase diagrams include a triple point known as the *eutectic point*, where, upon cooling, a liquid phase undergoes an isothermal, reversible transformation directly into a heterogeneous mixture of two solid phases. An example is shown in Fig. 6.8 for the Pb–Sn alloy system, where six distinct regions are present. In this phase diagram, three distinct phases are present, one liquid, l, and two solid solution phases, α and β, which are called *terminal solid solutions*. The solid α phase is Pb-rich and has the FCC crystal structure of Pb, while the solid β phase is Sn-rich and has the tetragonal crystal structure of metallic white Sn. In addition to three

[†] D. de Fontaine, *Solid State Phys.*, **34**, 73 (1979); **47**, 33 (1994).

single-phase regions, the three phases combine to yield three distinct two-phase regions: $\alpha + l$, $\beta + l$, and $\alpha + \beta$. Solidus and liquidus lines are present, separating $\alpha + l$ from single-phase α and single-phase l, as well as separating $\beta + l$ from single-phase β and from single-phase l. The lines separating α and β from the $\alpha + \beta$ two-phase region are called *solvus lines*. The relative amounts of the phases present in these two-phase regions can be determined by using the lever law given in Eqs. (6.28) and (6.29).

The eutectic point in the Pb–Sn phase diagram at 73.9 at % Sn (61.9 wt % Sn) and $T_e = 183°C$ is the only point where liquid and the two solid phases α and β are in equilibrium. This point occurs at the *eutectic temperature* T_e and corresponds to the lowest solidification temperature in the alloy system. When liquid is cooled to the eutectic point, it solidifies directly into the α and β phases, with compositions of 29 and 98.5 at % Sn, respectively. According to the Gibbs phase rule there are no degrees of freedom remaining at this triple point since $P = 3$, $C = 2$, and the pressure is fixed (i.e., $F = 2 - 3 + 1 = 0$). The compositions of all three phases are therefore fixed at the values given above. As the temperature is lowered further, the equilibrium solid phases become even more Pb- and Sn-rich via diffusion.

Figure 6.9 illustrates a schematic binary AB phase diagram with various labels superimposed on it. The labels refer to specific compositions (letters a, b, . . . , g) and specific temperatures (numbers 1, 2, 3, 4). Thus the sequence a1–a2–a3–a4 denotes a sample of pure A whose temperature is being lowered from the liquid phase, past the melting temperature and into the solid A phase. The sequence b1–b2–b3–b4 passes from the l phase, past the liquidus line into the $\alpha + l$ coexistence region, past the solidus line into the α region and finally past the solvus line into the eutectic region labeled $\alpha + \beta$. The c1–c2–c3–c4 sequence passes from l through the liquidus curve to $\alpha + l$, then through the eutectic line to $\alpha + \beta$. The sequence d1–d2–d3–d4, corresponding to the eutectic composition, passes from the l phase through the eutectic point directly into the $\alpha + \beta$ region. The sequence e1–e2–e3–e4 is analogous to c1–c2–c3–c4, with the role of the α phase being replaced by the β phase. Similarly, f1–f2–f3–f4 is analogous to b1–b2–b3–b4, and g1–g2–g3–g4 to a1–a2–a3–a4.

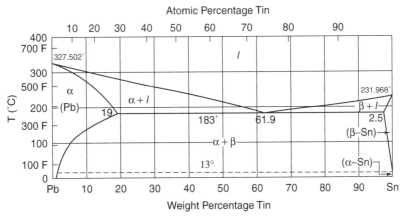

Figure 6.8. Pb–Sn eutectic phase diagram. (From *Metals Handbook*, 8th ed., Vol. 8, *Metallography, Structures and Phase Diagrams*, ASM International, Materials Park, Ohio, 1973, p. 330, Pb-Sn.)

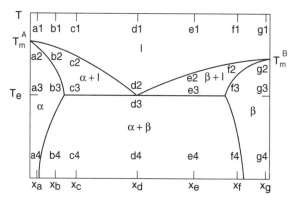

Figure 6.9. Points on a binary eutectic phase diagram corresponding to various compositions (a–g) and temperatures (1–4).

Figure 6.10 illustrates the phase compositions of alloys of the various compositions at different temperatures. The key to the phases is found on the left of the diagram. The symbols "α" and "β" represent the solid phases α and β. The liquid is represented by "l" and the eutectic composition by "e." In all cases a to g, the starting temperature corresponds to the liquid phase. Thus composition a1 starts with pure liquid A. When the melting temperature T_{mA} is passed, liquid A freezes into the α solid phase. Compositions a2, a3, and a4 are below the melting temperature. In case b the liquid with a minority component B is cooled starting at point b1. At b2 some α phase has solidified. When the solidus line is crossed, the remaining liquid solidifies into the α phase. Upon further cooling the $\alpha + \beta$ phase is entered.

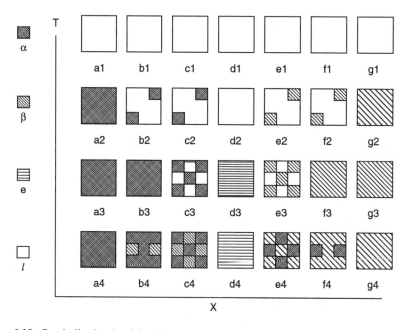

Figure 6.10. Symbolic sketch of the phase separation at various compositions and temperatures.

Case c begins with liquid c1, crosses the liquidus, and comes to the $\alpha + l$ region. Crystallites of the α phase form at c2. These grow in size and number as the temperature c3 is reached. When the eutectic line is crossed, the liquid freezes in the eutectic $\alpha + \beta$ phase, illustrated in c4. Case d is unique in that the liquid has precisely the eutectic composition. Points d1 and d2 lie in the liquid phase, whereas points d3 and d4 are the eutectic solid. Case e begins with the liquid e1, crosses the liquidus, and starts forming microcrystals of phase β along with l at point e2. These crystals grow until the eutectic line is about to be crossed, e3. Finally, in e4, the eutectic solid is formed. Case f is analogous to case b, with the roles of α and β interchanged. Similarly, case g is analogous to case a, with B replacing A.

The eutectic temperature T_e can be much lower than the melting points T_m of either of the pure components. The Pb–Sn system shown in Fig. 6.8 is one example, with $T_m(\text{Pb}) = 327°\text{C}$, $T_m(\text{Sn}) = 232°\text{C}$, and $T_e = 183°\text{C}$. Thus eutectic or near-eutectic Pb–Sn alloys are useful low-melting-point alloys which are commonly used as solders. An even more dramatic example is the Au–Si alloy system, which has a eutectic point at 18.6 at % Si and $T_e = 370°\text{C}$, as shown in the phase diagram presented in Chapter W12. The melting points T_m of pure Au and Si are 1063 and 1414°C, respectively. The instability of the solid phase in Au–Si alloys is likely to be connected to the relatively weak bonding between Si atoms and Au atoms since the former prefer covalent bonding and the latter prefer metallic bonding. Au-Si eutectic alloys, composed of heterogeneous mixtures of almost pure Au, with the FCC crystal structure, and almost pure Si, with the diamond crystal structure, are readily formed at low temperatures when Au wires are welded to Si, thus making this alloy useful in semiconductor technology.

In practice, the two-phase $\alpha + \beta$ eutectic region is rarely homogeneous: that is, the α and β solid phases are not completely mixed on an atomic or even on a microscopic level, for the same reasons as described for binary solid-solution alloys. In the $\alpha + \beta$ region the two phases can have a variety of microstructures, depending on how the phases nucleate and grow. In such a heterogeneous system the minimization of the Gibbs free energy associated with interfaces between the two phases will be critical. An interesting example of this phenomenon for Pb–Sn alloys is the lamellar microstructure shown in Fig. 6.11. This microstructure can result from atomic diffusion over relatively short distances. Another example of a lamellar microstructure that occurs in binary Fe–C alloys is pearlite, a mixture of alternating layers of cementite, Fe_3C, and ferrite (see Chapter 21, where the synthesis and processing of steels is discussed).

An important example of phase separation in a binary or multicomponent system is the phenomenon of *spinodal decomposition*, which can take place in a single-phase region when a solubility or miscibility gap appears as the temperature is lowered. The phase boundary of this two-phase region that lies within the single-phase region is called the *solvus*. In this case the two new phases have the same crystal structures but different compositions, then those of the single-phase alloy or matrix present above the miscibility gap. Since, in general, a surface energy barrier exists for nucleation of particles of the new phases, the single-phase alloy can remain metastable until these nucleation energy barriers are overcome. For a certain range of compositions within the spinodal region, however, phase separation can occur spontaneously as the temperature is lowered, without a nucleation barrier. This can occur because small fluctuations of composition within this range lower the Gibbs free energy of the system. The rate of

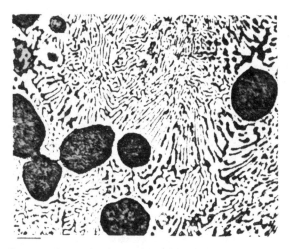

Figure 6.11. Lamellar eutectic matrix observed in a Pb-50 wt % Sn eutectic alloy, with Sn-rich regions (light) and Pb-rich regions, including dendritic grains (dark). The line at the lower left represents 25 μm. (From *ASM Handbook*, 9th ed., Vol. 9, *Metallography and Microstructures*, ASM International, Materials Park, Ohio, 1985, p. 422.)

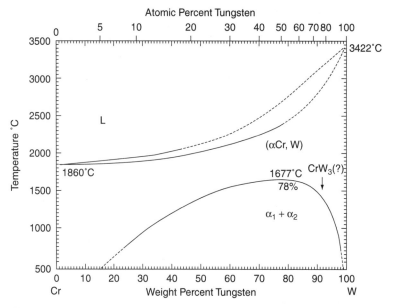

Figure 6.12. Phase diagram for Cr–W. The region of spinodal decomposition lies below the miscibility gap which opens up at 50 at % W and $T = 1677°C$. (From *ASM Handbook*, Vol. 3, *Alloy Phase Diagrams*, ASM International, Materials Park, Ohio, 1992, p. 2.162, Cr-W.)

the spinodal decomposition transformation can be quite rapid but is, of course, still controlled by interdiffusion of the components of the system.

Examples of systems exhibiting miscibility gaps and spinodal decomposition include the Au–Ni, Au–Pt, and Cr–W binary alloy systems, the ThO_2–ZrO_2, $Na_{1-x}K_xAlSi_3O_8$, and NaCl–LiCl systems, and also Cu–Ni alloys (e.g., $Cu_{0.7}Ni_{0.3}$), containing small amounts of Cr or Sn. In the latter case, the phase separation resulting from the addition of Cr to Cu–Ni solid-solution alloys increases their strength by impeding the motion of dislocations. In this case, the phase transition is referred to as *spinodal hardening*. The phase diagram for Cr–W is shown in Fig. 6.12, where it can be seen that the miscibility gap at 50 at % W opens up at $T = 1677°C$.

Additional examples of binary phase diagrams are presented in Chapters 11 to 18, where different classes of materials are discussed. In particular, the Fe–Fe$_3$C phase diagram in which a eutectoid point exists is discussed in detail in Chapter 21. Below the eutectoid temperature T_e, the austenitic γ-phase transforms to a mixture of the ferritic α-phase and cementite phase.

STRUCTURAL PHASE TRANSITIONS

In this section a survey of phase transitions that result in changes in the structure of solid materials is presented. The focus is on structural phase transitions such as melting and order–disorder transitions which are thermally activated (i.e., for which the driving force is the change of temperature). Structural phase transitions often occur during the synthesis and processing of materials, as described in Chapter 21. This section will therefore serve as an introduction to this important subject. The melting/solidification phase transition is discussed first, followed by a survey of structural phase transitions occurring in solids. The order–disorder transition in a binary alloy is then discussed and will serve as an example of a transition that involves changes in the microstructure rather than actual changes in the crystal structure or composition of the alloy.

6.6   Melting

Melting can be understood as a change of phase resulting from the complete breakdown of long-range order in a solid as the temperature is raised due to the increasing amplitude of vibration of the atoms about their equilibrium positions. In addition, changes in the short-range order (SRO) which reflect changes in chemical bonding can also occur at the melting temperature T_m. A dramatic example of a change in SRO at T_m is the "collapse" of the Si crystal structure with its fourfold coordination to a higher coordination of about 6 in the liquid state, reflecting a breakdown of the covalent bonding. This change of SRO results in an increased packing fraction in liquid Si and a higher rather than lower liquid density compared to solid Si. Another example of a solid whose volume contracts upon melting is ice.

Any discussion of melting should include both the macroscopic (i.e., thermodynamic), and microscopic aspects of the transformation. In general, T_m is an intrinsic, macroscopic property of a solid determined by the strength of the interaction potential (i.e., by the strength of the bonding). The melting temperatures of the elements are presented in Table 6.1. The initiation of melting and the rate at which the transition occurs are often controlled by extrinsic lattice defects such as surfaces, grain boundaries, and dislocations.

Thermodynamics of Melting. As for any phase transition occurring under equilibrium conditions, the Gibbs free energies of the solid, G_s, and liquid, G_l, phases

TABLE 6.1 Melting Points T_m (K) and Enthalpies of Fusion (or Melting) ΔH_m (kJ/mol) of the Elemental Solids at Atmospheric Pressure

1	2	3	4	5	6	7	8	9	10	11	12	13	14	15	16	17	18
1 H_2 13.9 0.12																	2 He
3 Li 453.7 3.00	4 Be 1560 7.90											5 B 2348 50.2	6 C diamond ≈3820 105	7 N_2 63.2 0.71	8 O_2 54.5 0.44	9 F_2 53.6 0.51	10 Ne 24.7 0.34
11 Na 371 2.60	12 Mg 923 8.48											13 Al 933.5 10.71	14 Si 1687 50.21	15 P^a 317 0.66	16 S 388 1.72	17 Cl_2 171.5 6.40	18 Ar 83.6 1.12
19 K 336.5 2.33	20 Ca 1115 8.54	21 Sc 1814 14.1	22 Ti 1941 14.15	23 V 2183 21.5	24 Cr 2180 21.0	25 Mn 1519 12.91	26 Fe 1811 13.81	27 Co 1768 16.2	28 Ni 1728 17.48	29 Cu 1358 13.26	30 Zn 693 7.32	31 Ga 302.9 5.59	32 Ge 1211 36.9	33 As^b 1090 24.44	34 Se 494 6.69	35 Br_2 266 10.57	36 Kr 115.9 1.37
37 Rb 312.5 2.19	38 Sr 1050 7.43	39 Y 1795 11.4	40 Zr 2128 21.0	41 Nb 2750 30	42 Mo 2896 37.48	43 Tc 2430 33.29	44 Ru 2607 38.59	45 Rh 2237 26.59	46 Pd 1828 16.74	47 Ag 1235 11.30	48 Cd 594 6.19	49 In 430 3.28	50 Sn 505 7.03	51 Sb 904 19.87	52 Te 723 17.49	53 I_2 350 15.52	54 Xe 161.5 1.81
55 Cs 301.6 2.09	56 Ba 1000 7.12	57 La 1191 6.20	72 Hf 2506 27.2	73 Ta 3290 36.57	74 W 3695 52.31	75 Re 3459 60.43	76 Os 3306 57.85	77 Ir 2719 41.12	78 Pt 2042 22.17	79 Au 1337 12.55	80 Hg 234.3 2.29	81 Tl 577 4.14	82 Pb 601 4.77	83 Bi 544.6 11.30	84 Po 527	85 At 575 —	86 Rd 202 —
87 Fr 300 —	88 Ra 973 7.15	89 Ac 1324 14.2															

58	59	60	61	62	63	64	65	66	67	68	69	70	71
Ce 1071 5.46	Pr 1204 6.89	Nd 1294 7.14	Pm 1441 12.6	Sm 1347 8.62	Eu 1095 9.21	Gd 1586 10.0	Tb 1629 10.79	Dy 1685 11.06	Ho 1747 17	Er 1802 19.9	Tm 1818 16.84	Yb 1092 7.66	Lu 1936 22

90	91	92	93	94	95	96	97	98	99	100	101	102	103
Th 2023 13.81	Pa 1845 12.34	U 1408 9.14	Np 917 3.20	Pu 913 2.82	Am 1449 14.39	Cm 1173 —	Bk 1323 —	Cf 1173 —	Es 1133 —	Fm 1800 —	Md 1100 —	No 1100 —	Lr 1900 —

Source: Most data from D. R. Lide, ed.-in-chief, *CRC Handbook of Chemistry and Physics*, 79th ed. (CRCnetBASE 1999), CRC Press, Boca Raton, Fla., 1999, pp. 6–116 to 6–121.

[a] White phosphorus.
[b] As under pressure (As sublimes at atmospheric pressure).

are equal at T_m. There are, however, discontinuous changes at T_m in the entropy S, volume V, and enthalpy H of the solid (Fig. 6.13). Melting and the reverse process of solidification are therefore first-order phase transitions. The solid-to-liquid transition corresponds to an increase in the entropy by an amount $\Delta S_m = S_l(T_m) - S_s(T_m)$ and an increase in the enthalpy by $\Delta H_m = H_l(T_m) - H_s(T_m)$. The quantity ΔH_m is the *latent heat* for the transition. Since $\Delta G_m = G_l(T_m) - G_s(T_m) = 0$, it follows that

$$\Delta H_m = T_m \, \Delta S_m. \tag{6.37}$$

The fact that $S_l(T_m) > S_s(T_m)$ is a result of increased disorder, both configurational and vibrational, in the liquid phase, while the increase in enthalpy results from the higher internal energy (but lower cohesive or bonding energy) in the liquid phase. It is clear that it is the $-TS$ term in the free energy (i.e., the entropy), that stabilizes the liquid phase above T_m.

The melting transition is an example of a *first-order phase transition*, defined as one in which the molar volume v_m and the entropy S of the material change discontinuously at the transition temperature or critical temperature T_c. According to the Ehrenfest classification of phase transitions, a phase transition has the same order as the derivatives of the Gibbs free energy, which show discontinuous changes at T_c. Thus the first derivatives of the Gibbs free energy $S = -(\partial G/\partial T)_P$ and $V = (\partial G/\partial P)_T$ have discontinuities at a first-order phase transition. First-order phase transitions also have latent heats $\Delta H = T_c \Delta S$. In a *second-order phase transition*, such as the transition between the superconducting and normal states, there are no discontinuities in the volume, the entropy, or in the order parameter of the new phase (i.e., the density of superconducting electrons). There are, however, discontinuities in second derivatives of the Gibbs free energy [e.g., in the specific heat $c_P = -T(\partial^2 G/\partial T^2)_P$], at a second-order phase transition. V. L. Ginzburg and L. Landau (G-L) developed a theory for the description of second-order phase transitions. The G-L theory is discussed briefly in Chapter 16, where its application to the superconducting phase transition is described, and also in Chapter W15, where it is applied to ferroelectric phase transitions.

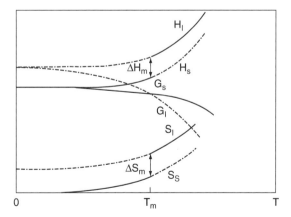

Figure 6.13. Schematic variations of the Gibbs free energy G, the entropy S, and the enthalpy H of the liquid and solid phases of a material above and below the melting temperature T_m. (From R. J. Borg and G. J. Dienes, *The Physical Chemistry of Solids*, copyright 1992. Reprinted by permission of Academic Press, Inc.)

Microscopic Aspects of Melting. An understanding of the microscopic or atomic-level processes involved in melting and solidification is still incomplete. Whereas quantitative, accurate models for the structure and dynamics of the solid state are available, the same cannot be said of the liquid state where the available models are at best qualitative. A useful phenomenological model for the melting transition has been provided by Lindemann.[†] According to Lindemann, melting is due to an intrinsic instability of the lattice which results when the amplitudes of the thermal vibrations of the atoms exceed a critical fraction f_c of the NN interatomic distance d (i.e., $\sqrt{\langle u^2 \rangle} = f_c d$).

If the lattice potential is assumed to be harmonic, the average potential energy of an atom vibrating with this critical amplitude is given by

$$\langle U \rangle = \frac{C(f_c d)^2}{4}. \tag{6.38}$$

Here $C = m\omega^2$ is an effective spring constant, m is the atomic mass, and ω is an angular frequency of vibration which can be identified with an average phonon frequency. It should be noted that this is just an estimate, since a truly harmonic crystal will never melt. The total vibrational energy $\langle E \rangle$ of the atom is the sum of the kinetic and potential contributions and is therefore given by

$$\langle E \rangle = \langle K \rangle + \langle U \rangle = 2\langle U \rangle = \frac{m\omega^2 (f_c d)^2}{2}. \tag{6.39}$$

Since T_m is always greater than the Debye temperature Θ_D, it is reasonable to use the classical result for a three-dimensional harmonic oscillator that $\langle E \rangle = 3k_B T_m$. It follows that

$$T_m = \frac{m\omega^2 (f_c d)^2}{6k_B} = \frac{m k_B \Theta_D^2 (f_c d)^2}{6\hbar^2} \tag{6.40}$$

where $\hbar\omega = k_B \Theta_D$ has been used.

A test of this prediction, known as the *Lindemann criterion*, for C (diamond), Si, and Ge is presented in Table 6.2, where $d = \sqrt{3}a/4$, a being the lattice constant. The values obtained for f_c for C, Si, and Ge are similar, as expected for materials with the same crystal structure and bonding. The critical amplitude $\sqrt{\langle u^2 \rangle}$ for these crystals is therefore predicted to be only 8 to 9% of the NN distance d.

TABLE 6.2 Test of the Lindemann Criterion

Crystal	T_m (at $P = 1$ atm) (K)	m (amu)	Θ_D (K)	a (nm)	f_c
C	≈3820	12	2340	0.357	0.084
Si	1687	28.1	645	0.543	0.087
Ge	1211	72.6	374	0.566	0.078

Source: Data from D. R. Lide, ed., *CRC Handbook of Chemistry and Physics*, 75th ed., CRC Press, Boca Raton, Fla., 1994, p. 12–87.

[†] F. A. Lindemann, *Phys. Z.*, **11**, 609 (1910).

When static disorder due to extrinsic defects is present in a material in addition to thermal disorder, it has been proposed that a crystalline-to-amorphous solid-state transition can take place at a temperature below the T_m value for the defect-free bulk material. The transition is proposed to occur in a disordered material when the sum of the dynamic (i.e., thermal), and static root-mean-square atomic displacements $\sqrt{\langle u^2 \rangle_{th}} + \sqrt{\langle u^2 \rangle_{st}}$ reaches a critical value $f_c d$ identical to that for melting of the perfect crystal in which only thermal disorder is present. This represents a generalization of the Lindemann criterion for the melting of perfect crystals and can apparently account for a wide range of observations of the solid-state amorphization of materials.

Other microscopic mechanisms that have been suggested for melting include:

1. The mechanical or elastic instability model of Born[†] in which the resistance of the crystal to shear stress vanishes at T_m
2. The spontaneous generation of intrinsic lattice defects such as vacancies and arrays of dislocations, as in the theories of Cahn[‡] and others

These models all assume that melting occurs homogeneously throughout the bulk of the solid as opposed to heterogeneous melting, which is initiated at a surface or other extended defect.

One useful current approach for understanding melting involves the use of computer-based molecular dynamics simulations which predict the thermally activated motions of atoms using realistic lattice potentials. Molecular dynamics simulations of the melting of Si and of Cu have investigated the effects of surfaces and interfaces on melting and on the kinetics of the melting process.[§] These simulations involve the solution of Newton's equations of motion for the atoms, thereby obtaining their positions and velocities as functions of time. Simulations for both Si and Cu show convincingly that melting at T_m is initiated at grain boundaries and free surfaces rather than in the bulk. The melted regions then propagate into the bulk of the crystal. Typical propagation velocities, on the order of 10 to 100 m/s, are found to increase with temperature, indicating that melting is a relatively slow process controlled by the thermally activated diffusion of atoms.

These simulations also show that, in the absence of extended defects, homogeneous melting in the bulk occurs at a temperature T_s that is about 40% above T_m for Si and about 20% above T_m for Cu. The melting that occurs at T_s is referred to as *mechanical melting*, as opposed to the *thermodynamic melting* that occurs at T_m. Mechanical melting apparently corresponds to the type of mechanical instability of the lattice proposed by Born.

The phenomena of *superheating* and *supercooling* can also occur during the melting–solidification process. Supercooling of liquids can readily be observed due to the kinetic limitations associated with the formation of a critical nucleus of the solid phase as the liquid is cooled below T_m. A *critical nucleus* of solid is one that is large enough to continue growing rather than decaying. The nucleation of a critical solid nucleus in liquids is inhibited by surface energy effects related to the fact that atoms

---

[†] M. Born, and K. Huang, *Dynamical Theory of Crystal Lattices*, Oxford University Press, London, 1962.
[‡] R. W. Cahn, *Nature*, **323**, 668 (1986).
[§] S. R. Philpotts, S. Yip, and D. Wolf, *Comput. Phys.*, Nov./Dec. 1989, p. 20.

near the surface of a solid nucleus are more weakly bound than those in the bulk. The nucleation and growth of solids is discussed in more detail in Chapter 21. Superheating of the solid above T_m, on the other hand, is rarely observed, since nucleation sites for the liquid phase are readily available at the free surfaces of the crystal or at other extended defects.

As indicated earlier, the melting of the bulk of a solid at T_m is often initiated at its surface. In fact, the surface of a bulk solid can undergo "melting" at temperatures below T_m for the bulk. This effect is consistent with the surface having a lower Debye temperature Θ_D than that of the bulk (see Chapter 19). *Surface melting* is an intrinsic effect, not related to impurities or temperature gradients, which corresponds to the appearance at the surface of a disordered or liquidlike film. This surface film wets the solid–vapor interface, and its thickness diverges at T_m. The details of surface melting are strongly dependent on the orientation of the external surface of the solid, with close-packed surfaces showing less of a tendency to become disordered. The criterion for surface melting is that the solid–vapor interface free energy per unit area σ_{sv} be greater than the sum $\sigma_{sl} + \sigma_{lv}$ of the solid–liquid and liquid–vapor interface free energies per unit area.

Melting and Cohesion. It is interesting to compare the enthalpy change associated with melting, ΔH_m, with the enthalpy of cohesion, ΔH_c, of a solid. For this purpose, selected values of T_m, $\Delta H_m(T_m)$, and ΔH_c (0 K) are presented in Table 6.3 for a wide variety of solids. Whereas the values of T_m, ΔH_m, and ΔH_c show considerable variation due to the wide range of bonding strengths found in different materials, the ratio $\Delta H_m/\Delta H_c$ is much more constant. The magnitude of this ratio indicates that ΔH_m is only a few percent of the total cohesive energy stored in the solid, thus indicating that a significant amount of cohesion or bonding remains in the liquid phase. The relatively high value, 0.111, of this ratio for Si indicates that a significant change of bonding occurs in the liquid state, as discussed earlier, while the low value, 0.0052, for SiO_2 indicates that the bonding between Si and O atoms in the liquid phase differs only slightly from the bonding found in solid SiO_2.

Values of the entropy change $\Delta S_m = \Delta H_m/T_m$ (in units of k_B, Boltzmann's constant) indicate that a significant increase in disorder occurs during the melting of Si, Ge, and GaSb ($\Delta S_m \approx 3$ to $4\ k_B$), whereas the reverse is true for the melting of SiO_2 ($\Delta S_m = 0.56\ k_B$). For the metals listed in Table 6.3, ΔS_m is in the range 0.8 to $1.7\ k_B$.

Also shown is the ratio $3N_A k_B T_m/\Delta H_m$, which shows that the thermal energy $3N_A k_B T_m$ associated with melting is higher than the enthalpy of melting ΔH_m by about a factor of 3 for metals. Here N_A is Avogadro's number.

6.7   Solid-State Phase Transitions

A survey of solid-state structural phase transitions, including some that involve changes in composition, is presented next. Since the interest here is in structural transitions occurring under equilibrium or near-equilibrium conditions, it is appropriate to focus initially on some general observations involving thermodynamics. At absolute zero, where entropy effects are absent, the stable structures of a solid and its surface will be the ones that have the lowest enthalpy or energy. Whether the stable bulk structure is an open one, such as the diamond crystal structure, or has a high packing fraction,

TABLE 6.3   Melting and Cohesion Data for Crystals

Crystal	T_m (at $P = 1$ atm) (K)	ΔH_m (at T_m) (kJ/mol)	ΔH_c (at $T = 0$ K) (kJ/mol)	$\dfrac{\Delta H_m}{\Delta H_c}$	$\dfrac{3N_A k_B T_m}{\Delta H_m}$
Na	371	2.60	107.8	0.024	3.56
Mg	923	8.48	145.9	0.058	2.71
Al	933	10.71	327.6	0.033	2.17
Cu	1358	13.26	336.2	0.039	2.55
Pt	2042	22.17	564.4	0.039	2.30
V	2183	21.5	511.0	0.042	2.53
Mo	2896	37.48	656.6	0.057	1.93
W	3695	52.31	848.1	0.062	1.76
Si	1687	50.21	451.3	0.111	0.84
Ge	1211	36.9	369.2	0.100	0.82
GaSb[a]	976	25.10	352.9	0.071	0.97
NaCl[b]	1074	28.16	638.3	0.044	0.95
SiO$_2$[c]	1996	9.58	≈1850	0.0052	5.19

Source: Data from NBS Tables of Chemical Thermodynamic Properties, *J. Phys. Chem. Ref. Data,* **11**, Suppl. 2 (1982); JANAF Thermochemical Tables, *J. Phys. Chem. Ref. Data,* **14**, Suppl. 1 (1985); and D. R. Lide and H. V. Kehiaian, eds., *CRC Handbook of Thermophysical and Thermochemical Data,* CRC Press, Boca Raton, Fla., 1994.

[a]Value of ΔH_c (0 K) is for solid GaSb relative to Ga atoms and Sb$_2$ dimers.
[b]Value of ΔH_c (0 K) is for solid NaCl relative to neutral Na and Cl atoms, while the value $\Delta H_c(0$ K$) =$ 786 kJ/mol given in Table 2.8 is relative to Na$^+$ and Cl$^-$ ions.
[c]β-Cristobalite.

such as the FCC or HCP crystal structures, depends on the nature of the bonding in the solid. In general, the lowest energy state is close packed when the bonding is metallic.

When the temperature is raised above absolute zero, entropy effects become significant and often lead to phase transitions to more open, less tightly bound, or more disordered structures. The temperature at which transitions occur and the structure of the new phase or phases depend critically on a wide range of factors which influence both the energy and the entropy of the system.

It is clearly not feasible to discuss or even list here all possible types of structural transitions that can occur in solids. A list of some of the important classes of transitions is presented in Table 6.4. *Martensitic (displacive) transitions* are nonthermally activated (i.e., athermal), transformations. The martensitic transition occurring in Fe–C alloys is discussed in Chapters W12 and W21. Spinodal decomposition transitions were mentioned briefly in Section 6.4. Structural transitions on surfaces involving reconstruction and roughening are discussed in Chapter 19.

To illustrate some of the principles that determine the nature of solid-state phase transitions, a description of the order–disorder transition in a binary alloy is given next.

Order–Disorder Transitions. *Order–disorder transitions* correspond to rearrangements of the positions of atoms, the orientations of molecules, or the directions of spins relative to the crystalline lattice or axes. They do not involve changes in crystal structure or composition. In general, an ordered state that exists at low temperatures

TABLE 6.4 **Characteristics of Solid-State Structural Phase Transitions**

Type of Transition	Change of Crystal Structure?	Change in Composition?	Examples
Due to change of temperature	Yes	No	Ti: HCP $\leftrightarrow$ BCC SiO_2: $\alpha \leftrightarrow \beta$
Pressure-induced	Yes	No	Fe: BCC $\leftrightarrow$ HCP Graphite $\leftrightarrow$ diamond
Order–disorder	No	No	CuZn (β-brass) NH_4Cl
Martensitic	Yes	No	Fe–Fe_3C
Spinodal decomposition	No	Yes	Au–Ni Cr–W ThO_2–ZrO_2 NaCl-LiCl
Phase segregation	Possible	Yes	Dispersion strengthening in steels, precipitation hardening in Al alloys

undergoes a transition to a disordered state as the temperature is raised and the effects of entropy become more important.

The specific order–disorder transition to be considered here involves a binary $A_{1-x}B_x$ alloy with a BCC crystal structure and the distribution of A and B atoms on the two simple-cubic sublattices of the BCC lattice. In the BCC crystal structure the sites on one cubic sublattice, α, have as NNs eight sites on the other sublattice, β, and vice versa. This is also the case for the CsCl crystal structure in which Cs^+ ions are surrounded by eight NN Cl^- ions, and vice versa (see Fig. 1.22). Order and disorder in solids are also discussed in Chapter 4 and are illustrated in Fig. 4.4. Only the equiatomic case with $x = \frac{1}{2}$ and with N atoms of type A and also N atoms of type B present are considered here. The expressions obtained for $x = \frac{1}{2}$ can be generalized readily to arbitrary composition x. The numbers of A atoms on the α and β sublattices are $N(A,\alpha)$ and $N(A,\beta)$, respectively. With analogous definitions for $N(B, \alpha)$ and $N(B, \beta)$, it follows that

$$N(A) = N = N(A,\alpha) + N(A,\beta),$$
$$N(B) = N = N(B, \alpha) + N(B, \beta). \tag{6.41}$$

It is important to distinguish between two types of order that may be present in the crystal: short range and long range. The LRO parameter s is zero when there are equal numbers of A and B atoms randomly arranged on each of the sublattices [i.e., when $N(A,\alpha) = N(B, \alpha) = N(A,\beta) = N(B, \beta) = N/2$]. This is the case in an ideal binary solid-solution alloy, as discussed in Section 6.5. Any fluctuation away from $s = 0$, (i.e., from a perfectly random arrangement of A and B atoms), signifies that some degree of LRO is present. In general, s is defined so that

$$N(\text{A},\alpha) = \frac{(1+s)N}{2},$$

$$N(\text{A},\beta) = \frac{(1-s)N}{2}, \tag{6.42}$$

from which it follows that

$$N(\text{B},\alpha) = N - N(\text{A},\alpha) = \frac{(1-s)N}{2},$$

$$N(\text{B},\beta) = \frac{(1+s)N}{2}. \tag{6.43}$$

Note that $N(\text{A},\beta) = N(\text{B},\alpha)$ and $N(\text{A},\alpha) = N(\text{B},\beta)$ for the situation under consideration here, in which $x = \frac{1}{2}$ and $N(\text{A}) = N(\text{B})$. When $s = \pm 1$, the LRO is perfect since each sublattice then contains only one type of atom.

Whether the alloy is in the perfectly ordered state or in a disordered state is determined by competition between the effects of enthalpy (i.e., internal energy), and entropy. The Gibbs free energy of mixing is given by

$$\Delta G_{\text{mix}} = G^S - G^M = \Delta H_{\text{mix}} - T\Delta S_{\text{mix}}, \tag{6.44}$$

where G^S and G^M are defined in Eq. (6.35) and the enthalpy of mixing is $\Delta H_{\text{mix}} = \Delta H^{\text{xs}}$. The standard approach for analyzing nonideal or regular solid solutions assumes that the enthalpy resides entirely in the bonds between NN atoms (i.e., results from NN interactions only). This approach is known as the *quasichemical model* (Gordon 1968, Chaps. 4 and 5), as noted earlier in Section 6.5. The enthalpy ΔH_{mix} will be minimized when the total energy associated with the NN A–A, A–B, and B–B bonds is maximized. When there are $N(\text{A–A})$ bonds with energy $E(\text{A–A})$, $N(\text{A–B})$ bonds with energy $E(\text{A–B})$, and $N(\text{B–B})$ bonds with energy $E(\text{B-B})$, the total bond energy is given by

$$E(\text{bonds}) = N(\text{A–A})E(\text{A–A}) + N(\text{A–B})E(\text{A–B}) + N(\text{B–B})E(\text{B–B}). \tag{6.45}$$

Since bond energies are defined here to be positive, the bonds make a negative contribution to the enthalpy [i.e., $\Delta H_{\text{mix}} = -E(\text{bonds})$].

The number $N(\text{A–A})$ of A–A bonds is determined by the product of three factors: the number $N(\text{A},\alpha)$ of A atoms on the α sublattice, the probability $P(\text{A},\beta) = N(\text{A},\beta)/N$ of finding an A atom on any given β site, and the number z of sites on the β sublattice that are NNs of each α site. Using similar arguments for $N(\text{A–B})$ and $N(\text{B–B})$ and also Eqs. (6.41) to (6.43), it follows that

$$N(\text{A–A}) = 2(1 - s^2)N,$$

$$N(\text{B–B}) = 2(1 - s^2)N, \tag{6.46}$$

$$N(\text{A–B}) = 4(1 + s^2)N.$$

The enthalpy of mixing is therefore

$$\Delta H_{\text{mix}} = -2N[E(\text{A–A}) + E(\text{B–B}) + 2E(\text{A–B})]$$

$$- 2Ns^2[2E(\text{A–B}) - E(\text{A–A}) - E(\text{B–B})] \tag{6.47}$$

The factor $2E(A-B) - E(A-A) - E(B-B)$ appearing above is defined to be the *interaction parameter* Ω. Note that Ω represents the "activation" energy for the bond reaction in which one A–A bond and one B–B bond are broken and two A–B bonds are formed (i.e., $A-A + B-B \leftrightarrow 2A-B$). If $\Omega > 0$, then A–B bonds are favored over A–A and B–B bonds and an ordered state will be formed at low T, corresponding to compound formation. If $\Omega < 0$, A–A and B–B bonds are preferred instead, leading to separation into A-rich and B-rich phases as T is lowered. If $\Omega = 0$, the alloy will be ideal and will in principle be disordered for all T since there is then no energetic preference for A–B bonds over A–A and B–B bonds.

The entropy of mixing (i.e., the configurational entropy associated with the distribution of A and B atoms on the two sublattices), is $\Delta S_{\text{mix}} = k_B \ln W$. Here W is the number of possible configurations of $N(A,\alpha)$ and $N(B, \alpha)$ atoms on the N sites of the α sublattice and of $N(A,\beta)$ and $N(B, \beta)$ atoms on the N sites of the β sublattice. Therefore,

$$\Delta S_{\text{mix}} = k_B \ln \left[\frac{N!}{N(A,\alpha)!N(B, \alpha)!} \frac{N!}{N(A,\beta)!N(B, \beta)!} \right]. \tag{6.48}$$

Taking $N \gg 1$ and using Stirling's approximation, it can be shown that

$$\Delta S_{\text{mix}} = 2N k_B \ln 2 - N k_B[(1 + s) \ln(1 + s) + (1 - s) \ln(1 - s)]. \tag{6.49}$$

Note that ΔS_{mix} as defined is zero and the alloy is completely ordered when $s = \pm 1$.

The equilibrium state of the AB alloy is that for which the Gibbs free energy of mixing ΔG_{mix} is minimized, in this case with respect to the LRO parameter s. Taking $\partial \Delta G_{\text{mix}} / \partial s = 0$, one obtains

$$-4Ns\Omega + Nk_B T \ln \frac{1 + s}{1 - s} = 0. \tag{6.50}$$

Note that if $\Omega = 0$, then s is also zero. The solution for s varies smoothly with T, indicating a second-order phase transition (Fig. 6.14), and goes to zero at the *critical*

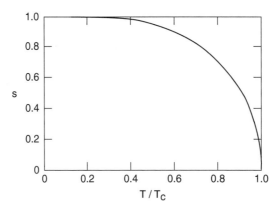

Figure 6.14. Variation of the long-range order parameter s with reduced temperature T/T_c according to Eq. (6.50).

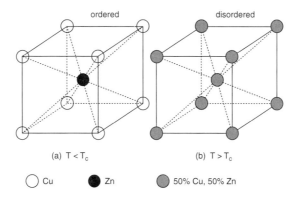

Figure 6.15. Distributions of atoms in both the ordered $(T < T_c)$ and disordered $(T > T_c)$ phases of a binary alloy such as CuZn.

temperature T_c, given by

$$T_c = \frac{2\Omega}{k_B}. \tag{6.51}$$

It is clear that Ω must be positive in order for a transition to an ordered alloy (i.e., compound), to occur at T_c. The distribution of the atoms between the two sublattices in both the ordered $(T < T_c)$ and disordered $(T > T_c)$ phases is indicated in Fig. 6.15. Examples of the type of equiatomic alloy considered here are CuZn (i.e., β-brass), and NiTi. The alloy CuZn undergoes a transition to an ordered CsCl crystal structure below $T_c \approx 460°C$ (Fig. 6.15a). Above this temperature the alloy is a random BCC solid solution (Fig. 6.15b).

Some interesting features of the order–disorder transition can occur for alloys with $x = 0.75$ (i.e., AB$_3$, an example of which is AuCu$_3$). For such alloys the transition is first order, involving discontinuous changes in both the enthalpy and entropy. As a result, the ordering transition for AuCu$_3$ has an energy barrier requiring thermal fluctuations in order to be crossed. The ordered phase of an equiatomic AB alloy like that discussed here, however, nucleates without an energy barrier since a second-order transition is involved.

Although LRO disappears and $s = 0$ for $T > T_c$, SRO can still be present. The SRO parameter σ is defined in terms of the average number $n(A-B) = N(A-B)/N$ of A–B bonds at any site. For a perfectly ordered equiatomic alloy in which every A atom is surrounded by eight B atoms, and vice versa, $n(A-B) = 8$. For the disordered case, however, $n(A-B) = 4$. The definition of σ is

$$\sigma = \frac{n(A-B) - n^r(A-B)}{n^o(A-B) - n^r(A-B)} = \frac{n(A-B) - 4}{4}, \tag{6.52}$$

where $n^r(A-B)$ and $n^o(A-B)$ correspond to completely random and completely ordered bonding, respectively. From this definition it can be seen that $\sigma = 1$ for perfect SRO while $\sigma = 0$ for complete short-range disorder. For the equiatomic alloy under consideration here, SRO can exist above T_c even though LRO is completely absent. The existence of SRO above T_c is not a static phenomenon, but rather, corresponds to a fluctuating NN configuration, which nevertheless, has a value of $n(A-B) > 4$.

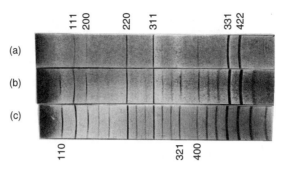

Figure 6.16. X-ray powder diffraction pattern for $AuCu_3$: (a) disordered (heated to $T = 425°C > T_c$ and then quenched; the so-called superlattice reflections disappear and the diffuse background scattering is high); (b) partially ordered (heated to $T = 375°C$); (c) highly ordered (heated at $T = 350°C$ for 50 h; the superlattice reflections due to the existence of LRO appear). [From C. Sykes et al, *J. Inst. Met.*, **58**, 255 (1936). Reprinted by permission of the Institute of Materials.]

The degree of LRO present in alloys such as those described here can be determined quantitatively using x-ray diffraction techniques, as described in Chapter W22.

Near T_c the mean-field type of approach used here to describe the order–disorder transition breaks down due to large fluctuations that occur in the LRO in the alloy. These fluctuations are typically correlated over a characteristic distance ξ, known as the *correlation length* (or *coherence length*). The fluctuations in LRO become increasingly important as $T \to T_c$ since ξ diverges [i.e., $\xi \propto 1/(T - T_c)^\nu$, $\nu > 0$]. Near T_c concepts known as scaling and the renormalization group are used to explore the physical properties of the system. These concepts and their implementation lie outside the scope of this book.

The structure factor of the unit cell depends on whether the alloy is ordered or disordered. Below T_c, superlattice reflections are observed in the x-ray powder diffraction pattern due to the existence of LRO. When heated above T_c and then quenched, the superlattice reflections disappear and the diffuse background scattering increases, as shown in Fig. 6.16 for $AuCu_3$.

REFERENCES

Borg, R. J., and G. J. Dienes, *The Physical Chemistry of Solids*, Academic Press, San Diego, Calif., 1991.

Gaskell, D. R., *Introduction to the Thermodynamics of Materials*, 3rd Ed., Taylor & Francis, Philadelphia, 1995.

Gordon, P., *Principles of Phase Diagrams in Material Systems*, McGraw-Hill, New York, 1968.

Grove, A. S., *Physics and Technology of Semiconductor Devices*, Wiley, New York, 1967.

Hansen, M., *Constitution of Binary Alloys*, 2nd ed., McGraw-Hill, New York, 1958. Elliott, R. P., *Constitution of Binary Alloys, First Supplement*, McGraw-Hill, New York, 1965. Shunk, F. A., *Constitution of Binary Alloys, Second Supplement*, McGraw-Hill, New York, 1969.

Kittel, C., and H. Kroemer, *Thermal Physics*, 2nd ed., W.H. Freeman, San Francisco, 1980.

Le Claire, A. D., *Diffusion*, in, N. B. Hannay, ed., *Treatise on Solid State Chemistry*, Vol. 4, Plenum Press, New York, 1976.

Massalski, T. B., *Metall. Trans.*, **20A**, 1295 (1989).

Rao, C. N. R., and K. J. Rao, *Phase Transitions in Solids*, McGraw-Hill, New York, 1978.

Zemansky, M. W., and R. H. Dittman, *Heat and Thermodynamics*, 6th ed., McGraw-Hill, New York, 1981.

PROBLEMS

6.1 Consider a slice of solid of thickness Δx in which the concentration gradient dC/dx for a given element is not constant in space. As a result of diffusion there is a net flux of atoms J_1 into the slice on one side and a net flux J_2 out of the slice on the opposite side. By conservation of matter the rate of change of the concentration $\partial C/\partial t$ in the slice is given by $(J_1 - J_2)/\Delta x$. Use Fick's first law, Eq. (6.3), to derive Fick's second law, Eq. (6.8).

6.2 Derive the equivalent of Eq. (6.13) for a BCC crystal structure and show that the result is the same as for an FCC crystal structure.

6.3 Given a simple-cubic lattice with a vacancy initially at $\mathbf{R} = 0$. Assume that after every time interval τ the vacancy hops randomly to a NN site. Find the mean-square displacement $\langle X^2 \rangle$ after time t, where $t \gg \tau$.

6.4 In a binary alloy $A_{1-x}B_x$ show that the weight fractions of the two components are given by

$$w_A = \frac{(1-x)W_A}{(1-x)W_A + xW_B} \quad \text{and} \quad w_B = \frac{xW_B}{(1-x)W_A + xW_B},$$

where W_A and W_B are the corresponding atomic weights.

6.5 For the case of completely random bonding in an $A_{1-x}B_x$ alloy, the number of distinct atomic configurations is $W = N!/N(A)!N(B)!$, with $N = N(A) + N(B)$, $N(A) = (1-x)N$, and $N(B) = xN$.

   (a) Show that $\Delta S_{mix}(x) = k_B \ln W = -k_B[x \ln x + (1-x) \ln(1-x)]$. (*Hint:* Use Stirling's approximation, $\ln N! \approx N \ln N - N$, valid for $N \gg 1$.)

   (b) Plot the term in the brackets in part (a) for $0 < x < 1$ and find the composition where ΔS_{mix} is a maximum.

6.6 Using the Gibbs phase rule, determine the number of degrees of freedom F available in the binary AB alloy system at each of the 28 points a1, a2, and so on, shown in Fig. 6.9.

6.7 For the solids listed in Table 6.3, calculate the entropy changes ΔS_m using the data given and Eq. (6.37). State which solids have the greatest and smallest changes in ΔS_m and explain why this may be so.

6.8 Derive the expressions for the numbers of bonds $N(A-A)$, $N(B-B)$, and $N(A-B)$ given in Eq. (6.46). Show that the total number of bonds is equal to $8N$ and explain why the factor of 8 appears here.

6.9 Using Eq. (6.42), derive the following expression for the long-range order parameter s,

$$s = \frac{N(A,\alpha) - N(A,\beta)}{N(A,\alpha) + N(A,\beta)}.$$

Describe the long-range order in the alloy when $s = +1$, 0, and -1.

Note: Additional problems are given in Chapter W6.

CHAPTER 7

Electrons in Solids: Electrical and Thermal Properties

7.1 Introduction

The electrical properties of materials are determined largely by the response of the electrons to external fields. The central theme of this chapter is electrical conductivity and how it varies across different classes of materials. The classical Drude model and how it attempts to explain conductivity and the Hall effect form the starting point. While providing an explanation of Ohm's law and Joule heating, Drude theory leaves open the question of why there is such a large difference in conductivity between different elements of the periodic table.

The Sommerfeld theory for the free-electron gas is then introduced. This treats the electrons quantum mechanically but ignores the presence of ions. It is applied to the calculation of the specific heat and the thermopower.

The focus then changes to the quantum theory of crystalline solids. Bloch's theorem is derived. A quantum-mechanical analysis is given in several limiting cases, such as the nearly free electron case and the tight-binding case. A qualitative discussion of the differences among metals, insulators, semiconductors, and semimetals is presented. Expressions for the density of electronic states are derived. The quantum theory of solids is then applied to a study of the temperature dependence of the resistivity of metals.

Following this is a section devoted to semiconductors. The valence and conduction energy bands are defined and the bandgap energy and effective-mass tensors are introduced. An application of band theory is made to calculation of the magnetoresistance of semiconductors.

Other phenomena occurring in semiconductors and insulators are considered. Variable-range hopping is discussed from the viewpoints of Mott and Efros-Shklovskii. Electrical conductivity in strong electrical fields is considered in the section on the Poole–Frenkel effect.

Some materials exhibit a radical change in conductivity as physical parameters are varied. In this chapter we discuss granular metals that are insulators until the concentration of the metallic inclusions is high enough to cause a percolation transition.

The remainder of the chapter concerns conduction in reduced-dimensional spaces. A discussion of conduction in carbon nanotubes based on the tight-binding method is presented. This is followed by the Landauer theory of one-dimensional conductance.

Additional topics covered at our Web site[†] include further material on the Onsager relations, the random tight-binding approximation, the Kronig–Penney model, the Hall effect in band theory, electron localization, and the evaluation of Fermi integrals.

CLASSICAL THEORY OF ELECTRICAL CONDUCTION

When a constant electric field is established in some materials, current flows. Assuming a local relationship, one writes $\mathbf{J}(\mathbf{r}) = \sigma \mathbf{E}(\mathbf{r})$, where $\mathbf{J}$ is the current per unit area (current density), σ the electrical conductivity and $\mathbf{E}$ the electric field. For an isotropic medium, σ is a scalar and the vectors $\mathbf{J}$ and $\mathbf{E}$ are parallel. For an anisotropic material σ becomes a tensor and these vectors need no longer be parallel. The microscopic form of Ohm's law states that σ is independent of the electric field. Its magnitude determines whether the material is a conductor (high σ), an insulator (low σ), or a semiconductor (intermediate σ). The difference in the value of σ among materials can be enormous, varying by more than 20 orders of magnitude in going from a good conductor to a good insulator. At first sight it is a mystery how, simply by changing the atomic number a few units in the periodic table, one can observe such a huge variation in a physical parameter. The answer to the mystery will involve understanding how tightly (relative to $k_B T$) the valence electrons are bound to the atoms and ions. We begin this section by studying metals and then proceed to semiconductors and insulators.

7.2   Drude Theory

An early attempt at a theory of Ohm's law was made by Drude, who applied Newtonian mechanics to study the motion of electrons through a metal. The electrons that are not tightly bound to atoms or ions (i.e., the "conduction" electrons), are accelerated by an applied electric field and are assumed to collide with "scatterers," which deflect them and randomize their velocities to a thermal distribution. Consider a sequence of collisions labeled by the index j. Let the velocity of an electron just after the jth collision be $\mathbf{v}_j$. Between collisions j and $j + 1$ the velocity is

$$\mathbf{v}(t) = \mathbf{v}_j - \frac{e\mathbf{E}}{m}(t - t_j), \tag{7.1}$$

since the acceleration is given by $-e\mathbf{E}/m$. The probability of surviving to time t without making a collision and then making a collision between times t and $t + dt$ is

$$dP = e^{-(t-t_j)/\tau}\frac{dt}{\tau}, \tag{7.2}$$

where τ is the mean time between collisions. The mean velocity between times t_j and t_{j+1} is

$$\langle \mathbf{v}_j(t) \rangle = \int \mathbf{v}(t)dP = \int_{t_j}^{\infty} \left[\mathbf{v}_j - \frac{e\mathbf{E}}{m}(t - t_j) \right] e^{-(t-t_j)/\tau}\frac{dt}{\tau}. \tag{7.3}$$

[†] Supplementary material for this textbook is included on the Web at the resource site (ftp://ftp.wiley.com/public/sci_tech_med/materials). Cross-references to elements of the Web material are prefixed by "W."

The direction of $\mathbf{v}_j$ is random and so will average to zero for all the conduction electrons. The term $-e\mathbf{E}\tau/m$ is the *drift velocity* $\langle \mathbf{v} \rangle$. The average current density is

$$\mathbf{J} = -ne\langle \mathbf{v}(t) \rangle = \frac{ne^2\tau}{m}\mathbf{E} = ne\mu\mathbf{E} \qquad (7.4)$$

where n is the number of conduction electrons per unit volume. The parameter $\mu = \langle v \rangle / E = e\tau/m$ is called the *mobility* of the electron. Thus the *Drude conductivity formula* is obtained:

$$\sigma = \frac{ne^2\tau}{m}. \qquad (7.5)$$

The *Joule heating formula* may be derived similarly by examining the kinetic energy loss in a typical collision:

$$\Delta K = \frac{m}{2}\mathbf{v}_{j+1}^2 - \frac{m}{2}\left[\mathbf{v}_j - \frac{e}{m}\mathbf{E}(t_{j+1} - t_j)\right]^2. \qquad (7.6)$$

On the average $\langle \mathbf{v}_{j+1}^2 \rangle = \langle \mathbf{v}_j^2 \rangle$ and $\langle \mathbf{v}_j \rangle = 0$, so

$$\langle \Delta K \rangle = -\frac{e^2\tau^2 E^2}{m}. \qquad (7.7)$$

Since the mean time between collisions is τ, the power produced per unit volume is

$$\mathscr{P} = -n\frac{\langle \Delta K \rangle}{\tau} = \sigma E^2. \qquad (7.8)$$

Use has been made of the integral $\int t^2\, dP = 2\tau^2$, with dP given by Eq. (7.2). This power is dissipated as heat through the production of phonons or other excitations of the metal.

Representative values for the dc electrical conductivity of metals are given in Table 7.1. The electron density, n, is computed from the relation $n = N_A v\rho_m/A$, where

TABLE 7.1   Parameters of Some Metals at $T = 295$ K

Metal	Atomic Number A	Valence z	Mass Density ρ_m $(10^3\ \text{kg/m}^3)$	Electron Density n $(10^{29}\ \text{m}^{-3})$	Conductivity σ $[10^6(\Omega \cdot \text{m})^{-1}]$	Collision Time τ $(10^{-15}\ \text{s})$
Ag	107.9	1	10.5	0.585	62.1	37.6
Al	26.98	3	2.70	1.81	36.5	7.17
Ba	137.3	2	3.59	0.315	2.6	2.93
Be	9.012	2	1.82	2.43	30.8	4.50
Ca	40.08	2	1.53	0.460	27.8	21.5
Cd	112.4	2	8.65	0.927	13.8	5.29
Cs	132.3	1	2.00	0.091	5.0	19.5
Cu	63.55	1	8.93	0.846	58.8	24.7
In	114.8	3	7.29	1.15	11.4	3.53
K	39.10	1	0.91	0.140	13.9	35.2
Li	6.939	1	0.54	0.469	10.7	8.11
Mg	24.31	2	1.74	0.862	22.3	9.18

N_A is Avogadro's number, z the valence, ρ_m the mass density, and A the atomic weight. Note that the SI unit for conductivity is $(\Omega \cdot m)^{-1}$. The values of the collision times computed from Eq. (7.5) are displayed. The ac conductivity of metals is covered in Section 8.3.

Without collisions σ and τ are infinite and the metal is a perfect conductor. This state may actually be achieved in superconductors, the subject of Chapter 16.

7.3   Hall Effect in Metals

A conductor that carries an electric current in the presence of a transverse magnetic field develops a potential difference across the sample perpendicular to both the current and the magnetic field. This is the *Hall effect*, and the potential difference produced is the Hall voltage, V_H. By measuring V_H it is possible to measure directly the product nq, the mobile carrier density multiplied by the charge of the carrier.

Figure 7.1 depicts a situation in which a voltage V is impressed across a rectangular parallelipiped of metal of dimensions L by w by h (assume that $L \gg w$ and $L \gg h$). At steady state a current I exists and a longitudinal current density $J = I/hw$ is established in response to the electric field $E = V/L$. Ohm's law relates the two quantities: $J = \sigma E$.

From Eq. (7.4) it follows that the magnitude of the current density is proportional to the drift velocity of the carriers, $\langle v \rangle$ (i.e., $J = nq\langle v \rangle$). Figure 7.1 is drawn for the case of carriers of negative charge, (e.g., electrons). A magnetic force on the carrier, $F_B = q\langle v \rangle B$, acts vertically downward. If this were the only force acting, negative charge would accumulate on the lower face, and to preserve charge neutrality, positive charge would collect on the upper face. These charge sheets create the Hall electric field E_H, which is directed downward. Hence there is an upward electrostatic force of magnitude $F_H = qE_H$ which will continue to grow until F_H and F_B equilibrate, giving $E_H = \langle v \rangle B$. The resulting Hall voltage is $V_H = E_H h$. Thus, finally, an expression for nq in terms of macroscopically measurable quantities is obtained:

$$nq = \frac{IB}{wV_H} \equiv -\frac{1}{R_H}, \tag{7.9}$$

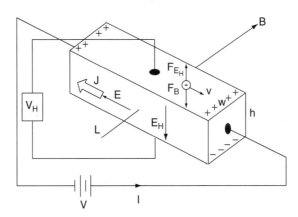

Figure 7.1. Geometry of a Hall effect measurement.

TABLE 7.2 **Comparison of Measured Hall Coefficients with the Free-Electron Theory Prediction**[a]

Metal	$-1/R_H ne$
Li	0.78
Na	0.99
K	1.00
Rb	1.08
Cs	1.11
Ag	1.19
Cu	1.37
Au	1.48
Al	$-1.00/3$
In	$-1.00/3$

[a]If the prediction were correct, $-1/R_H ne = 1$.

where R_H is called the *Hall coefficient*. Also, from the formulas $E = V/L$ and $I = V/R$, one obtains an expression for the resistance:

$$R = \frac{L}{\sigma h w},$$ (7.10)

which is independent of the strength of the magnetic field.

A comparison of measured values of R_H with the free-electron parameters is given in Table 7.2. For the alkali metals the quantitative agreement is reasonable. It is poorer for other good conductors and disagrees substantially in magnitude (and sign!) for In and Al. It is curious that if one allowed In and Al to have valence -1 rather than $+3$, the agreement with the free-electron theory would be restored. The theory is further frustrated by the observation that both R_H and ρ are often found to depend on the magnetic field. A proper accounting of these anomalies must await the quantum treatment of metals.

FREE-ELECTRON GASES

7.4 Sommerfeld Theory

Sommerfeld proposed a simple model treating a metal as a free-electron gas confined by the surfaces of the solid. The electrons, however, are treated using quantum mechanics. In the bulk of the material, the surface is disregarded altogether and the electrons are completely free. The free-electron model works best for low-valence metals, such as the alkalis (Li, Na, K, Rb, Cs), although it is often employed for others as well, including such valence 3 metals as Al. The ion-core potential is strongly screened by the valence electrons and the ions are unable to bind the valence electrons. (Note the self-consistency.) The valence electrons are free to wander about the solid. Recalling the fact that the average electric field in a conductor at equilibrium is zero, one can argue that the same screening effect acts to diminish the electron–electron interaction. Thus an independent-particle picture may be employed, and each electron is imagined to interact only with the constant background potential of the solid.

Quantum mechanics enters in two important ways. First, it defines the allowed eigenstates that the electrons can occupy. Second, since electrons obey Fermi–Dirac statistics, the Pauli exclusion principle applies (i.e., at most one electron may occupy a given eigenstate). These eigenstates are characterized by their wave vector and spin projection. The single-particle Hamiltonian is simply the free-electron kinetic energy operator, since the potential energy $V(\mathbf{r})$ is chosen by Sommerfeld to be zero,

$$H = \frac{p^2}{2m} = -\frac{\hbar^2}{2m}\nabla^2. \tag{7.11}$$

The eigenfunctions are plane waves multiplied by a two-element column vector specifying the spin state

$$\psi_{\mathbf{k},s}(\mathbf{r}) = e^{i\mathbf{k}\cdot\mathbf{r}}\chi_s. \tag{7.12}$$

The two independent spin states (up and down) are described by the column vectors (spinors)

$$\chi_+ = \begin{bmatrix} 1 \\ 0 \end{bmatrix}, \qquad \chi_- = \begin{bmatrix} 0 \\ 1 \end{bmatrix}. \tag{7.13}$$

The energy is independent of the spin and grows with increasing k:

$$E_{\mathbf{k},s} = \frac{\hbar^2 k^2}{2m}. \tag{7.14}$$

Suppose that there are N free electrons in a volume V. At $T = 0$ K, the ground state is obtained by filling the N lowest-lying energy levels. This implies a maximum value for occupied k's, called the *Fermi wave vector*, k_F. It is defined by

$$N = \sum_{\mathbf{k},s} \Theta(k_F - k) = 2 \int \frac{V d^3 k}{(2\pi)^3} \Theta(k_F - k)$$

$$= \frac{2V}{(2\pi)^3} \int_0^{k_F} 4\pi k^2 dk = V\frac{k_F^3}{3\pi^2}. \tag{7.15}$$

Here the unit step function $\Theta(k_F - k)$ imposes the restriction ($k < k_F$). Introducing the electron number density $n = N/V$, one finds that

$$k_F = (3\pi^2 n)^{1/3}. \tag{7.16}$$

The electrons have a range of energies extending from zero up to the Fermi energy, defined by

$$E_F = \frac{\hbar^2 k_F^2}{2m}. \tag{7.17}$$

The Fermi velocity is defined by $v_F = \hbar k_F/m$ and the Fermi temperature by $T_F = E_F/k_B$. The total electron energy is computed by integrating the kinetic energy over the Fermi sphere:

$$U = \sum_{\mathbf{k},s} E_{\mathbf{k},s}\Theta(k_F - k) = 2\int \frac{Vd^3k}{(2\pi)^3}\frac{\hbar^2k^2}{2m}\Theta(k_F - k) = \frac{3}{5}NE_F, \qquad (7.18)$$

from which it is seen that the average electron energy at $T = 0$ K is 60% of the Fermi energy. Some typical values of k_F, v_F, E_F, and T_F are presented in Table 7.3. Figure 7.2 depicts the Fermi sphere in **k** space.

At absolute zero the occupancy of any given state is either 1 or 0. At finite temperatures the occupancy is given by the Fermi–Dirac distribution $f(E_k, T) = [\exp(\beta(E_k - \mu)) + 1]^{-1}$ (see Appendix WB at our Web site) where $\beta = 1/k_BT$ and

TABLE 7.3   Free-Electron Parameters for Various Metals

Metal	k_F $(10^{10}\ \mathrm{m}^{-1})$	v_F $(10^6\ \mathrm{m/s})$	E_F (eV)	T_F $(10^3\ \mathrm{K})$
Ag	1.20	1.39	5.49	63.7
Al	1.75	2.03	11.7	135
Ba	0.977	1.13	3.64	42.2
Be	1.93	2.23	14.2	165
Ca	1.11	1.28	4.68	54.3
Cd	1.40	1.62	7.47	86.6
Cs	0.646	0.748	1.59	18.4
Cu	1.36	1.57	7.03	81.5
In	1.50	1.74	8.63	100
K	0.745	0.863	2.12	24.6
Li	1.12	1.29	4.74	55.0
Mg	1.37	1.58	7.11	82.5
Na	0.922	1.07	3.24	37.6
Rb	0.698	0.808	1.86	21.5
Sr	1.02	1.18	3.94	45.7
Zn	1.57	1.82	9.40	109

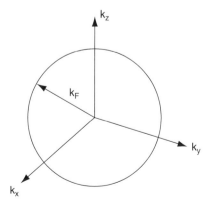

Figure 7.2. Fermi sphere in **k** space.

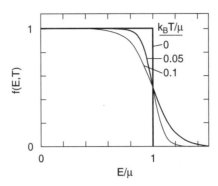

Figure 7.3. Fermi–Dirac distribution function $f(E, T)$ plotted for several values of T. Both $k_B T$ and E are given in units of μ, the chemical potential.

μ is the chemical potential of the electrons. A graph of this function is presented in Fig. 7.3 for several values of T. The size of the step width is seen to be on the order of $k_B T$. In place of Eq. (7.15), there is

$$N = \sum_{k,s} f(E_k, T) = 2 \int \frac{V d^3 k}{(2\pi)^3} \frac{1}{e^{\beta(E_k - \mu)} + 1}$$

$$= \frac{V}{2\pi^2} \left(\frac{2m}{\hbar^2} \right)^{3/2} \int_0^\infty \frac{E^{1/2}}{e^{\beta(E-\mu)} + 1} dE. \tag{7.19}$$

In place of Eq. (7.18), the internal energy is now given by

$$U = \sum_{k,s} E_{k,s} f(E_k) = 2 \int \frac{V d^3 k}{(2\pi)^3} \frac{\hbar^2 k^2}{2m} \frac{1}{e^{\beta(E_k - \mu)} + 1}$$

$$= \frac{V}{2\pi^2} \left(\frac{2m}{\hbar^2} \right)^{3/2} \int_0^\infty \frac{E^{3/2}}{e^{\beta(E-\mu)} + 1} dE. \tag{7.20}$$

One is interested in these formulas in the limit in which $E_F / k_B T \gg 1$, which is valid for metals, as may be seen from Table 7.3. The two required integrals are of the form

$$I_j(\beta, \beta\mu) = \int_0^\infty \frac{E^{j+1/2}}{e^{\beta(E-\mu)} + 1} dE, \tag{7.21}$$

with $j = 0$ and $j = 1$, respectively. One makes a power series in T (see Appendix W7A) and finds that

$$I_j(\beta, \beta\mu) = \frac{1}{(j + \frac{3}{2})\beta^{j+3/2}} \left[(\beta\mu)^{j+3/2} + \frac{\pi^2}{6} \left(j + \frac{3}{2} \right) \left(j + \frac{1}{2} \right) (\beta\mu)^{j-1/2} + \cdots \right].$$
$$\tag{7.22}$$

The Fermi energy may be written as

$$E_F = (\tfrac{3}{2} I_0)^{3/2}. \tag{7.23}$$

Solving for the chemical potential in terms of E_F and T gives

$$\mu = E_F \left[1 - \frac{\pi^2}{12} \frac{1}{(\beta E_F)^2} + \cdots \right]. \tag{7.24}$$

To order $(k_B T / E_F)^2$ the chemical potential and the Fermi energy are the same. The internal energy per unit volume, $u = U/V$, becomes

$$u = \frac{1}{5\pi^2} \left(\frac{2m}{\hbar^2} \right)^{3/2} E_F^{5/2} \left[1 + \frac{5\pi^2}{12} \left(\frac{k_B T}{E_F} \right)^2 + \cdots \right]. \tag{7.25}$$

The specific heat (at constant volume) per unit volume is

$$c_v = \frac{\partial u}{\partial T} = \frac{1}{6} \left(\frac{2m}{\hbar^2} \right)^{3/2} E_F^{1/2} k_B^2 T \equiv \frac{\pi^2 k_B^2}{3} \rho(E_F) T \equiv \gamma T. \tag{7.26}$$

[The significance of the quantity $\rho(E_F) = 3n/2E_F$ is discussed in Section 7.7.] Thus for metals there is an electronic contribution to the specific heat which is linear in T for $k_B T \ll E_F$. This is in addition to the lattice contribution discussed in Sections 5.5 and 5.6. The net result is that at low temperatures ($T < 10$ K) in metals,

$$c_v \approx \gamma T + A T^3. \tag{7.27}$$

A simple way of understanding the linear behavior is to recognize that most of the electrons in the Fermi sea cannot be thermally excited, since the states above them are occupied. Only those electrons within an energy band of width $\approx k_B T$ near the Fermi energy are capable of being excited to vacant states (see Fig. 7.3). Their number per unit volume is approximately $k_B T (\partial n / \partial E)|_{E_F} = 3n k_B T / 2E_F$. The typical excitation of electrons from these states will also be of order $k_B T$, so the electronic contribution to the internal energy has a term varying as $(k_B T)^2$, and hence one obtains a linear specific heat.

An example of the low-T specific heat of a metal is presented in Fig. 7.4 for Ag. The measured value of γ is approximately 10% larger than the free-electron prediction.

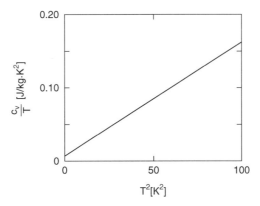

Figure 7.4. Ratio c_v/T plotted as a function of T^2 for Ag. (Data from CRC Handbook of Chemistry and Physics, 66th ed., R. C. Weast, ed., Boca Raton, Fla., 1985.)

This can be attributed to an enhancement of the free-electron mass due to interactions of the electron with the lattice and other electrons.

TRANSPORT THEORY

The description of conductivity in the Sommerfeld theory is facilitated by studying the motion of a collection of electrons in phase space, a six-dimensional space constructed from the three spatial and three momentum variables. The Boltzmann equation provides an alternative formulation for Newton's equations of motion used in Section 7.2. It also provides a framework in which other transport phenomena, such as thermal conductivity, may be studied. Details are presented in Section W7.1.

7.5   Onsager Relations

One may show that both an electric field $\mathbf{E}_0$ and a thermal gradient can produce an electric current density $\mathbf{J}$ and a heat current density $\mathbf{J}_Q$:

$$\mathbf{J} = \sigma \mathbf{E}_0 - \sigma S \nabla T, \tag{7.28}$$

$$\mathbf{J}_Q = \sigma S T \mathbf{E}_0 - \kappa' \nabla T. \tag{7.29}$$

These are the Onsager relations.

Integral expressions for the parameters are given in Section W7.1 and lead to

$$\sigma = \frac{ne^2 \tau(E_F)}{m}, \tag{7.30}$$

$$S = -\frac{\pi^2}{3} \frac{k_B^2 T}{e} \frac{\partial}{\partial E} \ln[E^{3/2} \tau(E)]|_{E=\mu}, \tag{7.31}$$

where $\tau(E)$ is the collision time for electrons with energy E and

$$\kappa' = \frac{\pi^2}{3} \frac{k_B^2}{e^2} \sigma T. \tag{7.32}$$

The parameter σ is the same as encountered in the Drude theory for the electrical conductivity, except that $\tau(E)$ is evaluated at the Fermi energy. Indeed, when T is constant, Eq. (7.28) reduces to Ohm's law, $\mathbf{J} = \sigma \mathbf{E}_0$. The electron mean free path is defined by $\lambda = v_F \tau(E_F)$ indicating that carriers at the Fermi energy dominate the electrical conductivity. Typically, v_F is much larger than the drift velocity by many orders of magnitude.

The parameter S is called the *thermopower*. When a thermal gradient is established in a material and there is no flow of current, $\mathbf{J} = 0$, Eq. (7.28) predicts that a thermoelectric field is induced, given by

$$\mathbf{E}_0 = S \nabla T. \tag{7.33}$$

Equation (7.31) shows that the thermopower increases linearly in the absolute temperature. For metals typical values of S are on the order of $1 \ \mu V/K$ at room temperature.

The thermopower can be an order of magnitude larger in transition metals, due to the contribution of d electrons to the conduction process. The sign and magnitude of S in a given metal are very sensitive to deviations from free-electron behavior.

The parameter κ' is related to the thermal conductivity, and Eq. (7.32) is known as the *Wiedemann–Franz law*. When no electrical current flows, Eqs. (7.28) and (7.29) may be combined to give

$$\mathbf{J_Q} = (\sigma S^2 T - \kappa')\nabla T \equiv -\kappa \nabla T \qquad (7.34)$$

where κ is the thermal conductivity. To a good approximation, $\kappa' \approx \kappa$. Equation (7.34) is Fourier's heat conduction formula. Equation (7.32) shows that the thermal conductivity grows linearly with T at low T. It also demonstrates why good electrical conductors will also be good thermal conductors. The same electrons responsible for transporting electrical charge also tranport thermal energy. The ratio $L = \kappa/\sigma T = \pi^2 k_B^2/3e^2$ called the *Lorenz number*, depends only on fundamental constants. Its numerical value is 2.44×10^{-8} W $\cdot$ Ω/K^2. The experimental values of L at $T = 273$ K for (Li, Na, K, Rb, Al, Ag, Au) are $(2.22, 2.12, 2.23, 2.42, 2.14, 2.31, 2.32) \times 10^{-8}$ W $\cdot$ Ω/K^2, in reasonable agreement with theoretical prediction.

For metals the thermal conductivity grows linearly with T at low T, peaks at intermediate T, and falls off at high T. The falloff at high T is due largely to the shortened mean free path of the electrons due to electron–phonon scattering.

THE QUANTUM THEORY OF SOLIDS

In going beyond the Sommerfeld theory one attempts to describe all aspects of a solid in quantum-mechanical terms. The scattering from the lattice ions must be treated properly. The wave character of the electrons inevitably leads to interference effects, since typical wavelengths of the electrons are on the order of the Fermi wavelength, $\lambda_F = 2\pi/k_F$ and are comparable to interatomic spacings. In crystalline solids the ordered array of ions can produce Bragg diffraction effects, which block the propagation of electrons in certain directions and/or with certain ranges of energies. In the following sections the quantum theory of solids is developed.

7.6 Bloch's Theorem

The treatment of a many-particle system such as a solid requires that a number of simplifying approximations be made before a concise mathematical description is possible. One begins with the independent-electron approximation, where it is assumed that each electron interacts with a potential $V(\mathbf{r})$. This potential includes the effect of the ion cores as well as the other electrons in the solid. The result is that one need only solve a one-electron Schrödinger equation to find a set of eigenvalues and eigenfunctions.

In this section attention is focused on the solution of the Schrödinger equation for a periodic solid, where the potential has translation symmetry

$$V(\mathbf{r} + \mathbf{R}) = V(\mathbf{r}), \qquad (7.35)$$

where the set of vectors $\{\mathbf{R}\}$ defines the Bravais lattice. Fourier-analyzing $V(\mathbf{r})$ gives

$$V(\mathbf{r}) = \sum_{\mathbf{G}} V_{\mathbf{G}} e^{i\mathbf{G}\cdot\mathbf{r}}, \tag{7.36}$$

where $\{\mathbf{G}\}$ are the reciprocal lattice vectors. The reality of $V(\mathbf{r})$ requires that the Fourier coefficients satisfy the "reality condition" $V_G^* = V_{-G}$, since

$$V^*(\mathbf{r}) = \sum_{\mathbf{G}} V_{\mathbf{G}}^* e^{-i\mathbf{G}\cdot\mathbf{r}} = \sum_{\mathbf{G}} V_{-\mathbf{G}}^* e^{i\mathbf{G}\cdot\mathbf{r}} = V(\mathbf{r}) = \sum_{\mathbf{G}} V_{\mathbf{G}} e^{i\mathbf{G}\cdot\mathbf{r}}. \tag{7.37}$$

The periodicity of $V(\mathbf{r})$ is obvious:

$$V(\mathbf{r} + \mathbf{R}) = \sum_{\mathbf{G}} V_{\mathbf{G}} e^{i\mathbf{G}\cdot(\mathbf{r}+\mathbf{R})} = \sum_{\mathbf{G}} V_{\mathbf{G}} e^{i\mathbf{G}\cdot\mathbf{r}} = V(\mathbf{r}), \tag{7.38}$$

where $\exp(i\mathbf{G}\cdot\mathbf{R}) = 1$ has been used.

The Schrödinger equation is

$$[H - E]\psi(\mathbf{r}) = \left[-\frac{\hbar^2}{2m}\nabla^2 + V(\mathbf{r}) - E \right]\psi(\mathbf{r}) = 0. \tag{7.39}$$

Note that H is unchanged if a translation through vector $\mathbf{R}$ is made. The wavefunction $\psi(\mathbf{r} + \mathbf{R})$ satisfies the same equation that $\psi(\mathbf{r})$ does, and so differs from it by at most a constant, that is,

$$\psi(\mathbf{r} + \mathbf{R}) = \tau_{\mathbf{R}}\psi(\mathbf{r}). \tag{7.40}$$

The quantity $\tau_{\mathbf{R}}$ must have magnitude 1. If it were greater than 1, repeated translations would make the wavefunction grow in magnitude in an exponential fashion, that is,

$$\psi(\mathbf{r} + N\mathbf{R}) = (\tau_{\mathbf{R}})^N \psi(\mathbf{r}), \tag{7.41}$$

and it would not be possible to normalize the wavefunction in the infinite solid limit. Similarly, if the modulus were less than 1, $\psi(\mathbf{r} - N\mathbf{R})$ would not be normalizable as $N \to \infty$. Thus $\tau_{\mathbf{R}} = \exp(i\theta_{\mathbf{R}})$ with $\theta_{\mathbf{R}}$ real. By compounding translations, one has

$$\tau_{\mathbf{R}_1}\tau_{\mathbf{R}_2} = \tau_{\mathbf{R}_1+\mathbf{R}_2}, \tag{7.42}$$

so

$$\theta_{\mathbf{R}_1} + \theta_{\mathbf{R}_2} = \theta_{\mathbf{R}_1+\mathbf{R}_2}, \tag{7.43}$$

which is satisfied by $\theta_{\mathbf{R}} = \mathbf{k}\cdot\mathbf{R}$ with $\mathbf{k}$ being a real vector. Thus

$$\psi(\mathbf{r} + \mathbf{R}) = e^{i\mathbf{k}\cdot\mathbf{R}}\psi(\mathbf{r}). \tag{7.44}$$

The function defined by

$$u(\mathbf{r}) = e^{-i\mathbf{k}\cdot\mathbf{r}}\psi(\mathbf{r}) \tag{7.45}$$

is seen to be a periodic function, since

$$u(\mathbf{r} + \mathbf{R}) = e^{-i\mathbf{k}\cdot(\mathbf{r}+\mathbf{R})}\psi(\mathbf{r} + \mathbf{R}) = e^{-i\mathbf{k}\cdot(\mathbf{r}+\mathbf{R})}e^{i\mathbf{k}\cdot\mathbf{R}}\psi(\mathbf{r}) = u(\mathbf{r}). \qquad (7.46)$$

Enlarging the notation, Bloch's theorem states that the solution of the Schrödinger equation may be factored into a plane wave multiplied by a periodic function $u_k(\mathbf{r})$ with the Bravais lattice periodicity:

$$\psi_\mathbf{k}(\mathbf{r}) = e^{i\mathbf{k}\cdot\mathbf{r}}u_\mathbf{k}(\mathbf{r}). \qquad (7.47)$$

The acceptable values of $\mathbf{k}$ may be determined by imposing periodic boundary conditions (see Section 5.1). For a solid of size N_1 by N_2 by N_3 atoms, a translation through $N_1\mathbf{u}_1$ should leave the wavefunction unchanged. Thus

$$e^{i\mathbf{k}\cdot\mathbf{u}_1 N_1} = 1, \qquad (7.48)$$

which has N_1 independent solutions of the form

$$\mathbf{k}_1 = \frac{j_1}{N_1}\mathbf{g}_1, \qquad j_1 = 0, 1, \ldots, N_1 - 1, \qquad (7.49)$$

where $\mathbf{g}_1 \cdot \mathbf{u}_1 = 2\pi$. More generally,

$$\mathbf{k} = \frac{j_1}{N_1}\mathbf{g}_1 + \frac{j_2}{N_2}\mathbf{g}_2 + \frac{j_3}{N_3}\mathbf{g}_3, \qquad j_n = 0, 1, \ldots, N_n - 1, \qquad (7.50)$$

where $\{\mathbf{g}_i\}$ are the primitive reciprocal lattice vectors [see Eq. (3.8).] Thus $\mathbf{k}$ is a point in the first Brillouin zone. The total number of such points is $N = N_1 N_2 N_3$, which is the number of lattice cells in the crystal.

Due to the periodicity of $u_\mathbf{k}(\mathbf{r})$ it may be expanded as a Fourier series:

$$u_\mathbf{k}(\mathbf{r}) = \sum_\mathbf{G} u_\mathbf{G}(\mathbf{k})e^{i\mathbf{G}\cdot\mathbf{r}}. \qquad (7.51)$$

Inserting this into the Schrödinger equation gives

$$\left[-\frac{\hbar^2}{2m}\nabla^2 + \sum_{\mathbf{G}'}V_{\mathbf{G}'}e^{i\mathbf{G}'\cdot\mathbf{r}} - E \right]\sum_\mathbf{G} u_\mathbf{G}(\mathbf{k})e^{i(\mathbf{G}+\mathbf{k})\cdot\mathbf{r}} = 0. \qquad (7.52)$$

Using

$$\sum_{\mathbf{G},\mathbf{G}'}V_{\mathbf{G}'}u_\mathbf{G} e^{i(\mathbf{G}+\mathbf{G}')\cdot\mathbf{r}} = \sum_{\mathbf{G},\mathbf{G}'}V_{\mathbf{G}'}u_{\mathbf{G}-\mathbf{G}'}e^{i\mathbf{G}\cdot\mathbf{r}}, \qquad (7.53)$$

this may be simplified to

$$\left[\frac{\hbar^2}{2m}(\mathbf{k}+\mathbf{G})^2 - E \right]u_\mathbf{G}(\mathbf{k}) + \sum_{\mathbf{G}'}V_{\mathbf{G}'}u_{\mathbf{G}-\mathbf{G}'}(\mathbf{k}) = 0, \qquad (7.54)$$

where use has been made of the linear independence of the functions $\exp(i\mathbf{G} \cdot \mathbf{r})$ in removing the $\mathbf{G}$ sum. This infinite set of coupled linear algebraic equations for $\{u_{\mathbf{G}}(\mathbf{k})\}$ may have many solutions, so the notation is expanded to $\{u_{n,\mathbf{G}}(\mathbf{k})\}$, where $n = 1, 2, 3, \ldots$ is called the *band-index*. The energy eigenvalues will also be labeled by this index, $E_n(\mathbf{k})$. The relations in Eq. (7.54) are referred to as *Bloch's difference equations*.

The condition for nontrivial solutions to Eq. (7.54) to exist is the vanishing of the determinant (called the *Hill determinant*):

$$\left\| \left[\frac{\hbar^2}{2m}(\mathbf{k} + \mathbf{G})^2 - E \right] \delta_{\mathbf{G},\mathbf{G}''} + V_{\mathbf{G}-\mathbf{G}''} \right\| = 0. \tag{7.55}$$

The roots of this equation determine the eigenvalue spectrum $\{E_n(\mathbf{k})\}$. This spectrum is invariant under the transformation $\mathbf{k} \rightarrow \mathbf{k} + \mathbf{K}$, where $\mathbf{K}$ is a reciprocal lattice vector. Making this substitution into the Hill determinant and letting $\mathbf{G}' = \mathbf{G} + \mathbf{K}$ and $\mathbf{G}'' = \mathbf{K}' - \mathbf{K}$ yields

$$\left\| \left[\frac{\hbar^2}{2m}(\mathbf{k} + \mathbf{G}')^2 - E \right] \delta_{\mathbf{G}',\mathbf{K}'} + V_{\mathbf{G}'-\mathbf{K}'} \right\| = 0. \tag{7.56}$$

The condition remains invariant under the transformation above. Hence the first Brillouin zone contains the entire energy spectrum. Other Brillouin zones simply contain replicas of this spectrum.

Truncating the Hill determinant by employing a finite set of $\mathbf{G}$ vectors and solving the resulting secular determinant for the roots $E = E_n(\mathbf{k})$ provides a method, in principle, for calculating band structures. Since the Fourier coefficients V_G fall off with increasing $\mathbf{G}$ for large $\mathbf{G}$, the energy eigenvalue spectrum will converge as the size of the determinant is increased. More practical methods exist, such as the Green's function [Korringa, Kohn, and Rostoker (KKR)] method, the augmented plane wave (APW) method, and the pseudopotential method, but these are beyond the scope of the present book. The books by Fletcher (1971) and Ashcroft and Mermin (1976) provide good introductions to modern band-structure computation techniques.

Figure 7.5 presents the results of an electronic band-structure calculation for diamond along various directions in the first Brillouin zone. The lowest band corresponds to $n = 1$, the next higher band to $n = 2$, and so on. One sees that there are two degenerate conduction bands and three degenerate valence bands at the Γ point. Diamond is an insulator. The Fermi level lies at *midgap*, so all states with $E < 0$ comprise the valence bands and are occupied. The conduction bands are empty. The bottom of the conduction band does not occur at the Γ point but rather, along the [100] direction. The bandgap is close to 6 eV, making diamond transparent to visible light.

7.7   Nearly Free Electron Approximation

In analyzing some metals the free-electron approximation forms a suitable starting point for explaining the band structure. In these materials the ion cores are largely screened by the valence electrons, leaving behind a weak potential that may be treated using perturbation theory. Examples of such materials are the alkali metals (Li, Na, K, Rb, Cs), Mg, and Al.

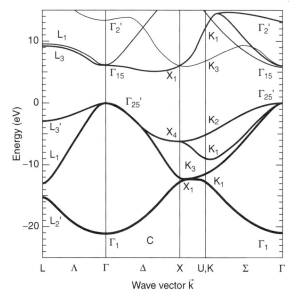

Figure 7.5. Electron band-structure calculation for diamond. [From J. R. Chelikowsky et al, *Phys. Rev. B*, **29**, 3470 (1984). Copyright 1984 by the American Physical Society.]

Begin with Bloch's difference equation [see Eq. (7.54)],

$$(\varepsilon_{\mathbf{k+G}} - E)u_{\mathbf{G}}(\mathbf{k}) + \sum_{\mathbf{G'}} V_{\mathbf{G'}}u_{\mathbf{G-G'}}(\mathbf{k}) = 0, \tag{7.57}$$

where a free-electron kinetic energy is defined by

$$\varepsilon_{\mathbf{k+G}} = \frac{\hbar^2}{2m}(\mathbf{k} + \mathbf{G})^2. \tag{7.58}$$

In the case where the $\{V_G\}$ all vanish, the solutions are

$$E = \varepsilon_{\mathbf{k+G}}. \tag{7.59}$$

In the extended-zone scheme the energy dispersion curve is a single parabola of revolution. In the reduced-zone scheme the information gets compressed into the first Brillouin zone (e.g., $-\pi/a < k < \pi/a$ for the one-dimensional case). In the periodic-zone scheme the information is repeated periodically in each Brillouin zone. Five parabolas are illustrated in the respective schemes in Fig. 7.6.

There is one case for which the free-electron approximation fails badly, and that is when energy bands intersect each other. At such degeneracies any perturbation will have a large effect. To see this in detail, suppose that the bands described by reciprocal lattice vectors $\mathbf{G}_1$ and $\mathbf{G}_2$ cross each other. An example of this is seen in the periodic-zone scheme in Fig. 7.6. The one-dimensional bands with $G = 0$ and $G = 2\pi/a$ intersect at $k = \pi/a$. Other couplings beside those between the two bands are neglected. Thus Eq. (7.57) becomes a pair of equations,

$$(\varepsilon_{\mathbf{k+G}_1} + V_0 - E)u_{\mathbf{G}_1} + V_{\mathbf{G}_1-\mathbf{G}_2}u_{\mathbf{G}_2} = 0, \tag{7.60a}$$

$$(\varepsilon_{\mathbf{k+G}_2} + V_0 - E)u_{\mathbf{G}_2} + V_{\mathbf{G}_2-\mathbf{G}_1}u_{\mathbf{G}_1} = 0. \tag{7.60b}$$

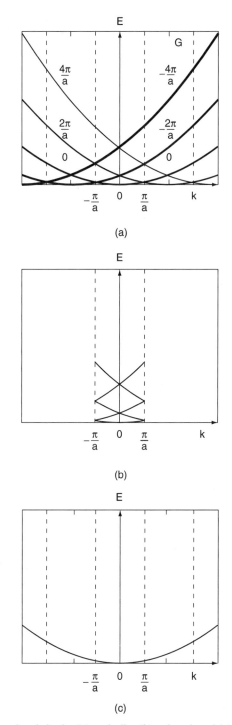

Figure 7.6. Free-electron bands in the (*a*) periodic, (*b*) reduced, and (*c*) extended-zone schemes. The first Brillouin zone extends from $-\pi/a$ to π/a.

Here V_0 describes a uniform background potential. The condition for the existence of solutions for this pair of linear equations is the vanishing of the secular determinant

$$\begin{vmatrix} \varepsilon_{\mathbf{k}+\mathbf{G}_1} + V_0 - E & V_{\mathbf{G}_1-\mathbf{G}_2} \\ V^*_{\mathbf{G}_1-\mathbf{G}_2} & \varepsilon_{\mathbf{k}+\mathbf{G}_2} + V_0 - E \end{vmatrix} = 0, \tag{7.61}$$

yielding

$$E^2 - E(\varepsilon_{\mathbf{k}+\mathbf{G}_1} + \varepsilon_{\mathbf{k}+\mathbf{G}_2} + 2V_0) + (\varepsilon_{\mathbf{k}+\mathbf{G}_1} + V_0)(\varepsilon_{\mathbf{k}+\mathbf{G}_2} + V_0) - |V_{\mathbf{G}_1-\mathbf{G}_2}|^2 = 0, \tag{7.62}$$

where $V_{\mathbf{G}_2-\mathbf{G}_1} = V^*_{\mathbf{G}_1-\mathbf{G}_2}$ has been used. The two roots are

$$E_\pm = V_0 + \frac{\varepsilon_{\mathbf{k}+\mathbf{G}_1} + \varepsilon_{\mathbf{k}+\mathbf{G}_2}}{2} \pm \sqrt{\left(\frac{\varepsilon_{\mathbf{k}+\mathbf{G}_1} - \varepsilon_{\mathbf{k}+\mathbf{G}_2}}{2}\right)^2 + |V_{\mathbf{G}_1-\mathbf{G}_2}|^2}. \tag{7.63}$$

At the point of degeneracy the two $\varepsilon_{\mathbf{k}+\mathbf{G}}$ factors are equal and

$$E_\pm = V_0 + \varepsilon_{\mathbf{k}+\mathbf{G}_1} \pm |V_{\mathbf{G}_1-\mathbf{G}_2}|. \tag{7.64}$$

The condition for the equality of the free-electron energies reduces to

$$G_1^2 + 2\mathbf{k} \cdot \mathbf{G}_1 = G_2^2 + 2\mathbf{k} \cdot \mathbf{G}_2. \tag{7.65}$$

This is precisely the von Laue condition for Bragg scattering of an electron. An energy gap is opened up between the two branches of the dispersion curve equal to twice the strength of the Fourier coefficient of the potential at that reciprocal lattice vector:

$$E_g = \Delta E = E_+ - E_- = 2|V_{\mathbf{G}_1-\mathbf{G}_2}|. \tag{7.66}$$

Typical graphs for the two branches of the dispersion curves are presented in Fig. 7.7 in the reduced-zone scheme.

As in the case of phonons, the spectral properties of the electronic states of a solid may be represented by the density of states, $\rho(E)$, the number of states per unit volume

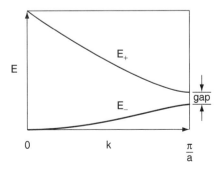

Figure 7.7. Typical dispersion curves for electrons in two interacting bands in the reduced-zone scheme.

per unit energy interval. It is defined by

$$\rho(E) = \frac{1}{V} \sum_{n,\mathbf{k},s} \delta(E - E_{n,\mathbf{k}}) = 2 \sum_n \int \frac{d^d k}{(2\pi)^d} \delta(E - E_{n,\mathbf{k}}), \qquad (7.67)$$

where d is the dimensionality of the solid and the sum extends over both electron spin projections, s, and the band index, n. For nonmagnetic materials the energy is independent of s. The $\mathbf{k}$ integration is limited to the first Brillouin zone. Note that $\rho(E)$ is a function of the form of the energy dispersion curves as well as the dimensionality of the solid. The quantity V is the crystal volume in three dimensions, area in two dimensions and length in one dimension.

For the one-dimensional free-electron gas one may use an extended-zone scheme and simply write

$$\rho(E) = 2 \int_{-\infty}^{\infty} \frac{dk}{2\pi} \delta\left(E - \frac{\hbar^2 k^2}{2m}\right) = \frac{1}{\pi\hbar} \sqrt{\frac{2m}{E}} \Theta(E), \qquad (7.68)$$

where $\Theta(E)$ is the unit step function. There is a threshold at $E = 0$ and an inverse-square-root behavior for $E > 0$. Note that for energies sufficiently far from the band edge, the limits of the $\mathbf{k}$ integration may be set to infinity. The corresponding result for a two-dimensional free-electron gas is

$$\rho(E) = 2 \int_{-\infty}^{\infty} \frac{d^2 k}{(2\pi)^2} \delta\left(E - \frac{\hbar^2 k^2}{2m}\right) = \frac{m}{\pi\hbar^2} \Theta(E), \qquad (7.69)$$

which has a threshold at $E = 0$ and is then constant for $E > 0$, For three dimensions the density of states is

$$\rho(E) = 2 \int \frac{d^3 k}{(2\pi)^3} \delta\left(E - \frac{\hbar^2 k^2}{2m}\right)$$
$$= \frac{1}{\pi^2} \int_0^E \frac{dk}{dE} k^2 \delta\left(E - \frac{\hbar^2 k^2}{2m}\right) dE = \frac{1}{2\pi^2} \left(\frac{2m}{\hbar^2}\right)^{3/2} \sqrt{E}\, \Theta(E), \quad (7.70)$$

and the density of states grows as $\sqrt{E}$ above the threshold at $E = 0$. Extending these results to the nearly free-electron model results in the introduction of van Hove singularities, which may occur at the zone boundaries or at any other extremal point of the energy band spectrum.

7.8   Tight-Binding Approximation in One Dimension

When an electron is bound in a deep-enough attractive potential well, the low-lying spectrum is a set of discrete energy levels. This is in contrast to the free-electron case, where the spectrum is a continuum. Just as the nearly free electron description of a solid in Section 7.7 was built around the free-electron model, it is possible to take as a starting point a solid treated as a collection of bound electrons. This is the idea behind the tight-binding approximation. We begin this section by reviewing the Hückel approximation from molecular physics. The approximation is then extended to a periodic solid.

The interaction of two atoms to form a molecule is most simply described in terms of the Hückel approximation. Molecular orbitals are constructed as a linear combination of atomic orbitals. A model Hamiltonian is introduced in which the atomic properties and the interaction between any pairs of atoms are parametrized in terms of numbers. Thus, for a two-atom situation, with one relevant state on each atom, the electronic states are denoted by $|1\rangle$ and $|2\rangle$ respectively, and the molecular wavefunction is written as a linear combination of these states

$$|\psi\rangle = a_1|1\rangle + a_2|2\rangle. \tag{7.71}$$

The on-site matrix elements of the Hamiltonian are $\langle 1|H|1\rangle = \varepsilon_1$ and $\langle 2|H|2\rangle = \varepsilon_2$ These are the energies of the electronic states of the individual atoms. The off-diagonal matrix elements, or *tunneling* (or *hopping*) *matrix elements*, are $\langle 1|H|2\rangle = t$ and $\langle 2|H|1\rangle = t^*$. They come about because the Coulomb interaction has matrix elements connecting the states on different atoms. The Hamiltonian matrix is

$$H = \begin{bmatrix} \varepsilon_1 & t \\ t^* & \varepsilon_2 \end{bmatrix}. \tag{7.72}$$

Note that H defines a Hermitian matrix (i.e., one whose transpose and complex conjugate have identical matrix elements). It is convenient to neglect the direct overlap of states $|1\rangle$ and $|2\rangle$ and assume the orthonormality condition $\langle i|j\rangle = \delta_{ij}$. This implies the normalization condition $|a_1|^2 + |a_2|^2 = 1$. The eigenfunctions and eigenvalues are determined from the Schrödinger equation $(H - E)|\psi\rangle = 0$, which leads to the following pair of equations:

$$(\varepsilon_1 - E)a_1 + t^*a_2 = 0, \tag{7.73a}$$

$$ta_1 + (\varepsilon_2 - E)a_2 = 0. \tag{7.73b}$$

The secular determinant vanishes, so

$$E^2 - E(\varepsilon_1 + \varepsilon_2) + \varepsilon_1\varepsilon_2 - |t|^2 = 0, \tag{7.74}$$

with the solutions

$$E_\pm = \frac{\varepsilon_1 + \varepsilon_2}{2} \pm \sqrt{\left(\frac{\varepsilon_1 - \varepsilon_2}{2}\right)^2 + |t|^2}. \tag{7.75}$$

Here E_- is the energy of the bonding state and E_+ is that of the antibonding state of the molecule.

This result is readily generalized to a chain of N atoms. For finite N it may represent a linear polymer. In the limit of large N it may be thought of as a one-dimensional solid. Thus

$$|\psi\rangle = \sum_{j=1}^{N} a_j|j\rangle. \tag{7.76}$$

In the simplest case, one makes the assumption of NN interactions only. The nonvanishing matrix elements of H are then

$$\langle j|H|j\rangle = \varepsilon_j, \qquad \langle j+1|H|j\rangle = t_j, \qquad \langle j|H|j+1\rangle = t_j^*. \qquad (7.77)$$

Periodic boundary conditions, introduced in Section 5.1, are imposed, so there is the subscript identification $N+1 \rightarrow 1$ and $0 \rightarrow N$. The Schrödinger equation leads to a set of N coupled linear algebraic equations:

$$(\varepsilon_j - E)a_j + t_{j-1}a_{j-1} + t_j^*a_{j+1} = 0. \qquad (7.78)$$

In the case where all the atoms are identical, $\varepsilon_j = \varepsilon$ and $t_j = t$, for all j. Let $a_j = Ac^j$, so that

$$\varepsilon - E + \frac{t}{c} + t^*c = 0. \qquad (7.79)$$

As in Section 7.6, it is found that the quantity c must be of modulus 1, so $c = \exp(i\theta)$. It is convenient to assume t to be real. Then

$$\varepsilon - E + 2t \cos\theta = 0. \qquad (7.80)$$

Imposing the periodic boundary conditions $a_{N+1} = a_1$ implies that $\exp(iN\theta) = 1$, so that $\theta = 2\pi n/N$, where $n = 0, 1, 2, \ldots, N-1$. Letting $k_n = 2\pi n/Na$, where a is the lattice constant, and suppressing the subscript n leads to the dispersion curve:

$$E(k) = \varepsilon + 2t \cos ka. \qquad (7.81)$$

Thus one finds a single allowed energy band of width $4|t|$. Depending on the sign of t, it could have a maximum or a minimum at $k = 0$. The allowed values of k extend over the first Brillouin zone, from $k = -\pi/a$ to π/a. For core levels it is safe to assume that $t = 0$, so there is no overlap or interaction.

For $t < 0$ there is a minimum at $k = 0$, and for small k it follows that

$$E \approx \varepsilon - 2|t| + |t|a^2k^2 = E_0 + \frac{\hbar^2k^2}{2m^*}, \qquad (7.82)$$

where $E_0 = \varepsilon - 2|t|$ and m^* denotes the effective mass of an electron,

$$m^* = \frac{\hbar^2}{2|t|a^2}. \qquad (7.83)$$

The inverse of m^* is proportional to the curvature of the energy dispersion curve at $k = 0$. One may define in a similar way the effective mass for the case in which there is a maximum at $k = 0$. In that case m^* will be negative. Equation (7.82) predicts that as $m^* \rightarrow \infty$, the energy E will be independent of k. The tight-binding bands are sketched in Fig. 7.8 for the cases $t > 0$ and $t < 0$.

The tight-binding approximation finds application in the description of semiconductors and the d bands of transition metals. It is also of use in finding the band

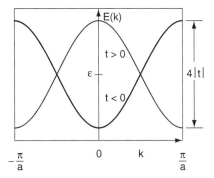

Figure 7.8. Tight-binding bands for the cases $t > 0$ and $t < 0$ in the reduced-zone scheme.

structure of polymers. It may be generalized to higher dimensions, to more than just NN interactions, and to the case where there is more than one state per site.

The density of states for the one-dimensional tight-binding model is obtained from Eq. (7.67) by integrating over the first Brillouin zone:

$$\rho(E) = 2 \int_{-\pi/a}^{\pi/a} \frac{dk}{2\pi} \delta(E - \varepsilon - 2t \cos ka) = \frac{2\Theta[4t^2 - (E - \varepsilon)^2]}{\pi a \sqrt{4t^2 - (E - \varepsilon)^2}}. \tag{7.84}$$

This gives a diverging density of states at the upper and lower band edges, $E = \varepsilon \pm 2t$ at $k = 0$ and at $k = \pm\pi/a$. The density of states for a random one-dimensional solid is discussed briefly in Section W7.2.

7.9   Tight-Binding Approximation in Two Dimensions

The tight-binding approximation will now be applied to calculation of the band structure of a two-dimensional crystal: a CuO_2 plane. As discussed in Chapter 11, these planes of atoms play a central role in the structure of ceramic high-temperature superconductors. Assume that there is a square lattice with three atoms per unit cell: a copper atom, an oxygen to the right of it (in the x direction), and an oxygen above it (in the y direction). The dsp^2 orbitals from the copper atoms interact with the p_x and p_y orbitals of the oxygen atoms to form bonds (Fig. 7.9). These bonds have mixed ionic and covalent character.

Let the amplitude of the wavefunction on the copper atom in cell (m, n) be denoted by $A_{m,n}$ and the corresponding amplitudes of the two oxygen atoms of the cell by $R_{m,n}$ and $U_{m,n}$, respectively. The tight-binding equations are

$$(E_{Cu} - E)A_{m,n} + t(R_{m,n} + U_{m,n} + R_{m-1,n} + U_{m,n-1}) = 0, \tag{7.85a}$$

$$(E_O - E)R_{m,n} + t(A_{m+1,n} + A_{m,n}) = 0, \tag{7.85b}$$

$$(E_O - E)U_{m,n} + t(A_{m,n} + A_{m,n+1}) = 0, \tag{7.85c}$$

where E_{Cu} and E_O are the on-site electron energies and t is the tunneling matrix element between the Cu and O atoms. Inserting the expressions

$$A_{m,n} = Ae^{i(mk_x + nk_y)a}, \qquad R_{m,n} = Re^{i(mk_x + nk_y)a}, \qquad U_{m,n} = Ue^{i(mk_x + nk_y)a}, \tag{7.86}$$

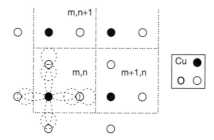

Figure 7.9. Copper atoms (solid circles) bonded to oxygen atoms (open circles) in the CuO_2 plane. The square unit cell is shown. The hybrid dsp^2-Cu orbitals combine with the p_x- and p_y-O orbitals to form σ-molecular orbitals.

where a is the lattice constant, leads to

$$(E_{Cu} - E)A + tR(1 + e^{-ik_x a}) + tU(1 + e^{-ik_y a}) = 0, \qquad (7.87)$$

with

$$R = \frac{tA}{E - E_O}(1 + e^{ik_x a}), \qquad (7.88)$$

$$U = \frac{tA}{E - E_O}(1 + e^{ik_y a}). \qquad (7.89)$$

The resulting secular equation is

$$E^2 - (E_O + E_{Cu})E + E_O E_{Cu} - 4t^2 \left(\cos^2 \frac{k_x a}{2} + \cos^2 \frac{k_y a}{2} \right) = 0. \qquad (7.90)$$

The two solutions are

$$E_{\pm}(k_x, k_y) = \frac{E_{Cu} + E_O}{2} \pm \sqrt{ \left(\frac{E_{Cu} - E_O}{2} \right)^2 + 4t^2 \left(\cos^2 \frac{k_x a}{2} + \cos^2 \frac{k_y a}{2} \right)}. \qquad (7.91)$$

The first Brillouin zone is a square occupying the space $|k_x| \leq \pi/a$, $|k_y| \leq \pi/a$. Along the line $k_x + k_y = \pi/a$, the energy is constant and has the value

$$E_{\pm} = \frac{E_{Cu} + E_O}{2} \pm \sqrt{ \left(\frac{E_{Cu} - E_O}{2} \right)^2 + 4t^2}, \qquad (7.92)$$

which is independent of **k**. The shape of the Fermi boundary is determined by the electron density in the plane, a factor that is often controlled by the nature of other atoms above or below the planes. For example, a half-filled energy band has all the states within the square defined by $|k_x + k_y| < \pi/a$ and $|k_x - k_y| < \pi/a$ occupied at $T = 0$ K. Several lines of constant energy are sketched in Fig. 7.10. Note that the free-electron model works well when k_x and k_y are near the zone center and the constant-energy contour is a circle.

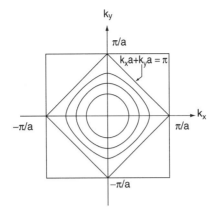

Figure 7.10. Lines of constant energy for the lower energy band in Eq. (7.91) for a typical case.

It is of some value to relate the bandwidth in the tight-binding approximation to the number of NNs in a lattice. The Schrödinger equation is of the form

$$(E_0 - E)A_0 + t \sum_{n=1}^{Z} A_n = 0, \tag{7.93}$$

where A_0 denotes the amplitude on a given site and A_n are the amplitudes on the Z neighboring sites. The latter amplitudes are related to A_0 through simple phase factors. At some points in the Brillouin zone all these phases are equal to 1 and the sum is maximized so that $E = E_0 + Zt$. On the other hand, at other points the sum equals -1 and is therefore minimized. Then $E = E_0 - Zt$. The bandwidth is therefore given by

$$B = 2Zt. \tag{7.94}$$

For fixed t, the larger the number of NNs, the larger the bandwidth.

7.10 Metals, Insulators, Semiconductors, and Semimetals

The energy-band picture provides a simple understanding for the wide variation in conductivity in going from material to material, at least for the case of crystalline solids. In a *metal* the Fermi level lies within an allowed energy band. There are unoccupied states with energies immediately above the Fermi energy and there are occupied states immediately below it. The application of even a weak electric field can elevate an electron in energy from an occupied state to a vacant state, where it can participate in the flow of a net current. In metals the concentration of charge carriers is essentially constant, independent of T, and determined by the atomic concentration and the valence of the ion. Referring to the Drude formula $\sigma = ne^2\tau/m$, the temperature variation of the conductivity may be understood. As the temperature of the material increases, σ decreases, due to the shortening of the collision time brought about by the emission, absorption, and scattering of phonons.

In an ideal *insulator* the Fermi level lies within an energy gap. All energy bands are either completely filled with electrons or completely vacant. (This is never true in real materials, due to defects, impurities, etc.). Since the number of states (with a given spin projection) in an energy band equals the number of unit cells in the crystal, a necessary condition for an insulator is that there be an even number of electrons per unit cell. This, however, is not a sufficient condition. For example, divalent Mg is a metal due to band overlap in different directions in **k** space. The occupied bands are called *valence bands*, and the unoccupied bands are called *conduction bands*. When a weak electric field is applied, there is no electric current in the filled bands since these electrons cannot be excited. The field is unable to provide the energy needed for making an interband transition to a conduction band. At finite temperatures, thermal excitation of electrons from the valence to the conduction bands is possible. The conductivity will be determined largely by the number of carriers produced. For wide-bandgap materials, this number will be very small. Free charge trapped in an insulator can remain there for very long periods of time without being conducted away, because of the very high electrical resistance. This forms the basis of solid-state CMOS memory devices, where a bit of information corresponds to a stored electric charge.

An intrinsic semiconductor is an insulator with a relatively small bandgap. Examples of elemental semiconductors include Ge and Si. For such materials thermal excitation of carriers is nonnegligible at room temperature. As the temperature rises the number of thermally generated carriers grows exponentially and the conductivity also increases exponentially. This effect more than offsets the shortening of the collision lifetime due to phonon interactions.

The conductivities of insulators and semiconductors are highly sensitive to the presence of impurities. Impure semiconductors are called *extrinsic semiconductors*: if the impurities are introduced in a controlled fashion, they are termed *doped semiconductors*. Semiconductors are discussed in detail in Section 7.12 and Chapter 11. The question of conduction in impure materials is considered later in this chapter.

Semimetals have a slight overlap between the valence and conduction bands. As a result, the valence band is nearly filled and the conduction band is nearly empty. The electrons in the conduction band can carry an electric current, as can the vacant electron states (holes) in the valence band. Examples of elemental semimetals are graphite, As, Bi, and Sb.

In Fig. 7.11 hypothetical band structures are sketched for a metal, an insulator (or semiconductor), and a semimetal.

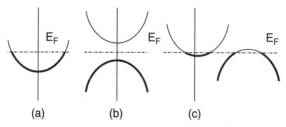

(a)   (b)   (c)

Figure 7.11. Hypothetical band structures for (*a*) a metal, (*b*) an insulator, and (*c*) a semimetal. The dashed line denotes the Fermi level. The heavy lines represent occupied states.

QUANTUM EFFECTS IN ELECTRICAL CONDUCTION

In the following section the temperature dependence of resistivity is analyzed and we will see that quantum-mechanical effects need to be introduced to describe the low-T resistivity. Then the Hall effect is reexamined from the vantage point of electron band theory in Section W7.4.

7.11   Temperature Dependence of Resistivity in Metals

Resistivity is defined as the inverse of conductivity: $\rho = 1/\sigma$. It has been seen that conductivity is determined by the average collision time, τ. A collision is any process that destroys the forward momentum of electrons. It may involve scattering from impurities or defects (time τ_i), or it may result from the emission or absorption of phonons or the scattering from phonons (time τ_{ph}). Since one may write the total scattering rate as the sum of the individual scattering rates, the resistivities will be additive:

$$\rho = \frac{m}{ne^2}\left(\frac{1}{\tau_i} + \frac{1}{\tau_{ph}}\right). \tag{7.95}$$

This separation, called *Matthiessen's rule*, is an approximation. The value of τ_i is controlled by the quality of the material and is essentially independent of the temperature, at least at low T. For a perfect material with no impurities (chemical or even isotopic) τ_i can be made arbitrarily long. The scattering time τ_{ph} depends on the temperature. In Fig. 7.12 data are presented on the variation of resistivity with temperature for aluminum.

Temperature dependence displays approximately linear behavior for $T > \Theta_D$. This may be understood simply in terms of the thermal occupancy of phonon modes in this temperature range. The Bose–Einstein distribution function for $\beta\hbar\omega \ll 1$ becomes

$$n(\omega, T) = \frac{1}{e^{\beta\hbar\omega} - 1} \approx \frac{k_B T}{\hbar\omega}, \tag{7.96}$$

so the number of phonons available (to be absorbed, to stimulate an emission process, or to scatter from) is linearly proportional to the temperature. These processes are

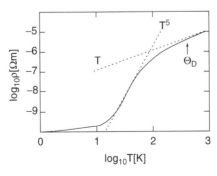

Figure 7.12. Plot of resistivity (in $\Omega\cdot$m) of Al versus temperature (in kelvin) on a log-log plot. The Debye temperature is denoted by $\Theta_D = 428$ K. Note that Al melts at $T = 934$ K. (Data from D. R. Lide, ed., *CRC Handbook of Chemistry and Physics*, 73rd ed., CRC Press, Boca Raton, Fla., 1991.)

likely to involve phonons with wave vectors anywhere in the Brillouin zone. The emission/absorption/scattering rate is proportional to the phonon mode occupancy, hence the linear behavior at high T.

At low temperatures, however, note that there is a range in which $\rho \propto T^5$. This comes about because of three effects. First, at low temperatures only low-energy phonons (with energies of approximately $k_B T$) can be absorbed. The phonon density of states grows quadratically with energy, so this introduces a factor proportional to T^2. Second, emission or absorption of these "soft" phonons are likely to result in only small deflections, and hence are not very effective in randomizing the electron's momentum direction. If θ is the scattering angle, the forward momentum will be reduced by the factor $1 - \cos\theta \sim \theta^2/2$. But $\theta \propto k/p$, where k and $p/\hbar$ are the wave vectors of the phonon and electron, respectively. This introduces an additional k^2 factor. Since ω and k are proportional for acoustic phonons and for a typical phonon $\omega \propto T$, one obtains an additional T^2 factor. The final factor of T arises from the behavior of the electron–phonon coupling constant. For example, in the deformation-potential interaction the phonon produces a transitory dilation or compression of the lattice. This, in turn, causes the energy bands to be shifted up or down locally and results in the conduction electron's wavefunction acquiring a phase shift. The result is that the electrons are scattered. The degree of dilation is measured by $\nabla \cdot \mathbf{u}$, $\mathbf{u}$ being the local displacement vector. Upon Fourier analysis this leads to a factor $\mathbf{k} \cdot \mathbf{u}$, and hence an additional T factor. The net result is the Bloch T^5 law for the resistivity. Other electron–phonon mechanisms produce a similar factor.

At very low temperatures (but above $T = 1.18$ K) one sees in Fig. 7.12 the residual effects of impurity scattering, and the resistivity tends to a constant value. Below $T = 1.18$ K the resistivity falls abruptly to zero. This is due to the onset of superconductivity, which is described in Chapter 16.

7.12 Semiconductors

Crystalline solids are characterized by the existence of allowed and forbidden electron energy bands. The distinction between insulators and semiconductors resides in the size of the bandgap and where the Fermi level sits relative to the band edges. In an insulator the Fermi level lies within a wide forbidden band, considerably removed from the filled valence band and the vacant conduction band. The gap is sufficiently wide that a typical applied electric field is unable to excite interband transitions, even if assisted by thermal fluctuations. (Very strong fields can lead to breakdown, however, due to interband tunneling.) In semiconductors there is generally a narrower gap. Two cases are distinguished: intrinsic and extrinsic. In the *intrinsic* case, the semiconductor is of high purity and has few defects. There are some thermally excited electrons that reside in the conduction band, and these leave behind holes in the valence band. These electrons and holes are each able to conduct an electrical current. In the *extrinsic* case, dopant atoms are added to the semiconductor, which may become thermally ionized and contribute either electrons to the conduction band or holes to the valence band.

The dynamics of carriers (electrons or holes) is often described in the semiclassical approximation by equations resembling the Hamilton equations of classical mechanics

$$\hbar \frac{d\mathbf{k}}{dt} = -e(\mathbf{E} + \mathbf{v}_n \times \mathbf{B}), \tag{7.97}$$

$$\mathbf{v}_n = \frac{1}{\hbar}\frac{\partial}{\partial\mathbf{k}}E_n(\mathbf{k}). \tag{7.98}$$

Each electron is assumed to reside in a particular band (fixed n), and interband transitions are neglected. This assumption puts an upper limit on the strength of the applied fields and the frequency at which they can change. Unlike the Hamilton equations, $\hbar\mathbf{k}$ is the crystal momentum (not the true momentum). Recalling that the energy bands are periodic functions in $\mathbf{k}$ space, $\mathbf{k}$ is effectively confined to the first Brillouin zone. Its value evolves under the influence of the Lorentz force.

Consider the case of an electron in a conduction band. Let E_c denote the minimum energy in the conduction band and assume that it occurs at the points $\mathbf{k} = \mathbf{k}^\nu$, $\nu = 1,\dots,N_\nu$. If $N_\nu = 1$, there is a unique minimum, whereas if $N_\nu > 1$, there is a degeneracy and the corresponding valleys are probably linked by a crystal-symmetry operation. In the neighborhood of the minimum of the energy band at $\mathbf{k} = \mathbf{k}^\nu$ (where $\nabla_\mathbf{k}E_n(\mathbf{k}) = 0$), one may expand the energy surface

$$E_n(\mathbf{k}) = E_c + \frac{1}{2}\sum_{ij}(k_i - k_i^\nu)(k_j - k_j^\nu)\frac{\partial^2 E_n}{\partial k_i\partial k_j} + \cdots$$

$$\equiv E_c + \frac{\hbar^2}{2}(\mathbf{k} - \mathbf{k}^\nu)\cdot\frac{1}{\overleftrightarrow{m}_e^*}\cdot(\mathbf{k} - \mathbf{k}^\nu) + \cdots, \tag{7.99}$$

where

$$\left(\frac{1}{\overleftrightarrow{m}_e^*}\right)_{ij} = \frac{1}{\hbar^2}\frac{\partial^2 E_n}{\partial k_i\partial k_j} \tag{7.100}$$

is called the *inverse* of the effective-mass tensor for the electrons.

In general, $\overleftrightarrow{m}_e^*$ is a symmetric matrix. It is positive definite since there is an absolute minimum in the band. The tensor may be diagonalized by an orthogonal matrix $\mathbf{R}$. This means that there exists a matrix $\mathbf{R}$ such that $\mathbf{R}\overleftrightarrow{m}_e^*\mathbf{R}^{-1}$ is a diagonal matrix. An orthogonal matrix is one whose transpose and inverse are the same. In the diagonal representation the mass eigenvalues are m_1^*, m_2^*, and m_3^*. In the case of twofold degeneracy, one sometimes writes $m_1^* = m_2^* = m_\perp^*$ (or m_t^*, the transverse mass) and $m_1^* = m_\parallel^*$ (or m_l^*, the longitudinal mass). The semiclassical equations for electrons may be written as

$$\overleftrightarrow{m}_e^*\cdot\frac{d\mathbf{v}_n}{dt} = -e(\mathbf{E} + \mathbf{v}_n\times\mathbf{B}). \tag{7.101}$$

In the case of the valence band, the maximum energy is denoted by E_v and the expansion of the energy surface around E_v leads to the introduction of a mass tensor defined with a negative sign:

$$\left(\frac{1}{\overleftrightarrow{m}_h^*}\right)_{ij} = -\frac{1}{\hbar^2}\frac{\partial^2 E_n}{\partial k_i\partial k_j}. \tag{7.102}$$

The equation for the holes in the valence band is

$$\overleftrightarrow{m}_h^*\cdot\frac{d\mathbf{v}_h}{dt} = +e(\mathbf{E} + \mathbf{v}_h\times\mathbf{B}). \tag{7.103}$$

Because of the downward curvature of the valence band, they behave effectively as if they had a positive charge.

The dynamical equations (7.101) and (7.103) for the velocity describe the motion of individual carriers in a band. They may be averaged over the band, however, and may then equally describe the motion of the average velocity of the carrier. One may introduce the effect of collisions phenomenologically, by means of a collision time matrix. The reason for introducing this quantity as a matrix is to take account of the possible anisotropy in the collision rates. Thus

$$\overset{\leftrightarrow}{m}^*_e \cdot \left(\frac{d\mathbf{v}_e}{dt} + \frac{1}{\overset{\leftrightarrow}{\tau}_e} \cdot \mathbf{v}_e \right) = -e(\mathbf{E} + \mathbf{v}_e \times \mathbf{B}), \tag{7.104a}$$

$$\overset{\leftrightarrow}{m}^*_h \cdot \left(\frac{d\mathbf{v}_h}{dt} + \frac{1}{\overset{\leftrightarrow}{\tau}_h} \cdot \mathbf{v}_h \right) = +e(\mathbf{E} + \mathbf{v}_h \times \mathbf{B}). \tag{7.104b}$$

In the absence of a magnetic field and for steady-state conditions, the electrical current may be written as

$$\mathbf{J} = \overset{\leftrightarrow}{\sigma} \cdot \mathbf{E}, \tag{7.105}$$

where the conductivity tensor is described in terms of contributions of electrons and holes by

$$\overset{\leftrightarrow}{\sigma} = e^2 \left(n_h \overset{\leftrightarrow}{\tau}_h \cdot \frac{1}{\overset{\leftrightarrow}{m}^*_h} + n_e \overset{\leftrightarrow}{\tau}_e \cdot \frac{1}{\overset{\leftrightarrow}{m}^*_e} \right). \tag{7.106}$$

Here n_h and n_e are the hole and electron concentrations, respectively.

7.13 Magnetoresistance

The Drude theory may be used to obtain a simple understanding of magnetoresistance (i.e., the effect of a magnetic field on the conductivity). Begin with the metallic case, where there is only one band, and then extend the analysis to semiconductors, where two or more bands may be involved. The equation of motion includes the magnetic force in addition to the electric force:

$$\frac{d\mathbf{v}}{dt} + \frac{\mathbf{v}}{\tau} = -\frac{e}{m^*}(\mathbf{E} + \mathbf{v} \times \mathbf{B}). \tag{7.107}$$

Take the magnetic field to be along the z direction and introduce the cyclotron frequency $\omega_c = eB/m^*$. In the absence of an electric field the electrons would move in cyclotron orbits, which are either planar circles or helices whose axes are along the direction of the magnetic field (z direction). The period around the circular orbit (or one turn of a helical orbit) is given by $2\pi/\omega_c$. In the presence of an electric field one may explore the solutions of Eq. (7.107) for the case in which the damping term dominates over the inertial term. This leads to the expressions

$$v_x + \omega_c \tau v_y = -\frac{e\tau}{m^*} E_x, \tag{7.108a}$$

$$v_y - \omega_c \tau v_x = -\frac{e\tau}{m^*} E_y, \tag{7.108b}$$

$$v_z = -\frac{e\tau}{m^*} E_z. \tag{7.108c}$$

Use $\mathbf{J} = ne\mathbf{v}$ to obtain

$$J_x = \sigma \frac{E_x - \omega_c\tau E_y}{1 + (\omega_c\tau)^2} = \sigma_{xx}E_x + \sigma_{xy}E_y, \tag{7.109a}$$

$$J_y = \sigma \frac{E_y + \omega_c\tau E_x}{1 + (\omega_c\tau)^2} = \sigma_{yx}E_x + \sigma_{yy}E_y, \tag{7.109b}$$

$$J_z = \sigma E_z = \sigma_{zz}E_z, \tag{7.109c}$$

where σ is the Drude conductivity. If $J_y = 0$, then $E_y = -\omega_c\tau E_x$ and $J_x = \sigma E_x$. There is no transverse magnetoresistance in this one-band model. As noted in Section 7.3, the Hall coefficient is given as $R_H = E_y/BJ_x = -1/ne$.

Next consider a semiconductor containing two bands, one with electrons and the other with holes. Let n_e and n_h denote the number of electrons and holes per unit volume, respectively. Their parameters are $(\tau_e, m_e^*, \sigma_e)$ and $(\tau_h, m_h^*, \sigma_h)$. The current density components are now

$$J_x = \sigma_e \frac{E_x - \omega_e\tau_e E_y}{1 + (\omega_e\tau_e)^2} + \sigma_h \frac{E_x + \omega_h\tau_h E_y}{1 + (\omega_h\tau_h)^2}, \tag{7.110a}$$

$$J_y = \sigma_e \frac{E_y + \omega_e\tau_e E_x}{1 + (\omega_e\tau_e)^2} + \sigma_h \frac{E_y - \omega_h\tau_h E_x}{1 + (\omega_h\tau_h)^2}, \tag{7.110b}$$

$$J_z = (\sigma_e + \sigma_h)E_z. \tag{7.110c}$$

For the case where $J_y = 0$ one now finds

$$E_y = \frac{\sigma_h Q_h/\Delta_h - \sigma_e Q_e/\Delta_e}{\sigma_h/\Delta_h + \sigma_e/\Delta_e} E_x, \tag{7.111}$$

where $Q = \omega_c\tau = \mu B$, where μ is the mobility and $\Delta = 1 + (\omega_c\tau)^2$. This leads to an expression for the magnetoconductivity:

$$J_x \left[\left(\frac{\sigma_e}{\Delta_e} + \frac{\sigma_h}{\Delta_h} \right) + \frac{(\sigma_e Q_e/\Delta_e - \sigma_h Q_h/\Delta_h)^2}{\sigma_e/\Delta_e + \sigma_h/\Delta_h} \right] E_x = \sigma_{xx}E_x. \tag{7.112}$$

In the limit of very large magnetic fields (i.e., $\mu_e B \gg 1$ and $\mu_h B \gg 1$), the first term in Eq. (7.112) becomes small and the B dependence of the second term drops out. Thus the magnetoconductivity is

$$\sigma_{xx} \approx \frac{(n_e - n_h)^2 e^2}{n_e m_e^*/\tau_e + n_h m_h^*/\tau_h}, \tag{7.113}$$

This, again, is independent of B (i.e., saturates). The high-field Hall coefficient approaches

$$R_H = -\frac{1}{(n_e - n_h)e}, \tag{7.114}$$

which displays the competition between electrons and holes in determining its sign.

In many metals, for example, the noble metals, things are more complicated. The starting assumption that the electrons are always characterized by a positive mass is not valid. The Fermi surface is such that it may contain necks connecting different Brillouin zones. The effective mass tensor then varies in magnitude and sign with location on the Fermi surface. A mass that varies in sign leads to a curvature of the orbit that also varies in sign. This leads to the possibility of open orbits instead of closed orbits perpendicular to the magnetic field. In such a situation the magnetoresistance can continue to grow with increasing magnetic field instead of saturating as in Eq. (7.113). The implications of open orbits are not considered further here.

CONDUCTION IN INSULATORS

A perfect insulator would, of course, block the flow of all charge. In practice, however, insulators do have some residual conductivity. There is always the possibility of activated conduction by transferring electrons to the conduction band or creating holes in the valence band. In the intrinsic case, this would require a thermal fluctuation to create an electron–hole pair, a process that has probability $\exp(-\beta E_g)$. For a wide-gap insulator, this would be negligible. For an insulator with impurity levels in the gap, the excitation energy need not be as large, but it may still be improbable. In this section it will be seen that one need not make transitions all the way to the conduction band for a carrier to move. All that is necessary is for a carrier from an occupied impurity level in the gap to find a vacant impurity level in the gap so that transitions can occur. The study of variable-range hopping explores this possibility. In addition, it will be shown that there is a possibility of nonohmic conduction in insulators, particularly at stronger electric fields, due to the Poole–Frenkel effect.

7.14   Variable-Range Hopping

Impurity levels in insulators or semiconductors may lie in the energy gap either below or above the Fermi level. At high temperatures these levels are likely to donate electrons to the conduction band (or accept electrons from the valence band) and the material will function as a doped semiconductor. At low temperatures, conduction occurs when an electron in an occupied impurity state is thermally excited and hops to a vacant state some distance R away (Fig. 7.13). Let the energy for this excitation be denoted by ΔE. The probability for being excited is $\exp(-\beta\Delta E)$. The probability that the electron will hop a distance R is also proportional to the overlap of the wavefunctions

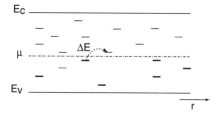

Figure 7.13. Set of impurity levels lying within the bandgap. Conduction occurs by hopping processes.

at the initial and final sites. This falls off exponentially with separation as $\exp(-2\alpha R)$. The parameter α is related to the binding energy E_B of the electron to the impurity ion (relative to the bottom of the conduction band) by $\alpha \approx \sqrt{2mE_B/\hbar^2}$. Thus the net hopping probability contains the factor $\exp(-\beta\Delta E - 2\alpha R)$.

An estimate for ΔE can be made by assuming an impurity concentration n_I, letting B denote the spread of energy values of impurity levels, and assuming a uniform spread of levels

$$\Delta E = \frac{B}{(4\pi R^3/3)n_I}. \tag{7.115}$$

Here the denominator represents the total number of impurities within a radius R of a given impurity. (More generally, one may replace n_I/B by the density of states at the Fermi level.) There is competition between the two terms in the exponent in the expression for the hopping probability. If $\beta\Delta E \gg 1$, electrons must hop long distances to find a state of the proper energy. If the hopping distance is to be small, the excitation energy can be large. This type of conduction is called *Mott variable-range hopping*.

The maximum hopping probability will be achieved when the magnitude of the exponent is minimized as a function of R:

$$\frac{\partial}{\partial R}\left(\frac{3\beta B}{4\pi n_I R^3} + 2\alpha R\right) = 0, \tag{7.116}$$

The most probable hopping distance is therefore

$$R_{\max} = \left(\frac{9\beta B}{8\pi n_I \alpha}\right)^{1/4}, \tag{7.117}$$

and the hopping probability and the conductivity are proportional to

$$\sigma = \sigma_0 \exp\left(-\frac{8}{3}\alpha R_{\max}\right) = \sigma_0 \exp\left[-\left(\frac{T_0}{T}\right)^{1/4}\right], \tag{7.118}$$

where $T_0 = 512 B\alpha^3/9\pi n_I k_B$. Here σ_0 contains more slowly varying factors of temperature. Typical mobilities in a localized state are small.

Efros and Shklovskii introduced modifications to this formula arising from the Coulomb attraction between an electron that undergoes hopping and the hole it leaves behind. This changes the T dependence of the conductivity. A qualitative derivation of the effect is presented. The Coulomb interaction energy is $\Delta E_c = -e^2/4\pi\epsilon R$, where ϵ is the static dielectric permittivity. Thus there is a minimum distance that an electron must hop to overcome the Coulomb attraction. For an electron at the Fermi level to hop to a site a distance R away requires an energy input of amount ΔE such that $\Delta E + \Delta E_c > 0$, or else the electron will not reach a state above the Fermi energy. This leads to a connection between the minimum hopping distance and the energy:

$$R = \frac{e^2}{4\pi\epsilon\Delta E}. \tag{7.119}$$

The optimum hopping probability selects a value for ΔE given by

$$\frac{\partial}{\partial \Delta E} \left(\beta \Delta E + \frac{\alpha e^2}{2\pi \epsilon \Delta E} \right) = 0, \tag{7.120}$$

which leads to

$$\Delta E = \sqrt{\frac{\alpha e^2}{2\pi \epsilon \beta}}. \tag{7.121}$$

The conductivity therefore varies with temperature as

$$\sigma = \sigma_0 \exp \left(-\sqrt{\frac{T_{\text{ES}}}{T}} \right), \tag{7.122}$$

where the characteristic temperature is $T_{\text{ES}} = 2\alpha e^2 / \pi \epsilon k_B$.

One expects the Efros–Shklovskii form for the temperature dependence to be valid at the lowest temperatures. At intermediate temperatures the Mott formula is more suitable. At high temperatures, if the energy is sufficient to excite the impurity electrons to the conduction band, the material will conduct as a doped semiconductor.

7.15   Poole–Frenkel Effect

The conductivity of an insulator is often not independent of the applied electric field and may be increased considerably by increasing the strength of the applied field. The long-range interaction of an electron with an impurity ion (with $z = 1$) in a host insulating or semiconducting crystal is given by the Coulomb interaction,

$$U_0(r) = -\frac{e^2}{4\pi \epsilon r}. \tag{7.123}$$

For ionization to occur, the electron must in principle be moved from the ion position to $r = \infty$, where U_0 vanishes. Let I_o denote the ionization potential of the impurity atom embedded in the host crystal. Now suppose that an electric field $\mathbf{E}$ is established in the crystal parallel to the z direction. The interaction becomes

$$U(r, \theta) = -\frac{e^2}{4\pi \epsilon r} - eEr \cos \theta, \tag{7.124}$$

where θ is the polar angle that $\mathbf{r}$ makes with the z direction. There is a saddle point in this function at the location

$$r = \sqrt{\frac{e}{4\pi \epsilon E}}, \qquad \theta = 0, \tag{7.125}$$

and the value of U there is

$$U = -\sqrt{\frac{e^3 E}{\pi \epsilon}}. \tag{7.126}$$

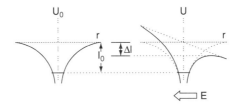

Figure 7.14. Potential energy without the electric field, U_0, and with the electric field, U.

For ionization to occur, one need only transport the electron to the location of the saddle point rather than all the way to infinity. The net result is that the ionization energy is lowered by an amount

$$\Delta I = -\sqrt{\frac{e^3 E}{\pi \epsilon}}. \tag{7.127}$$

The process of lowering the ionization barrier is illustrated in Fig. 7.14.

The ionization probability of the donor impurity, the number of ionized carriers, and the conductivity are all proportional to the Boltzmann factor $\exp(-\beta I_o)$. Hence the conductivity is enhanced and becomes

$$\sigma(E) = \sigma_0 \exp \left(\frac{1}{k_B T} \sqrt{\frac{e^3 E}{\pi \epsilon}} \right), \tag{7.128}$$

where σ_0 is the conductivity in the zero-field limit. This is called the *Poole–Frenkel effect*. It serves to show that nonohmic conductivity is likely to be important in describing charge transport in insulators at elevated field strengths. Dielectric breakdown is described in Chapter 15.

METAL–INSULATOR TRANSITION

There are materials that behave as metals for one range of parameters, but behave as insulators for another range. The separation between these two behaviors is often very sharp. Changing a physical parameter slightly can cause the material to undergo a metal–insulator transition. In this section several causes for this will be studied. The study begins with the phenomenon of percolation. One is concerned with an inhomogeneous system consisting of isolated conducting clusters (in three dimensions) or islands (in two dimensions). As their volume fraction increases, eventually they make contact with each other and the system can conduct electricity. This is followed by a discussion of the Mott transition, which is appropriate to homogeneous systems. In Section W7.5 the phenomenon of localization in solids is introduced, with attention being given to both weak and strong (Anderson) localization.

7.16 Percolation

Classical transport through a random medium may often be idealized by studying motion on a lattice. There are two conventional ways to introduce randomness: by

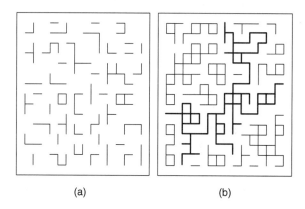

(a)   (b)

Figure 7.15. Random-bond model for a square lattice for the cases (a) $p < p_{cb}$ and (b) $p > p_{cb}$.

means of random-site or random-bond occupancy. In the *random-site* (RS) *model*, any site on the lattice is either occupied, with probability p, or vacant, with probability $q = 1 - p$. If NN sites are occupied, a path (bond) is drawn between these sites: otherwise, it is omitted. In the *random-bond* (RB) *model*, one distributes bonds randomly between all NN pairs of sites in the lattice, with probability p. In either case, connected clusters of bonds are formed. The distribution of cluster sizes is determined by the dimensionality of the lattice, d; the symmetry of the lattice; and the parameter p. If p is less than a critical probability (p_{cs} for RS, p_{cb} for RB), all clusters are finite in size. For $p > p_c$ (generic for p_{cs} or p_{cb}) there exists an infinite cluster spanning the lattice, in which case one says that *percolation* has occurred. Electrical conduction, for example, can occur only if there is percolation. In Fig. 7.15 a sketch is made of the RB model for prepercolation and percolation conditions on a square lattice. The percolated cluster is darkened in the figure.

Percolation theory is usually studied by means of simulations on a computer. Some crude insights into percolation theory can be obtained by studying a simple model that may be analyzed analytically. This is the tree structure called a *modified Bethe lattice* (Fig. 7.16). Each interior site (except the center) has Z nearest neighbors. There

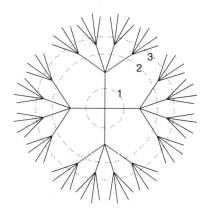

Figure 7.16. Modified Bethe lattice, illustrated for the case $Z = 5$.

are no closed loops in this structure. The dotted circles identify levels of increasing complexity in the lattice. The number of branches in zone n is $(Z-1)^n$. One wishes to know whether a particle starting at the center of the lattice will find its way to the outer perimeter by passing through occupied bonds. The main interest is in the limit as $n \to \infty$.

In the case $Z = 2$, the modified Bethe lattice is simply a single line passing through all the sites. The probability of reaching the nth level starting at the center is p^n. This will vanish as $n \to \infty$ for $p < 1$. For $Z = 2$, any break in the lattice is sufficient to prevent percolation.

The average number of paths that connect the center to the perimeter, N_p, is the product of the number of paths at level n, $(Z-1)^n$, multiplied by the probability of surviving the n steps to the perimeter, p^n:

$$N_p = [p(Z-1)]^n. \tag{7.129}$$

If $p(Z-1) < 1$, this will vanish in the limit $n \to \infty$. If $p(Z-1) > 1$, N_p will, in fact, be infinite. Therefore, $p_{cb} = 1/(Z-1)$ for the Bethe lattice.

For more familiar lattices in two and three dimensions, values for p_{cb} and p_{cs} are usually determined by computer simulations using Monte Carlo techniques. Values of these parameters are presented in Table 7.4. The critical cluster is found to be a fractal structure of noninteger dimension, D. An approximate formula for the RB critical probability is

$$p_{cb} \approx \frac{1}{Z} \frac{d}{d-1}. \tag{7.130}$$

One may think of the factor $d/(d-1)$ as a correction arising from the formation of closed loops in a finite-dimensional lattice.

A number of universal scaling relations pertain to the vicinity of the critical point, $p \approx p_c$. For example, the probability that a given bond belongs to the percolated cluster is $P \sim (p - p_c)^\beta$, with $\beta = (0.19, 0.41)$ for $d = (2, 3)$. The fraction of bonds that actually carry current (the *backbone*) is $P' \sim (p - p_c)^{\beta'}$, with $\beta' = (0.48, 1.05)$ for $d = (2, 3)$. For $p > p_c$, homogeneity of the lattice, on average, is maintained on distance scales larger than the *correlation length*, ξ. The correlation length obeys the scaling formula $\xi \sim |p - p_c|^{-\nu}$, with the critical exponent given by $\nu = (1.33, 0.88)$

TABLE 7.4 Percolation Thresholds for Some Lattices in Two and Three Dimensions

Lattic	Z	d	p_{cs}	D^a	p_{cb}
Honeycomb	3	2	0.653	91/48	0.696
Square	4	2	0.500	91/48	0.593
Triangular	6	2	0.347	91/48	0.500
Diamond	4	3	0.389	2.53	0.430
SC	6	3	0.249	2.53	0.312
BCC	8	3	0.180	2.53	0.246
FCC	12	3	0.198	2.53	0.119

Source: Data from M. Sahimi, *Applications of Percolation Theory*, Taylor & Francis, London, 1994.

[a]Fractal dimension of the critical percolation cluster.

for $d = (2, 3)$. For $p < p_c$, ξ also determines the size of a typical cluster. Note that the correlation length diverges at $p = p_c$.

Percolation is useful, for example, in analyzing the conductivity of a *cermet*, a ceramic material with embedded metallic clusters. For low cluster concentration the material is an insulator. As the concentration increases, one passes through the percolation threshold and the conductivity grows until it saturates at the value of appropriate to an amorphous metal.

7.17 Mott Metal–Insulator Transition

Imagine assembling a solid by starting with an array of atoms separated from each other by large distances and gradually decreasing the lattice constant until solid-state densities are achieved. When far apart, electrons are bound to their individual ions and the material is an insulator. When close together the interaction of an electron with its ion gets weakened by the presence of other electrons. If the other electrons were free to move around, for example, the ion potential will be shielded from its electron. For sufficiently strong shielding it is possible that the electron will no longer be bound to its ion and will delocalize over the solid. Therefore, a metal–insulator (M-I) transition may be expected to be seen. In more general situations a host insulator or semiconductor may be doped with impurity atoms and the conductivity studied as a function of increasing concentration of donors or acceptors.

The condition for the M-I transition density will be developed for the case of hydrogen atoms, although, strictly speaking, there is the need to consider atoms with an even number of electrons, so the bands have a possibility of being either full or empty. Begin by assuming that there is a concentration of free electrons, n, and look for the condition that the bound state is destroyed. In a metal the ion–electron interaction is given by the screened Coulomb interaction (again assuming that $Z = 1$)

$$V(r) = -\frac{e^2}{4\pi\epsilon r} \exp(-k_{\text{TF}}r), \qquad (7.131)$$

where $k_{\text{TF}} = (3ne^2/2\epsilon E_F)^{1/2}$ is the inverse Thomas–Fermi screening length. Here ϵ is the electric permittivity of the host material and E_F is the Fermi energy.

The derivation of Eq. (7.131) proceeds from a semiclassical expression for the chemical potential of an electron

$$\frac{\hbar^2 k_F^2}{2m_e^*} - e\phi = \mu, \qquad (7.132)$$

where $k_F = (3\pi^2 n)^{1/3}$ and ϕ is the potential due to the ion. Combining this with Poisson's equation,

$$\nabla^2\phi = \frac{e}{\epsilon}(n - n_0), \qquad (7.133)$$

with n_0 being the background ion density, and linearizing about $n = n_0$ gives the Yukawa equation:

$$(\nabla^2 - k_{\text{TF}}^2)\phi = -\frac{e}{\epsilon}\delta(\mathbf{r}), \qquad (7.134)$$

whose solution is the screened Coulomb potential given in Eq. (7.131).

A variational estimate of the energy of a bound state may be made. Assume a wavefunction of the form $\psi(r) = N \exp(-\alpha r)$ with the normalization constant given by $N = \alpha^{3/2}/\sqrt{\pi}$. The expectation value of the energy in this state is

$$E = \left\langle \psi \left| \left(\frac{p^2}{2m_e^*} + V \right) \right| \psi \right\rangle = \frac{\hbar^2 \alpha^2}{2m_e^*} - \frac{e^2 \alpha^3}{\pi\epsilon(2\alpha + k_{TF})^2}. \tag{7.135}$$

For small n, where k_{TF} is also small, there exists an absolute minimum in the E versus α curve for $\alpha > 0$ and the material is an insulator. For small α a minimum occurs at $\alpha = 0$ and the electron is delocalized over the solid so that the material is a metal. The metal–insulator transition occurs when a critical value for α is reached such that $E = 0$ and $\partial E/\partial \alpha = 0$ occur simultaneously, that is, when

$$\frac{4\pi\hbar^2 \epsilon k_{TF}}{m_e^* e^2} = k_{TF} a_d = 1, \tag{7.136}$$

where $a_d = 4\pi\hbar^2 \epsilon/m_e^* e^2$ is the Bohr radius for donors in the material.

CONDUCTIVITY OF REDUCED-DIMENSIONAL SYSTEMS

As integrated circuits become smaller and smaller, one begins to probe the intrinsic limits on the conduction process set by nature. Quantum wires refer to one-dimensional conductors whose length is comparable to or less than the elastic scattering length of the electrons passing through them. In this section two aspects of quantum wires are studied. First, a newly discovered system presented by nature, the carbon nanotube, is examined. Then the Landauer theory for conductivity will be developed and it will be shown that the resistance of a carbon nanotube is quantized.

7.18 Carbon Nanotubes

In graphite each carbon atom sits at a vertex of a planar hexagonal honeycomb lattice separated from its nearest neighbors by a bond distance $d = 0.142$ nm. The bonds are composed of sigmalike molecular orbitals arising from the interaction of sp^2 hybrid orbitals on the NN C atoms. There also delocalized π-like molecular orbitals. In the carbon nanotube most of the graphite sheet is wrapped into a seamless cylinder of finite diameter rather than extending over a plane. This requires an admixture of sp^3 hybrid orbitals, with the amount increasing as the radius of curvature decreases. The nanotubes currently fabricated in a carbon arc at a temperature of $T = 4000$ K are typically 1 μm in length, with diameters ranging from 1 to 20 nm. The ends of the tubes are often capped by an array of six carbon pentagons. In some cases one finds nested nanotubes with an intertubular separation of 0.34 nm, which is close to the distance between adjacent planes in graphite. Often, the nanotubes are bundled together to form filamentary fibers. Even closed rings have been found. The interest in these nanotubes stems from their high mechanical strength, light weight, and their potential for use as microscopic wires. When subjected to bending stresses, the nanotubes will deform, but they return to their original shape promptly when the stress is removed as long as no rupture occurs.

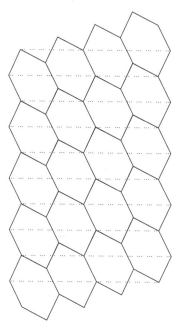

Figure 7.17. Section of graphite about to be wrapped into a chiral nanotube. The dotted lines denote points to be joined.

The nanotubes are observed to have varying degrees of chirality (i.e., one may think of rows of C atoms as if they were wire wound around a cylinder with various degrees of pitch). Consider the primitive honeycomb lattice with a row of parallel hexagons. Adjacent to this is another row of hexagons offset from the first row (Fig. 7.17), with the pattern repeated periodically. The hexagons are located by specifying the primitive vectors $\mathbf{u}_1 = \hat{i} d\sqrt{3}$, $\mathbf{u}_2 = d(\hat{i}\sqrt{3} + 3\hat{j})/2$, and writing $\mathbf{R}_{mn} = m\mathbf{u}_1 + n\mathbf{u}_2$, where m and n are integers. The chirality comes about during the folding into a cylinder when the hexagon at $\mathbf{R}_{00}$ is made to overlap the hexagon at $\mathbf{R}_{mn}$. The circumference of the cylinder is therefore $C = |m\mathbf{u}_1 + n\mathbf{u}_2|$ or $C = d\sqrt{3(m^2 + n^2 + mn)}$, and the diameter is $D = C/\pi$. This is illustrated in Fig. 7.17, where $m = 4$ and $n = 1$. In the limit where m and n become infinite, the nanotube becomes equivalent to a graphite sheet.

Depending on the numbers (m, n) determining chirality, the fiber can be either a semiconductor or a metal. This comes about as a consequence of the tight-binding model for graphite with NN interaction t. The honeycomb lattice has two inequivalent sites, with the π-electron wavefunction amplitudes on these sites being $A_{m,n}$ and $B_{m,n}$, in the unit cell designated by $\mathbf{R}_{m,n}$. The coupled tight-binding equations determining the energy eigenvalue, E, are

$$t(B_{m,n} + B_{m-1,n} + B_{m,n-1}) = EA_{m,n}, \qquad (7.137a)$$

$$t(A_{m,n} + A_{m-1,n} + A_{m,n-1}) = EB_{m,n}. \qquad (7.137b)$$

(see Sections 7.8 and 7.9). Letting

$$A_{m,n} = A \exp(i\mathbf{k} \cdot \mathbf{R}_{mn}) \qquad (7.138a)$$

and

$$B_{m,n} = B \exp\left[i\mathbf{k} \cdot \mathbf{R}_{mn} + i\mathbf{k} \cdot d\left(\hat{i}\frac{\sqrt{3}}{2} + \hat{j}\frac{3}{2}\right)\right], \qquad (7.138b)$$

one finds that

$$E^2(k) = t^2|1 + e^{i\mathbf{k}\cdot\mathbf{u}_1} + e^{i\mathbf{k}\cdot\mathbf{u}_2}|^2. \qquad (7.139)$$

The bonding and antibonding energy bands for the delocalized π-electron system are given by

$$E_\pm = \pm t\sqrt{1 + 4\cos\frac{3k_y d}{2}\cos\frac{k_x d\sqrt{3}}{2} + 4\cos^2\frac{k_x d\sqrt{3}}{2}}. \qquad (7.140)$$

The value of t is 2.66 eV, as determined by a fit to the graphite band structure.

Wrapping the structure into a chiral cylinder restricts the solutions to a one-dimensional band where the values of $\mathbf{k}$ are constrained by the periodicity condition $\mathbf{R}_{mn} \cdot \mathbf{k} = 2\pi j$, where j is an integer. Instead of having a two-dimensional band structure extending over the hexagonal first Brillouin zone, the band structure is defined on a series of parallel lines in the Brillouin zone given by this formula. In Fig. 7.18, electronic band structures are presented for the cases $(m, n) = (8, 0)$ and $(m, n) = (7, 1)$. Note that in the former case there is a gap between the bonding and antibonding states, so the nanotube is a semiconductor. In the latter case there is no gap and the material is metallic.

The bonding and antibonding bands will be degenerate at some point in $\mathbf{k}$ space when

$$1 + e^{i\mathbf{k}\cdot\mathbf{u}_1} + e^{i\mathbf{k}\cdot\mathbf{u}_2} = 0. \qquad (7.141)$$

If $\mathbf{k} \cdot \mathbf{u}_1 = 2\pi/3$ and $\mathbf{k} \cdot \mathbf{u}_2 = 4\pi/3$, this equation will be satisfied. (Another possibility is $\mathbf{k} \cdot \mathbf{u}_1 = 4\pi/3$ and $\mathbf{k} \cdot \mathbf{u}_2 = 2\pi/3$.) Coupling this with the periodicity condition, it is found that

$$\mathbf{k} \cdot \mathbf{R}_{mn} = \frac{2\pi m}{3} + \frac{4\pi n}{3} = 2\pi j. \qquad (7.142)$$

This will be satisfied for $m + 2n = 3j$. (In the other case it is $n + 2m = 3j$.) When this condition is obeyed, there is no gap between the bonding and antibonding bands, and the nanotube is a metal. When $m + 2n = 3j + 1$ or $m + 2n = 3j + 2$, there is a gap and it is a semiconductor.

It is possible to introduce defects into the nanotube in such a way that the number of edges, vertices, and faces is not changed. One way of doing this is to replace a pair of hexagons by a pentagon and a heptagon. Several other ways of accomplishing this are possible. When one introduces these defects, a bend is introduced, the diameter and helicity of the tubule changes, and the value of $m + 2n$ changes, since $m \to m \pm 1$ and $n \to n \mp 1$. It is thus possible to change metals to semiconductors and to fabricate metal–semiconductor junctions simply by inserting a pentagon–heptagon pair in place of a pair of hexagons. More generally, it should be possible to fabricate nanotube heterojunctions by controlling the defect locations. For example, the (7,1) tube is metallic, whereas the (8,0) tube is a semiconductor with a 1.2 eV energy gap. A sketch of the (8,0)/(7,1) junction is given in Fig. 7.19. Some additional features of carbon nanotubes may be found in Section W7.6.

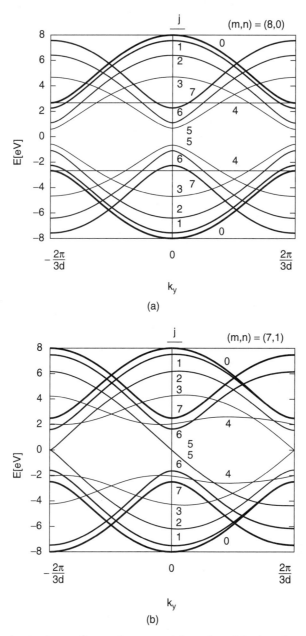

Figure 7.18. Band structures for carbon nanotubes for (a) $(m, n) = (8, 0)$ and (b) $(m, n) = (7, 1)$. The various curves correspond to values of j in the range 0 to 7.

There is evidence for *weak localization* in the electric conduction of nanotubes at low temperatures (see Section W7.5). This comes about because the probability that an electron entering a nanotube will backscatter is increased because of constructive interference between a given electron path and a time-reversed path. If backscattering is enhanced, it comes at the expense of transmission through the nanotube

Figure 7.19. The (8,0)/(7,1) junction between two carbon nanotubes. [From L. Chico et al, *Phys. Rev. Lett.*, **76**, 971 (1996). Copyright 1996 by the American Physical Society.]

(i.e., conduction). Four-probe resistance measurements have shown a marked variation of the resistance of a nanotube with diameter and helical pitch. Resistances per unit length are typically in the range 10^4 to 10^5 Ω/m for diameters in the range 5 to 20 nm. Current densities as large as 6×10^{10} A/m^2 could be passed through a nanotube without causing irreversible damage. The resistance grows considerably when there are curves or kinks in the tubes.

There is also evidence[†] that carbon nanotubes can serve as quantum-mechanical wires in the sense that conduction occurs by having electrons tunnel through delocalized quantum states. When idealized as a strictly one-dimensional system of length L, the states of the wire are given by $E_n = (\hbar n\pi)^2/2mL^2$, with n a positive integer. The energy separation between states, $\Delta E = \hbar\pi v_F/L$, can be substantial (i.e., on the order of 0.6 meV for a length $L = 3$ μm). Coherence lengths in excess of 140 nm and perhaps as large as L have been inferred. For a Fermi velocity of 0.8×10^6 m/s this implies a coherence time of several tenths of a picosecond.

It is possible both to fill and coat carbon nanotubes with molten materials, and thus have them serve as templates. This occurs if the surface tension of the adsorbate

[†] S. J. Tans et al., *Nature*, **386**, 474 (1997).

is sufficiently low ($\gamma < 0.2$ N/m). For example, such materials as Pb, Bi, V_2O_5, Se, S, Cs, and Rb as well as oxides of Ni, Co, Fe, and U have been taken up by the nanotubes. The formation of metallic nanowires seems to be related to the presence of an incomplete electronic shell of the metal ions.

Recently, it was shown that arrays of carbon nanotubes can function as field-emission electron sources and may therefore play a future role in the construction of flat-panel displays. The large enhancement of the local electric field due to the sharpness of the tube (lightning-rod effect) allows them to function with very high efficiency, despite the nearly 5 eV work function of graphite. Additional properties of carbon nanotubes are discussed Chapter W21.

7.19   Landauer Theory of Conduction

A quantum-mechanical description of the conduction process in one-dimensional wires was developed by Landauer[†]. Rather than attempt to derive the general formula, in this section attention is focused on an idealized model for the single-walled carbon nanotube, and a formula for its conductance is developed. It will be seen that this conductance is quantized (i.e., comes in multiples of a basic conductance).

The carbon nanotube will be treated as a hollow cylinder described by the coordinates (z, ϕ) and the electron is assumed to move freely on the surface of the tube. The Schrödinger equation governing the electrons is

$$\left[-\frac{\hbar^2}{2m^*} \left(\frac{\partial^2}{\partial z^2} + \frac{1}{a^2} \frac{\partial^2}{\partial \phi^2} \right) - E \right] \psi(z, \phi) = 0, \qquad (7.143)$$

where a is the radius of the tube. Here m^* is the effective mass of an electron in the conduction band. One writes the wavefunction in the form $\psi = \exp[i(kz + \mu\phi)]$, where μ must be an integer to maintain single valuedness in the azimuthal direction, and

$$k = \sqrt{\frac{2m^*E}{\hbar^2} - \frac{\mu^2}{a^2}}. \qquad (7.144)$$

Note that there are two values for μ, leading to the same k, as long as $\mu \neq 0$. Also note that for a given value of E, there is a maximum magnitude that μ can have, leading to a real value for k. Landauer viewed quantum currents as transmission processes. Let τ denote the transmission amplitude. The current is given by

$$I = -e \sum_s \sum_\mu \int_{-\infty}^{\infty} \frac{dk}{2\pi} v_z \Theta(k) f(E, T)[1 - f(E - eV, T)]|\tau|^2, \qquad (7.145)$$

where the unit step function $\Theta(k)$ ensures that there is only a forward current. The s-sum is over the two spin projections and the μ-sum is over allowable integers. Since the nanotube has a very long elastic-scattering length, it will be assumed that the

---

[†] R. Landauer, *Philos Mag.*, **21**, 863 (1970). See also D. S. Fisher and P. A. Lee, *Phys. Rev. B*, **23**, 6851 (1981).

transmission is perfect and $|\tau|^2 = 1$. One says that the conduction is ballistic. The voltage drop along the tube is denoted by V. For low-enough T, as in the case of metals, one may approximate the Fermi factor product as

$$f(E, T)[1 - f(E - eV, T)] \approx f(E, T)\Theta(E - eV - E_F)$$

$$\approx f(E, T)[\Theta(E - E_F) - eV\delta(E - E_F)]$$

$$\approx -\frac{eV}{2}\delta(E - E_F). \tag{7.146}$$

The current becomes

$$I = \frac{e^2 V}{2h} \sum_s \sum_\mu \int_0^\infty \delta\left(E_\mu + \frac{\hbar^2 \mu^2}{2m^* a^2} - E_F\right) dE_\mu$$

$$= \frac{e^2 V}{2h} \sum_s \sum_\mu \Theta\left(E_F - \frac{\hbar^2 \mu^2}{2m^* a^2}\right) \equiv GV, \tag{7.147}$$

where G is the conductance. Thus

$$G = \frac{2e^2}{h}\left(n + \frac{1}{2}\right). \tag{7.148}$$

Here n is the number of subbands, defined by $|\mu|$, that can conduct.

Experiments on carbon nanotube quantum resistors[†] find $G = 2e^2/h = (12.9 \text{ k}\Omega)^{-1}$, consistent with $n = \frac{1}{2}$. Although this is not yet fully understood, it may be that the Coulomb blockade effect prevents a larger number of electrons from traversing the quantum wire.

REFERENCES

Ashcroft, N. W., and N. D. Mermin, *Solid State Physics*, Holt, Rinehart and Winston, New York, 1976.

Fletcher, G. C., *The Electron Band Theory of Solids*, North-Holland, Amsterdam, 1971.

Rossiter, P. L., *The Electrical Resistivity of Metals and Alloys*, Cambridge University Press, New York, 1991.

Sahimi, M., *Applications of Percolation Theory*, Taylor & Francis, London, 1994.

PROBLEMS

7.1 Graphite is a hexagonal crystal with a four-atom unit cell. In one plane there is a honeycomb lattice with lattice constant a = 0.246 nm. The adjacent layer has atoms sitting over the centers of the hexagons of the honeycomb lattice. The interplanar spacing is large (0.337 nm) (see Fig. 1.27). This allows the interplanar

---

[†] S. Frank et al, *Science*, **280**, 1744 (1998).

coupling to be neglected as a first approximation. Use the two-dimensional tight-binding approximation to obtain the band structure for graphite.

7.2 Use the tight-binding approximation to obtain the band structure for linear poly-acetylene. You may assume that there is one mobile π-electron per carbon atom (refer to Fig. W14.7).

7.3 Electrons are confined to a two-dimensional sheet with areal density N. If a positive ion is placed on the sheet analyze the role played by the electrons in possibly screening the potential of the ion.

7.4 Derive the equations for J_x, J_y, and J_z in Section 7.13 for the magnetoresistance.

7.5 Show that the electronic density of states at E_F according to the free-electron model is $3z/2E_F$ per atom, $3N/4E_F$ per spin, and $3n/2E_F$ per unit volume (z = valence, N = number of electrons, n = density of electrons).

7.6 For energies E and temperatures T at which the Fermi–Dirac distribution $f(E, T) \ll 1$, show that $f(E, T)$ reduces to the classical Maxwell–Boltzmann distribution $\exp[-\beta(\mathbf{E} - \boldsymbol{\mu})]$. For Cu at $T = 300$ K, to what electron energies $\mathbf{E} - \boldsymbol{\mu}$ does this correspond? What fraction of the electrons in Cu at $T = 300$ K are excited to these, or higher, energies?

7.7 Calculate the mean free path and mobility of electrons in Cu at $T = 300$ K using the electrical conductivity and the free-electron theory.

7.8 Calculate the drift velocity, $\langle v \rangle$, of electrons in Cu at $T = 300$ K in an electric field of 100 V/m. Compare this with the Fermi velocity and explain the difference.

7.9 Compute the isothermal compressibility $K = -(1/V)(\partial V/\partial P)_T$ of a free-electron gas.

7.10 Use a computer to calculate the π-electron energy spectrum of a fullerene molecule, C_{60}, assuming a tight-binding model with nearest-neighbor interactions. The molecule is in the shape of a regular dodecahedron (see Fig. 4.8).

7.11 Show that in the tight-binding approximation [see Eq. (7.93)] $\langle (E_0 - E)^2 \rangle = Z|t|^2$, where the average is taken over the band.

7.12 Calculate the enhancement of the conductivity $\sigma(E)/\sigma_0$ at $T = 300$ K according to the Poole–Frenkel effect [Eq. (7.128)] for electric field strengths $E = 10^6$, 10^7, and 10^8 V/m in α-SiO_2 and in NaCl.

7.13 Consider the high-field, $\omega_c\tau \gg 1$ limit of Eq. (7.109). Show that $\sigma_{xx} \to 0$ and $\sigma_{xy} = -ne/B = -\sigma_{yx}$ in this limit. The quantity $\sigma_{xy} = 1/R_H B$ is known as the *Hall conductivity*.

CHAPTER 8

Optical Properties of Materials

8.1 Introduction

In this chapter the fundamental optical properties of materials are discussed, beginning with a brief review of the electromagnetic spectrum, and showing how the interaction of electromagnetic radiation with materials is governed by the complex index of refraction. An expression for the ac conductivity is derived using Drude theory. The reflectivity of solids is discussed, with particular emphasis on metals and ionic crystals. The optical properties of semiconductors are then considered and a comparison is made with those of metals and insulators. Band-structure effects are taken into account and a general expression for the linear dielectric tensor is obtained. The Kramers–Kronig relations, relating the real and imaginary parts of the dielectric function, are introduced. Consideration is then given to the index of refraction of composite media.

A quantum-mechanical exposition of the nonlinear optical polarizability is given, with attention limited to second- and third-harmonic generation. The chapter continues with a discussion of the role played by excitons and color centers. It concludes with an analysis of the emissivity of solids. Supplementary material concerning propagation in anisotropic media and the index ellipsoid, phase matching, polaritons, and the nonlocal dielectric function appears at the Web site.[†] A succinct review of Maxwell's equations also appears at the Web site.

Optical properties of materials are important for a number of reasons. First, they provide useful information concerning the electronic band structure and lattice modes of materials. They are also a probe of impurities and defects. The properties are important for technological applications, such as the design of lasers, light-emitting diodes, nonlinear-optical crystals, and so on, discussed in Chapter 18, where particular optical materials are studied.

Light interacts with materials in several different ways. It could simply be transmitted through matter, with little interaction other than refraction. It could be reflected, with the degree of reflectivity determined by optical properties of the material. It could also be absorbed in processes involving the excitation of electron–hole pairs or the excitation of lattice modes. There could be luminescence or phosphorescence. It could undergo Raman or Brillouin scattering. It could excite excitons or excitations associated with defects or impurities. It could damage the material, ablate material from the surface, or initiate photochemical changes.

[†] Supplementary material for this textbook is included on the Web at the resource site (ftp://ftp.wiley.com/public/sci_tech_med/materials). Cross-references to elements of the Web material are prefixed by "W."

8.2 The Electromagnetic Spectrum

A brief overview of the electromagnetic (EM) spectrum is provided in Table 8.1. The types of radiation are listed, along with their wavelength and frequency ranges, and typical methods of generating and detecting the radiation. The applications are well known. Much of the technology of the present century has been involved with employing EM radiation for such diverse applications as communications, unraveling the mysteries of nature, and medical instrumentation. Progress is still being made. The field of photonics (i.e., the application of light to technology) is now blossoming, supplementing traditional electronic components and devices with high-speed optical counterparts.

The classical theory of electromagnetism was formulated by Maxwell, Faraday, Ampère, and Gauss, among others. A brief review is presented in Appendix W8A. The quantum interpretation was provided by Planck and Einstein. In this chapter some of the physics relating to optical properties of matter are developed. In Chapter 18, we study optical materials.

Some general remarks are in order. Although always present, quantum effects become important when the photon energy, $\hbar\omega$, becomes comparable to or larger than characteristic excitation energies in matter. This accounts for the transition in methods for producing and detecting radiation as one progresses toward higher frequencies. At high frequencies quantum processes (absorption or emission of photons) characterize the interaction of radiation and matter. At low frequencies the number of photons involved is so large that a classical description suffices.

Use is made of some elementary formulas from electromagnetic theory. A monochromatic plane electromagnetic wave, propagating along the z direction, is described by its electric field,

$$\mathbf{E}(z, t) = \hat{i}E_0 e^{i(kz-\omega t)}, \tag{8.1a}$$

TABLE 8.1 Overview of the Electromagnetic Spectrum

Type of Radiation[a]	Wavelength λ (m)	Frequency f (Hz)	Methods of Generation	Methods of Detection
Radio wave	>0.1	$<3 \times 10^9$	Electronics	Electronics
Microwave	0.1 to 0.001	3×10^9 to 3×10^{11}	Klystron, magnetron, Gunn diode, maser	Diodes
Infrared radiation	0.001 to 7×10^{-7}	3×10^{11} to 4.3×10^{14}	LEDs, laser, thermal	Diodes, photoconductor
Visible light	7×10^{-7} to 4×10^{-7}	4.3×10^{14} to 7.5×10^{14}	Incandescence, arc lamp, vapor lamp, LEDs, fluorescence, laser	Photography, detectors, photoconductors, photocells
Ultraviolet light	4×10^{-7} to 10^{-9}	7.5×10^{14} to 3×10^{17}	Arc lamp, laser, synchrotron	Photography, detectors, photocells
X-ray	10^{-9} to 10^{-11}	3×10^{17} to 3×10^{19}	X-ray tube, radioactivity, synchrotron	Photography, detectors, scintillation counters
Gamma ray	$<10^{-11}$	$>3 \times 10^{19}$	Radioactivity	Photography, detectors, scintillation counters

[a]The boundaries between the various types of radiation are somewhat arbitrary.

and magnetic field intensity

$$\mathbf{H}(z, t) = \hat{j} H_0 e^{i(kz - \omega t)}. \tag{8.1b}$$

It is understood that the actual real fields are obtained by taking the real parts of these complex expressions. Here E_0 and H_0 denote the amplitudes of the respective fields, $k = 2\pi/\lambda$ is the wave vector, or propagation constant, expressed in terms of the wavelength, λ, and $\omega = 2\pi f$ is the angular frequency of the radiation, f being its frequency. The intensity (power per unit area) of the wave, I, is given by the magnitude of the Poynting vector [see Eq. (W8A.11) at our Web site]. Thus

$$\mathbf{S} = \tfrac{1}{2}\mathrm{Re}(\mathbf{E} \times \mathbf{H}^*), \tag{8.2}$$

where the real part [$\mathrm{Re}(\cdots)$] of the cross-product is taken. Recall that $\omega = k c_m$ and $c_m = 1/\sqrt{\epsilon\mu}$, where c_m is the speed of light in the material, with ϵ the permittivity and μ the permeability of the material. Using the formula connecting E_0 and H_0 [see Eq. (W8A.4)], $H_0 = \omega\epsilon E_0/k$, the intensity may be expressed as

$$I = \frac{1}{2}|E_0|^2 \, \mathrm{Re}\left(\sqrt{\frac{\epsilon}{\mu}}\right) e^{i(k - k^*)z}. \tag{8.3}$$

As will be seen shortly, ϵ is to be treated as a complex parameter.

In optics the primary quantity of interest is the complex index of refraction, $\tilde{n} = n + i\kappa$, where both n and κ are functions of frequency. The real part of $\tilde{n}$ determines the speed of propagation of light in the material: $c_m = c/n$. The imaginary part, κ, is the extinction coefficient and determines the attenuation of the light intensity as it propagates in a material (Beer's law):

$$I(z) = I_0 \exp\left(-2\frac{\kappa\omega}{c}z\right) = I_0 \exp(-\alpha z). \tag{8.4}$$

After traveling a distance z through the material, the intensity is attenuated from I_0 to $I(z)$. The quantity α is called the *absorption coefficient*. In general, it is frequency dependent. In isotropic materials or crystalline solids with cubic symmetry, α is independent of the direction of propagation. In anisotropic materials it is a function of the orientation of the polarization vector of the electric field of the light wave (i.e., the direction in which the electric field points) relative to the crystal axes. In linear materials α is independent of the intensity of the light. In nonlinear materials it depends on the intensity. In some materials it may also be nonlocal (i.e., the interaction of the light with the material at one point in space depends on the electric field at other neighboring points as well). In such a case, α depends on the wave vector as well as the frequency.

8.3 AC Conductivity of Metals

The Drude model for conductivity is readily generalized to include alternating-current (ac) effects. Newton's second law gives

$$m\left(\frac{d\mathbf{v}}{dt} + \frac{\mathbf{v}}{\tau}\right) = -e\mathbf{E}(t) = -e\mathbf{E}_0 e^{-i\omega t}, \tag{8.5}$$

where $-m\mathbf{v}/\tau$ is a velocity-dependent friction force and τ is a relaxation time. Although a complex notation is used, it is understood that only the real part of this expression has a direct physical meaning. There is a steady-state solution of the form $\mathbf{v} = \mathbf{v}_0 \exp(-i\omega t)$, where the amplitude is

$$\mathbf{v}_0 = -\frac{e}{m}\frac{\mathbf{E}_0}{-i\omega + 1/\tau}. \tag{8.6}$$

The current density vector is written similarly, as $\mathbf{J} = \mathbf{J}_0 \exp(-i\omega t)$. Combining this with the expressions $\mathbf{J}_0 = -ne\mathbf{v}_0$ and $\mathbf{J}_0 = \sigma(\omega)\mathbf{E}_0$ gives an expression for the frequency-dependent conductivity:

$$\sigma(\omega) = \frac{ne^2}{m}\frac{\tau}{1 - i\omega\tau} = \frac{ne^2\tau}{m}\frac{1 + i\omega\tau}{1 + \omega^2\tau^2}. \tag{8.7}$$

It is seen that as ω decreases, the real part, $\mathrm{Re}[\sigma(\omega)]$, increases and the formula reduces to the expression obtained previously, Eq. (7.5), for $\omega \to 0$. The imaginary part, $\mathrm{Im}[\sigma(\omega)]$, goes to zero as $\omega \to 0$. The current density may be combined with the free-space displacement current and expressed as

$$\mathbf{J} + \frac{\partial \epsilon_0 \mathbf{E}}{\partial t} = \frac{\partial \epsilon \mathbf{E}}{\partial t}, \tag{8.8}$$

which yields a formula for the frequency-dependent permittivity:

$$\epsilon(\omega) = \epsilon_0 + i\frac{\sigma(\omega)}{\omega}. \tag{8.9}$$

This leads to a formula for the frequency-dependent dielectric function:

$$\epsilon_r(\omega) = \frac{\epsilon(\omega)}{\epsilon_0} = 1 - \frac{\omega_p^2}{\omega(\omega + i/\tau)}, \tag{8.10}$$

where ω_p denotes the plasma frequency, defined through the relation

$$\omega_p^2 = \frac{ne^2}{m\epsilon_0}. \tag{8.11}$$

The significance of ω_p is that it is the natural frequency of oscillation of the electron gas.

In Table 8.2, values of the corresponding plasma energies, $E_p = \hbar\omega_p$, are presented for some metals. Theoretical values, based on Eq. (8.11), are compared with experimental data, obtained from the condition $\mathrm{Re}[\epsilon_r(\omega_p)] \approx 0$. Reasonably good agreement is found for the alkali metals and Al. Although good conductors, the noble metals (Cu, Ag, Au) are seen not to be free-electron metals. The agreement for Fe, a transition metal, is probably coincidental. For Si, although it is a semiconductor, good agreement is found if it is assumed that the four valence electrons respond as a plasma.

8.4 Reflectivity

Some materials are suitable for mirrors because they reflect all spectral components of white light with little absorption or transmission. Other materials appear colored, since

TABLE 8.2   Comparison of Calculated Plasma Energies, E_p, with Experimental Values[a]

Metal	E_p (eV)	
	Calculated	Experimental
Li	8.04	6.64
Na	6.05	5.4–5.94
K	4.40	3.79–4.15
Rb	3.98	3.45
Cs	3.54	2.82
Al	15.8	12.2–14.7
Cu	10.8	8.0
Ag	8.98	3.83

[a]Some data deduced from E. D. Palik, ed., *Handbook of Optical Constants of Solids*, Academic Press, San Diego, Calif., 1985, by looking for $\epsilon_r(\omega) = 0$.

they selectively absorb particular wavelength ranges. In addition, depending on the surface conditions, the reflection may be specular, with parallel incident rays mapping into parallel reflected rays, or diffuse, in which case the reflected rays go in different directions. Specular reflection occurs when the surface is smooth, whereas diffuse scattering occurs when the surface is rough. The characteristic roughness length scale l_r differentiating these two regimes is the wavelength of the light, with nonspecular scattering being enhanced when $\lambda < l_r$.

Consider the interface between two media, characterized by the electrical permittivity functions ϵ and ϵ'. In general, the interface could be either solid–solid, solid–liquid, or liquid–liquid. In Fig. 8.1 a planar interface is depicted with an incident wave of electric and magnetic field amplitudes (E_1, B_1) impinging at normal incidence. A reflected wave (E_2, B_2) and a transmitted wave (E_3, B_3) are produced. The wave vectors are k and k', respectively. It will be assumed that the materials are nonmagnetic, so $\mathbf{B} = \mu_0\mathbf{H}$ everywhere. The constitutive equations for the electric displacement vector are $\mathbf{D} = \epsilon\mathbf{E}$ or $\mathbf{D} = \epsilon'\mathbf{E}$ in the two materials, respectively. Note that for a perfect conductor, the field $\mathbf{E}_3$ would be zero, and the vector sum of $\mathbf{E}_1$ and $\mathbf{E}_2$ at the surface would correspondingly be zero.

From Faraday's law it follows that

$$kE_1 = \omega B_1, \quad kE_2 = \omega B_2, \quad k'E_3 = \omega B_3, \tag{8.12}$$

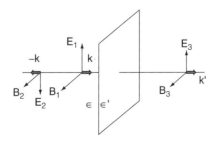

Figure 8.1. Incident, reflected, and transmitted wave at an interface.

while Maxwell's generalization of Ampère's law implies that

$$kB_1 = \omega\mu_0\epsilon E_1, \quad kB_2 = \omega\mu_0\epsilon E_2, \quad k'B_3 = \omega\mu_0\epsilon' E_3. \tag{8.13}$$

The continuity of the tangential **E** and **H** components implies that

$$E_1 - E_2 = E_3, \quad B_1 + B_2 = B_3. \tag{8.14}$$

Combining them gives

$$k = \frac{\omega}{c}\sqrt{\frac{\epsilon}{\epsilon_0}} = \frac{\omega}{c}\sqrt{\epsilon_r(\omega)}, \quad k' = \frac{\omega}{c}\sqrt{\frac{\epsilon'}{\epsilon_0}} = \frac{\omega}{c}\sqrt{\epsilon_r'(\omega)}. \tag{8.15}$$

where $\epsilon_r(\omega)$ and $\epsilon_r'(\omega)$ are the dielectric functions of the respective media. From these formulas it is seen that the (complex) index of refraction is given by $\tilde{n} = \sqrt{\epsilon_r}$. The intensities of the waves are determined by the respective Poynting vectors, given by Eq. (8.2). The reflection probability, or reflectivity, is

$$R = \left|\frac{E_2}{E_1}\right|^2 = \left|\frac{\sqrt{\epsilon_r'} - \sqrt{\epsilon_r}}{\sqrt{\epsilon_r'} + \sqrt{\epsilon_r}}\right|^2. \tag{8.16}$$

Reflection thus occurs due to the dielectric mismatch at the interface. Similarly, in the absence of absorption processes, the *transmission probability*, or *transmittance*, is determined by

$$T = \frac{k'}{k}\left|\frac{E_3}{E_1}\right|^2 = \frac{4\sqrt{\epsilon_r\epsilon_r'}}{\left(\sqrt{\epsilon_r} + \sqrt{\epsilon_r'}\right)^2}. \tag{8.17}$$

In the absence of absorption, $R + T = 1$, as required by the law of conservation of energy.

Reflectivity of Metals. Recall from Chapter 7 that in the Sommerfeld description of a metal, all the electron-energy bands are filled except for the conduction band, which is only partially filled. Electrons in the conduction band are free to move throughout the solid. Two factors govern the reflectivity of metals. One is the response of an essentially free electron gas to incident radiation (Drude theory). The other is determined by interband transitions.

Begin by studying the free-electron gas, let $\epsilon_r = 1$, and from Eq. (8.10),

$$\epsilon_r' = \epsilon_1' + i\epsilon_2' = n^2 - \kappa^2 + 2in\kappa = 1 - \frac{\omega_p^2}{\omega(\omega + i/\tau)}$$

$$= 1 - \frac{\omega_p^2\tau^2}{1 + \omega^2\tau^2} + i\frac{\omega_p^2\tau}{\omega(1 + \omega^2\tau^2)}. \tag{8.18}$$

The relative dielectric constant depends on two material parameters: ω_p^2, which by Eq. (8.11) is proportional to the electron concentration n, and the electron collision rate τ, which is determined by collisions with impurities, imperfections, and phonons. In

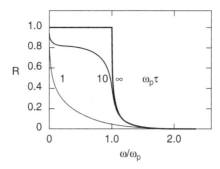

Figure 8.2. Reflectivity as a function of frequency (in units of the plasma frequency) for several values of $\omega_p \tau$.

Fig. 8.2 the behavior of the reflectivity R as a function of ω for several values of $\omega_p \tau$ is displayed. For $\omega < \omega_p$ there is high reflectivity, approaching perfect reflectivity as $\tau \to \infty$. Then $\epsilon' < 0$, $\sqrt{\epsilon'}$ becomes a pure imaginary number, the numerator and denominator in Eq. (8.16) have the same magnitude, and $R = 1$. The electric and magnetic fields are attenuated in amplitude as one progresses into the metal. The reason is that the electrons oscillate out of phase with the electric field when $\omega < \omega_p$, with amplitude depending on the electric field strength. The field set up by the electrons therefore acts to partially cancel the incident field. For finite collision times, τ, electron collisions extract energy from the field, but as $\tau \to \infty$, this effect becomes small and the light is reflected with high efficiency. For $\omega > \omega_p$, however, the electrons oscillate in phase with the field and the light is only partially reflected. At high frequencies little light is reflected but instead is eventually absorbed.

In addition to the free-electron contributions to the dielectric function, the effects of interband transitions must be included. For example, in the metals Cu, Ag, and Au, there are energy bands of d electrons that lie just below the Fermi level. If the energy of an incident photon is sufficiently high, electrons from these bands can be excited to the Fermi surface if the photon is absorbed. Since the d band has a finite width, typically on the order of several electron volts, this gives rise to a strong contribution to ϵ_2'. In addition, electrons from below the Fermi surface can be promoted to higher-energy bands above the Fermi surface. In Fig. 8.3 the electron-energy band structure is presented for Au. Direct transitions can occur at the X point, with a photon energy of 3.0 eV (labeled A). At the L point the corresponding photon has energy 3.7 eV (labeled B). The minimum photon energy occurs along the Γ–X direction and has an energy of 2.0 eV (labeled C). An examination of the graph of ϵ_2' versus E in Fig. 8.4 shows an onset of absorption occurring at the latter energy, and spectral features occurring at higher energies. The low-energy (<2 eV) absorption due to the conduction electrons is describable in terms of Drude theory [i.e., by Eq. (8.18)]. Figure 8.4 may be contrasted with Fig. 8.8, which gives ϵ_2 for Si. Since Si is a semiconductor and not a metal, the Drude contribution at low frequencies is missing. For photons with energies below the band gap, there is no interband absorption.

In Fig. 8.5 the real part of the dielectric function of Au is contrasted with that of a semiconductor, Si. The locations of the plasma frequencies are marked by arrows. In Au it is noted that ε_1 passes through zero at the plasma frequency (≈ 5.5 eV). For $\omega < \omega_p$, ϵ_1 becomes negative and diverges toward -∞ as $\omega \to 0$. This is not true for a

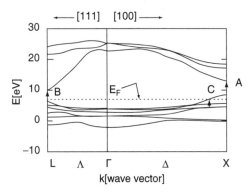

Figure 8.3. Calculated electron energy-band structure of Au. The low-lying flat bands are primarily those of d electrons. [From N. E. Christensen et al, *Phys. Rev. B*, **4**, 3321 (1971). Copyright 1971 by the American Physical Society].

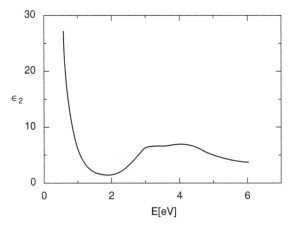

Figure 8.4. Experimental measurement of ϵ_2 as a function of photon energy for Au. [from M.-L. Thèye, *Phys. Rev. B*, **2**, 3060 (1970). Copyright 1970 by the American Physical Society].

semiconductor such as Si, where ϵ_1, and hence $n_1 = \sqrt{\epsilon_1}$, remain positive as $\omega \to 0$. The difference is due to the presence of electrons that are free to move around in metals but which are bound in semiconductors.

Reflectivity of Ionic Crystals. The reflectivity of metals is determined by the valence electrons. The characteristic frequencies are either the plasma frequency or those of interband transitions. Typically, they are on the order of several electron volts. In this section, attention is directed toward ionic crystals or to crystals that are partially ionic and partially covalent. In the absence of free electrons the ions play a more prominent role, particularly in the infrared, where the frequencies of the optic phonons lie.

Rather than develop a general theory, attention is limited to the linear-diatomic system considered in Section 5.2; more general conclusions are drawn by inference. The atoms, labeled A and B, are assumed to have charges Q_A and Q_B, respectively. Overall neutrality of the unit cell requires that $Q_A + Q_B = 0$. A macroscopic electric

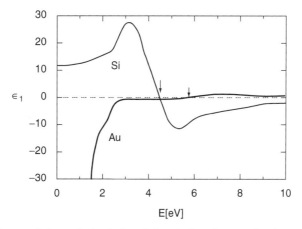

Figure 8.5. Real part of the optical relative dielectric function, ϵ_1, for Au and Si. (Data from E. D. Palik, ed., *Handbook of Optical Constants of Solids*, Vol. I, Academic Press, San Diego, Calif., 1991.)

field due to an electromagnetic wave will be assumed to exist in the crystal. In place of Eqs. (5.12*a*) and (5.12*b*), there is now

$$M_A \ddot{u}_n = K(v_n - u_n) - K'(u_n - v_{n-1}) + Q_A E_0 e^{i(kna - \omega t)}, \tag{8.19a}$$

$$M_B \ddot{v}_n = -K(v_n - u_n) + K'(u_{n+1} - v_n) + Q_B E_0 e^{i(k(na+b) - \omega t)}. \tag{8.19b}$$

Making the substitutions of Eqs. (5.13*a*) and (5.13*b*) yields

$$-M_A \omega^2 U = K(V - U) - K'(U - Ve^{-ika}) + Q_A E_0, \tag{8.20a}$$

$$-M_B \omega^2 V = -K(V - U) + K'(Ue^{ika} - V) + Q_B E_0 e^{ikb}. \tag{8.20b}$$

Since the wavelength of light is much larger than the size of a unit cell, it will be assumed that $k \approx 0$. The ions A and B oscillate $180°$ out of phase. Equations (8.20*a*) and (8.20*b*) may be solved to obtain the ionic contribution to the net electric dipole moment of the unit cell:

$$\mu_n = Q_A u_n + Q_B v_n = Q_B(V - U) = -\frac{Q_B^2 (1/M_A + 1/M_B) E_0}{\omega^2 - \omega_T^2}, \tag{8.21}$$

where

$$\omega_T = \sqrt{(K + K')\left(\frac{1}{M_A} + \frac{1}{M_B}\right)}. \tag{8.22}$$

From Eqs. (5.19) and (5.20) one identifies ω_T with the zone-center optic mode of the diatomic lattice. Note that μ_n is independent of n .

Having obtained this result, it may now be applied to a more general three-dimensional situation. Let there be n_c unit cells per unit volume. The ionic contribution to the electric polarization vector is $P = n_c \mu_n$. Assume that there is also a background permittivity ϵ_b due to possible electronic polarization contributions. Using $D = \epsilon_b E + P = \epsilon(\omega)E$, the permittivity is given by

$$\epsilon(\omega) = \epsilon_b \left(1 - \frac{\Omega^2}{\omega^2 - \omega_T^2}\right), \tag{8.23}$$

where

$$\Omega^2 \equiv \frac{n_c Q_B^2}{\epsilon_b} \left(\frac{1}{M_A} + \frac{1}{M_B}\right). \tag{8.24}$$

By comparison of Eq. (8.24) with Eq. (8.11), the quantity Ω is interpreted as the ionic-plasma frequency.

A plot of the dielectric function $\epsilon_r(\omega) = \epsilon(\omega)/\epsilon_0$ is given in Fig. 8.6a. Note that between frequencies ω_T and ω_L, where the frequency ω_L is given by $\omega_L = \sqrt{\omega_T^2 + \Omega^2}$, $\epsilon_r(\omega) < 0$, and no electromagnetic wave propagation is possible. This defines what is called the *reststrahlen band* (i.e., in this region the crystal acts as a reflector). From

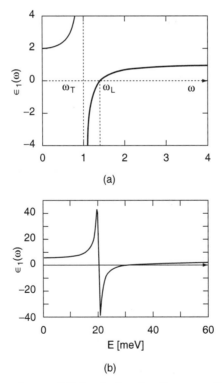

Figure 8.6. (*a*) Relative real part of dielectric function for a one-dimensional diatomic ionic crystal. Here $\omega_T = 1$ and $\omega_L = \sqrt{2}$, in arbitrary units. (*b*) Real part of dielectric function for NaCl. (Data from E. D. Palik, ed., *Handbook of Optical Constants of Solids*, Vol. I, Academic Press, San Diego, Calif., 1991.)

TABLE 8.3   Energies E_T and E_L for Representative Materials at $T = 300$ K

Material	E_T (meV)	E_L (meV)
NaCl	21.1	33.4
NaBr	16.8	26.2
NaI	14.5	21.8
AgBr	9.8	17.1
GaAs	33.4	36.2
GaP	45.6	50.1
InAs	26.9	29.6
AlAs	45.1	49.8
InSb	22.2	23.7
MgO	49.7	89.0
ZrO_2	43.9	84.3
BN	131	162
AlN	82.9	113.7
CdTe	14.4	17.3

Eq. (8.23) one may deduce the *Lydanne–Sachs–Teller formula*:

$$\frac{\epsilon_r(0)}{\epsilon_r(\infty)} = \frac{\epsilon(0)}{\epsilon(\infty)} = \left(\frac{\omega_L}{\omega_T}\right)^2. \tag{8.25}$$

At the frequency ω_L the real part of the dielectric function is zero and longitudinal modes may propagate. Equations (8.19a) and (8.19b) may be generalized to include a phenomenological damping for the optic modes. With damping included, the crystal no longer acts as a perfect reflector in the restrahlen band. In Fig. 8.6b experimental data are presented for the real part of the relative dielectric function, ϵ_1, for NaCl. The finite damping is responsible for ϵ_1 not diverging at ω_T. Values of the equivalent energies $E_T = \hbar\omega_T$ and $E_L = \hbar\omega_L$ are given in Table 8.3 for some representative materials. The frequencies are 10^{13} to 10^{14} rad/s (i.e., lie in the infrared). Note that the ratio E_T/E_L is close to 1 for crystals in which covalent bonding dominates over ionic bonding. The values are in good agreement with the predictions of the Lydanne–Sachs–Teller formula. For example, for NaI, $\epsilon_1(0) = 7.28$ and $\epsilon_1(\infty) = 2.93$. The left-hand side of Eq. (8.25) is 2.48 and the right-hand side is 2.26.

8.5   Optical Properties of Semiconductors

Semiconductors differ from metals in that there are no (or few) free electrons present. The only free electrons and/or holes present are those generated by doping (extrinsic case) or by thermal activation (intrinsic case). There is a set of filled valence bands separated from one or more empty conduction bands by an energy gap. Consider, for now, the electronic excitations. If a photon enters the crystal with an energy less than the gap (i.e., $\hbar\omega < E_g$), propagation is possible and the crystal is transparent. If the gap energy is exceeded, however, interband transitions and absorption begin to occur. Here the possibility of nonlinear processes involving two or more photons being absorbed is neglected.

It should be noted that the wave vector associated with a photon ($q = n\omega/c$) is small compared with the size of the first Brillouin zone ($\approx 2\pi/a$). The ratio is typically on the order of 0.1% for visible light. To a first approximation, one often simply sets $q = 0$. Two types of transitions can occur: direct and indirect. In a *direct transition*, an electron is promoted vertically on an energy-band diagram from the valence band to the conduction band, and hence there is little change in the electron's wave vector. The energies of the initial and final state are related by $E' = E + \hbar\omega$. In an *indirect transition*, phonon emission or absorption (of a thermally activated phonon) occurs, with both energy and momentum being taken up or supplied by the phonon. Thus $\mathbf{k}' = \mathbf{k} \pm \mathbf{q}_{ph}$ and $E' = E + \hbar\omega \pm \hbar\Omega(\mathbf{q}_{ph})$. Since a phonon perturbation is involved, indirect transitions are weaker than direct transitions. In some semiconductors (e.g., GaAs) the minimum of the conduction band and the maximum of the valence band both occur at $\mathbf{k} = 0$ (Γ point), so the threshold absorption is of the direct type. In other semiconductors, such as Si or Ge, the mimina and maxima occur at different points in wave vector space, so the threshold absorption is indirect. Only at frequencies well above the threshold can direct (i.e., vertical) transitions occur.

Figure 8.7 depicts the band structure of Si (see also Fig. 11.2). One notes the existence of an indirect transition from the top of the valence band at the Γ point to the bottom of the conduction band along the [100] direction. This is denoted by the dotted line connecting points a and c in the diagram. The photon alone cannot provide the crystal momentum to make this transition. There must be the emission or absorption of a phonon. The imaginary part of the dielectric function of Si is shown in Fig. 8.8. The small values of ϵ_2 immediately above an energy of $E_g = 1.1$ eV are associated with the threshold for indirect transitions in Si. The large increase in ϵ_2 at energies around 3 eV is due to the opening up of direct transition channels for absorption. This is denoted by the near-vertical dotted line connecting points a and b in Fig. 8.7. The real part of the dielectric function for Si is given in Fig. 8.5.

The optical properties of semiconductors are discussed in more detail in Chapter 11.

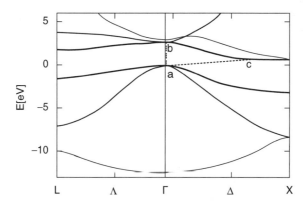

Figure 8.7. Band structure of Si along the [100] and [111] directions in wave-vector space. [From A. Zunger et al., *Phys. Rev. B*,**20**, 4082 (1979). Copyright 1979 by the American Physical Society.]

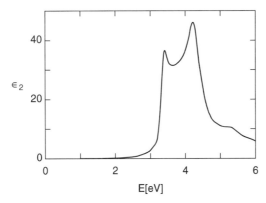

Figure 8.8. Imaginary part of the dielectric function for Si. (Data from E. D. Palik, ed., *Handbook of Optical Constants of Solids*, Vol. I, Academic Press, San Diego, Calif., 1991.)

8.6 Optical Dielectric Function

Since the propagation of light through matter is determined largely by the optical dielectric function, $\epsilon_r(\omega)$, it is worthwhile deriving a formula for it. For visible radiation as well as for ultraviolet and near-infrared radiation the dielectric function is determined primarily by the electronic excitations of the system. Since the wavelength of light, λ, is long compared with microscopic length scales in the solid, for many materials it is sufficient to treat the electromagnetic field as simply a harmonically time-varying electric field. This would not apply, however, to composite materials or materials with microstructure whose size is comparable with λ. In this section an expression for $\epsilon_r(\omega)$ is developed based on the Lorentz oscillator model for the electrons. In Appendix W8C a quantum-mechanical model is introduced to accomplish a similar goal, but is more precise. In the following discussion there is nothing that limits the discussion to crystalline materials. All classes of materials, including amorphous solids, polymers, and even liquids, could be described by the formalism.

In the Lorentz model the bound electrons are imagined to be bound in harmonic potentials. The equation of motion for the oscillator is

$$m \left(\frac{d^2\mathbf{r}}{dt^2} + \gamma_j \frac{d\mathbf{r}}{dt} + \omega_j^2 \mathbf{r} \right) = -e\mathbf{E}_0 e^{-i\omega t}, \tag{8.26}$$

where E_0 is the strength of an electric field of angular frequency ω, ω_j the oscillator frequency, and γ_j a damping constant. A steady-state solution is obtained by inserting $\mathbf{r} = \mathbf{r}_0 \exp(-i\omega t)$ and computing the electric-dipole moment of the oscillator. Thus

$$\boldsymbol{\mu} = -e\mathbf{r}_0 = \frac{e^2}{m} \frac{\mathbf{E}_0}{\omega_j^2 - i\omega\gamma_j - \omega^2}. \tag{8.27}$$

Let the fraction of electrons that are characterized by oscillator j be f_j, which is called the *oscillator strength* for the jth oscillator. The electric polarization vector is then given by $\mathbf{P} = n \sum f_j \boldsymbol{\mu}_j$, where n is the electron density and the sum extends over all types of oscillators. Using formulas from Appendix W8A: $\mathbf{D} = \epsilon_0 \mathbf{E} + \mathbf{P}$, $\mathbf{D} = \epsilon \mathbf{E}$, and $\epsilon_r(\omega) = \epsilon(\omega)/\epsilon_0$, one obtains

$$\epsilon_r(\omega) = 1 + \sum_j \frac{f_j \omega_p^2}{\omega_j^2 - i\omega\gamma_j - \omega^2}, \tag{8.28}$$

where Eq. (8.11) has been used. Note that $\sum f_j = 1$.

One shortfall of the Lorentz model is that it predicts an isotropic dielectric function. Although this is adequate for cubic crystals, other crystal classes have anisotropic physical properties, including the dielectric function. It is described by a three-dimensional matrix whose elements are denoted by $\epsilon_{r_{\alpha\beta}}(\omega)$. It is possible to empirically construct such a dielectric function by generalizing the concept of oscillator strength, as follows. One simply introduces a set of matrices with elements $\{f_{\alpha\beta}^j\}$ that obey the sum rule

$$\sum_j f_{\alpha\beta}^j = \delta_{\alpha\beta}, \tag{8.29}$$

where the right-hand side is the Kronecker delta. In place of Eq. (8.28), one writes

$$\epsilon_{r_{\alpha\beta}}(\omega) = \delta_{\alpha\beta} + \sum_j \frac{f_{\alpha\beta}^j \omega_p^2}{\omega_j^2 - i\omega\gamma_j - \omega^2}. \tag{8.30}$$

One may also write this in the equivalent dyadic notation

$$\overleftrightarrow{\epsilon}_r(\omega) = \overleftrightarrow{I} + \sum_j \frac{\omega_p^2}{\omega_j^2 - i\omega\gamma_j - \omega^2} \overleftrightarrow{f}^j. \tag{8.31}$$

where the unit dyadic $\overleftrightarrow{I}$ has 1 for the diagonal elements and 0 for the off-diagonal elements. At the level of the Lorentz treatment, the $\overleftrightarrow{f}^j$ elements are simply fitting parameters. The discussion of the dielectric function above assumes a local relation between the electric field and the polarization vector. It is possible to generalize the discussion to the nonlocal case, where the polarization at a given point in space is related to the electric field at neighboring points in the solid. The details are presented in Appendix W8B. In such a case one writes the dielectric function Fourier-transformed in both wave vector and frequency as $\epsilon_r(\mathbf{q}, \omega)$.

From Eqs. (8.28) or (8.31) one notes three important factors contributing to a large dielectric function. One element involves having large oscillator strengths. As seen in Appendix W8C, this will occur when the electric dipole matrix elements for a particular transition are large. Symmetry requires the two states linked by the electric dipole to have angular momentum quantum numbers, l, differing by $+1$ or -1 to be nonvanishing. In addition, a strong spatial overlap of the wavefunctions favors large matrix elements. A second element is the presence of resonance denominators, which could cause $\epsilon_r(\omega)$ to be large over a restricted frequency range. This will be particularly true if the damping constant, which is related to the lifetime of excited states, is small. The resonance occurs when a photon of frequency ω has sufficient energy to link the two quantum states. The lifetime includes both radiative and nonradiative channels of decay. The third element has to do with carrier occupancy. The more carriers that can participate in the polarization, the larger $\epsilon_r(\omega)$ will be.

It is worthwhile examining Eq. (8.28) in the limit of frequencies low compared with electronic frequencies but still high compared with phonon frequencies. In the case of semiconductors the typical electronic frequency scale is set by the energy gap (i.e., $\omega_j = E_g/\hbar$ for all j), so

$$\epsilon_1(0) \to 1 + \sum_j \frac{f_j \omega_p^2}{\omega_j^2} \to 1 + \sum_j \frac{f_j \omega_p^2}{(E_g/\hbar)^2} = 1 + \left(\frac{\hbar \omega_p}{E_g}\right)^2. \tag{8.32}$$

It should be noted that the low-frequency index of refraction, $n(0) = \sqrt{\epsilon_1(0)}$ is determined by ω_p and E_g in a semiconductor. In the limit of frequencies high compared with electronic frequencies, Eq. (8.28) reduces to the familiar formula

$$\epsilon_1(\omega) \to 1 - \sum_j \frac{f_j \omega_p^2}{\omega^2} \to 1 - \frac{\omega_p^2}{\omega^2}. \tag{8.33}$$

This is the same as Eq. (8.10) obtained for metals, in the high-frequency limit. One may argue that since ω is high compared with the frequencies of the electronic oscillators, the electrons respond as if they were free electrons, and therefore the free-electron dielectric function is obtained.

8.7 Kramers–Kronig Relations

The Kramers–Kronig relations are a pair of formulas that provide a connection between the real and imaginary parts of the dielectric function. By extension from the formula $\tilde{n}(\omega) = [\epsilon_r(\omega)]^{1/2}$, a relation is also established between the dispersive behavior of the index of refraction and the extinction coefficient of a material.

The relations are

$$\epsilon_1(\omega) - 1 = \frac{2}{\pi} P \int_0^\infty d\omega' \frac{\omega' \epsilon_2(\omega')}{\omega'^2 - \omega^2}, \tag{8.34}$$

$$\epsilon_2(\omega) = \frac{2\omega}{\pi} P \int_0^\infty d\omega' \frac{\epsilon_1(\omega') - 1}{\omega'^2 - \omega^2}. \tag{8.35}$$

Here P means "take the principal part" of the integral. Evaluation of the integrals in Eqs. (8.34) and (8.35) requires care because of the singularity at $\omega' = \omega$. The procedure implied by P is to perform the integrals as follows:

$$P \int_0^\infty d\omega' \frac{F(\omega)'}{\omega'^2 - \omega^2} = \lim_{\alpha \to 0} \left[\int_0^{\omega - \alpha} d\omega' \frac{F(\omega')}{\omega'^2 - \omega^2} + \int_{\omega + \alpha}^\infty d\omega' \frac{F(\omega')}{\omega'^2 - \omega^2} \right], \tag{8.36}$$

that is, to remove an infinitesimal slice of size 2α from the region of integration in order to avoid the singularity.

The Kramers–Kronig relations show that $\epsilon_1(\omega)$ and $\epsilon_2(\omega)$ are not independent functions. A knowledge of one is sufficient to obtain the other. Derivation of these formulas makes use of the complex-variable theory and Cauchy's theorem. Details are not presented here.

Since the index of refraction is related to the optical dielectric function (for a nonmagnetic material) by $\tilde{n}^2(\omega) = \epsilon_r(\omega)$, one may rewrite the Kramers–Kronig relations in terms of the real part of the index of refraction and the extinction coefficient.

8.8 Optical Properties of Composite Media

A homogeneous material is characterized by a single index of refraction n and extinction coefficient κ. The question arises how to generalize these concepts to the case of granular materials or composite structures. If the size of the particles and the separation between them is small compared with the wavelength of light, one may characterize the material by effective optical constants. These parameters depend on the total number density of all the particles, n_i, and the complex dielectric functions of each species, ϵ_{r_i}.

Let there be a average macroscopic electric field $\mathbf{E}$ existing in a material. Begin by isolating a spherical region and examining the local electric field, $\mathbf{E}_{\text{loc}}$, at its center. According to Lorentz, the local field consists of three contributions: the macroscopic field $\mathbf{E}$; the field due to the dipoles on the surface of the sphere, $\mathbf{P}/3\epsilon_0$; and the fields due to the dipoles inside the sphere, $\mathbf{E}_{\text{in}}$. This last term is zero for a random medium or a medium with cubic symmetry. Thus,

$$\mathbf{E}_{loc} = \mathbf{E} + \frac{\mathbf{P}}{3\epsilon_0}. \tag{8.37}$$

The polarization vector represents the average dipole moment per unit volume and may be written as

$$\mathbf{p} = \sum_i n_i \boldsymbol{\mu_i} = \epsilon_0 \sum_i n_i \alpha_i \mathbf{E}_{loc}, \tag{8.38}$$

where $\boldsymbol{\mu}_i = \epsilon_0 \alpha_i \mathbf{E}_{loc}$ is the dipole moment and α_i is the polarizability of the ith species. It is assumed that all the particles are spherical and isotropic, for simplicity's sake. Using the formula

$$\mathbf{p} = (\epsilon_r - 1)\epsilon_0 \mathbf{E}, \tag{8.39}$$

one may derive an expression for the effective dielectric function ϵ_r:

$$\frac{\epsilon_r - 1}{\epsilon_r + 2} = \frac{4\pi}{3} \sum_i n_i \alpha_i, \tag{8.40}$$

where $\sum n_i = n$. This represents a simple generalization of the Clausius–Mossotti formula, which would be obtained if only one type of particle were present. It was derived under the assumption that the particles are embedded in vacuum. If there is a background dielectric host material with dielectric function ϵ_{r_b}, this formula readily generalizes to the Maxwell–Garnett formula:

$$\frac{\epsilon_r - \epsilon_{r_b}}{\epsilon_r + 2\epsilon_{r_b}} = \sum_i f_i \frac{\epsilon_{r_i} - \epsilon_{r_b}}{\epsilon_{r_i} + 2\epsilon_{r_b}}. \tag{8.41}$$

This may be written for the case where there are two components, one of which (e.g., ϵ_{r_1}) is dominant. Taking as an approximation $\epsilon_{r_b} = \epsilon_{r_1}$, formula (8.41) may be written as

$$\frac{\epsilon_r - \epsilon_{r_1}}{\epsilon_r + 2\epsilon_{r_1}} = f_2 \frac{\epsilon_{r_2} - \epsilon_{r_1}}{\epsilon_{r_2} + 2\epsilon_{r_1}}. \tag{8.42}$$

(Note that ϵ_{r_1} here denotes the full dielectric function of material 1 and not the real part only.)

In the effective medium approximation (EMA) one adopts a more democratic approach and argues that there should be no difference between the host dielectric function and the effective dielectric function for a purely composite medium. Thus $\epsilon_{r_b} = \epsilon_r$ and

$$\sum_i f_i \frac{\epsilon_{r_i} - \epsilon_r}{\epsilon_{r_i} + 2\epsilon_r} = 0. \tag{8.43}$$

This is the *Bruggeman formula* for ϵ_r.

Equations (8.40), (8.42), and (8.43) often represent adequate interpolation formulas for the case where there is a strong majority of one material and a small minority of the others. In cases where the materials are of comparable proportions, fluctuations in the local environment are significant. This is particularly true near what would be the percolation threshold in the case of conductors.

8.9 Nonlinear Polarization

Consider the interaction of light with a molecule. The simplest interaction involves the elastic scattering of the light. The molecule is polarized by the electric field of the light, and the resulting dipole moment radiates photons of the same frequency in all directions. The microscopic parameters governing the coupling of light to the molecule are given by the polarizability tensor $\overset{\leftrightarrow}{\alpha}(\omega)$. For weak fields the polarization vector is given by

$$\mathbf{p} = \epsilon_0 \overset{\leftrightarrow}{\chi}(\omega) \cdot \mathbf{E}. \tag{8.44}$$

In a dilute medium consisting of n molecules per unit volume, the linear susceptibility tensor from Eqs. (8.38) and (8.30) is approximately $\overset{\leftrightarrow}{\chi}(\omega) = n\overset{\leftrightarrow}{\alpha}(\omega)$.

From a quantum-mechanical perspective, $\overset{\leftrightarrow}{\chi}(\omega)$ is calculated from time-dependent perturbation theory by including all second-order interactions with the electric field which leave the material unexcited. Let $|p\rangle$ and $|q\rangle$ represent energy eigenstates of the molecule and let α and β be polarization indices.

The susceptibility tensor [see Eq. (W8C.9)] may be rewritten as

$$\chi_{\alpha\beta}(\omega) = -\frac{n}{\hbar\epsilon_0} \sum_{pq} f(E_q, T) \left(\frac{\mu_{qp}^{\beta} \mu_{pq}^{\alpha}}{\omega - \omega_{pq}} - \frac{\mu_{qp}^{\alpha} \mu_{pq}^{\beta}}{\omega + \omega_{pq}} \right), \tag{8.45}$$

where μ is the dipole operator and ω_{pq} is the transition frequency from state p to state q. The matrix element of the dipole operator $\mu_{pq} = \langle p|\mu|q\rangle$ taken between two states is called the *transition dipole moment*. The Fermi factor for filled electronic states, f, is included in Eq. (8.45). The number of molecules per unit volume is denoted by n. It should be remarked that near resonance (e.g., $\omega = \omega_{pq}$) the susceptibility can become very large and is limited only by the natural lifetime of the excited state [not shown in formula (8.45) but which could be included by adding an imaginary term to the resonance frequency (i.e., $\omega_{pq} \to \omega_{pq} + i/2\tau$)]. The susceptibility also grows with increasing concentration, n, and increasing size of the transition dipole moment.

In a similar way, one may introduce the nonlinear susceptibility, which is used to describe the interaction of a material with a stronger electric field, such as that due to a laser. In addition to the linear polarization, there is a nonlinear polarization. To lowest order, it is quadratic in the incident electric field:

$$P_\alpha = \epsilon_0 \sum_{\beta\gamma} d^{(2)}_{\alpha\beta\gamma}(-2\omega; \omega, \omega) E_\beta E_\gamma. \tag{8.46}$$

The nonlinear optical coefficient $d^{(2)}$ is now a third-rank tensor. The formula for it is

$$d^{(2)}_{\gamma\alpha\beta}(-2\omega; \omega, \omega) = \frac{n}{2\epsilon_0 \hbar^2} \sum_{qlm} f(E_1, T)(M^1_{\gamma\alpha\beta} + M^2_{\gamma\alpha\beta} + M^3_{\gamma\alpha\beta}$$

$$+ M^1_{\gamma\beta\alpha} + M^2_{\gamma\beta\alpha} + M^3_{\gamma\beta\alpha}), \tag{8.47}$$

with

$$M^1_{\gamma\alpha\beta} = \frac{\mu^\gamma_{qm}\mu^\beta_{ml}\mu^\alpha_{lq}}{[2\omega - (\omega + \omega_{lq})](2\omega - \omega_{mq})}, \tag{8.48a}$$

$$M^2_{\gamma\alpha\beta} = \frac{\mu^\gamma_{qm}\mu^\beta_{ml}\mu^\alpha_{lq}}{[2\omega - (\omega + \omega_{lq})][2\omega - (3\omega + \omega_{mq})]}, \tag{8.48b}$$

$$M^3_{\gamma\alpha\beta} = \frac{\mu^\gamma_{qm}\mu^\beta_{ml}\mu^\alpha_{lq}}{[2\omega - (4\omega + \omega_{lq})][2\omega - (3\omega + \omega_{mq})]}. \tag{8.48c}$$

In this case the possible resonance structure is much richer. Resonances can occur when $\omega = \omega_{lq}$, and also $2\omega = \omega_{mq}$. A double resonance is possible if both $2\omega = \omega_{mq}$ and $\omega = \omega_{lq}$, and so on.

Still higher-order nonlinear optical coefficients are defined. Higher-order terms are important for the case of a centrosymmetric system (i.e., one having inversion symmetry) where $d^{(2)} \equiv 0$. This comes about because under a parity reversal the vector **E** reverses direction and so should **P**, but the formula $P_\alpha = \epsilon_0 \sum d^{(2)}_{\alpha\beta\gamma} E_\beta E_\gamma$ shows it not to reverse. For such media the nonlinear interaction is determined by $d^{(3)}$. Third-harmonic generation, in which photons at frequency ω combine to produce photons at frequency 3ω, is one consequence of such an interaction and is described by the fourth-rank tensor $d^{(3)}_{\delta\alpha\beta\gamma}(-3\omega; \omega, \omega, \omega)$. The third-harmonic polarization is then given by $P_\alpha = \epsilon_0 \sum d^{(3)}_{\alpha\beta\gamma\delta} E_\beta E_\gamma E_\delta$.

Some representative values of second-order nonlinear optical coefficients are given in Table 8.4. In the $d^{(2)}$ case one sees that the semiconductors GaAs and InSb have the largest values. This may be attributed to the small energy bandgap present in these materials. Equations (8.48) show that when the denominators become small, the nonlinear coefficient will be large.

Some values of third-order coefficients are provided in Table 8.5. Part of the nonlinearity may be attributed to the nonparabolic nature of the conduction band. When

TABLE 8.4 Nonlinear Optical Coefficients at $\lambda = 1.06$ μm

Material	Component	$d^{(2)}$ (pm/V)
α-Quartz	xxx	0.34
AlN	zzz	7.4
	zxx	0.2
ZnO	zzz	-7.0
	zxx	2.1
	xxz	2.3
GaAs	xyz	274
	zxy	249
InSb	xyz	520
PbTiO$_3$	xxz	2.1
	zxx	2.4
	zzz	0.48
BaTiO$_3$	zzz	-8
	zxx	-18
	xxz	19
LiNbO$_3$	zxx	-6.0
	yyy	2.9
	zzz	-34
NH$_4$H$_2$PO$_4$ (ADP)	zxy	0.56
	xyz	0.55
KH$_2$PO$_4$ (KDP)	zxy	0.47
	xyz	0.49
Ba$_2$NaNb$_5$O$_{15}$	zzz	13
	zyy	9.3
β-BaB$_2$O$_4$ (BBO)	yyy	1.8
	zxx	0.12
KBe$_2$BO$_3$F$_2$ (KBBF)	xxx	0.9
Sr$_2$Be$_2$B$_2$O$_7$	yyy	2.2
Polyvinylidene fluoride	zxx	0.18
	zyy	0
	zzz	0.36
C$_{60}$	zzz	0.88

Source: Data from A. Yariv, *Optical Electronics*, 3rd ed., Holt, Rinehart and Winston, New York, 1985; M. Bass *Handbook of Optics*, Vol. II, McGraw-Hill, New York, 1995; R. L. Sutherland, *Handbook of Nonlinear Optics*, Marcel Dekker, New York, 1996; and I. S. Grigoriev and E. Z. Meilikhov, eds., *CRC Handbook of Physical Quantities*, CRC Press, Boca Raton, Fla., 1997.

exposed to a pulse of laser light, multiphoton transitions will often excite electrons from the valence band to the conduction band. These carriers can respond further to the incident light pulse. They will acquire a crystal momentum proportional to the strength of the electric field of the light. For parabolic bands, the velocity, and hence the current, will be proportional to the crystal momentum, so the response frequency will be the same as the frequency of the light. For nonparabolic bands, however, the velocity will be nonlinear in the momentum and will contain harmonics of the light frequency. In noncentrosymmetric crystals this would include the second and

TABLE 8.5   Third-Order Susceptibilities

Material	λ (nm)	$d^{(3)}_{xxxx}$ (10^{-22} m^2/V^2)
BaF$_2$	575	1.2
CaF$_2$	575	0.6
Diamond	545	6.4
GaP	577	2.9×10^4
ZnSe	1064	2.5×10^2
V$_2$O$_5$	532	1.7×10^6
Cr$_2$O$_3$	532	1.1×10^5
Mn$_3$O$_4$	532	2.8×10^5
Fe$_2$O$_3$	532	5.9×10^5
Au-glass[a]	532	1.7×10^7
CS$_2$(l)	532	3.9×10^2
Poly(2,5-thiophene) PT	602	5.6×10^4
Poly(di-n-hexylsilylene) PDHS	1064	3.2×10^2

Source: Data from M. Ando et al., *Nature,* **324**, 625 (1995); R. L. Sutherland, *Handbook of Nonlinear Optics,* Marcel Dekker, New York, 1996; and elsewhere.

[a] $r = 2.9$ nm, $f = 0.076$. For colloidal Au in glass, r is the sphere radius and f is the fraction of the volume occupied by the Au.

higher harmonics. In centrosymmetric crystals the third and higher harmonics will be included.

Note again that narrow-energy-bandgap materials tend to have higher values for $d^{(3)}$ than wide-gap materials. In the case of some of the doped polymers the $d^{(3)}$ values can be very large, due to the large transition-dipole moments that may be obtained by having charge transfer occur over large distances.

Nonlinear optical materials provide scientists and engineers with the ability to extend the spectral range over which coherent radiation may be produced. Using the materials of Tables 8.4 and 8.5, powerful lasers in the infrared, visible, and ultraviolet can have their frequencies extended to still higher frequencies. Common examples of second-harmonic generation include the use of a KH$_2$PO$_4$ (KDP) crystal to halve the wavelength of the ruby laser, from 0.694 to 0.347 μm and use of a Ba$_2$NaNb$_5$O$_{15}$ crystal to halve the wavelength of a Nd^{3+}:YAG laser from 1.06 to 0.53 μm.

8.10   Excitons

The optical effects considered until now were based largely on the independent-electron approximation and involved perfect solids. Now effects that involve more than one electron or involve imperfect solids are considered. The study begins by looking at electron–hole pairs and the effect of the Coulomb interaction between them. This gives rise to what are called *excitons.*

Suppose that a photon is incident on a semiconductor or insulator with an energy greater than the gap energy, E_g. This is sufficient to create a free electron–hole pair (i.e., one in which the electron and hole are independent of each other). However, since there is a Coulomb attraction between the electron and hole, bound states also exist, and these may be produced with photons of energy below E_g. These states are excitons and give rise to sharp lines in the absorption spectra below the band edge. A description of excitons requires going beyond the independent-electron approximation and is an example of a correlation effect.

Wannier Excitons. There are various limiting cases in which the description of the exciton is simple. In one case, called the *Wannier exciton*, both the electron and hole are mobile but remain bound to each other in a hydrogenic system. The unperturbed Hamiltonian is based on the free-electron approximation, in which there is an electron in the conduction band and a missing electron (i.e., hole) in the valence band. The energy of the excitation is

$$\Delta E = E_c(\mathbf{p_e}) - E_v(\mathbf{p_h}). \tag{8.49}$$

The full Hamiltonian includes the Coulomb interaction. In the parabolic band approximation, it is

$$H = E_g + \frac{p_e^2}{2m_e^*} + \frac{p_h^2}{2m_h^*} - \frac{e^2}{4\pi\epsilon_r\epsilon_0|\mathbf{r}_e - \mathbf{r}_h|}, \tag{8.50}$$

where m_e^* and m_h^* are the band effective masses of the electron and hole, respectively, and ϵ_r is the dielectric function. [Since $\epsilon_r(\omega)$ is frequency dependent, its value must be determined in a self-consistent way once the orbital frequencies are calculated. If the binding energy is large compared with optical phonon frequencies, $\epsilon_r(\infty)$ should be used. In the other limit $\epsilon_r(0)$ is needed. More generally, coupled phonon–exciton modes exist.] The Hamiltonian may be rewritten in terms of center-of-mass coordinates $\mathbf{R}$ and relative coordinates $\mathbf{r}$:

$$\mathbf{R} = \frac{m_e^*\mathbf{r}_e + m_h^*\mathbf{r}_h}{m_e^* + m_h^*}, \quad \mathbf{r} = \mathbf{r}_e - \mathbf{r}_h, \tag{8.51}$$

so

$$H = E_g + \frac{P^2}{2M} + \frac{p^2}{2\mu^*} - \frac{e^2}{4\pi\epsilon_r\epsilon_0 r}, \tag{8.52}$$

where $M = m_e^* + m_h^*$ and $\mu^* = m_e^*m_h^*/(m_e^* + m_h^*)$ is the reduced effective mass. The vectors $\mathbf{P} = \hbar\mathbf{K}$ and $\mathbf{p}$ refer to the total momentum of the exciton and relative momentum of the electron and hole, respectively. With this Hamiltonian the eigenfunctions separate into products of plane waves and internal hydrogenic states:

$$\psi(\mathbf{r}, \mathbf{R}) = e^{i\mathbf{K}\cdot\mathbf{R}}\phi_{nlm}(\mathbf{r}). \tag{8.53}$$

The energy eigenvalues are

$$E_{nlm}(K) = E_g + \frac{\hbar^2K^2}{2M} - \left(\frac{\mu^*}{\epsilon_r^2 m}\right)\frac{e^2}{8\pi\epsilon_0 a_1 n^2}, \tag{8.54}$$

where $n = 1, 2, 3, \ldots$, and a_1 is the Bohr radius. Note that the bound states are given by the usual Rydberg formula with a correction factor in parentheses for the effect

TABLE 8.6 Exciton Parameters for Several Semiconductors

Semiconductor	E_g (eV)	$\epsilon_r(0)$	μ^*/m	E_B (meV)	a_{ex} (nm)
Ge	0.67	16	0.132	7.01	6.41
Si	1.11	11.8	0.190	18.6	3.28
GaAs	1.42	13.2	0.0616	4.81	11.3
InSb	0.163	17.7	0.0135	0.586	69.4

of the crystal. Since μ^*/m is usually small and ϵ_r is large, the binding energy of the exciton is typically measured in meV.

For example, in GaAs $m_e^*/m = 0.067$ and $m_h^*/m = 0.76$, respectively (see Table 11.4). The static dielectric constant is $\epsilon_1(0) = 13.2$. This gives a theoretical exciton binding energy of $E_B = 4.81$ meV. The experimental binding energy is 4.9 meV. Table 8.6 gives some typical exciton binding energies along with some other relevant band-structure data. The binding energy of the lowest-lying exciton is denoted by E_B. The parameter E_g is the gap energy.

The size of an exciton is given by $a_{ex} = a_1 \epsilon_r (m/\mu^*)$. When this is large compared with the lattice constant, it is a Wannier exciton. This occurs when the shielding effect due to the lattice polarization is strong, so the Coulomb interaction between the electron and the hole is weak. A small reduced effective mass also favors the formation of Wannier excitons. For GaAs $a_{ex} = 213a_1$.

The complexity of the band structure leaves its mark on the exciton spectrum. For example, in Cu_2O there are three low-lying conduction bands and three high-lying valence bands. At the Γ point they form a set of four energy levels (two doubly degenerate and two nondegenerate). Thus four direct transitions are possible. The frequencies for these transitions correspond to the colors yellow, green, blue, and violet (photon energies 2.2, 2.3, 2.7, and 2.9 eV). Associated with each of these transitions is a series of excitons.

Frenkel Excitons. Frenkel excitons occur when the Coulomb interaction is not so strongly shielded by polarization charge. Consequently, a_{ex} is comparable to the lattice constant, and the long-range Coulomb potential no longer describes the interaction. Suppose that there is a tightly bound electron–hole pair with a total energy $E_g - E_B$ on a given site. This pair can tunnel to neighboring sites and the system can be described by the tight-binding Hamiltonian:

$$H = \sum_{\mathbf{R}} (E_g - E_B)|\mathbf{R}\rangle\langle\mathbf{R}| + \sum_{\mathbf{R},\mathbf{u}} (t|\mathbf{R}+\mathbf{u}\rangle\langle\mathbf{R}| + t^*|\mathbf{R}\rangle\langle\mathbf{R}+\mathbf{u}|), \tag{8.55}$$

where $\mathbf{u}$ denotes a set of vectors pointing to the NNs and $|\mathbf{R}\rangle$ denotes the state of a pair bound to site $\mathbf{R}$. The empirical parameter t determines the hopping rate of excitons from site to site. Diagonalization of this Hamiltonian occurs by writing the eigenstate as the phased sum:

$$|\psi\rangle = \sum_{\mathbf{R}} e^{i\mathbf{k}\cdot\mathbf{R}}|R\rangle. \tag{8.56}$$

The energy eigenvalues are distributed in a band:

$$E(\mathbf{k}) = E_g - E_B + \sum_{\mathbf{u}} (te^{-i\mathbf{k}\cdot\mathbf{u}} + t^* e^{i\mathbf{k}\cdot\mathbf{u}}). \tag{8.57}$$

The Bloch-type states represent a set of excitons that are delocalized over the entire crystal. Thus whereas the electron and hole remain tightly bound to each other, with a size comparable to a lattice constant, the correlated pair can have a wavefunction that is spread out over the entire crystal. Frenkel excitons are typically seen in alkali halide crystals.

8.11   Color Centers

The presence of defects or impurities in crystals leaves their mark on the optical properties of solids. In this section we describe the color centers that are present in imperfect crystals.

In ionic crystals such as the alkali halides, electrons can be trapped at negative ion vacancies that have an effective positive charge. The trapping is due to the combined effect of the Coulomb field of the neighboring cations and the Pauli exclusion principle, which leads to effective repulsion of the electron from the nearby negative ions. The laws of electrostatics (Earnshaw's theorem) forbid an absolute minimum in the potential at a vacancy site, so that a complex saddle-point potential resides near the vacancy. There is also substantial relaxation of the neighboring cation positions toward the anion vacancy, locally distorting the perfect lattice. The electron can be trapped in a localized *s*-like ground state as well as in a more extended *p*-like excited state. Photoabsorption can link these states, with the photon's energy typically lying in the range 2 to 5 eV, depending on the material. The system of an electron trapped at a single vacancy is referred to as an *F center*. The absorption generally appears as a broad band, due to the large lifetime broadening associated with nonradiative (phonon) processes that damp the excited state.

The F centers can be formed by introducing excess alkali atoms or by replacing some alkalis by higher-valence cations. Exposure to ionizing radiation such as x-rays can also produce F centers. The spectral properties turn out to be independent of the nature of the agent that produced them. A schematic of an F center is given in Fig. 8.9.

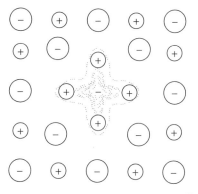

Figure 8.9. F center in an alkali halide crystal. Notice the lattice distortion in the neighborhood of the electron-occupied vacancy. The contours depict surfaces of constant potential.

Other types of color centers include *M centers* (pairs of adjacent F centers) and *R centers* (triplets of adjacent F centers). Trapped holes can also act as color centers. An example is the V_k center, in which a pair of adjacent halide ions constitute a singly negatively charged entity instead of existing as two negative ions. The F_2^+ center consists of a pair of adjacent vacancies with a trapped electron. It may be thought of as a solid-state analog of the hydrogen molecular ion, H_2^+. The energy levels are similar to those of the molecular ion but are scaled down by the factor $1/\epsilon_r^2$, where ϵ_r is the dielectric constant of the crystal. If there are foreign impurity ions present in the lattice, additional types of color centers are possible.

The F_2^+ center in alkali halide crystals is used to construct tunable color-center lasers. For example, a LiF mode-locked laser is tunable over the range 0.82 to 1.05 μm. The corresponding tuning ranges for NaF and KF are 0.99 to 1.22 μm and 1.22 to 1.50 μm, respectively.

In some circumstances, such as in gem stones, color centers are desirable, since they impart color to an otherwise transparent crystal. In other cases, such as in nonlinear optical crystals, they may be undesirable, since they produce unwanted absorption of electromagnetic radiation.

8.12   Polaritons

The chapter began by examining what happens when light couples with the internal degrees of freedom of the solid. Recall that pure light may be described as a set of harmonic oscillators involving the electric and magnetic fields. On the other hand, phonons in a solid are also harmonic oscillators involving the kinetic energy of the ions and the elastic energy of the crystal. When these oscillators coexist, coupled modes are formed, and it is often convenient to analyze the problem in terms of these modes. These are called *polaritons*. A discussion of polaritons can be found in Section W8.2.

8.13   Emissivity

The emissivity ε of a material is an optical property that provides information about its thermal radiative behavior (i.e., how the material behaves as a thermal source of radiant energy in the form of electromagnetic waves). Emissivity plays a particularly important role in the behavior of materials at elevated temperatures since the total radiated power is proportional to both ε and T^4, as described later. Although emissivity is an intrinsic property, it is nevertheless quite sensitive to the size, structure, and surface condition of the material.

The definition of ε can be given in terms of the optical absorptivity a of the material (i.e., the ratio of absorbed to incident power) by

$$\text{emissivity } \varepsilon = \text{absorptivity } a. \tag{8.58}$$

This expression indicates that emission and absorption are related processes and is effectively a statement of the conservation of energy for a solid object that is in thermal equilibrium with its surroundings. In other words, the thermal energy emitted or radiated by a solid is equal to the amount of energy that it absorbs under conditions of thermal equilibrium.

Equation (8.58) is, in fact, valid at every wavelength λ and also for every state of polarization and direction relative to the surface. It follows that

$$\varepsilon(\lambda, \theta, \psi, T) = a(\lambda, \theta, \psi, T) \tag{8.59a}$$

$$= 1 - \mathbf{R}(\lambda, \theta, \psi, T) - \mathbf{T}(\lambda, \theta, \psi, T) \tag{8.59b}$$

where θ defines the direction of emission or absorption, ψ is the state of polarization of the radiation, and $\mathbf{R}$ and $\mathbf{T}$ are the corresponding reflectivity and transmissivity of the object, respectively. Equation (8.59a) is known as *Kirchhoff's law* and $\varepsilon(\lambda)$ is known as the *spectral emissivity*.

There are several interesting and important limits of Eqs. (8.59):

1. In the idealized case when the absorptivity $a(\lambda)$ of a solid material is equal to 1 for all λ, the solid is called a *blackbody*. A blackbody is therefore also a "perfect" emitter, with $\varepsilon(\lambda) = 1$ for all λ.

2. When $a(\lambda)$ is a constant, but less than 1, for all λ, the solid is called a *graybody*. The emissivity $\varepsilon(\lambda)$ is therefore equal to the same constant for all λ.

3. When a solid is opaque for a given λ [i.e., $\mathbf{T}(\lambda) = 0$], $\varepsilon(\lambda) = 1 - \mathbf{R}(\lambda)$. In this case ε can be expressed in terms of the real and imaginary parts of the index of refraction $n(\lambda)$ and $\kappa(\lambda)$, respectively, by

$$\varepsilon(\lambda) = 1 - \frac{(n - n_0)^2 + \kappa^2}{(n + n_0)^2 + \kappa^2} = \frac{4nn_0}{(n + n_0)^2 + \kappa^2}, \tag{8.60}$$

where n_0 is the index of refraction of the surrounding medium.

4. When a solid is transparent (i.e., when it absorbs no radiation at a given λ), $a(\lambda) = 0$ and $\mathbf{R}(\lambda) + \mathbf{T}(\lambda) = 1$. In this case, the emissivity $\varepsilon(\lambda) = 0$.

It follows from these observations that ε can be very low (≈ 0) both for very highly reflecting ($\mathbf{R} \approx 1$) metals such as Cu or Au with polished surfaces and also for highly transparent ($a \approx 0$) solids such as quartz and diamond. The optical properties and hence the emissivities of metals are determined by the presence of free electrons (i.e., by free-carrier absorption) and at higher photon energies by interband transitions. The characteristic optical properties of metals are a high reflectivity $\mathbf{R}$ and also a high absorption coefficient α. The reflectivity of metals is discussed in Section 8.4.

The basis for understanding the thermal radiative properties of solid materials is *Planck's blackbody-radiation law*, which states that the power per unit area per unit wavelength radiated into the forward hemisphere by a blackbody is

$$I_{\text{bb}}(\lambda, T) = \frac{c_1}{\lambda^5[\exp(c_2/\lambda T) - 1]}. \tag{8.61}$$

The constants in this expression are $c_1 = 2\pi hc^2 = 3.7418 \times 10^{-16}$ J · m^2/s and $c_2 = hc/k_B = 1.4388 \times 10^{-2}$ m · K. When $I_{\text{bb}}(\lambda, T)$ is integrated over all wavelengths, the *Stefan–Boltzmann law* for the total power radiated by a blackbody per unit area is obtained:

$$I_{\text{tot,bb}} = \sigma T^4. \tag{8.62}$$

The Stefan–Boltzmann constant has a value $\sigma = 2\pi^5 k_B^4/15c^2h^3 = 5.6705 \times 10^{-8}$ J/s · m^2 · K^4.

The emissivity ε can also be defined as the ratio of the radiation emitted by a solid to that emitted under the same conditions by a blackbody. The radiation emitted by a nonblackbody can therefore be written as

$$I(\lambda, T) = \varepsilon(\lambda, T)I_{bb}(\lambda, T) = \frac{\varepsilon(\lambda, T)c_1}{\lambda^5[\exp(c_2/\lambda T) - 1]}, \tag{8.63}$$

and as

$$I_{tot} = \varepsilon(T)I_{tot,bb}(T) = \varepsilon_{tot}\sigma T^4. \tag{8.64}$$

where $\varepsilon_{tot}(T)$ is the total hemispherical emissivity.

The observed spectral and temperature dependencies of ε for Si will now be described to illustrate the behavior of the emissivity for a typical semiconductor. The normalized spectral emissivity $f_\varepsilon(\lambda)$ of lightly doped Si is shown as a function of λ for several temperatures in Fig. 8.10a, and the total hemispherical emissivities ε_{tot} for lightly and heavily doped Si are shown as functions of temperature up to $T \approx 800°C$ in Fig. 8.10b. The function $f_\varepsilon(\lambda)$ is defined to be the ratio of the spectral emissivity measured to that observed when the sample is opaque. Thus $f_\varepsilon(\lambda) = 1$ when the sample is opaque and $f_\varepsilon(\lambda) = 0$ when the sample is transparent.

It can be seen from Fig. 8.10a that $f_\varepsilon(\lambda)$ decreases from unity to zero with increasing λ, that is, for wavelengths corresponding to photon energies below the optical energy gap E_g(Si) (1.11 eV at $T = 300$ K) where Si is transparent. Note that λ_g(Si) $= hc/E_g$(Si) $= 1.12$ μm at $T = 300$ K. This decrease in $f_\varepsilon(\lambda)$ is no longer apparent at higher temperatures, as free carriers are generated in the Si conduction band via thermal excitation. As a result, both the absorptivity and emissivity of lightly doped Si increase with increasing T as the sample becomes opaque.

This behavior is also evident in Fig. 8.10b, where it can be seen that the total hemispherical emissivity ε_{tot} shows a strong dependence on doping level in Si. Thus ε_{tot} increases with T for lightly doped Si, in agreement with the results presented in Fig. 8.10a, whereas for heavily doped Si, ε_{tot} is high, ≈ 0.7, even at room temperature, due to the presence of free carriers generated in the conduction band as a result of the doping process. At higher temperatures, $T = 1000 - 1700$ K, the emissivity of Si at $\lambda = 0.65$ μm decreases with increasing T as a result of the increase in the reflectivity R, which is presumably due to an increase in the index of refraction n with increasing T.[†] It is clear that the emissivity of Si is determined by the existence of an optical energy gap and the effects of doping.

There are several ways in which the surface condition of a solid material can affect its emissivity. Surface roughness can decrease the reflectivity $\mathbf{R}(\lambda)$, thereby increasing the apparent emissivity $\varepsilon(\lambda)$ of a solid. The effect of roughness becomes more pronounced at shorter λ, where the scattering of light increases. Surface roughness has a much larger effect on the emissivity of highly absorbing samples such as metals where the optical penetration depth α^{-1} is small (α is the optical absorption coefficient). Another surface related effect results from the presence of a thin, transparent (or partially transparent) film with thickness $d \approx \lambda$ on the surface of a solid. The thin film can give rise to optical interference effects which can cause oscillations in both the reflectivity $\mathbf{R}$ and ε as λ or d changes.

[†] F. G. Allen, *J. Appl. Phys.*, **28**, 1510 (1957).

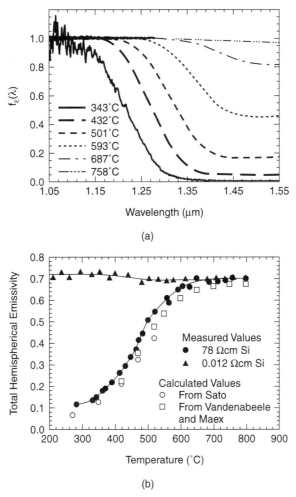

Figure 8.10. (*a*) Normalized spectral emissivity $f_\varepsilon(\lambda)$ of lightly doped Si as function of λ for several temperatures; (*b*) total hemispherical emissivity ε_{tot} for lightly and heavily doped Si as functions of temperature. [Reprinted with permission from P. J. Timans, *J. Appl. Phys.* **74**, 6353 (1993). Copyright 1993 by the American Institute of Physics.]

Practical applications involving the emissivity of materials include optical or radiation pyrometry, in which the temperature of an object is inferred from measurements of its thermal radiative properties. To determine the correct temperature, it is in general necessary to know the emissivity at the measured wavelength. Depending on how many wavelengths are probed at the same time, the technique is known as one-color pyrometry, two-color pyrometry, and so on. In two-color pyrometry, the temperature is determined from the measured intensity ratio $I(\lambda_1, T)/I(\lambda_2, T)$. Two-color pyrometry has the advantage that it is not necessary to know the spectral emissivity $\varepsilon(\lambda)$ of the object in order to determine T as long as $\varepsilon(\lambda_1) = \varepsilon(\lambda_2)$, (i.e., as long as the object is a graybody), since in this case the effects of emissivity cancel in the ratio $I(\lambda_1)/I(\lambda_2)$. Emissivity also plays a crucial role in rapid thermal processing (RTP), in which radiant energy from a lamp is used to rapidly heat a material such as Si during processing, leading to the fabrication of integrated circuits. In this case it is necessary to know the total hemispherical emissivity of the Si wafer in order to predict the equilibrium temperature reached during RTP. The emissivity of tungsten filaments is also obviously of critical importance to the operation and efficiency of incandescent lighting.

REFERENCES

Cohen, M. L., and J. R. Chelikowsky, *Electronic and Optical Properties of Semiconductors*, Springer-Verlag, Berlin, 1988.

Grigoriev, I. S., and E. Z. Meilikhov, eds., *CRC Handbook of Physical Quantities*, CRC Press, Boca Raton, Fla., 1997.

Palik, E. D., ed., *Handbook of Optical Constants of Solids*, Academic Press, San Diego, Calif., Vol. 1, 1995; Vol. 2, 1991; 5 vols., 1998.

Shen, Y. R., *The Principles of Nonlinear Optics*, Wiley, New York, 1984.

Singh, J., *Optoelectronics: An Introduction to Materials and Devices*, McGraw-Hill, New York, 1996.

Sutherland, R. L. ed., *Handbook of Nonlinear Optics*, Marcel Dekker, New York, 1996.

Yu, P. Y., and M. Cardona, *Fundamentals of Semiconductors*, Springer-Verlag, Berlin, 1996.

PROBLEMS

8.1 Derive Beer's law in Eq. (8.4) starting from the assumption that the rate of attenuation of the intensity (dI/dz) a given distance z into a material is proportional to the intensity $I(z)$ at that point.

8.2 Generalize the theory developed in Eqs. (8.19)–(8.25) to include a phenomenological lifetime for the optic modes, characterized by a damping constant τ.

8.3 Show that in the high-frequency limit the expression for $\overleftrightarrow{\epsilon}_r(\omega)$ reduces to the formula for the dielectric function of a plasma.

8.4 Derive the effective medium expressions in Eqs. (8.41) and (8.43).

8.5 Find the index of refraction for the case where the polar angles of $\mathbf{k}$ are given by (θ, ϕ) and the polarization is defined by a rotation angle ψ around the direction of the $\mathbf{k}$ vector.

8.6 Given a nonparabolic conduction band described by the formula $E(p) = [(E_g/2)^2 + E_g p^2/2m^*]^{1/2}$. Assuming that n carriers per unit volume populate the band, derive an expression for $d^{(3)}_{xxxx}$.

8.7 **(a)** In the limit that $c_2/\lambda T \gg 1$ show that Planck's radiation law, Eq. (8.61), becomes $I_{bb}(\lambda, T) \rightarrow c_1/\lambda^5 \exp(c_2/\lambda T)$. This form of Planck's law is known as *Wien's law*.

   (b) For $\lambda = 1 \, \mu m$, what is the range of temperatures for which Wien's law is valid?

   (c) The brightness temperature T_b of an object as measured by a one-color pyrometer is defined by $I_{bb}(\lambda, T_b) = \varepsilon(\lambda, T)I_{bb}(\lambda, T)$, where T is the true temperature. Using Wien's law, show that the true temperature of the object can be found from T_b and $\varepsilon(\lambda, T)$ using $T = T_b/[1 + (\lambda T_b/c_2) \ln \varepsilon]$.

(d) A one-color disappearing-filament pyrometer operating at $\lambda = 0.65$ μm measures $T_b = 1100$ K for a Si sample whose spectral emissivity is $\varepsilon(0.65$ μm, 1100 K$) \approx 0.62$. What is the true temperature T of the Si sample?

8.8 Verify the Kramers–Kronig relations for the single damped oscillator with dielectric function $\epsilon_r(\omega) = 1 + \omega_p^2/(\omega_0^2 - \omega^2 - i\omega\gamma)$ and the metal with dielectric function $\epsilon_r(\omega) = 1 - \omega_p^2/[\omega(\omega + i/\tau)]$.

CHAPTER 9

Magnetic Properties of Materials

9.1 Introduction

The subjects of magnetism and the magnetic properties of materials are of fundamental interest, due in part to the fact that magnetic materials can exhibit long-range cooperative phenomena such as ferromagnetism, antiferromagnetism, and ferrimagnetism. Magnetism is also of great practical importance due to the wide range of applications that magnetic materials find in magnetic recording, in electrical power generation and transmission, in communications, and so on. In this chapter the basic features of magnetism in solids which are needed to understand the magnetic properties of materials are presented. Additional characteristic properties and specific examples of magnetic materials and their applications are the subject of Chapter 17.

The outline of this chapter is as follows. The origins of magnetism in solids are discussed first, with the focus on free atoms and ions and also on atoms and ions with partially filled electron shells in solids. The important types of magnetism and magnetic behavior observed in materials, including paramagnetism, ferromagnetism, antiferromagnetism, ferrimagnetism, and diamagnetism, are then discussed. Throughout these discussions it will become clear that many fundamental questions concerning magnetism and the magnetic properties of materials remain unanswered.

ORIGINS OF MAGNETISM IN SOLIDS

The origins of magnetism and ultimately of the magnetic properties of materials essentially always result from the spin and orbital motions of electrons, the magnetic moments of electrons, and the resulting magnetic moments of atoms and ions. Magnetism is therefore essentially a quantum-mechanical phenomenon, depending on the charge and spin of the electron and the Pauli exclusion principle. The nuclei of atoms can also have magnetic moments and these can contribute to the magnetic properties of materials under special conditions (e.g., at very low temperatures). The magnetic moments of electrons, of atoms and ions, and of nuclei correspond in lowest order to magnetic-dipole moments since magnetic monopoles do not exist. The magnetic moments of free atoms and ions are discussed first, followed by a discussion of the magnetic moments of atoms and ions in solids.

9.2 Free Atoms and Ions

The magnetic properties and magnetic moments m of electrons in isolated or free atoms and ions are discussed first. A convenient unit for m is the magnetic moment of the electron due to its intrinsic spin. This is the *Bohr magneton* μ_B, which is given in standard international (SI) units by

$$\mu_B = \frac{e\hbar}{2m} = 9.274 \times 10^{-24} \text{ J/T}, \tag{9.1}$$

where $e = 1.602 \times 10^{-19}$ C is the unit of electrical charge, $\hbar = h/2\pi = 1.055 \times 10^{-34}$ J·s, where h is Planck's constant, and $m = 9.11 \times 10^{-31}$ kg is the mass of the electron.[†] The tesla is the unit of magnetic induction B in SI units and is equal to 1 N/A·m. Therefore, the units of J/T and A·m^2 are the same.

Electrons in free atoms and ions possess magnetic moments that are associated with both the *spin angular momentum* (or spin) $\mathbf{s}$ of the electron, due to its own intrinsic rotation, and the *orbital angular momentum* $\mathbf{l}$, due to its motion relative to the nucleus. The spin $\mathbf{s}$ is quantized and has the magnitude $\sqrt{s(s+1)}\hbar = \sqrt{3/4}\hbar$, where $s = \frac{1}{2}$ is the spin quantum number of the electron. The angular momentum $\mathbf{l}$ is also quantized with magnitude $\sqrt{l(l+1)}\hbar$, where l is the orbital angular momentum quantum number. The quantum number l can take on the integer values $0, 1, \ldots, n-1$, where n is the principal quantum number of the electron. Additional information on electrons and their quantum numbers in atomic orbitals is presented in Chapter W2 our Web site.[‡]

The relationship between angular momentum and the resulting *magnetic moment* of an electron in an atom is different for spin and orbital motion. The contribution of the spin of an electron to its magnetic moment is given by

$$\mathbf{m}_{\text{spin}} = -\frac{e}{m}\mathbf{s} = -\frac{g_e\mu_B}{\hbar}\mathbf{s}, \tag{9.2}$$

where g_e is the *Landé g factor*. The minus sign in this equation is due to the negative charge of the electron which causes $\mathbf{m}_{\text{spin}}$ and $\mathbf{s}$ to point in opposite directions. The corresponding contribution of the orbital motion of the electron is given by

$$\mathbf{m}_{\text{orb}} = -\frac{e}{2m}\mathbf{l} = -\frac{\mu_B}{\hbar}\mathbf{l}. \tag{9.3}$$

Note that $\mathbf{m}_{\text{orb}}$ and $\mathbf{l}$ also point in opposite directions for an electron. The *gyromagnetic ratio* γ is defined as the ratio of the magnetic moment to the corresponding angular momentum. With $\gamma = g_e(e/2m)$, it follows for electrons that the Landé g factor is $g_e = 2.0023 \approx 2$.

Consider now the total magnetic moment $\mathbf{m}$ of a free atom or ion which is the vector sum of the magnetic moments of its electrons. The electron configurations and energy levels in an atom play key roles in determining its magnetic moment. The electron configurations of atoms in their ground states are listed in Table W2.2. Electrons in filled shells do not contribute to the total angular momentum $\mathbf{J}$ or magnetic moment $\mathbf{m}$ of an atom since the net spin S and orbital angular momenta L of electrons in filled shells are both equal to zero. This follows from the *Pauli exclusion principle*, which states that two electrons in an atom cannot have the same set of quantum numbers.

[†] In cgs units the Bohr magneton is given by $\mu_B = e\hbar/2m_e c = 9.274 \times 10^{-21}$ erg/G.

[‡] Supplementary material for this textbook is included on the Web at the resource site (ftp://ftp.wiley.com/public/sci_tech_med/materials). Cross-references to elements of the Web material are prefixed by "W."

As a result, the spin momenta of the individual electrons in a filled shell cancel each other completely, as do the angular momenta.

When a shell is filled only partially, as can be the case for the $3d$ shell of the Fe-group transition elements, a net angular momentum and magnetic moment for the atom can result. For example, the net or total angular momentum $\mathbf{J}$ of the electrons corresponding to the lowest energy level of a partially filled $3d$ shell is usually given by the vector sum of their corresponding total orbital angular momentum $\mathbf{L}$ and total spin angular momentum $\mathbf{S}$ (i.e., $\mathbf{J} = \mathbf{L} + \mathbf{S}$). In this case $\mathbf{L} = \Sigma_i \mathbf{l}_i$ and $\mathbf{S} = \Sigma_i \mathbf{s}_i$ are the vector sums of the individual orbital momenta $\mathbf{l}_i$ and spin momenta $\mathbf{s}_i$, respectively, of the electrons in the $3d$ shell. This form of the coupling of $\mathbf{L}$ and $\mathbf{S}$ to give $\mathbf{J}$ is known as *Russell–Saunders* or *LS coupling*. Here $\mathbf{L}$ and $\mathbf{S}$ are coupled to each other through the *spin–orbit interaction*, which has the form $U(\text{spin–orbit}) = \lambda_{so}\mathbf{L} \cdot \mathbf{S}$. This interaction results from the fact that the spin of an electron in an orbit around the nucleus will experience a magnetic field due to the nuclear charge $+Ze$. The spin–orbit parameter λ_{so} is greater than 0 for a shell less than half-filled and less than 0 for a shell more than half-filled.

Russell–Saunders coupling has been shown to be valid for atoms in which the spin–orbit interaction is weak. This includes essentially all atoms with magnetic moments except for the heaviest atoms (e.g., the actinide elements).[†] The results of Russell–Saunders coupling in a two-electron atom are shown schematically in Fig. 9.1. Note that $\mathbf{l}_1$ and $\mathbf{l}_2$ combine to form $\mathbf{L}$, $\mathbf{s}_1$ and $\mathbf{s}_2$ combine to form $\mathbf{S}$, and the resulting $\mathbf{L}$ and $\mathbf{S}$ finally combine to form $\mathbf{J}$.

The results of the vector additions of the individual orbital and spin momenta of electrons in a partially filled shell to yield $\mathbf{L}$ and $\mathbf{S}$ and of $\mathbf{L}$ and $\mathbf{S}$ to yield the total angular momentum $\mathbf{J}$ are summarized by Hund's rules for the ground or lowest energy state of a free atom or ion. These rules also predict the relative orientations of the individual spins of the electrons in the partially filled shells.

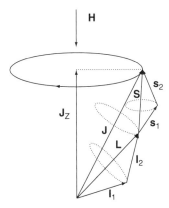

Figure 9.1. Results of Russell–Saunders (LS) coupling in a two-electron atom. Note that $\mathbf{l}_1$ and $\mathbf{l}_2$ combine to form $\mathbf{L}$, $\mathbf{s}_1$ and $\mathbf{s}_2$ combine to form $\mathbf{S}$, and $\mathbf{L}$ and $\mathbf{S}$ combine to form $\mathbf{J}$.

[†] In heavy atoms such as the actinides jj coupling rather than *LS* coupling occurs. In jj coupling $\mathbf{l}_i$ and $\mathbf{s}_i$ for each electron couple to form $\mathbf{j}_i = \mathbf{l}_i + \mathbf{s}_i$. The $\mathbf{j}_i$ then couple with each other to form the resultant $\mathbf{J} = \Sigma_i \mathbf{j}_i$.

Hund's Rules. For the ground state of a free atom or ion, *Hund's rules* state that:

1. The total spin **S** of the system is given by the vector sum $\mathbf{S} = \Sigma_i \mathbf{s}_i$, which yields the largest value of the magnitude of the spin angular momentum $S\hbar$ consistent with the Pauli exclusion principle. For example, when there are five electrons in the $3d$ ($n = 3$, $l = 2$) shell (i.e., when the $3d$ shell is half filled), the total spin quantum number S has its maximum possible value of $5s = \frac{5}{2}$, which corresponds to the spins of the five electrons all pointing in the same direction (e.g., ↑↑↑↑↑). Here ↑ represents the $s = \frac{1}{2}$ spin of an electron pointing "up" (i.e. with $m_s = +\frac{1}{2}$). In the general case of n electrons in the $3d$ shell, $S = n/2$ for $n \leq 5$, while $S = (10 - n)/2$ for $n \geq 6$. For example, for $n = 7$ the configuration of spins is given schematically by ↑↑↑↑↑↓↓, which corresponds to $S = \frac{3}{2}$.
2. The total orbital angular momentum **L** of the system is given by the vector sum $\mathbf{L} = \Sigma_i \mathbf{l}_i$, which yields the largest value of the magnitude of $L\hbar$ consistent with the Pauli exclusion principle and with rule 1. For example, for $n = 5$ electrons in the $3d$ ($l = 2$) shell, the total orbital quantum number $L = (2 + 1 + 0 - 1 - 2) = 0$. For the general case of n equivalent $3d$ electrons, $L = n(5 - n)/2$ for $n \leq 5$, while $L = (n - 5)(10 - n)/2$ for $n \geq 6$. For the case of n equivalent $4f$ electrons $L = n(7 - n)/2$ for $n \leq 7$, while $L = (n - 7)(14 - n)/2$ for $n \geq 8$.
3. The angular momenta **L** and **S** that result from the applications of rules 1 and 2 combine to yield the total angular momentum **J**, the magnitude of which (in units of $\hbar$) is $J = L - S$ for shells less than half-filled and $J = L + S$ for shells more than half-filled. For example, for $3d$ electrons $J = (3 - 1) = 2$ for $n = 2$ and $J = (2 + \frac{1}{2}) = \frac{5}{2}$ for $n = 9$.

Neglecting the nuclear spin I and the hyperfine interaction, the ground state of a free atom or ion corresponds to the values of **S**, **L**, and **J** that result from the application of Hund's rules. The ground state will have a degeneracy given by $(2J + 1)$, which can be lifted, for example, by the application of a weak external magnetic field. The lowest-lying excited states of the atom or ion will be characterized by the same values of S and L but with different values of J lying between $L - S$ and $L + S$. This fine-structure splitting of states with the same S and L but different J and J_z is due to the spin-orbit interaction.

The physical basis for rules 1 and 2 (i.e., the preferential occupation by the electrons in a partially filled shell of states with the same quantum number m_s but different quantum numbers m_l) is the stronger Coulomb repulsion (and resulting higher energy) that occurs between electrons in the same orbital (i.e., with the same value of m_l but with opposite spins s). Conversely, electrons in the same shell but with different values of m_l will reside in different orbitals. In this case their wavefunctions will not have as great a spatial overlap and the Coulomb repulsion between the electrons will be reduced. Rule 3 is consistent with the Russell–Saunders coupling mentioned above and results from the spin–orbit interaction between **L** and **S** and also from the fact that the spin–orbit parameter λ_{so} is > 0 for $n < 5$ and < 0 for $n \geq 5$. When $\lambda_{so} < 0$, **J** and **S** are antiparallel.

To summarize, the energy of a given electron configuration such as $3d^n$, which is characterized by the quantum numbers $n = 3$ and $l = 2$, is split into levels (or terms) by electron–electron Coulomb interactions, characterized in Russell–Saunders coupling by the values of **L** and **S**. These levels are split further via the spin–orbit interaction into levels (or multiplets) characterized by different values of $\mathbf{J}$, as mentioned previously. A multiplet with quantum numbers S, L, and J is referred to by the ground-state representation $^{2S+1}L_J$. The notation used for L in the ground-state representations follows standard usage, that is,

$$L = 0, \ 1, \ 2, \ 3, \ 4, \ 5, \ 6, \ \ldots$$
$$S, \ P, \ D, \ F, \ G, \ H, \ I, \ \ldots \ . \tag{9.4}$$

Therefore, $L = 0$ corresponds to an S state, $L = 1$ to a P state, and so on.

These splittings are illustrated for free ions of the $4f$ rare earths Eu^{3+} ($4f^6$), Gd^{3+} ($4f^7$), and Tb^{3+} ($4f^8$) in Fig. 9.2, where the corresponding energy-level diagrams are shown. The ground-state configurations for these three ions are, according to Hund's rules, $^7F_0$, $^8S_{7/2}$, and $^7F_6$, respectively. A series of excited-state energy levels corresponding to the ^7F configuration but with different values of J are shown for both Eu^{3+} and Tb^{3+}. The Gd^{3+} ion has no excited states within the ^8S configuration since $L = 0$ and $J = S$. A further discussion of the energy levels of the rare earth ions in solids is given in Section 9.3.

The total magnetic moment $\mathbf{m}$ of a free atom or ion is on the average parallel to $\mathbf{J}$ and, using Eqs. (9.2) and (9.3), is given by

$$\mathbf{m} = \mathbf{m}_{\text{spin}} + \mathbf{m}_{\text{orb}} = -\frac{\mu_B(\mathbf{L} + 2\mathbf{S})}{\hbar} = -\frac{g\mu_B\mathbf{J}}{\hbar}, \tag{9.5}$$

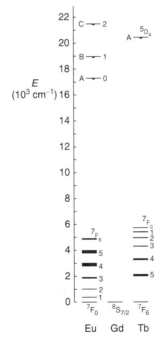

Figure 9.2. Energy-level diagrams for free ions of the $4f$ trivalent rare earths Eu^{3+} ($4f^6$), Gd^{3+} ($4f^7$), and Tb^{3+} ($4f^8$). The ground-state configurations for these three ions are $^7F_0$, $^8S_{7/2}$, and $^7F_6$, respectively. [From G. H. Dieke et al., *Appl. Opt.*, **2**, 675 (1963). Reprinted by permission of the Optical Society of America.]

where here the Landé g factor for the atom is defined by

$$g = 1 + \frac{J(J+1) + S(S+1) - L(L+1)}{2J(J+1)}. \tag{9.6}$$

The magnetic moment **m** can therefore be calculated using Hund's rules and Eqs. (9.5) and (9.6). Note that when **S** = 0, the moment is completely orbital in character. In this case, **J** = **L** and $g = 1$. Also, when **L** = 0, the moment is completely spin in character, so that **J** = **S** and $g = 2$. Thus measurement of the g factor for a magnetic ion using, for example, the electron spin resonance (ESR) techniques described in Chapter W22, can help to determine whether the magnetic moment of the ion in the solid arises from spin effects alone, from orbital effects alone, or from a combination of both. Even when **L** is *quenched* in a solid [i.e., when quantum effects average L to zero (so that L is no longer a good quantum number)], g can be greater or less than 2 due to spin–orbit and crystal field effects. Crystal field effects can also affect the magnetic behavior of an ion in a solid, as described briefly later.

It can be concluded from the discussion above that a net magnetic moment **m** will exist on a free atom or ion whenever unpaired electrons are present in a partially filled shell. Only for atoms or ions with all shells filled will **S**, **L**, and **J** each be equal to zero.

9.3 Atoms and Ions in Solids

Atoms and ions in solids can no longer be considered to be "free" since their outer electron orbitals and magnetic moments **m** are often strongly affected by interactions with their surroundings (i.e., with the crystal electric fields of neighboring ions and also with other magnetic moments in the material). In some cases both the orbital and spin parts of **m** are effectively destroyed or quenched by these interactions, so that no net atomic magnetic moment survives in the solid. In other cases only the orbital part of **m** is quenched while the spin part is either unaffected or only partially affected. If the interactions are sufficiently weak, the free-atom magnetic moment given by Eqs. (9.5) and (9.6) can be retained in the solid. Examples of these situations are given later. The total magnetic moment of the atom in the solid, if one exists, will often be referred to as its "spin" even if it contains an orbital component in addition to a spin component.

Atoms and ions with partially filled shells will now be discussed with regard to their magnetic behavior in solids. It should be pointed out once more that electrons in filled shells, including the core electrons, do not contribute to the magnetic moment of an atom. As a result, inert-gas atoms such as He, Ne, and Ar, apart from weak nuclear spin effects for He^3, do not have magnetic moments in solids and will therefore not be discussed here. Electrons in filled shells do contribute a diamagnetic response to external magnetic fields known as *Larmor diamagnetism*. This contribution is discussed in Sec. 9.9. Following the discussion of atoms and ions with partially filled shells, atomic nuclei and defects that have magnetic moments in solids are mentioned briefly.

Atoms and Ions with Partially Filled s and p Shells. The atoms whose only partially filled shells are s and p shells include all the elements of the periodic table with the exception of the inert-gas elements of group VIII, which have no unfilled shells; the 3d, 4d, and 5d transition elements, whose d shells are also only partially filled; and the $4f$ rare earth and $5f$ actinide elements, whose f shells are also only partially filled. The magnetic moments of free atoms and ions which are due solely to electrons in partially filled s and p shells are essentially always completely quenched and hence no longer present in the solid state. Examples include C $(2s^2 2p^2)$ and Si $(3s^2 3p^2)$ atoms in covalent crystals, where the s and p electrons enter covalent bonds in which pairs of electronic spins are directed antiparallel to each other; Na $(2s^2 2p^6 3s^1)$ and Cu $(3d^{10} 4s^1)$ atoms in metallic crystals, where Na$^+$ $(2s^2 2p^6)$ and Cu$^+$ $(3d^{10})$ ions with filled shells are formed and the s electrons act as delocalized conduction electrons; K $(3s^2 3p^6 4s^1)$ and Br $(4s^2 4p^5)$ atoms in the ionic solid KBr, where K$^+$ $(3s^2 3p^6)$ and Br$^-$ $(4p^2 4p^6)$ ions with filled s and p shells are formed. In all these cases both the orbital and spin parts of the magnetic moment **m** of the free atom are effectively destroyed or quenched by the strong interactions that occur with electrons on other atoms or ions in the solid. Therefore, no net magnetic moment survives.

An interesting exception to the behavior described above is Cu $(3d^{10} 4s^1)$. In the FCC crystal structure Cu is metallic and the Cu$^+$ $(3d^{10})$ ions have no net magnetic moment (i.e., $\mathbf{J} = \mathbf{S} = 0$). In solids such as CuO and YBa$_2$Cu$_3$O$_{7-x}$, however, Cu can exist as a Cu^{2+} $(3d^9)$ ion with a magnetic moment corresponding to $S = \frac{1}{2}$. It is apparently energetically favorable for one of the Cu $3d$ electrons to participate more directly in the bonding in the solid when strong covalent or ionic bonds can be formed.

Atoms and Ions with a Partially Filled d Shell. Atoms with partially filled $3d$, $4d$, or $5d$ shells belong to the corresponding transition element groups of the periodic table [i.e., to the $3d$ iron group beginning at scandium (Sc), to the $4d$ group beginning at yttrium (Y), or to the $5d$ group beginning at lanthanum (La)]. These elements are known in the solid state as the $3d$, $4d$, and $5d$ transition metals, respectively.

Consider first the important $3d$ iron group whose ionic $3d^n$ electron configurations, quantum numbers (S, L, and J), and ground-state representations $^{2S+1}L_J$ are given in Table 9.1 for the common valence states of these ions. Also presented are values of p, the *effective magneton number* calculated from $p = g\sqrt{S(S+1)}$ when $J = S$ and from $p = g\sqrt{J(J+1)}$ when $J = |L \pm S|$. Values for p obtained from measurements of the temperature dependence of the magnetic susceptibility χ for these $3d$ ions in paramagnetic ionic salts are also presented for comparison. The procedure for obtaining p from the measured $\chi(T)$ is explained in Section 9.4.

When the free atom or ion magnetic moments are retained in the solid state, the measured values of the effective magneton number p presented in the right-hand column of Table 9.1 agree with the calculated values $p = g\sqrt{J(J+1)}$ obtained from Hund's rules, which predict that $J = |L \pm S|$. The measured values of p for $3d$ ions in insulating solids are generally consistent instead with the values of p calculated under the assumption that $L = 0$, so that $J = S$ and $p = g\sqrt{S(S+1)}$. It can be concluded, therefore, for these $3d$ ions that the orbital angular momentum **L** is quenched (i.e., averaged to zero) by the effect of the crystal electric field due to the neighboring ions in the ionic compounds. This is a specific example of the general phenomenon known as *crystal field splitting* of energy levels in solids. The remaining magnetic moments are therefore essentially completely spinlike in character. Note that $L = 0$ for free ions with the $3d^5$ configuration (e.g., Mn^{2+}) even in the absence of crystal field effects. The effects of crystal fields on the electronic energy levels and spins of magnetic ions in materials are discussed next in more detail.

TABLE 9.1 $3d$ Iron-Group Transition Element Ions: Ground States of Free Ions and of Ions in Solids

Ion	Electron Configuration $3d^n$	Free-Ion Quantum Numbers S, L, J	Free-Ion Ground State $(^{2S+1}L_J)$	Calculated p		Measured p^a
				$g\sqrt{S(S+1)}$	$g\sqrt{J(J+1)}$	
Ca^{2+}	$3d^0$	0, 0, 0	$^1S_0$	0.0	0.0	0.0
Sc^{2+}	$3d^1$	$\frac{1}{2}, 2, \frac{3}{2}$	$^2D_{3/2}$	1.73	1.55	—
Ti^{3+}	$3d^1$	$\frac{1}{2}, 2, \frac{3}{2}$	$^2D_{3/2}$	1.73	1.55	—
V^{4+}	$3d^1$	$\frac{1}{2}, 2, \frac{3}{2}$	$^2D_{3/2}$	1.73	1.55	1.8
V^{3+}	$3d^2$	1, 3, 2	$^3F_2$	2.83	1.63	2.8
Cr^{4+}	$3d^2$	1, 3, 2	$^3F_2$	2.83	1.63	—
V^{2+}	$3d^3$	$\frac{3}{2}, 3, \frac{3}{2}$	$^4F_{3/2}$	3.87	0.77	3.8
Cr^{3+}	$3d^3$	$\frac{3}{2}, 3, \frac{3}{2}$	$^4F_{3/2}$	3.87	0.77	3.7
Mn^{4+}	$3d^3$	$\frac{3}{2}, 3, \frac{3}{2}$	$^4F_{3/2}$	3.87	0.77	4.0
Cr^{2+}	$3d^4$	2, 2, 0	$^5D_0$	4.90	0	4.8
Mn^{3+}	$3d^4$	2, 2, 0	$^5D_0$	4.90	0	5.0
Mn^{2+}	$3d^5$	$\frac{5}{2}, 0, \frac{5}{2}$	$^6S_{5/2}$	5.92	5.92	5.9
Fe^{3+}	$3d^5$	$\frac{5}{2}, 0, \frac{5}{2}$	$^6S_{5/2}$	5.92	5.92	5.9
Fe^{2+}	$3d^6$	2, 2, 4	$^5D_4$	4.90	6.70	5.4
Co^{2+}	$3d^7$	$\frac{3}{2}, 3, \frac{9}{2}$	$^4F_{9/2}$	3.87	6.54	4.8
Ni^{2+}	$3d^8$	1, 3, 4	$^3F_4$	2.83	5.59	3.2
Cu^{2+}	$3d^9$	$\frac{1}{2}, 2, \frac{5}{2}$	$^2D_{5/2}$	1.73	3.55	1.9
Cu^{1+}	$3d^{10}$	0, 0, 0	$^1S_0$	0.0	0.0	0.0

Source: Data from J. H. Van Vleck, *The Theory of Electric and Magnetic Susceptibilities*, Oxford University Press, Oxford, 1932; and R. Kubo and T. Nagamiya, eds., *Solid State Physics*, McGraw-Hill, New York, 1969, p. 453.

[a] —, Not observed or not known.

Crystal Fields. When an ion with an unfilled d or f shell is placed in a solid, it should not be surprising that its free-ion electronic energy levels and even its spin can be strongly affected by the interaction of electrons in its d or f orbitals with the electrostatic crystal field of neighboring ions. The basic features of the effects of the crystal field on the energy levels of $3d$ transition metal ions, for example, can be understood in a straightforward way as resulting from the shape and directionality of the $3d$ orbitals and from the electrostatic Coulomb repulsion between the electrons in these orbitals and the adjacent negative ions. An interesting and important example that is discussed in Chapter W9 is the Cr^{3+} ion, which is surrounded by six O^{2-} ions when present as an impurity in Al_2O_3. Also, crystal field effects determine the magnetocrystalline anisotropy in magnetic materials such as ferromagnets and ferrites, discussed in Section 9.6 and Chapter 17. The interactions responsible for the lifting of the degeneracy of the $3d$ levels are described in Chapter W22.

The important features of the effects of crystal fields on the energy levels of $3d$ ions and consequently on their magnetic behavior in materials are presented here in terms of *crystal field theory*. The physical basis for crystal field theory was first presented by H. Bethe.[†] This approach is oversimplified in that only the electrostatic interactions between the ions in the material are considered, and any possible covalent bonding between the transition metal cations and the neighboring anions is not taken into account. Extensions of crystal field theory that are important when the crystal field is strong take into account the mixing between the $3d$ orbitals and orbitals on neighboring anions or ligands. These extensions include ligand field theory and molecular orbital theory but are not discussed in detail here.

As an introduction to crystal field theory, consider the five possible d orbitals of a $3d$ transition metal ion which in the spherically symmetric potential of the free ion are degenerate in energy. In Fig. 9.3 these orbitals are shown for ions occupying sites

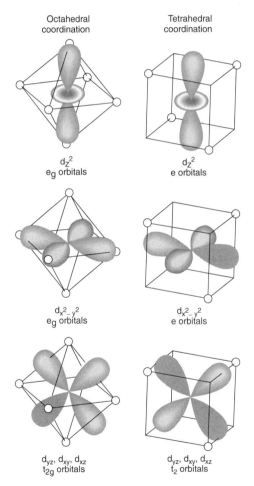

Figure 9.3. The d orbitals of a $3d$ transition metal ion such as Cr^{3+} in sites with octahedral (i.e., cubic) symmetry and with tetrahedral symmetry. (From R. J. Borg et al., *The Physical Chemistry of Solids*, copyright 1992. Reprinted by permission of Academic Press, Inc.)

[†] H. Bethe, *Ann. Phys.*, **3**, 133 (1929).

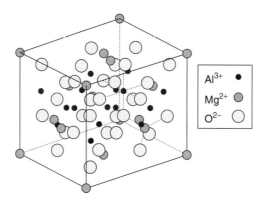

Figure 9.4. Crystal structure of spinel, $MgAl_2O_4$. In the cubic unit cell of the normal spinel structure shown, the oxygen O^{2-} anions form a close-packed FCC array, the Mg^{2+} ions are in tetrahedral coordination with four O^{2-} ions, and the Al^{3+} ions are in octahedral coordination with six O^{2-} ions.

where the electrostatic potential has octahedral (i.e., cubic) symmetry and also where it has tetrahedral symmetry. Both types of sites are found, for example, in crystals such as spinel, $MgAl_2O_4$, the prototype of the normal *spinel crystal structure* shown in Fig. 9.4. In normal spinel the O^{2-} anions form a close-packed FCC array, the Mg^{2+} ions are in tetrahedral coordination with four O^{2-} ions, and the Al^{3+} ions are in octahedral coordination with six O^{2-} ions. Electrons in d orbitals that are directed toward the surrounding negative ions will have energies higher than would be the case if the charge of the negative ions was distributed uniformly in a spherical shell at the same distance from the $3d$ ion. Conversely, electrons in d orbitals that do not point toward negative ions will have lower energies than in the spherical case. The fivefold degeneracy of the free-ion d orbitals will therefore be removed by the crystal field.

For the octahedral site with six NN anions shown in Fig 9.3, it can be seen that the $d_{x^2-y^2}$ and d_{z^2} orbitals are directed toward the negative ions and will therefore lie higher in energy than the d_{xy}, d_{yz}, and d_{xz} orbitals, which point toward regions between the negative ions. The energies of the d_{xy}, d_{yz}, and d_{xz} orbitals, also known in crystal field theory as t_{2g} orbitals (t_2 corresponds to triply degenerate), are still equal to each other in the tetrahedral field, due to their equivalent orientations with respect to the negative ions. The energies of the $d_{x^2-y^2}$ and d_{z^2} orbitals, also known as e_g orbitals (e corresponds to doubly degenerate), can also be shown to be equal to each other in the octahedral field. Therefore, the lower level in the crystal field will be threefold degenerate, and the upper level will be twofold degenerate. Using the same reasoning, it can be seen that for the tetrahedral site with four NN anions, also shown in Fig. 9.3, the $d_{x^2-y^2}$ and d_{z^2} orbitals will lie lower in energy than the d_{xy}, d_{yz}, and d_{xz} orbitals.

The corresponding energy-level diagrams for these two cases are shown schematically in Fig. 9.5, where it can be seen that the fivefold-degenerate $3d$ orbitals are split into two groups, with the energy splitting between the lower and upper groups of levels being Δ_o for the octahedral site and Δ_t for the tetrahedral site. These new levels in the crystal field are often referred to as *Stark levels*, since they arise from the effects of an internal electric field. Relative to the spherically symmetric degenerate level, the d_{xy}, d_{yz}, and d_{xz} orbitals are shifted lower in energy by $2\Delta_o/5$ for the octahedral case and higher in energy by $2\Delta_t/5$ for the tetrahedral case. The corresponding opposite shifts

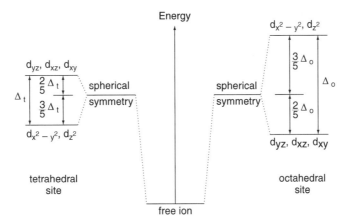

Figure 9.5. Schematic energy-level diagrams for $3d$ transition metal ions in octahedral and tetrahedral sites.

for the $d_{x^2-y^2}$ and d_{z^2} orbitals are by the amounts $3\Delta_o/5$ and $3\Delta_t/5$ for the octahedral and tetrahedral cases, respectively. These energy shifts are such that the total energy of the $3d^{10}$ configuration will be the same in both the spherically symmetric and the crystal field-split cases.

There can be additional splittings of these levels due to the crystal fields of next-NN cations.[†] For example, a $3d^7$ Co^{2+} cation in the spinel structure has six next-NN Co^{2+} cations, which produce an additional component of the crystal field having trigonal symmetry. In this case the lower level will be split into a singlet and a doublet, with the singlet having the lowest energy. The direction of the spin of the Co^{2+} ion can also be determined by the trigonal crystal field component. The *Jahn–Teller effect*, described in Chapter W9, can also result in distortions of the local structure which remove the ground-state degeneracy.

When placed in an octahedral (or tetrahedral) site in a crystal, an ion with a $3d^n$ electronic configuration will gain a *crystal field stabilization energy* (CFSE) relative to the spherically symmetric case. The magnitude of the CFSE depends on both the splitting Δ_o (or Δ_t) and the occupation of the crystal field split levels by the n d electrons. For example, the CFSE will be $2\Delta_o/5$ for a $3d^1$ ion such as Ti^{3+} in an octahedral site with its single $3d$ electron in one of the lower d_{xy}, d_{yz}, or d_{xz} orbitals. The values predicted for the CFSEs and spins S of the ground states of $3d^n$ ions in an octahedral site are given in Table 9.2 for the case where the crystal field is relatively weak. This weak-field limit is appropriate when Δ_o is small compared to the energy cost U of Coulomb repulsion in having two d electrons with opposite spins in the same orbital. Note that in this limit the spins S are just those predicted by Hund's rule 1. Predicted values of the CFSEs and spins S for the interesting strong-field limit in which $\Delta_o > U$ are also presented and are discussed later.

It can be seen that at least in the weak-field limit of an octahedral crystal field, the CFSE first increases as electrons are added to the d shell of a $3d$ ion, decreases to essentially zero at the half-filled shell, and then goes through the same cycle as the

---

[†] See Chapter W22 for the lifting of the degeneracy due to octahedral, tetragonal, and orthorhombic distortions of the spherically symmetric crystal field.

TABLE 9.2   Crystal Field Stabilization Energies and Spins S for $3d^n$ Ions in Octahedral Sites

Configuration $3d^n$	Weak Field ($\Delta_o < U$)		Strong Field ($\Delta_o > U$)	
	CFSE[a]	Spin S	CFSE[a]	Spin S
$3d^0$	0	0	0	0
$3d^1$	$2\Delta_o/5$	$\frac{1}{2}$	$2\Delta_o/5$	$\frac{1}{2}$
$3d^2$	$4\Delta_o/5$	1	$4\Delta_o/5$	1
$3d^3$	$6\Delta_o/5$	$\frac{3}{2}$	$6\Delta_o/5$	$\frac{3}{2}$
$3d^4$	$3\Delta_o/5$	2	$8\Delta_o/5 \; (-U)$	1
$3d^5$	0	$\frac{5}{2}$	$10\Delta_o/5(-2U)$	$\frac{1}{2}$
$3d^6$	$2\Delta_o/5 \; (-U)$	2	$12\Delta_o/5(-3U)$	0
$3d^7$	$4\Delta_o/5(-2U)$	$\frac{3}{2}$	$9\Delta_o/5 \; (-3U)$	$\frac{1}{2}$
$3d^8$	$6\Delta_o/5(-3U)$	1	$6\Delta_o/5 \; (-3U)$	1
$3d^9$	$3\Delta_o/5(-4U)$	$\frac{1}{2}$	$3\Delta_o/5 \; (-4U)$	$\frac{1}{2}$
$3d^{10}$	$0 \;\; (-5U)$	0	$0 \;\; (-5U)$	0

[a]CFSE is the crystal field stabilization energy and is given in units of the energy splitting Δ_o.

shell is filled completely. Experimental evidence supporting this predicted behavior can be found in the dependencies of the heats of formation and also the interionic separations on the occupancy n of the $3d$ shell for transition metal–based salts such as the MF_2 fluorides (where M is a $3d^n$ transition metal ion). It is found for these properties that a relative minimum occurs for Mn^{2+}, which has the $3d^5$ configuration, consistent with the predictions of crystal field theory presented in Table 9.2.

The strong-field limit for the crystal field is an important issue with regard to the magnetic behavior of $3d^n$ ions with $n = 4, 5, 6$, and 7 d electrons. Independent of the strength of the field and the ratio Δ_o/U, the first three d electrons associated with $3d$ ions in octahedral sites will always enter the lower crystal field energy level, corresponding to the d_{xy}, d_{yz}, and d_{xz} orbitals, where they gain the CFSE listed in Table 9.2. According to Hund's rules for free ions, the fourth d electron in the $3d$ shell would enter the upper energy level and the spin S of the $3d^4$ ion would then have its maximum allowed value of 2, corresponding to four unpaired electrons. This will actually occur in the weak-field limit only when the cost in energy Δ_o is less than the cost of energy U due to Coulomb repulsion of placing the fourth d electron with its spin reversed in one of the singly occupied d_{xy}, d_{yz}, or d_{xz} orbitals. In the opposite, strong-field limit, where $\Delta_o > U$, the fourth d electron will stay in the lower energy level, and the spin of the $3d^4$ ion will be $S = 1$, corresponding to only two unpaired electrons. The strong-field limit in which Hund's rule for the spin S breaks down is therefore also the limit of low spin. Note that the spin S of the $3d^6$ ion is actually predicted to be equal to zero in the strong-field limit, due to the occupation of each of the lower-energy d_{xy}, d_{yz}, and d_{xz} orbitals with pairs of electrons with opposite spins.

The magnitudes of the crystal field splittings Δ_o and Δ_t are proportional to the strengths of the corresponding crystal electric fields at the octahedral and tetrahedral sites, respectively. Electrostatic calculations show that for equivalent ionic charges and distances,

$$\Delta_t = \frac{4\Delta_o}{9}. \qquad (9.7)$$

That the crystal field is predicted to be weaker at tetrahedral sites is a reasonable result considering the fact that tetrahedral sites are surrounded by only four anions, while octahedral sites are surrounded by six anions. Theoretical estimates for Δ_o or Δ_t based solely on crystal field theory in which the interactions are purely ionic have not been very successful. To obtain reasonable estimates it is necessary to include the effects of covalent interactions between the $3d$ orbitals and orbitals on the neighboring anions. Typical values of Δ_o obtained from spectroscopic studies on $3d$ transition metal ions are ≈ 1 eV for divalent ions and ≈ 2 eV for trivalent ions.[†] The energy gained in the crystal field (i.e., the CFSE) represents only about 10% of the total binding energy that the free ion gains as a result of its attractive electrostatic interactions with the anions when it is placed in the material.

Other common site geometries for ions in crystals include tetragonal and square planar sites, whose relationships to octahedral sites with cubic symmetry are shown in Fig. 9.6. Here it can be seen that as the two anions on the z axis, for example, are moved farther away from the xy plane containing the other four anions, the geometry of the cation site changes from octahedral or cubic to tetragonal and then effectively to square planar as the two anions are moved far away. These distortions of the original octahedron can result from the Jahn–Teller effect. The energy-level diagrams for the d orbitals of a $3d$ ion on these three types of related sites are also shown schematically.

The effects of crystal fields on a Cr^{3+} ion in an octahedral site [e.g., as a dopant ion in ruby (Al_2O_3)] are described in Chapter W9.

As the nuclear charge $+Ze$ increases across the $3d$ series from $Z = 21$ for Sc to $Z = 28$ for Ni, the $3d$ electrons become more tightly bound by the nuclear potential. As a result, $3d$ ions such as Cr^{2+}, Mn^{2+}, Fe^{2+}, Co^{2+}, and Ni^{2+} toward the right-hand side of the $3d$ series retain more of their free-ion character and thus tend to retain a

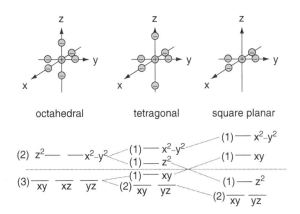

Figure 9.6. Relationships of tetragonal and square-planar sites to octahedral sites. As the two anions on the z axis are moved further away from the xy plane, the geometry of the cation site changes to tetragonal and then to square planar. The energy-level diagrams for the d orbitals of ions on the three types of related sites are also shown schematically, with degeneracies given in parentheses. (From A. L. Companion, *Chemical Bonding*, 2nd ed., McGraw-Hill, New York, 1979.)

[†] Y. Tanabe and S. Sugano, *J. Phys. Soc. Jpn.*, **9**, 766 (1954).

magnetic moment in solids. On the left-hand side the $3d$ ions Sc^{2+} and Ti^{3+} usually do not retain a magnetic moment in solids. One possible explanation for this behavior is that the $3d$ electrons of Sc and Ti interact strongly and therefore mix completely with the conduction electrons in the solid, thereby losing their magnetic character. This behavior can also occur for otherwise magnetic atoms or ions, such as Fe^{2+} and Mn^{2+}, when placed in metals such as Al with a high density of conduction electrons. (See Chapter W9 for further discussion of this effect.)

For the $4d$ ions ($Z = 39$ to 46, Y to Pd) and the $5d$ ions ($Z = 57$ and 72 to 78, La to Pt) the d electrons are more effectively screened from the attractive nuclear potential by the core electrons and thus are more weakly bound to the ion than is the case for the $3d$ ions. The $4d$ and $5d$ electrons therefore tend to become more delocalized in the solid and their energy states overlap the energy states of the $5s$ and $6s$ electrons, respectively, in the conduction band. This overlap and the resulting interaction and mixing usually occur to such an extent that no net magnetic moment can be associated with $4d$ or $5d$ ions in the transition metals. An alternative point of view is that both the orbital angular momentum **L** and the spin angular momentum **S** for $4d$ and $5d$ ions are quenched by crystal fields due to the strong interactions of the ions with their local atomic environments.

Atoms and Ions with a Partially Filled f Shell. The $4f$ rare earth *lanthanide* ions ($Z = 57$ to 71, La to Lu) and $5f$ *actinide* ions ($Z = 89$ to 103, Ac to Lr) have partially filled f shells which lie closer to the atomic nucleus and as a result tend to be more localized than the electrons in partially filled d shells. As a result of the greater localization of the $4f$ electrons, spin–orbit coupling is stronger for $4f$ ions than for $3d$ ions in solids. In addition, crystal field effects are weaker for $4f$ ions, due in part to the shielding of neighboring ions by the $5s$ and $5p$ electrons but due primarily to the larger size of $4f$ ions compared to $3d$ ions. Since the on-site spin–orbit interaction dominates over the effects of the crystal field, trivalent $4f$ ions, which have outer-electron configurations given by $4d^{10}4f^n5s^25p^6$, $n = 1$ to 13, often retain their free-ion magnetic moments in solids and to this extent can be treated as isolated ions. Nevertheless, the $4f$ electrons of certain rare earths such as Ce can interact and mix with the conduction electrons of the host material, resulting in *valence fluctuations* of the ions.

That $4f$ ions often retain their free-ion magnetic character in both insulating and conducting solids is clear from Table 9.3. Here the $4f$ ions are listed in their energetically preferred trivalent states, including La^{3+} with $4f^0$, along with some divalent $4f$ ions (in parentheses). Divalent ions such as Eu^{2+}, Sm^{2+}, and Yb^{2+} can be present as dopant ions in oxide glasses prepared under strongly reducing conditions. It can be seen that the measured effective magneton numbers p for most of the trivalent $4f$ ions tend to agree rather well with the values $p = g\sqrt{J(J + 1)}$ predicted by Hund's rules. Notable exceptions to this are Sm^{3+} and Eu^{3+}, for which the measured moments are even larger than those predicted by Hund's rules. For these ions excited energy levels lie not far above the Hund's rule ground-state level. This can be seen in Fig. 9.2 for Eu^{3+}, where the separation between the ground state and the first excited state is only 0.05 eV. As a result, Sm^{3+} and Eu^{3+} ions can readily be thermally excited out of their ground states where $J = L - S$ into states with higher values of J and hence higher magnetic moments. When the effects of thermal excitation are taken into account, the values of p predicted for Sm^{3+} and Eu^{3+} are in good agreement with experiment.

TABLE 9.3 **4f Rare Earth Element Ions: Ground States of Free Ions and of Ions in Solids**

Ion	Electron Configuration $4f^n$	Free-Ion Quantum Numbers S, L, J	Free-Ion Ground State $(^{2S+1}L_J)$	Calculated p $g\sqrt{J(J+1)}$	Measured p^a
La^{3+}	$4f^0$	0, 0, 0	$^1S_0$	0.0	0
Ce^{3+}	$4f^1$	$\frac{1}{2}$, 3, $\frac{5}{2}$	$^2F_{5/2}$	2.54	2.4
$(Ce^{2+}$	$4f^2$	1, 5, 4	$^3H_4$	2.54	—)
Pr^{3+}	$4f^2$	1, 5, 4	$^3H_4$	3.58	3.5
Nd^{3+}	$4f^3$	$\frac{3}{2}$, 6, $\frac{9}{2}$	$^4I_{9/2}$	3.62	3.5
Pm^{3+b}	$4f^4$	2, 6, 4	$^5I_4$	2.68	—
Sm^{3+}	$4f^5$	$\frac{5}{2}$, 5, $\frac{5}{2}$	$^6H_{5/2}$	0.84	1.5
$(Sm^{2+}$	$4f^6$	3, 3, 0	$^7F_0$	0.00	—)
Eu^{3+}	$4f^6$	3, 3, 0	$^7F_0$	0.00	3.4
$(Eu^{2+}$	$4f^7$	$\frac{7}{2}$, 0, $\frac{7}{2}$	$^8S_{7/2}$	7.94	—)
Gd^{3+}	$4f^7$	$\frac{7}{2}$, 0, $\frac{7}{2}$	$^8S_{7/2}$	7.94	8.0
Tb^{3+}	$4f^8$	3, 3, 6	$^7F_6$	9.72	9.5
Dy^{3+}	$4f^9$	$\frac{5}{2}$, 5, $\frac{15}{2}$	$^6H_{15/2}$	10.63	10.6
Ho^{3+}	$4f^{10}$	2, 6, 8	$^5I_8$	10.60	10.4
Er^{3+}	$4f^{11}$	$\frac{3}{2}$, 6, $\frac{15}{2}$	$^4I_{15/2}$	9.59	9.5
Tm^{3+}	$4f^{12}$	1, 5, 6	$^3H_6$	7.57	7.3
Yb^{3+}	$4f^{13}$	$\frac{1}{2}$, 3, $\frac{7}{2}$	$^2F_{7/2}$	4.54	4.5
$(Yb^{2+}$	$4f^{14}$	0, 0, 0	$^1S_0$	0.0	—)
Lu^{3+}	$4f^{14}$	0, 0, 0	$^1S_0$	0.0	0

Source: Date from J.H. Van Vleck, *The Theory of Electric and Magnetic Susceptibilities*, Oxford University Press, Oxford, 1932, p. 285; and R. Kubo and T. Nagamiya, eds., *Solid State Physics*, McGraw-Hill, New York, 1969, p. 453.

[a] —, Not observed or not known.

[b] Promethium, Pm, is a radioactive element with no stable isotopes. As a result, its properties cannot readily be determined.

Atomic Nuclei. The magnetic properties of atomic nuclei originate from the magnetic moments of the nuclear constituents (i.e., the protons and neutrons), both of which possess spin $s = \frac{1}{2}$. A convenient unit for nuclear magnetism is the magnetic moment of a proton, known as the *nuclear magneton* and given by

$$\mu_N = \frac{e\hbar}{2m_p} = 5.051 \times 10^{-27} J/T, \tag{9.8}$$

where $m_p = 1.673 \times 10^{-27}$ kg is the mass of the proton. Magnetic effects in solids due to the magnetic moments of nuclei are much weaker than those due to electrons because the nuclear magneton μ_N is smaller than the Bohr magneton μ_B by the factor $m_e/m_p = 1/1836$. As a result, the ordering of nuclear magnetic moments will occur at much lower temperatures, typically well below $T = 1$ K, than is the case for the ordering of magnetic moments due to electrons.

The magnetic moment of a nucleus can be expressed as follows:

$$\mathbf{m}_{\text{nucl}} = \frac{g_I \mu_N \mathbf{I}}{\hbar},\tag{9.9}$$

where g_I is the nuclear g factor and $\mathbf{I}$ is the total angular momentum of the nucleus. Some typical values for the magnitudes of $\mathbf{I}$ and $\mathbf{m}_{\text{nucl}}$ are $\hbar/2$ and $2.792\mu_N$ for ^{1}H, $\hbar$ and $0.857\mu_N$ for ^{2}H, and $\hbar/2$ and $0.090\mu_N$ for ^{57}Fe. Additional data on nuclear spins are given in Chapter W22.

Defects. Defects known as *dangling bonds* can be present when one of the two spin-paired electrons in a covalent bond is missing. As a result, the spin s of the remaining electron is unpaired and a localized magnetic moment can exist at the site of the dangling bond. Such defects often result from the inherent bonding disorder present in amorphous covalent solids such as a-Si or a-SiO$_2$ or from vacancies created in irradiated crystalline Si or SiO$_2$. Defects such as dangling bonds tend to be present in relatively low concentrations in solids and therefore do not contribute in a significant way to their magnetic properties. They can play a much more important role, however, in affecting the electronic properties of the material, as discussed in Chapter W11.

TYPES OF MAGNETISM AND MAGNETIC BEHAVIOR IN MATERIALS

The magnetic properties of materials are determined first and foremost by the magnetic moments of the electrons, atoms, and ions in the material, as discussed in Sections 9.2 and 9.3. The magnetic responses of electrons and of atoms and ions can exhibit a wide variety of behaviors in materials due to the wide range of interactions that can occur between the magnetic moments and their environment. It should not be surprising, therefore, that the subject of magnetism in materials is rich and complex.

The macroscopic constitutive relationships between the *magnetic field* $\mathbf{H}$, the *magnetic induction* or *flux density* $\mathbf{B}$ and the *magnetization* $\mathbf{M}$ are presented first. The magnetic behavior of materials due to well-defined atomic magnetic moments is then discussed on a microscopic level for the following types of magnetism: paramagnetism, ferromagnetism, antiferromagnetism, and ferrimagnetism. Finally, the diamagnetic and paramagnetic properties of materials due to electrons in filled shells and conduction electrons are described.

Independent of the particular type of magnetism in a material, its macroscopic response to an external magnetic field $\mathbf{H}$ can be characterized by the resulting magnetic induction $\mathbf{B}$ present in the material. This response is expressed using the following simple relation:

$$\mathbf{B} = \mu\mathbf{H}.\tag{9.10}$$

This expression is valid when the material is in the shape of a long, thin rod with $\mathbf{H}$ applied parallel to the long axis. For other sample shapes or field directions the *internal field* $\mathbf{H}_{\text{int}}$ within the material will not be equal to the applied $\mathbf{H}$, as described in detail in Chapter 17 in terms of the demagnetization coefficient of the sample. The *magnetic permeability* of the material,

$$\mu = \frac{B}{H},\tag{9.11}$$

can be dependent on both temperature and magnetic field. The magnetic response of the material can be expressed alternatively in terms of its magnetization or magnetic moment per unit volume **M** induced by the field **H**, as follows:

$$\mathbf{M} = \chi\mathbf{H}, \tag{9.12}$$

where

$$\chi = \frac{M}{H} \tag{9.13}$$

is the *magnetic susceptibility* of the material.[†]

In the currently preferred SI system of units, also known as the rationalized MKS system, **B**, **H**, and **M** are related by the following constitutive relationship:

$$\mathbf{B} = \mu_0(\mathbf{H} + \mathbf{M}) = (1 + \chi)\mu_0\mathbf{H} = \mu_r\mu_0\mathbf{H}, \tag{9.14}$$

where **H** is understood to be the internal magnetic field and μ, χ, and μ_r are related by

$$\mu = (1 + \chi)\mu_0 = \mu_r\mu_0. \tag{9.15}$$

Here $\mu_0 = 4\pi \times 10^{-7}$ N/A^2 is the permeability of free space and $\mu_r = 1 + \chi = \mu/\mu_0$ is the dimensionless relative permeability of the material. The magnetic induction B is expressed in units of tesla, where 1 T = 1 N/A·m = 1 Wb/m^2, when H and M are expressed in units of A/m. The units of *magnetic flux* Φ are webers, hence the term *magnetic flux density* for B.

The *cgs-emu system* of magnetic units, also known as the *Gaussian system*, is often used for these quantities, the relevant expressions being $\mathbf{B} = \mathbf{H} + 4\pi\mathbf{M} = (1 + 4\pi\chi)\mathbf{H}$ and $\mu = 1 + 4\pi\chi$. Note that while H, B, and M all have the same units in the cgs-emu system, the names of the units are different, with B and M expressed in gauss (G) when H is expressed in oersteds (Oe). The relationships between the SI and cgs-emu units for H, M, and B are as follows: for H: 1 A/m is equivalent to $4\pi \times 10^{-3}$ Oe (1 Oe = 79.6 A/m); for M: 1 A/m is equivalent to 10^{-3} G = 10^{-3} emu/cm^3; for B: 1 T is equivalent to 10^4 G. Note also that a magnetic induction B of 1 G is equivalent to a magnetic field H of 1 Oe. It should be noted that the magnetic permeability in SI units μ(SI) = μ(cgs-emu)μ_0, where μ(cgs-emu) is dimensionless. For the magnetic susceptibility χ(SI) = $4\pi\chi$(cgs-emu), where both χ(SI) and χ(cgs-emu) are dimensionless.[‡] The quantity χ(cgs-emu) can also be thought of as a volume susceptibility with units of emu/cm^3, which is in fact dimensionless.

[†] Both the magnetic permeability μ and susceptibility χ are in general frequency-dependent complex quantities [i.e., $\mu(\omega) = \mu_1(\omega) + i\mu_2(\omega)$, $\chi(\omega) = \chi_1(\omega) + i\chi_2(\omega)$]. The imaginary parts μ_2 and χ_2 are measures of the frequency-dependent magnetic losses in the material; see also the discussion of dynamic magnetic effects in Chapter 17 and of the ac bridge in Chapter W22.

[‡] Data for χ and M are often presented per unit mass or per mole. For example, in the cgs-emu system χ_ρ(cm^3/g or emu/g) = χ(cgs-emu)/ρ, where ρ is the mass density in g/cm^3 while χ_{mol}(cm^3/mol or emu/mol) = χ(cgs-emu)/ρ_{mol}, where ρ_{mol} is the molar density in mol/cm^3. Similarly, M_ρ(cm^3·G/g or emu·G/g) = M(G)/ρ and M_{mol}(cm^3·G/mol or emu·G/mol) = M(G)/ρ_{mol}. Note that emu in these expressions stands for cm^3. Similar expressions hold for the SI system of units.

The work w done per unit volume when a magnetic induction **B** is established in a material is expressed in SI units by

$$w = \frac{W}{V} = \int_0^B \mathbf{H} \cdot d\mathbf{B} = \mu_0 \int_0^H \mathbf{H} \cdot d\mathbf{H} + \mu_0 \int_0^M \mathbf{H} \cdot d\mathbf{M}$$

$$= \frac{\mu_0 H^2}{2} + \mu_0 \int_0^M \mathbf{H} \cdot d\mathbf{M} = u = \frac{U}{V}. \tag{9.16}$$

Note that w is also equal to the magnetic energy density u stored in the material. If χ is a constant, independent of H, and if the material is isotropic, then

$$w = u = \frac{\mu_0 H^2}{2} + \frac{\mu_0 MH}{2} = \frac{BH}{2} = \frac{\mu H^2}{2}. \tag{9.17}$$

The term $\mu_0 H^2/2$ is the energy density due solely to the magnetic field H, independent of the material. The term $\mu_0 MH/2$ represents the additional stored energy due to the magnetization M present in the material. The corresponding expressions in cgs-emu units are $u = H^2/8\pi + MH/2 = BH/8\pi$.

Note that the magnetization **M** corresponds to the magnetic response of the material itself (i.e., its own contribution to the magnetic flux density **B** within the material). If a material has no magnetic response of its own, both **M** and χ are zero. In this case, $\mathbf{B} = \mu_0\mathbf{H}$ and $\mu = \mu_0$ in SI units (while $\mathbf{B} = \mathbf{H}$ and $\mu = 1$ in cgs-emu units). It follows that the magnetic properties of a material are determined completely by the magnetization **M**, the dependence of **M** on temperature, magnetic field, pressure, and so on, and the spatial variation, if any, of **M** within the material.

The different types of magnetic behavior that can be found in materials will now be introduced and discussed on a microscopic level. When well-defined localized magnetic moments are present in a solid, the magnetic properties are determined to a large extent by the magnetic microstructure of the material (i.e., by the spatial arrangement and relative orientations of the magnetic moments) as well as by their interactions with each other and with external magnetic fields. The various types of magnetic behavior in solids due to the presence of well-defined magnetic moments (i.e., paramagnetism, ferromagnetism, antiferromagnetism, and ferrimagnetism) are characterized by magnetic microstructures with distinctive features. Examples of these magnetic microstructures are presented and discussed below for each type of magnetic behavior, beginning with paramagnetism.

9.4 Paramagnetism

Paramagnetism corresponds to the magnetic behavior found in materials in which localized magnetic moments are present but in which no net macroscopic magnetization or magnetic moment per unit volume **M** exists in zero applied field **H**. The simplest form of paramagnetism is found in materials in which the magnetic moments are present at sufficiently low concentrations so that they are well separated from each other. Under these conditions the spins do not interact with each other and are said to be free, responding only to thermal fluctuations and external magnetic fields.

Paramagnetism can also exist in materials even when there are interactions between the magnetic moments, as long as the interactions are sufficiently weak so that, as stated above, no net magnetization **M** appears in the material when **H** = 0. This type of behavior occurs, for example, in ferromagnetic materials above their critical temperatures. Paramagnetism corresponding to noninteracting or free spins and to interacting spins are discussed next in turn.

Free Spins: Curie Behavior. Consider first the paramagnetic behavior of a material containing a concentration $n = N/V$ of free spins. In the absence of an applied magnetic field **H**, the spins will be oriented completely randomly, as shown schematically in Fig. 9.7a. In addition, the direction of each spin will undergo rapid thermal fluctuations. It follows that the net magnetization **M** of the material will be zero. When a magnetic field **H** is applied, the resulting **M** will be in the same direction as **H**. It is generally true for paramagnetic materials in which the spins are free or only weakly interacting that the resulting magnetization M is much smaller in magnitude than H so that to a very good approximation the magnetic induction $\mathbf{B} = \mu_0\mathbf{H}$. This approximation will be made in this discussion of paramagnetism.

When a magnetic field **H** is applied and a magnetic induction **B** exists in the material, a torque $\boldsymbol{\tau} = \mathbf{m} \times \mathbf{B} = \mathbf{m} \times \mu_0\mathbf{H}$ of magnitude $\tau = mB\sin\theta$ will be exerted on each magnetic moment. Here θ is the angle between **m** and **B**. The magnetic moments will respond to this torque by tending to orient themselves in the direction of **B** (Fig. 9.7b). This reorientation results in a state of lower energy for the system of spins, which is, however, opposed by thermal fluctuations at finite T. The resulting potential energy U of a magnetic moment **m** in the presence of **B** is given by

$$U = -\mathbf{m} \cdot \mathbf{B} = -mB\cos\theta. \tag{9.18}$$

Note that U is minimized when θ is 0 (i.e., when **m** and **B** are parallel and pointing in the same direction).

It is useful to compare a typical value of U with the competing thermal energy k_BT. In a relatively large applied magnetic field $H = 10^6$ A/m, it follows that $B = \mu_0H = 1.26$ T and $\cos\theta \approx 1$. Therefore, $U \approx \mu_BB = (9.274 \times 10^{-24} \text{ J/T})(1.26 \text{ T}) = 1.17 \times 10^{-23}$ J. At room temperature, $T = 300$ K, the corresponding thermal energy $k_BT = (1.38 \times 10^{-23} \text{ J/K})(300 \text{ K}) = 4.14 \times 10^{-21}$ J is more than 300 times greater than U. As a result, free spins in a paramagnet will be essentially completely disordered at $T = 300$ K even in the presence of relatively strong magnetic fields.

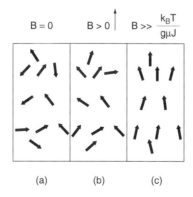

Figure 9.7. Two-dimensional array of magnetic moments in a paramagnetic solid.

Substituting $\mathbf{m} = -g\mu_B\mathbf{J}/\hbar$ from Eq. (9.5) into Eq. (9.18) yields the following expression for U, where the direction of $\mathbf{B}$ has now been taken to be along the z axis:

$$U = +\frac{g\mu_B\mathbf{J}\cdot\mathbf{B}}{\hbar} = +\frac{g\mu_B J_z B}{\hbar}. \qquad (9.19)$$

The total angular momentum vector $\mathbf{J}$ precesses about the direction of $\mathbf{B}$ with the $(2J+1)$ allowed values of its z-component given by $J_z = m_J\hbar$, with $m_J = J, J-1, \ldots, -J+1, -J$. From this expression for J_z, it can be seen that the allowed energies of the spin in the presence of B have the form $U(m_J) = +g\mu_B m_J B = +g\mu_0\mu_B m_J H$. The allowed directions for $\mathbf{J}$ relative to $\mathbf{B}$ and the resulting threefold splitting of the zero-field energy level of the spin by $\mathbf{B}$ for the case $J = 1$ are shown in Fig. 9.8. This splitting of the energy level in a magnetic field is known as the *Zeeman effect* and can be treated quantum mechanically by adding the term $-g\mu_B\mathbf{J}\cdot\mathbf{B}/\hbar$ to the Hamiltonian of the system. Note that adjacent energy levels of the spin in the presence of B are separated by an energy $\Delta E = g\mu_B B$. For $B = 1$ T, $J = 1$, and $g = 2$, a typical Zeeman splitting is $\Delta E = 1.85 \times 10^{-23}$ J (i.e., 1.2×10^{-4} eV).

The resulting magnetization $\mathbf{M}$ of a system of n free spins per unit volume in the presence of $\mathbf{B}$ is determined by the statistical occupation of the allowed energy levels by the spins. The magnetization $\mathbf{M}$ will be in the same direction as $\mathbf{B}$ and $\mathbf{H}$ and will be equal to $n\langle m_z\rangle$, where $\langle m_z\rangle = g\mu_B m_J$ is the average z component of an atomic magnetic moment $\mathbf{m}$. At thermal equilibrium the probability $P_J(B, T)$ of occupation of an energy level $g\mu_B m_J B$ is proportional to the Boltzmann factor:

$$P_J(B, T) \propto \exp\left(-\frac{g\mu_B m_J B}{k_B T}\right). \qquad (9.20)$$

The quantity $\langle m_z\rangle$ can be calculated as the thermal equilibrium average of $m_z = g\mu_B m_J$ as follows:

$$\langle m_z\rangle = \frac{\displaystyle\sum_{m_J=-J}^{+J} g\mu_B m_J P_J(B, T)}{\displaystyle\sum_{m_J=-J}^{+J} P_J(B, T)}. \qquad (9.21)$$

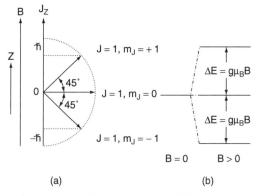

(a)   (b)

Figure 9.8. (*a*) Allowed directions for $\mathbf{J}$ relative to $\mathbf{B}$ and (*b*) resulting threefold splitting of the zero-field energy level of the spin by $\mathbf{B}$, for the case $J = 1$.

For the simplest possible case of a spin with $J = S = \frac{1}{2}$, $L = 0$, and $g = 2$, it follows that $m_J = \pm\frac{1}{2}$ and that

$$\langle m_z \rangle = \frac{+\mu_B e^{-\mu_B B} - \mu_B e^{+\mu_B B}}{e^{-\mu_B B} + e^{+\mu_B B}} = \mu_B \tanh\frac{\mu_B B}{k_B T}. \tag{9.22}$$

The general expression for the magnetization M for arbitrary J, L, and S is

$$M(B, T) = n\langle m_z \rangle = n g \mu_B J B_J \left(\frac{g\mu_B J B}{k_B T}\right) = M_{sat} B_J \left(\frac{g\mu_B J B}{k_B T}\right), \tag{9.23}$$

where the *Brillouin function* $B_J(x)$ is defined by

$$B_J(x) = \frac{2J+1}{2J} \coth\frac{(2J+1)x}{2J} - \frac{1}{2J}\coth\frac{x}{2J}. \tag{9.24}$$

The *saturation magnetization* is $M_{sat} = n g \mu_B J$ and corresponds to all the individual magnetic moments pointing in the direction of **B** (Fig. 9.7c). The variation of the magnetization with magnetic induction B and temperature T is shown in Fig. 9.9, where $M(B, T)/M_{sat}$ is plotted as a function of the reduced variable $x = g\mu_B J B/k_B T$ for $J = \frac{1}{2}$, $\frac{3}{2}$, and $\frac{5}{2}$. It can be seen that M initially increases linearly with x for $x \ll 1$ and then approaches a maximum value M_{sat} typically for $x \approx 4$. At $T = 300$ K, $x \approx 4$ corresponds to the extremely high value of $B \approx 4k_B T/\mu_B = 1800$ T. The magnetization curves shown here are reversible, with no net magnetization **M** remaining in a paramagnetic material at **B** = **H** = 0.

The behaviors of M and $\chi = M/H$ for low magnetic fields or high temperatures (i.e., for $x \ll 1$ and $B \ll k_B T/g\mu_B J$) are found by using the expansion $\coth\phi = 1/\phi + \phi/3 + \cdots$. The magnetic susceptibility χ is then given by

$$\chi = \frac{M}{H} = \frac{\mu_0 n g^2 \mu_B^2 J(J+1)}{3k_B T} = \frac{\mu_0 n p^2 \mu_B^2}{3k_B T} = \frac{C}{T}, \tag{9.25}$$

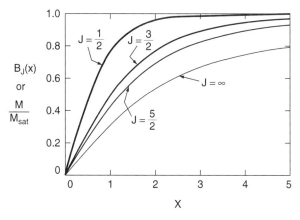

Figure 9.9. Variation with magnetic induction B and temperature T of the magnetization $M(B, T)/M_{sat}$ for free spins in a paramagnetic material as a function of $x = g\mu_B J B/k_B T$, thus illustrating the behavior of the Brillouin function $B_J(x)$.

where $p = g\sqrt{J(J+1)}$ is the effective magneton number and

$$C = \frac{\mu_0 n g^2 \mu_B^2 J(J+1)}{3k_B} \tag{9.26}$$

is the *Curie constant*, which, as defined, has units of temperature. Predicted and measured values for p have already been given in Tables 9.1 and 9.3 for $3d^n$ and $4f^n$ ions, respectively. The expression $\chi(T) = C/T$, known as the *Curie law*, is the low-field limit of the magnetic susceptibility of a paramagnetic material with n free spins per unit volume. In the opposite limit of high magnetic fields or low temperatures (i.e., for $x \gg 1$ and $B \gg k_B T/g\mu_B J$), $\coth \phi \to 1$, $\chi \to 0$, and $M \to ng\mu_B J = M_{\text{sat}}$. Note that when $T \gg C$, it follows that $\chi \ll 1$, $M \ll H$, and $B \approx \mu_0 H$. In the opposite limit, $T \ll C$, it follows that $\chi \gg 1$, $M \gg H$, and $B \approx \mu_0 M$.

The Curie law behavior for $\chi(T)$ expressed in Eq. (9.25) is illustrated schematically in Fig. 9.10, where both $\chi = C/T$ and $1/\chi = T/C$ are shown plotted as functions of T. The plot of χ versus T shows clearly that χ diverges (i.e., $\chi \to \infty$) as $T \to 0$ K, indicating that in the absence of the effects of thermal disorder, the spins can be aligned by an arbitrarily small applied field H. The linear plot of $1/\chi$ versus T allows both a sensitive test of the validity of the Curie law for a specific paramagnetic material and straightforward determination of the Curie constant C from the inverse of the slope of the linear plot. The effective magneton number $p = g\sqrt{J(J+1)} = \sqrt{3k_B C/\mu_0 n \mu_B^2}$ can be determined readily from C if the concentration n of spins is known.

A classical theory for the paramagnetism of free spins with magnetic moments **m** was developed by Langevin before the quantum-mechanical theory. Such a classical theory is strictly valid only in the limit $J \to \infty$. In this limit **J** acts like a classical spin which can have any orientation with respect to an applied field **H**. The magnetization is then given by $M = nmL(mB/k_B T)$, where $L(x) = \coth x - 1/x$ is the *Langevin function*. The function $L(x)$ and the Brillouin function $B_J(x)$ are in fact identical in the limit $J \to \infty$. The Langevin theory also predicts Curie law behavior for $\chi(T)$.

In metallic alloys such as $Fe_x Al_{1-x}$ and $Cu_x Ni_{1-x}$, a type of paramagnetic behavior associated with clusters of Fe or Ni spins is observed. These clusters of magnetic atoms have "giant" magnetic moments which are much larger than those of a single Fe or Ni atom. The resulting magnetic behavior of such alloys, termed *superparamagnetism*, corresponds to essentially independent magnetic clusters which interact more

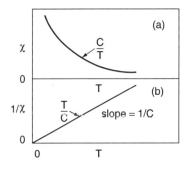

Figure 9.10. Curie law behavior for $\chi(T)$ as expressed in Eq. (9.25).

strongly with thermal fluctuations and external magnetic fields than with each other. The behavior of the spins within each cluster is clearly not free but rather, is determined by the strong intracluster interactions. Small ferromagnetic particles can also exhibit superparamagnetism, as discussed in Chapter 17.

Interacting Spins. While spins may be free and noninteracting when present in materials at sufficiently low concentrations and high temperatures, at higher concentrations and lower temperatures the spins may be near enough to interact with and exert forces on each other with energies of interaction $U(\text{int})$ comparable to or greater than $k_B T$. The condition $U(\text{int}) \geq k_B T$ indicates that the spin–spin interactions are strong enough not to be destroyed by thermal fluctuations of the directions of individual spins. When $U(\text{int}) \gg k_B T$, cooperative types of magnetism such as ferromagnetism, antiferromagnetism, and ferrimagnetism in which long-range magnetic order exists in the system of spins can be present in the material. These cooperative types of magnetic behavior are discussed later.

Van Vleck Paramagnetism. Ions with partially filled shells that have $J = 0$ in the ground state can still exhibit magnetic moments and paramagnetic behavior in solids when electrons in the unfilled shells are excited to higher-lying energy levels with $J > 0$. Examples are ions that fall one electron short of having a half-filled shell, including Cr^{2+} and Mn^{3+} with $3d^4$ electronic configurations as well as Eu^{3+} with a $4f^6$ configuration (see Tables 9.1 and 9.3). When the excited state of the ion with $J > 0$ lies at an energy Δ which is well above the ground state and when $\Delta \gg k_B T$, the system of n such ions per unit volume will have a temperature-independent paramagnetic susceptibility given by

$$\chi_{VV} = \frac{2n\mu_0 |\langle 1|m_z|0\rangle|^2}{\Delta}, \tag{9.27}$$

where $|\langle 1|m_z|0\rangle|^2$ is the matrix element of the magnetic moment between the ground state and the excited state. This contribution to the magnetic response is known as *Van Vleck paramagnetism.*

The ways in which spins or magnetic moments interact with each other depends not only on the properties of the material in which they reside but also on the degree with which the individual spins themselves interact with their surroundings. The direct and indirect interactions that typically occur between magnetic moments in materials are discussed next.

9.5 Interactions Between Magnetic Moments

The interactions between magnetic moments in a solid can be either direct or indirect. The difference between direct and indirect in this case is related to the extent to which the host material itself plays a role in the interaction.

Direct Interactions. Direct interactions between magnetic moments occur within the host material, but without the host itself playing a important role. Direct interactions can occur either via magnetic dipole–dipole interactions between the magnetic moments or via electrostatic interactions between the charge distributions of the magnetic ions.

Dipole–Dipole Interaction. Consider first the *magnetic dipole–dipole interaction* (i.e., the classical interaction occurring between two magnetic dipole moments $\mathbf{m}_1$ and $\mathbf{m}_2$). This interaction has the following form in free space:

$$U(\text{dipole}) = \frac{\mu_0[\mathbf{m}_1 \cdot \mathbf{m}_2 - 3(\mathbf{m}_1 \cdot \hat{\mathbf{r}})(\mathbf{m}_2 \cdot \hat{\mathbf{r}})]}{4\pi r^3}. \tag{9.28}$$

Here $U(\text{dipole})$ is the potential energy of interaction and $\mathbf{r} = \mathbf{r}_1 - \mathbf{r}_2$ is the vector connecting $\mathbf{m}_1$ and $\mathbf{m}_2$. This interaction has its origins in the magnetic fields of the spins and the resulting magnetic forces and torques which they exert on each other. It tends to align both $\mathbf{m}_1$ and $\mathbf{m}_2$ parallel to each other along $\mathbf{r}$.

An estimate for the strength of $U(\text{dipole})$ can be obtained by using the following approximation, which assumes that the spins are no closer than a typical interatomic separation $r \approx 0.2$ nm:

$$U(\text{dipole}) \approx \frac{\mu_0 m^2}{4\pi r^3} \approx \frac{\mu_0 \mu_B^2}{r^3} \approx 1.3 \times 10^{-23} \text{ J} \approx 10^{-4} \text{ eV}. \tag{9.29}$$

This estimated value is equivalent to a thermal energy $k_B T$ corresponding to $T \approx 1$ K. Due to the size of $U(\text{dipole})$, magnetic dipole–dipole interactions will be important only when they are the sole interaction occurring between the individual magnetic moments or when $\mathbf{m}_1$ and $\mathbf{m}_2$ represent the total moments of large group of spins as in the case of the magnetic domains found in ferromagnetic materials. Examples of direct dipole–dipole interactions between magnetic domains are discussed in Sections 17.2 and 17.3.

Heisenberg Exchange Interaction. Consider now a much stronger direct interaction between pairs of magnetic moments which has its origin in the direct Coulomb or electrostatic interaction between the overlapping charge distributions of the two magnetic ions in conjunction with the Pauli exclusion principle. This is the quantum-mechanical *Heisenberg exchange interaction*, which has the form

$$U(\text{exch}) = -J(\mathbf{r})\mathbf{S}_1 \cdot \mathbf{S}_2, \tag{9.30}$$

where $\mathbf{S}_1$ and $\mathbf{S}_2$ are two interacting spins separated by a distance $\mathbf{r} = \mathbf{r}_1 - \mathbf{r}_2$ and $J(\mathbf{r})$ is the *exchange integral* (or exchange constant), with units of energy.[†] Here $\mathbf{S}_1$ and $\mathbf{S}_2$ are dimensionless spins rather than angular momenta with units of $\hbar$. They represent in general the total angular momentum $\mathbf{J}$ of the spins, not just the intrinsic part. It should be emphasized that this direct coupling between $\mathbf{S}_1$ and $\mathbf{S}_2$ which depends on their magnitudes and their relative orientation is not actually of a magnetic nature. It is due instead to the electrostatic energy of interaction between the electrons which can be shown to depend on the relative orientation of their spins. The magnitude and range of $J(\mathbf{r})$ typically lead to significant exchange interactions between $\mathbf{S}_1$ and $\mathbf{S}_2$ for small values of r only (i.e., for NNs or next-NNs), where the electron clouds overlap appreciably. Note that the Heisenberg exchange interaction is an isotropic interaction since it depends only on the directions of the spins S_1 and S_2 relative to each other but

---

[†] A factor of 2 is often included in Eq. (9.30), so that $U(\text{exch}) = -2J(\mathbf{r})\mathbf{S}_1 \cdot \mathbf{S}_2$.

not to the crystalline axes. Anisotropic exchange which does depend on the directions of spins relative to the crystalline axes is discussed in Section 9.7, where magnetic anisotropy is discussed.

The sign of the exchange integral $J(\mathbf{r})$ can be either positive or negative and is the factor that determines whether a parallel ($\uparrow\uparrow$ or $\downarrow\downarrow$) or antiparallel ($\uparrow\downarrow$ or $\downarrow\uparrow$) alignment of the two interacting spins leads to a lowering of their energy [i.e., to $U(\text{exch}) < 0$]. This can be illustrated as follows:

$$
\begin{aligned}
J > 0: \quad & U(\text{exch}) \text{ is } < 0 \text{ when } \mathbf{S}_1 \text{ and } \mathbf{S}_2 \text{ are parallel } (\uparrow\uparrow \text{ or } \downarrow\downarrow) \\
J < 0: \quad & U(\text{exch}) \text{ is } < 0 \text{ when } \mathbf{S}_1 \text{ and } \mathbf{S}_2 \text{ are antiparallel } (\uparrow\downarrow \text{ or } \downarrow\uparrow).
\end{aligned}
\tag{9.31}
$$

Note that $J > 0$ corresponds to a ferromagnetic while $J < 0$ corresponds to an anti-ferromagnetic spin–spin interaction. The role of the direct Heisenberg interaction in ferromagnetic and antiferromagnetic materials is discussed in more detail later.

It can be shown for a simple two-electron system that J is equal to $E_s - E_t$, where E_s and E_t are the energies of the singlet ($S = 0$, $\uparrow\downarrow$ or $\downarrow\uparrow$) and triplet ($S = 1$, $\uparrow\uparrow$ or $\downarrow\downarrow$) states of the system. For many-electron systems the sign of J is determined simply by whether E_s or E_t is lower in energy. For example, $J < 0$ corresponds to $E_s < E_t$ and the singlet state in which the spins are antiparallel is preferred. The reverse is, of course, true for $J > 0$.

That an electrostatic interaction can have such a significant effect on the magnetic behavior of spins in a solid can be understood as arising from the behavior of electrons as stated in the Pauli exclusion principle. When two electrons occupy the same orbital, their individual spins $\mathbf{s}$ must point in opposite directions (i.e., if $m_s = +\frac{1}{2}$ for the first electron, then $m_s = -\frac{1}{2}$ for the second). This is the singlet state in which the Coulomb repulsion between the two electrons will be relatively strong due to the spatial overlap of their wavefunctions. This overlap and the resulting repulsion will be reduced in the triplet state, where the spins of the two electrons are parallel and where they in fact reside in different orbitals. Therefore, the overlap of their wavefunctions will decrease when the spins are parallel.

From this discussion it follows that the Coulomb energy of interaction of the two electrons and the relative orientation of their spins are correlated. The same argument holds even in the more general case when the electrons are not on the same atom or ion as long as there is an overlap of the electronic charge distributions and hence a mutual electrostatic interaction.

Indirect Interactions. When the interactions between spins are indirect, the host material itself participates in the interaction. Two types of indirect magnetic interactions are discussed here: the superexchange mechanism which can play an important role in insulators and the Ruderman, Kittel, Kasuya, and Yosida (RKKY) interaction in metals.

Superexchange. When ions with magnetic moments such as Mn^{2+}, Fe^{2+}, or Fe^{3+} are separated from each other by nonmagnetic ions such as O^{2-} or F^-, the direct overlap of their charge distributions will clearly be diminished as will be the direct coupling of their spins via the Heisenberg exchange interaction $U(\text{exch})$. The spin–spin interaction in this case can instead occur indirectly through the p orbitals of the intervening nonmagnetic ions whose electronic shells are filled completely. It is clear that covalent bonding plays an important role in this interaction since it occurs via the overlap of extended wavefunctions. This indirect interaction is known as *superexchange* and can have a profound effect on the nature of the spin–spin coupling. For example, the indirect superexchange mechanism favors an antiferromagnetic ground state for the spins in some insulators (e.g., FeO, MnO, and the high-temperature superconductor La_2CuO_4), whereas the direct exchange interaction in metallic Fe leads to a strongly ferromagnetic interaction. Additional discussion of the superexchange mechanism is presented in Sections 9.7 and 9.8.

RKKY Interaction. When a magnetic ion such as Mn^{2+} is present in a metal, its spin **S** can be strongly coupled to the spins σ of the conduction electrons via an exchange interaction of the form

$$U_{sd}(\text{exch}) = -J_{sd}(r)\boldsymbol{\sigma} \cdot \mathbf{S}, \tag{9.32}$$

where $J_{sd}(r)$ is the exchange integral corresponding to the $\boldsymbol{\sigma} \cdot \mathbf{S}$ spin–spin interaction. This interaction is often referred to as the *s–d interaction*, with *s* corresponding to the conduction electrons and *d* to the 3d electrons of the magnetic ion. The effects of this interaction on the conduction electrons are threefold:

1. The usual decrease of the electron mean free path due to scattering by the magnetic ion
2. The possibility that the spin σ of the conduction electron will change its direction (i.e., be flipped from up to down, or vice versa)
3. The establishment of a conduction electron spatial spin-density oscillation centered at the magnetic ion

When the spin of the conduction electron flips, the spin **S** of the magnetic ion will change correspondingly to conserve angular momentum during the scattering process.

If the scattered electron is then scattered again by the spin of a second magnetic ion in the vicinity of the first, an indirect interaction will have effectively occurred between the first and second spins. This indirect interaction between spins transmitted via the conduction electrons is known as the Ruderman–Kittel–Kasuya–Yosida (*RKKY*) interaction. The resulting oscillatory interaction between spins $\mathbf{S}_1$ and $\mathbf{S}_2$ has the form

$$U_{\text{RKKY}}(r) = -J_{\text{RKKY}}(r)\mathbf{S}_1 \cdot \mathbf{S}_2, \tag{9.33}$$

where

$$J_{\text{RKKY}}(r) \approx \frac{V_0 \cos(2k_F r)}{r^3}. \tag{9.34}$$

The factor V_0 is the strength of the RKKY interaction with units of $J \cdot m^3$ and k_F is the Fermi wave vector of the conduction electrons. The oscillatory nature of the RKKY interaction arises in the free-electron model from the presence of a sharp Fermi surface and its effect on the screening of the magnetic moment by the spins of the conduction electrons. The relationship between V_0 and J_{sd} is expressed by

$$V_0 = \frac{\rho(E_F)J_{sd}^2}{8\pi}, \tag{9.35}$$

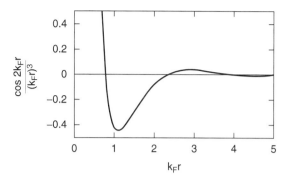

Figure 9.11. Oscillatory dependence of the RKKY interaction on the spatial separation r of the two spins.

where $\rho(E_F) = 3n/2E_F$ is the density of electron states at the Fermi energy per unit volume. It is not surprising that $V_o \propto J_{sd}^2$ since the s–d interaction of Eq. (9.32) occurs twice in the indirect RKKY interaction.

The oscillatory $\cos(2k_F r)$ dependence of the RKKY interaction on the spatial separation r of the two spins is shown in Fig. 9.11. It can be seen that the interaction is damped by the $1/r^3$ factor and that the sign of J_{RKKY} oscillates between ferromagnetic ($J_{\text{RKKY}} > 0$) and antiferromagnetic ($J_{\text{RKKY}} < 0$) couplings between $\mathbf{S}_1$ and $\mathbf{S}_2$ as their separation increases.

As described in Chapter W9, clear evidence for the existence of the RKKY interaction has been found from studies of the magnetic properties of dilute alloys (e.g., Mn in Au, Ag, Cu, and Zn).

One more interesting effect involving localized spins and the conduction electrons in metals can be mentioned. At sufficiently low temperatures the s–d or exchange interaction given in Eq. (9.32) can lead to a complicated many-body ground state of the system of the spin $\mathbf{S}$ and the conduction electrons of the metal. This behavior, known as the *Kondo effect*, is described in Chapter W9.

9.6   Ferromagnetism

Cooperative effects can appear in a magnetic material when the concentration of spins increases so that each spin begins to interact with several other spins in its vicinity. If the interactions are such that all the spins in extended regions of the sample point in the same direction, the material can have a nonzero spontaneous magnetization $\mathbf{M}$ even when $\mathbf{H} = 0$ as long as the temperature is below a critical temperature T_c. In this case $\chi \gg 1$ and $\mu \gg \mu_0$ below T_c. Such materials are called *ferromagnets* and the resulting magnetic behavior is known as *ferromagnetism*. Above T_c the material will be paramagnetic, with the spins pointing in random directions when $\mathbf{H} = 0$, as shown in Fig. 9.7a. Some residual short-range order can exist in the system of spins even above T_c, as is evidenced by the fact that the observed specific heat anomaly at T_c is not abrupt, but extends above the critical temperature. The transition at T_c is a second-order phase transition, with $M \to 0$ as $T \to T_c$. The types of magnetic ordering found in ferromagnets and in other types of magnetically ordered solids to be discussed later are shown schematically in Fig. 9.12.

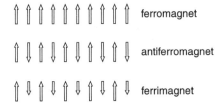

Figure 9.12. Types of magnetic ordering found in ferromagnets, antiferromagnets, and ferrimagnets below their respective critical temperatures.

TABLE 9.4   Magnetic Properties of Selected Ferromagnets, Antiferromagnets, and Ferrimagnets

Material (Magnetic Ions)	T_C or T_N[a] (K)	M_{sat}[b]$(T = 0$ K$)$ (10^6 A/m)	$n_B = gJ$[c] (μ_B per ion)
Ferromagnets			
Fe (Fe^{2+})	$T_C = 1042$	1.752	2.22
Co (Co^{2+})	1388	1.446	1.72 (HCP)
			1.75 (FCC)
Ni (Ni^{2+})	627	0.51	0.61
Gd (Gd^{3+})	293	2.06	7.63
Cu_2MnAl (Mn^{2+})	630	0.726	4.1
EuO (Eu^{2+})	77	1.91	6.8
$Nd_2Fe_{12}B$ (Nd^{3+}, Fe^{2+})	585	1.27	Nd^{3+}: 3.2, Fe^{2+}: 2.1
Antiferromagnets			
Cr (none)	$T_N = 311$	0	—
MnO (Mn^{2+})	122	0	4.6–4.8
FeO (Fe^{2+})	198	0	3.32
Ferrimagnets			
Fe_3O_4 (Fe^{2+}, Fe^{3+})	$T_C = 858$	0.51	Fe^{2+}: 4.1, Fe^{3+}: 5.0
$NiFe_2O_4$ (Ni^{2+}, Fe^{3+})	858	0.30	Ni^{2+}: 2.3, Fe^{3+}: 5.0
$Y_3Fe_5O_{12}$ (Fe^{3+})	560	0.195	5.0

Source: Data from F. Keffer, *Handbuch der Physik*, Vol. 18, Pt. 2, Springer-Verlag, New York, 1966; P. Heller, *Rep. Prog. Phys.* **30**, 731 (1967); S. Chikazumi, *Physics of Magnetism*, Wiley, New York, 1964; D. R. Lide, ed., *CRC Handbook of Chemistry and Physics*, 75th ed., CRC Press, Boca Raton, Fla., 1994; data for NdFe$_{12}$B from J. Herbst, *Rev. Mod. Phys.*, **63**, 819 (1991).

[a]T_C, Curie temperature, T_N, Néel temperature.
[b]Values of the saturation magnetization $M_{sat}(T = 0$ K$)$. When given in the cgs-emu unit of gauss, M_{sat} can be converted readily to the SI unit of A/m by multiplying by 10^3.
[c]$n_B = gJ = M_{sat}(T = 0$ K$)/n\mu_B$ is the number of Bohr magnetons μ_B per magnetic ion in ferromagnetic materials, where n is the concentration of magnetic ions. For example, n_B for metallic Fe with $n = 8.5 \times 10^{28}$ Fe atoms or spins per cubic meter is given by $(1.752 \times 10^6$ A/m$)/(8.5 \times 10^{28}$ m$^{-3})(9.274 \times 10^{-24}$ J/T$) = 2.22$. For antiferromagnetic materials n_B can be obtained from magnetic susceptibility data for $T \gg T_N$. For ferrimagnetic materials, neutron scattering results yield the values of n_B for the magnetic ions.

Ferromagnetism can exist in a wide variety of materials, including metals and alloys such as Fe, Ni, and Cu_xNi_{1-x}, intermetallic compounds such as Cu_2MnAl, and occasionally in oxides such as EuO. A short list of ferromagnets is presented in Table 9.4, along with their critical temperature T_C, also known as the Curie temperature, the saturation magnetization $M_{sat}(T = 0$ K$)$, and the effective number of Bohr magnetons per magnetic ion $n_B = gJ$, given by the product of the g factor and the total angular momentum quantum number J. In this section the focus is on elemental ferromagnetic metals such as Fe and Ni. Additional examples of ferromagnets with important technological applications are discussed in Chapter 17.

Cooperative effects due to spin–spin interactions in ferromagnets typically occur when spins occupy neighboring sites in the structure. In the case when spins are adjacent to each other with no intervening ions, the Heisenberg exchange interaction $U(\text{exch}) = -J(\mathbf{r})\mathbf{S}_1 \cdot \mathbf{S}_2$ is the dominant interaction. This interaction favors a ferromagnetic, parallel alignment of the spins when $J > 0$. The problem of predicting the allowed energy levels of the spins and the wavefunctions of the $3d$ or $4f$ electrons by solving the Schrödinger equation with $U(\text{exch})$ included as a potential energy term is extremely complex. For some materials models of ferromagnetism based not on localized but on delocalized or itinerant electrons have been proposed. Imbalances in the densities of up-spin ($\uparrow$) and down-spin ($\downarrow$) delocalized electrons and details of the electronic band structure play important roles in these models. An itinerant model of ferromagnetism for Ni is discussed later.

The RKKY interaction has been proposed to be the mechanism leading to ferromagnetism in transition metal compounds such as the Heusler "alloy" Cu_2MnAl, in which the Mn ions are believed to be too far apart for direct exchange mechanisms to apply. It is interesting to note that ferromagnetism is observed in Cu_2MnAl even though none of the pure constituents (i.e., Cu, Mn, or Al) is ferromagnetic.

A classical approach to the problem of explaining ferromagnetism was developed by P. Weiss in 1907,[†] before the advent of quantum mechanics. This phenomenological approach, known as the *molecular field theory of magnetism*, assumes that the interactions which align spins in ferromagnets do so by giving rise to an internal magnetic field, known as the molecular field. The Weiss molecular field theory of magnetism is capable of offering a qualitative explanation for many aspects of ferromagnetism.

Weiss Molecular Field Theory. Weiss proposed that the parallel alignment of spins in ferromagnets at temperatures that can be as high as several hundreds of kelvin is due to the presence within the material of the *molecular field*, a strong, effective internal magnetic field. The magnetic induction $\mathbf{B}_{\text{eff}}$ rather than the corresponding magnetic field $\mathbf{H}_{\text{eff}}$ is used to represent the molecular field since the origins of the field are internal (i.e., within the material), not external. Each spin is assumed to be acted on by the molecular field, which in turn is taken to be proportional to the local spontaneous magnetization $\mathbf{M}_s$, that is,

$$\mathbf{B}_{\text{eff}} = \lambda \mu_0 \mathbf{M}_s. \tag{9.36}$$

The proportionality factor λ is a measure of the strength of the molecular field. In a ferromagnet $\mathbf{B}_{\text{eff}}$ and the spontaneous magnetization $\mathbf{M}_s$, which is present even when $\mathbf{H} = 0$, are both assumed to be uniform (i.e., having the same direction throughout the material). The effect of $\mathbf{B}_{\text{eff}}$ is to align the spins, giving rise to the spontaneous magnetization $\mathbf{M}_s$.

Whether the effective field $\mathbf{B}_{\text{eff}}$ or the spontaneous magnetization $\mathbf{M}_s$ comes first is, in fact, a meaningless question since they are both ultimately due to the ferromagnetic exchange interaction $U(\text{exch})$ between spins. To estimate the magnitude of

---

[†] P. Weiss, *J. Phys.*, **6**, 661 (1907).

B_{eff}, consider the fact that the molecular field must overcome the randomizing effect of thermal fluctuations on the directions of the spins, so that $\mu_B B_{eff} \gg k_B T$. Taking $T = 300$ K requires that $B_{eff} \gg 447$ T. This is equivalent to 4.47×10^6 G in cgs-emu units and is clearly orders of magnitude too large to be due to magnetic dipole–dipole interactions.

Putting aside for the moment the question of the microscopic origins of such a strong internal field, consider how the macroscopic magnetic properties of a system of spins are affected by the presence of $\mathbf{B}_{eff}$. When an external field $\mathbf{H}$ is applied to the material with its direction parallel to that of $\mathbf{B}_{eff}$, the total magnetic induction tending to align the spins is given by

$$\mathbf{B} = \mu_0 \mathbf{H} + \mathbf{B}_{eff} = \mu_0 (\mathbf{H} + \lambda \mathbf{M}_s). \tag{9.37}$$

The resulting magnetization due to $\mathbf{B}$ is given by

$$M(B) = n g \mu_B J B_J \left(\frac{g \mu_B J B}{k_B T} \right) = n g \mu_B J B_J \left(\frac{g \mu_B \mu_0 J (H + \lambda M)}{k_B T} \right), \tag{9.38}$$

where here the Brillouin function $B_J(x)$ replaces the classical Langevin function $L(x)$ used by Weiss.

The expression for M in Eq. (9.38) cannot be solved analytically since M also appears in the argument of $B_J(x)$. A graphical approach to obtaining a solution has often been used but is not necessary since the stable, nonzero solution for $M(H, T)$ can be obtained through the use of computers to any desired degree of accuracy. The following solutions to Eq. (9.38) can be obtained, if necessary, by using a power series expansion and are valid as specified:

(a) $T = 0$ K $\qquad\qquad M = M_{sat}(T = 0 \text{ K}) = n g \mu_B J.$

(b) $T = T_C, M = 0 \qquad T_C = \dfrac{\mu_0 n g^2 \mu_B^2 J(J+1) \lambda}{3 k_B} = \lambda C.$ $\qquad$ (9.39)

(c) $T \gg T_C \qquad\qquad \chi(T) = \dfrac{\mu_0 n g^2 \mu_B^2 J(J+1)}{3 k_B (T - \lambda C)} = \dfrac{C}{T - T_C}.$

The saturation magnetization M_{sat} at $T = 0$ K corresponds to the complete alignment of the spins in the molecular field $\mathbf{B}_{eff}$ even when the external field $\mathbf{H} = 0$. The critical temperature at which $M_s \to 0$ is $T_C = \lambda C$, where C is the usual Curie constant. The fact that the *Curie temperature* T_C is directly proportional to the strength of the molecular field λ is a reasonable result. Finally, the Curie law $\chi(T) = C/T$ for free spins is modified here to the Curie–Weiss form $C/(T - T_C)$, so that χ is predicted to diverge at $T = T_C$ instead of at $T = 0$ K.

Next, the predictions of the Weiss molecular field theory are compared with experimental results for typical ferromagnets such as Fe. The measured temperature dependence of the spontaneous magnetization $M_s(T)$ for Fe, normalized to the saturation magnetization M_{sat}, is shown in Fig. 9.13 along with the result predicted using Eq. (9.38). The agreement between the Weiss prediction for $M_s(T)/M_{sat}$ and experiment for Fe is qualitatively satisfactory, although there are some important differences in detail, to be discussed later.

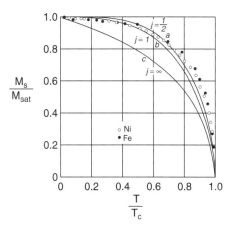

Figure 9.13. Measured temperature dependencies of the spontaneous magnetizations $M_s(T)$ for Fe and Ni, normalized to the saturation magnetization M_{sat}, as a function of T/T_C. The predicted results obtained from the Weiss molecular field theory using Eq. (9.38) for $J = \frac{1}{2}$ and for $J = 1$ (along with $g = 2$) are also shown. (From R. Becker et al., *Ferromagnetismus*, Springer-Verlag, Berlin, 1939.)

When the more realistic Heisenberg exchange interaction given in Eq. (9.30) is used for the calculation of $\chi(T)$ for $T \gg T_C$, the following result is obtained:

$$\chi(T) = \frac{C(1 + \theta/T + \cdots)}{T} \approx \frac{C}{T - \theta}, \tag{9.40}$$

where

$$\theta = \frac{S(S + 1)J_o}{3k_B} \quad \text{and} \quad J_o = \sum_{\mathbf{R}} J(\mathbf{R}). \tag{9.41}$$

Here the exchange integral $J(\mathbf{R})$ is summed over the sites $\mathbf{R}$ of all the interacting spins. Note that Eq. (9.40) is similar to the Curie–Weiss expression for $\chi(T)$ predicted in Eq. (9.39c) by the Weiss molecular field theory. The result given in Eq. (9.41) provides the relationship between the *Curie–Weiss temperature* θ, the spin S, and the parameter J_o determined by the sign and the strength of the exchange interaction $J(\mathbf{R})$. It follows that the sign of θ determined from measurements of $\chi(T)$ for $T \gg T_C$ can give the sign of J and hence can specify whether the low-temperature magnetically ordered state has ferromagnetic ($J > 0$) or antiferromagnetic ($J < 0$) coupling between spins.

An approximate value for the strength λ of the molecular field obtained from the measured values $T_C = 1043$ K and $gJ = 2.2$ for Fe and using Eq. (9.39b) is

$$\lambda = \frac{3k_B T_C}{\mu_0 n g^2 \mu_B^2 J(J + 1)} \approx 509. \tag{9.42}$$

This corresponds to a molecular field $B_{eff} = \lambda \mu_0 M_{sat}$ for Fe of about 1120 T. It should be noted that the molecular field can be determined experimentally in magnetic materials using the technique of ferromagnetic resonance, as described in Chapter W17. As stated earlier, the origin of this extremely strong field cannot be explained classically by the Weiss molecular field model and is, in fact, due to the Heisenberg exchange interaction. Since $T_C \approx \theta$, it follows from Eqs. (9.41) and (9.42) that

$$\lambda \approx \frac{zJ(\mathbf{R}_{NN})}{\mu_0 n g^2 \mu_B^2} \tag{9.43}$$

when each spins interacts only with its z NNs. This relationship confirms the expectation that the source of the molecular field is the Heisenberg exchange interaction and in fact that its strength λ is proportional to J. With this additional insight, the Weiss theory serves as a very useful qualitative model for ferromagnetism.

Itinerant and Localized Magnetism. As mentioned earlier, attempts have been made to explain the magnetism of ferromagnets such as metallic Ni and certain magnetic 3d transition metal alloys containing Fe, Ni, or Co using itinerant or free electrons within an energy band model, as opposed to the view taken earlier where the electrons responsible for magnetism are taken to be localized completely on a given atom or ion. In models based on itinerant magnetism, the wavefunctions of the 3d electrons on neighboring atoms are viewed as overlapping each other to the extent that a band of 3d electrons is formed which itself overlaps both the energy band of the 4s conduction electrons and the Fermi energy E_F. In this energy band model the 3d electrons lose some of their localized character and to this extent are no longer associated with a given transition metal ion.

In the case of Ni the free atom has eight 3d electrons and two 4s electrons. In the solid state the wavefunctions of these electrons overlap each other spatially, as stated earlier. For $T > T_C$, the densities $\rho(E)$ of occupied 3d and 4s electronic states in the corresponding overlapping energy bands are shown schematically in Fig. 9.14a. Here the density of states curves $\rho\uparrow(E)$ and $\rho\downarrow(E)$ for up- and down-spin electrons are shown separately, with the approximate number of electrons z per Ni ion of each type given in parentheses. Note that about 1.4 4s electrons per Ni ion have been shifted to the 3d band. With $z_{4s}\uparrow = z_{4s}\downarrow = 0.3$ and $z_{3d}\uparrow = z_{3d}\downarrow = 4.7$, the net magnetic moment per Ni ion for $T > T_C$ is $m = (z_{4s}\uparrow - z_{4s}\downarrow)\mu_B + (z_{3d}\uparrow - z_{3d}\downarrow)\mu_B = 0$. This is due to the fact that pairs of up- and down-spin 3d and 4s electrons occupy the available energy levels up to E_F.

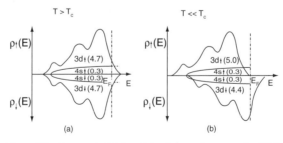

Figure 9.14. Densities $\rho(E)$ of occupied 3d and 4s electronic states for Ni both for $T > T_C$ and $T \ll T_C$. The densities of states curves $\rho_\uparrow(E)$ and $\rho_\downarrow(E)$ for up- and down-spin electrons are shown separately with the numbers of electrons z per Ni ion of each type given in parentheses.

The corresponding curves for $T \ll T_C$ are shown in Fig. 9.14b, where $z_{4s}\uparrow = z_{4s}\downarrow =$ 0.3 still holds but $z_{3d}\uparrow = 5.0$ (i.e., the majority-spin $3d\uparrow$ band is filled), while for the minority spins, $z_{3d}\downarrow = 4.4$. Now $m = (0.3 - 0.3)\mu_B + (5.0 - 4.4)\mu_B = +0.6\mu_B$ per Ni ion, essentially equal to the value measured for Ni and given in Table 9.4. This magnetic moment m arises from an upward shift of the $3d\downarrow$ energy band by an amount $\Delta E = 0.25$ to 0.4 eV relative to the $3d\uparrow$ energy band. This shift ΔE can be attributed to an exchange interaction between the electrons related to Coulomb interactions and to the Pauli exclusion principle, which tends to keep parallel-spin electrons spatially separated.

The itinerant model for ferromagnetism can thus explain the observed magnetic moment in Ni and can also offer an explanation for the fractional numbers n_B of Bohr magnetons observed per magnetic ion in ferromagnetic materials listed in Table 9.4. Whether the $3d$ electrons actually appear to be localized or itinerant in a magnetic material often depends on the experimental technique used as a probe. For example, in photoemission spectroscopy the time scale of the probe is so short that essentially all the electrons appear localized. Other techniques, such as Mossbauer spectroscopy, with longer time scales tend to see the $3d$ electrons as itinerant.

The Slater–Pauling curve for the dependence of the average atomic magnetic moment $\langle m \rangle$ on the average valence $\langle z \rangle$ per atom observed for $3d$ transition metals and their alloys is presented and discussed in Chapter 17. The observed dependence of $\langle m \rangle$ on $\langle z \rangle$ is explained there in terms of an itinerant model of magnetism for the $3d$ electrons.

In the *Stoner model* for itinerant electron or band ferromagnetism the splitting ΔE of the up- and down-spin energy bands is taken to be proportional to the magnetization [i.e., $\Delta E = I(z\uparrow - z\downarrow)\mu_B$]. The constant of proportionality I is the difference between the up- and down-spin exchange energies. The paramagnetic susceptibility of the itinerant electrons can be enhanced by these effects and is given by

$$\chi_{\text{para}} = \frac{\chi_P}{1 - I\rho(E_F)}, \tag{9.44}$$

where χ_P is the free-electron paramagnetic susceptibility defined in Section 9.9. The condition for itinerant electron ferromagnetism at $T = 0$ K is then $I\rho(E_F) \geq 1$ (i.e., $\chi_{\text{para}} \to \infty$), which shows that high values of the density of states $\rho(E_F)$ favor ferromagnetism. The factor $1/[1 - I\rho(E_F)]$ is known as the *Stoner enhancement factor*.

An approach that attempts to include both itinerant and localized effects and electron correlations within the same model is based on a proposal by Hubbard.[†] The Hubbard model is outlined in Chapter W9.

Spin Waves. At $T = 0$ K in a fully magnetized ferromagnetic material, all the spins are aligned to the maximum extent possible in the same direction, as shown in Fig. 9.7c. For $T > 0$ K, the spontaneous magnetization M_s will decrease below the saturation value M_{sat} when some of the spins are forced to change their directions as they are thermally excited to states of higher energy. The form of these thermal excitations could in principle correspond to a change in the z-component J_z of the total angular momentum **J** of a single spin by the amount $\Delta J_z = \hbar$ (assuming that the z axis is the

[†] J. Hubbard, *Proc. R. Soc.*, **A276**, 238 (1963); **A277**, 237 (1964); **A281**, 401 (1964).

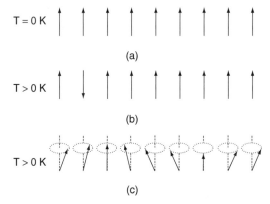

Figure 9.15. Spin waves in a ferromagnet: (*a*) $T = 0$ K: all spins point in the same direction; (*b*) $T > 0$ K: reversal of a single spin by the amount $\Delta J_z = -\hbar$; (*c*) $T > 0$ K: a spin wave in which the z component J_z of each spin is decreased by the same small amount and in which the directions of all spins deviate from the z direction.

direction along which the magnetization $\mathbf{M}_s$ is pointing). In fact, the lowest-energy thermal excitations are *spin waves*, illustrated schematically in Fig. 9.15. In a spin wave the component J_z of each spin is decreased on average by the same small amount and the directions of essentially all spins deviate from the z direction. A spin-wave excitation has a lower energy than does an excitation corresponding to the reversal of a single spin. Spin waves are the quantized thermal excitations of the system of spins and are in many ways analogous to phonons, which are the quantized thermal excitations of the lattice. This correspondence is indicated by the fact that spin waves are also known as *magnons*.

The decrease in magnetization M due to the excitation of spin waves for $T > 0$ K has been calculated on the basis of the Heisenberg exchange interaction and can be shown to have the form

$$M_s(T) = M_{\text{sat}}(1 - AT^{3/2}),\tag{9.45}$$

where A is a constant depending on the spin S and on the specific details of the exchange interaction. This dependence of M_s on T, known as the *Bloch $T^{3/2}$ law*, is obeyed experimentally by ferromagnetic materials. In contrast, the Weiss molecular field theory, which does not include spin waves, predicts that $M_s(T)$ deviates from M_{sat} as T increases above 0 K by a term depending exponentially on T [i.e., $M_s(T) \approx M_{\text{sat}}(1 - Ae^{-B/T})$, where A and B are constants].

The inelastic magnetic scattering of neutrons by spin waves allows the spin-wave energy dispersion curve $E(\mathbf{k})$ to be determined for ferromagnets in the same way that phonon dispersion curves can be mapped out by nonmagnetic neutron scattering. It has been found for $H = 0$ that $\omega(\mathbf{k}) \propto k^2$ for small k, as predicted by spin-wave theory. In particular, $\hbar\omega(k) = 2JSa^2k^2$ for a cubic lattice where a is the interspin spacing. Spin waves also exist in other types of ordered magnetic systems, such as antiferromagnets and ferrimagnets, to be discussed later.

It is interesting to note that the total magnetization $\mathbf{M}$ of a ferromagnetic material can be observed to be zero even for $T < T_C$, in contrast to statements made previously. This effect is related to the existence of macroscopic magnetic domains. In each domain the spins all point in the same direction, but the net spin and magnetization of each domain can point in different directions in the material, resulting in $\mathbf{M} = 0$. The properties of magnetic domains and their effect on the response of ferromagnets to externally applied magnetic fields is presented in Chapter 17.

A phenomenological discussion of the effects of anisotropy on the magnetic properties of materials, including the effect on the direction of the magnetization within the material, is presented next.

Magnetic Anisotropy. The magnetic properties of ferromagnets are anisotropic below T_C. The same is also true of antiferromagnetic and ferrimagnetic materials below their respective critical temperatures. In the high-T, disordered paramagnetic phase these materials are all magnetically isotropic since there are no preferred directions for the spins to point in the structure. The anisotropy of magnetic properties that results from the crystal structure or the local atomic environment of the magnetic moments, called *magnetocrystalline anisotropy*, is an intrinsic property of the material. Other types of extrinsic magnetic anisotropy can be induced by applying external stress to the material (magnetostrictive or magnetoelastic anisotropy) or by cooling the material in an applied magnetic field (induced anisotropy). Even the external shape of a magnetic material (e.g., a thin needle or a thin film) can induce anisotropy and hence determine the direction of the magnetization $\mathbf{M}$. Only the intrinsic form of magnetocrystalline anisotropy is described here, with additional discussions of magnetostrictive and shape-induced anisotropy presented in Section 17.5.

Magnetocrystalline anisotropy exists whenever the energy associated with the magnetization $\mathbf{M}$ of a magnetic material as expressed in Eq. (9.16) depends, at least in part, on the direction of $\mathbf{M}$ with respect to the crystal axes. The Heisenberg exchange interaction $U(\text{exch}) = -J(\mathbf{r})\mathbf{S}_1 \cdot \mathbf{S}_2$, which depends only on the relative directions of the two spins (and also their spatial separation $\mathbf{r}$), cannot be responsible for the anisotropy observed. The energy $U(\text{exch})$ is isotropic because it does not include the spin–orbit interaction $-\lambda_{so}\mathbf{L} \cdot \mathbf{S}$, which couples the spin to the electronic charge density. Since the charge density can be anisotropic, the energy of the spin can depend on its orientation with respect to the crystal axes when spin–orbit effects are important. Magnetic dipole–dipole interactions can also break rotational symmetry and can therefore give rise to anisotropic coupling between spins. The effects of magnetic anisotropy that can be observed in ferromagnets will now be discussed, while those found in antiferromagnets and ferrimagnets are described in Sections 9.7 and 9.8, respectively.

The type of uniaxial anisotropy that exists, for example, in the HCP ferromagnet Co is the simplest type of magnetocrystalline anisotropy and will serve here as an illustrative example. It is observed that the preferred direction of the spontaneous magnetization $\mathbf{M}_s$ in HCP Co below $T_C = 1388$ K is along the c axis (i.e., the [0001] direction). The *magnetic anisotropy energy density* E_a for a uniaxial material such as Co depends on the angle ϕ between the direction of $\mathbf{M}$ and the c axis as follows:

$$E_a(\phi) = K_{u1} \sin^2 \phi + K_{u2} \sin^4 \phi + \cdots, \tag{9.46}$$

where K_{u1} and K_{u2} are the first- and second-order uniaxial anisotropy coefficients, respectively. Uniaxial magnetocrystalline anisotropy can be included in the quantum-mechanical treatment of a spin system by adding the term $-DS_z^2$ to the Hamiltonian.

In this case the coefficients D and K_{u1} representing the strength of the anistropic interaction are proportional to each other, but with K_{u1} also incorporating the effects of the magnitude of S_z. When K_{u1} and K_{u2} are both positive, E_a has its minimum value $E_a(\text{min}) = 0$ for $\phi = 0$ or $180°$ (i.e., for **M** along the c axis), and its maximum value $E_a(\text{max}) = K_{u1} + K_{u2}$ for $\phi = 90°$ (i.e., for **M** perpendicular to the c axis).

The first- and second-order magnetocrystalline anisotropy coefficients for the ferromagnets Co, Fe, and Ni are listed in Table 9.5. For Co at room temperature both coefficients are positive. The parameter K_{u1} decreases with increasing temperature, changes sign near $T = 533$ K, and along with K_{u2} is equal to zero above T_C in the paramagnetic phase. To illustrate this anisotropy, the magnetization curves for Co for **H** $\parallel$ and $\perp$ to the c axis are shown in Fig. 9.16. While Co can be magnetized to saturation in either direction, the c axis is the easy direction, while directions in the basal plane perpendicular to the c axis are the hard directions. When K_{u1} and K_{u2} are both negative, the magnetization prefers to lie in the basal plane (i.e., the easy directions are perpendicular to the c axis). It can also be seen that $\chi_\parallel = M_\parallel / H$ is much larger than $\chi_\perp = M_\perp / H$ for Co. This is opposite to the behavior found for antiferromagnets, to be discussed later.

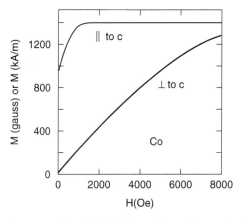

Figure 9.16. Magnetization curves for Co for **H** $\parallel$ and $\perp$ to the c axis, illustrating the uniaxial anisotropy of this HCP ferromagnet. (Note that 1 gauss $= 10^3$ A/m $= 1$ kA/m.) (From G. Burns, *Solid State Physics*, copyright 1985. Reprinted by permission of Academic Press, Inc)

TABLE 9.5   First- and Second-Order Magnetocrystalline Anisotropy Coefficients

Material	K_{u1}, K_1 $(10^4$ J/m$^3)$	K_{u2}, K_2 $(10^4$ J/m$^3)$
Co	40	20
Fe	4.8	0.5
Ni	−0.5	−0.2

Source: Data from J. H. Judy, *Mater. Res. Soc. Bull.*, Mar. 1990, p. 63.

In cubic ferromagnets such as Fe and Ni the magnetic anisotropy energy density can be shown to have the general form

$$E_a(\text{cubic}) = K_1(\alpha_1^2\alpha_2^2 + \alpha_2^2\alpha_3^2 + \alpha_3^2\alpha_1^2) + K_2\alpha_1^2\alpha_2^2\alpha_3^2 + \cdots, \qquad (9.47)$$

where K_1 and K_2 are the first- and second-order cubic anisotropy coefficients, respectively, and $\alpha_i = \cos\theta_i$, $i = 1, 2, 3$, are the direction cosines of $\mathbf{M}$ relative to the x, y, and z axes, respectively. There are now multiple directions corresponding to the minimum value of E_a (i.e., the easy directions of magnetization), and these depend on the sign of K_1. For BCC Fe, K_1 is positive and the easy directions are the six $\langle 100 \rangle$ directions (i.e., the cube axes); the eight $\langle 111 \rangle$ directions are the hard directions. For FCC Ni, K_1 is negative and the easy directions are the set of eight $\langle 111 \rangle$ directions. The microscopic origins of magnetocrystalline anisotropy are described in Chapter W9.

Magnetic anisotropy is expected to be small in crystals of high symmetry, such as the cubic ferromagnets Fe and Ni, but can be large in crystals of low symmetry or for ions in positions of low symmetry, such as Nd^{3+} in the ferromagnet $Nd_2Fe_{12}B$, as discussed in Chapter 17.

9.7 Antiferromagnetism

In an *antiferromagnet* there is no net spontaneous macroscopic magnetization $\mathbf{M}_s$ in zero magnetic field even below the critical temperature for ordering of the spins. Well above the critical temperature T_N, known as the *Néel temperature*, the spins are randomly oriented for $H = 0$ and the material behaves as a paramagnet. Néel proposed in 1932 that the behavior observed below T_N resulted from each spin in the material being surrounded by NN spins all pointing in the opposite direction, corresponding to a NN antiferromagnetic spin–spin interaction with $J < 0$.

The magnetic structures observed in antiferromagnetic (AF) materials (i.e., the spatial arrangements and orientations of the spins) can be quite complicated. All that is necessary is that the net macroscopic magnetic moment be zero. This can be accomplished in an essentially infinite number of ways. Several examples of antiferromagnets have been given in Table 9.4. The focus here will initially be on the particularly simple AF ceramic material MnO, which has the NaCl crystal structure based on magnetic Mn^{2+} and nonmagnetic O^{2-} ions (Fig. 9.17). MnO is actually very slightly distorted from a cubic to a trigonal crystal structure below T_N due to crystal field effects. In this crystal structure the two FCC lattices of Mn^{2+} and O^{2-} ions can be viewed as being displaced from each other by one half of the lattice constant a along a $\langle 100 \rangle$ direction. Each Mn^{2+} ion has six NN O^{2-} ions at a distance $a/2$ and 12 next-NN Mn^{2+} ions at a distance $\sqrt{2}\,a/2$. The magnetic structure of MnO is also shown, in which the spins of all the Mn^{2+} ions in a given (111) plane are parallel to each other, lie in the (111) plane, and are antiparallel to the spins of the Mn^{2+} ions in the two adjacent (111) planes. Note that there are no strong magnetic interactions between spins in a given (111) plane.

The antiferromagnetic interaction between the spin of each Mn^{2+} ion and the six second-NN Mn^{2+} spins at the distance a takes place indirectly through the intervening O^{2-} ions via the superexchange mechanism discussed in Section 9.5. The direct exchange interaction $U(\text{exch})$ between each Mn^{2+} ion and its 12 next-NN Mn^{2+} ions at the distance $\sqrt{2}\,a/2$ is weakened by the presence of the O^{2-} ions in the structure

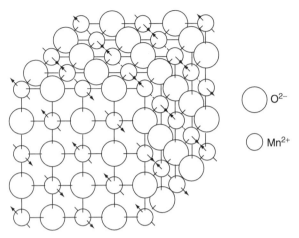

Figure 9.17. The antiferromagnetic material MnO whose crystal structure is NaCl is shown with the Mn^{2+} and O^{2-} ions occupying the two FCC sublattices. The AF magnetic structure of the Mn^{2+} ions is also shown in which the spins of all the Mn^{2+} ions in a given (111) plane are parallel to each other and antiparallel to the spins in the two adjacent (111) planes. (Reprinted by permission from S. Chikazumi, *Physics of Magnetism*, Wiley, New York, 1964.)

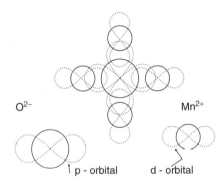

Figure 9.18. The indirect superexchange interaction between two Mn^{2+} ions in MnO involves the Mn^{2+} d orbitals and the p orbitals of the intervening O^{2-} ion, as shown here schematically.

and so does not play an important role in the magnetic structure of MnO. The superexchange interaction occurs via the p orbitals of the O^{2-} ions, as shown schematically in Fig. 9.18, and is particularly effective in the MnO structure, due to the fact that O^{2-} ions lie directly between pairs of interacting Mn^{2+} ions. It can be seen here that each O^{2-} ion is actually involved via its p_x, p_y, and p_z orbitals in the superexchange interaction between three pairs of Mn^{2+} ions, while each Mn^{2+} ion interacts indirectly through its six NN O^{2-} ions with its six Mn^{2+} second NNs at the distance a. This superexchange interaction is less effective between each Mn spin and its 12 next-NN Mn spins, since in this case the Mn^{2+}–O^{2-}–Mn^{2+} angle is equal to 90°.

Predictions for the magnetization M and magnetic susceptibility χ of an antiferromagnet such as MnO have been obtained from a phenomenological molecular field theory by Van Vleck in 1941. The model is more complicated than the corresponding

Weiss molecular field model presented earlier for ferromagnets since the molecular field in an antiferromagnet cannot be spatially uniform but must be directed antiparallel for the two sets of oppositely directed spins. An outline of the molecular field model for antiferromagnets is presented next. Details can be found in Chikazumi (1964).

In a magnetic material with N spins, let A denote the set of $N/2$ up spins and B the set of $N/2$ down spins, respectively, with spins of each type having the same magnetic moment $m = g\mu_B J$. It is assumed that each A spin is acted on by a molecular field $\mathbf{B}_{A,eff}$, consisting only of contributions from the A and B spins, which are its magnetic NNs; that is,

$$\mathbf{B}_{A,\text{eff}} = \lambda_{AA}\mu_0\mathbf{M}_{sA} + \lambda_{AB}\mu_0\mathbf{M}_{sB}. \tag{9.48}$$

The same is true for the molecular field $\mathbf{B}_{B,\text{eff}}$ acting on each B spin,

$$\mathbf{B}_{B,\text{eff}} = \lambda_{BA}\mu_0\mathbf{M}_{sA} + \lambda_{BB}\mu_0\mathbf{M}_{sB}. \tag{9.49}$$

Here the factors λ_{IJ}, with $\lambda_{AA} = \lambda_{BB} > 0$ and $\lambda_{AB} = \lambda_{BA} < 0$, represent the strengths of the molecular fields due to the spins J acting on the NN spins I, as shown schematically in Fig. 9.19. It is usually true in antiferromagnets that $|\lambda_{AB}| > |\lambda_{AA}|$ due to the dominance of the superexchange mechanism between A and B spins. The vectors $\mathbf{M}_{sA}$ and $\mathbf{M}_{sB}$ are the local spontaneous sublattice magnetizations at the sites of the A spins and B spins, respectively. In the absence of an external magnetic field, $\mathbf{M}_{sA} = -\mathbf{M}_{sB}$ (i.e., $\mathbf{M}_s = \mathbf{M}_{sA} + \mathbf{M}_{sB} = 0$) and the two sublattice magnetizations cancel each other. The antiferromagnetic nature of the interaction between A and B spins is consistent with both λ_{AB} and $\lambda_{BA} < 0$.

The temperature dependencies of the spontaneous sublattice magnetizations $\mathbf{M}_{sA}$ and $\mathbf{M}_{sB}$ can be determined in the same manner as presented earlier for the ferromagnetic or Weiss molecular field case. It is predicted that $\mathbf{M}_{sA}$ and $\mathbf{M}_{sB}$ both decrease with increasing T, falling to zero at $T = T_N$. The critical temperature T_N predicted is given in terms of the strengths of the molecular fields by

$$T_N = \frac{\mu_0 n g^2 \mu_B^2 J(J+1)(\lambda_{AA} - \lambda_{AB})}{6k_B} = \frac{C(\lambda_{AA} - \lambda_{AB})}{2} > 0, \tag{9.50}$$

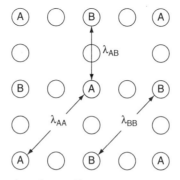

Figure 9.19. Schematic illustration of an antiferromagnet composed of A and B spins on different sublattices separated by oxygen ions. The λ_{IJ} represent the strengths of the molecular fields due to the spins J acting on the NN spins I.

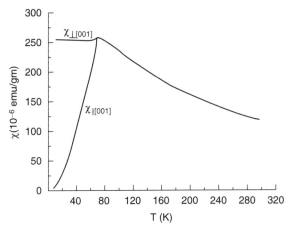

Figure 9.20. Measured $\chi(T)$ for the antiferromagnet MnF$_2$ ($T_N = 70$ K). (From S. Foner, in G. T. Rado and H. Suhl, eds., *Magnetism*, Vol. 1, Chap. 9, copyright 1963. Reprinted by permission of Academic Press Inc.)

where $n = N/V$ and C is the Curie constant. This can be compared with the ferromagnetic case, where the prediction for the Curie temperature is $T_C = \lambda C$ [see Eq. (9.39b)].

The magnetic susceptibility $\chi(T)$ for $T \gg T_N$ is predicted to have a Curie–Weiss form; that is,

$$\chi(T) = \frac{C}{T - \theta}. \tag{9.51}$$

The Curie–Weiss temperature $\theta = C(\lambda_{AA} + \lambda_{AB})/2$ is predicted to be negative since $|\lambda_{AB}| > |\lambda_{AA}|$. This is consistent with the prediction obtained using the Heisenberg exchange interaction as presented in Eq. (9.40) and illustrated schematically in Fig. 9.20, where the $\chi(T)$ measured for the antiferromagnet MnF$_2$ is shown. At $T = T_N$, χ reaches its predicted maximum value, $\chi_{\max} = -1/\lambda_{AB} > 0$. The antiferromagnetic ordering of the spins then limits their ability to respond to external magnetic fields for $T < T_N$.

As T decreases below T_N, the magnetic susceptibility χ will depend critically on the direction of the applied field **H** relative to the direction of the spins. If **H** is applied parallel to the spin direction (i.e., to **M**$_{sA}$ and **M**$_{sB}$), $\chi_\parallel(T)$ is predicted to decrease from the value $\chi_{\max}$ at T_N to 0 at $T = 0$ K (see Fig. 9.20). When **H** is applied perpendicular to **M**$_{sA}$ and **M**$_{sB}$, $\chi_\perp(T)$ is predicted to remain positive even at $T = 0$ K, since the spins can be partially rotated from their fully-magnetized directions into the direction of the resultant field, which is the vector sum of **H** and the molecular field **B**$_{\text{eff}}$.

The metal Cr is an interesting example of an AF material with $T_N = 311$ K and a second transition at $T = 120$ K. Above T_N there is no evidence for paramagnetic local moments in Cr, whereas for $T_N > T > 120$ K, a spin wave with a spatial period incommensurate with the lattice exists, apparently in a band of itinerant electrons. A spin–flop transition occurs as T is lowered to 120 K, at which point the polarization of the spin wave changes from transverse, where the directions of the spins are perpendicular to the spin-wave vector **k**, to longitudinal. In general, spin waves are also the lowest-energy spin excitations in antiferromagnets and have a linear dispersion relation $\omega(k) \approx 4|J|Ska$ for $ka \ll 1$.

9.8 Ferrimagnetism

Magnetic materials that have two spin sublattices interacting antiferromagnetically with each other, but in which the two spontaneous sublattice magnetizations $\mathbf{M}_{sA}$ and $\mathbf{M}_{sB}$ are not equal and opposite, are known as *ferrimagnets*. Below the *ferrimagnetic critical temperature* T_C, also known as the *Curie temperature*, ferrimagnetic materials have a net, nonzero spontaneous magnetization $\mathbf{M}_s = \mathbf{M}_{sA} + \mathbf{M}_{sB}$. Above T_C the spins are oriented randomly in zero magnetic field and the material behaves as a paramagnet. Some examples of ferrimagnets have been given in Table 9.4 and include the ceramic oxides known as *ferrites*, which have the chemical formula $MO \cdot Fe_2O_3$ (i.e., MFe_2O_4) and either the normal spinel or inverse spinel crystal structures, the former shown in Fig. 9.4. Here M is a divalent metal ion such as Fe^{2+}, Mn^{2+}, Cu^{2+}, or Mg^{2+}. Ferrites and other ferrimagnetic materials, such as garnets, have a variety of interesting magnetic applications and are discussed in more detail in Chapter 17.

The magnetic behavior of ferrimagnets is discussed next using a molecular field model similar to those applied earlier to ferromagnets and antiferromagnets. The key differences are that in ferrimagnets there can be different numbers of spins on the two sublattices, the spins on the two sublattices can experience different molecular fields, and the spins need not all have the same magnetic moment. As a result, it will be true in general that $\mathbf{M}_s = \mathbf{M}_{sA} + \mathbf{M}_{sB} \neq 0$ below T_C. For the sake of simplicity, the molecular field model will be developed here for ferrimagnets, in which the N_A spins on sublattice A and the N_B spins on sublattice B have the same magnetic moment, $m_A = m_B = g\mu_B J$. Each A or B spin can still make a different contribution to the magnetization if the local molecular fields at the A and B sites are different.

With a total of N spins in the ferrimagnet, let $N_A = xN$ and $N_B = (1-x)N$. As in the case of antiferromagnetism, it is assumed that each A spin is acted on by a molecular field $\mathbf{B}_{A,\text{eff}}$ consisting of contributions from its NN A and B spins, while the same is true for the molecular field $\mathbf{B}_{B,\text{eff}}$ acting on each B spin. The expressions for the molecular fields are in fact identical to those given in Eqs. (9.48) and (9.49) for the case of antiferromagnetism. The $A–B$ spin–spin interactions are antiferromagnetic in ferrites since they occur via superexchange, typically through the intervening O^{2-} ions. Hence $\lambda_{AB} = \lambda_{BA} < 0$. The like-spin $A–A$ interaction λ_{AA} and $B–B$ interaction λ_{BB} can be either ferromagnetic or antiferromagnetic. Letting $n = N/V$, then $\mathbf{M}_{sA} = xn\mathbf{m}_{Az}$ and $\mathbf{M}_B = (1-x)n\mathbf{m}_{Bz}$ are the local sublattice magnetizations at the sites of the A and B spins, respectively. Here $\mathbf{m}_{Az}$ and $\mathbf{m}_{Bz}$ are the components of the corresponding magnetic moments $\mathbf{m}_A$ and $\mathbf{m}_B$ in the direction of the applied field $\mathbf{H}$, which is taken to be along the z axis.

The temperature dependence of the spontaneous magnetization $\mathbf{M}_s$ of the ferrimagnet below T_C depends sensitively on the fractions x of A spins and $(1-x)$ of B spins, as well as on the ratios of the molecular field strengths $\lambda_{AA}/\lambda_{AB}$ and $\lambda_{BB}/\lambda_{AB}$. The magnetic susceptibility $\chi(T)$ has the following Curie–Weiss form in the paramagnetic state for $T \gg T_C$:

$$\chi(T) = \frac{C}{T - \theta}, \tag{9.52}$$

where $\theta = C[2x(1-x)\lambda_{AB} + x^2\lambda_{AA} + (1-x)^2\lambda_{BB}]$ is generally < 0 and C is the usual Curie constant with $n = n_A + n_B$. Note that this expression for θ reduces to the ferromagnetic result given in Eq. (9.39b) for $x = 0$ or $x = 1$. The susceptibility χ actually diverges at the transition temperature T_C, which is a function of x, $\lambda_{AA}/\lambda_{AB}$, and $\lambda_{BB}/\lambda_{AB}$. For ferrites with the spinel crystal structure, all three interactions are found to be antiferromagnetic (i.e., λ_{AA}, λ_{BB}, and λ_{AB} are all < 0). When in addition $\lambda_{AA}\lambda_{BB} > \lambda_{AB}^2$, T_C is predicted to be < 0 and the material will therefore be paramagnetic for all T. This occurs when the antiferromagnetic A–A and B–B like-spin interactions within each sublattice dominate the antiferromagnetic A–B interaction, which provides the tendency toward ferrimagnetism.

The magnetization $\mathbf{M}_s$ in ferrimagnets exhibits an interesting effect for $x < \frac{1}{2}$ (more B spins than A spins) when $\lambda_{BB} > \lambda_{AB}$ and $x(\lambda_{AB} - \lambda_{AA}) > (1 - x)(\lambda_{AB} - \lambda_{BB})$. Just below T_C, $\mathbf{M}_s$ will be positive (i.e., $|\mathbf{M}_{sA}| > |\mathbf{M}_{sB}|$), while at $T = 0$ K, $\mathbf{M}_s$ will be negative (i.e., $|\mathbf{M}_{sA}| < |\mathbf{M}_{sB}|$). The temperature below T_C at which $\mathbf{M}_s$ changes from positive to negative (i.e., at which $\mathbf{M}_s = 0$ and $\mathbf{M}_{sA} = -\mathbf{M}_{sB}$) is called the *compensation point*, T_{comp}. This behavior is illustrated in Fig. 9.21 for the higher-spin ferrimagnetic rare earth iron garnets $RE_3Fe_5O_{12}$, with RE = Gd, Tb, Dy, Ho, and Er. At T_{comp} the net magnetic moment of the RE ions cancels the net magnetic moment of the Fe ions. These RE iron garnets are discussed in more detail in Chapter W17.

In the ferrite Fe_3O_4 known as magnetite, one-third of the magnetic ions are present as Fe^{2+}, while the remaining two-thirds are present as Fe^{3+}. Fe_3O_4 is, in fact, considered

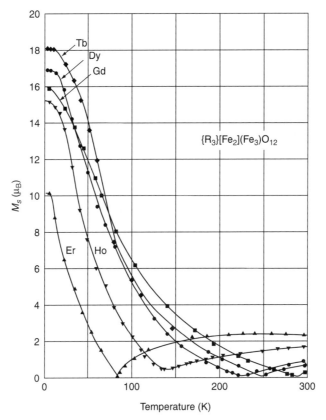

Figure 9.21. Saturation magnetic moments M_s in units of μ_B (per $RE_3Fe_5O_{12}$ formula unit) for the higher-spin rare earth iron garnets, from Gd to Er, as a function of temperature T. Note that the net magnetic moment M_s goes to zero at $T = T_{\text{comp}}$. [From S. Geller et al., *Phys. Rev.*, **137A**, 1034 (1965). Copyright 1965 by the American Physical Society.]

to be the archetype of a mixed-valent $3d$ transition metal compound. In the normal spinel structure the Fe^{2+} ions would occupy sites with tetrahedral symmetry having four O^{2-} ions as NNs, while the Fe^{3+} ions would occupy sites with octahedral symmetry having six O^{2-} ions as NNs. In fact, the crystal structure of Fe_3O_4 is inverse spinel in which the Fe^{2+} ions are in octahedral sites, where they gain a crystal field stabilization energy of $2\Delta_o/5$ (see Table 9.2). One half of the Fe^{3+} ions do occupy octahedral sites, as expected, while the remainder occupy tetrahedral sites. The spins of the Fe^{3+} ions in the tetrahedral sites are aligned antiferromagnetically with respect to the spins of the Fe^{3+} ions in the octahedral sites. As a result, the magnetic moment and ferrimagnetism of Fe_3O_4 are due solely to the spins of the Fe^{2+} ions, which all point in the same direction. The numbers n_B of Bohr magnetons for the Fe^{2+} and Fe^{3+} ions have been listed in Table 9.4.

Having discussed the three important types of long-range order found in magnetic materials (i.e., ferro-, antiferro-, and ferrimagnetism), it is interesting to note that Fe atoms have been observed to play dominant roles in all three types of magnetic behavior. This observation is summarized in Table 9.6, where several different crystals containing Fe are listed, along with the number n_B of Bohr magnetons per Fe ion and the type of magnetic order observed. These results show clearly that the local atomic environment has a strong effect on the magnetic moment of the Fe ion and on the type of magnetic order observed in the crystal.

The effect of the local atomic environment is seen very clearly in the Fe–Al compounds and alloys. In the ordered intermetallic compound Fe_3Al there are two types of sites for Fe ions, one having eight Fe^{2+} NNs (as in BCC Fe) and denoted Fe_I, while the second has four Fe^{2+} and four Al^{3+} NNs and is denoted Fe_{II}. The magnetic moment of Fe^{2+} ions in the Fe_I sites is found[†] to be $2.2\mu_B$ (again as in BCC

TABLE 9.6 Crystals Containing Magnetic Fe Ions

Crystal	Magnetic Ion	$n_B = gJ$	Type of Magnetic Order[a]
(Free ion	Fe^{2+}	4.0	—)
(Free ion	Fe^{3+}	5.0	—)
Fe (BCC)	Fe^{2+}	2.2	FM
FeO	Fe^{2+}	3.3	AFM
α-Fe_2O_3	Fe^{3+}	4.9	AFM
Fe_3O_4	Fe^{2+}, Fe^{3+}	Fe^{2+}: 4.1	FiM
		Fe^{3+}: 5.0	
FeAl	—	0	NM
Fe_3Al	Fe^{2+}	Fe_I: 2.2	FM
		Fe_{II} : $\approx$1.8	
$Fe_{1-x}Al_x$	Fe^{2+}	$2.2 \rightarrow 0$	FM $\rightarrow$ AFM $\rightarrow$ NM as x increases
$Nd_2Fe_{12}B$	Fe^{2+}	2.1	FM
Au:Fe	Fe^{2+}	2.4	Disordered (spin glass)
Mo:Fe	Fe^{2+}	1.8	Disordered (spin glass)
Al:Fe	—	0	NM

[a]FM, ferromagnetic; AFM, antiferromagnetic; FiM, ferrimagnetic; NM, nonmagnetic.

[†] P. A. Beck, *Metall. Trans.*, **2**, 2015 (1971).

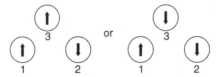

Figure 9.22. Cluster of three spins in which each spin prefers to interact antiferromagnetically with the other two, shown as a schematic illustration of magnetic frustration.

Fe), whereas in Fe_{II} sites the magnetic moment is only $\approx 1.8\mu_B$. The presence of the nonmagnetic Al^{3+} ions as NNs therefore reduces the magnetic moment of the Fe^{2+} ion. In the compound FeAl, each Fe^{2+} ion has eight Al^{3+} ions as NNs and is found to be nonmagnetic. Not surprisingly, dilute concentrations of Fe impurities in crystals of Al are also nonmagnetic. The effect is not as strong in crystals of Au and Mo, where, for example, Fe^{2+} ions have magnetic moments of 2.4 and $1.8\mu_B$, respectively.

Magnetism in Disordered Materials. Magnetism in disordered materials differs from the ordered forms of magnetism already discussed in that the magnetic material is either itself structurally amorphous or is crystalline but contains spins that are randomly distributed (i.e., spatially disordered). In either case no long-range magnetic order will be present in the material at any temperature, although short-range ferromagnetic or antiferromagnetic order may exist locally. Examples of the first type of material are amorphous metallic Fe and oxide glasses containing Fe; examples of the second type include crystalline Cu alloys known as spin glasses, which contain dilute concentrations of magnetic impurities such as Fe or Mn. Dilute magnetic semiconductors, such as crystalline $Zn_{1-x}Mn_xS$, are described in Chapter W17 and are also of the second type. Spin glasses are discussed in more detail in Chapter W9.

The term *magnetic frustration* is used to describe the magnetic behavior observed in amorphous magnetic materials when spins cannot simultaneously achieve the most energetically favorable alignment relative to all of their NN spins. Consider as an example a cluster of three spins, shown schematically in Fig. 9.22, in which each spin prefers to interact antiferromagnetically with its neighbors. If spins 1 and 2 are aligned antiferromagnetically as shown, then either spin 1 or spin 2, but not both, can be antiferromagnetically aligned with spin 3. Thus the spins are "frustrated" in their attempts to achieve favorable interactions with all of their NN spins. Magnetic frustration of this type will prevent long-range magnetic order from being established in the material, as observed experimentally.

9.9 Magnetic Behavior of Electrons in Closed Shells and of Conduction Electrons

In contrast to the magnetic behavior observed for materials containing well-defined magnetic moments that is determined by electrons in partially filled shells, the magnetic properties of solids in which no well-defined magnetic moments are present are determined primarily by the valence or bonding electrons and the conduction electrons. The diamagnetic response due to these electrons is discussed first, followed by a description of their paramagnetic response.

Diamagnetism is usually a relatively weak form of magnetism corresponding to a negative magnetic susceptibility $\chi < 0$ (and $\mu < \mu_0$), with typical values being

$\chi \approx -10^{-6}$ to -10^{-5}. The main exceptions to this weak diamagnetism are super-conductors, which can exist in a state in which they are perfect diamagnets (i.e., where $\mathbf{B} = 0$ even in the presence of a magnetic field $\mathbf{H}$) and for which $\chi = -1$. In the case of diamagnetism the magnetization $\mathbf{M}$ induced in the material is directed opposite to the applied magnetic field $\mathbf{H}$. Only a few materials have a net diamagnetic response, including simple metals such as Cu, Ag, and Au, inorganic solids such as Si, NaCl, Al_2O_3, and the inert-gas solids, as well as superconductors.

Larmor Diamagnetism. The contribution to the magnetic response of a material due to the orbital motion of electrons in filled shells is considered first. This response, called *Larmor diamagnetism*, is analogous to the diamagnetic response of any closed current loop in a changing magnetic field, in accord with Lenz's law. According to Hund's rules, electrons in filled shells have no net intrinsic spin $\mathbf{S}$, orbital spin $\mathbf{L}$, or total spin $\mathbf{J}$. They therefore do not contribute to any well-defined magnetic moment that the atom or ion may otherwise have in the solid.

The response to an external magnetic field H of an ion that has Z_{fs} electrons in filled shells involves an increase in the electronic orbital kinetic energy E_{orb}. A quantum-mechanical calculation of the resulting shift of the ground-state energy of the system begins by the addition to the Hamiltonian of the vector potential $\mathbf{A}$ of the magnetic field as follows: $\mathbf{p}^2/2m \rightarrow (\mathbf{p} + e\mathbf{A})^2/2m$. Here $\mathbf{p}$ is the momentum operator. Using $\mathbf{A} = \mu_0\mathbf{H} \times \mathbf{r}/2$ for a uniform field $\mathbf{H}$ leads to $A^2 = \mu_0^2 H^2 r^2/6$ and the following energy shift:

$$\Delta E_{orb} = \frac{Z_{fs}e^2\mu_0^2H^2\langle r^2 \rangle}{12m}. \tag{9.53}$$

Here $\langle r^2 \rangle$ is the mean-square ionic radius of an electron in a filled shell. An additional energy shift that is linear in H will not contribute to the magnetic susceptibility. It is assumed here that the electrons remain in their ground state with $\mathbf{J} = 0$.

The resulting diamagnetic response for such a solid composed of N ions in volume V is given by

$$\chi_{Larmor} = -\frac{N}{\mu_0 V}\frac{\partial^2 \Delta E_{orb}}{\partial H^2} = -\frac{\mu_0 N Z_{fs}e^2\langle r^2 \rangle}{6mV}. \tag{9.54}$$

The quantity χ_{Larmor}, known as the *Larmor* (sometimes as the *Langevin*) *diamagnetic susceptibility*, is a dimensionless quantity. Note that χ_{Larmor} is predicted to be independent of temperature.

This expression for the Larmor diamagnetic susceptibility can be expected to correspond to the total magnetic response of solids of inert-gas atoms such as solid Ar or to ionic crystals such as the alkali halides, which are composed of ions with filled (or nearly filled) shells. Typical values[†] for χ_{Larmor} are -10^{-5}. For a noble metal such as Au with $N/V = 5.9 \times 10^{28}$ m^{-3}, $Z_{fs} = 78$, and using the typical value $\langle r^2 \rangle = (0.5 \times 10^{-10}$ m$)^2$, Eq. (9.54) yields $\chi_{Larmor}(Au) = -6.7 \times 10^{-5}$. As will be seen later, there exist additional diamagnetic and paramagnetic contributions from the conduction electrons to the magnetic susceptibility of metals such as Au.

[†] Values for χ_{Larmor} for some filled-shell ions and inert-gas atoms are given in R. Kubo and T. Nagamiya, eds., *Solid State Physics*, McGraw-Hill, New York, 1969, p. 439.

Landau Diamagnetism. In addition to Larmor diamagnetism, there is a diamagnetic contribution to the magnetic susceptibility of metals from the orbital motion of the conduction electrons induced by applied magnetic fields. This is known as the *Landau diamagnetic susceptibility*, and a quantum-mechanical calculation gives the prediction for free electrons that

$$\chi_{\text{Landau}} = -\frac{\mu_0 \mu_B^2 \rho(E_F)}{3}, \tag{9.55}$$

where $\rho(E_F)$ is the density of electron states at E_F per unit volume and per unit energy. For metals, $\rho(E_F)$ is $\approx 10^{47}$ m^{-3}/J, and thus a typical value for χ_{Landau} is $\approx -10^{-5}$, of the same order of magnitude as the value predicted for χ_{Larmor}.

Pauli Paramagnetism. The conduction electrons contribute a paramagnetic component to the magnetic susceptibility of metals in addition to their diamagnetic χ_{Landau} contribution. This contribution, known as the *Pauli paramagnetic susceptibility* χ_P, results from an imbalance in the concentrations $n\uparrow$ of up-spin ($m_s = +\frac{1}{2}$) and $n\downarrow$ of down-spin ($m_s = -\frac{1}{2}$) conduction electrons in the presence of a magnetic field H. The total concentration of conduction electrons is $n = n\uparrow + n\downarrow$. The resulting magnetization M is given by

$$M(H) = n\uparrow(-\mu_B) + n\downarrow(+\mu_B) = -(n\uparrow - n\downarrow)\mu_B = -\Delta n \mu_B. \tag{9.56}$$

Note that up-spin (down-spin) electrons have their magnetic moments directed opposite (parallel) to the field **H**. In zero-field $n\uparrow = n\downarrow = n/2$ and **M** $= 0$, since, according to the Pauli exclusion principle, pairs of up- and down-spin electrons will occupy the available energy levels up to the Fermi level E_F, as shown schematically in Fig. 9.23a.

The following simple argument can be used to predict the correct order of magnitude for χ_P. When a field **H** is applied to a metal, a fraction $f(H)$ of the conduction electrons can lower their energies by an amount $\Delta E \approx -2\mu_0\mu_B H$ per electron by reorienting their magnetic moments **m** parallel (and spins antiparallel) to **H**. Only the fraction

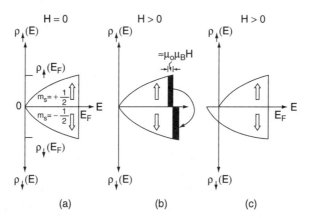

Figure 9.23. Densities of states $\rho\uparrow(E)$ and $\rho\downarrow(E)$ for up- and down-spin conduction electrons in a metal as functions of electron energy. (*a*) Pairs of up- and down-spin electrons occupy the available energy levels up to the Fermi level E_F, as shown here for $H = 0$. (*b*) In a field **H**, only electrons within an energy $\approx \mu_0\mu_B H$ of E_F can lower their energy by reorienting their magnetic moments **m** parallel (and spins antiparallel) to **H**. (*c*) Final distribution of electrons in a field **H** is shown.

$f(H) \approx \mu_0 \mu_B H / E_F$ of up-spin electrons within an energy $\approx \mu_0 \mu_B H$ of E_F can find available empty down-spin energy levels so that they can lower their energy in this way. This fraction f is quite small, $\approx 10^{-4}$, even for strong fields $H \approx 10^6$ A/m. The resulting shift in the energy distribution of electrons induced by **H** is shown schematically in Fig. 9.23b, with the final distribution shown in Fig. 9.23c.

As a result of the shift, there will now be more down-spin than up-spin electrons (i.e., $n\downarrow > n\uparrow$). The difference Δn will be given approximately by n times the fraction f, that is,

$$\Delta n(H) \approx n f(H) \approx -\frac{n \mu_0 \mu_B H}{E_F}. \tag{9.57}$$

The resulting magnetization is therefore given by $\Delta n \mu_B$:

$$M \approx \frac{n \mu_0 \mu_B H \mu_B}{E_F} = \frac{n \mu_0 \mu_B^2 H}{E_F}$$

$$\approx \mu_0 \mu_B^2 \rho(E_F) H, \tag{9.58}$$

where $\rho(E_F) = 3n/2E_F$ is the density of electron states at E_F for electrons of both spin directions. The Pauli paramagnetic susceptibility is therefore given by $\chi_P = M/H \approx \mu_0 \mu_B^2 \rho(E_F)$. An exact calculation in fact gives the same result, that is,

$$\chi_P = \mu_0 \mu_B^2 \rho(E_F). \tag{9.59}$$

Pauli paramagnetism is a weak form of magnetism, a typical value being $+10^{-5}$, and thus is similar in magnitude, but not in sign, to the Larmor and Landau diamagnetic susceptibilities.

This expression for χ_P is strictly valid only at $T = 0$ K. In fact, χ_P can be shown to be essentially independent of temperature and given by Eq. (9.59) up to $k_B T \approx 0.1 E_F$ or $T \approx 10^3$ K. That free electrons do not act as free spins with a resulting Curie law susceptibility is yet another result of the fact that electrons are fermions and so must obey Fermi–Dirac statistics and the Pauli exclusion principle.

The total magnetic susceptibility of the conduction electrons is the sum of their paramagnetic and diamagnetic contributions:

$$\chi_{el} = \chi_P + \chi_{Landau} = \mu_0 \mu_B^2 \rho(E_F) - \frac{\mu_0 \mu_B^2 \rho(E_F)}{3} = +\frac{2\mu_0 \mu_B^2 \rho(E_F)}{3}. \tag{9.60}$$

The conduction electrons therefore provide a net paramagnetic contribution to the magnetic susceptibility and magnetic moment of the metal.

Measured values of χ at room temperature are presented in the dimensionless SI units in Table 9.7 for selected diamagnetic and paramagnetic materials and in cgs-emu units of cm^3/mol in Fig. 9.24 for the elements up to Bi, $Z = 83$. Not shown in the figure are values for χ for the metals Fe, Co, and Ni, $Z = 26$ to 28, which are ferromagnetic at room temperature.

In addition to the more delocalized s and p electrons, for which the free-electron approximation is usually valid, many metals also contain d and f electrons, which are at least partially delocalized and which therefore interact with each other and with

TABLE 9.7 Magnetic Susceptibilities at Room Temperature for Selected Diamagnetic and Paramagnetic Materials

Diamagnetic	$\chi \ (10^{-5})^a$	Paramagnetic	$\chi(10^{-5})^a$
Ag	−2.39	Al	+2.07
Au	−3.45	Ce (β)	+150
Bi	−16.5	Er	+3270
C (diamond)	−2.16	Mo	+9.6
Cu	−0.96	Na	+0.85
Si	−0.325	Pd	+76.5
Zn	−1.26	Ti	+17.9
AgBr	−2.64	W	+7.0
Al_2O_3	−1.8	CuO	+24.1
Cu_2O	−1.05	FeO	+720
NaCl	−1.41	MnO	+468
PbS	−3.29	Ti_2O_3	+5.31
SiO_2	−1.42		

Source: Data from D. R. Lide, ed.-in-chief, *CRC Handbook of Chemistry and Physics*, 79th ed. (CRCnet-BASE 1999), CRC Press, Boca Raton, Fla., 1999, pp. 4-130 to 4-135.

[a]The quantity χ is the volume susceptibility defined in Eq. (9.13) and is given in dimensionless SI units. To obtain χ(cgs-emu) in units of emu/cm^3, divide χ(SI) in the table by 4π.

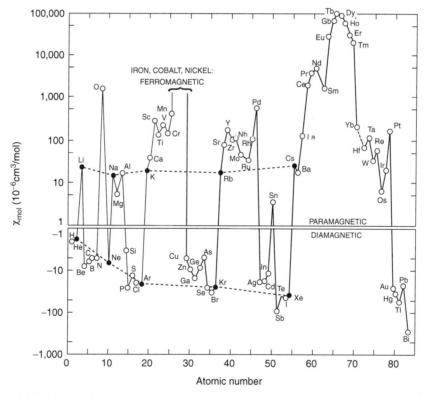

Figure 9.24. Measured values of the molar susceptibility χ_{mol} at room temperature are presented in cgs-emu units of cm^3/mol for the elements up to Bi, $Z = 83$. (From D. E. Gray, coord. ed., *American Institute of Physics Handbook*, 3rd ed., McGraw-Hill, New York, 1974. Reprinted by permission of the McGraw-Hill Companies.)

the s and p electrons. In such metals χ_P can be enhanced well above the predicted free-electron value, which, as shown above, is proportional to the density of electron states $\rho(E_F)$. The expression for this enhanced susceptibility has been given in Eq. (9.44). Examples of metals with exchange-enhanced paramagnetic susceptibilities include several listed in Table 9.7 and shown in Fig. 9.24.

The dominance of χ_P over χ_{Landau} can be reversed in metals such as Bi and in semiconductors that are doped with specific impurities. For semiconductors this reversal follows from the fact that $\rho(E_F)$ is proportional to the electron mass m for free electrons and to the effective mass m^* in general. Since $m^*/m \approx 0.1$ in semiconductors due to band-structure effects, χ_P will be reduced by this same factor. The susceptibility χ_{Landau}, on the other hand, originates from the orbital motion of the conduction electrons and is proportional to, and enhanced by, the factor $m/m^* \approx 10$. As a result, in doped semiconductors it is not unusual to have $\chi_{el} < 0$ since $|\chi_{\text{Landau}}| \gg \chi_P$. The relatively large diamagnetic susceptibility of Bi also arises from the low effective mass of electrons, which, in turn, is due to the high curvature of the energy bands in Bi.

REFERENCES

Ashcroft, N. W., and N. D. Mermin, *Solid State Physics*, Saunders College, Philadelphia, 1976.

Burns, G., *Solid State Physics*, Academic Press, San Diego, Calif., 1985.

Chen, C.-W., *Magnetism and Metallurgy of Soft Magnetic Materials*, Dover, Mineola, N.Y., 1986.

Chikazumi, S., *Physics of Magnetism*, Wiley, New York, 1964.

Companion, A. L., *Chemical Bonding*, 2nd ed., McGraw-Hill, New York, 1979.

Cotton, F. A., *Chemical Applications of Group Theory*, 3rd ed., Wiley-Interscience, New York, 1990.

Craik, D., *Magnetism: Principles and Applications*, Wiley, Chichester, West Sussex, England, 1995.

Herring, C., *Exchange Interactions Among Itinerant Electrons*, Vol. IV of G. T. Rado and H. Suhl, eds., *Magnetism*, Academic Press, San Diego, Calif., 1966.

Kittel, C., *Introduction to Solid State Physics*, 7th ed., Wiley, New York, 1996.

Reif, F., *Fundamentals of Statistical and Thermal Physics*, McGraw-Hill, New York, 1965.

Sugano, S., Y. Tanabe, and H. Kamimura, *Multiplets of Transition-Metal Ions in Crystals*, Academic Press, San Diego, Calif., 1970.

Taylor, J. R., and C. D. Zafiratos, *Modern Physics for Scientists and Engineers*, Prentice Hall, Upper Saddle River, N.J., 1991.

Van Vleck, J. H., *The Theory of Electric and Magnetic Susceptibilities*, Oxford University Press, Oxford, 1932.

White, R. M., and T. H. Geballe, *Long Range Order in Solids*, Suppl. 15 of H. Ehrenreich, F. Seitz, and D. Turnbull, eds., *Solid State Physics*, Academic Press, San Diego, Calif., 1979.

PROBLEMS

9.1 For n $3d$ electrons show that according to Hund's rules:

(a) $J = (L - S) = n(4 - n)/2$ for $n < 5$.

(b) $J = (L + S) = (n - 4)(10 - n)/2$ for $n > 5$.

9.2 Verify the calculated values of the effective magneton number $p = g\sqrt{J(J+1)}$ for the $3d$ iron group transition metal ions listed in Table 9.1 for the following two cases:

   (a) When the orbital angular momentum of the ion is quenched by its interaction with the local crystal field (i.e., when $\mathbf{L} = 0$ and $\mathbf{J} = \mathbf{S}$).

   (b) When L has its free-ion value and $J = |L \pm S|$.

9.3 Calculate the classical value $m = iA$ for the magnetic moment of a current loop of area A carrying a current $i = 1$ pA $= 10^{-12}$ A for the case where the loop is a circle of radius 0.2 nm. Express your answer for m in units of the Bohr magneton μ_B.

9.4 From the appropriate constitutive relations show that:

   (a) $\mu(\text{SI}) = \mu_0\mu(\text{cgs-emu})$ for the magnetic permeability.

   (b) $\chi(\text{SI}) = 4\pi\chi(\text{cgs-emu})$ for the magnetic susceptibility.

9.5 Derive Eq. (9.25) for the magnetic susceptibility $\chi(T) = M/H$ of free spins from the Brillouin function $B_J(x)$ in the limit $x \to 0$.

9.6 Find the magnetization M of a classical free spin J in a magnetic field H from the $J \to \infty$ limit of the Brillouin function $B_J(x)$ given in Eq. (9.24). Show that the result is equivalent to that predicted using the classical Langevin function $L(x) = \coth x - 1/x$.

9.7 The Heusler alloy Cu_2MnAl has the CsCl structure with Cu ions on one simple cubic sublattice and Mn and Al ions ordered alternately on the other simple cubic sublattice. If the Mn–Mn distance is 0.42 nm, show that $n_B = gJ = 4.1$ using the magnetic data provided in Table 9.4.

9.8 For cubic ferromagnets such as Fe and Ni, show that:

   (a) When $K_1 > 0$, as for BCC Fe, the anisotropy energy $E_a(\text{cubic})$ given in Eq. (9.47) is minimized when $\mathbf{M}$ points along any of the six $\langle 100 \rangle$ directions.

   (b) when $K_1 < 0$, as for FCC Ni, the anisotropy energy $E_a(\text{cubic})$ is minimized when $\mathbf{M}$ points along any of the eight $\langle 111 \rangle$ directions.

9.9 Show that $M_{\text{sat}} = 0.51 \times 10^6$ A/m (see Table 9.4) for Fe_3O_4 is consistent with 8 Fe^{2+} ions per cubic unit cell with $m = 4\mu_B$ per ion. The lattice constant is $a = 0.839$ nm.

9.10 In the calculation of the Pauli paramagnetic susceptibility χ_P, show that the gain in energy of the conduction electrons in a magnetic field H is given approximately by $\Delta E \approx (\mu_0\mu_B H/E_F) \cdot 2\mu_0\mu_B H$. Show also that $\chi_P \approx \mu_0\mu_b^2\rho(E_F)$. [*Hint*: see Eq. (9.58).]

9.11 Show that the Pauli paramagnetic susceptibility χ_P can be expressed in a Curie law form as $\chi = C/T_F$ with the Curie constant $C = 3n\mu_0\mu_B^2/2k_B$ and the Fermi temperature $T_F = E_F/k_B$ replacing the true temperature T. Show that $\chi_P/\chi_{\text{Curie}} \approx 10^{-3}$ for $T \approx 100$ K.

Note: Additional problems are given in Chapter W9.

CHAPTER 10

Mechanical Properties of Materials

10.1 Introduction

When external forces act on a solid material, the solid often responds by moving through space and by rotating about an axis, undergoing what is known as *rigid-body motion*. Since solids are not perfectly rigid, other types of response to external forces must also be considered. In particular, solids can undergo mechanical deformations or changes in dimensions or shape. In fact, the principal feature of solid materials that distinguishes them from other forms of matter, such as liquids and gases, is their strength (i.e., ability to withstand mechanical deformation). In this chapter we describe the mechanical properties of materials, specifically the nature of the deformations that result from application of external forces to materials and also the ways in which different classes of materials respond to external forces.

The mechanical properties of materials include elastic, anelastic, or inelastic behavior. A solid is said to be *elastic* when any deformations disappear "quickly" once the external forces are removed (i.e., the solid returns to its initial state). In the proportional elastic region the deformations are directly proportional to the external forces (i.e., the mechanical strain tensor components ε_{ij} are directly proportional to the stress tensor components σ_{ij}). The *anelastic* behavior of materials is a type of time-dependent mechanical behavior in which the applied stresses and the resulting strains are not uniquely related to each other due to relaxation effects. Other types of response, including permanent deformations, plasticity, viscoelasticity, fracture, and so on, correspond to *inelastic* behavior. Phase transitions corresponding to changes in the crystal structure of materials and due to the application of external pressure are a type of inelastic behavior that is not discussed here.

The definitions of stress and strain in materials are introduced first, along with the material properties known as *elastic constants*. A detailed discussion of the elastic properties will then be presented, followed by descriptions of the anelastic and inelastic properties of materials.

STRESS, STRAIN, AND ELASTIC CONSTANTS

The types and magnitudes of the deformations that can occur in a solid as a result of the application of external forces depend on how the forces are applied and on the magnitude of the forces in relation to the strength of the bonding forces between the atoms in the solid. Atoms in solids in general occupy positions of stable equilibrium as a result of their interactions with neighboring atoms. Atoms can undergo displacements away from their equilibrium positions in response to the application of external forces. A deformation results when changes in relative position occur for at least some of the atoms in the solid, corresponding to changes in the dimensions or shape or to the generation of defects.

External forces acting on solids are often mechanical in nature, resulting from physical contact with other solid objects. Deformations of solids can also be caused by external gravitational, electrical, and magnetic forces. *Piezoelectricity* corresponds to the generation of stresses and strains in materials due to the application of electric fields and is described in Chapter 15. The corresponding effect, known as *magnetostriction*, which results from the application of magnetic fields to materials, is described in Chapter 17. Changes in temperature leading to thermal expansion or contraction of a solid can also create stresses and thus cause deformations. The relationship between the forces acting on a solid and the stresses created in the solid is discussed next.

10.2 Stress

When external forces are applied to a solid material, stresses can be created within the solid. For a solid in mechanical equilibrium, the stresses result from the internal forces that are generated within the solid to counterbalance the externally applied forces. These internal forces are exerted by adjacent elements of the solid on each other and ultimately arise from the forces between atoms. The response of the solid to the stresses σ are the deformations known as strains ε.

To define the stresses σ in terms of the internal forces in a solid, the situation where an external force F_{ext} is applied to the end of a solid bar is examined. This is shown schematically in Fig. 10.1. Consider the forces acting on a small element of volume

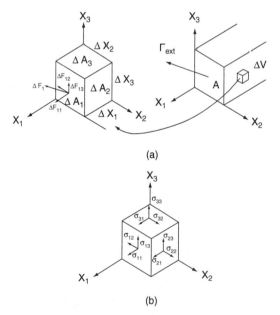

(a)

(b)

Figure 10.1. Internal forces and stresses σ in a solid bar resulting from the application of an external force F_{ext}: (a) internal force ΔF_1 acting on one face of a small element of volume $\Delta V = \Delta x_1 \Delta x_2 \Delta x_3$; (b) the three components σ_{11}, σ_{12}, and σ_{13} of the stress acting on this face.

$\Delta V = \Delta x_1 \Delta x_2 \Delta x_3$ within the solid. The internal force ΔF_1 acting on the face with area $\Delta A_1 = \Delta x_2 \Delta x_3$ is exerted by the adjacent material in the solid. The volume ΔV is chosen to be small enough so that the force ΔF_1 acts uniformly over the area ΔA_1. The components of the force ΔF_1 acting along the three coordinate axes x_1, x_2, and x_3 are ΔF_{11}, ΔF_{12}, and ΔF_{13}, respectively, as shown in Fig. 10.1a.

The three components σ_{ij} of the *stress* σ which act on the face shown have magnitudes given by

$$\sigma_{11} = \frac{\Delta F_{11}}{\Delta A_1}, \qquad \sigma_{12} = \frac{\Delta F_{12}}{\Delta A_1}, \qquad \sigma_{13} = \frac{\Delta F_{13}}{\Delta A_1}. \qquad (10.1)$$

The directions of σ_{11}, σ_{12}, and σ_{13} are shown in Fig. 10.1b and are the same as the directions of the corresponding force components (i.e., the first subscript indicates the direction of the normal of the face to which the stress is applied, and the second subscript indicates the direction in which the force component acts). For example, the stress component σ_{11} is a *normal (tensile) stress*, with units of N/m^2 or pascal (1 Pa = 1 N/m^2), which tends to cause an elongation of the volume element ΔV in the x_1 direction. If the direction of ΔF_{11} were reversed, σ_{11} would be a *normal compressive stress*, tending to cause a compression of ΔV.

The stress components σ_{12} and σ_{13} are *shear stresses*, also with units of pascal, which tend to deform the shape of ΔV. (In later sections the symbol τ will sometimes be used for shear stress.) The components σ_{21}, σ_{22}, and σ_{23} resulting from a force ΔF_2 acting on the face with area $\Delta A_2 = \Delta x_1 \Delta x_3$ and the components σ_{31}, σ_{32}, and σ_{33} resulting from a force ΔF_3 acting on the face with area $\Delta A_3 = \Delta x_1 \Delta x_2$ are shown in Fig. 10.1b and are defined using equations analogous to Eq. (10.1). These nine stress components σ_{ij} can be represented using the *stress tensor* σ, given by

$$\sigma = \begin{pmatrix} \sigma_{11} & \sigma_{12} & \sigma_{13} \\ \sigma_{21} & \sigma_{22} & \sigma_{23} \\ \sigma_{31} & \sigma_{32} & \sigma_{33} \end{pmatrix}. \qquad (10.2)$$

When the volume element ΔV is in mechanical equilibrium with regard to translation, the internal forces yielding the normal stress components σ_{11}, σ_{22}, and σ_{33} must be balanced by equal but oppositely directed forces acting on the other faces of ΔV. Similarly, for ΔV to be in rotational equilibrium, the couple $\Delta F_{12} \Delta x_1$ due to the two shear stress components corresponding to σ_{12} acting on the two opposite faces with areas ΔA_1 must be balanced by an opposing couple acting on the two opposite faces with areas ΔA_2. This opposing torque is given by $\Delta F_{21} \Delta x_2$ and results from the shear stress components corresponding to σ_{21} (see Figs. 10.1b and 10.2). It therefore follows that $\sigma_{12} = \sigma_{21}$ and similarly, that $\sigma_{13} = \sigma_{31}$ and $\sigma_{23} = \sigma_{32}$ for rotational equilibrium of the volume element ΔV.

In the limit $\Delta V \to 0$, the state of stress at a given point in a solid can therefore be specified by just six independent stress components instead of the nine possible σ_{ij}. These six stress components are often labeled using a single subscript, as follows:

$$\sigma_1 = \sigma_{11}, \qquad \sigma_2 = \sigma_{22}, \qquad \sigma_3 = \sigma_{33},$$
$$\sigma_4 = \sigma_{23} = \sigma_{32}, \qquad \sigma_5 = \sigma_{31} = \sigma_{13}, \qquad \sigma_6 = \sigma_{12} = \sigma_{21}. \qquad (10.3)$$

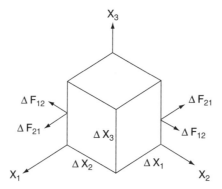

Figure 10.2. Torques acting on a volume element ΔV in a solid.

The stress tensor σ defined in Eq. (10.2) is therefore symmetric, with the form

$$\sigma = \begin{pmatrix} \sigma_{11} & \sigma_{12} & \sigma_{13} \\ \sigma_{12} & \sigma_{22} & \sigma_{23} \\ \sigma_{13} & \sigma_{23} & \sigma_{33} \end{pmatrix} = \begin{pmatrix} \sigma_1 & \sigma_6 & \sigma_5 \\ \sigma_6 & \sigma_2 & \sigma_4 \\ \sigma_5 & \sigma_4 & \sigma_3 \end{pmatrix}. \tag{10.4}$$

The relationships between the external forces and the resulting internal forces and stresses σ are simple only for special cases. For example, if external forces F_{ext} act along the axis of a long bar of original cross-sectional area A_0, as is usually the case in the tensile testing of materials, the internal forces F_{int} and normal tensile stresses also act along the axis of the bar. It follows that $\sigma = F_{\text{ext}}/A_0$ as long as changes in A_0 are small. When changes in A_0 cannot be neglected, the *nominal stress* $\sigma = F_{\text{ext}}/A_0$ no longer represents the true tensile stress in the material. The *true stress* is defined by $\sigma_{\text{true}} = F_{\text{ext}}/A$ where A is the actual cross sectional area, which can vary along the length of the bar.

Within the elastic limit of a material, the stresses can often be calculated analytically from the forces applied. In practice, however, the actual state of stress in a solid cannot be calculated easily or measured directly. The only general way to determine the stresses in a solid is through experimental study of the resulting strains.

10.3 Strain

Having defined the stresses σ that can exist within a solid due to the application of external forces, the next step is to describe the response of the solid (i.e., its deformation) in terms of the resulting *strains* ε created within the solid. There are in general six strain components ε_i, $i = 1$ to 6, analogous to the six stress components σ_i defined in Eq. (10.3). The ε_i are illustrated using Fig. 10.3, where a solid with initial volume $V_0 = x_{10}x_{20}x_{30}$ and with edges lying initially along the x_1, x_2, and x_3 coordinate axes is shown both before and after a general deformation. This general *deformation* can be described as follows:

$$\mathbf{x}_1 = \mathbf{x}_{10} + \mathbf{u}_1, \qquad \mathbf{x}_2 = \mathbf{x}_{20} + \mathbf{u}_2, \qquad \mathbf{x}_3 = \mathbf{x}_{30} + \mathbf{u}_3, \tag{10.5}$$

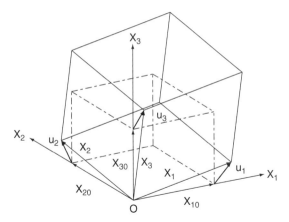

Figure 10.3. Solid with an initial volume $V_0 = x_{10}x_{20}x_{30}$ both before and after undergoing a general deformation. Also shown are the vector displacements $\mathbf{u}_1$, $\mathbf{u}_2$, and $\mathbf{u}_3$ of the three edges that are used to define the strains ε.

where the vectors $\mathbf{x}_{10}$, $\mathbf{x}_{20}$, and $\mathbf{x}_{30}$ and $\mathbf{x}_1$, $\mathbf{x}_2$, and $\mathbf{x}_3$ define the edges of the solid before and after the deformation, respectively. The vectors $\mathbf{u}_1$, $\mathbf{u}_2$, and $\mathbf{u}_3$ are therefore the displacements corresponding to the deformation, as shown. The components of $\mathbf{u}_1$ along the three coordinate axes are u_{11}, u_{12}, and u_{13}.

The three *normal strain* components $\varepsilon_i(= \varepsilon_{ii})$ are defined in terms of the displacements u_{ii} by

$$\varepsilon_1(= \varepsilon_{11}) = \frac{u_{11}}{x_{10}}, \qquad \varepsilon_2(= \varepsilon_{22}) = \frac{u_{22}}{x_{20}}, \qquad \varepsilon_3(= \varepsilon_{33}) = \frac{u_{33}}{x_{30}}. \qquad (10.6)$$

The strains ε_i as defined are fractional changes in length and so are dimensionless quantities. Deformations for which $\varepsilon_i > 0$ are *elongations*, while those for which $\varepsilon_i < 0$ are *compressions*.

An alternative definition for normal strain involves the changes in length relative to the instantaneous length l instead of relative to the initial length l_0. Following the definition given in Eq. (10.6), the *nominal strain* ε can be defined by $\Delta l/l_0 = (l - l_0)/l_0$. In terms of changes relative to the instantaneous length l, the *true strain* $\varepsilon_{\text{true}}$ is defined as follows:

$$\varepsilon_{\text{true}} = \int_{l_0}^{l} \frac{dl}{l_0} = \ln \frac{l}{l_0} = \ln \frac{l_0 + \Delta l}{l_0} = ln \left(1 + \frac{\Delta l}{l_0} \right) = \ln(1 + \varepsilon). \qquad (10.7)$$

In the limit of very small strains (i.e., $\varepsilon \ll 1$) it can be seen that the nominal strain ε and $\varepsilon_{\text{true}}$ are equivalent. This definition of true strain has the advantage that consecutive true strains $\varepsilon_{\text{true}}$ (1) and $\varepsilon_{\text{true}}$ (2) are additive; that is,

$$\varepsilon_{\text{true}} \text{ (total)} = \varepsilon_{\text{true}}(1) + \varepsilon_{\text{true}}(2) = \ln \frac{l_1}{l_0} + \ln \frac{l_2}{l_1} = \ln = \frac{l_2}{l_0}. \qquad (10.8)$$

The same is not true for the sum of the nominal strains $\varepsilon_1 + \varepsilon_2$, as defined in Eq. (10.6).

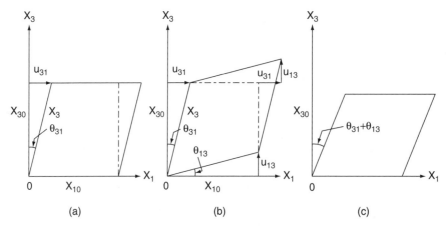

Figure 10.4. Components of shear strain in a solid defined using schematic diagrams. The deformation shown in (a) corresponds to the displacement u_{31} and is equivalent to a rotation of the side x_{30} through an angle θ_{31}. In (b) an additional displacement u_{13} is shown. For small strains, the total displacement shown in (b) is equivalent to a rotation of x_{30} (or x_{10}) through a total angle $\theta_{31} + \theta_{13}$, as shown in (c).

For the definitions of the *shear strain* components, consider Fig. 10.4a, in which the side view of a solid is shown in the x_1x_3 plane. The specific deformation shown here corresponds to a displacement u_{31} of the end of the side x_{30} in the x_1 direction. This deformation can be seen to be equivalent to a rotation of the side x_{30} through an angle θ_{31}. The corresponding shear strain component is defined by

$$\varepsilon_{31} = \frac{u_{31}}{x_{30}} = \tan\theta_{31} \approx \theta_{31}. \tag{10.9}$$

Note that $\varepsilon_{31} \approx \theta_{31}$ for small strains (i.e., for $u_{31}/x_{30} \ll 1$).

In Fig. 10.4b an additional displacement u_{13} is shown. For small strains, the total displacement is equivalent to a rotation of x_{30} (or x_{10}) through a total angle $\theta_{31} + \theta_{13}$, as shown in Fig. 10.4c. The resulting total shear strain component therefore can be defined as follows:

$$\varepsilon_5 (= \varepsilon_{31} = \varepsilon_{13}) = \frac{u_{31}}{x_{30}} + \frac{u_{13}}{x_{10}}. \tag{10.10}$$

The two remaining shear strain components ε_4 and ε_6 are defined similarly:

$$\varepsilon_4 (= \varepsilon_{32} = \varepsilon_{23}) = \frac{u_{32}}{x_{30}} + \frac{u_{23}}{x_{20}},$$
$$\varepsilon_6 (= \varepsilon_{21} = \varepsilon_{12}) = \frac{u_{21}}{x_{20}} + \frac{u_{12}}{x_{10}}. \tag{10.11}$$

It follows from these definitions that the *strain tensor* ε is also symmetric, with the form

$$\varepsilon = \begin{pmatrix} \varepsilon_{11} & \varepsilon_{12} & \varepsilon_{13} \\ \varepsilon_{12} & \varepsilon_{22} & \varepsilon_{23} \\ \varepsilon_{13} & \varepsilon_{23} & \varepsilon_{33} \end{pmatrix} = \begin{pmatrix} \varepsilon_1 & \varepsilon_6 & \varepsilon_5 \\ \varepsilon_6 & \varepsilon_2 & \varepsilon_4 \\ \varepsilon_5 & \varepsilon_4 & \varepsilon_3 \end{pmatrix}. \tag{10.12}$$

10.4 Relationships Between Stress and Strain: Elastic Constants

Linear elasticity refers in general to a relationship between two second-rank tensors, the stress tensor σ_{ij} and the strain tensor ε_{mn}, through the fourth-rank *elastic coefficient tensor* C_{ijmn}; that is,

$$\sigma_{ij} = \Sigma_{m,n} C_{ijmn} \varepsilon_{mn}. \tag{10.13}$$

Due to the symmetry mentioned earlier of σ_{ij} and of ε_{mn}, the number of independent elements of C_{ijmn} is reduced from 81 to 36 in general. Hence the following contracted notations are used:

$$\sigma_i = \Sigma_j C_{ij} \varepsilon_j, \tag{10.14a}$$

$$\varepsilon_i = \Sigma_j S_{ij} \sigma_j, \tag{10.14b}$$

with $i, j = 1$ to 6. The relationships between the stresses σ and the resulting strains ε in a material can thus be expressed using either the *elastic constants* C_{ij} or the *elastic compliances* S_{ij} of the material. This terminology is not unique since the S_{ij} have also been called the elastic constants, while the C_{ij} have also been called the *elastic coefficients*, the *elastic moduli*, or the *elastic stiffness constants*. In addition, the symbol C is sometimes used for the elastic compliances. It should be clear from the context which notation is being used.

Equations (10.14a) and (10.14b) are generalized versions of Hooke's law, discussed later. Specific components of the stress and strain (e.g., ε_1 and σ_1), are given by

$$\begin{aligned}
\varepsilon_1 &= S_{11}\sigma_1 + S_{12}\sigma_2 + S_{13}\sigma_3 + S_{14}\sigma_4 + S_{15}\sigma_5 + S_{16}\sigma_6, \\
\sigma_1 &= C_{11}\varepsilon_1 + C_{12}\varepsilon_2 + C_{13}\varepsilon_3 + C_{14}\varepsilon_4 + C_{15}\varepsilon_5 + C_{16}\varepsilon_6.
\end{aligned} \tag{10.15}$$

In Eq. (10.14a) the stresses are determined from the strains, whereas in Eq. (10.14b) the reverse is true. Although the second approach is more logical, the mechanical properties of materials are more often expressed using the elastic constants C_{ij}. It is clear that the two approaches are closely related since the matrices S_{ij} and C_{ij} are inverses of each other.

The region in which the strains σ and the stresses ε are directly proportional to each other [i.e., in which the S_{ij} are independent of the stresses σ_j and the C_{ij} are independent of the strains ε_j (and both S_{ij} and C_{ij} are independent of time)], is known as the *proportional elastic limit*. The mechanical response of the material in this limit obeys *Hooke's law*, which states that the response of the material (i.e., the deformation or strain) is directly proportional to the force or stress applied. When the S_{ij} depend on the σ_j or when the C_{ij} depend on the ε_j, the material is said to be *inelastic*. When the S_{ij} or C_{ij} are time dependent, the material is said to be *anelastic*.

Four different types of mechanical behavior observed for materials are illustrated in Fig. 10.5 in the usual form of curves of nominal stress versus nominal strain. These curves are determined using a procedure known as *tensile testing*, taken with the material in the form of a bar and under tension. In Fig. 10.5a the response of alumina, Al_2O_3, can be seen to be elastic initially, followed immediately by brittle fracture. In Fig. 10.5b the elastic region observed for a low-carbon steel is followed by an extensive region of plastic deformation and then finally, by ductile fracture. In Fig. 10.5c the type of elastomeric response found for a cross-linked natural rubber is shown, while in Fig. 10.5d the viscoelastic behavior of a plastic polymer, polymethylmethacrylate (PMMA), is shown.

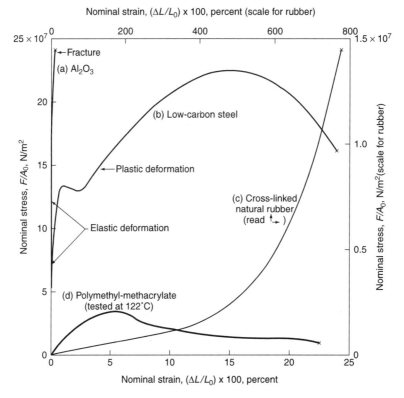

Figure 10.5. Four examples of the mechanical response of materials corresponding to (*a*) elastic, (*b*) plastic, (*c*) elastomeric, and (*d*) viscoelastic behavior are illustrated using nominal stress–nominal strain curves. (From A. G. Guy, *Introduction to Materials Science*, copyright 1972 by McGraw-Hill. Reprinted by permission of the McGraw-Hill Companies.)

It should be noted that within the elastic limit, typical strains in materials such as metals and ceramics extend up to about 10^{-3}, beyond which point inelasticity involving permanent deformations and ultimately, fracture can occur. In other materials, such as polymers, elastic strains up to unity or beyond are not unusual. Clearly, as illustrated for the materials shown here, the mechanical response of materials can be quite complicated. The elastomeric and viscoelastic behaviors of polymers are discussed in Chapter 14.

While a solid material in principle has 36 independent elastic compliances S_{ij} and also 36 independent elastic constants C_{ij}, in practice the actual number of independent S_{ij} and C_{ij} is always less than 36 (and can even be equal to 2 for isotropic solids). The 36 elastic constants C_{ij} of a single crystal of a uniform, homogeneous material are conveniently displayed in a 6×6 matrix, as follows:

$$C_{ij} = \begin{pmatrix} C_{11} & C_{12} & C_{13} & C_{14} & C_{15} & C_{16} \\ C_{21} & C_{22} & C_{23} & C_{24} & C_{25} & C_{26} \\ C_{31} & C_{32} & C_{33} & C_{34} & C_{35} & C_{36} \\ C_{41} & C_{42} & C_{43} & C_{44} & C_{45} & C_{46} \\ C_{51} & C_{52} & C_{53} & C_{54} & C_{55} & C_{56} \\ C_{61} & C_{62} & C_{63} & C_{64} & C_{65} & C_{66} \end{pmatrix} \tag{10.16}$$

There are several reasons why the actual number of independent elastic constants needed to describe the mechanical response of a material is reduced below the maximum possible number of 36.

1. The first reduction is independent of the structure of the material and results from a consideration of the elastic energy stored in the material due to the strains. This argument demonstrates that the C_{ij} matrix is symmetric; that is,

$$C_{ij} = \begin{pmatrix} C_{11} & C_{12} & C_{13} & C_{14} & C_{15} & C_{16} \\ C_{12} & C_{22} & C_{23} & C_{24} & C_{25} & C_{26} \\ C_{13} & C_{23} & C_{33} & C_{34} & C_{35} & C_{36} \\ C_{14} & C_{24} & C_{34} & C_{44} & C_{45} & C_{46} \\ C_{15} & C_{25} & C_{35} & C_{45} & C_{55} & C_{56} \\ C_{16} & C_{26} & C_{36} & C_{46} & C_{56} & C_{66} \end{pmatrix} \tag{10.17}$$

Thus only 21 independent elastic constants are sufficient in general for the description of the mechanical response of a material in the elastic limit.

2. Further reductions in the number of independent C_{ij} depend on the specific crystal structure and microstructure of the material in question. For materials in the form of single crystals, the number of independent C_{ij} decreases below 21 as the symmetry of the Bravais lattice of the crystal structure increases. This is due to restrictions placed on the elastic constants by symmetry relations.

The number of independent elastic constants for crystals with different Bravais lattices is indicated in Table 10.1. The essential symmetry for each lattice is also given.

Note that triclinic crystals with the most general crystal structure retain 21 independent elastic constants, while for isotropic solids such as fine-grained polycrystalline solids, amorphous solids, or glasses, only two elastic constants are sufficient. The elastic constant tensor for α-quartz, a trigonal crystal, is presented in Chapter W15 at our Web site[†] and can be seen to contain six independent elastic constants. For a

TABLE 10.1 Number of Independent Elastic Constants C_{ij} for Crystals with Different Bravais Lattices

Bravais Lattice	Essential Symmetry	Number of Independent C_{ij}
Triclinic	None	21
Monoclinic	One twofold axis	13
Orthorhombic	Two orthogonal twofold axes	9
Tetragonal	One fourfold axis	6 or 7
Trigonal (rhombohedral)	One threefold axis	6 or 7
Hexagonal	One sixfold axis	5
Cubic	Four threefold axes	3
(Isotropic)		2

[†] Supplementary material for this textbook is included on the Web at the resource site (ftp://ftp.wiley.com/public/sci_tech_med/materials). Cross-references to elements of the Web material are prefixed by "W."

cubic crystal only three elastic constants are needed and the matrix of elastic constants becomes

$$
C_{ij} = \begin{pmatrix}
C_{11} & C_{12} & C_{12} & 0 & 0 & 0 \\
C_{12} & C_{11} & C_{12} & 0 & 0 & 0 \\
C_{12} & C_{12} & C_{11} & 0 & 0 & 0 \\
0 & 0 & 0 & C_{44} & 0 & 0 \\
0 & 0 & 0 & 0 & C_{44} & 0 \\
0 & 0 & 0 & 0 & 0 & C_{44}
\end{pmatrix}
\tag{10.18}
$$

When only pairwise central forces exist in cubic crystals and when each atom is at a center of symmetry, it has been shown by Cauchy that the relation

$$
C_{12} = C_{44}
\tag{10.19}
$$

is valid. The interatomic forces will then be spherically symmetric and nondirectional. The *Cauchy relation* is usually not observed to be valid for cubic crystals, as illustrated in Section 10.5, where measured values of the elastic constants are presented.

For cubic solids that are isotropic, the relation

$$
C_{44} = \frac{C_{11} - C_{12}}{2}
\tag{10.20}
$$

is always valid. The number of independent elastic constants is therefore reduced to just two for these materials. This relation can be derived by requiring that the elastic constants of a cubic crystal be independent of any rotation.[†] It should be noted that a cubic crystal can be elastically isotropic without having central forces, and vice versa. If a cubic crystal is isotropic and has central forces, it follows that $C_{11} = 3C_{12} = 3C_{44}$ (i.e., there would be only one independent elastic constant).

ELASTIC PROPERTIES OF MATERIALS

The mechanical properties of materials in the elastic limit are discussed next. As stated previously, the response of a solid is said to be elastic when the deformations or strains due to external forces disappear completely (i.e., the solid returns to its initial state) as soon as the external forces are removed. The law describing the elastic behavior of materials, Hooke's law, is discussed in detail. The elastic properties of materials known as Young's modulus E, the shear modulus G, the bulk modulus B, and Poisson's ratio v are defined. The elastic behavior of isotropic solids is discussed as are elastic potential energy and elastic waves in solids.

10.5 Hooke's Law, Young's Modulus, and Shear Modulus

The laws governing the elastic behavior of metals in response to the application of external forces were investigated by Robert Hooke in 1678. His results for the behavior of metals in the elastic limit are summarized below.

[†] For details of the derivation of Eq. (10.20) for the case of a 45° rotation about the z axis, see Meyers and Chawla (1984, pp. 52–54).

The change in length Δl of a metal sample in a given direction increases in direct proportion to the external force F acting in that direction, in direct proportion to its initial length l_0, and in inverse proportion to its initial cross-sectional area A_0 (i.e., $\Delta l \propto Fl_0/A_0$). These observations are summarized in *Hooke's law*, which can be expressed as

$$\Delta l = \frac{SFl_0}{A_0} = \frac{Fl_0}{EA_0}. \tag{10.21}$$

The constant of proportionality in Hooke's law which determines the change in length Δl for a particular material is the elastic compliance S, defined in Eq. (10.14b). Note that $S = 1/E$, where E is known as *Young's modulus*, the *modulus of elasticity*, or the *elastic modulus* of the material in the direction in question. The simple relationship $E = 1/S$, or $E_{ii} = 1/S_{ii}$, between Young's modulus and the elastic compliance S_{ii} is valid for normal stresses and strains only under special conditions. These include the conditions normally used in tensile testing where the external forces are applied along the axis of a uniform bar (i.e., when only a single, uniaxial normal stress component is present in the material). In general, the relationship between E_{ii} and the elastic constant C_{ii} is not as simple. For example, Young's modulus for the set of $\langle 100 \rangle$ directions in a cubic crystal is

$$E_{11} = \frac{1}{S_{11}} = \frac{(C_{11} - C_{12})(C_{11} + 2C_{12})}{C_{11} + C_{12}}. \tag{10.22}$$

Since the nominal normal strain and stress can be defined by $\varepsilon = \Delta l/l_0$ and $\sigma = F/A_0$, respectively, it follows that Hooke's law can also be written in the form

$$\varepsilon_i = \frac{\sigma_i}{E} \quad \text{or} \quad \sigma_i = E\varepsilon_i \quad (i = 1, 2, 3) \tag{10.23}$$

when only a single stress component is acting.[†] Note that Young's modulus E for a material has the same units as stress, Pa, and is equal to the slope of the initial, proportional region of the corresponding stress–strain curve, the region in which Hooke's law is valid (see Fig. 10.5). Evidence that elastic behavior is not always linear has been presented in Fig. 10.5c where the elastomeric response of cross-linked natural rubber is shown.

The form of Hooke's law appropriate for shear stresses and strains is the same as that for normal stresses and strains given in Eq. (10.23), that is,

$$\varepsilon_i = \frac{\sigma_i}{G} \quad \text{or} \quad \sigma_i = G\varepsilon_i, \quad (i = 4, 5, 6) \tag{10.24}$$

where the shear strain ε_i, the shear stress σ_i, and the *shear modulus* G all refer to the same direction in the material. Hooke's law as expressed in Eqs. (10.23) and (10.24) is valid only for normal or shear stresses and strains acting in the same direction. In general, however, a deformation in a solid can also occur in a direction different from that of the stress creating it. This type of behavior has been expressed in the generalized versions of Hooke's law given in Eq. (10.14).

[†] The form of Hooke's law usually given for the change in length Δl of an elastic spring is $F = k\Delta l = EA_0\Delta l/l_0$. The "spring" or stiffness constant $k = EA_0/l_0$ can thus be seen to depend not only on the stiffness of the material through E but also on the dimensions and shape of the spring through A_0 and l_0.

TABLE 10.2   Elastic Moduli and Poisson's Ratio for Polycrystalline Cubic Metals at Room Temperature

Material	Young's Modulus E (GPa)[a]	Shear Modulus G (GPa)[a]	Bulk Modulus B (GPa)[a]	Poisson's Ratio v
Al	70.3	26.1	76.0	0.345
Cr	279	115	194	0.210
Cu	130	48.3	134	0.343
Fe	211	81.6	172	0.293
Nb	105	37.5	167	0.397
Ta	186	69.2	211	0.342
W	411	161	319	0.280

Source: Data from M. A. Meyers and K. K. Chawla, *Mechanical Metallurgy*, Prentice Hall, Upper Saddle River, N. J., 1984, p. 58, Table 1.5; N. W. Ashcroft and N. D. Mermin, *Solid State Physics*, Saunders College, Philadelphia, 1976, p. 39, Table 2.2; and M. L. Bernstein and V. A. Zaimovsky, *Mechanical Properties of Metals*, Mir Publishers, Moscow, 1988, p. 45, Table 6.

[a] 1 GPa = 10^9 N/m^2.

Values of the elastic moduli and of Poisson's ratio v, to be defined later, are given for several polycrystalline cubic metals in Table 10.2. It is interesting to note that the ratio E/G of Young's modulus to the shear modulus for the metals listed is essentially constant, $E/G \approx 2.6 \pm 0.2$, and that Poisson's ratio $v \approx 0.3 \pm 0.1$ for these metals. Finally, the bulk modulus B is roughly proportional to the cohesive energy ΔH_c and to the melting temperature T_m of the metal (see Table 6.3).

The elastic constants for various single crystals are given in Table 10.3. Also listed in this table are values of the *anisotropy ratio* $A = 2C_{44}/(C_{11} - C_{12})$ for crystals with cubic crystal structures.

It should be noted that $C_{11} > C_{12}$ for all the materials listed in Table 10.3, indicating that it is in general harder to generate a tensile strain than a shear strain in these materials. For the hexagonal materials listed, it is found that $C_{33} > C_{11}$, indicating a somewhat higher stiffness parallel to the c axis than in the basal plane.

The Cauchy relation $C_{12} = C_{44}$ can be seen from Table 10.3 to hold reasonably well only for ionic crystals (e.g., the alkali halides LiCl and NaCl). This is reasonable since the electrostatic forces in ionic crystals can be expected to be Coulomb-like and hence central. The fact that the Cauchy relation fails for the metallic and covalent cubic crystals listed above indicates that the interatomic bonding interactions in these materials are nonspherical. This is certainly the case for the directional bonding found in covalent crystals. The best examples from Table 10.3 of cubic crystals which are both isotropic and have central forces, for which $C_{11} = 3C_{12} = 3C_{44}$, seem to be metallic W with $C_{11} = 2.53C_{12} = 3.32C_{44}$ and ionic NaCl with $C_{11} = 3.93C_{12} = 3.87C_{44}$.

Diamond is a unique material for many reasons, including its mechanical properties. Young's modulus for diamond is $E \approx 1100$ GPa, its bulk modulus is $B = 443$ GPa, its shear modulus is $G = 535$ GPa, and its Poisson's ratio is only $v = 0.07$, values that are much higher and much lower, respectively, than the values given for the polycrystalline metals in Table 10.2. The high value of E can be attributed to the rigidity of diamond, specifically to the very high values of C_{11} and C_{44} (see Table 10.3).

TABLE 10.3   Elastic Constants for Various Single Crystals at Room Temperature

Material	Crystal Structure	Elastic Constant (GPa)						A^a
		C_{11}	C_{12}	C_{13}	C_{33}	C_{44}	C_{66}	
Na	BCC	7.0	6.1	—	—	4.5	—	10
K	BCC	3.7	3.14	—	—	1.88	—	6.7
Be	HCP	292	26.7	14	336	163	—	—
Mg	HCP	59.7	26.2	21.7	61.7	16.7	—	—
Al	FCC	107	61	—	—	28	—	1.22
In	Tetragonal	44.5	39.5	40.5	44.4	6.6	12.2	—
C	Diamond	1076	125	—	—	576	—	1.19
Si	Diamond	166	64	—	—	80	—	1.57
Cu	FCC	168	121	—	—	75	—	3.2
Au	FCC	186	157	—	—	42	—	2.9
V	BCC	229	119	—	—	43	—	0.78
Ta	BCC	267	161	—	—	82	—	1.55
W	BCC	501	198	—	—	151	—	1.02
Fe	BCC	234	136	—	—	118	—	2.41
Co	HCP	307	165	103	358	75.3	—	—
LiCl	NaCl	49.4	22.8	—	—	24.6	—	1.80
NaCl	NaCl	48.7	12.4	—	—	12.6	—	0.88

Source: Data from M. A. Meyers and K. K. Chawla, *Mechanical Metallurgy*, Prentice Hall, Upper Saddle River, N. J., 1984, p. 57, Table 1.3; N. W. Ashcroft and N. D. Mermin, *Solid State Physics*, Saunders College, Philadelphia, 1976, p. 447, Table 22.2; and C. Kittel, *Introduction to Solid State Physics*, 7th ed., Wiley, New York, 1996, p. 91, Table 11.

[a]A is the anisotropy ratio for cubic crystals, defined by $A = 2C_{44}/(C_{11} - C_{12})$.

Since the macroscopic deformation of a solid reflects the displacements of individual atoms from their equilibrium positions, it should not be surprising that the elastic response of a solid is determined by the nature of the interactions between neighboring atoms. A discussion of the relationship between Hooke's law and the potential energy of interaction $U(r)$ for a pair of atoms is presented in Chapter W10.

10.6   Compressibility and Bulk Modulus

A solid material that is subjected to an inward-directed hydrostatic pressure P (i.e., to an external pressure that is uniform on every surface of the solid) will be compressed. The change in volume $\Delta V = V - V_0$ is negative. In the elastic limit the original volume V_0 and the shape of the solid will be recovered soon after external pressure is removed. The response of a material to hydrostatic pressure in the elastic limit is determined by its *isothermal bulk modulus B* or, alternatively, by its *isothermal compressibility* $K = 1/B$, where

$$K = -\frac{1}{V}\left(\frac{\partial V}{\partial P}\right)_T. \tag{10.25}$$

The bulk modulus[†] of a cubic material is given in terms of its elastic constants by

$$B = \frac{C_{11} + 2C_{12}}{3}. \tag{10.26}$$

[†] The symbol K is often used for the bulk modulus.

Values of the bulk modulus B for several polycrystalline cubic metals are given in Table 10.2. For metals much of the bulk modulus arises from the resistance to compression of the conduction electrons due to Coulomb effects. The expression $B = 2nE_F/3$ can be derived for free electrons (see Chapter 12). Here n is the conduction electron concentration and E_F is the Fermi energy. When typical values for n and E_F are used, values of $B \approx 10$ to 200 GPa are obtained which are of the same order of magnitude as the values observed (see Tables 10.2 and 12.1).

10.7 Poisson's Ratio

When a solid is in tension (or compression) due to a stress acting along a given direction, the dimensions of the solid in general will change not only along the direction in question but also in the two orthogonal directions. As an example, consider a solid material with a cubic, tetragonal, or orthorhombic crystal structure and with a tensile stress acting along a crystal axis that is taken to be the z axis (Fig. 10.6). The changes in the dimensions of the solid along the three axes are $2\Delta x$, $2\Delta y$, and $2\Delta z$, with Δx and Δy both usually < 0 and $\Delta z > 0$.

The ratios of the transverse strains $\varepsilon_x = 2\Delta x/x_0$ or $\varepsilon_y = 2\Delta y/y_0$ to the longitudinal strain $\varepsilon_z = 2\Delta z/z_0$ in the solid are known as *Poisson's ratios*, given here by

$$v(z, x) = -\frac{\varepsilon_x}{\varepsilon_z} \quad \text{and} \quad v(z, y) = -\frac{\varepsilon_y}{\varepsilon_z}. \tag{10.27}$$

Note that v will in general depend on the specific longitudinal and transverse directions chosen. For isotropic solids that experience no change in volume ΔV under the application of stress, Poisson's ratio is $v = \frac{1}{2}$. Similarly, $v = -1$ for isotropic solids when there is no change of shape. In fact, $-1 < v < \frac{1}{2}$ for isotropic solids. Values of Poisson's ratio for several polycrystalline metals are given in Table 10.2, where it can

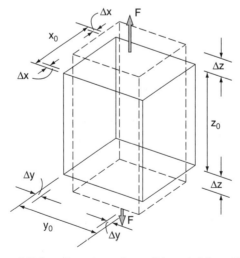

Figure 10.6. Definition of Poisson's ratio v for solid materials with cubic, tetragonal, or orthorhombic structures. The changes in the dimensions of the solid along the three axes are shown.

be seen that ν is typically in the range 0.2 to 0.4. Thus $\Delta V > 0$ for these isotropic metals.

As stated earlier, the values of Poisson's ratio for a mechanically anisotropic single crystal can depend on the direction along which the tensile stress is applied. In the case of BCC-like CuZn (i.e., β-brass), $\nu(100) = +0.39$ while $\nu(111) = -0.39$. Thus a tensile stress in one of the $\langle 111 \rangle$ directions in a single crystal of β-brass actually causes an expansion of the crystal in the transverse direction. Negative Poisson's ratios for $\langle 110 \rangle$ directions have been predicted, on the basis of the measured elastic constants, to be common in mechanically anisotropic elemental cubic metals. In particular, a hard-sphere model for solids with BCC crystal structures indicates that $\nu(110, 1\bar{1}0) = -1$ and $\nu(110, 001) = +2$. Such large negative or positive values for ν are not, however, expected on the basis of hard-sphere models for FCC and HCP solids.

10.8 Isotropic Solids: Relationships Between the Elastic Moduli

Many solid materials are isotropic on a macroscopic scale (i.e., their physical properties are the same in all directions), even though they are anisotropic on a microscopic scale. A typical example of an isotropic material is a fine-grained polycrystalline solid in which the individual crystallites are randomly oriented. The crystallites are individually anisotropic since they retain the symmetry of the crystal. Other examples of isotropic solids include glasses and other amorphous materials.

A single crystal of a cubic material can be isotropic with respect to its mechanical properties if its elastic constants obey the relation $C_{44} = (C_{11} - C_{12})/2$ given in Eq. (10.20). Single crystals of cubic solids such as Al, C, W, and NaCl are nearly isotropic, as can be seen in Table 10.3, where values of the anisotropy ratio $A = 2C_{44}/(C_{11} - C_{12})$ are given. These values of A have been calculated from the elastic constants listed in the same table. Note that a cubic crystal with $A = 1$ will be elastically isotropic.

Parameters such as Young's modulus E and the shear modulus G ordinarily used to characterize the elastic properties of isotropic solids are simply related to the elastic constants C_{ij} of the corresponding single crystal and to each other. Some useful relationships are listed in Table 10.4. It should be noted that any two parameters (e.g., C_{11} and C_{12} or E and ν) are sufficient for the specification of the elastic properties of isotropic solids. Another pair of parameters that are sometimes used to describe the elastic properties of isotropic solids are Lamé's constants, $\mu = C_{44} = (C_{11} - C_{12})/2$ and $\lambda = C_{12}$.

TABLE 10.4   Relationships Between the Elastic Moduli of Isotropic Solids

Young's modulus:	$E = \dfrac{1}{S_{11}} = \dfrac{(C_{11} - C_{12})(C_{11} + 2C_{12})}{C_{11} + C_{12}} = \dfrac{\mu(2\mu + 3\lambda)}{\mu + \lambda}$
Shear modulus:	$G = \dfrac{1}{2(S_{11} - S_{12})} = \dfrac{C_{11} - C_{12}}{2} = C_{44} = \dfrac{E}{2(1 + \nu)} = \mu$
Bulk modulus:	$B = \dfrac{C_{11} + 2C_{12}}{3} = \dfrac{E}{3(1 - 2\nu)} = \lambda + \dfrac{2\mu}{3}$
Poisson's ratio:	$\nu = -\dfrac{S_{12}}{S_{11}} = \dfrac{C_{12}}{C_{11} + C_{12}} = \dfrac{E}{2G} - 1 = \dfrac{\lambda}{2(\mu + \lambda)}$

The values of E, G, and ν given in Table 10.2 are consistent with the expressions listed in Table 10.4. The same cannot be said for the values of B also given in Table 10.2, which were obtained from different sources. From the expression given in Table 10.4 for the bulk modulus B in terms of E and ν, it can be seen that $B \to \infty$ for $\nu \to \frac{1}{2}$. In this case the solid is incompressible, which is consistent with the fact that $\Delta V = 0$ when $\nu = \frac{1}{2}$. From the expressions given in Table 10.4 for B and the shear modulus G, it can be seen that $-1 \leq \nu \leq \frac{1}{2}$ for isotropic solids. Also, when $\nu \approx \frac{1}{3}$, then $G \approx 3E/8$ and $B \approx E$. These results are consistent with the data presented for the polycrystalline metals in Table 10.2.

The expressions listed in Table 10.4 do not always yield the elastic moduli actually measured for polycrystalline metals when the elastic constants of the corresponding single crystals from Table 10.3 are used. This is illustrated by the fact that the relationship $G = C_{44}$ is not always obeyed when values of G for the polycrystalline cubic metals listed in Table 10.2 are compared with the values of C_{44} for single crystals of the same metals from Table 10.3. The relationship $G = C_{44}$ does work reasonably well, however, for Al and W due to the fact that even single crystals of these metals are nearly isotropic. A possible explanation for these discrepancies is that many polycrystalline materials are not truly isotropic but instead are textured (i.e., they contain regions in which the crystallites have preferred orientations).

10.9   Elastic Potential Energy

In the elastic limit the work done by external forces in the deformation of a solid is completely converted into elastic energy [i.e., the potential energy of interaction $U(r)$ of the atoms]. As long as the elastic limit is not exceeded, this elastic energy can be completely recovered when the external forces are removed. In the inelastic region most of the work done by external forces cannot be recovered and is instead converted to heat or sound (i.e., vibrations of the atoms or ions).

To determine the dependence of the elastic potential energy or strain energy $U(\varepsilon)$ on the strains present in a material, consider the work W done by an external force during an elastic deformation in the proportional region. The deformation consists of an elongation by an amount Δx of a solid bar of original cross-sectional area A_0 and original length x_0. Using Hooke's law in the form $F = EA_0\Delta x/x_0$ from Eq. (10.21) and the relationship $\Delta x = \varepsilon x_0$, the work can be calculated as follows:

$$W = \int_0^{\Delta x} F\, d(\Delta x) = \int_0^{\varepsilon} Fx_0\, d\varepsilon = EA_0x_0 \int_0^{\varepsilon} \varepsilon d\varepsilon = \frac{E\varepsilon^2 V_0}{2}, \qquad (10.28)$$

where $V_0 = A_0x_0$ is the initial volume of the bar. It follows that the work done per unit volume $w = W/V_0$ and hence the *elastic energy density* $u_{\text{el}}(\varepsilon) = U_{\text{el}}(\varepsilon)/V_0$ are given by

$$w = u_{\text{el}}(\varepsilon) = \frac{E\varepsilon^2}{2} = \frac{\sigma\varepsilon}{2}. \qquad (10.29)$$

The expression for $u_{\text{el}}(\varepsilon)$ when all six strain components $\sigma_i = E_i\varepsilon_i$ are involved is

$$u_{\text{el}}(\varepsilon) = \frac{\Sigma_i E_i\varepsilon_i^2}{2} = \frac{\Sigma_i \sigma_i\varepsilon_i}{2}. \qquad (10.30)$$

The general expression for $u_{el}(\varepsilon)$ in the elastic limit when stresses and strains in different directions may be coupled is

$$u_{el}(\varepsilon) = \frac{\Sigma_i \Sigma_j C_{ij}\varepsilon_j \varepsilon_i}{2} = \frac{\Sigma_i \sigma_i \varepsilon_i}{2},$$

(10.31)

where the expression $\sigma_i = \Sigma_i C_{ij}\varepsilon_j$ has been used. For a cubic crystal

$$u_{el}(\varepsilon) = \frac{C_{11}(\varepsilon_1^2 + \varepsilon_2^2 + \varepsilon_3^2)}{2} + C_{12}(\varepsilon_1\varepsilon_2 + \varepsilon_1\varepsilon_3 + \varepsilon_2\varepsilon_3)$$

$$+ \frac{C_{44}(\varepsilon_4^2 + \varepsilon_5^2 + \varepsilon_6^2)}{2}.$$

(10.32)

10.10 Elastic Waves

The propagation of *elastic waves* in a solid is another example of the dependence of the physical properties of a material on its mechanical properties. The quantized excitations of the lattice known as phonons are discussed in detail in Chapter 5. The elastic waves in question correspond to the long-wavelength ($\lambda \gg d$, the interatomic separation) sound waves which propagate through a solid due to the action of time-varying forces. In this long-wavelength limit the solid can be treated as a continuous medium rather than as a discrete set of atoms.

Consider the situation where a net force F acts on a volume element ΔV in a solid (Fig. 10.7). In this case the stress is taken to be inhomogeneous (i.e., it varies from point to point in the solid and therefore has components σ_i with different magnitudes on opposite faces of ΔV). As a result, the strain in the solid will also be inhomogeneous.

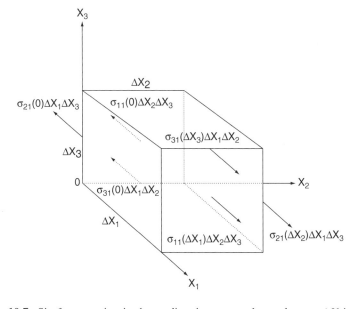

Figure 10.7. Six forces acting in the x_1 direction on a volume element ΔV in a solid.

The equation of motion in the x_1 direction for the center of mass of the element ΔV can be written as

$$F_1 = m \frac{\partial^2 u_1}{\partial t^2} = \Delta \sigma_1 \Delta x_2 \Delta x_3 + \Delta \sigma_5 \Delta x_1 \Delta x_2 + \Delta \sigma_6 \Delta x_1 \Delta x_3 \qquad (10.33)$$

where $m = \rho \Delta V$ is the mass of the volume element and u_1 is the displacement of the center of mass of ΔV resulting from the net force F_1 acting in the x_1 direction. The quantity $\Delta \sigma_1 \, (= \Delta \sigma_{11})$ is the net normal stress acting in the x_1 direction; $\Delta \sigma_5 (= \Delta \sigma_{31})$ and $\Delta \sigma_6 (= \Delta \sigma_{21})$ are the net shear stresses acting in the same direction. The net force F_1 thus includes contributions from forces acting on all six faces of ΔV. After dividing Eq. (10.33) by $\Delta V = \Delta x_1 \Delta x_2 \Delta x_3$ and setting $\Delta \sigma_1 / \Delta x_1 = \partial \sigma_1 / \partial x_1$, and so on, the equation of motion becomes

$$\rho \frac{\partial^2 u_1}{\partial t^2} = \frac{\partial \sigma_1}{\partial x_1} + \frac{\partial \sigma_5}{\partial x_3} + \frac{\partial \sigma_6}{\partial x_2}. \qquad (10.34)$$

Similar equations of motion for the center of mass of ΔV can be derived for the x_2 and x_3 directions.

For a cubic crystal it follows from Eqs. (10.18) and (10.34) that

$$\rho \frac{\partial^2 u_x}{\partial t^2} = C_{11} \frac{\partial \varepsilon_1}{\partial x} + C_{12} \left(\frac{\partial \varepsilon_2}{\partial x} + \frac{\partial \varepsilon_3}{\partial x} \right) + C_{44} \left(\frac{\partial \varepsilon_5}{\partial z} + \frac{\partial \varepsilon_6}{\partial y} \right) \qquad (10.35)$$

for motion in the x or [100] direction. It can be shown[†] that the general definitions of strains have the form $\varepsilon_1 = \partial u_x / \partial x$ and $\varepsilon_5 = \partial u_x / \partial z + \partial u_z / \partial x$, and so on. It follows that Eq. (10.35) can be written as

$$\rho \frac{\partial^2 u_x}{\partial t^2} = C_{11} \frac{\partial^2 u_x}{\partial x^2} + C_{44} \left(\frac{\partial^2 u_x}{\partial y^2} + \frac{\partial^2 u_x}{\partial z^2} \right) + (C_{12} + C_{44}) \left(\frac{\partial^2 u_y}{\partial x \partial y} + \frac{\partial^2 u_z}{\partial x \partial z} \right). \qquad (10.36)$$

This equation of motion for an elastic wave in a cubic crystal now has the form of the wave equation for the propagation of a deformation resulting from the inhomogeneity of the stress and strain. The analogous equations for the propagation of elastic waves in the y and z directions in cubic crystals are related to Eq. (10.36) by symmetry.

Simple solutions to this wave equation can be found for elastic waves traveling in the $\langle 100 \rangle$, $\langle 110 \rangle$, or $\langle 111 \rangle$ directions in a cubic crystal. First, consider elastic plane waves traveling in the [100] or positive x direction. A longitudinal wave will have both the direction of propagation and the atom displacements along the x direction, as follows:

$$u_x(t) = u_x(0) \exp[i(k_x x - \omega t)], \qquad (10.37)$$

where $k_x = 2\pi / \lambda$ is the wave vector. Substituting $u_x(t)$ in Eq. (10.36) yields the simple result that

$$\rho \omega^2 = C_{11} k_x^2. \qquad (10.38)$$

[†] See, for example, the derivation in Meyers and Chawla (1984, Chap. 1).

The velocity of this longitudinal wave is therefore $v_l(100) = \omega/k_x = \sqrt{C_{11}/\rho}$. Essentially the same result for the dispersion relation $\omega(k)$ has been obtained in Chapter 5 for the long-wavelength acoustic modes in the case of the hypothetical one-dimensional lattice. In that case the speed of sound is given by $c_s = \sqrt{Ka^2/M}$, where a is the NN distance, K the effective spring constant between NN ions, and M the ionic mass.

Transverse or shear elastic waves traveling in the [100] direction and with the atom displacements in the y or z direction have the forms

$$u_y(t) = u_y(0)\exp[i(k_x x - \omega t)], \tag{10.39}$$

or

$$u_z(t) = u_z(0)\exp[i(k_x x - \omega t)].$$

Substitution of either of these expressions into Eq. (10.36) yields

$$\rho\omega^2 = C_{44}k_x^2, \tag{10.40}$$

from which it follows that $v_t(100) = \sqrt{C_{44}/\rho}$ for shear waves polarized in the y or z direction.

Solutions of the wave equation for elastic waves traveling in the $\langle 110 \rangle$ and $\langle 111 \rangle$ directions in cubic crystals also correspond to the propagation of a single longitudinal wave and two transverse waves, with velocities listed in Table 10.5. These wave velocities for elastic waves correspond to the speed of sound in the elastic medium.

Based on the expressions listed in Table 10.5, typical values of the velocities of elastic waves in solids are found to lie in the range $(0.2 \text{ to } 2) \times 10^4$ m/s. Note that

TABLE 10.5   Velocities of Propagation of Elastic Waves in Cubic Crystals

Direction	Modes	Wave Velocity v
$\langle 100 \rangle$	Longitudinal	$\sqrt{\dfrac{C_{11}}{\rho}}$
	Transverse (2)	$\sqrt{\dfrac{C_{44}}{\rho}}$
$\langle 110 \rangle$	Longitudinal	$\sqrt{\dfrac{C_{11} + C_{12} + 2C_{44}}{2\rho}}$
	Transverse (polarized along [001])	$\sqrt{\dfrac{C_{44}}{\rho}}$
	Transverse (polarized along [$1\bar{1}0$])	$\sqrt{\dfrac{C_{11} - C_{12}}{2\rho}}$
$\langle 111 \rangle$	Longitudinal	$\sqrt{\dfrac{C_{11} + 2C_{12} + 4C_{44}}{3\rho}}$
	Transverse (2)	$\sqrt{\dfrac{C_{11} - C_{12} + C_{44}}{3\rho}}$

for cubic crystals with elastic constants which satisfy the isotropy condition $C_{44} = (C_{11} - C_{12})/2$, the longitudinal wave velocity v_l will be $\sqrt{C_{11}/\rho}$ and the transverse wave velocity v_t will be $\sqrt{C_{44}/\rho} = \sqrt{G/\rho}$ for the $\langle 100 \rangle$, $\langle 110 \rangle$, and $\langle 111 \rangle$ directions. Such crystals are therefore elastically isotropic (i.e., the wave velocities are the same in all directions). Accurate measurements of the velocities of elastic waves are quite useful as a dynamic method of determining the elastic constants C_{11} and C_{44} for cubic crystals and isotropic solids.

ANELASTIC PROPERTIES OF MATERIALS

Anelasticity refers to a type of mechanical behavior of materials in which time-dependent effects occur and in which dissipation of energy can also take place. In addition, the applied stress and the resulting strain are not uniquely related to each other, due to relaxation effects. For example, anelastic behavior can correspond to a situation in which a deformation in a material does not disappear instantaneously when the applied stress is removed. Anelasticity is thus a form of time-dependent elasticity (or reversible plasticity) since the deformations involved will eventually disappear. All other types of irreversible mechanical behavior corresponding to the formation of permanent deformations are referred to as inelasticity and are discussed later. It should be noted that inelastic effects can also be time dependent, as is the case for creep, discussed in Section 10.13, and for the viscoelastic behavior of polymers discussed in Chapter 14. Anelastic effects are also observed in magnetic materials when couplings between the elastic and magnetic properties occur, as is the case in magnetic relaxation and magnetomechanical damping. These effects are discussed in Chapter W17.

Anelasticity is described first from the macroscopic or continuum point of view, where the focus is on stress–strain curves, hysteresis loops, internal friction, and general features of the relaxation processes involved in anelastic behavior. The Zener model for anelasticity is presented in Chapter W10. Microscopic aspects of anelasticity are the subject of Section 10.12, where specific atomic-level mechanisms leading to anelastic behavior such as the reversible motions of impurities, grain boundaries, and dislocations are discussed.

10.11 Macroscopic Aspects of Anelasticity

The schematic stress–strain curve shown in Fig. 10.8a serves to illustrate some of the important macroscopic aspects of time-dependent anelastic behavior. This curve corresponds to the type of mechanical response observed for a sample of Fe containing interstitial C, to be discussed in Section 10.12. The material is first taken within the elastic proportional limit from point A to point B. When the applied stress σ is then removed, the strain in the material does not disappear immediately but instead retains a nonzero value ε_c at point C. The magnitude of the strain ε_c will depend on how rapidly the applied stress is removed. The strain will then continue to decrease slowly, reaching the starting point A on a time scale determined by the nature of the specific microscopic relaxation processes occurring within the material.

If the stress applied to the material is cycled periodically between positive and negative values within the elastic limit, a *hysteresis curve* similar to that shown in Fig. 10.8b will be traced out. Under these circumstances the strain in the material is out of phase with the applied stress due to time-dependent relaxation processes related,

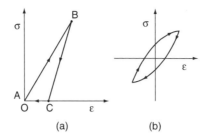

(a)   (b)

Figure 10.8. Schematic stress–strain curve and hysteresis loop for an anelastic material: (*a*) the stress–strain curve shown illustrates time-dependent, non reversible behavior; (*b*) Schematic hysteresis loop obtained when the stress applied to the material is cycled periodically between positive and negative values within the elastic limit.

for example, to the reversible motion of impurities or dislocations. As a result, some of the work done on the material is converted to heat, the amount of elastic energy ΔU_{el} dissipated per cycle being given by the area contained within the hysteresis loop. This area is shown greatly exaggerated in Fig. 10.8*b*. The ratio $\Delta U_{el}/U_{el}$, where U_{el} is the total elastic energy stored in the material in one cycle, is known as the *damping capacity*.

This dissipation of energy is related to the presence of *internal friction* in the material and results from the occurrence of time-dependent microscopic processes. The term internal friction can also correspond to processes leading to the dissipation of energy that occur outside the elastic limit. Materials with low internal friction and with $\Delta U_{el}/U_{el} < 10^{-2}$ find applications in springs, bells, piezoelectric devices, and so on, where dissipation of elastic energy is to be avoided. The opposite limit of high internal friction, corresponding to $10^{-2} < \Delta U_{el}/U_{el} < 1$, is desirable in materials such as gray cast iron and polymers used in shock-absorbing applications.

In addition to irreversible stress–strain curves and hysteresis loops, other evidence for the occurrence of anelastic effects in materials include elastic after effects and stress relaxation (Fig. 10.9). Figure 10.9*a* shows the time dependence of the strain ε in an anelastic solid resulting from the application of a constant stress σ_0 for a finite period of

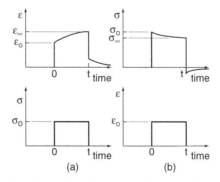

(a)   (b)

Figure 10.9. (a) Elastic after effect. The time dependence of the strain ε resulting from the application of a constant stress σ_0 for a finite period of time is shown. (*b*) Stress relaxation. The time dependence of the stress σ resulting from the existence of a strain ε_0 in a solid for a finite period of time is shown.

time. Following the instantaneous application of the stress σ_0 at time $t = 0$, the strain jumps to a value $\varepsilon_0 = \sigma_0/E$, which is the elastic strain corresponding to σ_0. Subsequent increase in the strain to a limiting value ε_∞ is the *elastic after effect* and corresponds to a type of creep, discussed in Section 10.13. The component of the strain given by $\varepsilon_\infty - \varepsilon_0$ is the reversible plastic strain. When the stress σ_0 is removed instantaneously at time t, the elastic component ε_0 of the strain also disappears instantaneously, followed by the elastic after effect corresponding to gradual disappearance of the remaining reversible plastic strain.

Stress relaxation is illustrated in Fig. 10.9*b*, where the time dependence of the stress σ required to generate a constant strain ε_0 in a solid is shown. The initial stress σ_0 decreases slowly to a value σ_∞ as additional anelastic deformation processes occur within the solid. This results in a corresponding decrease of the stress needed to maintain the strain at the value ε_0. When the strain is taken to zero at time t, the stress σ falls to a negative value (needed to maintain an elastic strain in the material opposing the reversible plastic strain component). The stress then decreases slowly to zero as the reversible plastic strain in the solid disappears. The time dependencies of ε and σ are predicted on the basis of the Zener model for anelasticity described in Chapter W10.

10.12   Microscopic Aspects of Anelasticity

The anelastic behavior of a material results from microscopic processes that take place in a solid on various time scales. These processes are important not only for the mechanical properties of materials but can also influence their electrical, optical, and magnetic properties as well. Examples of some relaxation processes and their corresponding characteristic times or relaxation times τ are presented in Chapter W10.

If the relaxation time τ for a certain process is of the same order of magnitude as the testing time t_{test} or the time of actual use of the material in service, it can be expected that significant effects associated with the process will be observed. On the other hand, if $\tau \gg t_{\text{test}}$, the relaxation process is "slow" and will not reach equilibrium during the time of the test. If $\tau \ll t_{\text{test}}$, the relaxation process is "fast" and will reach equilibrium.

Relaxation times τ for dislocation motion and the diffusion of impurities depend very strongly on temperature and stress since these processes involve the motion of atoms via thermally activated jumps. For such processes the dependence of τ on temperature can be expressed by

$$\tau(T) = \tau_0 e^{E_a/k_B T} \tag{10.41}$$

where E_a is the activation energy for the relaxation process (e.g., for the jump of an atom between sites in a solid) and $\tau_0 = 1/f_{\text{vib}}$, where f_{vib} is a typical atomic vibrational frequency. Such relaxation times thus decrease exponentially as the temperature increases. This temperature dependence is the physical basis for the increasing effectiveness at higher temperatures of annealing processes which are often used to remove stress- or radiation-induced defects or damage in materials. Measurements of internal friction as a function of temperature are often used to determine values of E_a for the relaxation processes involved. An example is $E_a = 1.1$ eV obtained for the diffusion of interstitial C in Ta. The activated process of diffusion is discussed in more detail in Chapter 6.

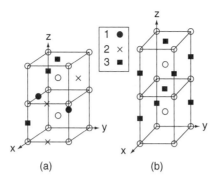

Figure 10.10. C atoms in BCC α-Fe. Labeling of C atoms: 1, on x axis; 2, on y axis; 3, on z axis. (*a*) Random distribution of C atoms in two types of octahedral interstitial sites under zero applied stress; (*b*) when a tensile stress is applied along the z axis, the C atoms lower their energies by occupying the interstitial sites on the cube edges lying along the z axis.

A classic example of an anelastic effect in a metal is the *Snoek effect*, involving a lattice distortion and the corresponding lattice strain arising from the redistribution under stress of interstitial C or N atoms in BCC α-Fe. The Snoek effect is also mentioned in Chapter W17 in regard to its role in magnetic relaxation in α-Fe. Figure 10.10*a* shows the random distribution of C atoms in two types of interstitial sites, with octahedral symmetry in the BCC lattice of α-Fe (i.e., face-centered sites and edge-centered sites) under the condition of zero applied stress. The actual concentration of C atoms would be much less than that shown here. The C atoms create local distortions or strains within the lattice, but the crystal structure remains cubic since the distribution of C atoms is random and there is no preferred direction for the local strain.

When a tensile stress is applied to a crystal of α-Fe along the z axis, the C atoms can lower their energies by occupying the interstitial sites on the cube edges lying along the z axis (Fig. 10.10*b*). Thus, if the tensile stress is applied essentially instantaneously, the usual elastic strain will be followed by a gradual buildup of an additional anelastic strain as the C atoms move slowly into the lower-energy interstitial sites. This is just the type of elastic after effect shown in Fig. 10.9*a*. As a result of the strains, the crystal structure becomes tetragonal under stress.

When the stress is removed, an additional elastic after effect, similar to that shown in Fig. 10.9*b*, occurs as the C atoms gradually jump away from the z axis sites and return to a random distribution within the Fe lattice. Stress relaxation under conditions of constant strain can also be observed in this system since the motion of C atoms to the preferred interstitial sites on the z axis causes a reduction in the applied stress needed to maintain the strain at its initial, completely elastic value. When the C atoms have again achieved a random distribution, the crystal again becomes cubic.

Measurements of the Snoek effect (i.e., of internal friction) in α-Fe as a function of temperature at a given frequency allow the diffusion coefficient $D(T)$ and the concentration of free, interstitial C atoms to be determined. The height of the *Snoek peak* is proportional to the free C-atom concentration (i.e., those not tied up at dislocations or grain boundaries). The relationship between D and the experimentally-determined relaxation time τ for the thermally activated jumps of C atoms is $\tau = a^2/36D$, where a is the lattice constant of α-Fe. For T slightly above 300 K, $\tau \approx 100$ s, which corresponds to $D \approx 10^{-24}$ m^2/s for C in α-Fe.

INELASTIC PROPERTIES OF MATERIALS

When the stress applied to a solid material exceeds the elastic limit or when the stress is applied for a sufficiently long time, a part of the resulting deformation of the material will be permanent (i.e., it will remain indefinitely after the applied stress is removed). This type of inelastic mechanical response of a material is known as *plasticity*, and the deformations involved are known as plastic deformations, as opposed to the elastic and anelastic deformations discussed earlier. When even higher stresses are applied, the material can undergo mechanical failure or *fracture* into two or more parts. In both plasticity and fracture, the internal structure of the material changes and permanent atomic displacements occur.

It should be noted that the plastic deformation of a material is determined not only by the applied stress σ but also by the strain rate $\partial \varepsilon / \partial t$, the temperature T, and the microstructure of the material. As a result, various mechanisms of plastic deformation can exist for a given material, and the resulting inelastic behavior is often quite complicated.

One of the most important mechanical properties of solid materials, in particular metals, is their *ductility* (i.e., their ability to undergo a considerable amount of plastic deformation before fracture occurs). Brittle materials, on the other hand, often undergo fracture before significant plastic deformation has taken place. The mechanical processing of ductile metals and alloys via plastic deformation using a variety of techniques, such as forging, rolling, and so on, not only helps to achieve the desired final shape of the material but also often contributes additional mechanical strength via the important mechanism of work hardening, to be discussed later. The inelastic region is also important due to the current trend of allowing limited plastic deformation of certain materials in some applications. These include applications where the weight factor is critical, such as space vehicles and rockets.

Another key mechanical property of solids is their high strength. *Strength* is a general term used to describe the ability of a material to resist deformation or fracture. Structural materials, in particular, must have both sufficiently high ductility to allow significant deformation to take place before failure and also sufficiently high strength so that deformation takes place only at relatively high loads. Metals can possess both of these properties and can also absorb large amounts of mechanical energy before failure occurs. Although typical metals are not as strong as glass fibers, are less hard than quartz crystals, and are less elastic than rubber, these nonmetallic materials possess either poor ductility or low strength and thus do not have the advantageous combination of both mechanical properties possessed by metals.

The discussion of inelasticity presented here will focus on plasticity and fracture, with phenomena occurring on the macroscopic level discussed in Section 10.13 and microscopic or atomic-level phenomena the subject of Section 10.14. Structural defects such as dislocations play critical roles in the inelastic behavior of materials and are described in Chapter 4. Since the plastic response and failure of materials are very structure- and material-specific, the presentation here deals only with important general concepts and some key details. The discussion of some specific types of inelastic mechanical behavior is deferred to Chapters 11 to 18, where various classes of materials are considered. For example, the time-dependent type of inelasticity known as viscoelasticity is discussed in Chapter 14 since viscous behavior is usually not observed for classes of materials other than polymers (except possibly at high temperatures). Strengthening procedures are an important aspect of the fabrication of metals and alloys and are also discussed in Chapter W21.

10.13 Macroscopic Aspects of Plasticity and Fracture

From the macroscopic point of view, a material is a continuum and no consideration is given to its structure on the atomic level. Most of the important macroscopic aspects of inelasticity, including the parameters describing the inelastic behavior of materials, are conveniently illustrated using *stress–strain* (σ–ε) *curves*.

Stress–Strain Curves. Several σ–ε curves have been presented earlier in this chapter to illustrate the elastic and anelastic behaviors of materials. The tensile σ–ε curve shown in Fig. 10.11 will now be used to illustrate various aspects of inelastic behavior, specifically for the case of a ductile material exhibiting a considerable amount of plastic deformation prior to fracture. True stress and strain, as opposed to the nominal values, are shown plotted in this figure. The schematic σ–ε curve shown can be divided into three distinct regions: region I, extending from the origin 0 to point A: region II, from A to B; and region III, from B to C, with the terminal point C corresponding to the point at which the material undergoes mechanical failure.

Region I is the reversible *elastic region* discussed in Section 10.5, which gradually merges at point A with region II in which irreversible *plastic deformation* occurs. The σ–ε curve in region I can consist of a proportional regime in which σ and ε are directly proportional to each other and also a nonlinear elastic region at higher stress. Point A is known as the *yield point* and σ_y and ε_y are the corresponding *yield stress* (or strength) and *yield strain*, respectively. An alternative definition of σ_y is the stress at which the plastic strain reaches 0.002. The yield stress will in general depend on the direction in which an external stress is applied to a single crystal. Analogous yield points exist for shear σ–ε curves, where the corresponding parameters are known as the *shear yield stress* τ_y and *strain* ε_y.

The material undergoes uniform plastic deformation in region II [i.e., plastic deformation, which occurs essentially uniformly along the length of the material (although

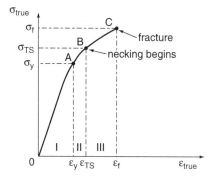

Figure 10.11. Stress–strain curve illustrating inelastic behavior, specifically for the case of a ductile material exhibiting a considerable amount of plastic deformation prior to fracture. True stress and strain are plotted here.

certainly not uniformly on a microscopic scale)]. At point B the cross-sectional area normal to the applied stress begins to decrease in a localized region due to a weakness or instability occurring at some flaw or defect in the material. This process is known as *necking*. On a stress–strain curve where nominal stress and strain are plotted (e.g., Fig. 10.5b), point B corresponds to the point on the curve where the slope is zero. Beyond point B in region II the cross-sectional area decreases as necking occurs and therefore the applied force F and the nominal stress $\sigma = F/A_0$ actually decrease. The applied stress at point B is σ_{TS}, the *tensile strength* of the material. The quantity σ_{TS} is the maximum stress that can be sustained by the material under tension before nonuniform deformation occurs. The corresponding strain is ε_{TS}.

Following the onset of necking, plastic deformation in region III proceeds nonuniformly along the length of the material with the cross-sectional area of the necked region continuing to shrink. This shrinking continues until fracture finally occurs at a weak point or flaw within the bulk or on the surface of the material in the necked region. The *fracture stress* (or *strength*) and *strain* at point C of region III are σ_f and ε_f, respectively.

Not all materials have $\sigma-\varepsilon$ curves which exhibit all three regions shown here. For example, brittle materials undergo little if any plastic deformation before failure. In this case the $\sigma-\varepsilon$ curve would terminate with fracture at or just beyond point A and regions II and III would be absent. An example of this behavior has already been shown in Fig. 10.5a for alumina, Al_2O_3, a brittle ceramic.

As mentioned earlier, the microstructure of a material can undergo significant changes in regions II and III as plastic deformation occurs. In particular, the microstructure can become nonuniform with a variety of extended defects, such as dislocations and twins being generated in the material, in addition to those that may already have been present at zero stress. The nature and structure of these extended defects, as well as their interactions with each other and with point defects such as vacancies and impurities, are discussed in Section 10.14.

The existence of these extended defects in the material as the applied stress exceeds the elastic limit can be understood from the point of view of the work done on the material by the external forces and the resulting energy stored in the material. In region I, the region of elastic deformation, the work done by external forces is stored in the potential energy of interaction between the atoms and is completely recoverable when the applied stress is removed. In regions II and III, 90 to 95% of the work done in the process of plastic deformation can be dissipated as heat and so is not recoverable. Therefore, only 5 to 10% may be used to create additional extended defects. It is energetically favorable to increase the number of defects in the material rather than to continue to increase the elastic strain. This follows from the fact that dislocations and twins allow a high level of strain to be accommodated in localized regions, while the bulk of the material remains at a lower level of strain.

The volume of a solid material is observed to remain constant during plastic deformation, in contrast to the generally observed increase (decrease) in volume that occurs in the elastic limit when the material is under tension (compression). This indicates that tensile stresses in a material are not the source of plastic strain, which must therefore be attributed to shear stresses. Since plastic deformations occur in localized regions, the bulk of the material, and hence its volume, are unaffected. It follows that Poisson's ratio ν increases during plastic deformation.

As the level of strain in a ductile material increases in region III, failure via fracture eventually occurs. Although fracture in a material most often results from the shear component of the applied stress, it can also occur due to the tensile component when cleavage occurs (i.e., when the fracture path is perpendicular to the applied stress). Fracture is often initiated at a weak point or flaw and so is critically dependent on microstructure. From a macroscopic point of view an estimate for the stress and corresponding strain that lead to mechanical failure can be obtained by setting the cohesive energy per unit volume Δh_c of the material equal to the stored elastic energy density $u_{el} = E\varepsilon^2/2$. The maximum elastic strain and applied stress allowed in the material would then be given, respectively, by

$$\varepsilon_{max} = \sqrt{\frac{2\Delta h_c}{E}} \quad \text{and} \quad \sigma_{max} = \sqrt{2E\,\Delta h_c}. \tag{10.42}$$

For the refractory transition metal tungsten with $E = 411$ GPa (see Table 10.2) and $\Delta h_c = 8.9 \times 10^{10}$ J/m^3, Eq. (10.42) yields $\varepsilon_{max} = 0.66$ and $\sigma_{max} = 270$ GPa. Similarly, for Al with $E = 70.3$ GPa and $\Delta h_c = 3.25 \times 10^{10}$ J/m^3, ε_{max} and σ_{max} are predicted to be 0.96 and 68 GPa, respectively. These predicted values are far above the typical yield and tensile strengths observed experimentally for these and other materials, as listed in Table 10.6. These values of σ_y and σ_{TS} can be enhanced by various hardening procedures, as mentioned in the following sections. Even thin whiskers, which are the strongest forms of pure materials, have strengths well below those predicted by Eq. (10.42). In real materials, defects that are present limit the stress and strain to values well below those predicted by Eq. (10.42). Values of the Vickers hardness, VHN, are also presented in Table 10.6. Hardness as a mechanical property of materials is discussed briefly later in this section.

It is interesting to note that both the weakest and strongest forms of some metals (e.g., Fe) are single crystals. Dislocations can move very freely in large single crystals of Fe since no other defects such as grain boundaries are present to impede their motion. As a result, the yield strength σ_y is relatively low. In whiskers of Fe, which are also defect-free single crystals, dislocations are pinned by the surfaces, so this material has a high yield strength.

The effect of increasing temperature on the inelastic behavior of materials is to greatly reduce the yield stress σ_y and also to limit the process of work hardening, to be defined later. Materials therefore become softer at higher temperature as a result of the enhanced motion of dislocations due to thermal activation, which allows them to move past obstacles that would otherwise have pinned them. The mechanical processing of materials (e.g., rolling) is therefore easier as T increases since less energy is required. Brittle materials such as glass become softer at high T and as a result become increasingly ductile. Conversely, ductile materials such as rubber often become brittle as the temperature is lowered. This is known as the *brittle-to-ductile transformation*, which can occur in materials when the motion of dislocations is essentially frozen out as the temperature is lowered.

Work Hardening. The shape of the $\sigma-\varepsilon$ curve shown in Fig. 10.11 in regions II and III exhibits the inelastic behavior known as *work hardening* or *strain hardening*. This behavior is also illustrated in Fig. 10.5 for a low-carbon steel. Work hardening, one of the most important mechanisms leading to the strengthening of metals, is the

TABLE 10.6 Mechanical Properties of Selected Materials at Room Temperature

Material	Yield Strength σ_y (MPa)	Tensile Strength σ_{TS} (MPa)	Vickers Hardness VHN[a] (kg/mm^2)
Bulk			
Pb	—	—	4.2
Sn	—	—	9
Au	—	103	22
Al	17	55	24
Zn	—	—	35
Ag	55	125	35
Pt	—	—	40
Cu	69	220	50
Fe	130	260	—
Ni	138	483	130
Cr	—	282	200
Ti	240	330	200
Mo	565	655	—
W	150	380	330
C (diamond)			9,800
Thin whiskers			
Cu	—	2,000	—
Fe	—	12,600	—
Si	—	7,000	—

	Young's Modulus E (GPa)	Fracture Toughness K_{1c} (MPa$\sqrt{m}$)	VHN[a] (kg/mm^2)
Bulk			
BN (cubic)	—	—	6,430
WC (3% Co)	641	—	3,060
β-SiC	415–441	5–7	2,860
Al$_2$O$_3$	434	2.7–4.2	2,240
TiC	430	—	1,840
SiO$_2$	74	—	920–1120

Source: Data from A. G. Guy, *Introduction to Materials Science*, McGraw-Hill, New York, 1972, p. 416, Table 9-3; W. D. Callister, *Materials Science and Engineering: An Introduction*, 2nd ed., Wiley, New York, 1991, pp. 738–739, Tables C1 and C2; M. A. Meyers and K. K. Chawla, *Mechanical Metallurgy*, Prentice Hall, Upper Saddle River, N.J., 1984, p. 197, Table 4.5; and D. M. Teter, *Mater. Res. Soc. Bull.*, Jan. 1998, p. 22; hardness: L. E. Samuels, *Metallographic Polishing by Mechanical Methods*, 3rd ed., ASM, Metals Park, Ohio, 1982; ceramics: *ASM Engineered Materials Handbook*, Vol. 4, AMS International, Materials Park, Ohio, 1990, p. 30; metals: *ASM Engineered Materials Handbook*, Vol. 2, ASM International, Material Park, Ohio, 1990.

[a]A Vickers hardness of VHN = 1 kg/mm^2 is equivalent to a hardness of VHN = 9.8 MPa in SI units.

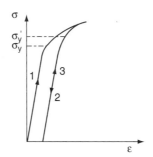

Figure 10.12. Stress–strain curve illustrating work hardening. Upon repeated loading, the stress can be increased to $\sigma_y' > \sigma_y$, the initial yield stress, before plastic deformation begins.

process by which the plastic deformation of a ductile material causes it to become more resistant to further plastic deformation (i.e., once the yield stress σ_y has been exceeded in a material, the next time the material is stressed it will be observed that σ_y has increased). As a result, the material is stronger. Work hardening is illustrated in Fig. 10.12, where it is observed that upon repeated loading, the stress can be increased to $\sigma_y' > \sigma_y$ before plastic deformation begins. An example is a bar of copper, which can be bent readily into a horseshoe shape but cannot then be straightened, due to the work hardening that has occurred. The mechanical work hardening of steels to increase their strength is discussed in Chapter W21, where the processing of metals is described. Brittle materials cannot undergo significant work hardening, due to their inability to deform plastically.

To understand the process of work hardening, consider the case in which plastic deformation increases the size and number of dislocations in the material. Dislocation generation can occur via the Frank–Read mechanism, as described in Section 10.14, and can result in increases of the density of dislocations from 10^{10} to 10^{12} up to 10^{14} to 10^{16} m^{-2}. Work hardening will occur if the work or cost in strain energy of creating additional dislocations increases as the density of dislocations in the material increases. This increasing energy cost can result from interaction of dislocations with each other and also with other defects, such as grain boundaries, twins, and so on. These interactions pin the dislocations so that they cannot move freely through the material. As a result of work hardening, the applied stress must be increased continually in regions II and III in order to create additional dislocations. Work hardening also increases the tensile strength σ_{TS}, since a work-hardened material can withstand greater applied stresses before the plastic deformation becomes nonuniform, leading eventually to failure.

The phenomenon of work hardening is difficult to treat theoretically, the most difficult aspect being to predict how the density and distribution of dislocations vary with the strain in the material. A further discussion of work hardening is presented in Chapter W10.

If no work hardening were observed in a given ductile material, the σ–ε curve would be horizontal for applied stresses exceeding the elastic limit (i.e., for $\sigma > \sigma_y$ or $\tau > \tau_y$). In this case the cost in energy to create additional dislocations would be constant, independent of the number of dislocations already present. The material would then be *perfectly plastic* since the strain would continue to increase as additional dislocations were created but without the requirement of any further increase in the applied stress.

In addition to cold work resulting from forging, rolling, and so on, other mechanisms that can lead to hardening and increased strength in materials include dispersion strengthening, precipitation hardening, and solid-solution strengthening. These mechanisms are described briefly in Chapter W10. The use of these mechanisms in the processing of metals and alloys is also described in Chapter W21.

Materials can undergo mechanical failure not only due to the application of high stresses but also because of the phenomena known as *creep* or *fatigue*. Creep is the time-dependent, gradual increase of plastic deformation at a constant applied stress, often observed at high temperatures $T > 0.5T_m$, where T_m is the melting temperature. Fatigue is the failure at relatively low stress levels of materials that are subjected to fluctuating or cyclic stresses.

Creep. Creep is a form of mechanical degradation that occurs at an accelerated rate at high temperatures where thermally activated diffusion becomes more important. A material under a constant tensile stress at high T will undergo plastic deformation resulting in an increase in length. A corresponding decrease in cross-sectional area usually also occurs, leading to an increase in the true stress. When the increase in stress reaches the tensile strength σ_{TS}, failure will eventually occur. The actual mechanisms contributing to the plastic deformation that is involved in creep can involve the self-diffusion of vacancies, dislocation glide, and grain-boundary sliding. These mechanisms are typically dominant only within certain ranges of applied stress and temperature. The plastic deformation involved in creep can also lead to some work hardening. These competing effects of weakening and strengthening, both due to the same plastic deformation, make the phenomenon of creep difficult to analyze or predict. A typical creep test is discussed in Chapter W10.

Fatigue. Fatigue is the failure at relatively low stress levels of materials subjected to fluctuating or cyclic stresses. Examples range from the failure by repeated bending of a paper clip to the weakening of aircraft wings by repeated stressing. As a material is cycled back and forth along the stress–strain curve, dislocations may coalesce to form microscopic cracks. As discussed in Section 10.14, these cracks have the ability to concentrate stress in their neighborhood and thus may continue to grow, leading eventually to brittle fracture.

In general, the higher the stress, the fewer the number of cycles before the material fails. This is illustrated in the $\sigma-N_f$ curves presented in Fig. 10.13, where the number

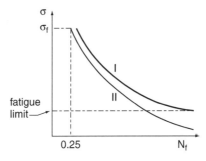

Figure 10.13. Fatigue-induced failure of a material illustrated using $\sigma-N_f$ curves. The curves indicate the number of cycles N_f at which failure occurs for a given applied stress amplitude.

of cycles N_f at which fatigue-induced failure occurs in a material for a given applied stress amplitude is shown. Note that when the stress amplitude $\sigma = \sigma_f$, then $N_f = \frac{1}{4}$ (i.e., failure occurs the first time that the stress reaches its full amplitude). Below a certain stress, called the *fatigue limit*, materials such as steels can withstand essentially any number of stress cycles. This is the case for curve I in Fig. 10.13. The existence of a fatigue limit in steels may be due to the phenomenon of strain aging, associated with the diffusion of C and N to dislocations and the resulting pinning of the dislocations. Strain aging is discussed in Chapter W21, where solid-solution strengthening is described.

The stress corresponding to the fatigue limit is usually less than the yield stress σ_y and thus falls within the elastic region. No such fatigue limit exists for nonferrous metals, which will fail eventually no matter how low the stress. This corresponds to curve II. The fatigue limit is temperature dependent, being lower for higher temperatures. Fatigue is particularly structure sensitive since defects permit plastic deformation at stresses below the bulk yield strength σ_y in areas of local weakness. A further discussion of fatigue appears in Chapter W10.

Hardness. An important although not uniquely defined mechanical property of materials is *hardness*. Hardness can be defined generally as the mechanical resistance of a solid object to a permanent change. More specifically, hardness is a measure of the minimum stress needed to produce an irreversible plastic deformation into the surface of a solid. Other definitions are possible, due to the many different hardness tests that are in use. See Table 10.6 for values of Vickers hardness for some materials. The different hardness tests have in common local application of an external load or stress by direct contact with the surface of the material. Hardness therefore corresponds to various aspects of the resistance of a material to a local deformation concentrated in a small volume at an external surface. For various classes of materials it has been observed that hardness and tensile strength σ_{TS} are proportional to each other. Thus the nondestructive measurement of hardness for a material serves as an indirect measure of σ_{TS}. Details of the measurement of hardness are presented in Chapter W10.

10.14   Microscopic Aspects of Plasticity and Fracture

Most of the recent progress in understanding inelasticity has been derived from efforts that have explored the microscopic or atomic-level mechanisms involved in plasticity and fracture. In this section important microscopic processes such as the generation, motion, and pinning of dislocations, slip, twinning, and crack development and growth are discussed briefly. In addition, models that attempt to predict the magnitudes of yield stresses and fracture strengths from a microscopic point of view are described.

Microscopic Inelastic Processes. Plastic deformation in a crystalline material usually begins either with the generation of dislocations or with their motion if they are already present at zero stress. Dislocations are generated by shear stresses when one section of a solid is displaced relative to another, as discussed in Chapter 4 and shown here in Fig. 10.14. The dislocation line is the path of the region in the crystal which is distorted from the original atomic structure. In Fig. 10.14 the dislocation line is curved, terminating in an edge dislocation on one face and in a screw dislocation on the adjacent face.

Dislocations can be generated in the bulk of a perfect crystal via homogeneous nucleation, a process involving the breaking of bonds between adjacent planes of

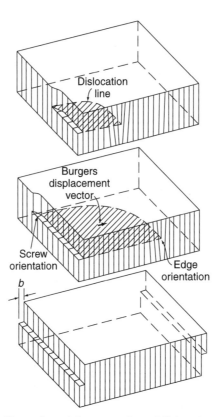

Figure 10.14. Schematic illustration of the generation of dislocations by shear stresses. (From J. J. Gilman, *Micromechanics of Flow in Solids*, copyright 1969 by McGraw-Hill. Reprinted by permission of the McGraw-Hill Companies.)

atoms. To illustrate this, imagine having a deck of cards lying on a table and applying both a shear stress τ_x and a normal stress σ_z to the top card. The shear stress needed to make the cards begin to slide is determined by elementary statics, the result being $\tau_x > \mu_s \sigma_z$ (neglecting the weight of the cards), where μ_s is the coefficient of static friction. Once the cards have moved and the stresses are removed, the deck has been permanently deformed.

Now suppose that a shear stress is applied parallel to a set of lattice planes in a perfect crystal. To determine the stress required to cause one plane to slide past an adjacent plane, consider a simple cubic array of atoms with lattice constant a subjected to the shear stress τ_x. One layer will be displaced by distance x parallel to an adjacent layer, and the shear strain will be $\varepsilon_x = x/a$ (Fig. 10.15). The stress and strain are related (for small strains) by the shear modulus (i.e., $\tau_x = G\varepsilon_x = Gx/a$). When the strain x exceeds a value on the order of half a lattice constant, one plane will slide past the next. Thus the theoretical maximum shear stress is given approximately by $\tau_f = G/2$. Note that unlike the case of the playing cards, the lattice planes are elastically linked to each other, so even in the absence of a normal stress there is a maximum shear stress.

In practice, this estimate grossly overestimates the strength of real materials by a factor on the order of 100 to 1000, and homogeneous nucleation is rarely observed. The reason for this discrepancy, as will be seen, lies in the faulty assumption that the

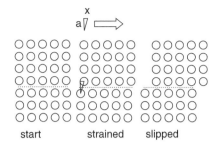

Figure 10.15. A shear force applied to an ideal crystal causes a shear strain and then slippage.

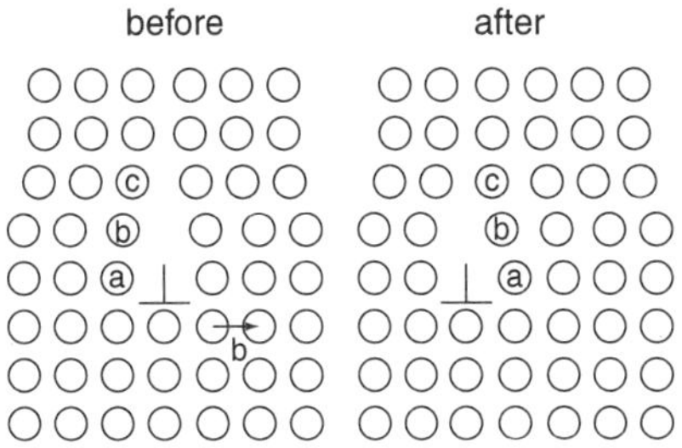

Figure 10.16. The motion of only a few atoms in a plane produces a shift of the edge dislocation. The Burgers vector **b** of the dislocation is shown.

crystal is perfect. Real crystals contain dislocations and twin boundaries, and these play a crucial role in limiting the strength. The basic reason is the following. In the ideal case every atom in the upper plane had to move at once in order to slide this plane past the lower plane, as in Fig. 10.15. In the case where a dislocation is already present, only a few atoms have to move at a time to displace the dislocation. This is illustrated in Fig. 10.16 for the case of an edge dislocation. Notice that in the drawing only the atoms labeled *a*, *b*, and *c* in the plane of the paper have moved in order to displace the dislocation one lattice constant to the left. Of course, in parallel planes, corresponding atoms must also move. In the course of time the dislocation line sweeps across the plane and every atom is displaced. It is very much like trying to open one's jacket by grabbing the two sides with both hands and pulling them apart rather than simply opening the jacket with the zipper. In the theoretical case, there is a large stress that need be applied for only a short period of time to produce the slippage, whereas in the case of a real solid, there is only a need to supply a small stress but for a longer period of time. The energy required, however, is the same in either case.

Grain boundaries are currently thought to be important sources of dislocations in polycrystalline materials, with dislocations being emitted from steps or ledges on the interfaces between the grains. Steps on external surfaces of solids can act as stress concentration sites where dislocations can be formed and emitted into the bulk. It should be noted that materials can be strengthened by surface-hardening procedures that prevent the surfaces from acting as dislocation sources. Surface treatments of metals for improving wear resistance and surface hardness, including surface carburizing and nitriding and ion-beam treatments, are described in Chapter W21.

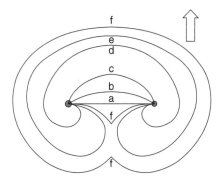

Figure 10.17. Frank–Read dislocation source. A shear stress stretches a dislocation line *a* through various elongations, *b*–*e*, before it eventually fissions into two dislocations, *f*.

The *Frank–Read mechanism* for the generation of dislocation loops is recognized as being important for the multiplication of existing dislocations. When existing dislocations are under stress in a material but are unable to move, segments of the dislocation can be forced to expand and break away from the original dislocation, forming a dislocation loop. This is illustrated in Fig. 10.17, where an edge dislocation depicted by line *a* is bound or pinned initially by two impurities. This line represents the intersection of the edge dislocation with a slip plane, to be described later. As a result of an applied shear stress, the dislocation grows, and therefore this line stretches. Lines *b*, *c*, *d*, and *e* depict various elongated states of the edge dislocation. Finally, line *f* shows the dislocation splitting into two separate, disjoint segments. One portion assumes the configuration held by *a* while the other is free to move. This process may be repeated many times, with the result that many new dislocations can be produced via the Frank–Read mechanism. Frank–Read sources may be the dominant mechanism of dislocation generation after the first few percent of plastic strain.

Once generated, dislocations can move through a material as the applied stress continues to be increased. The two main types of dislocation motion are *glide*, illustrated in Fig. 10.14, and *climb*, which can occur when glide is not possible. The glide of a dislocation occurs on the plane (i.e., the slip plane) in which the dislocation line lies. Cross slip occurs, for example, in BCC α-Fe, when a dislocation moves from one slip plane to another. Glide of dislocations often occurs in region II of the $\sigma-\varepsilon$ curve shown in Fig. 10.11. In climb, edge dislocations move from one slip plane to an adjacent one via the addition or removal of atoms from the incomplete plane of atoms at the dislocation core.

Another inelastic process associated with plastic deformation is *slip*. Slip corresponds to the relative motion of adjacent regions of the material and results from the motion of a dislocation on a *slip plane*. This relative motion can occur over many multiples of a lattice constant, with the crystal structure of the two regions remaining unchanged. Slip is the primary mechanism by which very large strains are generated in materials and is illustrated on the macroscopic level in Fig. 10.18. The dislocation motion corresponding to slip, as illustrated in Fig. 10.15, occurs preferentially only in certain directions on certain planes in a crystal structure. The general rule is that slip is observed to occur on the planes with the greatest area density of atoms [e.g., the close-packed {111} planes of the FCC crystal structure or the (0001) planes of HCP (at least for $c/a > 1.633$)] and in directions in those planes that have the greatest linear

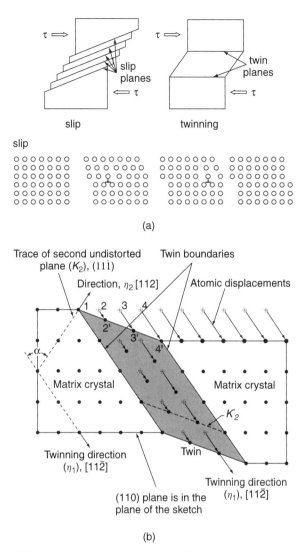

Figure 10.18. (*a*) Differences between the effects of slip and twinning illustrated on the macroscopic level; (*b*) corresponding arrangements of atoms. [(*a*) From W. D. Callister, Jr., *Materials Science and Engineering*, 2nd ed., copyright 1991 by John Wiley & Sons, Inc. Reprinted by permission of John Wiley & Sons, Inc. (*b*) From A. G. Guy, *Introduction to Materials Science*, copyright 1972 by McGraw-Hill. Reprinted by permission of the McGraw-Hill Companies.]

density of atoms [e.g., the $\langle 101 \rangle$ directions of the $\{111\}$ FCC planes]. These observations can be understood qualitatively as follows: (1) the spacing d between adjacent close-packed planes is larger than for other planes, leading to weaker bonding between atoms in adjacent planes, and (2) the distance between neighboring atoms is smallest in those directions with the greatest linear density of atoms, allowing slip by one atomic spacing to occur at a lower applied stress. These directions in the slip planes are also the directions of the Burgers vector **b** for the dislocation undergoing slip. The slip planes and directions observed for several crystal structures are listed in Table 10.7.

TABLE 10.7 Slip Planes and Directions

Crystal Structure	Slip Planes	Slip Directions	Number of Slip Systems
FCC (Cu)	$\{111\}$	$\langle 101 \rangle$	12
BCC (Fe)	$\{110\}^a$	$\langle 111 \rangle$	12
	$\{112\}$	$\langle 111 \rangle$	12
	$\{123\}$	$\langle 111 \rangle$	12
HCP (Cd)	(0001)	$\langle 11\bar{2}0 \rangle$	3^b
Diamond (Si)	$\{110\}$	$\langle 111 \rangle$	12
NaCl	$\{110\}$	$\langle 101 \rangle$	12
Hexagonal (Al_2O_3)	(0001)	$\langle 11\bar{2}0 \rangle$	3^b
Polyethylene^c $(CH_2)_n$	$\{hk0\}$	$\langle 001 \rangle$	—

^{a}The $\{110\}$ set of planes is active at lower temperatures and the sets $\{112\}$ and $\{123\}$ at higher temperatures.
^{b}There are only two linearly-independent Burgers vectors in the (0001) plane.
^{c}Polyethylene has an orthorhombic lattice.

Each combination of a given slip direction in a slip plane is known as a *slip system*. For example, slip can occur in three directions in the (111) plane of FCC Cu (i.e., the $[1\bar{1}0]$, $[10\bar{1}]$, and $[01\bar{1}]$ directions). Three slip directions also exist in each of the $(11\bar{1})$, $(1\bar{1}1)$, and $(\bar{1}11)$ planes, for a total of 12 slip systems in the FCC crystal structure. The number of active slip systems in various crystal structures is also listed in Table 10.7. The more slip systems that a material has, the more readily can the conditions for slip be satisfied, and thus the greater the ductility of the material. For example, materials such as Cu with FCC crystal structures will be more ductile than HCP materials such as Cd, which have only three active slip systems.

It has been shown that materials with at least five independent slip systems such as Cu can undergo any arbitrary change in shape due to plastic deformation, including necking. Materials with less than five slip systems are inherently brittle. For example, since slip is absent in Al_2O_3 for $T < 1000°C$, the fracture observed is brittle.

The ease with which slip occurs is strongly influenced by the bonding in the material. For example, slip is difficult in materials such as diamond and Si with strong covalent bonds, since the process of slip would involve the breaking of the strong bonds between atoms in adjacent planes. Such materials therefore tend to be brittle. Instead of slip, twinning often occurs in covalently bonded materials since the process of twinning, discussed later, can occur without the breaking of bonds. Slip in metals often occurs readily since the bonding in metals is usually nondirectional. As far as slip and ductility are concerned, ionic crystals are intermediate between covalent and metallic materials, with NaCl being brittle while AgCl and LiF are ductile. Another illustrative example of the effect of bonding on slip and ductility occurs for the two crystalline forms of Sn. White Sn is metallic and fairly ductile; gray Sn is a covalently bonded semiconductor and is brittle. The process of slip can influence and be influenced by other defects in the material. Grain boundaries, for example, can interfere with slip by blocking the relative motions involved in slip.

There is a *critical shear stress* τ_y for the occurrence of slip on a given slip system. Even when a material is under a tensile load, shear stresses will in general exist along every direction in every plane, with magnitudes determined by the orientations of both the direction and the plane in question relative to the applied stress. Consider a slip plane in a cylindrical single crystal whose normal makes an angle ϕ with respect to the direction of the applied uniaxial stress. The tensile stress acting on this plane is $\sigma \cos^2 \phi$, where one factor of $\cos \phi$ accounts for the component of the force F acting normal to the slip plane, and the other factor accounts for the increase in the area on which the force component acts. The resulting shear stress $\tau(\theta, \phi)$ acting in a given direction in this plane, which makes an angle θ with respect to the cylinder axis, is given by $\tau(\theta, \phi) = \sigma \cos \phi \cos \theta$. The factor $\cos \phi \cos \theta$ is known as the *Schmid factor*. Slip will occur first on the primary slip system (i.e., the slip system on which the critical shear stress is reached first).

Twinning is the inelastic process leading to the formation of the extended defects known as *twins*. Twins result from only small displacements of atoms and correspond to the relative motion of adjacent regions in a material by only a fraction of a lattice constant. The difference between the effects of slip and twinning in a material can be observed on the macroscopic level and are shown clearly in Fig. 10.18a. The corresponding arrangements of atoms on the microscopic level are shown in Fig. 10.18b. Although not as common as slip due to the requirement of a higher applied stress, twinning may nevertheless be favored at low temperatures, at high strain rates, or when the stress is applied in a direction perpendicular to the active slip plane [e.g., the (0001) plane in an HCP metal]. Once twinning has occurred, however, the process of slip may then occur more readily.

Cracks are often the ultimate cause of failure in materials. They result when the strength of the material is exceeded locally due either to the coalescence of dislocations or to voids that are formed in heavily deformed regions (e.g., in the central, necked region of a ductile material). Once formed, cracks further enhance or concentrate the local applied stress and can lead to rapid failure via fracture. Surface cracks often control the strength of brittle materials such as glass. Some very effective methods for increasing the strength of materials are surface treatments such as etching which remove cracks, the application of hard coatings that prevent cracks from spreading, or the prestressing of a thin surface layer in compression, which tends to "seal" the cracks. Having a surface layer in compression will assist the surface in resisting external tensile stresses. The microscopic model leading to the Griffith criterion for the critical stress for crack propagation is presented in the following section.

Microscopic Models of Inelasticity. Although it is difficult to estimate the yield stress σ_y of even a defect-free material, there are useful relationships that express the effect of grain size on σ_y in polycrystalline metals and ceramics. One commonly used expression is the *Hall–Petch relation*, which has the form

$$\sigma_y = \sigma_0 + \frac{k_y}{\sqrt{d}}, \tag{10.43}$$

where d is the average grain size in the material and σ_0 and k_y are assumed to be constants, with k_y proportional to the strength of the material. The quantity σ_0, called the *friction stress*, corresponds to the yield stress in a single crystal (i.e., the yield stress in the limit $d \to \infty$). According to this expression, the yield stress is predicted to increase as the average grain size d decreases (i.e., as the mean free path for dislocation motion decreases). It is assumed that dislocations are pinned by grain boundaries. Equation (10.43) appears to be valid for a wide range of materials under a wide range of conditions. Its use in describing the effect of grain boundaries on the strengthening of steels is described in Chapter W21. There are, however, uncertainties concerning its theoretical justification and whether the observed value of the exponent of d is always $-\frac{1}{2}$. Further discussion of the Hall–Petch relation appears in Chapter W10.

Theoretical estimates for the stress at which fracture occurs in a material can be obtained in several different ways. An example is the macroscopic estimate given in Eq. (10.42) for the energy required to overcome the bonding between atoms in the material (i.e., $\sigma_{max} = \sqrt{2E\Delta h_c}$). On the microscopic level, estimates have been proposed that involve the stress needed to separate adjacent planes of atoms or the stress required for the propagation of a crack (i.e., the *Griffith criterion* for the effect of cracks on the brittle fracture of materials).

Consider a crack passing completely through a plate of a material, as shown in Fig. 10.19. The crack is modeled as an elliptical cavity with dimensions $2a$, the major axis, and $2b$, the minor axis. Under application of a stress σ to the plate, as shown, the maximum local stress σ_{max} will occur at the tip of the crack, where the radius of curvature is $\rho = b^2/a$. This stress can be shown to be given by

$$\sigma_{max} = \sigma \left(1 + 2\sqrt{\frac{a}{\rho}} \right). \qquad (10.44)$$

The stress within the material is indicated schematically in the figure by "force" lines, analogous to electric field lines, whose density is proportional to the magnitude of the stress. Note that σ_{max} increases as the semimajor axis a increases and, hence, ρ decreases (i.e., for long, narrow cracks with $b \ll a$).

Griffith calculated the change in elastic energy due to the presence of an elliptical crack in a thin plate of thickness d and found it to be given by

$$\Delta U_{el} = -\frac{\pi \sigma^2 a^2 d}{E}, \qquad (10.45)$$

thus corresponding to a net decrease in elastic energy. Here E is Young's modulus. The other important contribution is the energy E_s needed to separate two planes of

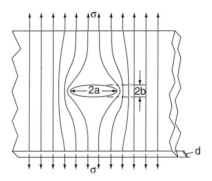

Figure 10.19. Crack passing through a material modeled schematically as an elliptical cavity with dimensions $2a$ and $2b$.

atoms in order to form the crack. The energy E_s is positive and is proportional to the energy of bonding per unit area between planes of atoms (i.e. to the surface energy γ_s). Since E_s is equal to $2(2ad)\gamma_s$, the total energy associated with the crack is

$$E = \Delta U_{\text{el}} + E_s = -\frac{\pi\sigma^2 a^2 d}{E} + 4a\gamma_s d. \tag{10.46}$$

The crack will grow when the resulting gain in elastic energy ΔU_{el} overcomes the energy E_s needed to create the new internal crack surfaces. The criterion for the critical stress σ_c required for crack propagation can be obtained by setting $\partial E/\partial a$ equal to zero (and confirming that $\partial^2 E/\partial a^2 < 0$). This yields

$$\frac{\partial E}{\partial a} = 0 = -\frac{2\pi\sigma_c^2 ad}{E} + 4\gamma_s d. \tag{10.47}$$

The *critical fracture stress* is therefore

$$\sigma_c = \sqrt{\frac{2\gamma_s E}{\pi a}}. \tag{10.48}$$

Thus the Griffith criterion for brittle fracture is obtained by considering both the need to provide the surface energy γ_s associated with the free, internal surfaces of the crack and the availability of strain energy in the vicinity of the crack for this purpose.

The Griffith criterion indicates that when $\sigma > \sigma_c$ and the crack starts to grow, the critical stress σ_c actually drops due to the increase in the length of the crack $2a$. Thus the crack will continue to grow indefinitely, leading ultimately to catastrophic failure. These important ideas have formed the basis of the field of study known as *fracture mechanics*. In order to have a high resistance to crack propagation and therefore a high fracture strength, a brittle material should have a high Young's modulus E and a high surface energy γ_s. Typical values for metals are $E = 100$ to 200 GPa (see Table 10.2) and $\gamma_s = 1$ to 2 J/m^2. When used in Eq. (10.48) along with $a = 1$ to 10 μm, these values for γ_s and E yield typical fracture stresses in the range 300 to 500 MPa.

The *fracture toughness* K_c of a material in the form of a thin plate is a measure of its resistance to failure via fracture due to crack propagation. The quantity K_c is defined by

$$K_c = \sigma_c\sqrt{\pi a} = \sqrt{(2\gamma_s + \gamma_p)E}, \tag{10.49}$$

where γ_p is the effective surface energy associated with the generation of plastic deformation near the crack tip. Note that K_c is also known as the *critical stress intensity factor for plane stress* (when the material is completely constrained in the two directions in the plane of the plate) and is usually quoted in units of MPa$\sqrt{\text{m}}$. In practice, the fracture toughness of a material is observed to depend on its thickness. The fracture toughness will be higher in ductile materials where $\gamma_p \gg \gamma_s$.

For the case of plane strain, when the material is constrained completely in the direction perpendicular to the plate, the appropriate fracture toughness is K_{1c}, where

$$K_{1c} = \sqrt{\frac{2\gamma_s E}{1 - \nu^2}}. \tag{10.50}$$

Here ν is Poisson's ratio. The quantity K_{1c} is also known as the *critical stress intensity factor for plane strain* and is a material constant, independent of the dimensions of the plate. Thus K_{1c} should be used to characterize the fracture toughness of a material. Typical values of K_{1c} for metals are 20 to 200 MPa$\sqrt{m}$, while ceramics, polymers, and plastics have K_{1c} typically in the range 1 to 10 MPa$\sqrt{m}$. Thus ceramics, although typically of high strength, tend to be more brittle (i.e., less resistant to fracture) than metals. The high strength of ceramics actually makes them more susceptible to flaws since plastic deformation is limited in ceramics. Values of the fracture toughness K_{1c} for some metals, ceramics, polymers, and plastics are presented in Table 10.8.

Materials exposed to hydrostatic pressure alone do not undergo plastic deformation since no shear stresses are generated. The combination of hydrostatic pressure and shear stresses can, however, generate much higher levels of plastic deformation before failure occurs. This results from the tendency of hydrostatic pressure to oppose the generation and propagation of cracks rather than to any change of the yield stress of the material. The generation of higher levels of deformation by the use of hydrostatic pressure can be very useful in mechanical-processing procedures such as rolling. The closing of cracks is also evident during the compressive loading of materials. As a

TABLE 10.8   Fracture Toughness K_{1c} of Various Materials

Material	K_{1c} (MPa$\sqrt{m}$)
Metals	
Al–4% Cu	26
Ti–6% Al–1% V	60
Stainless steel (18–8)	200
12% Cr steel	50
Cast iron	20
Ceramics	
Pyrex glass	0.75
Al$_2$O$_3$	2.7–4.2
Mullite	2.2
TiB$_2$	6–8
β-SiC	5–7
Polymers and plastics	
Polymethylmethacrylate (PMMA)	0.7–1.6
Polystyrene (PS)	0.7–1.1
Polycarbonate (PC)	2.2
Polyvinyl chloride (PVC)	2–4
Polyethylene (PE)	1–6
Epoxy	0.6
Polyethylene terephthalate (PET)	$\approx$5

Source: Metals: M. A. Meyers and K. K. Chawla, *Mechanical Metallurgy*, Prentice Hall, Upper Saddle River, N.J., 1984, p. 167; ceramics: *ASM Engineered Materials Handbook*, Vol. 4, ASM International, Material Park, Ohio, 1990, p. 30; polymers and plastics: *ASM Engineered Materials Handbook*, Vol. 2, ASM International, Materials Park, Ohio, 1990, p. 739.

result, the compressive strengths of brittle materials such as cast iron and concrete are often an order of magnitude higher than their tensile strengths. Thus, when possible, these materials are put under compressive loads in use.

REFERENCES

Ashcroft, N. W., and N. D. Mermin, *Solid State Physics*, Saunders College, Philadelphia, 1976.

Bernstein, M. L., and V. A. Zaimovsky, *Mechanical Properties of Metals*, Mir Publishers, Moscow, 1983 (English translation).

Callister, W. D., *Materials Science and Engineering: An Introduction*, 2nd ed., Wiley, New York, 1991.

Gilman, J. J., *Micromechanics of Flow in Solids*, McGraw-Hill, New York, 1969.

Guy, A. G., *Introduction to Materials Science*, McGraw-Hill, New York, 1972.

Haasen, P., *Physical Metallurgy*, 3rd ed., Cambridge University Press, Cambridge, 1996.

Houwink, R., and H. K. de Decker, eds., *Elasticity, Plasticity, and Structure of Matter*, 3rd ed., Cambridge University Press, Cambridge, 1971.

Kittel, C., *Introduction to Solid State Physics*, 7th ed., Wiley, New York, 1996, Chap. 3.

Meyers, M. A., and K. K. Chawla, *Mechanical Metallurgy*, Prentice Hall, Upper Saddle River, N.J., 1984.

Nabarro, F. R. N., *Theory of Crystal Dislocations*, Dover, Mineola, N.Y., 1987.

Read, W. T., Jr., *Dislocations in Crystals*, McGraw-Hill, New York, 1953.

Tegart, W. J. M., *Elements of Mechanical Metallurgy*, Macmillan, New York, 1966.

Zener, C., *Elasticity and Anelasticity of Metals*, University of Chicago Press, Chicago, 1948.

PROBLEMS

10.1 Show that if a volume element ΔV in a solid (see Fig. 10.2) is in rotational equilibrium, the shear stress components σ_{12} and σ_{21} must be equal to each other. (*Hint*: Begin by using the fact that the corresponding torques acting on ΔV must be equal in magnitude and acting in opposite directions.)

10.2 Consider a single crystal with a cubic crystal structure and also with the external shape of a cube.

    (a) If this crystal is exposed to a uniform external hydrostatic pressure P, show that the fractional change in volume is given in terms of the strain components ε_i by $\Delta V/V_0 = 3\varepsilon$ where $\varepsilon_1 = \varepsilon_2 = \varepsilon_3 = \varepsilon$.

    (b) From the stress–strain relationships expressed by $\sigma_i = \Sigma_i C_{ij}\varepsilon_j$, show that $\sigma_1 = \sigma_2 = \sigma_3 = \sigma = (C_{11} + 2C_{12})\varepsilon$.

    (c) Show that the bulk modulus is given by $B = (C_{11} + 2C_{12})/3$. (*Hint*: Note that in the case of hydrostatic pressure $\sigma = -P$.)

10.3 Consider a solid material with a cubic, tetragonal, or orthorhombic crystal structure and with a tensile stress σ acting along the z axis, as shown in Fig. 10.6. Show that Poisson's ratio ν defined in Eq. (10.27) is equal to $\frac{1}{2}$ when the volume V of the solid does not change under the action of the tensile stress.

10.4 Derive the expression $v_t(110) = \sqrt{C_{44}/\rho}$ given in Table 10.5 for the wave velocity of a transverse elastic wave propagating in a $\langle 110 \rangle$ direction in a cubic crystal. [*Hint*: Take $u_z(t) = u_z(0)\exp[i(k_x x + k_y y) - i\omega t]$. Note that the wave equation for $u_z(t)$ is related to Eq. (10.36) for $u_x(t)$ by symmetry.]

10.5 For a cubic crystal with $C_{44} = (C_{11} - C_{12})/2$:

(a) Prove that the longitudinal wave velocity will be $v_l = \sqrt{C_{11}/\rho}$ and the transverse wave velocity will be $v_t = \sqrt{C_{44}/\rho}$ for the $\langle 100 \rangle$, $\langle 110 \rangle$, and $\langle 111 \rangle$ directions. Use the expressions for the wave velocities given in Table 10.5.

(b) Calculate the numerical values of v_t and v_l in the $\langle 100 \rangle$, $\langle 110 \rangle$, and $\langle 111 \rangle$ directions for a diamond crystal. Use the elastic constants listed in Table 10.3 and the mass density ρ for diamond given in Table 1.11.

10.6 Obtain values of the wave velocities for diamond from Fig. 5.7 and use them to determine the elastic constants C_{11}, C_{12}, and C_{44}. Compare your results with those presented in Table 10.3. You may use $\rho = 3520$ kg/m^3.

10.7 A polycrystalline, isotropic bar of Fe is placed under uniaxial tension.

(a) If the resulting longitudinal strain ε_z is 10^{-3}, calculate the elastic energy per unit volume $u(\varepsilon)$ stored in the bar. (*Hint*: Use the values of E and v for polycrystalline Fe given in Table 10.2.)

(b) If this energy were instantaneously converted to heat within the bar, estimate the resulting rise in temperature ΔT. Assume that the bar is initially at $T = 300$ K, that no heat leaves the bar, and that the Debye and free-electron theories of the heat capacity are valid.

10.8 Consider a cylindrical bar of polycrystalline Fe of cross-sectional area $A_0 = 10^{-4}$ m^2, length $l_0 = 0.1$ m, initially unstrained, and at a temperature of $T = 300$ K. The bar is held so that no elongation is possible. The bar is then heated to $T = 700$ K.

(a) Derive the expression $\varepsilon_T = \alpha\Delta T$ for the thermal strain that would be generated in the bar if it were free to elongate. What would ε_T be when the bar described above is at $T = 700$ K? (The coefficient of linear expansion for Fe is $\alpha = 12.2 \times 10^{-6}$ K^{-1}.)

(b) Calculate the resulting thermal stress σ_T in the bar if it is not allowed to elongate. (Young's modulus for polycrystalline Fe is $E = 211$ GPa.)

(c) What will be the actual elastic and plastic deformation of the bar under the foregoing circumstances?

10.9 Consider the (100), (110), and (111) planes of FCC Cu as possible slip planes.

(a) Calculate the d spacings for these three planes.

(b) Calculate the area densities of Cu atoms in these planes. Which planes are likely to be the preferred slip planes in FCC Cu? Explain your reasoning.

(c) Show in a sketch the three $\langle 101 \rangle$ directions in a (111) plane and calculate the linear density of Cu atoms in these directions. Are there any other directions in a (111) plane that have greater linear densities of Cu atoms?

10.10 Consider a cylindrical single crystal that is subjected to a tensile stress σ applied along its axis of symmetry.

(a) A slip plane in the crystal has a normal that makes an angle ϕ with respect to the axis of the cylinder (i.e., to the direction of the applied stress). Show that the tensile stress acting on this plane is $\sigma\cos^2\phi$.

(b) Show that the resulting shear stress $\tau(\theta, \phi)$ acting in a direction in this plane which makes an angle θ with respect to the cylinder axis is given by $\tau(\theta, \phi) = \sigma \cos \phi \cos \theta$.

(c) Explain why the (0001) plane in a hexagonal crystal is not an active slip plane when stress is applied along the c axis of the crystal.

Note: Additional problems are given in Chapter W10.

CLASSES OF MATERIALS

CHAPTER 11

Semiconductors

11.1 Introduction

Semiconductors are materials whose electrical conductivities are intermediate between those of metals and insulators. A more important attribute of semiconductors is that their electrical properties can be controlled in a variety of ways [e.g., by the addition of impurities (doping), by adjusting the size and dimensionality of the semiconductor and the physical structure into which it is placed, and also by the use of light and external electric fields]. The semiconductor revolution that has occurred since the invention of the transistor over 50 years ago and that is still in progress is a direct result of our ability to control the electrical properties of semiconductors. As an indication of the growing importance of semiconductors in the U.S. economy, semiconductor products have exceeded the value of steel mill products in recent years.

Investigations of new semiconducting materials and of new configurations and applications of existing semiconductors are very active areas of materials research. In addition to their obvious technological importance, interesting new physical phenomena in semiconductors such as the quantum Hall effect have been discovered in recent years.

In this chapter the properties and applications of semiconductors are described. The characteristic properties of homogeneous semiconductors are discussed first, followed by a description of the important classes of semiconducting materials. Inhomogeneous semiconductors and some of their significant applications are then discussed.

CHARACTERISTIC PROPERTIES OF SEMICONDUCTORS

A *semiconductor* can be defined in several related ways as a solid material which, when extremely pure, (i.e., containing less than $\approx 10^{20}$ impurities/m^3) has the following properties:

1. A filled valence band and an empty conduction band at $T = 0$ K and an energy gap E_g greater than zero but less than about 3 to 4 eV for the excitation of electrons from the highest normally filled energy band (the valence band) to the lowest normally empty energy band (the conduction band).

2. An electrical conductivity σ (300 K), typically in the range 10^{-8} to 10^3 ($\Omega \cdot$m)$^{-1}$. Typical *metals* such as Na and Cu have $\sigma(300$ K$) > 10^6$ ($\Omega \cdot$m)$^{-1}$; typical *semimetals* such as As, Bi, and graphite have $\sigma(300$ K$)$ in the range 10^3 to 10^6 ($\Omega \cdot$m)$^{-1}$; and typical *insulators* such as diamond, NaCl, and SiO$_2$ have $\sigma(300$ K$) < 10^{-8}(\Omega \cdot$m$)^{-1}$.

3. An electrical conductivity that is zero at $T = 0$ K and increases exponentially with increasing T, according to $\sigma(T) \propto \exp(-E_g/k_B T)$, due to increasing concentrations of charge carriers (electrons in the conduction band and holes in the valence band) resulting from the thermal activation of electrons across the energy gap.

4. A concentration of electrons in the conduction band at $T = 300$ K in the range 10^{10} m$^{-3} < n < 10^{20}$ m^{-3}. Typical metals have $n \approx 10^{28}$ to 10^{29} m^{-3}, typical semimetals have $n \approx 10^{23}$ to 10^{26} m^{-3}, and typical insulators have $n \lesssim 10^{10}$ m^{-3}.

Although the boundaries indicated above between the properties of metals, semimetals, semiconductors, and insulators involving σ and n are not well defined or even necessarily widely accepted, they are, nevertheless, useful for the purposes of providing a practical classification of materials. The energy gaps E_g and the intrinsic electron concentrations n at $T = 300$ K for the group IV elemental semiconductors, including diamond as an example of an insulator, as well as for some alloy and compound semiconductors, are presented in Table 11.1. Note that the listing of these materials in order of decreasing energy gap also corresponds to the order of increasing electron concentration.

According to the classification scheme just outlined, the elements Si and Ge, the alloy Si$_{0.5}$Ge$_{0.5}$, and the compounds GaAs and cubic β-SiC are semiconductors, the element C (diamond) is an insulator, and the element α-Sn (gray Sn) is a semimetal. A group of materials including diamond, cubic BN, and the group III–V nitrides based on Ga, In, and Al are ordinarily considered to be insulators due to their wider bandgaps, $E_g \approx 3$ to 6 eV. Significant concentrations of charge carriers are usually not available in these materials at $T = 300$ K but can be generated at $T \approx 500$ to 800 K as thermal excitation of electrons becomes possible. As a result, these materials are often classified as high-temperature semiconductors and have potential applications in high-temperature, high-power devices.

TABLE 11.1 Energy Gaps E_g and Electron Concentrations n of Selected Semiconductors at $T = 300$ K

	$E_g{}^a$ (eV)	n^b (m^{-3})
C (diamond)	5.4	$\approx 10^{-21}$
β-SiC (cubic)	2.3	$\approx 4 \times 10^5$
GaAs	1.42	3×10^{12}
Si	1.11	8×10^{15}
Si$_{0.5}$Ge$_{0.5}$ (alloy)	0.92	$\approx 1 \times 10^{17}$
Ge	0.67	2×10^{19}
α-Sn (gray Sn)	0.08	$\approx 10^{24}$

[a]Most values are from D. R. Lide, ed., *CRC Handbook of Chemistry and Physics*, 75th ed., CRC Press, Boca Raton, Fla., 1994, pp. 12–87 to 12–96.

[b]GaAs, Si, and Ge are calculated from Eq. (11.29) using energy gaps from this table and band-structure effective masses from Table 11.8. For other materials, estimates are based on effective masses in the range 0.1 to 1m.

The sensitivity of the properties of semiconductors to very low concentrations, $\approx 10^{21}$ m^{-3}, of electrically active impurities actually delayed an understanding of the intrinsic properties of these materials until sophisticated techniques had been developed for synthesizing sufficiently pure samples. *Intrinsic semiconductors* are now among the purest materials available. *Extrinsic semiconductors* are materials whose electrical properties are controlled by the addition of electrically active dopant atoms. Semiconductors can also be classified as homogeneous or inhomogeneous according to whether their properties are uniform or nonuniform in space, respectively. Nonuniform properties can result from spatially varying distributions of dopant atoms, from interfaces with other materials such as metals, from the application of external fields, or from the absorption of electromagnetic radiation. The nonuniform properties of inhomogeneous semiconductors and some of their important applications are discussed in Sections 11.12 and 11.13.

The microscopic properties of homogeneous semiconductors which are associated with the energy levels of the electrons, with the dynamics of the electrons, with excited electron states, with doping and defects, and with effects due to the dimensionality of a semiconductor and quantum confinement are discussed next. The important and useful concept of the current carriers known as holes, arising from empty electron states in the valence band, is also introduced. Following the description of microscopic properties, the macroscopic properties of homogeneous semiconductors associated with the response of electrons and holes to external electric and magnetic fields and to light are described. In addition, the thermoelectric effects that appear in semiconductors when temperature gradients are present are discussed in Chapter W11 at our Web site.[†]

MICROSCOPIC PROPERTIES

The energy levels of the valence electrons in semiconductors are determined by their interaction with atoms or ions (i.e., with the lattice potential and with the potential of the valence electrons themselves). In addition, the effective masses of electrons and of holes and their motions are also determined by their interactions with the lattice potential and with each other. This discussion of the microscopic properties of homogeneous semiconductors begins with a description of the energy band structure of the electrons and the resulting energy gaps.

11.2 Energy-Band Structure and Energy Gaps

The *energy bands* of the electrons in a solid can be thought of as arising from the overlap of the atomic orbitals (AOs) of the outermost or valence electrons. As atoms approach each other and their valence electrons begin to interact, the electronic wavefunctions which are the appropriate solutions of the Schrödinger equation for the interacting atoms are no longer the free-atom AOs but instead can be thought of as corresponding to overlapping, hybridized orbitals. The degree of overlap and the resulting energy-band structure of the valence electrons determine whether the solid

---

[†] Supplementary material for this textbook is included on the Web at the resource site (ftp://ftp.wiley.com/public/sci_tech_med/materials). Cross-references to elements of the Web material are prefixed by "W."

will be a metal, semimetal, semiconductor, or insulator. The greater the overlap in a given material, in general the stronger will be the tendency for metallic behavior.

The development of the energy bands of the valence electrons in a covalent solid such as diamond or Si is presented schematically in Fig. 11.1, where the highest-lying valence electron energy levels of a system of $N \approx 10^{23}$ atoms are shown. For $r \gg r_0$, the equilibrium separation of atoms in the solid, these levels are just the degenerate atomic levels. The interacting atomic levels then broaden into the essentially continuous energy bands of the solid as $r \to r_0$. At $r = r_0$ in the case shown, an *energy gap* E_g exists between the highest filled energy band, the *valence band*, and the lowest empty energy band, the *conduction band*. Note that electrons in the higher-lying energy levels start to interact with each other at larger atomic separations than do electrons in the lower-lying levels. In the covalent bonding limit the valence electrons occupy orbitals corresponding roughly to σ molecular orbitals (MOs). The electrons involved in covalent bonding between pairs of atoms in solids are not as localized, however, as the electrons in the covalent bonds in molecules. The wavefunctions of the bonding electrons, instead, have probability densities that extend throughout the entire solid. An increase in the external pressure on a semiconductor such as Si can result in a transition to a more closely packed structure in which the atoms are closer together, the number of NN atoms is greater than four, and the energy gap decreases to zero (i.e., the material becomes a metal).

In a crystalline solid the *quantum numbers* for a conduction electron with wavefunction $\psi_{n,\mathbf{k}}(\mathbf{r})$ and energy $E_n(\mathbf{k})$ correspond to the three components of the wave vector $\mathbf{k}$ (i.e., k_x, k_y, and k_z) and the spin quantum number $m_s = \pm \frac{1}{2}$. Another useful parameter is the integer *band index* n which is needed to specify $\psi_{n,\mathbf{k}}(\mathbf{r})$ and $E_n(\mathbf{k})$ completely in the reduced-zone scheme, where the wave vector $\mathbf{k}$ of the electron is restricted to lie within the first Brillouin zone.

In addition to providing a graphical representation of the effect of the lattice potential on the electronic energy levels, the energy-band structure $E_n(\mathbf{k})$ of a semiconductor also provides the information that is necessary for calculating the dynamical properties of the electrons (and holes) and also the density of electron states as a function of energy. Although knowledge of the band structure over a wide range of energies is interesting, under most conditions the properties of semiconductors depend primarily on the electron states near the top of the valence band and the bottom of the conduction band.

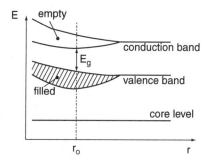

Figure 11.1. The development of the valence electron energy bands in a covalent solid such as diamond or Si is presented schematically.

All of the band-structure calculational methods that attempt to find solutions for the wavefunctions $\psi_{n,\mathbf{k}}(\mathbf{r})$ and energies $E_n(\mathbf{k})$ of the electrons begin with the Schrödinger equation (see Appendix WC),

$$H\psi_{n,\mathbf{k}}(\mathbf{r}) = \left[\frac{p^2}{2m} + eV(\mathbf{r})\right]\psi_{n,\mathbf{k}}(\mathbf{r}) = E_n(\mathbf{k})\psi_{n,\mathbf{k}}(\mathbf{r}), \tag{11.1}$$

where $V(\mathbf{r})$ is the potential seen by the electrons in the solid. The energy band structures of semiconductors are typically determined from model calculations which are based in part on first principles and in part on experimental values for the energies of characteristic electronic transitions in the material of interest. Two important calculational methods are the *tight-binding approximation* or *model* [also known as the *LCAO* (linear combination of atomic orbitals) *method*] and the *nearly free electron* (NFE) *approximation*. Both of these approaches have been outlined in Chapter 7 and are capable of predicting the existence of continuous energy bands separated by energy gaps, with details that depend on the particular parameters chosen for a given semiconductor.

The tight-binding approximation starts by using a basis of linear combinations of *atomic orbitals* $\phi_{ml}(\mathbf{r})$ to describe the electron wavefunction, that is,

$$\psi_{n,\mathbf{k}}(\mathbf{r}) = \Sigma_{m,l} c_{ml} \phi_{ml}(\mathbf{r}). \tag{11.2}$$

Here l and m are the quantum numbers for the orbital angular momentum of the electron and its z component, respectively. In this model the energy gaps are included from the beginning since "gaps" occur naturally between the atomic energy levels. The energy bands arise from the overlap of the atomic orbitals on neighboring atoms. The tight-binding approach focuses on the chemical bonding aspects of the formation of energy levels in the solid and is therefore more useful for predicting the structure of the valence band than of the conduction band.

The NFE approximation or model starts with a basis of *free-electron plane waves*, that is,

$$\psi_{n,\mathbf{k}}(\mathbf{r}) = \Sigma_k a_k e^{i\mathbf{k}\cdot\mathbf{r}}. \tag{11.3}$$

In this approach the energy bands are included from the beginning, whereas the energy gaps between bands arise from interactions of the electrons with the ions or, more formally, with the lattice potential $V(\mathbf{r})$. In the NFE model the gap in the electronic density of states at a given Brillouin zone boundary is predicted to be small compared to the Fermi energy E_F and also to be proportional to the strength of the Fourier component of the lattice potential for the zone boundary in question. The tight-binding and NFE models each provide their own useful insights into development of the electron energy bands and energy gaps in solids.

Si is the prototypical crystalline semiconductor with an *indirect energy gap* (i.e., the state with the lowest energy in the conduction band does not lie directly above, at the same wave vector $\mathbf{k}$, the electron state with the highest energy in the valence band). The energy-band structure for Si in the reduced-zone scheme as calculated via the pseudopotential method is shown in Fig. 11.2a. Here $E_n(\mathbf{k})$ is presented along three different directions in $\mathbf{k}$ space. These directions are indicated in the Brillouin zone of Si which is shown in Fig. 11.2b. It should be noted that the energy bands in $\mathbf{k}$ space are symmetric under inversion (i.e., they are the same in the $+\mathbf{k}$ and $-\mathbf{k}$

directions). Semiconducting crystals with the diamond, zincblende, and NaCl crystal structures (e.g., Si, GaAs, and PbS, respectively) all possess FCC Bravais lattices and all therefore have the Brillouin zone shown in Fig. 11.2*b*.

The valence bands of semiconductors such as Si, GaAs, and ZnS all correspond to two atoms and eight valence electrons per primitive unit cell and thus occupy a total of four Brillouin zones ($n = 1$ to 4) in **k** space. In the reduced-zone scheme shown in

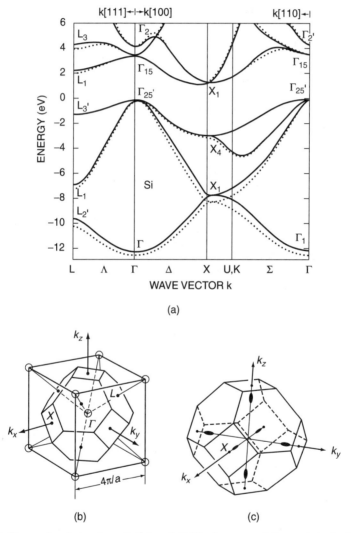

(a)

(b)   (c)

Figure 11.2. Energy-band structure $E_n(\mathbf{k})$ and Brillouin zone of the prototypical crystalline semiconductor Si. (*a*) $E_n(\mathbf{k})$ along the directions **k** [100], **k** [111], and **k** [110] in **k** space. (From M. L. Cohen and J. R. Chelikowsky, *Electronic Structure and Optical Properties of Semiconductors*, Springer-Verlag, New York, 1989.) (*b*) Brillouin zone of Si and some important directions in **k** space. (From B. Sapoval et al., *Physics of Semiconductors*, Springer-Verlag, New York, 1993.) (*c*) Brillouin zone of Si containing the six ellipsoidal regions corresponding to the conduction band minima in the six equivalent ⟨100⟩ directions. (From B. Sapoval et al., *Physics of Semiconductors*, Springer-Verlag, New York, 1993.)

Fig. 11.2a the four branches of $E_n(\mathbf{k})$ in the valence band correspond to $n = 1$ to 4. For Si the energy levels at the top of the valence band at $\mathbf{k} = 0$ are doubly degenerate but exhibit different curvatures and hence different effective masses, to be defined later. These two branches of $E_n(\mathbf{k})$ correspond to the *heavy holes* (low curvature) and *light holes* (high curvature). There is a third branch lying below these two which is split off as a result of the spin–orbit interaction.

The energy at the bottom of the conduction band of Si does not lie at $\mathbf{k} = 0$ but rather at the six symmetric points in $\mathbf{k}$ space given by $k(100) \approx 0.8(2\pi/a)$. The resulting indirect energy gap in Si corresponding to transitions between the valence band at $\mathbf{k} = 0$ and the bottom of the conduction band is $E_g = 1.11$ eV at $T = 300$ K. Because photons with energies $\hbar\omega = 1$ eV have wave vectors $k_{photon} = \omega/c \approx 5 \times 10^6$ m^{-1}, these indirect transitions require the participation of a phonon with $k_{phonon} \approx \pi/a \approx 10^{10}$ m^{-1} to provide the nonzero momentum, $\approx 1.6\pi(\hbar/a)$, which the electron must have at the conduction-band minimum. The locations of the ellipsoidal regions corresponding to the six conduction-band minima are shown in the Brillouin zone of Si in Fig. 11.2c. The ellipsoid in the [001] (i.e., $+k_z$) direction has a surface of constant energy defined by

$$E_n(\mathbf{k}) = \frac{\hbar^2}{2}\left[\frac{k_x^2 + k_y^2}{m_{Te}^*} + \frac{(k_z - k_{z0})^2}{m_{Le}^*}\right], \tag{11.4}$$

where $k_{z0} \approx 1.6\pi/a$. Similar expressions exist for the other ellipsoids. This equation provides the definition of the *transverse* and *longitudinal conduction-band effective masses* m_{Te}^* and m_{Le}^* of the electron, respectively. Values of these effective masses for Si and Ge are given in Table 11.8.

While the band structures of semiconductors such as Si shown in Fig. 11.2 appear to be rather complicated, the energy bands, in fact, resemble qualitatively those of the corresponding free-electron gas in two respects. First, the energy gaps between the valence and conduction bands are small compared to the overall valence bandwidth, and second, away from the BZ boundaries the energy bands are essentially parabolic, a fact that is obscured when they are all plotted in the first Brillouin zone. Since it is sufficient for most practical purposes to focus only on the electron states near the edges of the bandgap, it is often useful, at least to a first approximation, to consider the simpler, schematic "universal" semiconductor band structure shown in Fig. 11.3, where only the important regions near the band edges at $\mathbf{k} = 0$ are presented. Direct (i.e., vertical) and indirect optical transitions associated with the absorption of light are shown.

A useful semiempirical method for calculating $E_n(\mathbf{k})$ is the $\mathbf{k} \cdot \mathbf{p}$ perturbation theory, which uses experimentally determined energy gaps (or other relevant optical energies) and optical matrix elements as input in order to determine the energy bands throughout the Brillouin zone.

The valence-band densities of electron states $\rho_e(E)$ for Si and Ge measured using x-ray photoemission spectroscopy are presented in Fig. 11.4, where they are compared with the predicted densities of states from both local and nonlocal pseudopotential calculations. These curves correspond to averages over all directions in the Brillouin zone. The density of states $\rho_e(E)$ is predicted to be parabolic at the top of the valence band and also at the bottom of the conduction band [i.e., $\rho_e(E) \propto |E|^{1/2}$ in the valence band for $E < E_v = 0$ and $\rho_e(E) \propto |E - E_c|^{1/2}$ in the conduction band for $E > E_c = E_g$]. Here E_v and E_c are the electron energies corresponding to the top of the valence

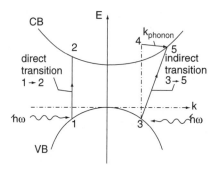

Figure 11.3. Schematic "universal" semiconductor band structure.

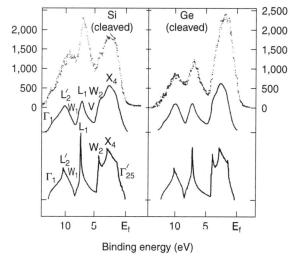

Binding energy (eV)

Figure 11.4. Measured valence-band electron density of states $\rho_e(E)$ for crystalline Si and Ge as obtained from x-ray photoemission spectroscopy along with the predicted $\rho_e(E)$. [From L. Ley et al., *Phys. Rev. Lett.* **29**, 1088 (1972). Copyright 1972 by the American Physical Society.]

band and the bottom of the conduction band, respectively. It follows that the energy gap $E_g = E_c - E_v$. The measured and calculated results for $\rho_e(E)$ can be seen to be in qualitative agreement with regard to its shape and in quantitative agreement with regard to the width of the valence band, ≈ 12 to 13 eV for Si. This width is close to the calculated Fermi energy, $E_F(\mathrm{Si}) = 7.85$ eV, which is the bandwidth according to the free-electron model. Note that $E_F(\mathrm{Si})$ is calculated on the basis of four valence electrons per Si atom. In both metals and semiconductors the width of the valence band results from the spatial delocalization of the valence electrons.

The prototypical crystalline semiconductor with a *direct gap* is GaAs. Its energy band structure as calculated by the pseudopotential method is presented in Fig. 11.5. The GaAs band structure is qualitatively similar to that of Si except at the very bottom of the valence band and in the region of the energy gap. In GaAs the lowest valence band has s-like character and is split off from the the rest of the valence band. The electrons occupying this lowest valence band originate from the As $4s$

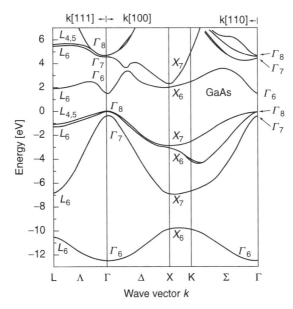

Figure 11.5. Energy-band structure of the prototypical direct-gap semiconductor GaAs. (From M. L. Cohen and J. R. Chelikowsky, et al., *Electronic Structure and Optical Properties of Semiconductors*, Springer-Verlag, New York, 1989.)

electrons and are partially localized in space near the more electronegative As ions. Also, in GaAs the minimum energy in the conduction band lies at the center of the Brillouin zone (i.e. at $\mathbf{k} = 0$ in this reduced-zone scheme). Direct gap semiconductors such as GaAs have the advantage that injected electrons and holes can recombine readily, thereby emitting light, without transferring any energy to the lattice via phonon emission. Devices making use of this property are described in Chapter 18. As the electronegativity differences between the group III and group V elements in the III–V zincblende semiconductors increase (i.e., as the semiconductor becomes more ionic), the width of the valence bands decreases while the width of the energy gap increases.

Both the magnitude of the energy gap and its nature (i.e., whether direct or indirect) can be affected by changes in temperature, pressure, or the composition of the semiconductor through doping or alloying. The energy gap E_g also varies with direction in the crystal and typically has weak, linear dependencies on temperature and pressure, assuming that no structural transformations occur. The effect of temperature on E_g arises both from the thermal expansion of the lattice and from electron–phonon interactions. In most semiconductors E_g decreases with increasing T. Tuning of the energy gap by alloying is common in group III–V semiconductor alloys such as $Ga_{1-x}Al_xAs$, where the energy gap at $T = 300$ K can be varied continuously from $E_g(GaAs) = 1.42$ eV to $E_g(AlAs) = 2.16$ eV. In some cases the nature of the energy gap can be changed from indirect to direct with alloying as in $GaP_{1-x}As_x$ and also by the application of external pressure. The properties of semiconductor alloys are discussed further in Section 11.11. The increase in E_g with increasing free-carrier concentration that is observed in degenerate semiconductors is known as the *Burstein–Moss shift*. This shift occurs in n-type semiconductors since electrons must make transitions from the valence band to empty states that lie higher in the conduction band as a result of n-type doping, discussed in Section 11.5.

11.3 Dynamics of Electron Motion

The motion of electrons through a material is determined by their interactions with the potential of the ions, with other electrons, and also by the presence of any external electric or magnetic fields. The dynamics of electrons are described in general by the Schrödinger equation, as discussed in Chapter 7. In semiconductors it is possible to focus only on the energy states that are occupied by mobile charge carriers under normal conditions (i.e., the energy states near the top of the valence band and the bottom of the conduction band). The energy-band structure in the vicinity of the energy gap will therefore play a critical role in the dynamics of electrons in semiconductors.

From the semiclassical point of view presented in Section 7.12, variation of the *crystal momentum* $\hbar\mathbf{k}_e$ of an electron in a solid with time (i.e., its equation of motion in $\mathbf{k}$ space), is determined by the external electric and magnetic fields through the *Lorentz force*,

$$\hbar\frac{d\mathbf{k}_e}{dt} = -e(\mathbf{E} + \mathbf{v}_{en} \times \mathbf{B}) = \mathbf{F}_{ext}. \tag{11.5}$$

Here $\mathbf{v}_{en}$ is the velocity of the electron in the nth energy band. Note that the external force $\mathbf{F}_{ext}$ on an electron in a solid as defined above is not in general equal to the time rate of change of its momentum $d\mathbf{p}_{en}/dt$, as expected from Newton's second law. The second law for an electron in a solid can be expressed in the form

$$\mathbf{F}_{tot} = \mathbf{F}_{ext} + \mathbf{F}_{latt} = \frac{d\mathbf{p}_{en}}{dt}, \tag{11.6}$$

where the effects of the lattice potential and of interactions with other electrons are all included in $\mathbf{F}_{latt}$. It should be remembered that in the free-electron model $\mathbf{F}_{latt} \equiv 0$ and $m_e\mathbf{v}_e = \hbar\mathbf{k}_e$. If $\mathbf{F}_{ext} = 0$, then by Eq. (11.5) the crystal momentum $\hbar\mathbf{k}_e$ of the electron will not change unless the electron is scattered by a phonon or a defect.

The true momentum $\mathbf{p}_{en} = m\mathbf{v}_{en}$ of the electron can change as it moves through the lattice and need not even be in the same direction as its crystal momentum $\hbar\mathbf{k}_e$ in $\mathbf{k}$ space. In fact, if the electron's effective mass, defined below, is negative (i.e., $m_e^* < 0$), $\mathbf{v}_{en}$ and $\mathbf{k}_e$ can even point in opposite directions. It is true, however, that in the absence of external fields or forces the magnitude of the electron's velocity v_{en} will not change as it moves through the periodic lattice potential. The electron's velocity can be expressed in terms of the band structure $E_n(\mathbf{k})$ by

$$\mathbf{v}_{en}(\mathbf{k}) = \frac{1}{\hbar}\frac{\partial E_n(\mathbf{k})}{\partial\mathbf{k}}. \tag{11.7}$$

In certain simple situations, for example, when the effective mass m_e^* is isotropic and $\mathbf{v}_e$ and $\mathbf{k}_e$ point in the same direction, the electron's crystal momentum can be written as

$$\hbar\mathbf{k}_e = m_e^*\mathbf{v}_{en}. \tag{11.8}$$

An electron interacting with the lattice potential $V(\mathbf{r})$ thus often responds to external electric and magnetic fields as if it had an *effective mass* m_e^* which can be very different in both magnitude and sign from its free-electron mass m. The effective mass tensor for an electron with wave vector $\mathbf{k}$ in band n is defined in Section 7.12 in terms of $E_n(\mathbf{k})$ for a crystalline material by

$$\left(\frac{1}{\widetilde{m}_e^*}\right)_{ij} = \frac{1}{\hbar^2}\frac{\partial^2 E_n(\mathbf{k})}{\partial \mathbf{k}_i \partial \mathbf{k}_j}. \tag{11.9}$$

This equation serves as the definition of the electron's *band-structure effective mass*, which results from the interaction of the electron with the periodic lattice potential. The effective mass m_e^* for motion in a certain direction in the crystal thus is inversely proportional to the curvature of the energy bands in the same direction in $\mathbf{k}$ space. An external force $\mathbf{F}_x$ acting in the x direction can cause an acceleration of the electron in the y direction due to the fact that the lattice can constrain the motion of the electron. Thus, as defined by Eq. (11.9), the effective mass is in general a tensor quantity. This constraint due to the lattice acts in the same way as, for example, a string acts when it constrains the motion of a pendulum ball to a circular arc under the influence of a gravitational field.

At energy-band minima where the curvature of $E_n(\mathbf{k})$ is positive (i.e., concave upward), electrons have positive effective masses, while at energy band maxima where the curvature is negative (i.e., concave downward), electrons have negative effective masses. To illustrate this, consider the energy-band diagram for Si shown in Fig. 11.2. At the conduction-band minimum at $k[100] \approx 1.6\pi/a$, the effective mass for electrons is both positive and anisotropic, with the longitudinal effective mass $m_{Le}^* = 0.90m$ and the transverse effective mass $m_{Te}^* = 0.19m$, as measured by the cyclotron resonance technique. For $\mathbf{k} = 0$ at the top of the valence band in Si there are two degenerate electron energy bands, both with negative curvature. As a result, there are two different negative electron effective masses. As discussed later, the light and heavy holes at $\mathbf{k} = 0$ in the valence band of Si have positive, isotropic effective masses $m_{lh}^* = 0.16m$ and $m_{hh}^* = 0.52m$, respectively. The light holes correspond to the energy band with the higher curvature, and the heavy holes correspond to the band with lower curvature.

Other effective masses resulting from many-body effects (e.g., the electron–phonon interaction) can also be defined and appear in expressions for the electronic specific heat (see Chapter 7). In addition, it will be convenient later to define density-of-states and conductivity effective masses.

11.4 Excited States of Electrons

Only at $T = 0$ K will the electrons in a semiconductor occupy the lowest set of available energy levels. At all higher temperatures, some electrons will be thermally excited across the energy gap into the conduction band (CB). As a result, empty electron states or *holes* will be left behind in the valence band (VB). For $T > 0$ K, equilibrium concentrations of excited electrons and holes will therefore exist in a semiconductor. These excited states are discussed next for the case of a pure, homogeneous semiconductor, starting with the definition of holes in the valence band and a description of their properties.

Holes in the Valence Band. The concept of holes in the valence band acting as charge carriers plays a key role in our understanding of both the microscopic and macroscopic properties of semiconductors. The electrons in the conduction band behave as a gas of free particles with effective mass m_e^*, charge $-e$, and spin $s = \frac{1}{2}$. The holes in the valence band whose origins can be understood in terms of the energy-band structure of the material are much more difficult to comprehend and visualize. In simple terms, holes correspond to empty electron states near the top of an otherwise-filled energy band. Although the concept of holes is introduced in Chapter 7, a more detailed discussion is presented here.

The concept of holes was introduced by A. H. Wilson in 1931 in his theory of semiconductors as materials with narrow energy gaps. The usefulness of this concept is due to the fact that a single hole can in principle represent the behavior of an entire band of electrons. For example, if only a single electron orbital is empty in an energy band consisting of $2N$ available energy states, the properties of the resulting hole are chosen to be the same as those of the remaining $2N - 1$ electrons. It is clearly easier to focus on the properties of a single hole, or of even 10^{20} holes, in a band than on the properties of the 10^{23} electrons which normally occupy the band. The concept of holes is most useful when dealing with a relatively few empty electron states near the top of an energy band.

The schematic diagram presented in Fig. 11.6a illustrates an excited electron and the resulting hole in the lattice of a semiconductor such as Si. The covalent bonds between NN Si atoms, normally occupied by pairs of electrons with opposite spins, are shown as pairs of parallel lines. Near the center of the diagram one electron is missing from a bond. This missing (i.e., excited) electron is shown moving through the lattice at the lower right. The motion of the hole in a given direction actually corresponds to that of an electron moving in the opposite direction into the empty orbital associated with the hole, as shown, either spontaneously or as a result of an applied electric field. The motion of the hole through the lattice corresponds to a sequence of these steps.

The creation of an *electron–hole* (e–h) *pair* via the absorption by an electron in the valence band of a photon with energy $\hbar\omega > E_g$ can be represented by the following process:

$$e(\text{VB}) + \text{photon}(\hbar\omega > E_g) \longrightarrow e(\text{CB}) + h(\text{VB}). \tag{11.10}$$

The wave vectors and energies of the following particles involved in this process are shown in Fig. 11.6b (note that the zero of energy for both electrons and holes has been chosen to lie at the top of the valence band):

1. The excited electron in the CB (electron I)
2. Electron I when it was originally in the VB
3. The "extra" unpaired electron left behind in the VB (electron II)
4. The hole in the VB

The properties of the hole will now be related to the properties that electron I possessed when it was in the valence band.

The "extra" electron II is the only electron in the VB whose wave vector $\mathbf{k}$ is now unpaired. Therefore, electron II carries the entire momentum of the remaining $2N - 1$ electrons. This follows from the fact that a filled energy band with $2N$ electrons has a total wave vector $\mathbf{k}_{\text{band}} = \Sigma_i \mathbf{k}_i = 0$ since, by symmetry, for every electron with wave

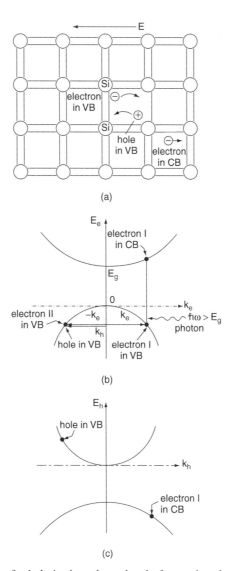

Figure 11.6. Properties of a hole in the valence band of a semiconductor: (*a*) excited electron and the resulting hole in the lattice of a semiconductor such as Si; (*b*) creation of an electron–hole pair via the absorption of a photon by an electron in the valence band. (*c*) When the electron energy diagram of part (*b*) is inverted, the hole energy $E_h = -E_e$ becomes positive in the upward direction.

vector $+\mathbf{k}_e$ in the band there exists an electron with wave vector $-\mathbf{k}_e$. The hole that represents the entire band of $2N - 1$ electrons therefore has a wave vector

$$\mathbf{k}_h = -\mathbf{k}_e, \qquad (11.11)$$

where $\mathbf{k}_e$ is the wave vector of electron I. Note that the hole in Fig. 11.6*b* is thus located in $\mathbf{k}$ space at $-\mathbf{k}_e$, the position of the "extra" electron II.

From conservation of energy it can be seen that

$$E_e(\text{VB}) + \hbar\omega = E_e(\text{CB}), \tag{11.12}$$

where $E_e(\text{VB}) < 0$ is the initial energy of electron I in the valence band and $E_e(\text{CB})$ is its final energy in the conduction band. An equally valid way of expressing energy conservation can be written as follows:

$$\hbar\omega = E_e(\text{CB}) + E_h(\text{VB}), \tag{11.13}$$

where $E_h(\text{VB})$ is the energy of the hole created in the valence band. From Eqs. (11.12) and (11.13) it follows that

$$E_h(\text{VB}) = -E_e(\text{VB}). \tag{11.14}$$

Thus the energy $E_h(\text{VB})$ of the hole in the valence band is a positive quantity that increases in the downward direction from $E_h(\text{VB}) = 0$ at the top of the valence band.

When the electron energy-level diagram is inverted as shown in Fig. 11.6c, the hole energy $E_h(\text{VB})$ is now positive in the upward direction, with $\mathbf{k}_h$ taken to be positive to the right. The definition of the *hole velocity* $\mathbf{v}_h$ is analogous to that of $\mathbf{v}_e$, that is,

$$\mathbf{v}_h(\mathbf{k}_h) = \frac{1}{\hbar} \frac{\partial E_h(\mathbf{k}_h)}{\partial \mathbf{k}_h}, \tag{11.15}$$

while that of the *hole effective mass* $\overleftrightarrow{m}_h^*$ tensor is

$$\left(\frac{1}{\overleftrightarrow{m}_h^*}\right)_{ij} = \frac{1}{\hbar^2} \frac{\partial^2 E_h(\mathbf{k}_h)}{\partial \mathbf{k}_{hi} \partial \mathbf{k}_{hj}}. \tag{11.16}$$

Using these definitions, it can be seen from the slopes and curvatures of the electron and hole energy bands shown in Fig. 11.6c that

$$\mathbf{v}_h(\mathbf{k}_h) = \mathbf{v}_e(\mathbf{k}_e) \tag{11.17}$$

and that

$$m_h^* = -m_e^*. \tag{11.18}$$

Here $\mathbf{v}_e(\mathbf{k}_e)$ and $m_e^* < 0$ are the velocity and effective mass of electron I in the valence band. Thus holes near the top of the valence band have positive effective masses.

The equation of motion of the hole in $\mathbf{k}$ space is

$$\hbar \frac{d\mathbf{k}_h}{dt} = q_h(\mathbf{E} + \mathbf{v}_h \times \mathbf{B}). \tag{11.19}$$

When $\mathbf{k}_h$ is replaced by $-\mathbf{k}_e$ and $\mathbf{v}_h$ by $\mathbf{v}_e$, the following equation is obtained:

$$-\hbar \frac{d\mathbf{k}_e}{dt} = q_h(\mathbf{E} + \mathbf{v}_e \times \mathbf{B}). \tag{11.20}$$

When this expression is compared with the corresponding expression for an electron given in Eq. (11.5), it can be seen that

$$q_h = -q_e = +e. \tag{11.21}$$

Thus holes behave as positively charged particles. This surprising property of holes results from the negative effective masses of the electrons missing from the valence band. The same result for q_h can be obtained by equating the current $\mathbf{j}_e = (-e)(-\mathbf{v}_e) = +e\mathbf{v}_e$ carried by the "extra" electron II in the valence band with the current $\mathbf{j}_h = q_h\mathbf{v}_h = +e\mathbf{v}_h$ carried by the hole.

Table 11.2 compares the properties of the electrons and the hole shown in Fig. 11.6b. Note that the properties of the hole are not simply those of electron I in the valence band but instead correspond to those of the remaining $2N - 1$ electrons. The contributions of holes to the macroscopic properties of semiconductors are discussed throughout the remainder of this chapter.

Equilibrium Carrier Concentrations. Any source of energy (thermal, electromagnetic, etc.) can in principle result in the excitation of pairs of electrons and holes in a semiconductor. The concentrations of electrons and holes in equilibrium at temperature T will now be derived, first for the case of intrinsic, defect-free semiconductors and later for the case of extrinsic or doped materials. It will be seen that the magnitude of the energy gap E_g plays a crucial role in determining these concentrations.

Consider the electron densities of states $\rho_e(E)$ near the band edges (Fig. 11.7) where the electron energy E is plotted vertically, with $E = 0$ at the top of the valence band. At $T = 0$ K the Fermi energy E_F lies in the center of the bandgap at $E = E_g/2$ and is equal to the electron's *chemical potential* μ. The probability of occupation of a given electron state of energy E at temperature T is given by the *Fermi–Dirac distribution function*

$$f_e(E, T) = \frac{1}{e^{\beta(E-\mu)} + 1}, \tag{11.22}$$

TABLE 11.2 Properties of Electrons and Holes

Property	Electron I Originally in VB	Electron I in CB	"Extra" Electron II in VB	Hole in VB
Wave vector **k**	$\mathbf{k}_e$	$\mathbf{k}_e$	$-\mathbf{k}_e$	$\mathbf{k}_h = -\mathbf{k}_e$
Energy E	$E_e(\text{VB}) < 0$	$E_e(\text{CB}) > 0$	$E_e(\text{VB}) < 0$	$E_h(\text{VB}) = -E_e(\text{VB}) > 0$
Velocity **v**	$\mathbf{v}_e(\text{VB})$	$\mathbf{v}_e(\text{CB})$	$-\mathbf{v}_e(\text{VB})$	$\mathbf{v}_h = \mathbf{v}_e(\text{VB})$
Effective mass m^*	$m_e^*(\text{VB}) < 0$	$m_e^*(\text{CB}) > 0$	$m_e^*(\text{VB}) < 0$	$m_h^* = -m_e^*(\text{VB}) > 0$
Charge q	$q_e = -e$	$q_e = -e$	$q_e = -e$	$q_h = -q_e = +e$
Equation of motion (in **k** space)^a	$\hbar\dfrac{d\mathbf{k}_e}{dt} = -e(\mathbf{E} + \mathbf{v}_e\times\mathbf{B})$			$\hbar\dfrac{d\mathbf{k}_h}{dt} = +e(\mathbf{E} + \mathbf{v}_h\times\mathbf{B})$

^{a}The electrons all have the same equation of motion.

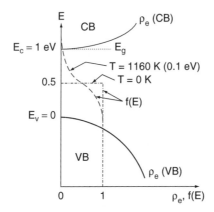

Figure 11.7. Electron densities of states $\rho_e(E)$ near the band edges with electron energy plotted vertically. The Fermi–Dirac distribution function $f_e(E, T)$ for electrons is also shown at $T = 0$ K and also for $T \approx E_g/10k_B = 1160$ K when $E_g = 1$ eV.

where $\beta = 1/k_B T$. At $T = 0$ K this is a step function with $f_e = 1$ for $E < E_F = \mu$ and $f_e = 0$ for $E > \mu$, as shown in Fig. 11.7. Thus at $T = 0$ K all electron states in the valence band are occupied and all those in the conduction band are empty. The electron energy E corresponding to the chemical potential is defined by $f_e(E = \mu, T) = 1/2$.

The concentration $n(T)$ of electrons excited into the conduction band is given by the following integral over this band:

$$n(T) = \int_{E_c}^{E_{\max}} \rho_e(E) f_e(E, T) \, dE, \tag{11.23}$$

where $E_{\max}$ is the maximum energy in the conduction band. Similarly, the concentration $p(T)$ of holes (i.e., of empty electron states in the valence band) is

$$p(T) = \int_{-E_{\min}}^{0} \rho_e(E)[1 - f_e(E, T)] \, dE, \tag{11.24}$$

where $-E_{\min}$ is the minimum energy in the valence band and $1 - f_e(E, T) \equiv f_h(E, T)$ is the probability that a given electron state at energy E is empty (i.e., occupied by a hole). In intrinsic semiconductors with $E_g > 0.15$ eV, under normal circumstances the quantity $|E - \mu| \gg k_B T$ for electron energies in both the conduction and valence bands. Thus the Fermi–Dirac function can be approximated quite accurately in the conduction band for $E > E_g$ by

$$f_e(E, T) \approx e^{\beta(\mu - E)} \ll 1 \tag{11.25a}$$

and in the valence band for $E < 0$ by

$$f_h(E, T) = 1 - f_e(E, T) \approx e^{\beta(E - \mu)} \ll 1. \tag{11.25b}$$

The fact that these Maxwell–Boltzmann-like exponential tails of f_e and f_h extend into the conduction and valence bands for any finite temperature indicates that some electrons have indeed been excited across the energy gap.

The electron densities of states per unit energy and per unit volume are given in the nearly free electron approximation by

$$
\rho_e(E) = \begin{cases}
\dfrac{1}{2\pi^2}\left(\dfrac{2m^*_{eds}}{\hbar^2}\right)^{3/2}(E-E_g)^{1/2} & \text{for } E > E_g \\[4mm]
\dfrac{1}{2\pi^2}\left(\dfrac{2m^*_{hds}}{\hbar^2}\right)^{3/2}|E|^{1/2} & \text{for } E < 0.
\end{cases}
\tag{11.26}
$$

Here m^*_{eds} and m^*_{hds} are the appropriate density-of-states effective masses at the bottom of the conduction band and the top of the valence band, respectively, and it is assumed that the conduction-band minimum is at $\mathbf{k} = 0$. In indirect-bandgap semiconductors such as Si, the sixfold degeneracy of the conduction-band minimum, the anisotropy of the effective mass m^*_e, and the degeneracy at the top of the valence band must be taken into account explicitly. The conduction-band density-of-states effective mass is given in terms of the band-structure effective masses by $m^*_{eds} = Z^{2/3}(m^*_{Le}m^{*2}_{Te})^{1/3}$, where Z is the number of equivalent minima in the conduction band. With this definition of m^*_{eds}, it follows that in the conduction band, $\rho_e(E)$ is proportional to Z, as expected. For Si with $Z = 6$, one obtains $m^*_{eds} = 1.05m$ using $m^*_{Le} = 0.90m$ and $m^*_{Te} = 0.19m$. The appropriate valence-band density-of-states effective mass is given by $(m^*_{hds})^{3/2} = (m^*_{lh})^{3/2} + (m^*_{hh})^{3/2}$. For Si one obtains $m^*_{hds} = 0.58m$ using the values $m^*_{lh} = 0.16m$ and $m^*_{hh} = 0.52m$ given in Section 11.3.

When Eqs. (11.25) and (11.26) are substituted into Eqs. (11.23) and (11.24), the resulting integrals can be evaluated using a change of variables to $\varepsilon = \beta(E - E_g)$. The following results are obtained in the intrinsic limit, using $E_g = E_c - E_v$ (and $E_v = 0$),

$$
n(T) = n_i(T) = 2\left(\frac{m^*_{eds}k_BT}{2\pi\hbar^2}\right)^{3/2}e^{-\beta(E_c-\mu)} = N_c(T)e^{-\beta(E_c-\mu)}
$$
$$
\tag{11.27}
$$
$$
p(T) = p_i(T) = 2\left(\frac{m^*_{hds}k_BT}{2\pi\hbar^2}\right)^{3/2}e^{-\beta(\mu-E_v)} = N_v(T)e^{-\beta(\mu-E_v)}.
$$

The quantities $N_c(T)$ and $N_v(T)$ as defined above are the *effective densities of states* in the conduction and valence bands, respectively. The concentrations of electrons and holes both increase exponentially with increasing temperature in intrinsic semiconductors, proportional to $\exp(-\beta E_g/2)$, since $\mu \approx E_g/2$.

The product $n(T)p(T) = n_i(T)p_i(T)$ obtained from Eq. (11.27) is given by

$$
n_i(T)p_i(T) = N_c(T)N_v(T)e^{-\beta E_g}.
\tag{11.28}
$$

This product is therefore independent of the position of the Fermi energy in the bandgap, a result that is of great importance in doped or extrinsic semiconductors. Equation (11.28) is an example of the law of mass action for the equilibrium between electrons and holes. Since $n_i = p_i$ in electrically neutral intrinsic semiconductors, one can obtain the following results:

$$n_i(T) = p_i(T) = 2 \left(\frac{k_B T}{2\pi\hbar^2} \right)^{3/2} (m_{eds}^* m_{hds}^*)^{3/4} e^{-\beta E_g/2}, \tag{11.29}$$

$$\mu(T) = \frac{E_g}{2} + \frac{3k_B T}{4} \ln \frac{m_{hds}^*}{m_{eds}^*}. \tag{11.30}$$

Equation (11.30) indicates that $\mu(T) \approx E_g/2$ (i.e., that μ always lies near the middle of the energy gap) in intrinsic semiconductors. Whether μ moves upward or downward in the energy gap as T increases is determined by whether the ratio m_{hds}^*/m_{eds}^* is greater or less than 1. This ratio is less than 1 in Si and Ge but is greater than 1 in group III–V and II–VI semiconductors such as GaAs and CdTe, as can be seen in Table 11.8. The more significant dependence of μ on T in extrinsic or doped semiconductors is discussed in the following section.

11.5 Doping and Defects

The deliberate introduction of controlled concentrations of impurity atoms into an intrinsic semiconductor such as highly purified Si is a critical step in the fabrication of semiconductor devices. The most common dopant impurity atoms enter the Si lattice substitutionally and are either *donors* like P, which "donate" electrons to the conduction band, or *acceptors* like B, which "accept" electrons from the valence band. Equivalently, acceptors "donate" holes to the valence band. The electrical properties of the resulting extrinsic semiconductor can be controlled accurately over a wide range. In the case of the Si, the intrinsic resistivity $\rho \approx 3000$ $\Omega\cdot$m at $T = 300$ K can be controllably reduced to $\approx 10^{-6}\Omega\cdot$m or lower through appropriate doping.

To be effective, the dopant atoms must be electronically active; that is, they must have electron energy levels that are in the energy gap and close enough to one of the band edges so that transitions of electrons between the bands and the dopant levels can be thermally activated with high probability at $T \approx 300$ K. Such energy levels are called *shallow*. To illustrate the doping of Si (Ge will be similar), consider the schematic diagram of the Si lattice shown in Fig. 11.8a. Here both a substitutional P donor atom with five valence electrons and a substitutional B acceptor atom with three valence electrons are indicated. These P and B valence electrons form Si–P and Si–B covalent bonds with neighboring Si atoms which are similar to the Si–Si bonds and whose energies lie in the Si valence band.

The additional electron energy levels associated with these dopant atoms are shown schematically in Fig. 11.8b. At $T = 0$ K the extra electron associated with the P atom occupies the donor level shown in the gap at an energy E_d below the conduction-band edge. The extra electron has not yet been donated or thermally excited into the Si conduction band, where it would be able to move through the lattice with high mobility. Also, at $T = 0$ K, the acceptor level associated with the B atom and located in the gap at an energy E_a above the valence-band edge is empty. An electron has not yet been thermally excited into this level from the valence band, a process that would create a mobile hole in the valence band. Thus doped Si remains an insulator at $T = 0$ K, at least for dopant concentrations less than about 4×10^{24} m^{-3}. Other dopant atoms in Si include the donors As and Sb from group V and the acceptors Al,

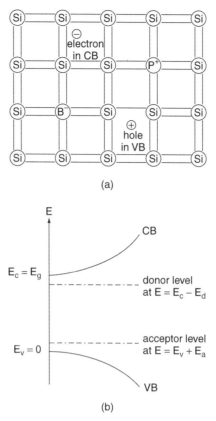

Figure 11.8. The n- and p-type doping of Si. (a) A phosphorus substitutional donor atom and a boron substitutional acceptor atom have been introduced into the Si lattice. The covalent bonds between NN atoms are shown as pairs of parallel lines. (b) The electron energy levels in the energy gap associated with these P and B dopant atoms are shown.

In, and Ga from group III of the periodic table. An important criterion for an impurity atom to have a sufficiently high solubility in solid Si so that it can function as a useful substitutional dopant is that its covalent radius be not too different from that of Si, $r_{cov} = 0.118$ nm. This criterion is met by only some of the elements from groups III and V of the periodic table (see Table 2.12).

A simple *hydrogenic* model based on the Bohr theory of the hydrogen atom can provide good estimates for the binding energies of electrons to donor ions[†] such as P^+ and of holes to acceptor ions such as B^-. These donor and acceptor binding or activation energies, E_d and E_a, respectively, are shown in Fig. 11.8b. Consider first the P donor impurity atom indicated in Fig. 11.8a. The Coulomb potential of the P^+ ion is more attractive to electrons than that of the Si ions, due to the extra nuclear charge $+e$ of P. Therefore, the extra electron is bound in the potential of the P^+ ion.

[†] The ionic charges of the P^+ and B^- ions are measured relative to fully bonded Si^o, which is taken in this case to be neutral. If Si is instead taken to be an ion with charge $+4e$ (i.e., Si^{4+}), the corresponding dopant ionic charges would be P^{5+} and B^{3+}.

In an analogous model the hole is bound in the Coulomb potential of the negatively charged acceptor ion (i.e., B^-). An acceptor ion such as B^- has an effective charge $-e$ because it has attracted an electron from the valence band (i.e., from a Si–Si bond) into its acceptor level. This acceptor level corresponds to a Si–B covalent bond with an energy only slightly less than that of a normal Si–Si bond.

An additional effect associated with dopant atoms is that once ionized, the resulting P^+ and B^- ions disturb the periodic potential of the Si lattice and can be effective in scattering the charge carriers they have produced. It is very desirable to separate these two effects, and this can be accomplished by the technique known as *modulation doping*, discussed in Chapter W11. This procedure can lead to very high carrier mobilities in certain semiconductor devices.

A comparison of the predictions of this hydrogenic model with those of the Bohr theory for the H atom is presented in Table 11.3. The derivation follows that of the Bohr theory, with replacements of the permittivity of free space ϵ_o by the static permittivity $\epsilon = \epsilon_r \epsilon_o$ of the polarizable host crystal and of the free-electron mass m by the appropriate average effective mass m_e^* for electrons at the conduction-band minimum or the average effective mass m_h^* for holes at the valence-band maximum. The details of the Bohr theory and the derivation of its predictions can be found in introductory modern physics texts. In the case of Si the appropriate effective masses are

$$m_e^* = (m_{Le}^* m_{Te}^{*2})^{1/3} = 0.32m,$$

$$m_h^* = \frac{m_{hh}^*(m_{hh}^*)^{3/2} + m_{lh}^*(m_{lh}^*)^{3/2}}{(m_{hh}^*)^{3/2} + (m_{lh}^*)^{3/2}} = 0.47m, \tag{11.31}$$

where m_{hh}^* and m_{lh}^* are weighted according to the corresponding densities of states in the two hole subbands [see Eq. (11.26)]. This approach is known as the *effective-mass approximation* and is valid for impurities that give rise to shallow energy levels in the bandgap.

TABLE 11.3   Comparison of Predictions of the Bohr Theory for the H Atom and of the Hydrogenic Dopant Impurity Model

	Bohr Theory for H Atom	Dopant Impurity Model
Potential $V(r)$ seen by electron (hole)	$\dfrac{+e}{4\pi\epsilon_0 r}$	$\dfrac{+e}{4\pi\epsilon r}\left(\dfrac{-e}{4\pi\epsilon r}\right)$
Ground-state binding energy of electron (hole)	$E_1 = \dfrac{-me^4}{2(4\pi\epsilon_0)^2\hbar^2} = -13.6$ eV	$E_{d(a)} = \dfrac{-m_{e(h)}^* e^4}{2(4\pi\epsilon)^2\hbar^2}$ $= \dfrac{m_{e(h)}^*}{m}\dfrac{\epsilon_0^2}{\epsilon^2}E_1$
Excited states of electron (hole)	$E_n = -\dfrac{E_1}{n^2}, \quad n = 2, 3, \ldots$	$E_n = -\dfrac{E_{d(a)}}{n^2}, \quad n = 2, 3, \ldots$
Bohr radius of electron (hole):	$a_1 = \dfrac{4\pi\epsilon_0\hbar^2}{me^2} = 0.0529$ nm	$a_{d(a)} = \dfrac{4\pi\epsilon\hbar^2}{m_{e(h)}^* e^2}$ $= \dfrac{m}{m_{e(h)}^*}\dfrac{\epsilon}{\epsilon_0}a_1$

Predicted and measured activation energies for dopant atoms in Si and in some other elemental and compound semiconductors are presented in Table 11.4. These activation energies are typically measured via optical absorption spectroscopy in the infrared region or from the measured temperature dependence of the electrical conductivity $\sigma(T)$ in the range where thermal activation of the dopant atoms occurs. Given the simplicity of the hydrogenic model, the agreement obtained between the predicted and measured energies can be considered to be satisfactory.

The large differences between the electron's ground-state binding energy of 13.6 eV in the H atom and the values of E_d and E_a for dopant ions in semiconductors result from the following:

1. Electrons in conduction-band minima in semiconductors have average effective masses $m_e^* \approx (0.1 \text{ to } 0.5)m$, while the free-electron mass m is appropriate for the H atom.

2. The potential of the dopant ion in a semiconductor is weaker, due to screening by the other valence electrons, an effect that is absent in the one-electron H atom. Since the frequencies associated with the orbiting electron are low compared

TABLE 11.4 Predicted and Measured Dopant Activation Energies

Host	Predicted[b]	$E_d{}^a$ (Donors)		
		P	As	Sb
C ($\epsilon = 5.7\epsilon_0$, $m_e^* = 0.57m$)	240	(no known donors)		
Si ($\epsilon = 11.8\epsilon_0$, $m_e^* = 0.32m$)	31.3	44	49	39
Ge ($\epsilon = 16\epsilon_0$, $m_e^* = 0.22m$)	11.7	12	12.7	9.6
		Si(Ga)[c]	Ge(Ga)	S(As) Se(As)
GaAs[d] ($\epsilon = 13.2\epsilon_0$, $m_e^* = 0.067m$)	5.23	5.84	5.88	5.87 5.79
		N(C)[e]		
6H–SiC ($\epsilon = 10.2\epsilon_0$, $m_e^* \approx m$?)	130[f]	85,125(2)		
		$E_a{}^a$ (Acceptors)		
		B	Al	Ga In
C ($\epsilon = 5.7\epsilon_0$, $m_h^* \approx m$?)	420[f]	370	—	— —
Si ($\epsilon = 11.8\epsilon_0$, $m_h^* = 0.47m$)	45.7	46	57	65 160
Ge ($\epsilon = 16\epsilon_0$, $m_h^* = 0.33m$)	17.5	10.4	10.2	10.8 11.2
		Zn(Ga)		
GaAs ($\epsilon = 13.2\epsilon_0$, $m_h^* = 0.76m$)	59.3	31		
		Al(Si)		
6H–SiC ($\epsilon = 10.2\epsilon_0$, $m_h^* \approx m$?)	130[f]	200		

[a]The units for E_d and E_a are meV. Values for dopants in Si and Ge obtained from B. Sapoval and C. Hermann, *Physics of Semiconductors*, Springer-Verlag, New York, 1993, p. 73.
[b]The predictions are based on the expressions for E_d and E_a given in Table 11.3.
[c]Si(Ga) corresponds to a Si atom occupying a Ga site.
[d]Data from R. K. Watts, *Point Defects in Crystals*, Wiley-Interscience, New York, 1977.
[e]There are three inequivalent carbon sites, one with hexagonal symmetry and two cubic symmetry, for substitutional N donors in 6H–SiC. Since one type of N donor has $E_d = 85$ meV while the other two types have $E_d = 125$ meV, it seems likely that the former corresponds to the hexagonal site and that the latter two correspond to the two cubic sites.
[f]Calculated using m_e^* or $m_h^* \approx m$.

with the characteristic frequencies of the permittivity $\epsilon(\omega)$, the static permittivity ϵ of the semiconductor is employed instead of the permittivity of free space ϵ_0.

Since $\epsilon = 11.8\epsilon_0$ and $m_e^* = 0.32m$ in Si, the measured donor activation energy $E_d \approx 40$ meV, while the donor radius $a_d \approx 2$ nm. Thus in Si, donor electrons are weakly bound to the donor ions in orbits whose radii exceed the lattice constant of Si, $a = 0.543$ nm. The energies E_d and E_a are larger in wide-bandgap semiconductors 6H–SiC and diamond due to their lower static dielectric constants $\epsilon_r = \epsilon/\epsilon_0$ which can be correlated with their higher energy gaps (see the dielectric model for bonding described in Chapter W11).

At $T = 300$ K the thermal energy $k_B T \approx 25$ meV is sufficient to ionize all or most of the common donor and acceptor dopant atoms in Si and Ge where typical activation energies are in the range 10 to 50 meV. Since the intrinsic concentration of electrons in Si at $T = 300$ K is only $n_i \approx 8 \times 10^{15}$ m^{-3}, even the unintentional presence of donors or acceptors in Si at the level of 1 part per million will in practice greatly exceed n_i. Therefore, the carriers in Si at $T = 300$ K will usually come from the impurity dopant atoms, whose activation energies are much lower than the energy E_g required for the excitation of electrons across the bandgap. Doping is a somewhat more complicated process in compound semiconductors such as GaAs, where, for example, a substitutional Si atom can act as a donor when on a Ga site or as an acceptor when on an As site. This issue is discussed further in Section 11.11.

The fact that the measured activation energies E_d for the common donor impurities in Si vary between 39 and 49 meV is evidence that the hydrogenic model is, in fact, oversimplified and that the impurity potentials seen by the extra donor electrons in their ground states are not strictly coulombic in nature. This breakdown of the effective-mass approximation which occurs when the dopant ground-state radii are only about four times greater than the NN atom distance is due to chemical effects associated with differences between the electronic structures of the dopant and host atoms. The necessary modifications to the simple effective-mass approximation are known as *central-cell corrections*, the central cell being the lattice site occupied by the dopant atom.

Defects other than substitutional impurity atoms from groups III and V can also affect the electrical properties of semiconductors. For example, the interstitial Li$^+$ ion can act as an n-type dopant in Si and Ge. In addition, defects such as vacancies or interstitials can either act as dopants when their energy levels are shallow or as *traps* for electrons and holes when their energy levels lie deep within the energy gap. Impurity atoms from groups I, II, VI, and VII and also transition metal impurity atoms such as interstitial Fe will typically have levels that are deeper in the energy gap in Si and Ge than those of atoms from groups III and V. This is usually a result of the perturbing potential of the defect being highly localized near the defect site. In this case the bound electron or hole will also be highly localized near the defect, with a correspondingly high activation energy for release from the trap into either the conduction or valence band. The deep-level defects can exhibit charge states of 0, $-e$, or $+e$, depending on whether they are neutral or have trapped an electron or hole.

The solubilities of atoms that generate deep levels will typically be much less than those of shallow impurity atoms, due to the fact that the former cause a greater disturbance to the crystal structure of the host. These disturbances often involve the scattering of electrons and holes by the defect. Defects with deep energy levels can play important roles in the recombination of electrons with holes. Extended defects such as dislocations are very detrimental to the electronic properties of semiconductors, due to the large number of trap states they introduce into the energy gap. Some structural (as opposed to chemical) defects are found to be electronically active in semiconductors. Examples discussed in Section 11.11 include doubly charged Zn vacancies such as $V^{2-}(Zn)$, which can act as acceptors in ZnTe. Defects in materials are discussed in more detail in Chapter 4.

The actual concentrations of electrons and holes present in a semiconductor will depend on the temperature T and on the dopant concentrations N_d of donors and N_a of acceptors. The excitation of carriers by external sources of energy such as electromagnetic radiation and the injection of electrons and holes into semiconductors via their contact to external current sources can lead to inhomogeneous distributions of carriers, a topic that is discussed in Section 11.12. No matter what the source of the carriers, the np product at equilibrium at temperature T will remain a constant, given by Eq. (11.28).

The combined effects of doping and of temperature on the concentrations of electrons and holes in an extrinsic semiconductor and on the position of the Fermi energy E_F in the energy gap are discussed next for the case of n-type doping of a semiconductor with a concentration N_d of donors and with $N_a = 0$. Details of the derivation are presented in Chapter W11. The case of p-type doping will be mentioned only briefly since essentially all of the important concepts carry over from the n-type case. These two cases correspond to *uncompensated* semiconductors. The more general case of *partially compensated* semiconductors containing both donors and acceptors will also be mentioned.

Consider first n-type doping with a concentration N_d of donors and $N_a = 0$. At any temperature the donor concentration can be expressed as

$$N_d = N_d^+(T) + N_d^0(T), \tag{11.32}$$

where N_d^+ and N_d^0 are the concentrations of ionized and neutral donor atoms, respectively. Note that $N_d^-(T)$, corresponding to the occupancy of the donor level by a pair of spin-up and spin-down electrons, is negligibly small due to the resulting large Coulomb repulsion energy in this case. The fractional occupation of the neutral ground state of the donor level at an energy E_d below the conduction-band edge is given by

$$\frac{N_d^0(T)}{N_d} = \frac{1}{\frac{1}{2}e^{\beta[E_g - E_d - \mu(T)]} + 1}. \tag{11.33}$$

The factor of $\frac{1}{2}$ appears in the denominator because, as stated earlier, only a single electron with spin up or spin down will ordinarily occupy the donor ground state. Only at high temperatures such that $k_B T \gg E_d$ will essentially all of the donors be ionized with $N_d^+(T) \approx N_d$.

The requirement of electrical neutrality in an n-type semiconductor can be expressed in a general way by

$$n(T) + N_a^-(T) = p(T) + N_d^+(T), \tag{11.34}$$

where the concentrations of electrons, holes, and ionized donors and acceptors will all vary with temperature. In the high-temperature limit when $\beta[(E_g - E_d) - \mu(T)] \approx 2$ (or greater), the electron concentration is given by

$$n(T) = \frac{N_d}{2} + \sqrt{\frac{N_d^2}{4} + n_i(T)p_i(T)}. \tag{11.35}$$

When the intrinsic, high-temperature limit where $n_i(T)p_i(T) \gg N_d^2/4$ is reached,

$$n(T) = \sqrt{n_i(T)p_i(T)} = \sqrt{N_c(T)N_v(T)} \exp\left(-\frac{\beta E_g}{2}\right). \tag{11.36}$$

This equation is equivalent to Eq. (11.29), which had been derived for the intrinsic case (i.e., in the absence of both donors and acceptors).

In the $T \to 0$ K limit, the following result is obtained:

$$n(T) = \sqrt{\frac{N_c(T)N_d}{2}} \exp\left(-\frac{\beta E_d}{2}\right). \tag{11.37}$$

In this limit only a small fraction of the donors are ionized and the chemical potential μ lies just above the donor level. Measurements of $n(T)$ in this temperature range using the Hall effect can be employed to determine the activation energies E_d or E_a.

In the intermediate temperature region where $E_g > 4[E_g - \mu(T)] > 8E_d$ and $\exp(-\beta E_g) \ll (N_d/2N_c)\exp(-\beta E_d)$, the electron concentration is

$$n(T) = N_d. \tag{11.38}$$

In this region all the donors are ionized, each having donated an electron to the conduction band, and thermal excitation across the energy gap is unimportant. The hole concentration, given by $p(T) = n_i(T)p_i(T)/N_d \ll p_i(T)$, has been lowered well below the intrinsic value $p_i(T)$, due to the fact that electrons from the donor levels have fallen into the normally empty hole states in the valence band. For example, if $N_d = 2 \times 10^{24}$ m^{-3}, then in Si at $T = 300$ K where $n_i p_i = (8 \times 10^{15})^2$ m^{-6}, it follows that $p = 3.2 \times 10^7$ m$^{-3} \ll p_i$. In an *n-type* semiconductor the electrons are called the *majority carriers* since $n(T) \gg p(T)$ except at very high T when $n_i(T) > N_d$. The holes are therefore known as the *minority carriers* in an *n*-type semiconductor.

These three temperature regions (high, intermediate, and low) are illustrated in Fig. 11.9, where $n(T)$, obtained from Eq. (W11.4), for *n*-type Si with a concentration of donors $N_d = 10^{22}$, 10^{23}, and 10^{24} m^{-3} is plotted against $1/T$ in a semilogarithmic plot. These three regions are also referred to as *exhaustion* or *intrinsic* (high T), *ionization* or *saturation* (intermediate T), and *freeze-out* (low T). The intrinsic electron concentration $n_i(T)$ is included in the figure for comparison. It should be noted that it has been assumed here that E_g, m_e^*, and E_d are all independent of temperature.

When a concentration $N_a(< N_d)$ of acceptors is also present in an *n*-type semiconductor, the acceptors will be able to accept electrons from the donor atoms even at $T = 0$ K, thereby compensating a fraction N_a/N_d of the donor atoms (i.e., rendering this fraction ineffective in donating electrons to the conduction band). In this case,

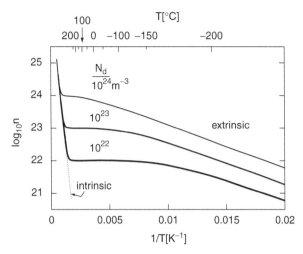

Figure 11.9. Electron concentration $n(T)$ in n-type Si with $N_d = 10^{22}$, 10^{23}, and 10^{24} m^{-3} plotted against $1/T$ on a semilogarithmic plot. The three temperature limits or regions discussed in the text are illustrated here. The intrinsic concentration of electrons, $n_i(T)$, is included for comparison.

corresponding to a partially compensated n-type semiconductor, the concentration N_d of donors should be replaced by the effective donor concentration $(N_d - N_a)$ in the formulas presented earlier.

The dependencies of the chemical potential $\mu(T)$ on temperature and indirectly on dopant concentration can be obtained from $n(T)$ by rearranging Eq. (11.27):

$$\mu(T) = E_g + k_B T \ln \frac{n(T)}{N_c(T)}. \tag{11.39}$$

The quantity $\mu(T)$ obtained in this way is presented in Fig. 11.10. Note that μ lies above the donor level at $T = 0$ K when all the donor levels are occupied in the uncompensated case. In addition, $\mu(T)$ decreases as T increases and the electrons in the donor levels are excited into the conduction band. Also, μ approaches midgap, $E_g/2$, at very high T when the concentration of carriers $n_i(T)$ thermally excited across the gap is much greater than N_d.

Expressions analogous to those derived for the n-type case just discussed can be obtained for p-$type$ doping with a concentration N_a of donors and with $N_d = 0$. In this case the holes will be the majority carriers and the electrons the minority carriers. In a partially compensated p-type semiconductor the effective acceptor concentration will be $(N_a - N_d)$. A useful plot that illustrates the effects of doping is presented in Fig. 11.11, where log n, log p, and μ are plotted versus $\log[|N_d - N_a|/(N_c N_v)^{1/2}]$ for Si at $T = 300$ K. Note that μ lies near midgap when the donor and acceptor concentrations are approximately equal (i.e., $N_d \approx N_a$).

At high concentrations of donors or acceptors, N_d or $N_a \approx 10^{25}$ m^{-3}, the dopant-related energy levels in the bandgap broaden into an energy band as a result of the spatial overlap of the electron or hole wavefunctions. At even higher concentrations, the donor energy band will merge with the bottom of the conduction band and the

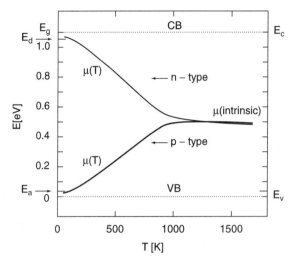

Figure 11.10. Temperature dependence of the chemical potential $\mu(T)$ for n-type Si with $N_d = 10^{22}$ m^{-3}. The chemical potential $\mu(T)$ for p-type Si with $N_a = 10^{22}$ m^{-3} is also shown.

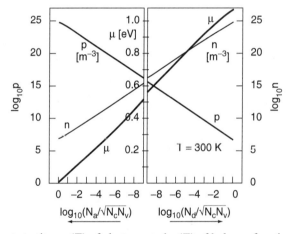

Figure 11.11. Concentrations $n(T)$ of electrons and $p(T)$ of holes as functions of the net donor and acceptor concentrations in Si at $T = 300$ K on a logarithmic plot. The position of the chemical potential μ is also shown.

acceptor energy band will merge with the top of the valence band. This occurs at $N_d \approx 2 \times 10^{25}$ m^{-3} for P in Si. In this case the Fermi level will lie in a region of extended states and Eq. (11.25) will no longer be valid. The material is then said to be *degenerate* and will behave in some ways more like a metal than a semiconductor.

11.6   Dimensionality and Quantum Confinement

The size and effective dimensionality of a solid can have important effects on its physical properties. For example, in a semiconductor the electronic wavefunctions, energy levels, and densities of states are all dependent on whether the semiconductor is a bulk, macroscopic solid (i.e., three-dimensional) or a "microscopic" solid (i.e., two-, one-, or zero-dimensional). Here the terms *two-*, *one-*, and *zero-dimensional* refer, respectively, to structures in which one, two, or all three physical dimensions are comparable to some characteristic length L, with the remaining dimensions much greater than L. Under certain circumstances L could be the electron mean free path for inelastic scattering. These structures are often referred to as *quasi-two-*, *quasi-one-*, and *quasi-zero-dimensional structures*. In the quasi-two-dimensional case, for example, this acknowledges the fact that the structure is not infinitesimally thin.

Materials in structures that have at least one dimension on the order of tens to hundreds of nanometers are said to be *mesoscopic*. Mesoscopic materials often have properties which differ from those of the same material in structures that are *macroscopic* (dimensions all typically greater than hundreds of nanometers) or *microscopic* (at least one dimension typically less than about 10 nm).

Materials that have at least one dimension on the scale of hundreds of nanometers or less are also known as *nanostructures*. It is now possible to fabricate materials such as semiconductors and metals into nanostructures with low dimensionality d in the form of *quantum wells* ($d = 2$), *quantum wires* ($d = 1$), and *quantum dots* ($d = 0$). A quantum dot can, in principle, correspond to a cluster of as few as ≈ 10 atoms or to a solid cube 10 nm on a side containing $\approx 10^5$ atoms.

The quantum nature of these low-dimensional structures can become evident if the separations of the electronic energy levels become comparable to the energies themselves, in contrast to the case of a bulk solid where the allowed energy levels within an energy band are essentially continuous. This situation corresponds to *quantum confinement* of electrons within the structure. The fabrication and study of semiconductor nanostructures is currently a very active area of research. Important applications of reduced or low-dimensional semiconductors are discussed in Sections 11.12 and 11.13 and also in Chapter 20, and their fabrication is discussed in Chapter W21.

The dependence on dimensionality of the wavefunctions, energy levels, densities of states, and Fermi wave vectors k_F for spatially confined electrons in $d = 3, 2, 1$, and 0 dimensions are summarized in Table 11.5. The three-dimensional (i.e., bulk) case has been discussed in Chapter 7. Here the electrons are taken to be confined within the regions shown in Fig. 11.12, inside which the potential $V = 0$ and at whose boundaries V increases to an infinite value. The boundaries are therefore infinite barriers. The derivation of the results for the infinite quantum well is presented in textbooks on modern physics and its extension to quantum wires and dots is straightforward. Since it is assumed that $V = 0$ within these structures, the free-electron model is appropriate for the description of the properties of these electrons. In real structures, electron-electron interactions which are not included in the free-electron model, for example, Coulomb repulsion, can modify the observed electron energy levels and densities of states. It is assumed that the electrons occupy a parabolic band with effective mass m_e^*. Similar expressions with m_e^* replaced by m_h^* apply to the confinement of holes in low-dimensional semiconductors.

The quantum numbers for electrons in quantum wells, wires, and dots will be different from their quantum numbers in bulk solids. For example, in a quantum well where the electron is confined in a region along the x axis, the appropriate quantum numbers will be the k_y and k_z components of the wave vector, the integer index n_x for the energy subbands defined in Table 11.5, and the usual spin quantum number

TABLE 11.5 Properties of Electrons in Solids of Reduced Dimensionality

	Dimensionality			
	$d = 0$ (Quantum Dot)	$d = 1$ (Quantum Wire)	$d = 2$ (Quantum Well)	$d = 3$ (Bulk)
$\psi_{\mathbf{k}}(\mathbf{r})^a$	$A\sin k_x x(\sin k_y y)(\sin k_z z)$	$A\sin k_x x(\sin k_y y)e^{ik_z z}$	$A(\sin k_x x)e^{i(k_y y+k_z z)}$	$Ae^{i(k_x x+k_y y+k_z z)}$
$E(\mathbf{k})^b = E(k_x)+$ $E(k_y)+E(k_z)$; n_x, n_y, $n_z = 1, 2, 3, \ldots$	$\dfrac{h^2}{8m_e^*}\left(\dfrac{n_x^2}{L_x^2}+\dfrac{n_y^2}{L_y^2}+\dfrac{n_z^2}{L_z^2}\right)$	$\dfrac{h^2}{8m_e^*}\left(\dfrac{n_x^2}{L_x^2}+\dfrac{n_y^2}{L_y^2}\right)+\dfrac{\hbar^2 k_z^2}{2m_e^*}$	$\dfrac{h^2 n_x^2}{8m_e^* L_x^2}+\dfrac{\hbar^2(k_y^2+k_z^2)}{2m_e^*}$	$\dfrac{\hbar^2(k_x^2+k_y^2+k_z^2)}{2m_e^*}$
$\rho_e(E)^{b,c}$	Discrete states	$\dfrac{\sqrt{2m_e^*}}{\pi\hbar L_x L_y}E^{-1/2}$	$\dfrac{m_e^*}{\pi\hbar^2 L_x}$	$\dfrac{1}{2\pi^2}\left(\dfrac{2m_e^*}{\hbar^2}\right)^{3/2}E^{1/2}$
k_F^d	$—^e$	$\dfrac{\pi n L_x L_y}{2}$	$\sqrt{2\pi n L_x}$	$(3\pi^2 n)^{1/3}$

[a]The components of the wave vector $\mathbf{k}$ of the electron are given by $k_x = n_x\pi/L_x$, $n_x = 1, 2, \ldots$.

[b]The electron (or hole) effective mass or masses appropriate to the direction or the plane of motion should be used in $E(\mathbf{k})$ and $\rho_e(E)$.

[c]Density of electron states per unit energy and unit volume.

[d]n = electron concentration = $N/L_x L_y L_z$, where N is the number of electrons confined in the region. Note that $n_{3d} = n$, $n_{2d} = nL_x$, and $n_{1d} = nL_x L_y$.

[e]For $d = 0$, $k_F = k_{max}$, where k_{max} is the maximum value of $k = \pi(n_x^2/L_x^2 + n_y^2/L_y^2 + n_z^2/L_z^2)^{1/2}$ for any electron in the quantum dot.

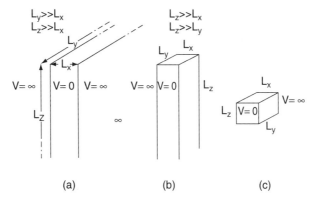

Figure 11.12. Materials with low dimensionality d in which electrons are confined: (a) a quantum well ($d = 2$); (b) a quantum wire ($d = 1$); (c) a quantum dot ($d = 0$). The boundaries of these regions are infinite potential barriers.

$m_s = \pm\frac{1}{2}$. Thus in this case the component k_x of the wave vector, while quantized in the sense that it can only take on discrete values, is not a good quantum number and is not the crystal momentum for the confined electrons.

Note that when an electron is spatially confined within a region of width L_x, its ground-state energy corresponding to $n_x = 1$ is increased above the unconfined value by an amount equal to $h^2/8m_e^*L_x^2$ due to the quantum confinement. The amount of this increase grows rapidly as L_x decreases, increasing from 1.5×10^{-12} eV to 1.5×10^{-6} eV, which is well below typical thermal energies, and then to 1.50 eV as L_x decreases from 1 mm to 1 μm, and then to 1 nm. This increase in energy is a result of the *Heisenberg uncertainty principle*, which states that a particle confined within a region of size L_x will have a resulting minimum uncertainty in its momentum given by $\Delta p_x \approx \hbar/L_x$. The corresponding minimum uncertainty in its energy, the *zero-point energy*, will therefore be $\Delta E \approx (\Delta p_x)^2/2m_e^*$, which corresponds to the ground-state energy given above. This energy enhancement can also be expected to occur for the binding energies of donors and acceptors atoms in low-dimensional doped semiconductors.

The discrete energy levels of electrons and holes in a quantum well with infinite barriers and width L_x are shown in Fig. 11.13a. These energy levels in $d = 2$ dimensions are called *energy subbands* to distinguish them from the continuous energy bands of a three-dimensional bulk solid, as illustrated in Fig. 11.3. It should be stressed that this case of electrons and holes confined in a *two-dimensional* solid structure corresponds to their confinement in a *one-dimensional* infinite quantum well. The wave vector $k_\perp$ in the plane of the quantum well is given by $k_\perp^2 = k_y^2 + k_z^2$. While in principle an infinite number of bound energy levels exist in such a quantum well with infinite potential barriers at its boundaries, in the physically attainable quantum wells with finite barriers there can exist only a finite number of bound energy levels.

Examples of two-dimensional semiconductors include quantum wells, thin films, inversion layers at the surface of a semiconductor, single-wall carbon nanotubes, and superlattices with artificial rather than natural periodicity. Some of these structures are discussed in Chapter 20. Note that in a quantum well, the effective optical energy gap

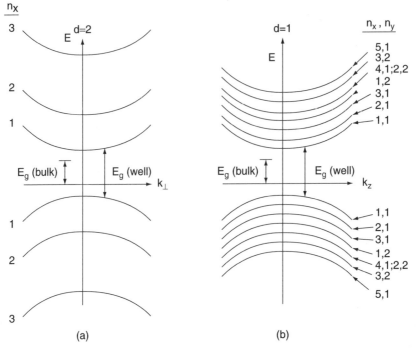

Figure 11.13. Energy levels of quantum-confined electrons and holes: (*a*) electron and hole energy levels in a quantum well with infinite barriers and width L_x as a function of the wave vector $k_\perp$ in the plane of the well; (*b*) Electron energy levels in a rectangular quantum wire with infinite potential barriers and dimensions $L_x = 2L_y \ll L_z$ as a function of the wave vector k_z along the length of the wire.

of the semiconductor depends on the width L_x of the well and is given by

$$E_g(L_x) = E_{g0} + \frac{h^2}{8L_x^2} \left(\frac{1}{m_e^*} + \frac{1}{m_h^*} \right) \qquad (11.40)$$

since the electron and hole ground-state energies are both shifted "higher" in energy due to the quantum confinement. Here E_{g0} is the energy gap in the bulk limit ($L_x \to \infty$) and m_e^* and m_h^* are the effective masses appropriate to the plane in which the electrons and holes move.

The electron energy levels for a rectangular-cross-section quantum wire with infinite barriers and dimensions $L_x = 2L_y \ll L_z$ are shown in Fig. 11.13*b* as a function of the wave vector k_z for motion along the wire. In this case the confinement of electrons in a *one-dimensional* solid corresponds to their confinement in a *two-dimensional* infinite quantum well. In practice the energy levels in quantum wells and wires shown in Fig. 11.13 will be broadened due to the scattering of electrons by phonons and lattice defects.

The densities of electron states $\rho_e(E)$ in $d = 0$, 1, 2, and 3 dimensions are presented in Fig. 11.14, where infinite barrier heights are assumed. Note that $\rho_e(E)$ is nonzero for all energies greater than the ground-state energies E_{11} in $d = 1$, E_1 in $d = 2$, and E_c in $d = 3$ dimensions. For $d = 0$, $\rho_e(E)$ consists of discrete states since in this case none of

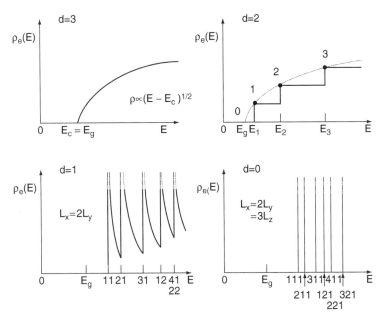

Figure 11.14. Electron densities of states $\rho_e(E)$ in $d = 0$, 1, 2, and 3 dimensions. The energy gap E_g for $d = 3$ is shown in each case.

the components of the electron's wave vector can take on a continuous range of values. Note that $\rho_e(E)$ for $d = 1$ and 2 are given by the superposition of the densities of states of the individual energy subbands. For the $d = 2$ quantum well, the resulting steplike $\rho_e(E)$ is constant within each energy subband, with the locus of points corresponding to the edges of the steps lying on the continuous $\rho_e(E)$ curve for $d = 3$, as shown. Due to the coupling between the quantum wells in a $d = 2$ superlattice, such as that formed when GaAs wells are separated by finite $Al_xGa_{1-x}As$ barriers, $\rho_e(E)$ will be intermediate between the densities of states of $d = 2$ and $d = 3$. In addition, the steps in $\rho_e(E)$ for a superlattice will not be as sharp as those shown in Fig. 11.14 for $d = 2$. It can be seen from Figs. 11.13 and 11.14 that the ground states of electrons and holes increase in energy as the dimensionality d decreases from 3 to 2 to 1 to 0 (i.e., as the degree of confinement increases).

Quantum confinement and the corresponding quantized energy levels of electrons and holes in semiconductors are more readily observed in materials at low temperatures when the electron mean free path l due to scattering from phonons increases and becomes comparable to or greater than the dimensions of the structure in which the electron is confined. For shorter values of l and corresponding shorter average scattering times $\langle \tau \rangle$, the broadening of the energy levels becomes comparable to their separation, and the effects of quantum confinement are not as obvious.

MACROSCOPIC PROPERTIES

The macroscopic properties of homogeneous semiconductors which are associated with the response of uniform distributions of electrons and holes to external electric and magnetic fields and to light are described next. Thermoelectric effects in semiconductors are described in Chapter W11. The microscopic properties of semiconductors involving the electron energy-band structure and energy gaps, the dynamics of the motions of electrons and holes, the equilibrium carrier concentrations, and the effects of doping play important roles in these discussions.

11.7 Electrical Conductivity and Mobility

The electrical properties of semiconductors involve the motions of electrons and holes in response to applied electric fields. This response can be described in terms of the electrical conductivities and mobilities of the carriers. The electrical properties of semiconductors are also quite sensitive to light for a variety of reasons. These light-induced effects are discussed later when the optical properties of semiconductors are described.

In the presence of an electric field $\mathbf{E}$ the total electrical current density $\mathbf{J}$ in a semiconductor will be the sum of the currents of the unpaired charge carriers in both the conduction and valence bands (i.e., the sum of the electron and hole currents). The contribution of holes to the electrical conductivity can be understood from Fig. 11.6. In an electric field the change in the wave vector $\mathbf{k}_h$ of a hole in the valence band will occur in the same direction as $\mathbf{E}$, while the change in $\mathbf{k}_e$ for an electron in the valence band will occur opposite to $\mathbf{E}$. In terms of the transport of electric charge the motion of the hole in the direction of $\mathbf{E}$ is thus completely equivalent to the motion of the electron opposite to $\mathbf{E}$ as it moves into the empty electron state associated with the hole, as illustrated schematically in Fig. 11.6a.

The current density $\mathbf{J}$ in a semiconductor is given by

$$\mathbf{J}(\mathbf{E}) = \sigma\mathbf{E} = (\sigma_e + \sigma_h)\mathbf{E}. \tag{11.41}$$

Here σ_e and σ_h are the electrical conductivities of the electrons and holes, respectively, given by

$$\sigma_e(T) = \frac{n(T)e^2\langle\tau_e(T)\rangle}{m_{ec}^*} = n(T)e\mu_e(T),$$

$$\sigma_h(T) = \frac{p(T)e^2\langle\tau_h(T)\rangle}{m_{hc}^*} = p(T)e\mu_h(T). \tag{11.42}$$

Note that m_{ec}^* and m_{hc}^* are the appropriate average effective masses, known as the *conductivity effective masses*,[†] and $\langle\tau_e\rangle$ and $\langle\tau_h\rangle$ are the average electron and hole *scattering times* (i.e., mean free times) for electrons at the bottom of the conduction band and holes at the top of the valence band. The times $\langle\tau_e\rangle$ and $\langle\tau_h\rangle$ are the momentum relaxation times. The *carrier mobilities* μ_e for electrons and μ_h for holes defined in Eq. (11.42) are discussed in more detail later. The general form of these expressions for σ_e and σ_h has been derived using the classical Drude model in Chapter 7.

[†] The conductivity effective masses for electrons and holes in Si correspond to the appropriate averages over the six conduction-band minima for m_{ec}^* and over the two hole valence bands for m_{hc}^*. These averages are $1/m_{ec}^* = 1/(3m_{Le}^*) + 2/(3m_{Te}^*)$ and $1/m_{hc}^* = 1/m_{hh}^* + 1/m_{lh}^*$. These yield $m_{ec}^* = 0.26m$ and $m_{hc}^* = 0.12m$ for Si when the effective masses listed in Table 11.8 are used.

The approach used to obtain these expressions for σ involves equating the rate at which energy is gained by the charge carriers from the electric field to the rate at which they lose energy to the lattice via inelastic scattering processes involving phonons. Ohm's law in the form $\mathbf{J} = \sigma\mathbf{E}$ is valid in the low-field limit where σ is independent of E. Deviations from Ohm's law can occur in the bulk of a semiconductor at high electric fields and also at its surface when electrical contacts made to the semiconductor are nonohmic.

The *drift velocity* $\langle v \rangle$ of a charge carrier in the presence of an electric field $\mathbf{E}$ is defined in terms of the current density by $\mathbf{J} = ne\langle\mathbf{v}\rangle$, where n is the concentration of charge carriers. Using Eqs. (11.41) and (11.42), the drift velocities of electrons and holes can be expressed in terms of the corresponding average scattering times and effective masses by

$$\langle v_e \rangle = \frac{e\langle \tau_e \rangle E}{m_{ec}^*} \quad \text{and} \quad \langle v_h \rangle = \frac{e\langle \tau_h \rangle E}{m_{hc}^*}, \tag{11.43}$$

with both drift velocities defined to be positive quantities. The electron and hole mobilities μ_e and μ_h may also be defined as the drift velocities per unit electric field and can therefore be expressed in terms of the average scattering times and effective masses using either Eq. (11.41) or (11.43) by

$$\mu_e = \frac{\langle v_e \rangle}{E} = \frac{e\langle \tau_e \rangle}{m_{ec}^*} \quad \text{and} \quad \mu_h = \frac{\langle v_h \rangle}{E} = \frac{e\langle \tau_h \rangle}{m_{hc}^*}. \tag{11.44}$$

It can be seen that carriers with low effective masses or long scattering times will have high mobilities. Mobilities and scattering times will in general depend not only on temperature and on the concentration of defects in the semiconductor but also on both the energy of the electron or hole and on the magnitude and direction of the electric field $\mathbf{E}$.

The important processes that shorten the average scattering times and therefore limit the carrier mobilities μ_e and μ_h are the scattering of electrons and holes by phonons and also by defects such as ionized impurities (e.g., by donor and acceptor ions). Elastic scattering from defects dominates at low temperatures, and phonon scattering dominates at higher temperatures as more phonons are created in the material. Inelastic scattering of charge carriers by polar optical phonons is particularly effective at high temperatures. When scattering from ionized defects dominates, the mobility will actually increase with T, with $\mu(T)$ predicted to be proportional to $T^{3/2}$. The opposite behavior is expected when scattering from phonons dominates, with $\mu(T) \propto T^{-3/2}$ predicted for nondegenerate semiconductors. In heavily doped semiconductors the mobility is often observed to go through a maximum as T increases and then to decrease as phonon scattering begins to dominate at higher T.

Values of μ_e and μ_h and the electrical conductivity σ are given Table 11.6 at $T = 300$ K for several intrinsic semiconductors. Note that the mobilities of carriers in semiconductors are typically in the range 0.01 to 1 $m^2/V \cdot s$. At low temperatures, $T \approx 1$ to 10 K, mobilities in semiconductors can be hundreds of times greater than these values at $T = 300$ K. Carrier mobilities in semiconductors are typically much higher than those in metals [e.g., $\mu_e(Cu) \approx 0.0035$ $m^2/V \cdot s$]. This is due primarily to

TABLE 11.6 Electron and Hole Mobilities and Electrical Conductivities of Several Intrinsic Semiconductors at $T = 300$ K

	$\mu_e(m^2/V{\cdot}s)^a$	$\mu_h(m^2/V{\cdot}s)^a$	$\sigma(\Omega^{-1}m^{-1})^b$
Si	0.190	0.050	$\approx 3 \times 10^{-4}$
Ge	0.380	0.182	≈ 1.8
β-SiC	0.10	0.004	$\approx 6.6 \times 10^{-15}$
GaAs	0.90	0.050	$\approx 4.6 \times 10^{-7}$

^{a}For Si, Ge, and GaAs: data from D. R. Lide, ed., *CRC Handbook of Chemistry and Physics*, 75th ed., CRC Press, Boca Raton, Fla., 1994, pp. 12–87 to 12–96. For β-SiC: data from H. Morkoc et al., *J. Appl. Phys.*, **76**, 1363 (1994).
^{b}Calculated from $\sigma = \sigma_e + \sigma_h$ and Eq. (11.42) using the mobilities given in this table and the carrier concentrations given in Table 11.1.

the very low effective masses for carriers in semiconductors compared to metals, where $m^* \approx m$.

The dependencies of the electron and hole mobilities on dopant atom concentrations in Si, Ge, and GaAs at $T = 300$ K are shown in Fig. 11.15. It can be seen that μ_e and μ_h are essentially independent of the dopant concentrations N_d or N_a up to 10^{22} m^{-3} and then decrease by up to an order of magnitude as N_d or N_a increase to $\approx 10^{25}$ m^{-3}. These observed decreases in mobility are due to the increasing scattering of the carriers from the ionized dopant atoms.

Carrier mobilities in semiconductors are typically determined indirectly from $\mu = \sigma/ne$ (for electrons) by using the results of measurements of the electrical conductivity σ and of the carrier concentration n or p from the Hall effect. The mobility obtained in this way is called the *Hall mobility* μ_H. The Hall mobility may not be the same as the mobility defined by $\mu = e\langle\tau\rangle/m_c^*$ when the scattering time depends on electron energy. Average scattering times can be determined from $\langle\tau\rangle = \mu m_c^*/e$, where m_c^* is the appropriate conductivity effective mass. Some typical values obtained in this way for intrinsic Si at $T = 300$ K are $\langle\tau_e\rangle = 2.8 \times 10^{-13}$ s and $\langle\tau_h\rangle = 3.5 \times 10^{-14}$ s.

The temperature dependence of $\sigma(T)$ in intrinsic semiconductors is dominated by the effects of the energy gap E_g (i.e., by the exponential temperature dependencies of the carrier concentrations n and p). Thus $\sigma(T)$ in this case increases exponentially with increasing T due to the thermal excitation of electrons and holes, in contrast to the behavior of metals, where the carrier concentration is constant and where σ decreases with increasing T due to increased electron–phonon scattering. In nondegenerate doped semiconductors $\sigma(T)$ is dominated at low T by thermal excitation of the donor and acceptor atoms and only at high T by thermal excitation of electrons across the bandgap. The weak power law temperature dependencies of the mobilities described earlier play relatively minor roles in the temperature dependence of $\sigma(T)$.

The electrical resistivity $\rho(T) = 1/\sigma(T)$ for n-type Si is shown in Fig. 11.16 from low temperatures, where most of the donor atoms are neutral, through the region where they are all ionized, and finally to high temperatures, where $\rho(T)$ is dominated by the thermal excitation of electrons across the energy gap. The resistivity $\rho(T)$ is therefore dominated by $n(T)$ at low T and again at high T, with the effects of $\mu(T)$ being observable only for intermediate T, where $n(T)$ is essentially constant and equal to

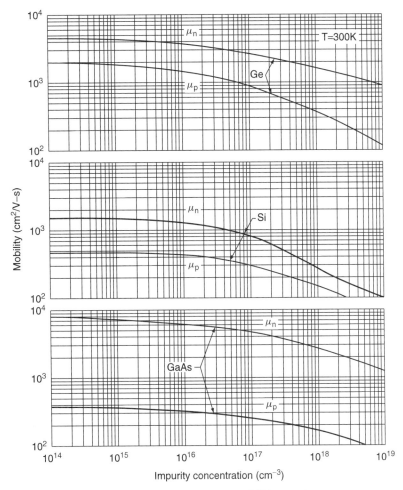

Figure 11.15. Dependencies of the electron and hole mobilities on dopant atom concentrations in Si, Ge, and GaAs at $T = 300$ K. (From S. M. Sze, *Physics of Semiconductor Devices*, 2nd ed., copyright 1981 by John Wiley & Sons Inc. Reprinted by permission of John Wiley & Sons, Inc.)

N_d, the concentration of donor atoms. At high T the temperature dependence of $\sigma(T)$ can be expressed by

$$\sigma(T) \propto T^{-3/2} \exp\left(\frac{-\beta E_g}{2}\right) \tag{11.45}$$

where the $T^{-3/2}$ prefactor originates from $\mu(T)$. The slope of a plot of the logarithm of $\sigma(T) \cdot T^s$ versus $1/T$ is a simple, direct way of determining E_g for a semiconductor. The exponent s can be used as a fitting parameter, with the value $s = \frac{3}{2}$ expected to apply when phonon scattering dominates.

In disordered or amorphous semiconductors such as a-Si the electrical conduction at low temperatures is dominated by *variable-range hopping* and obeys the $T^{1/4}$ *law* (i.e., $\sigma(T) = \sigma_0(T) \exp[-(T_0/T)^{1/4}]$). The charge carriers in this case occupy and

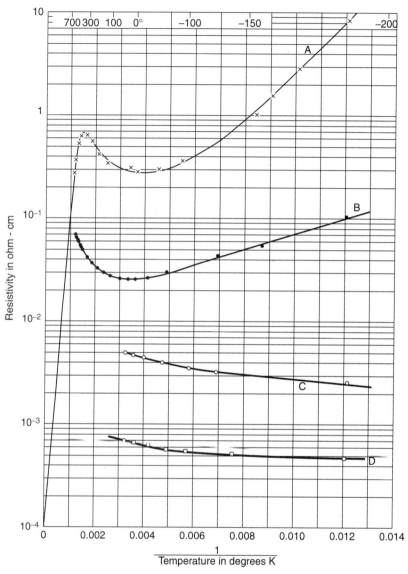

Figure 11.16. Electrical resistivity $\rho(T)$ for n-type phosphorus-doped Si from low temperatures where most of the donor atoms are neutral, through the region where they are all ionized, and finally, to high temperatures where $\rho(T)$ is dominated by the thermal excitation of electrons across the energy gap. P concentrations (in cm^{-3}): A = 4.7×10^{17}, B = 2.7×10^{18}, C = 4.7×10^{19}, D = 4.7×10^{20}. [From G. L. Pearson et al., *Phys. Rev.*, **75**, 865 (1949). Copyright 1949 by the American Physical Society.]

move through localized states and as a result have very low mobilities $\mu \approx 10^{-6}$ to 10^{-3} m^2/V·s. Their motion in the presence of an electric field occurs via thermally activated hopping from one localized state to another, which need not necessarily be the nearest state but one that is close to the first state in energy. Variable-range hopping is discussed in more detail and the $T^{1/4}$ law is derived in Chapter 7.

The flow of electrical currents resulting from the net diffusion of carriers in semiconductors can occur when gradients of carrier concentration exist. This situation is important in semiconductor devices and is discussed in Section 11.12.

High-Field Effects. Electrons or holes can gain energy from high electric fields at rates which are greater than the rates at which they lose energy to the lattice via phonon emission. Under these high-field conditions the electrons and holes will no longer be in thermal equilibrium with the lattice, although they can remain in thermal equilibrium with each other. This is a transient situation that lasts less than about 10^{-12} s after the external field is removed. Electrons and holes are referred to as *hot carriers* when their temperatures become higher than that of the lattice. Some examples of effects related to hot carriers that are important in semiconductor devices such as the junction field-effect transistor are described in Chapter W11.

At high electric fields, $E \approx 10^6$ V/m, the electrical current in a semiconductor is in general no longer simply proportional to the electric field E. In this case the electrical conductivity σ becomes field-dependent through the mobility μ and the drift velocity $\langle v \rangle$. This behavior of the measured drift velocities of electrons and holes in Si, Ge, and GaAs at $T = 300$ K is illustrated in Fig. 11.17. It can be seen that Ohm's law with $\langle v \rangle$ proportional to E is valid only up to 10^5 to 10^6 V/m for these semiconductors. At higher fields the drift velocities no longer increase linearly with E and eventually reach essentially constant values of about 10^5 m/s, known as the *saturation velocity*. The saturation of the drift velocity occurs when the energy of the carriers no longer increases with increasing E but instead, remains constant as a result of inelastic scattering processes involving the emission of optical phonons. High values of the electron saturation velocity are desirable in some semiconductor devices in order to decrease the transit time of the charge carriers through the active region of the device.

The observed decrease in $\langle v \rangle$ for electrons in GaAs at fields above about 2×10^5 V/m is termed *negative differential resistance* (NDR). This decrease in a direct-gap semiconductor such as GaAs results from the scattering of electrons at high electric fields into conduction bands which lie only a few tenths of an electron volt above the bottom of the conduction band at $\mathbf{k} = 0$. Since the electrons in the higher-conduction-band valleys have lower mobilities, the net effect is a decrease in the average mobility and drift velocity. The current oscillations in GaAs which can result from the NDR effect are known as the *Gunn effect*.[†]

Another important aspect of high-field behavior involves dielectric breakdown, also discussed in Chapter 15. Consider a semiconductor or insulator with an energy gap E_g and lattice constant a. An electric field of magnitude $E \approx E_g/ea$ is therefore required to provide the energy needed to promote a valence-band electron directly across the energy gap. The transition of the electron from the valence band to the conduction band can be viewed as a quantum-mechanical tunneling process since in a very high electric field the energy bands will be position dependent in the material. Using typical values of $E_g = 1$ eV and lattice constant $a = 0.3$ nm, the electric field required for dielectric breakdown is $E \approx 3 \times 10^9$ V/m, much higher than the electric fields at which field-dependent effects begin to be observed.

[†] J. B. Gunn, *Solid State Commun.*, **1**, 88 (1963).

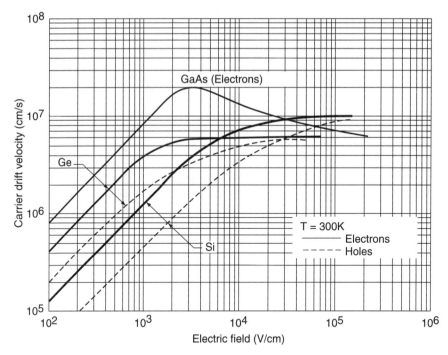

Figure 11.17. Dependencies of the drift velocities $\langle v \rangle$ of electrons and holes in Si, Ge, and GaAs on applied electric field at $T = 300$ K. (From S. M. Sze, *Physics of Semiconductor Devices*, 2nd ed., copyright 1981 by John Wiley & Sons Inc. Reprinted by permission of John Wiley & Sons, Inc.)

11.8   Effects of Magnetic Fields

The application of a magnetic field to a semiconductor is generally used for diagnostic purposes, such as measurements of the Hall effect or of cyclotron resonance, which yield, respectively, the sign and net concentration of the charge carriers and the carrier effective mass, m_e^* or m_h^*.

Hall Effect.   The effect of the magnetic induction or flux density $\mathbf{B} = \mu\mathbf{H}$ on the charge carriers in a conductor (i.e., a semiconductor or a metal) carrying a current density $\mathbf{J}$ is described in Chapter 7. Here μ is the magnetic permeability of the material and $\mathbf{H}$ is the applied magnetic field. When the current flow $\mathbf{J}_x$ is perpendicular to $\mathbf{B}_z$, a transverse electric field $\mathbf{E}_y$, the *Hall field*, is developed. For the simple case in which charge carriers of only one sign and all having the same isotropic effective mass and scattering time are present, the determination of the Hall field E_y and the *Hall coefficient* R_H allow both the sign and concentration of the charge carriers to be obtained using the expression

$$R_H = \frac{E_y}{J_x B_z} = \frac{1}{nq}, \qquad (11.46)$$

where $q = \pm e$. As mentioned earlier, in this case the Hall mobility μ_H of the charge carriers can be determined from $\mu_H = \sigma|R_H|$. The ratio of the transverse Hall field E_y

to the longitudinal electric field E_x determines the *Hall angle* θ_H through

$$\tan \theta_H = \frac{E_y}{E_x} = \frac{J_x B_z}{E_x n q} = \frac{\sigma_x B_z}{n q} = \frac{q \mu_x B_z}{|q|}, \tag{11.47}$$

where the definition of R_H and standard relationships between J, E, σ, and mobility μ have been used. It can be seen that the sign of the charge carrier in this case also determines the sign of the Hall angle θ_H.

Initially unexplained results obtained from Hall effect measurements in certain semiconductors indicated that the charge carriers could have positive charges and, in combination with cyclotron resonance results, that these positive charge carriers could also have high mobilities μ, comparable to those typically observed for electrons in semiconductors. These mobilities seemed too high to be associated with positively charged cations. It is now clear that these positive charge carriers are just the holes described earlier.

As described in Chapter W7, the analysis of the Hall effect becomes considerably more complicated in the more general case when charge carriers with different charges q, concentrations n, or mobilities μ are present. The results for R_H are presented here for the case when only two different types of charge carriers are present, with charges q_1 and q_2, concentrations n_1 and n_2, and mobilities μ_1 and μ_2, respectively. This case can correspond to the presence of electrons and holes, of two types of electrons, or of two types of holes. The analysis begins by adding the currents of the two types of carriers since, according to Eq. (11.41), the current densities of two types of charge carriers are additive even when they are electrons and holes.

It is assumed in this case that the transverse Hall voltage $V_y = E_y L_y$ is small compared to the longitudinal voltage $V_x = E_x L_x$ (i.e., that the dimensionless parameter $\mu_x B_z \ll 1$). It follows that the Hall angle $\theta_H \ll 1$ and $\tan \theta_H \approx \theta_H$. The following expression for R_H is then obtained:

$$R_H = \frac{q_1 n_1 \mu_1^2 + q_2 n_2 \mu_2^2}{q^2 (n_1 \mu_1 + n_2 \mu_2)^2}. \tag{11.48}$$

This equation reduces to the single-carrier case expressed by Eq. (11.46) when $n_1 = n$ and $n_2 = 0$. Even if the charges q_1 and q_2 are known, measurements of the Hall coefficient R_H, of $\sigma = n_1 q_1 \mu_1 + n_2 q_2 \mu_2$, and of the magnetoresistance, which also depends on these parameters, will not in general be sufficient to determine the four unknowns n_1, μ_1, n_2, and μ_2.

In an intrinsic semiconductor where the two types of charge carriers have equal concentrations (i.e., where $n_i = p_i$), the expression for R_H given in Eq. (11.48) becomes

$$R_H = \frac{\mu_h - \mu_e}{n e (\mu_h + \mu_e)}. \tag{11.49}$$

Here the sign of R_H will be determined by the sign of the charge carrier with the higher mobility. If electrons have much higher mobilities than holes (i.e., $\mu_e^2 \gg \mu_h^2$), as is often the case, this equation reduces to the simple expression $R_H = 1/nq$.

When the mobility μ and average scattering time $\langle \tau \rangle$ of the charge carriers depend on their energy, the Boltzmann equation must be used to analyze the Hall effect. In this case correction factors such as $\langle \tau^2 \rangle / \langle \tau \rangle^2$ involving different averages of τ will appear in R_H. In addition, in high magnetic fields such that $\omega_c \tau = \mu_e B \approx 1$, where ω_c is the cyclotron frequency, the Hall coefficient R_H becomes dependent on the applied magnetic field. In very high magnetic fields R_H approaches the limiting value $1/(p-n)e$ given in Eq. (7.114). See Ferry (1991) for a useful discussion of the Hall effect.

Cyclotron Resonance. The behavior of charge carriers in a conductor due to the presence of a magnetic flux density **B** is described in Chapter 7. When a frequency-dependent electric field $\mathbf{E}(\omega)$ in a direction perpendicular to **B** is also present, the charge carriers can absorb energy from the electric field when ω is equal to the *cyclotron frequency* $\omega_c = qB/m^*$. Thus a measurement of ω_c via *cyclotron resonance* allows the determination of the effective mass m^*. This band-structure effective mass is in general anisotropic, and its values in different directions in the material can be mapped out in this way. In addition, the measured width of the cyclotron resonance line can be used to obtain the scattering time τ of the charge carriers.

When cyclotron resonance is measured for electrons in semiconductors such as Si, the details of the anisotropic valleys at the conduction-band minima play an important role in the resonances observed. In addition, the presence of heavy and light holes at the top of the valence band also increases the number of resonant frequencies observed for a given direction of **B**. The analysis of the experimentally observed resonances, although complicated,[†] can be carried out in order to give, in the case of Si, the electron effective masses m^*_{Le} and m^*_{Te} as well as the hole effective masses m^*_{hh} and m^*_{lh} listed in Table 11.8.

11.9 Optical Properties

The optical response of semiconductors in the infrared (≈ 0.01 to 1.7 eV), visible (1.7 to 3 eV), and near-ultraviolet (3 to 10 eV) regions can involve the excitation of phonons, pairs of electrons and holes, and other species, such as excitons and polaritons. Phenomena associated with the absorption of light include photoconductivity, photoluminescence, the Raman and Brillouin effects, and other electric field–related effects such as electroluminescence and the electro-optic effect. The recombination of electrons in the conduction band with holes in the valence band can result in the emission of light via the radiative process:

$$\text{electron} + \text{hole} \longrightarrow \text{photon}. \tag{11.50}$$

Nonradiative processes are also possible in which phonons are emitted as a result either of electron–hole recombination or of electrons or holes deep within their respective energy bands relaxing or "falling" into lower-lying, vacant energy levels, thereby losing energy to the lattice which ultimately winds up in the form of heat.

Time-dependent effects associated with the absorption of light in a semiconductor, with the generation of nonequilibrium concentrations of electrons and holes, and with the recombination of electrons and holes are discussed in Section 11.12. The

---

[†] For a useful description of this procedure, see Sapoval and Hermann (1993, App. 3.1).

optical properties of semiconductors are critical for a variety of device applications, including light-emitting diodes, lasers, and photodetectors. The electrical properties of semiconductors are also quite sensitive to electromagnetic radiation with frequency $\omega > \omega_g = E_g/\hbar$. When the electric field $\mathbf{E}$ in Eq. (11.5) varies with a sufficiently high frequency $\omega \approx \omega_g \approx 10^{14}$ rad/s, the motion of electrons in a material can no longer be described in terms of the evolution of $\mathbf{k}_e$ within a given energy band, and transitions to higher energy bands must also be considered.

The topic of photoconductivity is discussed briefly in this section. Nonlinear optical effects which are dependent not only on the intensity I of the light but also on higher powers of I are discussed in Chapters 8 and 18. The optical properties of semiconductors, including excitons and polaritons, are discussed in Chapter 8; photoconductivity, the electro-optic effect, and semiconductor lasers and light-emitting diodes are discussed in Chapter 18.

The most obvious optical effect occurring in semiconductors is the transition from optical transparency for photons with energies $\hbar\omega < E_g$ to the strong absorption of photons with $\hbar\omega > E_g$. This transition occurs in the infrared for semiconductors such as Si, Ge, GaAs, HgTe, and InAs, in the visible for AlAs, GaP, β-SiC, and ZnSe, and in the near-ultraviolet for GaN, ZnS, and diamond. As described earlier, the absorption of light for $\hbar\omega > E_g$ can occur via direct or indirect (i.e., phonon-assisted) processes. The shape of the optical absorption edge of a semiconductor depends on the type of processes that can occur at or near $\hbar\omega = E_g$ and is illustrated later for both direct- and indirect-bandgap semiconductors.

Optical spectroscopy plays an important role in the determination of the energy levels associated with the elementary excitations in semiconductors [i.e., the electron energy-band structure $E_n(\mathbf{k})$ and also the energies associated with phonons and excitons]. A description of the various optical spectroscopic methods is presented in Chapter W22. Important information obtained from some of these techniques is presented next, beginning with measurements of optical absorption.

Optical Absorption. The *optical absorption coefficient* α is defined to be the fractional decrease per unit length of the intensity of light I in a material:

$$\alpha = -\frac{1}{I}\left(\frac{dI}{dx}\right). \tag{11.51}$$

In terms of the wavelength λ of the light in vacuum and the *complex index of refraction*, defined by $\tilde{n}(\lambda) = n(\lambda) + i\kappa(\lambda)$, where n is the *index of refraction* and κ is the *extinction coefficient*, the absorption coefficient is given by

$$\alpha(\lambda) = \frac{4\pi\kappa(\lambda)}{\lambda} = \frac{2\pi\epsilon_2(\lambda)}{n(\lambda)\lambda}. \tag{11.52}$$

Here ϵ_2 is the imaginary part of the *relative permittivity* or *dielectric function* $\epsilon_r = \epsilon/\epsilon_0 = \epsilon_1 + i\epsilon_2$. The inverse of the absorption coefficient, $1/\alpha(\lambda)$, can be considered to be the mean free path or absorption depth for a photon of wavelength λ/n in a medium of index of refraction n. For example, when $\alpha = 10^6$ m^{-1} at the band edge of a semiconductor, the optical absorption depth is 1 μm.

Various expressions have been derived for the dependence of α on photon energy near the absorption edge of a crystalline semiconductor. The form of these expressions depends on whether the semiconductor has a direct or indirect bandgap:

$$\text{Direct gap:} \quad \alpha(\omega) = A_d \frac{(2\mu)^{3/2}(\hbar\omega - E_g)^{1/2}}{n\omega}. \tag{11.53}$$

$$\text{Indirect gap:} \quad \alpha(\omega) = A_i \frac{(\hbar\omega - E_g \pm \hbar\omega_{\text{phonon}})^2}{\omega}. \tag{11.54}$$

In Eq. (11.53), μ is the reduced effective mass of the electron–hole pair created in the transition, that is,

$$\frac{1}{\mu} = \frac{1}{m_e^*} + \frac{1}{m_h^*}. \tag{11.55}$$

In Eq. (11.54), $\hbar\omega_{\text{phonon}}$ is the energy of the phonon that participates in the indirect absorption process. The prefactors A_d and A_i include the square of the momentum matrix elements and fundamental constants. In these expressions for $\alpha(\omega)$, the electron densities of states at the band edges have been taken to be parabolic [i.e., $\rho_e(E) \propto (E - E_g)^{1/2}$ for the conduction band, as in the free-electron model]. Thus a measurement of the dependence of α on photon energy $\hbar\omega$ near the absorption edge allows the determination of whether a semiconductor has a direct or indirect bandgap.

The absorption edge for the direct-bandgap semiconductor GaAs at $T = 300$ K and at 77 K is shown in Fig. 11.18, where the logarithm of α is plotted as a function of

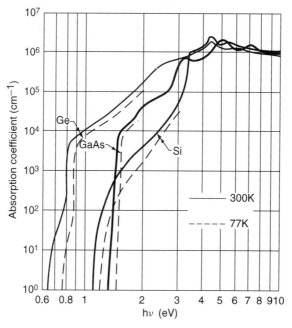

Figure 11.18. Optical absorption edges for the direct-bandgap semiconductor GaAs and for the indirect-bandgap semiconductors Ge and Si at $T = 300$ K and 77 K, with the logarithm of the absorption coefficient α plotted versus the photon energy $\hbar\omega$. (From S. M. Sze, *Physics of Semiconductor Devices*, 2nd ed., copyright 1981 by John Wiley & Sons, Inc. Reprinted by permission of John Wiley & Sons, Inc.)

the photon energy. It can be seen that α increases rapidly at the band edge, by about four orders of magnitude, thus allowing accurate determination of E_g. The predicted dependence of $\alpha(\omega)$ expressed in Eq. (11.53) is obeyed rather well as long as the photon energy is not much greater than E_g.

The absorption edges of the indirect-bandgap semiconductors Ge and Si are also shown in Fig. 11.18. It can be seen that the onset of absorption at E_g is rather weak, with strong absorption setting in only when the lowest-energy direct transition between the valence and the conduction bands is possible. The decrease of E_g with increasing T which has been mentioned earlier can be clearly observed here for all three semiconductors.

For bulk samples that are too thick for a measurable amount of light to be transmitted even for $\hbar\omega < E_g$, it is necessary to measure the reflectance R of the material and then to extract the real and imaginary parts of the optical dielectric function using the Kramers–Kronig relations, which are described in Chapter 8.

The real and imaginary parts of the complex index of refraction, n and κ, respectively, are shown for the group IV elements C (diamond), Si, and Ge as functions of photon energy $\hbar\omega$ in Fig. 11.19. Note that n approaches a constant value for $\hbar\omega \ll E_g$ and that κ increases from zero as $\hbar\omega$ approaches E_g. The quantity κ increases more rapidly for Ge and Si at 1 to 2 eV above E_g as direct transitions to higher-lying conduction bands at $\mathbf{k} = 0$ become possible.

Optical absorption is also a very useful tool for studying the effects of quantum confinement on the energy levels of electrons and holes. The primary effect is a "blue shift" in the absorption edge to higher energies, due to a widening of the effective energy gap. This increase in E_g can readily be observed in quantum wells and results from increases in the binding energies of both the electrons and the holes in the well, as expressed in Eq. (11.40). Experimental evidence that quantum confinement has occurred and that quantum size effects are observed can be obtained by studying the absorption spectrum as a function of sample size. If the energy levels and energy gaps for quantum wells of decreasing width L_x increase as $1/L_x^2$, this is a strong indication that quantum confinement has occurred.

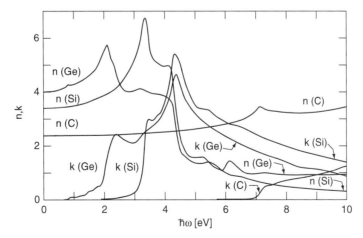

Figure 11.19. Real and imaginary parts, n and κ, of the complex index of refraction, for the group IV elements C (diamond), Si, and Ge as functions of photon energy $\hbar\omega$. (Data from E. D. Palik, *Handbook of Optical Constants of Solids*, Vol. 1, Academic Press, 1985.)

Excitons. Structure in the form of discrete absorption peaks is often observed in direct-bandgap semiconductors such as GaAs at photon energies just below E_g. These absorption peaks are due to the creation of *excitons* (i.e., the bound electron–hole pairs described in Chapter 8) and become more pronounced as the temperature is lowered. The absorption of a photon with energy $\hbar\omega \geq E_g$ provides enough energy for the creation of an electron in the conduction band and a hole in the valence band which are free to move independently through the material. Excitons can be created by the absorption of photons with energies $\hbar\omega < E_g$ when the electron and hole form a bound system due to their mutual Coulomb attraction. This is analogous to the case of an electron bound to a donor atom or a hole bound to an acceptor atom in a semiconductor, and a similar model can be developed to describe the properties of excitons. When generated by photons with energies $\hbar\omega > E_g$, a bound exciton can move through the semiconductor with kinetic energy which can be as large as $\hbar\omega - E_g$. Excitons can therefore either be free to move through the semiconductor or can be localized or bound (e.g., to neutral donor or acceptor impurity atoms and to strongly electronegative substitutional impurities such as N in GaP).

As derived in Chapter 8, the exciton ground-state binding energy in a simple hydrogenic model in which the valence and conduction bands are assumed to be isotropic and parabolic is predicted to be

$$E_B = -\frac{\mu e^4}{2(4\pi\epsilon)^2 \hbar^2} = -\frac{\mu}{m}\left(\frac{\epsilon_0}{\epsilon}\right)^2 E_1, \tag{11.56}$$

where μ is the reduced effective mass of the electron–hole pair defined in Eq. (11.55), and the other parameters are the same as in the dopant impurity model described earlier (i.e., $E_1 = 13.6$ eV). The exciton binding energy for GaAs, calculated using the information provided in Table 11.4, is $E_B = 4.81$ meV, slightly smaller than the corresponding donor electron binding energy of 5.23 meV. To this minimum binding energy should be added the kinetic energy of the center-of-mass motion of the exciton, if any. This prediction, based on the simple isotropic, hydrogenic model, has been found to be in good agreement with the results of optical absorption studies in GaAs which find that $E_B = 4.9$ meV. The size of the exciton in GaAs in its ground state is predicted to be $a_{ex} = 11.4$ nm, in surprisingly good agreement with experiment for such a relatively simple model. Since a_{ex} is much greater than the lattice constant, excitons of this type in a semiconductor such as GaAs where the Coulomb interaction between the electron and the hole is strongly screened can be classified as *Wannier excitons*, as described in Chapter 8.

Photoconductivity. Photoconductivity corresponds to the increase in the electrical conductivity σ of a semiconductor above the *dark conductivity* σ_0 when light is absorbed. This increase is easy to understand in terms of the resulting increase in the number of charges carriers, both electrons and holes. Steady-state effects are discussed here, with time-dependent effects described in Section 11.12.

The photo induced increase in σ can be written as

$$\Delta\sigma = \Delta n\, e\mu_e + \Delta p\, e\mu_h, \tag{11.57}$$

where Δn and Δp are the increases in the electron and hole concentrations, respectively, resulting from the absorption of above-bandgap light. When both electrons and holes are created, $\Delta n = \Delta p$ and the photoconductivity is called *intrinsic*. When either electrons or holes are excited from defect levels in the energy gap into the conduction or valence band, respectively, only one type of carrier may be responsible for the resulting *extrinsic* photoconductivity.

The dimensionless *quantum efficiency* $\eta(\omega)$ is defined to be the fraction of absorbed photons with energy $\hbar\omega$ which generate electron–hole pairs. The frequency-dependent generation rate $G_L(\omega)$ of carriers, per unit volume, in a uniformly illuminated semi-conductor will therefore be given by

$$G_L(\omega) = \frac{\eta(\omega)\alpha(\omega)I_0}{\hbar\omega}. \tag{11.58}$$

Here I_0 is the intensity of the incident light and α is the absorption coefficient. The resulting increases in the electron and hole concentrations in an n-type semiconductor under steady-state conditions are given by

$$\Delta n = \Delta p = G_L(\omega)\tau_p = \frac{\eta\alpha I_0\tau_p}{\hbar\omega}, \tag{11.59}$$

where τ_p is the recombination time or lifetime for the minority-carrier holes, as defined in Section 11.12. Measurements of both $\Delta\sigma$ and τ_p under steady-state conditions allow the quantum efficiency η to be determined. In a p-type semiconductor the minority-carrier lifetime τ_n for electrons in the conduction band will control both Δn and Δp.

Luminescence. *Luminescence* in semiconductors refers to optical processes in which energy in the form of incident light, injected carriers, bombarding electrons, or thermal energy excites electrons and holes, which then relax, leading ultimately to the emission of optical energy in the form of photons when electron–hole recombination occurs. These four types of luminescence are classified according to the source of the exciting energy and are known as *photoluminescence, electroluminescence, cathodo-luminescence*, and *thermoluminescence*, respectively. The emphasis here will be on photoluminescence. Except for the initial excitation step, the other types of lumines-cence behave similarly. The role that electroluminescence plays in the operation of semiconductor junction lasers and light-emitting diodes is described in Chapter 18.

Photoluminescence (PL) is the optical process that corresponds to the absorption and subsequent spontaneous emission of light at a lower frequency or energy. Since the principles of luminescence are presented in Chapter W22, the focus here is on the nature and spectrum of the light emitted and also on the applications of PL as a diagnostic tool for obtaining information about semiconductors (e.g., about the energy levels and concentrations of defects, etc.).

Photoluminescence begins with the absorption of above-bandgap light, which results in the creation of a nonequilibrium distribution of electron–hole pairs. These photoex-cited carriers typically lose energy and relax to the appropriate band edges via the nonradiative emission of phonons. This process is known as *thermalization*. As a result, when electrons finally return to the valence band, the spectrum of the emitted light will provide information not only about the absorption edge but also about the energy levels of excitons, shallow donor and acceptor impurities, and also impurities whose energy levels lie in or near the middle of the gap, away from the band edges.

For the case of band-to-band luminescence under low-level excitation in a direct-bandgap semiconductor such as GaAs, the dependence of the luminescence spectrum on emitted photon energy $\hbar\omega$ is given by

$$I_{PL}(\omega) \propto (\hbar\omega - E_g)^{1/2} \exp(-\beta\hbar\omega), \tag{11.60}$$

where the factor $(\hbar\omega - E_g)^{1/2}$ represents the joint density of states for the conduction and valence bands. The luminescence spectrum will therefore "turn on" at $\hbar\omega = E_g$ and, after an initial rise, will fall off exponentially with increasing $\hbar\omega$. At higher excitation levels, deviations from the exponential falloff are observed.

Radiative luminescent transitions can take place in n- or p-type semiconductors from an electron in a donor level to a hole in the valence band, emitting photons of energy $\hbar\omega = E_g - E_d$, or from an electron in the conduction band to a hole in an acceptor level, emitting photons of energy $\hbar\omega = E_g - E_a$, respectively. These free-to-bound luminescent transitions will dominate the band-to-band transitions at low temperatures and the observed spectra can be used to determine E_d and E_a.

Deep levels associated with defects can also participate in radiative luminescent transitions. Examples include free-to-bound transitions involving carriers in one of the bands and a deep level and also bound-to-bound transitions involving pairs of deep levels. These types of transitions can be generated by bombarding electrons as in the semiconducting phosphors used in cathode ray tubes and are thus an example of cathodoluminescence.

An example of a luminescence spectrum from a GaP light-emitting diode is shown in Fig. 11.20. The emission of light in both the red and green regions of the visible spectrum involves defect levels associated with the intentional doping of Cd, O, and S impurities into this indirect-bandgap semiconductor with $E_g = 2.24$ eV. An energy-level diagram showing the defect levels in the bandgap of GaP is also shown.

Raman and Brillouin Effects. Light penetrating into a semiconductor can undergo a variety of inelastic scattering processes with dynamic species such as phonons, plasmons, excitons, and so on, and also light scattering from static defects such as dislocations. These light-scattering effects are in general much weaker than those associated with the absorption and reflection of light. The inelastic scattering of photons in solids, known as *Raman scattering*, occurs through their interactions with optical phonons and plasmons, while inelastic *Brillouin scattering* is the light-scattering process involving interactions with acoustic phonons. Since the principles of Raman and Brillouin scattering are presented in Chapter W22, the focus here is on their applications to the characterization of semiconductors.

Raman scattering can correspond to either the emission or absorption of an excitation such as a phonon. The frequency of the Raman-scattered light is therefore given by

$$\omega = \omega_i \pm \omega_{phonon}, \tag{11.61}$$

where ω_i is the frequency of the incident light, the plus sign corresponds to phonon absorption, known as *anti-Stokes scattering*, and the minus sign to phonon emission,

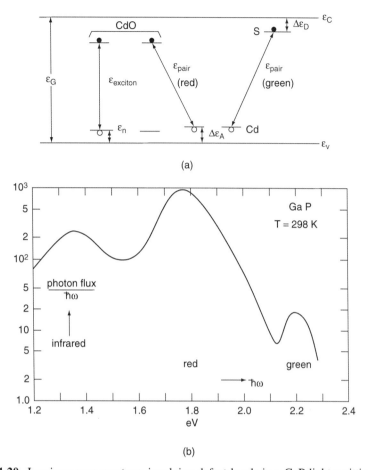

Figure 11.20. Luminescence spectrum involving defect levels in a GaP light-emitting diode in both the red and green regions of the visible spectrum. An energy-level diagram showing the defect levels in the GaP bandgap is also shown. (From K. Seeger, *Semiconductor Physics*, 7th ed., Springer-Verlag, New York, 1999.)

known as *Stokes scattering*. Phonon emission is the dominant process at low temperatures where few phonons are available. The Raman shift is $\Delta\omega = \omega_i - \omega$ and its measurement allows the determination of the phonon frequency. Although the scattering process described here is a one-phonon process, processes involving two or more phonons are also possible but with lower intensities.

Since both the total energy and the total momentum of the system are conserved in light scattering, the phonons participating in one-phonon scattering processes must have wave vectors with magnitudes very close to zero (i.e., to the center of the Brillouin zone), due to the fact that the wave vectors $k = \omega/c \approx 10^7$ m^{-1} of typical visible photons are much smaller than the maximum phonon wave vector $k_{BZ} \approx 1/a \approx 10^{10}$ m^{-1}, where a is the lattice constant. Thus one-phonon Raman scattering involves and provides information only about zone-center phonons.

Raman and Brillouin scattering are very powerful tools for determining the frequencies and symmetries of phonons in semiconductors. Raman scattering can also be used

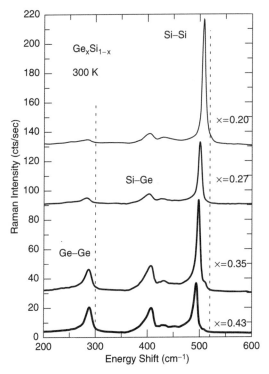

Figure 11.21. Raman spectra of several $Si_{1-x}Ge_x$ epitaxial layers grown on Si (100) substrates. The peaks correspond to the **k** = 0 LO phonon modes which result from the motions of adjacent Si–Si, Si–Ge, and Ge–Ge atom pairs. [Reprinted with permission from P. M. Mooney et al., *Appl. Phys. Lett.*, **62**, 2069 (1993). Copyright 1993 by American Institute of Physics.]

to study the local atomic structure in semiconductor alloys. An example is shown in Fig. 11.21, where the Raman spectra of several epitaxial $Si_{1-x}Ge_x$ layers grown on Si (100) substrates are shown. The peaks correspond to the **k** = 0 LO phonon modes, which result from the motions of adjacent Si–Si, Si–Ge, and Ge–Ge pairs. The peak energy depends on both the composition and the strain present in the epitaxial layer. The analysis of these results indicates that the atomic bonding in these Si:Ge alloys is random.

EXAMPLES OF SEMICONDUCTORS

Although only a small fraction of the elements can be classified as semiconductors, the number of semiconducting compounds and alloys is essentially unlimited. All that is necessary for a material to be a semiconductor is that it have an energy gap E_g, which is nonzero but also not too large, typically less than about 3 to 4 eV. Even materials commonly classified as insulators such as SiC ($E_g \approx 2.3$ eV when cubic and ≈ 3.0 eV when hexagonal), diamond ($E_g \approx 5.4$ eV), GaN ($E_g \approx 3.3$ eV), and cubic BN ($E_g > 6$ eV) are now often considered to be wide-bandgap semiconductors, which may have useful high-power, high-frequency, and high-temperature applications in electrical and optical devices.

The emphasis in the following sections is on describing the different classes of semiconductors in terms of their structure, bonding, and composition along with some of their important electrical and optical properties. These different classes will include the standard elemental, compound, and alloy semiconductors. Standard semiconductors, such as Si, GaAs, and ZnS, share several important attributes: They are typically composed of elements with valences 2 through 6 from groups II through VI of the periodic table and they possess either fourfold tetrahedral or sixfold octahedral local atomic coordination. In addition, several types of unusual or nonstandard semiconducting materials, including amorphous, oxide, organic, and magnetic semiconductors and porous Si, are described in Chapter W11.

Crystalline order and purity are critical issues in semiconductors since the presence of unintentional impurities and defects can make it impossible to dope these materials successfully. It is often necessary to control the concentrations of impurity atoms and defects such as dislocations which give rise to energy levels deep within the energy gap at the level of 10^{21} m^{-3} or $\approx 10^{-6}$ at %. These imperfections can act as traps that increase the recombination rates of electrons and holes.

11.10 Elemental Semiconductors and Their Compounds and Alloys

The chemical bonding in the elemental semiconductors C (diamond), Si, Ge, and α-Sn (gray Sn) is almost completely covalent since the electrons are shared equally in covalent bonds between pairs of identical NN atoms. The degree of localization of the valence electrons in the covalent bonds is greatest in diamond and decreases through Si and Ge to gray Sn, which is essentially a semimetal. It should be remembered, however, that the translational periodicity of these crystals and the resulting Bloch or delocalized nature of the electron wavefunctions result in the valence electrons actually being itinerant (i.e., delocalized throughout the solid). It is often difficult in practice to reconcile these two apparently conflicting viewpoints concerning the degree of localization of the valence electrons. The strength of the covalent bonds and the size of the energy gaps is highest for diamond and decreases down group IV to gray Sn. These four elemental semiconductors all have the cubic diamond crystal structure with covalent bonding based on the tetrahedral A$-$A$_4$ local bonding unit. The stable forms of carbon and tin are, respectively, hexagonal graphite, a semimetal with zero bandgap, and tetragonal β-Sn or white Sn, a metal.

Some of the important structural, optical, and thermal properties of these elemental semiconductors, of some important polytypes of SiC, and of the alloy Si$_{0.5}$Ge$_{0.5}$ are compared in Table 11.7.

The index of refraction increases from diamond to gray Sn as these materials become more polarizable. The increase in the concentration of free carriers n in the conduction band reflects the size of the energy gap E_g, as expressed by Eq. (11.29). The decrease of the melting temperature T_m from diamond to gray Sn is a reflection of the decreasing strength of the covalent bonding down group IV.

The real and imaginary parts n and κ, respectively, of the complex index of refraction for diamond, Si, and Ge have been presented as functions of photon energy in Fig. 11.19. These spectra can be seen to have characteristic shapes, with κ first increasing above zero at the fundamental absorption edge at E_g, which increases in energy from Ge to diamond. At higher energies, peaks and shoulders are observed in both n and κ and also in $\epsilon_1 = n^2 - \kappa^2$ and $\epsilon_2 = 2n\kappa$, the real and imaginary parts

TABLE 11.7 Properties of Semiconductors Based on the Group IV Elements at $T = 300$ K[a]

Semiconductor	a (nm)	$n(0)$	E_g (eV)	n (m^{-3})	T_m (K)
C (diamond)	0.357	2.4	5.4 (5.2[b])	$\approx 10^{-21}$	$\rightarrow$ graphite at $T > 980$ K
Si	0.543	3.4	1.11	8×10^{15}	1685
Ge	0.5658	4.0	0.67	2×10^{19}	1211
α-Sn (gray Sn)	0.649	≈ 4.3	0.08	$\approx 10^{24}$	$\rightarrow \beta$-Sn at $T \approx 286$ K
3C–SiC (β-SiC, cubic)	0.435	3.1[b,c]	2.3	$\approx 4 \times 10^5$	—[d]
2H–SiC (α-SiC, hex.)	0.308, 0.505[e]		3.33	$\approx 1 \times 10^{-3}$	2973
4H–SiC (hex.)	0.301, 1.005[e]	$\approx 2.6^c$	3.27	$\approx 3 \times 10^{-3}$	2973
6H–SiC (hex.)	0.308, 1.51[e]	$\approx 2.6^c$	3.02	$\approx 3 \times 10^{-1}$	2973
Si$_{0.5}$Ge$_{0.5}$ (alloy)	0.554	$\approx 3.6^c$	0.92	$\approx 1 \times 10^{17}$	≈ 1380 (sol. $\rightarrow$ sol. + liq.)

Source: Most data from D. R. Lide, ed., *CRC Handbook of Chemistry and Physics*, 75th ed., CRC Press, Boca Raton, Fla., 1994, pp. 12–87 to 12–96.

[a] a, lattice constant; $n(0)$, static index of refraction $= \sqrt{\epsilon_1(0)/\epsilon_0}$; E_g, energy gap; n, concentration of electrons (and holes) in the intrinsic material, all at $T = 300$ K; T_m, melting temperature.

[b] From E. D. Palik, ed., *Handbook of Optical Constants of Solids*, Vol. II, Academic Press, San Diego, Calif., 1991, Chap. 14.

[c] For compound and alloy semiconductors, the value of the static index of refraction $n(0)$ corresponds to photon energies below the bandgap E_g and above any strong absorption bands in the infrared.

[d] β-SiC transforms to 2H–SiC at $T = 2100°$C.

[e] lattice constants a and c of crystals with hexagonal unit cells.

of the optical dielectric function or relative permittivity $\epsilon_r = \epsilon/\epsilon_0$, respectively. These peaks and shoulders are related to features in the electronic band structure.

The effective masses, where available, and mobilities of the charge carriers in some of the technologically important semiconductors are presented in Table 11.8. The definitions of the various effective masses are given in Sections 11.2 and 11.3. Additional effective mass values can be found in Table 18.1.

Note that the highest mobilities for charge carriers listed in the Table 11.8 (e.g., for electrons in GaAs and InAs), are associated with the carriers that have the lowest effective masses, in agreement with Eq. (11.44). Differences between m_{Le}^* and m_{Te}^* and between m_{hh}^* and m_{lh}^* are due, respectively, to the curvatures of the energy bands in different directions in **k** space at the bottom of the conduction band and to the curvatures of the two degenerate energy bands at **k** $= 0$ at the top of the valence band.

Of the elemental semiconductors, Si has by far the widest range of current applications in electronics, micromechanics, and so on, while diamond may find a wide range of high-temperature semiconducting applications if a suitable n-type dopant can be found to complement the successful p-type dopant B. Typical dopants and their activation energies for some elemental and compound semiconductors are given in Table 11.4.

An energy-level diagram for deep impurities in the energy gap of Si is presented in Fig. 11.22. The energy values indicated are measured from the nearest band edge and the symbols $+$ and $-$ refer to levels with donorlike and acceptorlike character, respectively. Transition metal impurities such as Fe, Cr, and Mn and the noble metal impurities Cu, Ag, and Au have energy levels in Si which lie deep within the energy

TABLE 11.8 Effective Masses and Mobilities of Charge Carriers at $T = 300$ K^a

Semiconductor	Effective Mass (in units of m)						Mobility ($m^2/V{\cdot}s$)	
	m^*_{Le}	m^*_{Te}	m^*_{hh}	m^*_{lh}	m^*_{eds}	m^*_{hds}	μ_e	μ^a_h
Diamond	1.4	0.36	—	—	1.87	—	—	0.1^b
Si^c	0.90	0.19	0.52	0.16	1.05	0.58	0.19	0.050
Ge^c	1.59	0.082	0.34	0.043	0.56	0.35	0.38	0.182
3C–SiC (β-SiC)^d	0.68	0.25	—	—	—	—	0.10^d	0.0040^d
4H–SiC^e	0.33^f	0.42^f	—	—	0.61	—	0.050^g	$\approx$0.002–0.005^g
6H–SiC^e	2.0^h	0.42^h	—	—	0.89	—	0.025^g	$\approx$0.002–0.005^g
Si$_{0.5}$Ge$^b_{0.5}$	—	—	—	—	—	—	0.017	0.013
GaAs^c	—	0.067^i	0.80	0.12	0.067	0.83	0.90	0.050
InAs^c	—	0.022^i	0.41	0.025	0.022	0.41	3.3	0.046
CdTe^c (zincblende)	—	0.11^i	0.35	—	0.14	0.35	0.12	0.0050
GaN^j (wurtzite)	—	—	—	0.26	—	—	0.10	0.030
PbSe^c (NaCl)	0.095	0.047	—	—	0.3	0.34	0.15	0.15

^{a}The quantities m^*_{Le} and m^*_{Te} are the longitudinal and doubly degenerate transverse electron effective masses in the ellipsoidal conduction-band minima; m^*_{hh} and m^*_{lh} are the heavy- and light-hole effective masses at the top of the valence band; m^*_{eds} and m^*_{hds} are the density-of-states effective masses defined in Section 11.4. The quantity μ_h is the mobility of the light hole.

^{b}Data from L. I. Berger, *Semiconductor Materials*, CRC Press, Boca Raton, Fla., 1997.

^{c}Data from D. R. Lide, ed., *CRC Handbook of Chemistry and Physics*, 75th ed., CRC Press, Boca Raton, Fla., 1994, pp. 12–87 to 12–96.

^{d}Data from H. Morkoc et al., *J. Appl. Phys.*, **76**, 1363 (1994).

^{e} For 4H–SiC and 6H–SiC the parallel and the average perpendicular effective masses with respect to the c axis, respectively, are listed.

^{f} Data from D. Volm et al., *Phys. Rev. B*, **53**, 15,409 (1996).

^{g}Mobilities at $\approx 10^{23}$ m^{-3} doping levels. Data from K. Moore and R. J. Trew, *Mater. Res. Soc. Bull.*, Mar. 1997, p. 50.

^{h}Data from N. T. Son et al., *Appl. Phys. Lett.*, **65**, 3209 (1994).

^{i}Note that $m^*_{Le} = m^*_{Te}$ for single-valley, direct-gap semiconductors with symmetric conduction-band minima.

^{j}M. S. Shur and M. A. Khan, *Mater. Res. Soc. Bull.*, Feb. 1997, p. 44.

gap. These impurities can be neutral or charged positively or negatively, depending on whether they have trapped an electron or a hole from a donor or acceptor impurity. As charged traps, they can deactivate electrically active dopant atoms and thus interfere with the doping process. Whereas Si and Ge are easily doped during growth, doping of SiC is often achieved via ion implantation and subsequent annealing to remove the radiation-induced damage.

The chalcogenide elements Se and Te from group VI of the periodic table are semiconductors with energy gaps of 1.5 and 0.33 eV, respectively. Se and Te possess crystal structures composed of spiral chains oriented along the c axis and can readily be prepared as glasses. As, Sb, and Bi from group V are semimetals and although not significant as semiconductors themselves, these elements along with the nonmetals P and S often appear in other compound semiconductors, such as GaAs, InSb, Bi$_2$Te$_3$, GaP, and PbS. They are also useful as dopant atoms in semiconductors such as Si. Elemental B has a complicated crystal structure based on B$_{12}$ icosahedra with an energy gap $E_g = 1.55$ eV and a low electron mobility, $\mu_e \approx 0.001$ m^2/V·s.

The chemical bonding in compounds and alloys of the elemental semiconductors remains primarily covalent, with a ionic component whose magnitude depends on

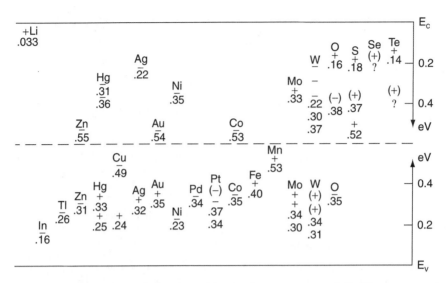

Figure 11.22. Energy levels of deep impurities in the energy gap of Si. The energy values indicated are measured from the nearest band edge and the symbols + and − refer to donors and acceptors, respectively. (Reprinted from E. Schibli et al., *Materials Science Engineering*, Vol. 2, p. 173, copyright 1967, with permission from Elsevier Science.)

the relative electronegativities of the atoms involved [i.e., C (2.5), Si (1.8), Ge (1.8), and Sn (1.8) (see Table 2.11)]. $Si_{1-x}Ge_x$ binary alloys exhibit complete mutual solid solubility from $x = 0$ to 1 and no stable compounds are found. The random bonding in $Si_{1-x}Ge_x$ alloys can be understood in terms of the relative strengths of the covalent bonds involved. The Si–Ge bond energy of 2.14 eV is nearly equal to the average of the Si–Si (2.34 eV) and Ge–Ge (1.95 eV) bonds. There is therefore no energetic preference for the formation of Si–Ge bonds.

In $Si_{1-x}Ge_x$ alloys the lattice constant a of the diamond crystal structure increases smoothly from $a(\text{Si}) = 0.543$ nm to $a(\text{Ge}) = 0.5658$ nm. The energy gap decreases smoothly from $E_g(\text{Si}) = 1.11$ eV to $E_g \approx 0.92$ eV for $Si_{0.15}Ge_{0.85}$ and then much more rapidly to $E_g(\text{Ge}) = 0.67$ eV. Some of the properties of the equiatomic $Si_{0.5}Ge_{0.5}$ alloy are presented in Tables 11.7 and 11.8. The low electron and hole mobilities in $Si_{0.5}Ge_{0.5}$ are due to the short scattering times $\langle \tau_e \rangle$ and $\langle \tau_h \rangle$ resulting from the chemical disorder in the lattice. The lattice constant in ternary $Si_{1-x-y}Ge_xC_y$ alloys can be adjusted at Ge levels of about 30 at % and C levels of about 4 at % to match that of crystalline Si, an important requirement for achieving defect-free epitaxial growth on Si substrates. In these alloys the smaller C atoms help to compensate for the introduction of the larger Ge atoms.

While the solid solubilities of Si in C and of C in Si are both low, less than 1 at %, the stoichiometric compound β-SiC with the cubic zincblende crystal structure is stable. The ordered bonding in β-SiC results from the fact that the Si–C bond energy of 3.21 eV is significantly greater than the average of the Si–Si (2.34 eV) and the C–C (3.70 eV) bonds. This occurs because the electronegativity of C, 2.5, is higher than that of Si, 1.8. Thus Si–C bonds and compound formation are energetically favored.

In addition to cubic β-SiC there also exists a sequence of hexagonal SiC poly-types starting with α-SiC or 2H–SiC which has the wurtzite crystal structure with $a = 0.308$ nm, $c = 0.505$ nm, and $E_g = 3.33$ eV. The stacking sequence of (0001) planes perpendicular to the c axis in 2H–SiC is $\mathbf{A}$(Si)$\mathbf{A}$(C)$\mathbf{B}$(Si)$\mathbf{B}$(C)$\mathbf{A}$(Si)$\mathbf{A}$(C)$\mathbf{B}$(Si)$\mathbf{B}$(C)$\ldots$. Other hexagonal polytypes, such as 4H–SiC ($E_g = 3.27$ eV) and 6H–SiC ($E_g = 3.02$ eV), exist in which the size of the unit cell is two and three times larger along the c axis, respectively. The stacking of planes within the unit cell of the 4H–SiC structure is $\mathbf{A}$(Si)$\mathbf{A}$(C)$\mathbf{B}$(Si)$\mathbf{B}$(C)$\mathbf{A}$(Si)$\mathbf{A}$(C)$\mathbf{C}$(Si)$\mathbf{C}$(C)$\ldots$. The 4H, 6H, and higher polytypes are all mixtures of the fundamental cubic 3C zincblende and hexagonal 2H wurtzite structures. All of the SiC polytypes are indirect-bandgap semiconductors. The possibility of both n- and p-type doping of the 6H–SiC polytype with N and Al, respectively (see Table 11.4), along with its chemical and thermal stability, high thermal conductivity, and high electron saturation velocity, make this material ideal for high-power operation in electronic devices.

The GeC compound is not stable, and in fact $Ge_{1-x}C_x$ alloys are observed to undergo phase separation into pure Ge and pure C. This behavior results from the fact that the Ge–C bond with energy ≈ 2.74 eV is weaker than the average of the Ge–Ge (1.95 eV) and C–C (3.70 eV) bonds. The determination of bond energies from thermochemical data is discussed in Chapter 2.

11.11 Compound Semiconductors and Their Alloys

In binary compound semiconductors such as SiC, GaAs, and ZnS the chemical bonding is of a mixed ionic–covalent type, due to the partial transfer of electrons from the less to the more electronegative atoms. These semiconductors typically have the zincblende (cubic ZnS) or wurtzite (hexagonal ZnS) crystal structures with tetrahedral A–B_4 local bonding units involving the participation of a total of eight valence electrons per formula unit. They are often referred to as group IV–IV (SiC), III–V (GaAs), or II–VI (ZnS) compounds. The group I–VII compounds, such as CuCl and AgBr, which contain noble metal cations have predominantly ionic bonding and are found with both the zincblende and NaCl or CsCl crystal structures. These I–VII materials are usually classified as insulators and have energy gaps in the range 2 to 3 eV. In addition to ordered binary compounds such as GaAs and GaP, ternary and even quaternary compounds and alloys such as $ZnSiP_2$, $GaAs_xP_{1-x}$, and $Ga_{1-y}Al_yAs_{1-x}P_x$, respectively, are also interesting semiconductors. The compound semiconductors and their alloys discussed here are classified as binary, ternary, or quaternary according to the number of elemental constituents. The group IV–IV compound SiC and IV–IV alloys have been discussed in Section 11.10.

Much of the interest in semiconductors such as GaAs stems from the fact that they have direct bandgaps. Direct-bandgap materials have the important advantage over indirect-bandgap elemental semiconductors such as Si and Ge of being compatible with the efficient generation and detection of light. Other advantages of GaAs and related materials include the high carrier mobilities and the bandgap flexibility that can be achieved via alloying. For example, alloying GaAs with Al increases E_g, while alloying with In accomplishes the reverse. A serious drawback to the device applications of GaAs, however, is the lack of a stable native oxide for surface passivation.

A model for the standard semiconducting elements and compounds that provides a useful framework for understanding trends in crystal structure, bonding, and related properties such as ionicity and dielectric response has been developed by Phillips and by Van Vechten[†] This model is presented in Chapter W11.

Binary, ternary, and quaternary semiconducting compounds and alloys are discussed next.

Binary Semiconductors. The common group III–V, II–VI, and IV–VI binary compound semiconductors and some of their important properties at or near $T = 300$ K are presented in matrix form in Table 11.9. The intersections of the rows and columns in this table refer to the corresponding stoichiometric compounds (e.g., BN lies at the intersection of the horizontal row containing B and the vertical column containing N). Note that the chemical bonding in these compounds becomes more metallic as one proceeds to the right and down each table. This occurs because the outer valence electrons become more weakly bound due to the increased shielding of the nuclei by the core electrons. The valence electron configuration of the neutral atom and the covalent radius r_{cov} are also listed under each element.

It can be seen from Table 11.9 that in the second row of the periodic table E_g increases in the isoelectronic sequence from Si (1.11 eV), to AlP (2.45 eV), to MgS (≈ 4.4 eV), and finally to NaCl (8.6 eV) as the ionic contribution to the bonding increases. The same trend for E_g also occurs in the third row: Ge (0.67 eV), GaAs (1.42 eV), ZnSe (2.58 eV), and KBr (7.8 eV). Note also that the lattice constants are nearly the same when the component elements belong to the same row of the periodic table [e.g., in the isoelectronic sequence from the third row $a(\text{Ge}) = 0.566$ nm, $a(\text{GaAs}) = 0.565$ nm, and $a(\text{ZnSe}) = 0.567$ nm]. The fact that energy gaps decrease to the right in each row and downward in each column confirms the tendency for metallic bonding to increase in these directions. In the lower right-hand corner of the table for the group II–VI compounds it can be seen that $E_g < 0$ for HgTe, indicating that band overlap and semimetallic behavior are present. All of the group III–V compounds except AlP, AlAs, AlSb, and GaP, all of the II–VIs, and all of the IV–VIs except GeSe, SnS, SnSe, and PbO are direct-bandgap semiconductors.

The most common crystal structures for the groups III–V and II–VI semiconductors are the cubic zincblende and hexagonal wurtzite structures. An AB compound with the wurtzite crystal structure consists of two interpenetrating HCP lattices, one consisting typically of group IIA atoms and the other of group VIB atoms, displaced from each other by $3c/8$ along the c axis. When compounds with the wurtzite crystal structure have the ideal c/a ratio of $\sqrt{8/3} = 1.633$, the local bonding units are tetrahedral A–B$_4$ and B–A$_4$ units. The ideal hexagonal wurtzite and cubic zincblende crystal structures both have tetrahedral bonding with the same NN configuration. The differences in bonding between these two structures are found only in the positions of the second-NN atoms. This similarity in structure and bonding can lead to very similar cohesive energies for some group II–VI compounds. An example is ZnS, which can crystallize in either the zincblende or wurtzite crystal structures.

The wurtzite crystal structure occurs preferentially in the more ionic group II–VI compounds, while the more covalent group III–V compounds prefer the zincblende crystal structure. In the wurtzite structure third-NN ions of opposite charge are closer to each other than in the zincblende structure, thereby increasing the ionic bonding energy.

[†] J. C. Phillips, *Phys. Rev. Lett.*, **20**, 550 (1968); *Rev. Mod. Phys.*, **42**, 317 (1970); J. A. Van Vechten, *Phys. Rev.*, **182**, 891 (1969); **187**, 1007 (1969).

TABLE 11.9 Binary Group III–V, II–VI, and IV–VI Compound Semiconductors

Data format, for each entry:
Lattice constant(s) a or a, c (nm); ± 0.005 nm
Energy gap E_g (eV): ± 0.03 eV (i = indirect bandgap)
Crystal structure [zb, zincblende; wz, wurtzite; rs, rocksalt (NaCl); rb, rhombohedral; hex, hexagonal; orth, orthorhombic]
Melting point T_m(K)

	More Ionic	← Chemical Bonding →		More Metallic
	N[a] $2s^2 2p^3$ 0.070 nm	P $3s^2 3p^3$ 0.110 nm	As $4s^2 4p^3$ 0.118 nm	Sb $5s^2 5p^3$ 0.136 nm
III–V Semiconductors				
B[a] $2s^2 2p$ 0.088 nm	0.362; 0.250,0.666[b] **6.2[b]; 5.8[b]** zb; hex $\approx$3300	0.454; 0.356,0.590 $\approx$**2.1;** — zb; wz $\approx$2800	0.478 $\approx$**1.5** zb $\approx$2300	
Al $3s^2 3p$ 0.126 nm	0.438[d]; 0.311,0.498 —; **6.02 (6.2)[c]** zb; wz $\approx$2500 (3273[c])	0.545 **2.45(i)** zb $\approx$2100	0.566 **2.16(i)** zb 2013	0.614 **1.60(i)** zb 1330
Ga $4s^2 4p$ 0.126 nm	0.452[d]; 0.319,0.519 **3.2–3.3[d];3.34** zb; wz 1500, (>1973[c])	0.545 **2.24(i)** zb 1750	0.565 **1.42[e]** zb 1510	0.610 **0.67 (0.75)[e]** zb 980

(Continued overleaf)

411

TABLE 11.9 (*Continued*)

	More Ionic	← Chemical Bonding →		More Metallic
III–V Semiconductors (Continued)				
	N[a]	P	As	Sb
	$2s^2 2p^3$	$3s^2 3p^3$	$4s^2 4p^3$	$5s^2 5p^3$
	0.070 nm	0.110 nm	0.118 nm	0.136 nm
In	0.498[d]; 0.353,0.569	0.587	0.606	0.648
$5s^2 5p$	—; **2.0 (1.89)**[c]	**1.27**	**0.36**	**0.163**
0.144 nm	zb; wz	zb	zb	zb
	1200 (≈1373[c])	1330	1215	798
II–VI Semiconductors				
	O[a]	S	Se	Te
	$2s^2 2p^4$	$3s^2 3p^4$	$4s^2 4p^4$	$5s^2 5p^4$
	0.066 nm	0.104 nm	0.114 nm	0.132 nm
Be[a]	0.270,0.438	0.487	0.514	0.563
$2s^2$	**9.4**[e]			
0.106 nm	wz	zb	zb	zb
	2800			
Mg	0.422[e]	≈0.565[d]	≈0.587[d]	0.454,0.739
$3s^2$	**7.8**[e]	**≈4.4**[d]	**≈3.6**[d]	
0.140 nm	rs	rs	rs	wz
	3099	>2473		≈2800

	O^a $2s^2 2p^4$ 0.066 nm	S $3s^2 3p^4$ 0.104 nm	Se $4s^2 4p^4$ 0.114 nm	Te $5s^2 5p^4$ 0.132 nm
Zn $4s^2$ 0.131 nm	0.463; 0.325,0.521 —; **3.2** zb; wz 2250	0.541; 0.381,0.626 **3.54(3.68)**^{c};**3.67(3.91)**^{e} zb; wz 2100	0.567 **2.58 (2.83)**^{e} zb 1790	0.610; 0.427,0.629 **2.26**; — zb; wz 1568
Cd $5s^2$ 0.148 nm	0.470 **2.5** rs 1700	0.583; 0.414,0.675 $\approx$**2.5**^{d}, **2.42 (2.55)**^{e} zb; wz 1748	0.605; 0.430,0.701 $\approx$**1.8**^{d}; **1.74** zb; wz 1512	0.648; 0.457,0.747 **1.44**; **1.50** zb; wz 1365
Hg $6s^2$ 0.148 nm		0.585; 0.415,0.950^b **<0; 2.1**^{b} zb; hex 2023	0.608 **2.12** zb 1070	0.646 **−0.15** zb 943

IV–VI Semiconductors

Ge^a $4s^2 4p^2$ 0.123 nm	1.04,0.365,0.43^b **1.65**^{b} orth 938^b	**1.08 (i)**^{b} 948^b	0.597; 0.601 **0.73–0.95**^{b}; **0.1–0.2** rb; rs 998^b

(Continued overleaf)

TABLE 11.9 (*Continued*)

	More Ionic	← Chemical Bonding →		More Metallic
IV–VI Semiconductors (*Continued*)	O[a] 2s²2p⁴ 0.066 nm	S 3s²3p⁴ 0.104 nm	Se 4s²4p⁴ 0.114 nm	Te 5s²5p⁴ 0.132 nm
Sn 5s²5p² 0.140 nm		1.10 **(i)**[b] 1153[b]	0.602 **0.9 (i)**[b] rs 1133	0.631 **0.5 (0.19)**[e] rs 1080
Pb 6s²6p² 0.146 nm	$0.398, 0.502$[b]; — **1.9 (i); 2.7 (i)**[b] tetr; orth 1163[b]	0.594 **0.37 (0.41)**[e] rs 1390	0.612 **0.26 (0.28)**[e] rs 1340	0.645 **0.25 (0.31)**[e] rs 1180

Source: (except when otherwise noted): D. R. Lide, ed., *CRC Handbook of Chemistry and Physics*, 75th ed., CRC Press, Boca Raton, Fla., 1994, pp. 12–87 to 12–96.

[a] The valence electron configuration of the neutral atom and the covalent radius r_{cov} are given for each element. The lattice constant $a(AB)$ of compound AB with the zincblende crystal structure is given approximately by the corresponding covalent radii as follows: $\sqrt{3}a(AB)/4 = r_{cov}(A) + r_{cov}(B)$.

[b] L. I. Berger, *Semiconductor Materials*, CRC Press, Boca Raton, Fla., 1997.

[c] M. S. Shur and M. A. Khan, *Mater. Res. Soc. Bull.*, Feb. 1997, p. 44.

[d] H. Morkoc et al., *J. Appl. Phys.*, **76**, 1363 (1994).

[e] E. D. Palik, ed., *Handbook of Optical Constants of Solids*, vol. II, Academic Press, San Diego, Calif., 1991, Chap. 14.

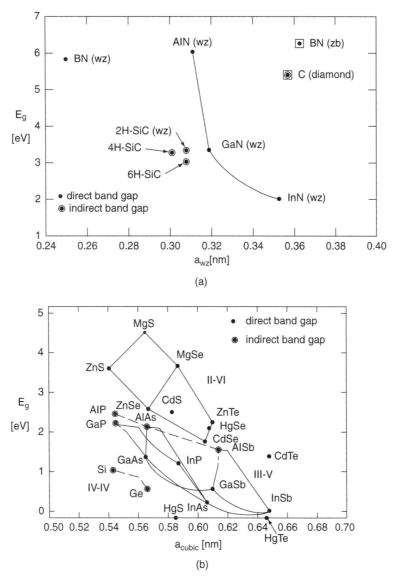

Figure 11.23. Plots of energy gap E_g versus lattice constant a are shown for semiconducting elements, binary and ternary compounds, and alloys: (*a*) E_g versus wurtzite lattice constant a_{wz}; (*b*) E_g versus cubic lattice constant a_{cubic} for crystals with diamond or zincblende crystal structures. The dashed lines correspond to indirect energy gaps.

A useful plot of E_g versus lattice constant a for the semiconducting elements, compounds, and alloys with the diamond and zincblende crystal structures is presented in Fig. 11.23. The lines connecting pairs of binary compounds such as GaAs and AlAs indicate the lattice constants and energy gaps either measured or expected for the corresponding pseudobinary alloys (e.g., $Al_xGa_{1-x}As$). Changes in slope of these lines at certain lattice constants signal changes from direct to indirect bandgaps at the corresponding compositions.

The group III nitrides, including AlN, GaN, InN, and their alloys, have direct bandgaps in the range 1.9 to 6.2 eV (650 to 200 nm). These nitrides are becoming important for applications where large bandgaps are needed (e.g., for semiconductor lasers operating at short wavelengths in the blue and ultraviolet regions and for high-temperature, high-power electronic devices). The electronic properties of these materials are potentially superior to those of SiC. Both n- and p-type doping have been achieved in GaN using Si donors and Mg acceptors, respectively, on the Ga sites. The insensitivity of GaN to defects such as dislocations introduced during growth and associated with the columnar microstructure of this material is important for potential optoelectronic applications. Apparently, the highly ionic bonding in GaN allows internal surfaces such as dislocations to be ineffective as sites for minority-carrier recombination.

The group IV-VI semiconductors PbS, PbSe, and SnTe have the NaCl crystal structure with each atom in an octahedral site with 6 NNs of the other type. The two $6s$ electrons of Pb occupy nonbonding orbitals and therefore do not participate in bonding in PbS and PbTe. The technological interest in these semiconductors is due to their small, direct bandgaps which lie in the infrared region. Large spin–orbit splittings play important roles in the energy-band structures of these group IV–VI compounds.

Binary semiconductors not shown in Table 11.9 include the group III–VI compound GaSe with $E_g = 2.05$ eV. GaSe has a complex layered crystal structure in which the local bonding units are Ga–GaSe$_3$ and Se–Ga$_2$ (i.e., each Ga atom is at the center of a tetrahedon of three Se atoms and one Ga atom, while Se atoms are bonded to pairs of Ga atoms). Binary compounds of the type V$_2$–VI$_3$ (e.g., As$_2$Se$_3$) form two-dimensional layered crystal structures in which the local bonding units are As–Se$_3$ and Se–As$_2$ (see Fig. 4.12 for a schematic representation of this two-dimensional network). Nonbonding orbitals consisting of lone-pair orbitals, one per As atom and two per Se atom, are also present in As$_2$Se$_3$, where they occupy the highest-lying valence band.

Doping in compound semiconductors is more complicated than in Si and Ge, due to the presence in the structure of atoms with different valences (e.g., Ga with valence 3 and As with valence 5 in GaAs). As a result, deviations from stoichiometry alone can give rise to levels in the energy gap that are shallow enough to provide doping. In very pure, stoichiometric GaAs, the concentration of antisite defects (e.g., Ga on an As site) can be less than 10^{20} m^{-3}, or less than one atom in 10^9 on the wrong site. Carbon in the configuration C$^-$(As) on an As site is the primary residual acceptor in "undoped" GaAs. The EL2 defect in GaAs gives rise to a deep donor level that can compensate shallow acceptors such as residual C, thereby producing the high-resistivity material used as substrates for MBE growth. The EL2 defect is an As antisite defect [i.e., As$^+$(Ga)] and is usually found near dislocations where the Ga vacancy concentrations are high. Impurities that act as acceptors in GaAs include both Zn$^-$(Ga) from group II and Si$^-$(As) from group IV; donors include Si$^+$(Ga) from group IV and Se$^+$(As) from group VI. Here the notation Zn$^-$(Ga) indicates that Zn substitutes for Ga and acts as an acceptor when ionized to form a negative ion. Thus Si can act as either a donor or acceptor in a III–V compound such as GaAs, depending on which atom it replaces. N and Bi from group V can act as isoelectronic acceptors and donors when they replace P in GaP due to their electronegativities, which are much greater and much less than that of P, respectively. A common problem in p-type doping of group III–V semiconductors is the passivation of acceptors by atomic hydrogen (e.g., by the formation of an Si–H complex instead of an Si$^-$ acceptor ion when Si replaces As in GaAs).

A type of metastable or bistable defect in $Al_xGa_{1-x}As$ is the deep-level *DX center* associated with donors such as $Si^+(Ga)$ or $S^+(As)$. Extensive research has shown that the isolated $Si^+(Ga)$ donor can exist in either of two states, the usual tetra-hedrally coordinated shallow donor configuration or the deep-donor DX configuration, which involves the breaking of one Si–As bond, leaving the Si and As atoms both threefold coordinated. The DX center is stabilized by lattice relaxation and the binding of an additional electron. For $x > 0.22$ in $Al_xGa_{1-x}As$ alloys the DX center is the stable configuration and the alloys become insulating. Thus the donors are self-compensated.

While most elemental and group III–V semiconductors can be doped either *n*- or *p*-type, the phenomenon of *self-compensation* often occurs in group II–VI semicon-ductors, which, as a result, can usually be doped only *n*- or *p*-type. As an example, consider the case of ZnTe, in which $I^+(Te)$ donors can exist and in which Zn vacancies can act as acceptors by generating pairs of holes [i.e., $V^{2-}(Zn)$ which corresponds to a vacancy V on a Zn site with charge $-2e$]. It is actually energetically favorable for Zn vacancies to be created in I-doped ZnTe, which can then each "accept" two electrons from the I donors, thereby removing the donated electrons from the conduction band. Thus *n*-type ZnTe is self-compensated via the formation of Zn vacancies and as a result can usually only be doped *p*-type. The energy required to create the Zn vacancy is regained when two electrons drop from the conduction band down to the Zn vacancy acceptor levels. The energetics of the self-compensation effect therefore depends both on the strength of the bonds that must be broken to create the defect and also on the magnitude of the energy gap in the semiconductor. The group II–VI semiconductors CdS, CdSe, ZnSe, and ZnO can all be doped *n*-type due to the energetically favorable formation of anion vacancies [e.g., S vacancies, $V^{2+}(S)$, which act as donors].

Ternary Semiconductors.   When ternary alloys such as $Al_xGa_{1-x}As$ (AlGaAs) and $Hg_xCd_{1-x}Te$ are formed from pairs of binary group III–V or II–VI compounds, their properties vary essentially continuously between those of the endpoint compounds (e.g., between those of AlAs and GaAs or of HgTe and CdTe). The energy gaps can be tuned in these *pseudobinary* alloys [e.g., $GaAs_{1-x}P_x$ alloys will have energy gaps lying between $E_g(GaAs) = 1.42$ eV and $E_g(GaP) = 2.24$ eV, while $Hg_{1-x}Cd_xTe$ alloys will have energy gaps between $E_g(HgTe) = -0.15$ eV and $E_g(CdTe) = +1.44$ eV (see Fig. 11.23)]. The $Hg_{1-x}Cd_xTe$ alloys have important applications as infrared detec-tors. Other pairs of binary compounds with nearly equal lattice constants include AlSb and GaSb, AlN and GaN, AlP and GaP, HgS and CdS, and HgSe and CdSe. These pairs can thus all readily form pseudobinary alloys.

In those cases where the lattice constant and energy gap vary continuously with composition, the variation is often approximated by *Vegard's law*. In pseudobinary alloys $A_{1-x}B_xC$ with the zincblende or wurtzite structure, Vegard's law proposes a strictly linear variation of the lattice constant a_{alloy} with composition according to the following composition-weighted average:

$$a_{alloy}(x) = (1 - x)a(AC) + xa(BC). \qquad (11.62)$$

Thus $a_{alloy}(x)$ is proposed to vary linearly between the endpoint values $a(AC)$ and $a(BC)$, which are the lattice constants of the III–V or II–VI compounds AC and BC, respectively. This is essentially a *virtual-crystal* approximation in which it is assumed that the bond lengths in the alloy are equal to a single, average bond length d_{alloy} given by

$$d_{\text{alloy}}(x) = (1 - x)d(A-C) + xd(B-C). \tag{11.63}$$

Note that there are no like-atom A–A, B–B, or C–C bonds in the ideal case. In this approximation all the atoms are assumed to lie on the ideal lattice whose lattice constant is given by Eq. (11.62). There is an analogous form of Vegard's law for the energy gap $E_g(x)$:

$$E_g(x, \text{alloy}) = (1 - x)E_g(\text{AC}) + xE_g(\text{BC}), \tag{11.64}$$

where $E_g(\text{AC})$ and $E_g(\text{BC})$ are the energy gaps of the AC and BC compounds, respectively.

An alternative approach to predicting the average bond length $d_{\text{alloy}}(x)$ is to assume that individual bond lengths are preserved in the alloy. This is in accord with Pauling's observation that bond lengths tend to be preserved in molecules and solids and therefore do not depend on the environment. For an $A_{1-x}B_xC$ alloy this approach in fact yields the same equations, (11.62) and (11.63), for $a_{\text{alloy}}(x)$ and $d_{\text{alloy}}(x)$, with their linear dependencies on composition. EXAFS studies have shown clearly that the NN bond lengths $d(A-C)$ and $d(B-C)$ to a first approximation tend to be preserved in both crystalline and amorphous alloys and thus are not strongly dependent on the local environment. In this case the average lattice constant of the alloy can still follow Vegard's law even though the average bond length $d_{\text{alloy}}(x)$ does not correspond to any actual bond length in the alloy. Applications of Vegard's law for predicting the average lattice constants and energy gaps in quaternary semiconducting alloys such as $In_{1-x}Ga_xAs_yP_{1-y}$ are discussed in Chapter 18.

Ternary compounds with tetrahedral bonding include the II–IV–V_2 and I–III–VI_2 *chalcopyrites* such as $ZnGeAs_2$ ($E_g = 0.85$ eV) and $CuInSe_2$ ($E_g = 1.04$ eV), respectively. $ZnGeAs_2$ can be thought of as being based on GaAs but with one-half of the group III Ga atoms in Ga–As_4 tetrahedral bonding units having been replaced by group II Zn atoms and the other half by group IV Ge atoms, thus maintaining the average valence of 3 on the former Ga FCC sublattice. The group II and IV atoms can, in principle, occupy the Ga sublattice in either an ordered or a random manner, while essentially all of the As atoms occupy the other FCC sublattice. The driving force for ordering is the minimization of the strain energy in these covalent materials, composed of atoms with different covalent radii and pairs of atoms with different bond lengths. When the occupation of the sites is ordered, the local bonding units are Zn–As_4, Ge–As_4, and As–Zn_2Ge_2 tetrahedra, the latter tetrahedron being more distorted than the former two, due to the differing As–Zn (≈ 0.249 nm) and As–Ge (≈ 0.240 nm) bond lengths. The resulting crystal structures are tetragonal with lattice constants $c \approx 2a$. $CuInSe_2$ and $CuIn_{1-x}Ga_xSe_2$ are employed in thin-film form in photovoltaic solar cells. The endpoint compound $CuGaSe_2$ has $E_g = 1.68$ eV.

Other ternary compounds with tetrahedral coordination include the following, with general formulas and examples provided: I_2–IV–VI_3 (Cu_2SiTe_3), I_3–V–VI_4 (Cu_3PS_4), and I–IV_2–V_3 ($CuSi_2P_3$). In each case there is an average of exactly four bonding electrons per atom, as required in tetrahedrally bonded structures.

Quaternary Semiconductors. Quaternary alloys such as $Ga_{1-x}Al_xAs_{1-y}P_y$ can be considered to be pseudobinary alloys of the binary compounds GaAs and AlP, a statement that is at least correct when $x = y$. The energy gaps and lattice constants for $Ga_{1-x}Al_xAs_{1-y}P_y$ alloys can be obtained from Fig. 11.23 and are limited by $E_g = 2.45$ eV and $a = 0.545$ nm for AlP and 1.42 eV and 0.565 nm for GaAs. It is usually found that separate FCC sublattices are present for the group III and V elements (e.g., Ga and Al atoms occupy one sublattice with As and P atoms on the other). Thus bonds tend to be formed only between group III and V atoms. In quaternary alloys such as $Al_{1-x-y}In_xGa_yP$ and $In_{1-x}Ga_xAs_{1-y}P_y$, the energy gap and the lattice constant can in principle be varied over wide ranges, as is evident from Fig. 11.23. In the group II–VI alloy system, $Zn_xCd_yMg_{1-x-y}Se$, alloys can be obtained with energy gaps that span the visible region. These alloys can have lattice constants which match that of InP, which can therefore be used as a substrate for their deposition via MBE.

No important quaternary semiconductor compounds of the standard variety are known, although quaternary oxides such as $YBa_2Cu_3O_7$ which are based on the perovskite structure and are semiconducting can be doped to the point of becoming metals and even superconductors, as described in Chapter 16.

In addition to the standard semiconductors discussed here, which typically have the diamond, zincblende, wurtzite, or NaCl crystal structures, there also exist nonstandard semiconducting materials with a variety of other structures and properties, including disordered or amorphous semiconductors, oxide, organic, and magnetic semiconductors, and porous Si. Some interesting and technologically important examples of these semiconductors are discussed in Chapter W11.

APPLICATIONS OF SEMICONDUCTORS

The fact that semiconductors play crucial roles in a wide variety of applications in electronics, microelectronics, and optoelectronics is due in large part to our ability to create semiconductor devices in which the motions of electrons and holes can be controlled rapidly and reliably. The widespread use of semiconductors in technology may be said to have originated with the Nobel prize–winning development of the transistor in 1947 by Bardeen, Brattain, and Shockley, which was based on the principles of the physics of semiconductors. Subsequent development of semiconductor technology based on Si has been remarkably rapid, steady, and continues today. It has been stated correctly that Si is not only an element, it is also an industry. The contribution that the development of semiconductor devices has made to information technology (i.e., the ability to store and process information) cannot be underestimated.

In this section several issues critical for the applications of semiconductors are discussed. In the following section some specific applications of semiconductors in electronic and optical devices are described briefly. Additional applications of semiconductors in transistors, in photovoltaic solar cells, in thermoelectric devices, and in the quantum Hall effect are described in Chapter W11.

11.12   Critical Issues

The semiconductors that are currently used in applications are essentially never intrinsic, bulk materials. They are, instead, extrinsic materials containing controlled amounts of dopant atoms and requiring elaborate processing to achieve the structures with the desired physical and chemical properties. Some critical issues related to these applications are addressed next. These include nonequilibrium effects and recombination, effects of spatial inhomogeneities (i.e., nonuniform spatial distributions of electrons and holes), and effects related to interfaces involving semiconductors.

Nonequilibrium Effects and Recombination. Time-dependent or nonequilibrium effects in both intrinsic and extrinsic semiconductors are typically the result of disturbances of the carrier concentrations away from their equilibrium values as a result of external influences such as the absorption of light, the injection of carriers through an interface, and so on. Nonequilibrium effects in homogeneous semiconductors related to deviations of the np product from its equilibrium value $n_i p_i$ are discussed next. Related nonequilibrium effects occurring in inhomogeneous semiconductors are discussed later. These effects play important roles in semiconductor devices such as p-n junction diodes, transistors, and light-emitting diodes, since all such devices operate under nonequilibrium conditions.

Two competing types of processes determine the steady-state concentrations of electrons and holes in semiconductors [i.e., the generation of carriers via thermal, optical, or other forms of excitation and their recombination or decay via the emission of energy in the form of photons (*radiative decay*) and phonons (*nonradiative decay*)]. To illustrate the effects of the generation or injection of carriers, a sample of n-type Si at $T = 300$ K with a donor concentration $N_d = 2 \times 10^{24}$ m^{-3} and with $N_a = 0$ will be considered. At this temperature essentially all the donors will be ionized, so the equilibrium concentrations are $n_0 \approx N_d = 2 \times 10^{24}$ m^{-3} for the majority-carrier electrons and $p_0 = n_i p_i / n_0 \approx (8 \times 10^{15} \ m^{-3})^2 / (2 \times 10^{24} \ m^{-3}) = 3.2 \times 10^7$ m^{-3} for the minority-carrier holes. Note that as a result of the n-type doping, p_0 is now 17 orders of magnitude smaller than n_0. The changes of n_0 and p_0 resulting from carrier injection will be denoted by Δn and Δp, and it will be required that $\Delta n = \Delta p$ to preserve overall charge neutrality in the semiconductor.

Low-level injection of carriers into an n-type semiconductor corresponds to changes in carrier concentrations such that $\Delta n \ll n_0$, while in general $\Delta p \gg p_0$. An example of low-level injection for the n-type Si sample described above is $\Delta n = \Delta p = 10^{16}$ m^{-3}, with the result that the new carrier concentrations are $n = n_0 + \Delta n \approx n_0$ and $p = p_0 + \Delta p \approx \Delta p$. These values of n and p clearly satisfy the criteria stated above for low-level injection. Note that the value of np is now $(2 \times 10^{24}$ m$^{-3})(10^{16}$ m$^{-3}) = 2 \times 10^{40}$m^{-6} and is no longer equal to the equilibrium value given by $n_i p_i = 6.4 \times 10^{31}$ m^{-6}. *High-level injection* corresponds to $\Delta p \gg p_0$ and to $\Delta n \gtrsim n_0$ (i.e., the change in the majority-carrier concentration n can no longer be neglected). For example, with $\Delta n = \Delta p = 10^{26}$ m^{-3}, it follows that $n \approx \Delta n$, $p = \Delta p$, and $np \approx 10^{52}$ m^{-6}. Although high-level injection is a common occurrence in semiconductor devices, the focus here is on the simpler case of low-level injection.

The changes Δn and Δp resulting from a constant low level of injection into the n-type Si sample described above are described next. The time rate of change of the minority-carrier hole concentration $p_n(t)$ in an n-type semiconductor is given by the following continuity equation:

$$\frac{dp_n}{dt} = G_{\text{T}} + G_{\text{I}} - R. \tag{11.65}$$

Here G_{T} is the temperature-dependent thermal generation rate of electrons and holes, G_{I} is the constant injection or generation rate of electrons and holes above the equilibrium

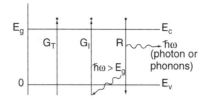

Figure 11.24. Rates listed in the continuity equation for the hole concentration, Eq. (11.65).

values, and R is the electron–hole recombination rate. These three rates are all assumed here to be spatially uniform and are shown schematically in Fig. 11.24.

Several types of recombination processes are possible in semiconductors, three important ones being band-to-band recombination, defect-mediated recombination, and surface recombination. Band-to-band recombination is considered here. Defect-mediated and surface recombination are described in Chapter W11.

In *band-to-band recombination* an electron near the bottom of the conduction band falls into an empty electron state (i.e., a hole state) near the top of the valence band. The band-to-band radiative recombination in this case corresponds to the spontaneous emission rate of light, and its rate will be proportional to the product of the concentrations of electrons in the conduction band and of holes in the valence band. This rate can therefore be written as

$$R_{\text{band}} = k_1 n\, p_n, \tag{11.66}$$

where k_1 is a proportionality constant that is independent of n and p_n but which does depend on the semiconductor. It is clear from this expression that one way of reducing the radiative recombination rate is to physically separate the electrons and the holes. This can be accomplished through the use of electric fields, either applied externally or existing within the semiconductor, as in the depletion region of a $p - n$ junction.

When band-to-band recombination dominates over both defect-mediated and surface recombination, the continuity equation (11.65) becomes

$$\frac{dp_n}{dt} = G_{\text{T}} + G_{\text{I}} - k_1 n\, p_n. \tag{11.67}$$

To focus initially on the thermal generation of carriers, consider the case when carrier injection is absent (i.e., when $G_{\text{I}} = 0$). At equilibrium when p_n and n have the values p_0 and n_0, respectively, it follows that

$$\frac{dp_n}{dt} = 0 = G_{\text{T}} - k_1 n_0 p_0 \tag{11.68}$$

and that

$$n_0 p_0 = \frac{G_{\text{T}}}{k_1}. \tag{11.69}$$

Thus in the absence of carrier injection the np product will be proportional to the thermal generation rate G_{T}. From Eq. (11.28) for the $n_i p_i$ product it can be seen that $G_{\text{T}} \propto \exp(-E_g/k_B T)$, as expected for thermally activated electron–hole generation.

In the general case when both thermal generation and injection of carriers are important, Eq. (11.67) can be written with the help of Eqs. (11.66) and (11.69) as

$$\frac{dp_n}{dt} = G_I + k_1 n_0 p_0 - k_1 n p_n. \tag{11.70}$$

Since $n \approx n_0$ for low-level injection in an n-type semiconductor, this equation can be also written as

$$\frac{dp_n}{dt} = G_I - k_1 n(p_n - p_0) = G_I - \frac{p_n - p_0}{\tau_p}. \tag{11.71}$$

This equation serves to define the band-to-band *recombination time* or *lifetime* τ_p for minority-carrier holes in an n-type semiconductor:

$$\tau_p = \frac{1}{k_1 n}. \tag{11.72}$$

It follows from Eqs. (11.66) and (11.69) that $G_T = p_0/\tau_p$ and that $R_{band} = p_n/\tau_p$.

The fact that τ_p is inversely proportional to the electron concentration n is physically reasonable since the minority-carrier holes will have a greater tendency to recombine (i.e., will have a shorter lifetime) the greater the concentration of majority-carrier electrons available to participate in this recombination process. The equation for minority-carrier electrons in a p-type semiconductor that corresponds to Eq. (11.71) is

$$\frac{dn_p}{dt} = G_I - \frac{n_p - n_0}{\tau_n}. \tag{11.73}$$

The lifetime $\tau_n = 1/k_1 p$ for minority-carrier electrons in a p-type semiconductor is inversely proportional to the hole concentration p.

The buildup of the hole concentration $p_n(t)$ from its initial value p_0 at the beginning of low-level injection can be obtained by integrating Eq. (11.71), with the result that

$$p_n(t) = p_0 + G_I \tau_p (1 - e^{-t/\tau_p}). \tag{11.74}$$

The steady-state concentration of holes is obtained from this equation in the limit $t \to \infty$:

$$p_n = p_0 + G_I \tau_p \approx G_I \tau_p \tag{11.75}$$

since $\Delta p \gg p_0$.

When the source of carrier injection is removed (i.e., when $G_I = 0$) the decay of the hole concentration $p_n(t)$ to its equilibrium value p_0 can also be obtained from Eq. (11.71), which now can be written as

$$\frac{dp_n}{dt} = -\frac{p_n - p_0}{\tau_p}. \tag{11.76}$$

Integration of this equation yields

$$p_n(t) = p_0 + G_I \tau_p e^{-t/\tau_p}. \tag{11.77}$$

The time dependence of n for this case of low-level injection into an n-type semiconductor is of little interest since $n(t) \approx n_0$.

Band-to-band recombination, including a table of minority-carrier lifetimes, is discussed further in Chapter W11.

Effects of Spatial Inhomogeneities and Diffusion. Spatially nonuniform distributions of electrons and holes resulting from the injection of carriers or from spatial variations in doping, and so on, are an important characteristic of most semiconductor devices. The diffusion of carriers resulting from these concentrations gradients must be included in any analysis of the characteristics of these devices. The focus in this section is on spatial inhomogeneities in otherwise homogeneous semiconductors which result from charge injection. Inhomogeneities that result from spatial variations in doping level and type are discussed later when the effects of interfaces are considered, specifically for the case of the $p - n$ junction.

The particle flux of electrons in a semiconductor resulting from a carrier concentration gradient $n(x)$ is given by the usual Fick's first law of diffusion, derived in Chapter 6:

$$J_e(x) = -D_e \frac{\partial n(x)}{\partial x}, \tag{11.78}$$

where the diffusion coefficient D_e for electrons is given in terms of the *electron thermal velocity* v_{nth}, average scattering time or momentum relaxation time $\langle \tau_e \rangle$, and effective mass m_e^* by

$$D_e(T) = \frac{v_{nth}^2 \langle \tau_e \rangle}{3} = \frac{k_B T \langle \tau_e \rangle}{m_e^*}. \tag{11.79}$$

An analogous equation is valid for holes:

$$D_h(T) = \frac{v_{pth}^2 \langle \tau_h \rangle}{3} = \frac{k_B T \langle \tau_h \rangle}{m_h^*}. \tag{11.80}$$

By comparison with Eq. (11.44) for the electron and hole mobilities, the following general expression can be shown to be valid for both electrons and holes in semiconductors:

$$D(T) = \frac{k_B T}{e} \mu(T). \tag{11.81}$$

This expression, the *Einstein relation*, is valid when the average scattering times $\langle \tau \rangle$ for charge transport and diffusion are the same.

When concentration gradients and electric fields E exist together in a semiconductor, as is often the case, their effects on the motion of charges can be added. The resulting electron current density in one dimension is

$$J_{ex} = ne\mu_e E_x + eD_e \frac{\partial n(x)}{\partial x}, \tag{11.82a}$$

and the hole current density is

$$J_{hx} = pe\mu_h E_x - eD_h \frac{\partial p(x)}{\partial x}. \tag{11.82b}$$

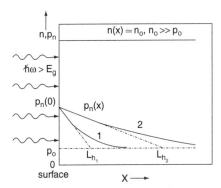

Figure 11.25. Injection of carriers via the absorption of light at the surface of an n-type semiconductor. The resulting spatial distributions of electrons and holes within the semiconductor are also shown for two values of the hole diffusion length L_h.

In these equations the first term is called the *drift term* (charges drift in electric fields) and the second term is the *diffusion term*. The total current density is, of course, given by $J_x = J_{ex} + J_{hx}$.

To illustrate the importance of diffusion and its effect on the spatial distributions $n(x)$ and $p(x)$ of charge carriers in a semiconductor, the low-level injection of carriers into the surface of a semiconductor via the absorption of light will now be described. The physical situation is shown schematically in Fig. 11.25, where, in the case considered, a uniformly doped (i.e. homogeneous) n-type semiconductor is illuminated by light with energy $\hbar\omega > E_g$. The resulting steady-state spatial distributions of electrons and holes within the semiconductor are also shown.

Electrons and holes are created near the surface in equal concentrations and in general have different diffusion coefficients. Strict charge neutrality in the semiconductor will therefore no longer hold near its illuminated surface and, as a result, an electric field E_x will be present there. Under normal conditions of low-level illumination the effect of this electric field on the minority carriers can be neglected, and so approximate electrical neutrality will still be valid in a homogeneous semiconductor. These statements will not be true, however, for high-level injection of carriers or in an inhomogeneous semiconductor (e.g., at a $p - n$ junction), as described later. Since the electric field E_x can usually be neglected, the diffusion term in Eq. (11.82b) will determine the net motion of the holes in an illuminated n-type semiconductor.

The continuity equation (11.71) for holes now has the form

$$\frac{\partial p_n(x)}{\partial t} = G_{\mathrm{I}} - \frac{p_n(x) - p_0}{\tau_p} - \frac{1}{e}\frac{\partial J_{hx}}{\partial x} = 0, \tag{11.83}$$

where G_{I} is the generation or injection rate of holes due to the absorption of light, the second term corresponds to the recombination of holes with electrons, and diffusion effects related to the hole current J_{hx} given in Eq. (11.82b) appear in the third term. The simple case where the light is strongly absorbed in the semiconductor is considered here. As a result, electron–hole pairs will be effectively created only at the surface at $x = 0$. This approximation will be valid when the optical absorption coefficient of the semiconductor is large, $\alpha \approx 10^8$ m^{-1}, so that most of the light is absorbed within $1/\alpha = 10$ nm of the surface. Under steady-state conditions and with $G_\mathrm{I} = 0$ and $E_x = 0$ within the bulk of the semiconductor away from its illuminated surface, the hole current will be a diffusion current and the continuity equation becomes

$$\frac{\partial p_n}{\partial t} = -\frac{p_n - p_0}{\tau_p} + D_h \frac{\partial^2 p_n}{\partial x^2} = 0. \tag{11.84}$$

This equation can be integrated using the boundary conditions $p_n(x) = p_n(0)$ at $x = 0$ and $p_n(x) \to p_0$ for $x \to \infty$. The solution corresponds to an exponential decay of p_n with distance x from the surface as shown in Fig. 11.25:

$$p_n(x) = p_0 + [p_n(0) - p_0]e^{-x/L_h}. \tag{11.85}$$

Here the hole (i.e., minority-carrier) *diffusion length* L_h is given by

$$L_h = \sqrt{D_h \tau_p}. \tag{11.86}$$

The quantity L_h can also be thought of as the minority-carrier recombination length. For Si at $T = 300$ K using the values $D_h = 1.3 \times 10^{-3}$ m^2/s and $\tau = 1$ μs, the result $L_h = 36$ μm is obtained. Thus L_h is much greater than $\alpha^{-1} = 10$ nm, which is consistent with the initial assumption that essentially all of the optically excited holes are created at the surface of the semiconductor. Note the analogy between Eq. (11.77) for the time dependence and Eq. (11.85) for the spatial dependence of the minority-carrier hole concentration p_n. The exponential falloff of p_n from its value at $t = 0$ is determined by the recombination time τ_p, while the exponential falloff of p_n from its value at $x = 0$ is determined by the product of τ_p and the diffusion coefficient D_h through the recombination length L_h. In both cases p_n approaches its equilibrium value p_0 for $t \gg \tau_p$ or for $x \gg L_h$.

The diffusion current of holes away from the illuminated surface through the n-type semiconductor is given by

$$J_h(x) = -eD_h \frac{\partial p_n}{\partial x} = e[p_n(0) - p_0]\frac{D_h}{L_h} e^{-x/L_h}. \tag{11.87}$$

The quantity $D_h/L_h = \sqrt{D_h/\tau_p}$ is known as the *diffusion velocity* for holes and has the value 36 m/s for Si at $T = 300$ K when the values of D_h and τ_p given earlier are used. There will, of course, be no net current flowing in the semiconductor under steady-state conditions in this case.

Under illumination, the photogenerated holes will therefore redistribute themselves spatially via diffusion to maintain approximate electrical neutrality in the semiconductor. The physical explanation for the exponential fall-off of the hole concentration in this case is the rapid recombination of the minority-carrier holes with the majority-carrier electrons. In addition, since $n \gg p$ in an n-type semiconductor, under low-level illumination the electron concentration $n(x)$ will not deviate appreciably from its value n_0 in the dark. The corresponding analysis of the creation of electron–hole pairs at the surface of a p-type semiconductor is straightforward†

Semiconductor Interfaces and $p - n$ Junctions. Nonuniform distributions of electrons and holes in a semiconductor which result from variations in space of doping levels and types are discussed next, specifically for the case of the $p - n$ *junction* consisting of an interface between two regions of opposite doping type. The term $p - n$ junction can refer both to the interface itself and to the entire device. The $p - n$ junction is a vital component of many semiconductor devices and illustrates concepts that are critical for an understanding of the operation of transistors, light-emitting diodes, and so on. The related topics of nonuniform charge distributions that occur when a semiconductor is placed in contact with a metal, insulator, or another semiconductor are discussed in Chapter 20.

Consider an interface within a nondegenerate semiconductor, separating a region with p-type doping on the left from one with n-type doping on the right. This type of interface is known as a *homojunction*. An idealized one-dimensional version of a $p - n$ junction is shown schematically in Fig. 11.26 both before and after diffusion of carriers occurs across the interface. The edges E_c and E_v of the conduction and valence bands, the position of the chemical potential μ in the energy gap, and the electron and hole concentrations are shown on both sides of the junction. The majority-carrier concentrations are p_p and n_n, while the minority-carrier concentrations are n_p and p_n, with the subscripts denoting the dominant doping type of the material. Thus p_n is the minority-carrier concentration of holes on the n-type side of the interface. The equilibrium properties of the $p - n$ junction are considered first, followed by the nonequilibrium case corresponding to the flow of current through the junction when an external voltage is applied.

Equilibrium Properties (External Voltage $V_{ext} = 0$). For the sake of illustration, discontinuities in electron energy levels and carrier concentrations are assumed to exist between the two sides of the interface initially, as shown in Fig. 11.26. Diffusion of carriers will then occur, with a net flux of holes from left to right and a net flux of electrons from right to left. Electrons and holes must also move across the junction in the same directions in which they diffuse in order to maintain the chemical potential μ at the same level throughout the semiconductor under equilibrium conditions (i.e., when no external voltage is applied and no net electrical current is flowing). Since initially, the chemical potential on the right-hand side of the junction lies above the chemical potential on the left-hand side, electrons on the right can lower their energies by moving to the left. In the same way, holes on the left can lower their energies by moving to the right.

The net result of the motions of the carriers is shown in the lower diagram, where the chemical potential μ is now at the same level throughout the semiconductor. As a result, the electron energy bands are no longer constant in space but instead, shift downward from left to right, with the shift occurring in a narrow region on both sides of the interface. The carrier concentrations are no longer discontinuous but instead, vary continuously though the junction, as shown. The bending of the energy bands

---

†For further discussions of the surface injection of carriers via illumination, see Sapoval and Hermann (1993, Chap. 7).

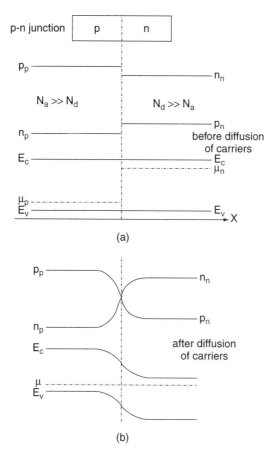

Figure 11.26. Interface between two sides of a semiconductor with opposite doping types. The interface, known as a $p - n$ junction, is shown both before (a) and after (b) diffusion of carriers occurs across the interface. The electron energy bands are also shown.

is a result of an internal electric field E_x which exists in the region of the interface. The existence of this electric field is also the reason that diffusion of the carriers does not continue until n and p reach uniform (i.e., constant) values throughout the entire semiconductor.

At equilibrium with $V_{\text{ext}} = 0$, the net flux of electrons across the junction is zero, as is the net flux of holes. This occurs when the diffusion flux of electrons (holes) down the electron (hole) concentration gradient is exactly canceled by a flux of electrons (holes) drifting in the internal electric field E_x in the opposite direction. Since the net current densities J_{ex} and J_{hx} are both equal to zero, it follows from Eq. (11.82) that

$$ ne\mu_e E_x = -eD_e \frac{\partial n}{\partial x} \quad \text{and} \quad pe\mu_h E_x = eD_h \frac{\partial p}{\partial x}. \tag{11.88} $$

These equations can also be expressed as

$$ J_{e,\text{drift}} = -J_{e,\text{diff}} \quad \text{and} \quad J_{h,\text{drift}} = -J_{e,\text{diff}}. \tag{11.89} $$

The *diffusion currents* J_{diff} are also known as *recombination currents* because the diffusing majority carriers recombine with carriers of the opposite type after diffusing across the junction. The *drift currents* J_{drift} are also known as *generation currents* because carriers that are thermally generated near the junction drift in the electric field found there.

The source of the electric field E_x in the junction region is the net, nonzero charge density $\rho(x)$ due to the uncompensated ionized donors and acceptors located there. As the electrons diffusing from right to left enter the p-type region, they recombine with the majority-carrier holes at a rapid rate. The same is true of the holes diffusing from left to right; they also recombine rapidly with the majority-carrier electrons. Thus near the junction a *depletion region* of approximate width $d = d_p + d_n$ is formed which is depleted of both electrons and holes. This depletion region with its nonzero net charge is also known as a *space-charge region*. The nonzero charge density on the left-hand side of the depletion region is due to uncompensated negative acceptor ions, while on the right-hand side it is due to uncompensated positive donor ions. The depletion region, the uncompensated dopant ions, and the resulting electric field E_x are all shown schematically in Fig. 11.27. Note that the depletion region is much wider than the junction itself where the doping changes from p-type to n-type. Since the entire semiconductor must be electrically neutral, the total negative ion charge $-Q$ on the left of the junction must be equal in magnitude to the total positive ion charge $+Q$ to the right. Note that the direction of the electric field E_x is such as to tend to keep holes on the p-type side, electrons on the n-type side, and thus to oppose the diffusion of these charges across the junction.

The net charge density $\rho(x)$, the electric field E_x, and the electrostatic potential $\phi(x)$ in the semiconductor are all related to each other. The electric field E_x and $\phi(x)$ are related in one dimension by

$$E_x = -\frac{dV}{dx}.$$ (11.90)

In addition, *Poisson's equation* is

$$\frac{d^2\phi}{dx^2} = -\frac{\rho(x)}{\epsilon} = -\frac{dE_x}{dx},$$ (11.91)

where ϵ is the static permittivity of the semiconductor. The electric field E_x can therefore be obtained by integrating the net charge density $\rho(x)$ across the depletion region.

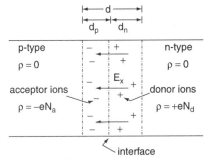

Figure 11.27. Depletion region, uncompensated dopant ions, and resulting electric field E_x for a $p - n$ junction.

As an example of the method used to obtain E_x, consider the idealized $p - n$ *step junction* in which the dopant concentrations N_d and N_a change discontinuously at the interface. This type of $p - n$ junction is also known as an *abrupt junction*. Other dopant distributions, such as the shallow (i.e., one-sided) step junction and the deep (i.e., linearly graded) junction can be analyzed in the same way. The total width of the depletion region is $d = d_p + d_n$ (i.e., the sum of the widths of the regions where uncompensated acceptor and donor ions reside). On the left-hand side of the junction, $\rho_a = -Q/Ad_p$, and on the right-hand side, $\rho_d = +Q/Ad_n$, where Q is the magnitude of the uncompensated ionic charge and A is the cross-sectional area of the junction. If the depletion region is essentially completely depleted of electrons and holes (i.e., if $n \ll N_d$ and $p \ll N_a$), $Q = eN_d d_n A = eN_a d_p A$, $\rho_a = -eN_a$, and $\rho_d = eN_d$ (see Fig. 11.27). As shown in Fig. 11.26, where $p_p > n_n$, N_a on the p-type side has been chosen to be greater than N_d on the n-type side. The integration of $\rho(x)$ is simple in this case, with the details left for a problem. The resulting electric field distribution is given in Fig. 11.28, where it can be seen that the sign of E_x is negative (i.e., E_x is directed from right to left), as expected. The maximum value of the electric field at the interface is $E_x = -Q/\epsilon A$.

The total shift of the energy bands across the $p - n$ junction can be seen from Fig. 11.26 to be given by

$$\Delta E = \mu_n - \mu_p \tag{11.92}$$

(i.e., by the difference in the chemical potentials before diffusion of carriers occurs). The chemical potential μ_n on the n-type side is given by Eq. (11.39) with $n = n_n \approx N_d$. A similar expression for μ_p can be obtained from Eq. (11.27) for $p(T)$ with $p = p_p \approx N_a$. When these expressions for μ_n and μ_p are substituted in Eq. (11.92), the following result can be obtained:

$$\Delta E = \frac{E_g}{2} + k_B T \ln \frac{N_d N_a}{N_c N_v} = k_B T \ln \frac{N_d N_a}{n_i^2}. \tag{11.93}$$

The bending of the energy bands is equivalent to a *built-in electric potential* or *voltage* $V_B = \Delta E/e$ at the $p - n$ junction, which is therefore given by

$$V_B = \frac{k_B T}{e} \ln \frac{N_d N_a}{n_i^2}. \tag{11.94}$$

For a Si $p - n$ junction at $T = 300$ K and with $N_d = N_a = 2 \times 10^{24}$ m^{-3}, the built-in voltage is calculated to be $V_B = 0.95$ V.

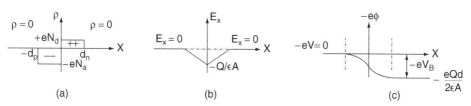

Figure 11.28. Idealized $p - n$ step junction: (*a*) on the left-hand side of the junction $\rho_a = -Q/Ad_p = -eN_a$, while on the right-hand side $\rho_d = +Q/Ad_n = +eN_d$; (*b*) resulting electric field distribution; (*c*) variation of the electron potential energy, given by $-e\phi(x)$.

The actual spatial variation of the band edges across the junction can be obtained by integrating the electric field E_x in Eq. (11.90) to obtain the electric potential $\phi(x)$. This macroscopic potential must be added to the microscopic potential seen by the electrons due to the lattice and the other electrons. The electron energy-band shift is therefore given by $-e\phi(x)$ and is shown in Fig. 11.28c.

The width d of the depletion or space-charge region can be shown to be given by

$$d = \sqrt{\frac{2\epsilon}{e} \frac{N_a + N_d}{N_a N_d} V_B}. \tag{11.95}$$

Thus d becomes smaller as the doping level increases. Using the values of N_d and N_a given earlier, the depletion region width for a Si $p - n$ junction at $T = 300$ K is $d = 35$ nm. For a one-sided step junction with $N_d \gg N_a$ (i.e., for a $p - n^+$ junction), the depletion width becomes

$$d = \sqrt{\frac{2\epsilon}{e} \frac{1}{N_a} V_B}. \tag{11.96}$$

The entire width of the depletion region in this case appears on the p-type side (i.e., on the side with the much lower doping level). An analogous expression exists for a one-sided step junction with $N_a \gg N_d$.

Nonequilibrium Properties (External Voltage $V_{ext} \neq 0$). The most important characteristic of a $p - n$ junction is that it allows the passage of electrical current in only one direction (i.e., it exhibits rectifying behavior). A brief description of the nonequilibrium properties of a $p - n$ junction under reverse and forward bias is presented next. Detailed discussions of these properties can be found in many texts.

In simple terms, when an external voltage V_{ext} is applied to a $p - n$ junction, the shift of the electron energy bands either decreases or increases in magnitude, depending on the polarity of V_{ext}. If a positive voltage is applied to the n-type side and a negative voltage to the p-type side, the energy bands will shift even more across the junction, the depletion region widens, and the total voltage appearing across the junction increases by approximately the applied voltage V_{ext}, that is,

$$V_{total} \approx V_B + V_{ext}. \tag{11.97}$$

The applied voltage appears across the depletion region due to the absence of carriers there (i.e., because the resistivity in this region is much higher than in the rest of the semiconductor). The electric field in the depletion region also increases. In this case corresponding to *reverse bias* the current flowing through the $p - n$ junction is vanishingly small as long as V_{ext} is less than the breakdown voltage, discussed later.

In the case of *forward bias* when a positive voltage is applied to the p-type side and a negative voltage to the n-type side, the energy bands shift less than at zero bias, the depletion region narrows, and the total built-in voltage decreases to the approximate value:

$$V_{total} \cong V_B - V_{ext}. \tag{11.98}$$

In forward bias larger currents, which are many orders of magnitude greater than those observed for the reverse-bias case, can flow through the $p - n$ junction. These forward-bias currents are, however, orders of magnitude smaller than the drift and diffusion currents, which exactly cancel each other at zero bias. The forward-bias currents thus result from a small imbalance between the drift and diffusion currents that occurs in the presence of an applied voltage.

The results of a simplified analysis of the biased $p - n$ junction is given next. A careful analysis of the spatial distributions of the minority-carrier concentrations n_p and p_n as functions of the applied voltage V_{ext} indicates that these minority carriers have either been injected into or extracted from the side of the junction of the opposite doping type, depending on whether the $p - n$ junction is forward or reverse biased, respectively. The injection of minority carriers (e.g., of electrons from the n-type side into the p-type side) occurs under forward bias due to the lowering of the potential barrier V_B at the junction according to Eq. (11.98). This small, excess electron diffusion current thus provides an imbalance between the drift and diffusion currents of electrons. The same imbalance occurs in the hole drift and diffusion currents. The drift currents of electrons and holes are, to a first approximation, unaffected by the applied voltage and the resulting changes in the electric field in the depletion region.

The excess minority-carrier diffusion currents are determined by changes in the potential barrier height V_B and so are proportional to the Boltzmann factor $\exp(eV_{ext}/k_BT)$, where $V_{ext} > 0$ for forward bias. The concentration of minority-carrier electrons n_p on the p-type side resulting from the application of the voltage V_{ext} can thus be expressed by

$$n_p = n_{p0} \exp\left(\frac{eV}{k_BT}\right), \tag{11.99}$$

where the subscript "ext" has been dropped from the voltage for simplicity. Here n_{p0} is the electron concentration on the p-type side for zero bias. Thus n_p is increased or decreased according to whether V is > 0 (forward bias) or < 0 (reverse bias). The excess electron diffusion current due to forward bias is proportional to $\Delta n_p = n_p - n_{p0}$ through the following equation:

$$J_e(V) = J_{e,\text{diff}}(V) + J_{e,\text{drift}}(V = 0) = e\Delta n_p \frac{D_e}{L_e}, \tag{11.100}$$

which is similar to the result expressed in Eq. (11.87) for the diffusion current of carriers injected at the surface of a semiconductor due to external illumination. Using Eq. (11.99), $J_e(V)$ can be written as

$$J_e(V) = en_{p0}\frac{D_e}{L_e}\left[\exp\left(\frac{eV}{k_BT}\right) - 1\right]. \tag{11.101}$$

The analogous equation for the additional hole diffusion current is

$$J_h(V) = ep_{n0}\frac{D_h}{L_h}\left[\exp\left(\frac{eV}{k_BT}\right) - 1\right], \tag{11.102}$$

where p_{n0} is the hole concentration on the n-type side of the junction under zero bias. The net current flowing through the junction is therefore

$$J(V) = J_e(V) + J_h(V) = e\left(n_{p0}\frac{D_e}{L_e} + p_{n0}\frac{D_h}{L_h}\right)\left[\exp\left(\frac{eV}{k_BT}\right) - 1\right]$$

$$= J_s\left[\exp\left(\frac{eV}{k_BT}\right) - 1\right]. \tag{11.103}$$

This result is valid for voltages such that $eV < E_g$ and also in the absence of recombination of the injected electrons and holes in the depletion region, corresponding physically to long minority-carrier recombination times τ_n and τ_p. This approximation may no longer be valid under high forward-bias conditions when the injected currents are large.

The prefactor J_s defined in Eq. (11.103) is given by

$$J_s = e\left(n_{p0}\frac{D_e}{L_e} + p_{n0}\frac{D_h}{L_h}\right) = en_i^2\left(\frac{D_e}{L_eN_d} + \frac{D_h}{L_hN_a}\right) \tag{11.104}$$

and is known as the *saturation current* since $J(V)$ saturates at the value $-J_s$ in reverse bias for large, negative voltages. The drift current flowing under reverse-bias conditions arises primarily from electrons and holes which have been thermally generated in the depletion region and is therefore often referred to as a *generation current*. This current is essentially independent of the applied voltage. The diffusion current under reverse bias becomes less important with increasing voltage, due to the higher potential barrier that the diffusing carriers must pass over. The saturation current J_s increases exponentially with T and is larger for semiconductors with low bandgaps, due to the presence of the factor n_i^2.

A typical value of J_s for Si at $T = 300$ K calculated for dopant concentrations of 10^{23} m^{-3} on both the p- and n-type sides of the junction is $J_s = 6.7 \times 10^{-8}$ A/m^2. In this calculation the values $D_e = 4.9 \times 10^{-3}$ m^2/s, $D_h = 1.3 \times 10^{-3}$ m^2/s, $L_e = 70$ μm, and $L_h = 36$ μm (valid for minority-carrier lifetimes of 1 μs) have been used.

There is no saturation of the current observed in forward bias, as can be seen from Fig. 11.29, where the prediction of Eq. (11.103) for $J(V)$ is presented. This current–voltage characteristic shows that the $p-n$ junction acts as a rectifier and hence is called a $p-n$ *diode* in practice. Deviations from the idealized picture of junction behavior presented here and from the expression for $J(V)$ given in Eq. (11.103)

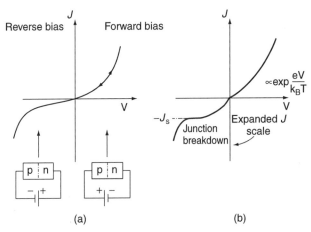

Figure 11.29. Current–voltage, $J-V$, characteristics for a $p-n$ junction, as predicted by Eq. (11.103): (*a*) current versus voltage for a wide range of voltages, illustrating rectifying behavior and also junction breakdown under high reverse bias; (*b*) current versus voltage near the origin, with an expanded current scale for reverse bias (i.e., $J < 0$). The saturation current J_s is shown for reverse bias.

can occur in real diodes for a variety of reasons. These include the following effects, which have been neglected here: the recombination of electrons and holes in the depletion region due to short minority-carrier lifetimes, surface recombination effects at the sides of the junction, the nonzero resistances of the p and n regions outside the junction, and effects due to high-level injection that occur at high forward bias. In particular, measurements of the forward current in $p - n$ junctions indicate that the exponential term has the form $\exp(eV/k_BT)$ given in Eq. (11.103), when the diffusion current dominates at high forward bias and the form $\exp(eV/2k_BT)$ when drift current dominates at low forward bias. When both diffusion and drift currents participate, the observed dependence is of the form $\exp(eV/mk_BT)$, with $1 \leq m \leq 2$.

In summary, the observed net currents flowing through $p - n$ junctions under conditions of both forward and reverse bias are minority-carrier currents. These currents originate as majority carriers which are then injected as minority carriers into the opposite side of the junction.

The electron energy-level diagrams for the cases of zero, forward, and reverse bias are shown in Fig. 11.30. Here the energy bands shift less in forward bias but shift even more, by approximately the amount $\Delta E = e(V_B + V_{\text{ext}})$, in reverse bias. The chemical potential is no longer constant throughout the semiconductor when $V_{\text{ext}} \neq 0$.

As illustrated in Fig. 11.29, very large, uncontrolled currents can flow through $p - n$ junctions at high reverse-bias voltages in a phenomenon known as *junction breakdown*.

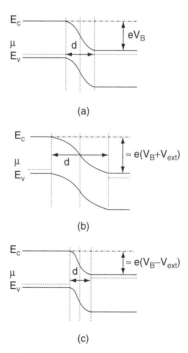

Figure 11.30. Electron energy bands for the cases of zero, reverse, and forward bias. Note the changes in width of the depletion regions with bias and that the chemical potential is no longer constant throughout the semiconductor. (*a*) In zero bias ($V_{\text{ext}} = 0$) the energy bands shift by $\Delta E = -eV_B$. (*b*) In reverse bias ($V_{\text{ext}} < 0$) the energy bands shift by approximately $\Delta E = e(V_B + |V_{\text{ext}}|)$. (*c*) In forward bias ($V_{\text{ext}} > 0$) the energy bands shift less across the junction.

Breakdown results from the presence of very high electric fields ($\approx 10^7$ to 10^8 V/m) in the depletion region and can occur via two distinct mechanisms, impact ionization and the Zener effect.

Impact ionization or the *avalanche effect* is, as the latter name implies, an uncontrolled increase in the electrical current in a semiconductor or insulator which occurs when electrons are accelerated by electric fields to high velocities (i.e., when they become "hot"). When the kinetic energy of an electron exceeds the bandgap energy E_g, an inelastic collision or impact with the lattice (i.e., an ion) or with an electron in the valence band can generate an electron–hole pair as the hot electron gives up the necessary energy. Since the newly excited electron can now also be accelerated, the numbers of excited carriers can increase exponentially, leading to breakdown.

The *Zener effect* is a tunneling process that occurs in very high electric fields E when the electron energy bands vary rapidly in space. An electron can tunnel directly from the valence band to an empty state at the same or lower energy in the conduction band. The electron tunneling current will depend exponentially on the width of the barrier, which itself decreases with increasing electric field. The electric fields that are required for Zener breakdown are $\approx 10^8$ V/m and, except at high doping levels, are higher than the fields at which impact ionization occurs in semiconductors such as Si. In both impact ionization and Zener breakdown, electrons receive enough energy from the electric field to effectively break covalent bonds between pairs of atoms.

The differential capacitance of a $p - n$ junction under reverse-bias conditions is given by $C = dQ/dV = \epsilon A/d$, where A is the cross-sectional area of the junction and d is the width of the depletion region. The appropriate width for the junction capacitance is d because all of the net uncompensated charge in the semiconductor is contained there. The junction capacitance is of interest for two reasons. The first results from the fact that C can be controlled by an applied reverse-bias voltage since the width d depends on the voltage drop across the depletion region; see Eqs. (11.95) and (11.97). A practical application of this dependence is the use of a $p - n$ junction as a voltage-controlled variable capacitor or *varactor* (variable reactor). The second reason is due to the fact that the measurement of $C(V)$ under reverse-bias conditions can provide information about the doping profile of the junction.

In addition to the semiconductor $p - n$ junction described here, other interesting and important systems involving an interface between a semiconductor and another semiconductor, a metal, an insulator, a liquid, or the vacuum are discussed elsewhere. Their location and a brief description of the structure in which the interface exists are presented in Table 11.10.

11.13 Specific Applications

Semiconductors have a wide range of applications in electronics, optics, optoelectronics, as thermoelectric devices, sensors, and so on. Selected applications of semiconductors in electronic devices (i.e., transistors), in photovoltaic solar cells, and in thermoelectric generators and refrigerators are discussed here and on our home page. Additional important applications of semiconductors which are discussed in other chapters include photoconductivity, light-emitting diodes (LEDs), semiconductor lasers, and bandgap engineering of semiconductors in Chapter 18 and semiconductor heterostructure superlattices in Chapter 20.

TABLE 11.10   Interfaces Between Semiconductors and Other Materials

Type of Interface[a]	Description of Structure	Location
S–S	$p - n$ junction (homojunction)	This section
	LEDs and semiconductor lasers	Chapter 18
S1–S2–S1	Quantum well	Section 11.6
	Buried heterostructure laser	Chapter 18
$(S1-S2)_n$	Superlattice or MQW (multiple quantum well) with n pairs of layers	Chapter 20
	Cascade laser	Chapter 18
S–M	Schottky barrier	Chapter 20
S–L	Semiconductor–electrolyte	Chapter W20
S–V or S–I	Band bending at semiconductor surfaces	Chapter 20

[a]S, semiconductor; M, metal; L, liquid; V, vacuum; I, insulator.

Electronic Devices. Electronic devices based on semiconductors typically have as building blocks the $p - n$ junctions described in Section 11.12. In addition to the many diode devices consisting of essentially a single $p - n$ junction (e.g., tunnel diodes, avalanche diodes, *p-i-n* diodes, rectifiers, and Gunn oscillators), there exist devices consisting of more than one $p - n$ junction (e.g., transistors with pairs of $p - n$ junctions). Since the operation of $p - n$ junctions has already been described, the focus here and on our home page is on descriptions of the important types of transistors and their operation.

Although the nature of the fundamental components of semiconducting electronic devices have not changed in recent years, their size has continued to decrease at an exponential rate. The typical feature size corresponding to the length of an MOS transistor gate or channel has decreased by a factor of 2 approximately every five to six years, leading to a doubling of the density of transistors on a chip every two to three years. Gordon Moore observed in 1964 that the number of transistors fabricated on a single Si chip had been doubling every year without an increase in cost. He predicted correctly that this rate of increase would continue at least into the near future. The doubling is now occurring approximately every 18 months. Details of some of the materials requirements for the fabrication of Si-based electronic devices are given in Chapter W21, where the synthesis and processing of semiconductors are discussed.

The use of *planar technology* for device fabrication was introduced at the beginning of the 1960s. This development made possible the *integration* of circuits onto a single substrate, as described briefly in Chapter W21, and opened up the continuing miniaturization of semiconductor devices. The area of a typical transistor in a semiconductor memory device is currently about 0.2 by 0.2 µm. According to recent predictions of the 1999 International Technology Roadmap for Semiconductors, feature sizes in semiconductor devices will reach 70 nm by the year 2008. There is some justification for a change in terminology from *microelectronics* to *nanoelectronics* as device dimensions shrink from the range of microns to hundreds and eventually tens of nanometers. The term *nanoscale devices* can have two meanings, either the scaling down of current devices to dimensions of hundreds of nanometers (or less) or the fabrication of new types of devices (e.g., devices based on quantum effects with dimensions on the order of the electron de Broglie wavelength $\lambda = h/p = h/\sqrt{2meV}$, where V is the accelerating voltage). When $V = 1$ V, $\lambda = 1.24$ nm. The present cost of one transistor on a Si chip is less than the cost of a single staple.

The current physical limits on semiconductor electronic or digital device applications and the challenges for future developments involve the continuing shrinking of the physical dimensions of the devices and the attainment of higher speeds. While the advantages of smaller dimensions are clear (e.g., decreasing carrier transit times and increasing speeds), the challenges that will arise as dimensions shrink are significant. These challenges include the following:

1. Device fabrication will become more difficult. Significant advances will be required in lithography and in chemical etching techniques. Processing limitations associated with decreasing device dimensions are discussed in Chapter W21.

2. Power dissipation in W/m^3 will go up dramatically and the thermal conductivity of the materials employed will become an important issue. Devices can now generate an energy flux from their surfaces of up to 3×10^5 W/m^2, a flux equivalent to that of a blackbody at $T = 1240°$C.

3. When transistors reach dimensions of tens of nanometers, the presence or absence of a single atom may become important. For example, if a cube of Si with dimensions of $L = 100$ nm is doped at the level of 10^{23} dopant atoms per cubic meter, the average number of dopant atoms in the cube will be $N_{dopant} = 10^{22} \times (10^{-7})^3 = 10$ atoms. The standard deviation of this number will therefore be about 30% of the average value, a level of variation that would have important effects on the operation of a device containing an active volume of doped Si of this size.

4. As device dimensions L shrink, the ratio V/L of applied voltage to size will increase rapidly and electrical breakdown will be much more likely. High-field transport and hot carriers will become increasingly important in small devices.

5. Two problems associated with decreasing device dimensions will involve the resistances and capacitances of elements in the device. As the widths w of metal interconnect lines of length l and thickness t decrease, the line resistance $R = \rho l/wt$ will increase. As their spacing d decreases, their capacitance $C \approx \epsilon A/d$ will increase. Both of these effects will lead to increases of the time constant $\tau = RC$ for changes of current in these devices. Possible materials solutions to these problems involve the reduction of the resistivity ρ by using Cu instead of Al and finding an insulator with a smaller permittivity ϵ than a-SiO$_2$.

Future electronic applications will involve devices in which quantum effects will play important roles. Interesting physical effects associated with shrinking device dimensions involve the quantized nature of the electron energy levels. These effects will become apparent and increasingly important in small MOSFETs or MODFETs. Other future developments may involve the applications of single-electron devices and possible applications of wide-bandgap semiconductors in high-power, high-temperature, and high-speed electronic devices.

The first semiconductor device exhibiting transistor action was developed in 1947 and was based on Ge. Even though Ge has an electron mobility twice that of Si, the lack of a stable native oxide (GeO$_2$ has a low melting point and is susceptible to attack by moisture) and the presence of a low bandgap have prevented Ge from playing an important role in semiconductor technology. Si remains the dominant material for most electronic device applications because no other material or technology has been shown to possess as much gain and to be as reliable. Other materials can, however, play important roles in certain applications. For example, although indirect wide-bandgap semiconductors such as SiC and diamond are not suitable for optoelectronic applications, they are strong candidates for power applications and high-temperature applications where Si cannot be used due to its relatively small energy gap of 1.1 eV. Low energy gaps lead to transitions in doped semiconductors to intrinsic behavior at relatively low temperatures and thus a loss of the dominant n- or p-type behavior.

The relative suitability of semiconductors for given types of applications is often evaluated on the basis of relevant *figures of merit* (FOMs), which are specific functions of the properties of the semiconductors. These are discussed in Chapter W11.

Transistors. Transistors are semiconductor electronic devices with at least three electrodes. A wide variety of structures are employed for transistors depending on the application (e.g., amplification or switching involving high frequency, high power, high speed, etc.). A brief outline of transistor action and several important transistor structures are described in Chapter W11. As an illustration, the simple structure known as a *metal–oxide–semiconductor field-effect transistor* (MOSFET) is described here. More detailed discussions can be found in the texts listed in the references.

A class of transistors whose operation involves only majority carriers is that of *field-effect transistors* (FETs). These devices are simpler than bipolar junction transistors and correspond in practice to a resistor whose resistance is controlled by an applied voltage and the resulting electric field in the semiconductor. They therefore operate on a completely different physical mechanism than that of bipolar junction transistors. Instead of having an emitter, collector, and base, FETs consist of a *source* and a *drain* for electrons and a *gate* that is used either to control or create a *conducting channel* in the semiconductor. FETs can be viewed as electronic switches which are either in an "on" or an "off" state. As a result, an FET corresponds in a real sense to a single bit (i.e., a binary unit of information).

The MOSFET is a relatively simple structure in which a gate voltage provides an electric field at the surface of the semiconductor, which creates a conducting channel between source and drain. A MOSFET is shown schematically in Fig. 11.31, where it can be seen that the metallic or doped polycrystalline Si (polySi) gate is isolated from the p-type region of the Si by an insulating oxide, in this case a thin layer of a-SiO$_2$. PolySi gates are used extensively in Si MOS devices because they can withstand high processing temperatures. The conducting gate–oxide–semiconductor MOS structure can be viewed here as a capacitor. The nature of the Si/a-SiO$_2$ interface is critical for operation of the MOSFET. This interface can be prepared so that the surface states which are present on the bare Si surface are eliminated through the formation of strong Si–O bonds. As a result, the Fermi level at the surface is no longer pinned by the surface-state traps and therefore can be shifted relative to the electron energy bands by the applied gate voltage.

In the absence of a gate voltage V_g, essentially no current can flow from the n-type source to the n-type drain since either one or the other of the two $p - n$ junctions in the current path will be reverse biased when a source-to-drain voltage V_d is applied. When a large enough positive gate voltage $V_g > 0$ is applied, electrons in the p-type

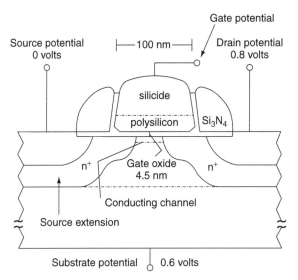

Figure 11.31. Metal–oxide–semiconductor FET (MOSFET). The metallic gate is isolated from the *p*-type region of the Si by an insulating oxide. The *n*-type conducting channel created when an inversion layer is formed under positive gate bias at the oxide–semiconductor interface is also shown. [Copyright 1988 by International Business Machine Corporation. Reprinted with permission from *IBM J. Res. Dev.*, **32**, (4).]

region will be drawn to the oxide–semiconductor interface and holes will be repelled. This bending of the electron energy bands will create an *n*-type inversion layer at the interface. The resulting layer of electrons is the active region of the MOSFET and will provide the conducting channel for the flow of majority-carrier electrons from source to drain. Since the active region in the MOSFET shown in Fig. 11.31 is an *n*-type inversion layer or *n*-channel, this transistor is referred to as *n*-MOS.

An advantage of the MOSFET over the bipolar junction transistor is that no current flows through the gate itself in the MOSFET due to the presence of the insulating oxide (i.e., no carriers are injected and $I_g = 0$). In addition, the MOSFET requires fewer processing steps. As can be seen in Fig. 11.31, the MOSFET is a four-terminal device, if one counts the voltage applied to the Si substrate.

An important property of MOSFETs is that the minority-carrier electrons or holes traveling through the inversion layer do not experience strong scattering by ionized dopant atoms since the channel itself can be lightly doped. As a result, interesting physical effects can be observed and studied in MOSFETs (e.g., effects related to the quasi-two-dimensional electron gas with variable density and Fermi level that exists in the inversion layer). The quantum Hall effect, observed in a high-mobility MOSFET, is described in Chapter W11.

The fundamental element of electronic logic circuits is the complementary MOS inverter or CMOS inverter. The CMOS inverter consists of an *n*-channel MOSFET and a *p*-channel MOSFET fabricated on the same Si substrate and in series with each other. CMOS uses much less power than bipolar transistors because when the *n*-channel MOSFET is on, the *p*-channel MOSFET is off, and vice versa. Therefore, power is dissipated in CMOS only during switching from one state to another. Due to their lower dissipation and ease of fabrication, FETs have important advantages over bipolar junction transistors for most integrated-circuit applications. Bipolar transistors, however, have the advantage of higher speeds than MOSFETs, due to the large capacitance of the MOS structure.

The MOSFET is an integral part of the dynamic random-access memory (DRAM) cells, which can consist of a single transistor and a single capacitor. Data in the form of a "0" or "1" are stored in the capacitor in its charged or uncharged state, respectively. The MOSFET acts as a switch, transferring data to and from the capacitor. DRAMs play a dominant role in computer main memories, due to their high density, one cell per 0.25 μm^2, and low cost, about 10^{-6} U.S. dollar per cell.

Optical and Optoelectronic Devices. Semiconductors have a wide range of applications involving the detection, absorption, and emission of light and also the conversion of light to other forms of energy (e.g., electrical energy), and vice versa. The fields in which these applications take place include optical communications, optical data storage, xerography, energy conversion, and so on. Possible future applications include areas such as optical computing. The topic of photovoltaic solar cells is discussed in Chapter W11. Other applications of semiconductors involving light are discussed elsewhere, including light-emitting diodes, semiconducting lasers, and xerography in Chapter 18. In addition, the topic of bandgap engineering in semiconductors for the design of interesting and potentially-useful optical materials and structures is discussed in Chapter 18.

The field of *optoelectronics* involves the use of electrons and photons in the same device and includes the use of light-emitting and detecting devices on the same chip (e.g., LEDs and photodiodes, respectively) for optical computing applications. One of the goals in this area is the replacement of electron currents by photon fluxes. An important advantage of photons is that at least at low fluxes, they can pass through each other in a material without interacting, in strong contrast to the behavior of electrons. An important materials parameter for these applications is the nonlinearity of the optical properties. Challenges for future developments include developing shorter-wavelength lasers (i.e., green, blue, and ultraviolet lasers).

REFERENCES

Berger, L. I., *Semiconductor Materials*, CRC Press, Boca Raton, Fla., 1997.

Blakemore, J. S., *Semiconductor Statistics*, Dover, Mineola, N.Y., 1987.

Brodsky, M. H., ed., *Amorphous Semiconductors*, 2nd ed., Springer-Verlag, New York, 1985.

Burns, G., *Solid State Physics*, Academic Press, San Diego, Calif., 1985.

Cohen, M. L., and J. R. Chelikowsky, *Electronic Structure and Optical Properties of Semiconductors*, Springer-Verlag, Berlin, 1989.

Ferry, D. K., *Semiconductors*, Macmillan, New York, 1991.

Grove, A. S., *Physics and Technology of Semiconductor Devices*, Wiley, New York, 1967.

Hannay, N. B., ed., *Semiconductors*, Reinhold, New York, 1959.

Milnes, A. G., *Deep Impurities in Semiconductors*, Wiley, New York, 1973.

Neamen, D. A., *Semiconductor Physics and Devices*, Richard D. Irwin, Homewood, Ill., 1992.

Phillips, J. C., *Bonds and Bands in Semiconductors*, Academic Press, San Diego, Calif., 1973.

Sapoval, B., and C. Hermann, *Physics of Semiconductors*, Springer-Verlag, New York, 1993.

Seeger, K., *Semiconductor Physics*, 7th ed., Springer-Verlag, New York, 1999.

Sze, S. M., *Physics of Semiconductor Devices*, 2nd ed., Wiley, New York, 1981.

Sze, S. M., ed., *Modern Semiconductor Device Physics*, Wiley, New York, 1998.

Yu, P. Y., and M. Cardona, *Fundamentals of Semiconductors: Physics and Materials Properties*, Springer-Verlag, Berlin, 1996.

PROBLEMS

11.1 (a) For a semiconductor, show that the np product obtained from Eq. (11.27) is proportional to $\exp(-\beta E_g)$ and thus is independent of the position of the chemical potential μ in the bandgap.

(b) The law of mass action in semiconductors for reactions creating pairs of electrons and holes [e.g., Eq. (11.28)] has the form $n(T)p(T) \propto \exp(-\beta E_g)$. Explain the significance of this law. (*Hint*: The law of mass action is described in Section 4.6.)

(c) Evaluate the np product at $T = 300$ K for Si with $E_g = 1.11$ eV and $m_{eds}^* = 1.05m$ and $m_{hds}^* = 0.58m$.

11.2 Using Eq. (11.30) and $m_{eds}^* = 1.05m$ and $m_{hds}^* = 0.58m$ for Si, calculate the change in the position of the chemical potential μ in the energy gap of intrinsic Si between $T = 0$ and 300 K.

11.3 Estimate the concentration of neutral phosphorus donors N_d^0 in Si at which overlap of the ground-state donor electron wavefunctions begins to occur. This is the concentration at which metallic behavior appears in P-doped Si. (*Hint*: Calculate the radius a_d of the donor orbit using the expression given in Table 11.3.)

11.4 Calculate the values of N_c and N_v as defined in Eq. (11.27) for Si at $T = 300$ K. The appropriate density-of-states effective masses for Si are $m_{eds}* = 1.05m$ and $m_{hds}* = 0.58m$.

11.5 Consider a semiconductor with a bulk energy gap $E_g = 1.5$ eV and with $m_e^* = m_h^* = 0.1m$. Calculate the increase in the energy gap of this semiconductor when it is incorporated into the following structures:

(a) A quantum well ($d = 2$) with $L_x = 10$ nm.

(b) A quantum wire ($d = 1$) with $L_x = L_y = 10$ nm.

(c) A quantum dot ($d = 0$) with $L_x = L_y = L_z = 10$ nm.

11.6 A Hall effect measurement is carried out on a rectangular bar of Si with dimensions $L_x = 0.04$ m (the direction of current flow) and $L_y = L_z = 0.002$ m. When a current $I_x = 5$ mA flows in the $+x$ direction and a magnetic field $B_z = 0.2$ T is applied in the $+z$ direction, the following voltages are measured: $V_x = 6$ V and $V_y = +0.3$ mV (i.e., increasing in the $+y$ direction). Determine the following properties of the Si bar from these data:

(a) The sign of the dominant charge carriers.

(b) The concentration of the dominant charge carriers.

(c) The electrical conductivity σ.

(d) The mobility μ of the dominant charge carriers.

11.7 Derive the results given in Table 11.5 for the density of states $\rho_e(E)$ and the Fermi wave vector k_F for $d = 1$ and $d = 2$ dimensions.

11.8 Using Eq. (11.59), estimate the increase Δn in the electron concentration in an n-type semiconductor due to the uniform absorption of light with $\alpha = 10^5$ m^{-1},

$I_0 = 1$ W/m^2, and $\hbar\omega = 1$ eV, a quantum efficiency $\eta = 1$, and a minority-carrier lifetime $\tau_p = 10^{-3}$ s.

11.9 Using the definition of the Hall mobility $\mu_H = |\sigma R_H|$ and the expression for R_H for an intrinsic semiconductor given in Eq. (11.49), show that $\mu_H = |\mu_h - \mu_e|$.

11.10 Consider the structural transformation of a binary crystal AB from the hexagonal wurtzite crystal structure to the cubic zincblende crystal structure in which the density of atoms remains constant. Find the lattice constant of the resulting cubic crystal if the lattice constants of the initial wurtzite crystal are $a = 0.3400$ nm and $c = 0.5552$ nm.

11.11 List all of the local tetrahedral bonding units, A–B$_4$, which are present in the ternary semiconducting compounds Cu$_2$SiTe$_3$, Cu$_3$PS$_4$, and CuSi$_2$P$_3$. Note that each tetrahedron must contain an average of four bonding electrons per atom.

11.12 Consider the idealized $p - n$ step junction shown in Fig. 11.28. If the charge densities are $\rho_a = -Q/Ad_p$ on the left-hand side and $\rho_d = +Q/Ad_n$ on the right-hand side of the junction, integrate $\rho(x)$ to obtain the electric field E_x.

11.13 Derive the expression for the shift ΔE of the electron energy bands from one side of a $p - n$ junction to the other under zero bias as given in Eq. (11.93). Calculate the magnitude of the built-in electric potential $V_B = \Delta E/e$ for Si at $T = 300$ K for $N_d = N_a = 2 \times 10^{24}$ m^{-3}. Using these same parameters, calculate the depletion width d and the maximum electric field $Q/\epsilon A$ for a Si $p - n$ junction at $T = 300$ K.

Note: Additional problems are given in Chapter W11.

CHAPTER 12

Metals and Alloys

12.1 Introduction

Metals share a number of characteristic properties which define them as a distinct class of materials. They tend to be good electrical and thermal conductors. This is attributed to the ease with which electrons may be excited within the partially filled conduction band. The electrical and thermal resistances grow with increasing temperature, since phonon scattering becomes more prevalent at higher temperatures, as described in Chapter 5. The specific heat grows linearly with temperature at low temperatures. This arises from the existence of a sharp Fermi surface separating occupied and vacant conduction electron states, and the fact that only electrons in the neighborhood of the Fermi level are excitable, as discussed in Chapter 7. Some metals, such as Al or Ag, are shiny because of their high electron plasma frequency, which causes them to be highly reflecting below that frequency. Other metals such as Au have a distinct color. This is attributed to interband transitions that lie in the visible spectral range. From the bonding viewpoint, many metals bind by having as many NN ions as possible simultaneously sharing a common pool of electrons.

Most of the solid-state physics studies of metals have been devoted to analyzing the properties of the metallic elements. These studies usually begin with the s- and p-bonded *free-electron metals*, such as Li, Na, K, Rb, Cs, Mg, and Al, then progress to studying more complicated metals such as Pb, Sn, and Bi, and culminate in a study of the transition metals (with their partially filled d shells) and rare earth metals (with their partially filled f shells). Band structures are calculated, densities of states are computed and measured, and Fermi surfaces are mapped out by ingenious magnetic resonance techniques. Collective excitations, such as plasmons, are identified and more exotic structures, such as charge density waves, are studied.

However, the types of metals of interest in materials science are often complicated alloys. In this chapter a primary focus is on these materials. For example, if one looks at the constitution of a typical aircraft alloy (Alcoa CW67) the following composition (wt %) is found: Cu(1.5), Mg(2.5), Ni(0.1), O(0.35), Zn(9.0), Zr(0.14), and the balance Al. Even the common "aluminum" beverage can is found to be a complex alloy consisting of Cu(0.15), Fe(0.4), Mn(1.0), Mg(1.0), Si(0.2), and the balance Al. Some of the added elements, such as Mn and Mg, serve to distort the crystal structure. The distorted lattice is found to resist deformation, as described in Chapter 10. The reason for the additives lies both with a finished product with improved physical and chemical properties as well as the need to find an easier fabrication process. In addition to the chemical composition, the properties of the material depend strongly on the methods used to prepare it (i.e., the synthesis and processing). This has to do with the management of both microstructure and defects, which are found to dominate the physical properties of the product.

Some aspects of metal and alloy physics have been considered in previous chapters. Electrical and thermal properties of free-electron metals were studied in Chapter 7. Thermal properties of binary alloys and the order–disorder transition were studied in Chapter 6. Mechanical properties of metals were considered in Chapter 10. Other aspects of metal and alloy physics are covered in later chapters. For example, annealing is discussed at the Web site[†] in Chapter W21. Metals are also discussed in Chapters 16 and 17 since metals and metallic alloys have proven to have interesting and technologically important magnetic and superconducting properties.

The chapter begins with a general review of properties of elemental metals. The discussion is broken into three parts: sp-bonded metals, transition metals, and rare earth metals. The formation of the metallic bond is studied and electronic screening is described. The Friedel model for the transition metals is introduced. Various theoretical tools that have proven to be valuable in the study of metals are discussed in Sections W12.1 and W12.2. These include the density-functional formalism and the embedded-atom method. The Peierls instability of a one-dimensional metal, discussed in Section W12.3, is given as an example of the interplay of the electron gas and elastic forces.

Alloy formation is then considered. The Hume–Rothery rules are presented. A brief introduction is given to electrical properties of metals. The chemistry of oxidation and corrosion is studied in Section W12.4. Then a review is made of a selection of metallic alloys and their applications to both traditional and modern metallurgy. The material in the text includes steel, intermetallic compounds, superalloys, and electromigration. Supplementary material appears on our home page relating to coatings, shape-memory alloys, metallic glasses, metal hydrides, solder joints, and porous metals (see Sections W12.5 to W12.10). The list is not meant to be exhaustive but rather representative. It is hoped that the student will obtain some insight into how the metallurgist and materials scientist confronts modern technology with the physics and chemistry of metals.

THREE CLASSES OF METALS

Scanning through the periodic table, one finds that most elements are metallic under standard conditions. They may roughly be classified into three groups. The first are the sp-bonded metals. In these metals the valence electrons are the s and p electrons surrounding inert cores with filled shells of electrons. The second group consists of the transition metals in which the d shells are partially filled. The third group are the rare earth metals with partially filled f shells.

12.2 *sp*-Bonded Metals

The sp-bonded metals include the elements from groups IA, IIA, and IIIA of the periodic table, as well as several elements from groups IVA and VA. Typical material properties are listed in Table 12.1. The atoms freely donate their z valence s

---

[†] Supplementary material for this textbook is included on the Web at the resource site (ftp://ftp.wiley.com/public/sci_tech_med/materials). Cross-references to elements of the Web material are prefixed by "W."

and p electrons to the Fermi sea. To a first approximation, the electrons comprise a homogeneous free-electron gas described by the Sommerfeld model, as discussed in Section 7.4.

The kinetic-energy contribution to the internal energy of the electron gas is given, at $T = 0$, by Eq. (7.18):

$$U_{\mathrm{KE}} = \frac{3}{5}NE_F = \frac{3}{10}\frac{N\hbar^2}{m}\left(\frac{3\pi^2 N}{V}\right)^{2/3}, \tag{12.1}$$

where N is the total number of valence electrons in the metal and V is the volume of the metal. The contribution of this term to the adiabatic bulk modulus, B, is readily evaluated from the formula

$$B = -V\left(\frac{\partial P}{\partial V}\right)_S = V\left(\frac{\partial^2 U}{\partial V^2}\right)_S, \tag{12.2}$$

so

$$B_{\mathrm{KE}} = \frac{\hbar^2 k_F^5}{9\pi^2 m} = \frac{2nE_F}{3}. \tag{12.3}$$

Inserting typical values for k_F from Table 7.3 yields results of the same order of magnitude as the values in Table 12.1, indicating that the kinetic energy of the electron gas plays an important role in determining the bulk modulus. However, the state of lowest kinetic energy, U_{KE}, is one in which V is infinite, since there is no attractive energy yet included in the model. Consequently, the cohesive energy, E_{coh}, is zero.

The jellium model idealizes the ions as a static uniform distribution of charge density $n_0 e$. The electrons are treated quantum mechanically. To ensure Fermi–Dirac statistics, the total electronic wavefunction must be symmetric under interchange of any two electrons. This property may be incorporated by having the wavefunction approximated by a determinant formed from the individual orbitals. In terms of these orbitals the internal energy of the system is

$$U = \sum_{k,s}{}' \int d\mathbf{r}\,\psi_k^*(\mathbf{r})\left[-\frac{\hbar^2}{2m}\nabla^2 - \int \frac{n_0 e^2}{4\pi\epsilon_0|\mathbf{r}-\mathbf{r}'|}d\mathbf{r}'\right]\psi_k(\mathbf{r})$$

$$+ \frac{1}{2}\int d\mathbf{r}\,d\mathbf{r}'\frac{(n_0 e)^2}{4\pi\epsilon_0|\mathbf{r}-\mathbf{r}'|} + \frac{1}{2}\sum_{k,s}{}'\sum_{k',s'}{}'\frac{e^2}{4\pi\epsilon_0|\mathbf{r}-\mathbf{r}'|}|\psi_k(\mathbf{r})|^2|\psi_{k'}(\mathbf{r}')|^2$$

$$- \frac{1}{2}\sum_{k,s}{}'\sum_{k',s'}{}'\frac{e^2}{4\pi\epsilon_0|\mathbf{r}-\mathbf{r}'|}\psi_k^*(\mathbf{r})\psi_k(\mathbf{r}')\psi_{k'}^*(\mathbf{r}')\psi_{k'}(\mathbf{r}')\delta_{s,s'}. \tag{12.4}$$

The $\mathbf{k}$ and $\mathbf{k}'$ sums are restricted to the interior of their respective Fermi spheres. The first two terms are the kinetic energy of the electrons and the energy of interaction of the electrons with the ions, respectively. The third and fourth terms describe ion–ion and the electron–electron coulombic interactions. The last term is the exchange correction to the coulombic interaction and is a consequence of the antisymmetry of the total wavefunction for the electrons.

In the spirit of the variation principle of quantum mechanics, a trial wavefunction is chosen and an estimate of the internal energy is obtained. The trial wavefunction represents the orbitals by plane waves:

$$\psi_{\mathbf{k}}(\mathbf{r}) = \frac{1}{\sqrt{V}} e^{i\mathbf{k}\cdot\mathbf{r}}. \tag{12.5}$$

Inserting this into the expression for U leads to

$$U = \sum_{\mathbf{k},s}' \int \frac{d\mathbf{r}}{V} \frac{\hbar^2 k^2}{2m} - \frac{e^2}{8\pi\epsilon_0 V^2} \sum_{\mathbf{k}}' \sum_{\mathbf{k}'}' \sum_s \int d\mathbf{r} \int d\mathbf{r}' \frac{e^{i(\mathbf{k}-\mathbf{k}')\cdot(\mathbf{r}-\mathbf{r}')}}{|\mathbf{r}-\mathbf{r}'|}. \tag{12.6}$$

The first integral is just the expression for the kinetic energy given by the Sommerfeld model. The second term, representing the exchange energy, may be evaluated analytically, with the result

$$U = N\left(\frac{3}{5}\frac{\hbar^2 k_F^2}{2m} - \frac{3e^2 k_F}{16\pi^2\epsilon_0}\right). \tag{12.7}$$

Note that the direct Coulomb energy is zero, since the uniform electron gas and the smeared ions occupy the same space and yield local electrical neutrality. The exchange energy comes about because an electron will exclude other electrons of the same spin projection from its environment. An electron residing at the center of a spherical hole of radius r in a conducting medium will have its energy lowered by $\Delta U = -e^2/8\pi\epsilon_0 r$. The expected size for r is of the order of the Fermi wavelength. A comparison of this expression with the exchange energy gives $r = \lambda_F/3$. This is referred to as the *Fermi hole*.

This expression leads to a unique prediction for k_F, E_{coh}, and B. Minimizing U with respect to k_F yields

$$k_F = \frac{5}{4\pi a_1} = 0.752 \times 10^{10} \text{ m}^{-1}, \tag{12.8}$$

where $a_1 = 4\pi\epsilon_0\hbar^2/me^2$ is the Bohr radius. Similarly,

$$E_{\mathrm{coh}} = -\frac{U}{2N} = \frac{15}{64\pi^2}\frac{\hbar^2}{ma_1^2} = 0.67 \text{ eV}, \tag{12.9}$$

$$B = \frac{1}{45\pi^2}\frac{\hbar^2 k_F^5}{m} = 0.661 \text{ GPa}. \tag{12.10}$$

The next correction to the internal energy arises from the electron–electron interaction and is called the *correlation energy*, U_{corr}. It originates because electrons tend to avoid each other due to their Coulomb repulsion. Therefore, the electrostatic energy is lower than would be found assuming just a uniform electron-charge distribution with no fluctuations. Low and Pines, using many-body theory, derived the approximate formula

$$U_{\mathrm{corr}} = -\frac{Ne^2}{8\pi\epsilon_0 a_1}(0.115 - 0.31\ln r_s), \tag{12.11}$$

where the dimensionless parameter r_s is defined through the relation

$$\frac{4\pi(r_s a_1)^3 n}{3} = 1.$$
(12.12)

However, inclusion of U_{corr} into the theory still fails to explain the element-to-element variation of the data in Table 12.1. Note that the valence z is absent from the formula for U.

The lattice is important, but not as important as it may seem at first sight. Two effects are operative. First, there is screening of the ions by the Fermi sea of electrons. Around each ion of charge $+ze$ embedded in the electron gas there will be a cloud of screening charge $-ze$, with a typical size for the cloud being the Thomas–Fermi screening length, k_{TF}^{-1}, as shown in Section 7.17. A second effect arises from the Pauli exclusion principle. The electrons of the Fermi sea avoid the filled electronic shells of the ionic-core electrons and are thus are effectively excluded from the ions by a distance r_c, the core radius. Ashcroft introduced a pseudopotential of the form

$$\Phi(r) = \frac{ze}{4\pi\epsilon_0 r}\Theta(r - r_c),$$
(12.13)

where the theta function cuts off the potential in the core region.

The Coulomb energy, which is zero in the jellium model, must be recomputed. The solid is partitioned into Wigner–Seitz cells. In evaluating the Coulomb energy there are terms involving intercellular interactions and terms involving intracellular interactions. Since each cell is electrically neutral, the former give zero contribution to the Coulomb energy. Within each cell there is an interaction of the ion with the electrons and an interaction of the z electrons with each other. The Wigner–Seitz cell may be approximated as a sphere of radius r_{WS}. This simplification should be good for close-packed lattices such as the FCC or HCP lattice. The Coulomb energy is therefore

$$U_{\text{Coul}} = N\left[\frac{1}{2}\int d\mathbf{r}\int d\mathbf{r}'\frac{(ne)^2}{4\pi\epsilon_0|\mathbf{r} - \mathbf{r}'|} - \int d\mathbf{r}\, ne\Phi(r)\right],$$
(12.14)

where the first term is the electron–electron interaction and the second term is the interaction of the electrons with the Ashcroft pseudopotential. The electron density is assumed to be uniformly spread over the Wigner–Seitz cell, so $n = 3z/4\pi r_{\text{ws}}^3$. Evaluation of the integrals leads to

$$U_{\text{Coul}} = \frac{Nze^2}{8\pi\epsilon_0 r_{\text{WS}}}\left[\frac{6}{5} - 3\left(1 - \frac{r_c^2}{r_{\text{WS}}^2}\right)\right].$$
(12.15)

The total energy is given by

$$U = N\left\{\frac{3}{5}\frac{\hbar^2 k_F^2}{2m} - \frac{3e^2 k_F}{16\pi^2\epsilon_0} - \frac{e^2}{8\pi\epsilon_0 a_1}\left[0.115 - 0.31\ln r_s - \frac{Za_1}{r_{\text{WS}}}\left(-\frac{9}{5} + \frac{3r_c^2}{r_{\text{WS}}^2}\right)\right]\right\}.$$
(12.16)

Minimization of this with respect to k_F leads to expressions for k_F, E_{coh}, and B in reasonably good agreement with experiment. The core radius r_c is taken to be a fitting parameter.

TABLE 12.1 Properties of *sp*-Bonded Metals

Element	Valence z	Crystal	Cohesive Energy E_{coh} (eV/atom)	Bulk Modulus B (GPa)	Melting Temperature T_m (K)	Debye Temperature Θ_D (K)	Coefficient of Thermal Expansion α (10^{-6} K^{-1})	Cutoff Radius[a] r_c (nm)	Calculated Bulk Modulus[b] B_{th} (GPa)
Li	1	BCC	1.66	11.6	454	352	45	0.070	15
Na	1	BCC	1.13	6.81	371	157	71	0.093	7.5
K	1	BCC	0.94	3.18	337	89	83	0.117	3.4
Rb	1	BCC	0.88	3.14	312	54	88	0.131	2.6
Cs	1	BCC	0.83	2.03	302	40	97	0.146	2.1
Be	2	HCP	3.34	100	1560	1160	12	0.040	170
Mg	2	HCP	1.53	35.4	923	396	26	0.069	47
Ca	2	FCC	1.83	15.2	1115	234	22	0.092	22
Sr	2	FCC	1.70	11.6	1050	147	20	0.102	16
Ba	2	BCC	1.86	10.3	1000	111	19	0.107	14
Al	3	FCC	3.34	72.1	933	423	23	0.059	160
Ga	3	ORT	2.81	56.1	303	317	18	0.063	130
In	3	TET	2.49	41.1	430	109	31	0.072	90
Tl	3	HCP	1.88	35.9	577	—	29	0.076	79
β-Sn	4	TET	3.12	111	505	236	5.3	—	—
Pb	4	FCC	2.04	42.9	601	102	29	—	—
Sb	5	TRIG	2.70	2511	904	150	11	—	—
Bi	5	TRIG	2.17	3054	545	119	13	—	—

Source: Data from K. A. Gschneider, Jr., Physical properties and interrelationships of metallic and semimetallic elements, in F. Seitz and D. Turnbull, eds. *Solid State Physics, Vol. 16*, Academic Press, San Diego, Calif., 1964. Theoretical results are adapted from D. Pettifor, *Bonding and Structure of Molecules and Solids*, Clarendon Press, Oxford, 1995.

[a]Defined in Eq. (12.13).
[b]See Eqs. (12.2) and (12.16).

12.3 Transition Metals

The transition metals are extremely important because of their high strength and hardness, large elastic constants, high melting temperatures, and their magnetic properties. The three series of transition metals are the set of 30 elements characterized by the filling of the $3d$, $4d$, and $5d$ electron shells, respectively. Each d shell can accommodate up to 10 electrons. Melting temperatures for the $3d$ series range from 693 K for Zn to 2183 K for V. For the $4d$ series, T_m is as high as 2896 K for Mo. For the $5d$ series the highest T_m occurs for W at 3695 K. A plot of the cohesive energies (Fig. 12.1) shows that as more electrons are added to a shell, the energy first rises and then falls, in an approximately parabolic fashion. Similar remarks apply to the variation of the bulk modulus (Fig. 12.2). The melting temperature and elastic constants follow a similar pattern. Friedel proposed a band-structure theory explaining this trend.

The d bands are relatively narrow, since the wavefunctions fall off rapidly with distance and there is little overlap between d electrons on NN sites. If the atoms were infinitely far apart from each other the energy of N_d electrons in the d shell of an atom would be $N_d E_d$, neglecting the fine-structure splitting. As the atoms are assembled into their lattice positions the energy levels broaden symmetrically into a band of width Δ. The density of states is taken to be a constant of magnitude $\rho_0(E) = 10/\Delta$ over this band. If N_d electrons are placed in the band, the Fermi energy is given by

$$N_d = \int_{E_d - \Delta/2}^{E_F} \rho_0 \, dE = \frac{10}{\Delta}\left(E_F - E_d + \frac{\Delta}{2}\right). \tag{12.17}$$

The energy of all the electrons in the d band is

$$E_{\text{band}} = \int_{E_d - \Delta/2}^{E_F} \rho_0 E \, dE = \frac{5}{\Delta}\left[E_F^2 - \left(E_d - \frac{\Delta}{2}\right)^2\right] = N_d\left(E_d - \frac{\Delta}{2} + \frac{N_d \Delta}{20}\right). \tag{12.18}$$

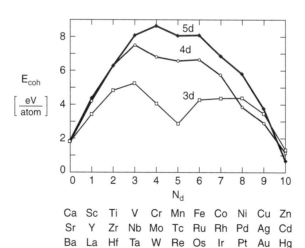

Figure 12.1. Cohesive energy, E_{coh}, for the transition metals. N_d is the number of d electrons. (After K. A. Gschneider, Jr., in F. Seitz and D. Turnbull, eds., *Solid State Physics*, Vol. 16, copyright 1964. Reprinted by permission of Academic Press, Inc.)

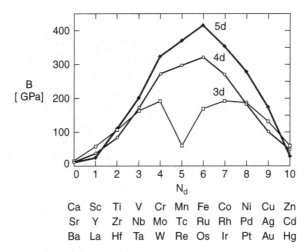

Figure 12.2. Bulk modulus, B, for the transition metals. (After K. A. Gschneider, Jr., in F. Seitz and D. Turnbull, eds., *Solid State Physics*, Vol. 16, copyright 1964. Reprinted by permission of Academic Press, Inc.)

The net change in the band energy upon assembly is therefore

$$\delta E_{\text{band}} = E_{\text{band}} - N_d E_d = -\frac{N_d(10 - N_d)}{20}\Delta. \tag{12.19}$$

Equation (7.93) may be used to determine the root-mean-square (RMS) bandwidth. Square both sides of the equation and average it over the band. Use the fact that $\langle A_n A_{n'}^* \rangle = |A_0|^2 \delta_{n,n'}$ to obtain

$$\langle (E - E_d)^2 \rangle = |t|^2 \sum_{n,n'} \delta_{n,n'} = Z|t|^2, \tag{12.20}$$

where Z is the number of NN. From the model band structure one finds that

$$\langle E^2 \rangle - \langle E \rangle^2 = \frac{1}{10} \int_{E_d - \Delta/2}^{E_d + \Delta/2} \rho_0 E^2 \, dE - E_d^2 = \frac{\Delta^2}{12}. \tag{12.21}$$

Therefore the bandwidth is proportional to the tunneling matrix element:

$$\Delta = \sqrt{12Z}|t|. \tag{12.22}$$

The energy shift of the $3s^2$, $4s^2$, or $5s^2$ electrons, δE_s, is added to the d-band energy shift and is treated as a constant. A repulsive energy of the form $U = U_0 \exp(-\gamma R)$, due to electron–electron repulsion, is also added. Here R is the NN distance. The tunneling integral is assumed to fall off exponentially with distance [i.e., $t = t_0 \exp(-\alpha R)$]. It is not unreasonable to assume that $\gamma = 2\alpha$. The total atomic energy is therefore

$$\delta E_{\text{total}} = \delta E_s - V_0 e^{-\alpha R} + U_0 e^{-\gamma R}, \tag{12.23}$$

with

$$V_0 = \frac{N_d(10 - N_d)}{20}\sqrt{12Z}|t_0|. \tag{12.24}$$

The minimum of the total atomic energy determines the equilibrium separation:

$$R_0 = \frac{1}{\gamma - \alpha} \ln \frac{\gamma U_0}{\alpha V_0}. \tag{12.25}$$

The total atomic energy is

$$\delta E_{\text{total}} = -2E_{\text{coh}} = E_s - \frac{3N^2(10 - N_d)^2 Z|t_0|^2}{400 U_0}, \tag{12.26}$$

which is twice the cohesive energy, E_{coh}, the factor 2 being included to avoid double counting.

The adiabatic bulk modulus is [see Eq. (WA.11a)] given by

$$B_S = -V \left(\frac{\partial P}{\partial V} \right)_{N,S} = V \left(\frac{\partial^2 U}{\partial V^2} \right)_{N,S} = v \left(\frac{\partial^2 U}{\partial v^2} \right)_{N,S}, \tag{12.27}$$

where v is the volume per atom. For an FCC lattice $v = R^3/\sqrt{2}$, where $R = a\sqrt{2}/2$ is the NN distance, and a is the side of the cubic cell. Therefore,

$$B_S = \frac{4\sqrt{2}\,\alpha^2}{9R_0} \left(E_{\text{coh}} + \frac{E_s}{2} \right). \tag{12.28}$$

Equations (12.26) and (12.28) account for the form of the data in Figs. 12.1 and 12.2.

12.4 Rare Earth Metals

The rare earths, comprising the lanthanide series, extending from La to Lu, and the actinide series, extending from Ac to Lr, are characterized by the filling of the $4f$ and $5f$ shells of the atoms, respectively. In their metallic phases the three outer electrons are donated to the conduction band and the trivalent ion is left with its f electrons. The f electrons are confined largely to the interior of the ions and are not strongly perturbed by coupling to the nearest neighbors. A measure of this coupling may be obtained by comparing the melting temperatures of the rare earth (RE) metals with those of the transition metals. One finds that the RE metals have lower values for T_m. The reduced interatomic coupling also implies that the electronic energy levels are well defined. Some optical applications exploit this feature.

RE ions (Nd, Sm, Eu, Tb, Dy, and Yb) are used as dopants in laser crystals and glasses, and provide colors to carbon arc lamps (La, Ce, Pr, and Sm). They are used to color glasses (Ce, Nd, Pr, and Er) and as phosphors (Eu, Gd, and Tb), as well as phosphor sensitizers and activators (Sm and Eu). They find utility as additives to alloys to refine various mechanical properties (La, Ce, Pr, Nd, Gd, Ho, Er, Tm, Yb, Lu, and Th). They are employed to make permanent magnets (Nd and Sm) and are found in magnetic-storage disks (Gd and Dy). Some function as catalysts (Ce, Sm, and Lu). Elements with higher atomic numbers have found use in the nuclear industry (Eu, Dy, Er, Tm, Ac, U, Np, Pu, and Am).

Some typical physical properties of the RE metals are given in Table 12.2. Most of the RE metals are trivalent, although Eu and Yb are divalent, and Ce is often of mixed valence. The conduction-electron concentration, n, is comparable in magnitude to those of trivalent sp-bonded metals.

The electrical resistivity is typically more than an order of magnitude larger than in the sp-bonded metals. As one progresses through the lanthanide series one sees that the resistance rises to a maximum at Gd and then decreases with increasing atomic number. This suggests that magnetic scattering contributes significantly to the resistance, since the magnetic moments of the ions also have this trend (see Table 9.3). The cross section for magnetic spin-flip scattering, referred to as *Kondo scattering*, can be expected to vary as S^2, where S is the spin of the ion, and this can approximately account for the rise and fall of the resistivity. Magnetic scattering becomes increasingly more important as the temperature is lowered and, in fact, leads to an increase of the resistivity as $T \to 0$, a phenomenon referred to as the *Kondo effect*. The Kondo temperature, T_K, may be defined as the temperature below which magnetic scattering dominates the thermal properties of the metal.

The melting temperatures are higher than those of sp-bonded metals but not as high as those of some of the transition metals. In the transition metals much of the bond strength arises from the tunneling of d electrons. The tunneling of f electrons is not as strong in the RE metals since the size of the ions is much less than the typical interatomic spacing.

In Table 12.2 the experimental values for the linear specific heat constant, γ, are given. The theoretical values are given by Eqs. (7.26) and (7.70). One may use these data to deduce an effective mass for the valence electrons, $m^*/m = 3\hbar^2 \gamma / m k_B^2 k_F$. (Note that one must first convert γ to the units of J/electron $\cdot$ K^2 by multiplying by n/N_A, where N_A is Avogadro's number). Values of m^*/m are one to two orders of magnitude larger than what they are for free-electron metals (e.g., $m^*/m = 1.3$ for Na). Heavy-fermion metals refer to intermetallic compounds with $m^*/m \sim 10^2 - 10^3$.

The effects of the heavy fermions are only felt at low temperatures, where only those electrons that are close to the Fermi level are thermally excited. The conduction electrons are strongly coupled to the f-band electrons by magnetic spin-flip interactions, particularly when the Fermi level lies near the f-band levels and the possibility for strong state mixing occurs. A many-body resonance state may form between the f electrons and the conduction electrons. It is believed that this resonance is a singlet state, so that the magnetic moment of the ion is shielded by those of the nearby conduction electrons, with the net result that the magnetic moment of the ion is quenched. The physics of this highly correlated system is a topic of current research.

An estimate of the specific heat at low temperatures may be made for a spin J moment by noting that the entropy associated with its degeneracy is $k_B \ln(2J + 1)$. Since the characteristic temperature is T_K,

$$c_v = T \left(\frac{\partial S}{\partial T} \right)_v = \frac{k_B \ln(2J + 1)}{T_K} T = \gamma T. \tag{12.29}$$

Low values of T_K account for high values of γ and hence large apparent effective masses for the electrons.

ALLOYS

Alloying different elements together greatly extends the number of metallic materials and allows the metallurgist the ability to optimize one or more physical or chemical

TABLE 12.2   Selected Properties of the Rare Earth Metals

Element	Crystal	E_{coh} (eV/atom)	T_m (K)	Θ_D (K)	n (10^{28} m^{-3})	ρ (10^{-8} $\Omega \cdot$ m)	γ (J/mol·K^2)	m^*/m
La	DHCP	4.4	1191	142	8.0	61	0.010	19
Ce	DHCP	4.3	1071	146	11	74	0.058	137
Pr	DHCP	3.7	1204	85	8.7	70	0.022	45
Nd	DHCP	3.3	1294	159	8.8	64	0.0089	18
Pm	DHCP	2.8	1315	158	8.9	75	0.010	21
Sm	TRIG	2.2	1347	116	9.1	94	0.011	23
Eu	BCC	1.9	1095	127	4.1	90	0.0028	3.4
Gd	HCP	3.6	1586	170	9.1	120	0.01	21
Tb	HCP	3.9	1629	150	9.4	120	0.0091	19
Dy	HCP	2.9	1685	172	9.6	91	0.0093	20
Ho	HCP	3.1	1747	114	9.6	94	0.026	56
Er	HCP	3.1	1802	134	9.8	86	0.013	29
Tm	HCP	2.5	1818	127	10	70	0.020	44
Yb	FCC	1.7	1092	118	4.9	28	0.0029	4.0
Lu	HCP	4.3	1936	210	10	56	0.010	22
Ac	FCC	4.5	1470	124	8.0	—	0.0096	18
Th	FCC	5.9	2028	170	9.1	15	0.0047	10
Pa	BCT	5.7	1845	159	11	18	0.007	17
U	ORTH	5.4	1408	200	14	28	0.011	31
Np	ORTH	4.9	912	121	14	120	0.011	31
Pu	MONO	4.0	913	171	15	150	0.049	143
Am	DHCP	—	1449					
Cm	—	—	1173					
Bk	—	—	1323					
Cf	—	—	1173					
Es	—	—	1133					
Fm								
Md								
No								
Lr								

Source: Data largely from K. A. Gschneidner, Jr., Physical properties and interrelationships of metallic and semimetallic elements, in F. Seitz and D. Turnbull, eds., *Solid State Physics*, Vol. 16, Academic Press, San Diego, Calif., 1964, p. 275.

properties. Alloys may be substitutional, in which the overall lattice structure is preserved, but the sites are occupied by more than one type of atom. They may be interstitial, in which case the added atoms do not occupy the elemental sites. There may be a transformation of the lattice to a completely new structure, with the formation of an intermetallic compound, or to an amorphous solid. The order–disorder transition was considered in Chapter 6. Here the formation a solid solution is considered.

This is followed by a study of the electrical resistance of alloys. Then the chemical stability of alloys is examined in Section W12.4, with emphasis on the oxidation and corrosion processes.

12.5   Hume–Rothery Rules

In a solid solution there is an underlying invariant lattice whose sites may be shared by two or more different types of atoms. In some instances the solid solution may only be formed over a specific range of stoichiometric compositions. In other cases any proportion is possible, and one says that there is complete solid solubility (e.g., in Cu–Au at high T). Hume–Rothery formulated a set of three empirical rules to help identify when solid solutions would be unlikely to occur and when they would be likely to occur.

The first rule involves the relative size of the atoms. If the atomic radii differ by more than 15%, solid solutions are unlikely to form, due to the lattice strain energy resulting from the mismatch. If too large an atom were placed in a lattice populated by smaller atoms, a compression field would be set up in the neighborhood of the larger atom. Conversely, if a small atom replaced a larger atom in a lattice, an expansion field would be produced. Either case involves an increase in strain energy.

The second rule concerns the electronegativity difference of the atoms in an alloy. A large difference indicates a proclivity for forming stable intermediate chemical compounds. This destroys the bonding of a pair of atoms with the rest of the lattice and disrupts the ordered lattice.

Darken and Gurry introduced a map that provides a graphic illustration of the two rules above. The abscissa gives the atomic radius r_{met}, assuming a fixed coordination number, and the ordinate gives the atomic electronegativity. Various atoms are plotted as points on the plane. One selects a given atom, A, and draws two ellipses around it. The inner ellipse has an axis parallel to the abscissa corresponding to a $\pm 7.5\%$ difference in radii. The outer ellipse has a corresponding axis corresponding to $\pm 15\%$. Similarly, the inner ellipse has an axis parallel to the ordinate corresponding to a ± 0.2 electronegativity difference. The corresponding value for the outer ellipse is ± 0.4. If a second atom, B, lies within both ellipses, there is a substantial probability for high solubility of B in A. This solubility could be expected to be in excess of 5 at %. If B lies between the ellipses, the solubility is lower. If B is outside the ellipses, solubility is unlikely. The equations for the two ellipses are

$$\left(\frac{X - X_A}{0.2}\right)^2 + \left(\frac{r - r_A}{0.075 r_A}\right)^2 = 1, \tag{12.30}$$

$$\left(\frac{X - X_A}{0.4}\right)^2 + \left(\frac{r - r_A}{0.15 r_A}\right)^2 = 1. \tag{12.31}$$

An example of a Darken–Gurry map for Al is presented in Fig. 12.3. The map predicts that solid solution is likely over some range of stoichiometry of Al with V, Ga, and Ta, possible solid solutions with Mn, Cr, W, Pa, Zr, Hf, and Sn, and unlikely solid solutions with Be, Ni, Cu, Ag, Pb, Mg, Y, Gd, and Ca.

The third rule relates to n, the valence electron concentration (VEC) in the metal. One computes the radius of the Fermi sphere, k_F, assuming the metal to be free-electron-like. A lowering of the internal energy may be expected for those metals whose Fermi sphere is tangent to a bounding face of a Brillouin zone. The reason has to do with electronic band structure. When an energy band intersects a face of the Brillouin zone, in most cases an energy gap will open up and the energy surfaces will be perpendicular to the Brillouin zone face. The bending of the lower band in a concave downward direction results in a lowering of the electron energies at those points of the Brillouin zone in the neighborhood of the point of tangency.

The computation of the VEC is made using empirical valences. Metals such as the alkalis, Cu, Ag, and Au have valence 1. The metals Mg, Ca, Sr, and Ba have

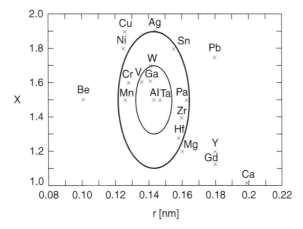

Figure 12.3. Darken–Gurry map for Al. Electronegativity is plotted as a function of metallic radius (for coordination number 12) for various elements. See Tables 2.11 and 2.12 for data.

valence 2. The metals Al, Tl, and In have valence 3. The elements Si, Ge, and Sn have valence 4, and Sb has valence 5. The average number of valence electrons per atom, $z = e/a$, is computed. Thus for CuZn (β-brass), $e/a = \frac{3}{2}$, for Cu_5Zn_8 (γ-brass), $e/a = 21/13$, and for $CuZn_3$ (ε-brass), $e/a = \frac{7}{4}$. The corresponding crystal structures are BCC, complex cubic, and HCP. Recall that for the BCC lattice, the primitive reciprocal lattice vectors are $\mathbf{g}_1 = 2\pi(\hat{j} + \hat{k})/a$, and so on, so tangency of the Fermi sphere and the zone face will occur when $k_F = g_1/2 = \pi\sqrt{2}/a$. Since $k_F = (3\pi^2 n)^{1/3}$, this gives $n = 2\pi\sqrt{2}/3a^3$. There are two atoms per conventional unit cell of volume a^3. If each atom provides, on the average, z valence electrons, the tangency condition is $2z/a^3 = 2\pi\sqrt{2}/3a^3$, so $z = \pi\sqrt{2}/3 = 1.481$, close to the value $\frac{3}{2}$ found for β-brass. Values of e/a in the range 1.36 to 1.59 are likely to form the BCC structure if the crystal has cubic symmetry. If e/a is in the range 1.54 to 1.70, the γ-brass structure is favored.

The various brass compounds have analogous structures involving other elements. For example, there is Cu_3Al, FeAl, Ag_3Al, and so on. The spherical Fermi surface assumption is obviously a weakness in rule 3, since only free-electron metals may be expected to possess a spherical surface. Nevertheless, rule 3 points to a correlation between the VEC and the crystal structure.

12.6   Electrical Resistance of Metallic Alloys

In the discussion of the electrical resistance of metals, two sources of resistance are identified. One, which is temperature dependent, is the scattering of electrons from phonons, which diminishes at low temperatures. The second is largely temperature independent and is due to scattering from impurities, defects, dislocations, and so on. If the metal were a perfect crystal without such imperfections, then according to the quantum theory of solids, it would be a perfect conductor at $T = 0$. In this section binary alloys of the form $A_x B_{1-x}$ are studied and the dependence of the low-temperature resistivity on composition is determined.

A plot of resistivity as a function of composition for $Cu_x Au_{1-x}$ is presented in Fig. 12.4. Two sets of data are given, one for tempered alloys and the other for annealed

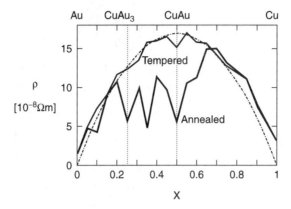

Figure 12.4. Electrical resistivity of Cu_xAu_{1-x} as a function of composition for tempered and annealed alloys at $T = 300$ K. The dashed curve is the theoretical prediction for the random alloy. The jaggedness of the curves is probably due to a sparsity of data points. (Data adapted from E. W. Washburn, ed., *International Critical Tables of Numerical Data*, Vol. 6, McGraw-Hill, New York, 1929.)

alloys. In the tempered samples the temperature is rapidly reduced from the melt and the alloys are unable to reach thermal equilibrium. In the annealed samples the temperature lowering is gradual and various intermetallic compounds have time to nucleate and grow. One notes sharp dips in the resistivity corresponding to the intermetallic compounds CuAu, $CuAu_2$, and $CuAu_3$. Due to the crystalline order, these compounds present diminished resistance. The data for the tempered alloys show the presence of only small numbers of crystallites, probably distributed inhomogeneously throughout the metal. Tempered alloys better approximate random alloys than do annealed alloys.

The resistivity is given by the Drude formula, Eq. (7.30),

$$\rho = \frac{m}{ne^2\tau}, \tag{12.32}$$

where n is the number density of electrons and τ is the collision time. The collision time is determined by the quantum-mechanical interaction of the electrons with the scatterers. The quantum-mechanical scattering rate is given by Fermi's golden rule (see Eq. (WC32)):

$$\frac{1}{\tau} = \frac{2\pi}{\hbar}|M|^2\rho(E_F)\langle 1 - \cos\theta\rangle, \tag{12.33}$$

where $\rho(E_F)$ is the density of electron states at the Fermi level and θ is the scattering angle. The quantity M is the matrix element of the potential fluctuation between the initial and final states. Mott argued that a disordered alloy may be treated as an effective medium in which there is an average background potential:

$$\langle V \rangle = xV_A + (1 - x)V_B. \tag{12.34}$$

The presence of an A atom causes a potential fluctuation

$$U_A = V_A - \langle V \rangle = (1 - x)(V_A - V_B), \tag{12.35}$$

and the B atom causes a fluctuation

$$U_B = V_B - \langle V \rangle = x(V_B - V_A). \tag{12.36}$$

Since the relative numbers of A and B atoms are in the proportion x to $1 - x$, the total temperature-independent resistivity is given by the sum of weighted contributions from the respective scatterings:

$$\rho = x\rho_A + (1 - x)\rho_B \propto x(1 - x)^2 + (1 - x)x^2 \propto x(1 - x). \tag{12.37}$$

The dashed curve in Fig. 12.4 is a plot of the theoretical prediction $\rho(x) = 4\rho_{max}x(1 - x)$. The prediction is in good agreement with the results for the tempered alloy.

EXAMPLES AND APPLICATIONS OF METALLIC ALLOYS

Having considered some general principles relating to metals and alloys, the discussion now proceeds to specific cases.

12.7   Steel

In this section the most important alloy, steel, is introduced. More details concerning steel are presented in Chapter W21. Steel is primarily an alloy of iron and carbon of the form $Fe_{1-x}C_x$. There are, however, many types of steels, and these have tertiary elements mixed in. Typically, these can be Cr, Mn, Mo, Ni, or Si. Trace amounts of other elements may also be present. The $Fe_{1-x}C_x$ phase diagram is rather complex, as may be seen in Fig. 12.5. At high temperatures there is the molten alloy. The α phase (ferrite) has a BCC crystal structure with a low concentration of C as interstitial impurity atoms. Since there is little change of the lattice constant with C content, there is a very low solubility of C in Fe for this phase. Carbon atoms occupy the octahedral sites on the faces or in the middle of the edges of the cubic unit cell. The γ phase (austenite) has the FCC crystal structure and tolerates a higher concentration of C atoms. The C atoms reside at the octahedral site at the center of the conventional unit cell. The lattice constant increases with increasing C concentration. At most, one site in 12 unit cells is occupied, due to the low solubility of C in Fe. There is also a δ phase (δ-ferrite) at higher temperatures for low x. The compound iron carbide, Fe_3C, has an orthorhombic structure and is called *cementite*. The lattice constants are $a = 0.509$ nm, $b = 0.674$ nm, and $c = 0.452$ nm. Not shown is the magnetic phase transition that occurs for pure iron at the Curie temperature, $T_C = 769°C$.

The physical properties of steel depend very much on the history of the preparation treatment, which is discussed in greater detail in Chapter W21. In particular, it is sensitive to the quench rate (i.e., how rapidly the alloy is cooled from the melt). If it is done very slowly, graphitization occurs and the phase diagram is then slightly altered, with phase boundary lines denoted by dotted lines in Fig. 12.5. If the quench rate is rapid, the system goes far from thermal equilibrium and the phase diagram is not directly applicable. Figure 12.5 is applicable to intermediate quench rates.

Notice a eutectoid transition at $T_e = 727°C$. This transition is analogous to the eutectic transition, but involves three solid phases rather than two solid phases and

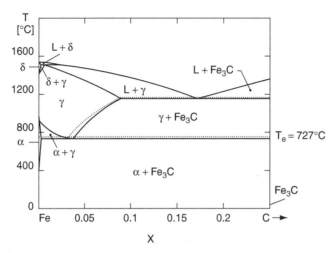

Figure 12.5. Binary phase diagram for steel. (From *Metals Handbook*, 8th ed., Vol. 8, *Metallography, Structures and Phase Diagrams*, ASM International, Materials Park, Ohio, 1973, p. 275, C–Fe.)

a liquid. The solid phases are α, γ, and Fe_3C. The eutectoid composition is at $x_e = 0.0361$. Hypoeutectoid steel has $x < x_e$, whereas hypereutectoid steel has $x > x_e$.

When hypoeutectoid steel is cooled from the melt, austenite (γ) begins nucleating as the liquidus is crossed. These nuclei continue to grow in size until the sample is completely austenite. As the temperature is lowered further, ferrite (α) particles nucleate and grow amidst the austenite. When the eutectoid temperature is passed, the remaining austenite transforms to cementite (Fe_3C) and ferrite, producing a eutectoid mixture of ferrite and cementite which is called *pearlite*.

For hypereutectoid steel, austenite (γ) nucleates as before. However, now cementite particles form amidst the austenite. Below the eutectoid temperature there is both pearlite as well as cementite present.

In all of these processes the texture depends, to a large extent, on the cooling rate. If hypoeutectoid steel is rapidly quenched, it is possible to prevent the transformation from austenite (γ) to cementite (Fe_3C) and to form a supersaturated mixture of interstitial carbon in the ferrite (α) phase. This material is called *martensite*. Martensitic transitions are considered in more detail in Section W12.6.

12.8 Intermetallic Compounds and Superalloys

For an alloy with the molar concentrations of the elements in integer ratios, it is possible to create intermetallic compounds. In Fig. 12.6 a typical binary phase diagram is sketched for the $La_{1-x}Cu_x$ alloy system. Note that for specific compositions, stoichiometric compounds form. These correspond to the vertical lines in the diagram at $x = \frac{1}{2}$, $\frac{2}{3}$, $\frac{4}{5}$, $\frac{5}{6}$, and $\frac{6}{7}$ and correspond to $CuLa$, Cu_2La, Cu_4La, Cu_5La, and Cu_6La, respectively.

As a metal is quenched from the melt, crystallites nucleate, grow, and eventually come into contact with each other. The common interfaces of adjacent crystallites are called *grain boundaries*. When subjected to stress, these grains can slip past each other. This limits the hardness of the metal. When the metal is subject to stress, atoms on

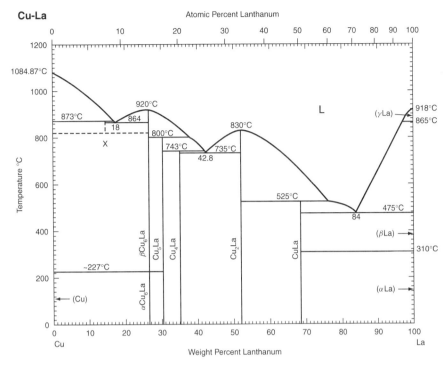

Figure 12.6. Binary phase diagram for La$_x$Cu$_{1-x}$. (From *ASM Handbook*, Vol. 3, *Alloy Phase Diagrams*, ASM International, Materials Park, Ohio, 1992, p. 2.171, Cu–La.)

the grains can undergo thermally activated translation which leads to slippage and a slow plastic creep of the material. Obviously, any mechanism that would tend to lock the grains together would strengthen the metal.

In a binary alloy, AB, the process of atomic rearrangement facilitates the microscopic slippage of one grain past another. Assume for the moment that the atoms in the grain are disordered; that is, even though there may be a lattice present, the site occupancy is random, being determined only by the overall stoichiometry. A given A grain-boundary atom is likely to be surrounded by a random mixture of A and B atoms. For this atom to exchange places with a neighboring B atom within the same grain is relatively simple, since the energy of the new configuration is likely to differ only slightly from the old one. This situation would change, however, if the alloy were ordered. Then an atomic interchange would radically change the environment and cost more energy. Thus, in the ordered phase an A atom may have only B neighbors, and vice versa. An interchange of A and B would now give A mostly A neighbors and B mostly B neighbors.

From the remarks concerning the order–disorder transition in Chapter 6, alloyed metals often have the tendency to form intermetallic compounds. These compounds include CuZn, NiAl, FeCo, Cu$_3$Zn, Ni$_3$Fe, Mg$_3$Cd, and Au$_3$Cu as well as many others. This ordering usually takes place considerably below the melting temperature T_m at some critical temperature, T_c. As the metal is quenched below T_c, some portion of the grains become atomically ordered. These ordered subdomains are likely to nucleate at the grain boundaries. Since it is no longer a simple matter to rearrange the A and B atoms, the ability of grains to slide past each other via atomic reshuffling is curtailed. The grain-boundary mobility may be reduced by about two orders of magnitude. Effectively, the grains become locked together.

At still higher degrees of ordering, other mechanisms come into play which tend to promote mobility, such as pairs of dislocations, connected by stacking faults, which become free to move. Thus, in a material like Ni_3Al, the stress needed to induce flow peaks at intermediate values of ordering.

A number of intermetallic compounds have found useful applications. The compound Fe_3Al, although brittle and of low strength, may be hardened by alloying with Cr, Mo, or TiB_2. This process, in addition, makes the material much more corrosion resistant. Presumably, the added atoms help pin the dislocations and grain boundaries to increase its strength. Any tendency toward more positive electronegativity would make an alloy more noble-metal-like and increase its corrosion resistance.

Ni_3Al, although normally brittle, becomes ductile when alloyed with B. This may be due to the tendency for the B atoms to migrate to the grain boundaries, where they form a disordered layer. However, for the reasons mentioned above, this also reduces its resistance to creep.

With the advent of the aerospace age, the need for alloys that could maintain their mechanical integrity at high temperatures became necessary. The term *super-alloy* was coined to describe such materials. The problem in designing such alloys was that there were many variables that had to be optimized simultaneously. These included castability, corrosion resistance, resistance to creep, ductility, the number of stress cycles until rupture, the ultimate tensile stress, workability, and the yield stress. The parameters at one's disposal consisted mainly of alloy composition, quench rates, and heat treatments. In Table 12.3 a selection of such superalloys is presented along with their elemental compositions. It is quickly noted that superalloys contain a list of elements comparable in size to that on a bottle of vitamin pills. Since a detailed theory of such materials does not exist, empirical approaches based on intuition are used.

Grain boundaries act as sites that can begin the failure processes. Elements such as B, C, Hf, and Zr form partially covalent bonds which promote the binding together of grains, thereby neutralizing them, and mechanically strengthen the alloy. However, the covalent bonds obtain their strength at the expense of the metallic bonds. The metallic grains now tend to melt at a lower temperature. In a single crystal, where grain boundaries are not relevant, it would be desirable to eliminate these elements. This may be seen by examining the composition of the Ni-based superalloy PWA 1480, which is manufactured in single crystals, and comparing it with the polycrystalline Waspaloy (see Table 12.3). Single-crystal superalloys are the best materials for high-temperature applications, such as jet-engine turbines.

There are two phases in the NiX alloys. The majority γ-phase is based on the Ni FCC crystal structure and the γ'-phase is based on the Ni_3Al crystal structure (the $L1_2$ structure), as in Fig. 12.7. The γ' phase tends to strengthen the material because the precipitated particles pin dislocations. (More precisely, there are pairs of screw dislocations that are pinned by {111} planes that cross-slip into {100} planes of the γ' phase.) One generally would like to keep the solvus temperature as high as possible (but below T_m) and thereby retain this phase during high-temperature applications. By testing Ni-based alloys with varying compositions, empirical formulas for the solvus temperature (in °C) as a function of chemical composition have been

TABLE 12.3   Chemical Composition (wt %) of Various Superalloys

Element	Alloy[a]							
	A	B	C	D	E	F	G	H
Al	—	—	—	1.35	—	0.2	—	—
B	—	—	—	0.005	—	—	—	—
C	0.03	0.12	0.58	0.036	—	0.05	0.1	—
Co	—	—	—	13.1	—	—	39.0	5.0
Cr	—	—	—	19.7	19.0	15.0	22.0	10.0
Fe	—	—	—	1.03	—	53.0	3.0	—
Hf	0.9	—	1.94	—	—	—	—	—
Mo	—	98.4	—	4.16	3.0	1.25	—	—
Nb	—	—	69.2	—	5.0	0.5	—	1.0
Ni	—	—	—	58.2	73.0	26.0	22.0	62.5
Si	—	—	—	0.07	—	—	—	—
Ta	89.8	—	—	—	—	—	—	12.0
Ti	1.1	1.2	—	2.95	—	2.15	—	1.5
V	—	—	—	—	—	0.2	—	—
W	8.2	—	28.3	—	—	—	14.0	4.0

[a]A, ASTAR 811C; B, TZC; C, B88; D, Waspaloy; E, Inconel 718; F, A286; G, HS-188; H, PWA 1480.

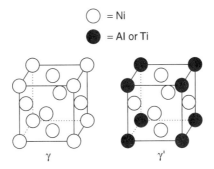

Figure 12.7. Unit cells for the γ phase of Ni and the strengthening γ' phase: Ni$_3$Al or Ni$_3$Ti.

obtained, for example,

$$T_{\gamma'} = 1152 - 358(\text{wt \% C}) - 1.26(\text{wt \% Cr}) - 6.9(\text{wt \% Co}) + 2.73(\text{wt \% Mo})$$
$$+ 8.65(\text{wt \% Nb}) + 47.5(\text{wt \% Al}) + 51.3(\text{wt \% Ti}) + 218(\text{wt \% B})$$
$$- 265(\text{wt \% Zr}) + (6.19\text{wt \% W}) \tag{12.38}$$

Note that C, Cr, Co, and Zr decrease $T_{\gamma'}$ whereas Mo, Nb, Al, Ti, B, and W increase $T_{\gamma'}$. Another such empirical formula may be developed for T_m. Such formulas assist the metallurgist in formulating recipes for alloy design.

The tensile strength varies inversely with the size of the γ'-phase grains, as described by the Hall–Petch relation, Eq. (10.43). Since grain size is to be kept small, one wants to cool the melt sufficiently rapidly so that, after nucleation, not much grain growth occurs. Dislocations can climb over small grains, whereas larger grains tend to make it easier for dislocations to propagate via the Frank–Read mechanism. There is an optimum size for the γ'-phase grains for producing the strongest material.

The hot workability of Ni-based materials is found to be enhanced by such elements as Ca, Hf, B, C, Si, Al, Ta, Ti, and Zr, and to be reduced by Ag, Au, Cu, Na, O, S, P, Sn, and Zn. Certain elements, such as Co, Cr, Ni, and Zr, assist in dispersing crystallites and therefore retard recrystallization. Oxidation protection may be provided by a thin layer of alumina on the surface, which acts as an obstacle to oxygen diffusion. The addition of Cr aids in providing corrosion resistance against hot salts. Adding Nb, Mo, Ta, and W strengthen the solid solution. Admixtures of B, Nb, Cr, Mo, Ta, Ti, and W strengthen the grain boundaries. Low concentrations of Hf and Zr assist in making the material more ductile. Certain elements (e.g., Bi, Te, and Tl) can "poison" the alloy by migrating to the grain boundaries. These elements melt at low temperatures and essentially lubricate the grains, thereby reducing strength. These, and similar considerations, go into the design of a superalloy.

12.9 Electromigration

The construction of very large scale integrated (VLSI) chips requires wires of microscopic dimensions to make the multitude of electrical connections within each device. The current densities in these interconnects can be rather substantial (e.g., up to 4×10^9 A/m^2), due to the small cross-sectional areas involved. It is found that ultimately these polycrystalline wires break apart and the chips fail, due to a process called *electromigration* (EM). The EM process involves a movement of metal ions in the direction of electron flow and the corresponding movement of vacancies in the opposite direction. The pile-up of vacancies at one end of the wire leads to voids and ultimately to the opening of the circuit. The large-scale separation of vacancies and ions along the wire, which is usually in contact with a rigid oxide insulator, leads to the buildup of mechanical stresses along the wire that can fracture it. Typical interconnect metals include Ag, Al, Au, and Cu. These metals are in contact with Si, SiO_2, Si_3N_4, or Si_2ON_2.

Consider the conduction process in the Drude model discussed in Section 7.2. The average velocity attained by an electron prior to a collision is $\langle v_i \rangle = -e\mathbf{E}\tau/m$, whereas just after a collision the velocity is randomized, so its average is zero. Thus the average momentum transferred to the lattice per collision is $\Delta \mathbf{p} = -e\mathbf{E}\tau$. This momentum transfer gives rise to a force in the direction of the electron flow. Competing with this is the electric force on the ion, $ze\mathbf{E}$, which is in the opposite direction. If the ions are associated with a nearby vacancy, the two forces will not cancel. The net driving force on the ions may be expressed as

$$\mathbf{F} = -Z^* e \mathbf{E}, \tag{12.39}$$

where Z^* is an empirical effective charge.

If there is a vacancy downwind from an ion, the ion will be driven to fill the vacancy. The vacancy will then occupy the previous position of the ion and will be ready to accommodate another drifting ion. The net result is that the vacancies (and ions) drift with a velocity

$$\langle \mathbf{v} \rangle = \frac{D\mathbf{F}}{k_B T} = -\frac{D|Z^*|e\mathbf{J}}{k_B T \sigma}, \tag{12.40}$$

where D is the diffusivity of the vacancy and use has been made of Ohm's law.

The buildup of an ion-vacancy gradient along the wire induces a backward diffusion. Using Fick's law, the net vacancy flux density is given by

$$\mathbf{J}_d = -n_d \langle \mathbf{v} \rangle - D \nabla n_d, \tag{12.41}$$

where n_d is the vacancy number density. The evolution of n_d is governed by the continuity equation $\nabla \cdot \mathbf{J}_d + \partial n_d / \partial t = 0$, so this leads to the diffusion equation with a drift term

$$\frac{\partial n_d}{\partial t} = D \nabla^2 n_d - \langle \mathbf{v} \rangle \cdot \nabla n_d. \tag{12.42}$$

A solution to this equation for $n_d(x, t)$ for $(-\infty < x \leq 0)$ subject to the initial condition $n_d(x, 0) = n_0$, and the boundary conditions $n_d(-\infty, t) = n_0$ and $J_d(0, t) = 0$ may be found. Note that the diffusion equation defines a characteristic length, $L = D/v$, and a characteristic time, $t_c = D/v^2$, v being the drift speed. It may be shown for long times that $n_d(0, t)/n_0 \sim v^2 t/D$. The mean time to failure, t_f, is the time at which $n_d(0, t_f)$ reaches the vacancy density needed to produce failure, n_f. Therefore,

$$t_f = \frac{D n_f}{n_0} \left(\frac{k_B T \sigma}{D |Z^*| e J} \right)^2. \tag{12.43}$$

The diffusion constant may be expressed in terms of the enthalpy needed to activate vacancy diffusion, $D = D_0 \exp(-\Delta H / k_B T)$, to obtain the formula[†]

$$t_f = B \left(\frac{T}{J} \right)^2 e^{\Delta H / k_B T}, \tag{12.44}$$

where B is a constant for a given material (i.e., a given n_0, σ, and Z^*). Experiment shows that $\Delta H = 0.66$ eV for Al.

Electromigration leads to a compactification of the metal as the vacancies are expelled to the ends of the wire. The void space per unit volume is given by $\Delta V / V = n_d \Omega$, where Ω is the atomic volume. If the wire is connected to fixed external components, stresses will develop in the wire. A rough measure of the stress is given by the hydrostatic pressure $P = -B n_d \Omega$, where B is the bulk modulus. The stress is tensile rather than compressive, so P is negative. Correspondingly, compressive stresses are induced in surrounding passivation materials.

A full understanding of EM requires including the vacancy–vacancy interactions and the production of voids. Experimental observations show these to be complex phenomena. It is found that voids nucleate at heterogeneous sites on the sidewalls of a wire, migrate inward, and then travel along the wire opposite to the electron's motion. The nucleation, referred to as *stress-induced voiding*, results from the expulsion of vacancies from the interior of the metal due to the hydrostatic tension.

REFERENCES

Fletcher, G. C., *The Electron Band Theory of Solids*, North-Holland, Amsterdam, 1971.

Guy, A. G., *Elements of Physical Metallurgy*, 2nd ed., Addison-Wesley, Reading, Mass., 1960.

[†] J. R. Black, *IEEE Trans. Electron Devices*, **16**, 338 (1969).

Hansen, M., *Constitution of Binary Alloys*, 2nd ed., McGraw-Hill, New York, 1958.

Hume-Rothery, W., *Electrons, Atoms, Metals and Alloys*, 3rd ed., Dover, Mineola, N.Y., 1963.

Kossowsky, R., *Surface Modification Engineering*, 2 vols, CRC Press, Boca Raton, Fla., 1989.

Mott, N. F., *The Theory of the Properties of Metals and Alloys*, Dover, Mineola, N.Y., 1958.

Pettifor, D., *Bonding and Structure of Molecules and Solids*, Oxford University Press, New York, 1995.

Porter, D. A., and K. E. Easterling, *Phase Transformations in Metals and Alloys*, 2nd ed., Chapman & Hall, London, 1992.

Richman, M. H., *An Introduction to the Science of Metals*, Blaisdell, Waltham, Mass., 1967.

Thompson, C. V., and J. R. Lloyd, Electromigration and IC interconnects, Mater. Res. Soc. Bull., Dec. 1993, p. 19.

PROBLEMS

12.1 Referring to Section 12.5, show that the condition for the tangency of the Fermi sphere to the Brillouin zone boundary for the FCC lattice is $N = 1.36$.

12.2 Derive Eq. (12.7).

12.3 Derive Eq. (12.15).

12.4 Draw the Darken–Gurry map for Fe, using the data from Tables 2.11 and 2.12. Compare the predictions of this map with measured phase diagrams, for example, those presented in Hansen (1958) and supplements, or *ASM Handbook*, Vol. 3, *Alloy Phase Diagrams*, ASM International, Materials Park, Ohio, 1992.

CHAPTER 13

Ceramics

13.1 Introduction

Ceramics are inorganic solids held together by bonds that are either ionic or partially ionic and partially covalent. Typically, they include oxides, borides, carbides, silicides, silicates, and nitrides. They maintain their mechanical strength to high temperatures and are resistant to wear, but are brittle. They are utilized both in crystalline form and as glasses.

Ceramics find application in a wide variety of fields and are considered, along with metals and polymers, one of the three primary classes of materials. Traditional uses include pottery, dinnerware, and other functional whiteware. Ceramics are widely used for structural applications in the construction industry in the form of bricks, cement, and tiles. The ability of ceramics to withstand high temperatures in often hostile environments and to serve as thermal insulators make them suitable for refractory containers in the steel, glass, and chemical industries. Typical refractory materials include the various oxides, such as magnesia (MgO), silica (SiO_2), alumina (Al_2O_3), and zirconia (ZrO_2). Ceramics such as SiC and Si_3N_4 are used as components in heat engines. The hardness of SiC and Al_2O_3 makes them appropriate for cutting tools and abrasives. The wear resistance and strength at high temperatures of Si_3N_4 makes it suitable for moving parts. Ceramics have found high-technology applications because of their versatile electrical and optical properties. Typically, they are used as insulators, resistors, capacitors, electronic-packaging materials, piezoelectric crystals, magnetic-memory devices, laser materials, nonlinear-optical materials, and high-temperature superconductors. The chemical industry makes broad use of ceramics as catalysts and filters. Natural gemstones are crystalline ceramics and techniques exist for creating synthetic gems.

The goal of the present chapter is twofold. The first is to introduce the reader to theoretical concepts which are particularly useful in ceramics. The second is to present a range of applications of ceramics that typify their application to technology. The discussion begins by presenting a set of bonding rules formulated by Pauling. Then the use of refractories in industrial processes is reviewed. The material silicon nitride, and the closely related SIALONs, are then introduced, as these provide examples of strong and tough ceramics. The section on zeolites reveals the importance of these materials as catalysts in the chemical industry.

The scope of phase diagrams introduced in Chapter 6 is enlarged to include ternary phase diagrams (see Section W13.1 at our Web site[†]). This is followed by a brief

---

[†] Supplementary material for this textbook is included on the Web at the resource site (ftp://ftp.wiley.com/public/sci_tech_med/materials). Cross-references to elements of the Web material are prefixed by "W."

discussion of the silicates (Section W13.2), which are the most common ceramic materials found in nature. A discussion of clay (Section W13.3) is followed by a section on cement (Section W13.4).

Glasses, an important class of amorphous ceramic materials, are studied next. Electronic applications of ceramic materials are considered in Chapter 15. Ceramic high-temperature superconductors and magnets are studied in Chapters 16 and 17, respectively. Optical applications of ceramic materials appear in Chapter 18.

In Table 13.1 some general properties of ceramic materials are listed. One notes that ceramic crystals adopt a number of crystal structures, some of which may be rather complex. Examples of some of these crystal structures appear in Fig. 13.1, where the rutile (e.g., TiO_2) and perovskite (e.g., $SrTiO_3$) crystal structures are sketched, and Fig. 13.2, where the fluorite crystal structure (e.g., ZrO_2) is illustrated. The large number of crystal structures available may be attributed to the nondirectional nature of the ionic forces responsible for binding the crystal. The fact that the forces are long-range allows complex unit cells to be built as the crystal searches for its state of lowest free energy. As the temperature is increased, the competition between internal energy and entropy contributions is increased and structural phase transitions often occur. These are important for some of the unique ferroelectric properties of ceramics (and possibly for the high-temperature superconductive properties of some ceramics, as well).

Table 13.1 also shows the ceramic materials to have high melting temperatures, making them suitable for use as refractory materials. The elastic constants are moderately

TABLE 13.1   Selected Properties of Ceramic Materials

Ceramic	Crystal Structure	Melting Temperature T_m (K)	Young's Modulus E (GPa)	Thermal Expansion Coefficient α (10^{-6} K^{-1})	Thermal Conductivity κ (W m^{-1} K^{-1})	Knoop Hardness (GPa)
MgO	Rocksalt	3125	270	12.8	41	8
CaO	Rocksalt	3200	—	—	—	—
ZnO	Wurtzite	2250	124	2.9	23	2
Al_2O_3	Corundum	2330	390	8.1	43	20
Cr_2O_3	Corundum	2600	103	7.5	10–33	29
SiO_2	Hexagonal	1700	94	—	1.0	12
Y_2O_3	Bixbyite	2680	175	9.3	2.3	8
TiO_2	Rutile	2130	290	9.4	6.7	10
ZrO_2	Fluorite	2950	220	10	2.1	12
SiC	Zincblende	3110	455	10	135	27
TiC	Rocksalt	3340	430	8	33	31
WC	Hexagonal	3050	700	5.2	—	29
B_4C	Trigonal	2740	455	5.5	16	29
AlN	Hexagonal	2500	330	5.0	100	11
BN	Hexagonal	3300		4.4	20	—
TiN	Rocksalt	3220	250	9	24	18
Si_3N_4	Hexagonal α, Hexagonal β	2150[a]	317	3.4	30	15
TiB_2	Hexagonal	2900	537	8.2	75	26
$MgAl_2O_4$	Spinel	2170		7.5	12	—

[a]Sublimates.

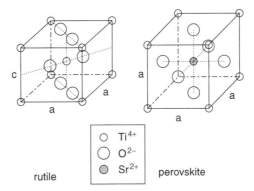

Figure 13.1. Rutile and perovskite cubic crystal structures, adopted by TiO_2 and $SrTiO_3$, respectively.

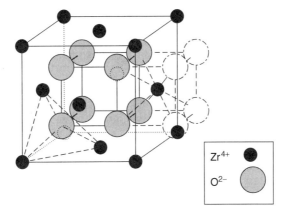

Figure 13.2. Fluorite crystal structure adopted by ZrO_2 for $T > 1400°C$.

strong, as is evidenced by the size of the Young's modulus, E. The coefficients of linear thermal expansion, α, are of moderate size, reflecting the sizable nonlinearities associated with ionic bonding. The thermal conductivities, κ, are also of moderate size. The large values of the Knoop hardness make ceramic materials useful for cutting tools and abrasives.

13.2 Pauling Bonding Rules

In most ceramic materials the bonding is partially ionic and partially covalent. There is electron transfer from the cation to the anion and there is also some sharing of electrons between the ions. By examining many such materials, Linus Pauling was able to formulate a set of rules which, although not always 100% reliable, are useful in summarizing the systematics of bonding. These rules work best for ionic crystals and progressively become less dependable when covalent bonding becomes significant. This underscores the competition between the tendency to minimize the electrostatic energy in the system and the tendency to populate the lowest molecular orbitals.

Pauling's first rule involves interionic distances. Associated with each ion in the crystal is an ionic radius (see Table 2.12). The ionic charges are taken to be integral multiples of the electronic charge, e. The nearest-neighbor (NN) interatomic distances are the sum of the corresponding ionic radii. The cation (of radius r_c) sits at the center of a polyhedral cluster of anions (of radius r_a). It is assumed that the NN anions are in intimate contact with the cation.[†] To the extent that covalent forces may be neglected, the repulsion between the ions (caused largely by the Pauli exclusion principle) is needed for stability, and therefore the anions have to "press" against the cations. The anions may or may not also be in contact with each other. Depending on the ratio $R = r_c/r_a$ various polyhedra are possible, based on the geometry. The determination as to which polyhedra are possible is made in Appendix W13A. As noted in Table 2.3, for $R < 0.155$ only a linear (twofold) coordination is possible. For the range 0.155 to 0.225 a triangular (threefold) coordination can also occur. For the range 0.225 to 0.414 the possibility of a tetrahedral (fourfold) coordination opens up. For the range 0.414 to 0.732 one may also have a (sixfold) octahedral coordination or a square (fourfold) coordination. For $R > 0.732$ the (eightfold) cubic coordination is feasible. The preference is often for the highest-possible coordination number, as this maximizes the number of ionic bonds. Anion radii are generally larger than cation radii, since the anions contain an excess of electrons. Thus, for the coordination of cations around an anion, the ratio $R' = r_a/r_c$ will exceed 1.

It is possible to introduce a more general set of ionic radii that are coordination-number dependent. For example, this would make $\{r\}$ different for a tetrahedral and an octahedral environment. This refinement is omitted here, for the take of simplicity.

Consider the layered crystal, brucite, $Mg_3(OH)_6$ (Fig. 13.3). Each layer consists of a sheet of Mg^{2+} ions clad with two layers of OH^- ions, one above and one below. The crystal is hexagonal with $a = 0.313$ nm and $c = 0.474$ nm. The ionic radius of the Mg^{2+} ion in the filled-shell configuration is $r_c = 0.065$ nm (see Table 2.12). The

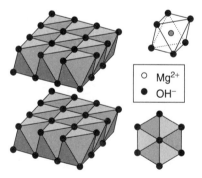

Figure 13.3. Brucite sheets consisting of a hexagonal array of Mg^{2+} cations octahedrally coordinated with OH^- ions. (For simplicity the OH^- ion is represented simply as a black circle.) A view along the c axis, shown in the lower right, displays the hexagonal symmetry.

[†] If this were not the case, there could not be stable equilibrium. Earnshaw's theorem states that under the influence of electrostatic forces alone, there is no stable equilibrium. This is because Laplace's equation, $\nabla^2 \Phi = 0$, implies a saddle-point behavior in the potential field rather than an absolute maximum or minimum.

Mg^{2+} ion is octahedrally coordinated, the distance from the ion to the $(OH)^-$ ion being $r = a/\sqrt{2} = 0.221$ nm. Thus the effective ionic radius of $(OH)^-$ is $r_a = r - r_c = 0.156$ nm. The ionic radius ratio is $R = 0.65/1.56 = 0.416$, consistent with the octahedral coordination.

Pauling's second rule requires "local" charge neutrality and is closely related to Gauss's law. Local charge neutrality is desirable to prevent the positive and compensating negative charges from being too far apart. This would cause an electric field to persist over a large volume and a large electrostatic energy would result. Pauling defined the ionic bond strength, S, as the ratio of the cation charge, q_c (in units of e), to its coordination number, N_c. This definition is equivalent to Faraday's "line of force" concept, in which the electric flux is represented by a line emanating from a cation, with each charge producing one line, according to Gauss's law. Since these lines must terminate on an anion, the sum of the bond strengths connecting to an anion must be equal in magnitude but opposite in sign to the charge of the anion (in units of e).

For the case of brucite, each of the six bonds radiating from a Mg^{2+} cation has a bond strength $S = \frac{2}{6} = \frac{1}{3}$. Therefore, each OH^- must be coordinated to three Mg^{2+} ions, in accordance with Pauling's second rule.

The *third Pauling rule* relates to how the polyhedra combine to form a crystal. Joining of adjacent polyhedra at one corner is more favorable than edge joining (at two adjacent corners). Edge joining is more favorable than face joining (three or more corners). Looking at the octahedra drawn for brucite, one may see an example of edge joining. An instance of corner joining may be found in Section W13.2. As one follows the progression from corner joining to edge joining to face joining, the ions are brought closer together and the electrostatic repulsive energy increases.

The *fourth Pauling rule* enforces the corner-joining prerogative when the bond strength becomes too high. Then the local electrostatic interactions (which vary as the square of the charges) preclude the other possibilities.

Finally, the *fifth Pauling rule* states that structures tend to be formed from only a few component elements. Among the many possible combinations, some are likely to result in significantly lower free energies, and these are the ones that will be more likely to nucleate and grow as the material is cooled from the melt. Note that increasing the number of components makes the formation of crystals more difficult. This concept was encountered in the discussion of metallic glasses in Chapter W12. The fifth rule, sometimes called the *rule of parsimony*, is the least significant of the rules.

13.3 Ionic Interactions

The interactions among the ions comprising ceramic materials are often modeled in terms of the Coulomb potential energy, which accounts for around 80% of the cohesive energy, and the short-range repulsive potential energy, caused by the Pauli exclusion principle when the electronic clouds overlap. This was introduced in Section 2.5, when ionic bonding was first discussed. Here the scope of the analysis is expanded somewhat. Consider the case of crystalline solids. Let ions of charge $z_i e$ be located at positions $\mathbf{R} + \mathbf{s}_i$, where $\{\mathbf{R}\}$ is a set of lattice vectors and $i = 1, 2, \ldots, n_s$, n_s being the number of ions in a unit cell. The interaction energy may be written as

$$U = \frac{1}{2} \sum_{\mathbf{R},\mathbf{R}'} \sum_{i,j} {}' \left[\frac{F_{ij}}{|\mathbf{R} + \mathbf{s}_i - \mathbf{R}' - \mathbf{s}_j|^{m_{ij}}} + \frac{z_i z_j e^2}{4\pi\epsilon_0 |\mathbf{R} + \mathbf{s}_i - \mathbf{R}' - \mathbf{s}_j|} \right], \qquad (13.1)$$

where the first term is the repulsive interaction and the second is the Coulomb attraction. In principle, all terms in the sum are included, except the ones corresponding to an ion interacting with itself. The factor $\frac{1}{2}$ is present to avoid double counting. The numbers F_{ij} and m_{ij} are particular to the given pair of ions, i and j. The numbers m_{ij} are positive and sufficiently large so as to make the short-range repulsion fall off rapidly with distance. Charge neutrality demands that

$$\sum_{i=1}^{n_s} z_i = 0. \tag{13.2}$$

The choice of the expression for the short-term repulsive interaction is somewhat arbitrary. Often, an exponential variation with distance is assumed instead.

The expression for U may be separated into intracell and intercell contributions:

$$U = \frac{1}{2} \sum_{\mathbf{R}} \sum_{i,j}{}' \left(\frac{F_{ij}}{|\mathbf{s}_i + \mathbf{s}_j|^{m_{ij}}} + \frac{z_i z_j e^2}{4\pi\epsilon_0 |\mathbf{s}_i - \mathbf{s}_j|} \right)$$

$$+ \frac{1}{2} \sum_{\mathbf{R},\mathbf{R}'}{}' \sum_{i,j} \left(\frac{F_{ij}}{|\mathbf{R} + \mathbf{s}_i - \mathbf{R}' - \mathbf{s}_j|^{m_{ij}}} + \frac{z_i z_j e^2}{4\pi\epsilon_0 |\mathbf{R} + \mathbf{s}_i - \mathbf{R}' - \mathbf{s}_j|} \right). \tag{13.3}$$

By making use of translational symmetry, this reduces to

$$U = \frac{N}{2} \sum_{i,j}{}' \left(\frac{F_{ij}}{s_{ij}^{m_{ij}}} + \frac{z_i z_j e^2}{4\pi\epsilon_0 s_{ij}} \right) + \frac{N}{2} \sum_{\mathbf{R}}{}' \sum_{i,j} \left(\frac{F_{ij}}{|\mathbf{R} + \mathbf{s}_{ij}|^{m_{ij}}} + \frac{z_i z_j e^2}{4\pi\epsilon_0 |\mathbf{R} + \mathbf{s}_{ij}|} \right), \tag{13.4}$$

where N is the number of unit cells in the crystal.

It is convenient to express the vectors $\{\mathbf{R}\}$ and $\{\mathbf{s}_{ij}\}$ in terms of dimensionless variables, $\mathbf{R} = a\boldsymbol{\rho}$ and $\mathbf{s}_{ij} = \mathbf{s}_i - \mathbf{s}_j = a\boldsymbol{\sigma}_{ij}$, where a is a characteristic cell dimension. Then

$$U = \frac{N}{2} \sum_{i,j}{}' \left(\frac{F_{ij}}{a^{m_{ij}}\sigma_{ij}^{m_{ij}}} + \frac{z_i z_j e^2}{4\pi\epsilon_0 a\sigma_{ij}} \right) + \frac{N}{2} \sum_{i,j} \left[\frac{F_{ij} W_{ij}(m_{ij})}{a^{m_{ij}}} + \frac{z_i z_j e^2 W_{ij}(1)}{4\pi\epsilon_0 a} \right], \tag{13.5}$$

where

$$W_{ij}(m) = \sum_{\boldsymbol{\rho}}{}' \frac{1}{|\boldsymbol{\rho} + \boldsymbol{\sigma}_{ij}|^m}. \tag{13.6}$$

The dimensionless numbers $W_{ij}(m)$ are called *Madelung sums*. These numbers will not be finite for $m \leq 3$. For the short-range repulsion, m is typically in the range 6 to 10, so there is no difficulty. For the Coulomb interaction, however, $m = 1$. Due to overall electrical neutrality, when the contributions from the ions of a unit cell are combined, the convergence is improved. Intercell Coulomb interactions occur by means of their electric multipoles. If there is no net electric dipole moment to the unit cell, the Coulomb interactions will converge. For the case where there is a net dipole moment within the unit cell, as occurs in ferroelectrics, special care must be taken to include the contributions arising from the surfaces of the crystal.

TABLE 13.2   Madelung Constants for Various Binary Ionic Solids

Structure	A
Sodium chloride	1.7476
Cesium chloride	1.7627
Zincblende	1.6381
Wurtzite	1.6407
Fluorite	2.5194
Rutile	2.385

The combined Madelung sums depend on the crystal structure. For a cubic crystal they yield pure numbers. For tetragonal (or orthorhombic) crystals they will also depend on the cell dimension ratios c/a (and b/a). For other crystal classes they may depend on the characteristic angles of the unit cell as well.

In the special case of an ionic solid with just two types of ions, one may express U in terms of the NN distance as

$$U = N \left(-\frac{z_1 z_2 e^2}{4\pi\epsilon_0} \frac{A}{r} + \frac{C}{r^m} \right), \tag{13.7}$$

where only the NN repulsive interaction is retained. The Madelung constants A are tabulated for simple crystals in Table 13.2. As the value of A increases, the stability of the crystal increases. However, the preferred crystal structure may be influenced by nonionic forces as well.

The expression for U above may be used to derive the equilibrium lattice constants, the cohesive energy, and the elastic moduli of the crystal. One may also generalize the expressions above to include small displacements about equilibrium and obtain predictions for the phonon spectra.

In addition to the static ion contributions considered above, one must include the effect of the electronic polarization of the ions. This is often modeled by imagining that surrounding the nuclear core is a massless shell of charge Ye connected to it by a spring of spring constant k. The spring constant is chosen to give the right dc electric polarizability. In the presence of an electric field E, at equilibrium, the Hooke's law force must balance the electric force on the shell (i.e., $kx = YeE$). The induced electric dipole moment is $\mu = \epsilon_0 \alpha E = Yex$, so $k = (Ye)^2/\alpha\epsilon_0$. The short-range interaction between neighboring shells is also modeled as a Hooke's law force with spring constant k'. Using this model it has been possible to obtain good fits to the phonon dispersion curves for the ionic solid, as well as the frequency dependence of the dielectric function.

APPLICATIONS

13.4 Refractories

Refractories are materials with high melting temperatures and suitable mechanical and chemical properties that may be utilized as linings and roofs for kilns, ovens,

TABLE 13.3 Melting Temperatures of Some Refractory Materials

Material	T_m (°C)	Material	T_m (°C)
C	>3550	HfC	3890
Mo	2610	SiC	2840
W	3410	B_4N	2470
BeO	2570	Si_3N_4	1880
CaO	2930	Zr_3N_4	2980
MgO	2850	SiO_2	1710
Al_2O_3	2060	$MoSi_2$	1680
ZrO_2	2680	TiO_2	1857

crucibles, driers, and high-temperature utensils. Typically, they find application in steel mills, glass manufacture, chemical reactors, and the aerospace industry. A list of some elemental and compound refractories, along with their melting temperatures, is given in Table 13.3.

These materials are characterized by a number of physical properties. The first is called *refractoriness*, which is the temperature at which a sample of solid softens and loses much of its mechanical strength. This may be due to a brittle-to-ductile transition. There is a standard test in which a cone of material, with its base down and its vertex upward, is heated until the cone distorts and the vertex slumps over to the side. The degree of slump may be compared with a standard set of shapes to get the *pyrometric cone equivalent* (PCE), a measure of the refractoriness at a given temperature. Closely related is the concept of refractoriness under load, in which case the system is put under stress and then heated. Often, materials fail at a lower temperature when stressed.

Thermal shock (spalling) resistance is the ability of a refractory material to undergo a sudden temperature change without cracking. It depends on several factors, including the thermal conductivity, the coefficient of thermal expansion, and the existence or nonexistence of phase transitions involving crystal distortion or volume changes. If a plate of material is suddenly heated on its outer faces, a thermal gradient will be established and heat will flow into or out of the material. Thermal expansion will cause a stress to form which can cause cracking. The higher the thermal conductivity, the more rapidly the temperature will equilibrate and the quicker the stress will be relieved. One may imagine that if a hot part of the material underwent a volume-changing phase transition while a cooler part didn't, this too could crack the object.

For example, consider an infinite slab of material of thickness w at an initial temperature T_0 whose surface temperature is suddenly changed to T_s at time $t = 0$. The temperature profile will be written as $T(z, t)$, where $-w/2 < z < w/2$. The heat flux is given by Eq. (5.64), $\mathbf{J_Q} = -\kappa \nabla T$, where κ is the thermal conductivity. The continuity equation is $\nabla \cdot \mathbf{J_Q} + \partial h / \partial t = 0$, where the enthalpy per unit volume, h, may be expressed in terms of the specific heat per unit volume, c_p, by the formula $h = h_0 + c_p T$. This results in the diffusion equation

$$\nabla^2 T = \frac{c_p}{\kappa} \frac{\partial T}{\partial t}.$$

(13.8)

The temperature field is given by the Fourier series,

$$T(z, t) = T_s + \frac{2(T_0 - T_s)}{\pi} \sum_{n=1}^{\infty} [1 - (-)^n] \sin\left[\frac{n\pi}{w}\left(z + \frac{w}{2}\right)\right] e^{-n^2 t/\tau}, \qquad (13.9)$$

where the thermal relaxation time is given by $\tau = w^2 c_p / \pi^2 \kappa$. The average temperature over the width of the bar, $\overline{T}$, is

$$\overline{T} = T_s + \frac{2(T_0 - T_s)}{\pi^2} \sum_{n=1}^{\infty} \frac{[1 - (-)n]^2}{n^2} e^{-n^2 t/\tau}. \qquad (13.10)$$

A plot of the temperature evolution as a function of position is given in Fig. 13.4.

If a point in the slab has temperature T there would be a local thermal expansion per unit length $\alpha(T - \overline{T})$, where α is the linear coefficient of thermal expansion. If the slab were constrained so as not to move in the x or y directions, the local strain would be $\epsilon_{xx} = \epsilon_{yy} = -\alpha(T - \overline{T})$. This would produce a plane stress

$$\sigma_{xx} = \frac{E}{1 - \nu^2}(\epsilon_{xx} + \nu\epsilon_{yy}) = -\frac{E\alpha}{1 - \nu}(T - \overline{T}), \qquad (13.11)$$

where E is Young's modulus and ν is Poisson's ratio. The same result is obtained for σ_{yy}. If the stress exceeds the fracture stress, σ_f, the slab will fracture.

In some articles in the literature the thermal resistance factor is defined by $K = 3\kappa\sigma_f(1 - \nu)/\alpha c_p E$, which is proportional to $w^2(T - \overline{T})/\tau$. Aside from the factor w^2, this is the rate of cooling per thermal relaxation time. A large value of K means that the material is more capable of withstanding thermal shock.

Porosity plays an important role in refractories. Molten metal can penetrate the pores and cause corrosion, so in this respect pores are a negative factor. Low-porosity material is also generally stronger. Pores also serve to limit the propagation of cracks and therefore help improve the thermal shock resistance. Porosity is also of value in creating good insulating materials.

The ability of the refractory to be wet by molten slag is also a factor in determining how effective the porosity will be. If the material is easily wet, penetration into the pores may be expected to be a problem.

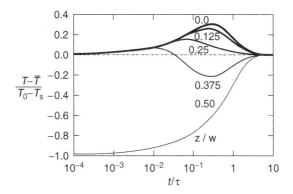

Figure 13.4. Plot of $[T(z, t) - \overline{T}]/(T_0 - T_s)$ as a function of t/τ for various z/w values.

Certain refractories behave as acids (e.g., if they contain NH_4^+ ions), others as bases (e.g., if they contain Na^+, K^+, or Mg^{2+} ions), and others are neutral. It is not good to have an acidic material in contact with a basic slag, or vice versa, as this could lead to considerable chemical corrosion. The result would be a deterioration of the physical structure of the refractory form.

Consider some typical refractory ceramic materials. Common oxides in use are alumina, beryllia, calcia, magnesia, silica, and zirconia. Alumina (Al_2O_3) is both mechanically and chemically stable and has low porosity. It is used for temperatures up to $T = 1850°C$. Its ability to withstand thermal shock is compromised by the fact that its thermal conductivity falls at high temperatures, as it does for all materials. The compressive strength falls by over a factor of 50 when the temperature is raised to $T = 1600°C$. It finds application in crucibles for refining metals, for furnace and kiln linings, as containers for molten glass, and as sheaths for thermocouples.

Beryllia (BeO) has a high melting temperature (2570°C), high mechanical strength, and high thermal conductivity. It finds application in crucibles for refining uranium and thorium.

Calcia (CaO) has a very high melting temperature, 2930°C. However, it is very hygroscopic (water-absorbing) and therefore cannot be used with water or water vapor. It is used for refining some metals, such as platinum. Pitch-bonded dolomite is a refractory brick consisting of 56% CaO, 40% MgO, and trace amounts of alumina, silica, chromia, and iron oxide.

Magnesia (MgO) also has a high melting temperature, 2850°C. It is used for linings of high-temperature ovens. It is also used in crucibles for smelting such metals as Al and U. Periclase is a refractory brick consisting of 90% MgO, 3% SiO_2, 3% Fe_2O_3, 2.5% CaO, 1% Al_2O_3, and 0.3% Cr_2O_3. A variant of forsterite (Mg_2SiO_4) is used as a refractory material. It is a composite consisting of 54.5% MgO, 33.3% SiO_2 and 9.1% Fe_2O_3, with trace amounts of alumina, chromia, and calcia. Periclase-chrome ore consists of 73% MgO, 9.0% Al_2O_3, 8.2% Cr_2O_3, 5.0% SiO_2, 2% Fe_2O_3 and 2.2% CaO. It is possible to calculate the lattice energy, a measure of stability, of MgO doped with various oxides using Madelung summation techniques (see Section 13.3). One predicts the following hierarchy of lattice energies: $TiO_2 > B_2O_3 > Al_2O_3 > Fe_2O_3 > Cr_2O_3 > MgO > NiO > CoO > FeO > CaO > Li_2O > Na_2O$.

Silica (SiO_2) is commonly used for furnaces and kilns, where the temperature requirements are not too demanding. The melting temperature is 1710°C. At a temperature of approximately $T = 1140°C$, the glass begins to anneal. At temperatures above $T = 1670°C$, silica softens, meaning that the viscosity drops below the value of 4×10^7 poise. Silica is acidic and therefore can be used with acidic slags. It has a high thermal shock resistance despite the fact that its thermal conductivity is relatively low.

Zirconia (ZrO_2) has very high mechanical strength below 1500°C. It finds use in the steel and glass industries and in crucibles for metallurgy. Generally, pure zirconia is not that useful because of its several crystalline forms. One needs to use additives, such as Ca^{2+} or Mg^{2+} ions, provided by CaO or MgO, respectively, to stabilize it. Its low thermal conductivity makes it suitable as a thermal insulator.

Refractory bricks are often composed of large-grained particles held together by fine-grained particles that serve as a binder. Fire clay is a common refractory consisting of both silica and alumina. An examination of the binary phase diagram of silica and alumina in Fig. 13.5 allows one to identify useful compositions. Mullite is a nonstoichiometric compound (solid solution) with the approximate formula $2SiO_2 \cdot 3Al_2O_3$.

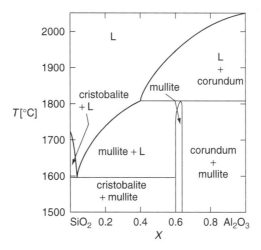

Figure 13.5. Binary phase diagram for silica–alumina. [Reprinted by permission from I. A. Aksay and J. A. Pask, *Science*, **183**, 69 (1974). Copyright 1974 by the American Association for the Advancement of Science.]

There is a eutectic point at a composition of about 4 at % alumina with a temperature of $T_e = 1595°C$. Obviously, it is important to stay away from the eutectic to improve the refractoriness. The melting point of pure silica is 1713°C and of pure alumina is 2060°C. Compositions to the right of the eutectic, up to about 60% alumina, are called *fire clays*. They are useful, in principle, up to $T = 1600°C$, where liquid starts to form. Fire clays offer the advantage that they maintain their strength almost to the temperature at which they melt. This makes them suitable for the roofs of kilns and ovens. Alumina refractories have still higher alumina concentrations, often exceeding 95%, and can be used up to $T = 1800°C$ before liquid begins to form.

Additional refractories include SiC, Si_3N_4, C, B_4N, as well as other borides and nitrides. The mechanical strength of these materials is often determined by the microstructure. For example, it is found that needle-shaped crystallites in Si_3N_4 act as "nails" that help hold the material together when it is subjected to stresses at high temperatures. Specialized refractories have been developed for particular purposes. For example, the material ACZS (7% zirconia, 4% silica, 70% alumina, 16% chromia, and 0.3% iron oxide) is useful in the glass-making industry.

13.5 Silicon Nitride

The strength of ceramics is determined largely by the nature of the chemical bond. Since there is a fair degree of covalency involved, the bonds in most ceramics tend to be strong. In this section the focus of attention is on one ceramic, silicon nitride (Si_3N_4), which is almost completely covalent, and has shown great promise as a structural material. The material has a low density (3200 kg/m^3), so that if an engine were to be constructed from it, the weight would be much less than if it were made from steel (density ≈ 7800 kg/m^3). It sublimates at $T = 2150$ K, higher than the melting temperature of steel (≈ 1810 K). It is already in use as a turbocharger rotor. It maintains mechanical strength to high temperatures and is thus superior in many ways to superalloys. It is also resistant to corrosion, oxidation, and wear. It has a low coefficient of thermal expansion (3.4×10^{-6} K^{-1} as opposed to steel with $\alpha = 12 \times 10^{-6}$ K^{-1}), so it is able to withstand sudden changes in temperature (thermal shock) without breaking.

Silicon nitride consists of trigonally corner-joined tetrahedra with N atoms at the vertices and Si atoms at the centers. There are two forms for the crystal, both of which have hexagonal lattices. The material α-Si$_3$N$_4$ has a basis consisting of 12 Si atoms and 16 N atoms. The lattice constants are $a = 0.775$ nm and $c = 0.562$ nm. The material β-Si$_3$N$_4$ has only six Si atoms and eight N atoms. The locations of the atoms for both forms of Si$_3$N$_4$ are given in Fig. 13.6. There are four arrangements of the atoms in the layers, denoted by A, B, C, and D. In α-Si$_3$N$_4$ the stacking arrangement is ABCDABCD..., whereas in β-Si$_3$N$_4$ it is ABAB.... The crystal β-Si$_3$N$_4$ has long empty channels parallel to the c axis, while α-Si$_3$N$_4$ has void spaces which are not interconnected. Because α-Si$_3$N$_4$ is more cross-linked, it is harder than β-Si$_3$N$_4$. The crystal β-Si$_3$N$_4$ tends to grow as needlelike structures, which serves to toughen the material in two respects. First, the needles tend to act as obstacles for crack propagation, thereby limiting the range of damage caused by brittle fracture. Second, the needlelike microstructure acts the same way that nails act in holding boards together. This limits the possible motion of grains.

Growth of Si$_3$N$_4$ takes place by nitridization of Si at high temperatures (1300 to 1500°C), and both the α and β forms crystallize. There is a kinetic competition between them which may be controlled by varying the cooling rate. The structure α-Si$_3$N$_4$ tends to nucleate and grow at the early stages if the temperature is kept low. It is believed to involve the addition of molecular N$_2$ to a surface Si atom. The structure β-Si$_3$N$_4$ grows at higher temperatures and probably involves atomic N forming a bond with a surface Si atom.

Despite its near-ideal properties, a serious technological problem is caused by the fact that Si$_3$N$_4$ decomposes at $T = 1877$°C before it densifies. It cannot simply be fired to densify it. A flux must be used to lower the melting temperature. Typical fluxing agents are alumina, magnesia, or yttria. Alternatively, the N$_2$ pressure in the furnace must be elevated to prevent decomposition.

SIALONs are a class of materials based on silicon nitride in which Al and O are added. Oxygen substitutes for a N atom and simultaneously, Al substitutes for

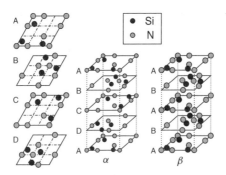

Figure 13.6. Layers of the unit cell for Si$_3$N$_4$. Dark circles are Si atoms, light circles are N atoms. The c axis is perpendicular to the planes. The atomic positions in a hexagonal unit cell are shown in perspective for $\alpha - $ Si$_3$N$_4$ and $\beta - $ Si$_3$N$_4$. [Adapted from K. H. Jack, *J. Mater. Ed.*, **13**, 1 (1991); and R. W. G. Wyckoff, *Crystal Structures*, Vol. II, Interscience, New York, 1964.]

a Si, while maintaining the tetrahedral coordination. The bond lengths tend to be short (and hence strong), with typical values for the bond lengths and bond energies being $(0.19, 0.17, 0.19, 0.17, 0.16)$ nm and $(2.9, 5.3, 3.2, 3.5, 4.8)$ eV for (Al–N, Al–O, Si–C, Si–N, Si–O), respectively. These numbers may be compared with 0.155 nm and 3.7 eV for diamond. SIALONs also exist in the α and β forms. The typical chemical composition (for β) is $Si_{6-x}Al_xO_xN_{8-x}$, with $0 < x < 4.2$. It should be noted that there is also a stable silicon oxynitride, Si_2N_2O.

SIALONs have found application as cutting tools because of their hardness and ability to withstand high temperatures. Their fracture toughness, as well as that of many other materials, may be further enhanced by embedding small zirconia particles in them as part of the microstructure. When zirconia is subject to intense stress, it undergoes a martensitic transition from a tetragonal crystal to a monoclinic crystal. Near the tip of a crack there is an enhanced stress. The structural phase transition of the ZrO_2 particle serves to extract energy from the stress field near the crack and to relieve it, somewhat. Such materials are referred to as *zirconia-toughened materials*.

13.6 Zeolites

The discussion now turns to a set of ceramics that form molecular sieves. These are porous materials which permit the "free" passage of small molecules through the pores. Natural zeolites are a class of porous minerals composed primarily of Al^{3+}, Si^{4+}, and O^{2-}, and monovalent or divalent cations such as Na^+, K^+, Ca^{2+}, and Mg^{2+}. They include such minerals as chabazite ($CaAl_2Si_4O_{12} \cdot 6H_2O$), faujasite ($Na_{13}Ca_{11}Mg_9K_2Al_{55}Si_{137}O_{384} \cdot 235H_2O$), heulandite ($CaAl_2Si_7O_{18} \cdot 6H_2O$), mordenite ($Na_8Al_8Si_{40}O_{96} \cdot 24H_2O$), natrolite ($Na_2Al_2Si_3O_{10} \cdot 2H_2O$), and stilbite ($NaCa_2Al_5Si_{13}O_{36} \cdot 14H_2O$). There are closely related minerals, such as sodalite ($Na_6Al_6Si_6O_{24} \cdot 2NaCl$) and cancrinite ($Na_6Al_6Si_6O_{24} \cdot CaCO_3 \cdot 2H_2O$), where the pores are clogged by the cations. There are also numerous synthetic zeolites such as Linde-A, Linde-X, Linde-Y, zeolite-β, ZSM-5 ($Na_nAl_nSi_{96-n}O_{192} \cdot 16H_2O$, with $n < 27$), and ZSM-11. When zeolites are heated, the water "boils" out and the network structure remains intact. (The word *zeolite* stems from the Greek words for "weeping stone".) All zeolites are crystalline, with the unit cell commonly being monoclinic, orthorhombic, or hexagonal.

In zeolites the Al and Si ions are tetrahedrally coordinated with O^{2-} ions. The tetrahedra are joined at the corners so that each O^{2-} ion is NN-connected either to two Si^{4+} or one Si^{4+} and one Al^{3+} ion. These tetrahedra form a three-dimensional network which contains void spaces or channels that run throughout the crystal. To satisfy the Pauling bonding rules, there must also be one monovalent cation (such as Na^+, K^+, or $CaOH^+$) per Al ion, or one-half of a divalent cation (such as Mg^{2+} or Ca^{2+}) per Al ion. These cations are not part of the skeletal structure but reside in the void spaces. It is possible to satisfy all valence requirements in a pure silicate structure, since the ionic bond strength of each Si^{4+} bond is $+1$ and the valence of O^{2-} matches two Si^{4+} that it is connected to. One never finds two adjacent tetrahedra containing Al^{3+} ions (Lowenstein's rule), although this can happen in cases of octahedral coordination, as in alumina.

In representing zeolite structures it is customary to show only the Al or Si ions (called T atoms) and not the tetrahedron of O ions. Often, the compensating cations are also not shown. Thus the emphasis is put on depicting the underlying skeleton. This

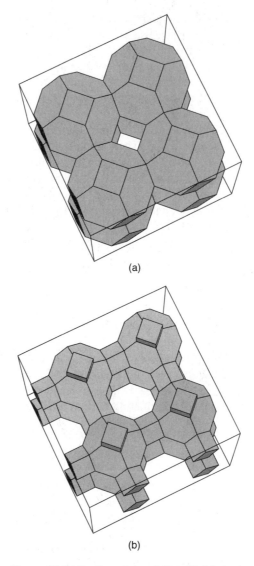

Figure 13.7. Zeolites: (*a*) sodalite; (*b*) Linde-A.

is illustrated in Fig. 13.7, where the skeletons for sodalite and Linde-A are drawn. The polygonal framework structure has been shaded in for aesthetic reasons. It is important to notice the regular network of void spaces that permeate the structure. It is these voids that provide the key to the utility of zeolites. They permit small molecules to enter while they block larger molecules and thus form molecular sieves. Depending on the particular zeolite, the channel diameter can range from 0.25 nm for sodalite to approximately 1 nm for some of the synthetic zeolites. Obviously, the types of molecules that the sieve will accept are determined by the channel dimensions and, to some extent, by the shape of the pores.

One of the early applications of zeolites was as an ion exchanger. For example, hard water, containing Ca^{2+} or Mg^{2+} ions, can be passed through a zeolite containing

Na$^+$ ions, and one divalent ion can replace two sodium ions. Thus the divalent ions are removed from water and are replaced by sodium ions. When the sieve becomes permeated with divalent ions, it may be rejuvenated by soaking it in hot brine, which causes the reaction to run in the reverse direction, and the Ca and Mg are replaced by Na.

The zeolite may also be employed as a desiccant. As mentioned, natural zeolites contain water, which may be driven out by heating. The dry zeolite is hydrophilic and will reabsorb water when exposed to damp air. The desiccant may be reactivated by subsequently reheating the material. The ability for zeolites to accept some molecules and to reject others make them suitable as chemical sensors. When cooled to liquid nitrogen temperature (77 K) they act as molecular sieve "pumps" and are used to help create a vacuum. Some research has been directed toward the production of zeolite membranes which could be employed for chemical separation.

The most important applications of zeolites are in the field of catalysis. Since it is possible to replace the channel ion by ions containing protons (e.g., NH$_4$$^+$), the zeolite can be made into an acid. This acid may then promote reactions involving the adsorbed molecules. This is effective for several reasons. First, the channels permeate the bulk of the material. Thus the entire bulk of the material is chemically active. This is in contrast to many heterogeneous catalysts for which only the surface is chemically active. Second, many zeolites can be heated to high temperatures ($\approx 500°$C) without destroying the material. The effectiveness of the acid as a proton donor increases dramatically with temperature. Third, and perhaps most important, there is a high degree of selectivity of products that occurs when reagents react. Some potential products simply may not fit inside the pore volume or the pore shape, or may have great difficulty leaving the bulk. On the other hand, other products may be able to leave readily. Finally, the very fact that the reagent molecules adsorb throughout the solid and come into proximity with each other, provides a concentrating effect that increases the rate of bimolecular reactions (proportional to the product of the concentrations). It is equivalent to having run the corresponding gas-phase reaction at a very high pressure.

For example, suppose that toluene (C$_7$H$_8$) and methanol (CH$_3$OH) enter the catalyst (in this case, ZSM-5). The reaction C$_7$H$_8$ + CH$_3$OH $\rightarrow$ H$_2$O + C$_8$H$_{10}$ (water + xylene) can produce three isomers of xylene: *ortho*-xylene, *meta*-xylene, and *para*-xylene. Alternatively, one may have the disproportionation reaction C$_7$H$_8$ + C$_7$H$_8$ $\rightarrow$ C$_6$H$_6$ + C$_8$H$_{10}$ (benzene + xylene). (The precise reaction pathways may be more complex, but are not important for now.) Certain side reactions can be blocked by the steric constraints of the pores, such as the disproportionation reaction C$_8$H$_{10}$ + C$_8$H$_{10}$ $\rightarrow$ C$_7$H$_8$ + C$_9$H$_{12}$ or the formation and polymerization of H-deficient hydrocarbons (coking). *para*-Xylene has important commercial value in the manufacture of polyester fibers. The various molecules are depicted in Fig. 13.8. Because of their different sizes they have greatly different diffusion constants through the pores, with *para*-xylene being the most mobile and *ortho*-xylene the least. In fact, before the larger molecules can leave the zeolite, there is a high probability that the reverse reaction will occur and replenish the starting material. Effectively, the catalyst produces only *para*-xylene. The material behaves as a *shape-dependent catalyst*. A sketch of the porous structure of ZSM-5 is provided in Fig. 13.9.

Zeolites provide efficient catalysts for the cracking of large hydrocarbons to produce gasoline. The zeolite Linde-Y permits the entry of *n*-paraffins into its pores. The Al^{3+}

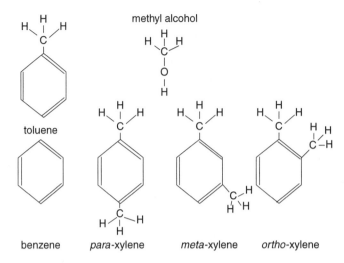

Figure 13.8. Organic molecules involved in the production of *para*-xylene.

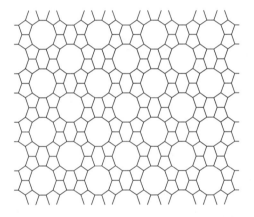

Figure 13.9. Porous structure of the zeolite ZSM-5.

sites, being highly charged, are effective in attracting polar molecules such as H_2O toward them. Some OH^- ions from H_2O become bound, leaving their protons relatively free. These protons may be loaned to a chemical reaction, and thus the zeolite functions as an acid. These are effective in breaking the carbon–carbon bonds and can crack large hydrocarbons to produce small molecules such as ethylene ($CH_2=CH_2$) and propene ($CH_3CH=CH_2$). While a high concentration of Al sites would increase the reactivity of the catalyst, it also serves to destabilize it both thermally and chemically. Thus the Al concentration cannot be too high. By operating at higher temperatures the reactivity of the catalyst is restored. A concentration of Al sites must be sought which is a compromise between these competing trends. The Si/Al ratio in Linde-Y varies between 1.5 and 3.

Zeolites are usually prepared from aqueous solutions. Current research is centering on the use of organic molecules to act as templates upon which the zeolites crystallize (see Section W21.18). When the crystal grows to the desired size the organic molecule may be removed by thermal or chemical means. For example, the zeolite DAF-1, containing Mg and PO_4, has two sets of independent channels, of diameters 0.61 and 0.74 nm, respectively. The latter channels interconnect pockets of 1.6 nm diameter that can accommodate rather large intermediate organic molecules. Since this extends the range of organic molecules amenable to catalytic preparation, it has attracted considerable interest. Much of the recent research has centered on properties of aluminophosphates (ALPOs).

The tools for studying zeolites and for resolving their structure include magnetic resonance, x-ray diffraction, high-resolution electron microscopy, and neutron scattering. They will be discussed in Chapter W22.

The synthesis of zeolites is more of an art than a science, in that no reliable models exist for predicting precisely how the crystal growth relates to the aqueous concentrations of materials and other physical parameters. Nevertheless, it has been possible to engineer materials with large pore spaces by using a clever interplay between micelles and crystal growth. A detailed discussion of micelles is given in Chapter 19. For now, all that needs to be known is that the micelles are long organic molecules with a hydrophilic polar group at one end and a hydrophobic alkyl chain at the other. Placed in water, some micelles self-organize into cylinders with the molecules oriented so that the hydrophobic groups are on the axis and the hydrophylic groups are on the surface, in contact with the water. The cylinders align themselves parallel to each other, forming a hexagonal structure, with water permeating the intercylinder regions. The silica salts are dissolved in the water and crystallize in the intercylinder region. Ultimately, the organic molecules are removed. In this way one creates a material such as MC-41, a hexagonal mesoporous material with void spaces of sizes 2.0 to 10.0 nm (Fig. 13.10). By protonating the walls one can create an acid catalyst in much the same way as is done in a conventional zeolite.

13.7 Glasses

If a liquid is cooled slowly enough, it solidifies and forms a crystal. The temperature at which this occurs, called the *melting temperature*, is denoted by T_m. If it is cooled sufficiently rapidly, it may fail to solidify at temperature T_m and instead may become a supercooled viscous liquid. At some lower temperature T_g, called the *glass-transition temperature*, the liquid solidifies and forms an amorphous solid. The

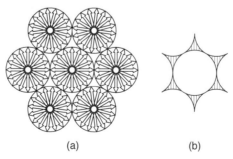

(a)   (b)

Figure 13.10. (*a*) Micelles arranged in a hexagonal array of cylinders; (*b*) crystallized silica formed in the intercylinder region produces a mesoporous material.

TABLE 13.4 Melting Temperatures of Glass-Forming Materials

Glass	T_m (°C)
SiO_2	1710
B_2O_3	450
GeO_2	1116
P_2O_5	580
V_2O_5	670
$Na_2O \cdot 2SiO_2$	878
$K_2O \cdot 2SiO_2$	1040
$BaO \cdot 2B_2O_3$	910
$PbO \cdot 2B_2O_5$	774

Source: Based on R. H. Doremus, *Glass Science*, 2nd ed., Wiley, New York, 1994.

tendency for the glass to crystallize is frustrated by the steric constraints imposed by the solidification. The melting temperatures of some materials that can form glasses are given in Table 13.4. Unlike T_m, T_g is not unique but is dependent on the cooling rate. The crystal represents a state of thermodynamic equilibrium, whereas the glass is frozen into a metastable state of higher free energy. Kauzmann compared the entropy of a supercooled liquid (SL) to that of a crystalline solid and found that the entropy of the SL falls more rapidly with decreasing temperature. In fact, his extrapolation predicted that the entropy of the SL would go to zero at a finite temperature (called the *Kauzmann temperature*, T_k), in violation of the third law of thermodynamics. This *Kauzmann paradox* is averted by having the SL solidify as a glass before T_k is reached.

Most glasses in common use are optically transparent. They are based on oxides which, in their crystalline state, have an even number of electrons per unit cell. Thus the crystal consists of filled valence band(s) and empty conduction band(s). The bandgap is wide, corresponding to ultraviolet photon energies. In the amorphous state, where the concept of bandgap loses its precise meaning, there nevertheless exists a minimum photon energy needed to excite the system. Another family of glasses, based on chalcogenides, are semiconductors in their crystalline state. These glasses absorb at the high-frequency end of the visible spectrum and thus are colored.

On the scale of the wavelength of light, glass appears homogeneous, so there is not much light scattering. (A notable exception is As_2Se_3, which exhibits hexagonal domain structure on a length scale of 100 nm.) The net result is that light passes through a glass with little extinction (i.e., losses due to absorption and/or scattering). More will be said about these effects later, when optical fibers are studied in greater detail in Chapter 18.

The structures of glasses bear some similarities and some important differences to those of crystals. Unlike crystals, there is no long-range order. In a crystal the coordination numbers, atomic bond distances, and bond angles all take on precise values, the last two to be modified only by thermal fluctuations. The atoms in a glass form a continuous random network, as is discussed in Section 4.3. There is, however, considerable short-range order. In analogy to the crystal, the coordination number and the NN bond distances are largely fixed, but with some additional spread in the latter. Some bond angles are also fixed, whereas others assume a broad range of possible values. This is illustrated in Fig. 4.12a for a two-dimensional glass with the composition A_2B_3.

In this section the primary concern is with what are called oxide glasses. The primary glasses of concern are based on silicates (SiO_2), borates (B_2O_3), phosphates (P_2O_5), and germanates (GeO_2). These glasses consist of a continuous random network in which the Si, B, P, and Ge ions, along with the O ions, form an integral part of the network. For this reason Si, B, P, and Ge are called *network formers*. Other elements are utilized in glasses and they fall into three major categories. The first are *network modifiers*, such as Ba, Ca, K, Li, Na, and Mg, which are often introduced in their oxide forms: BaO, CaO, K_2O, Li_2O, Na_2O, and MgO. The alkali or alkaline earth ions reside in the interior void spaces but do not become members of the primary network. The modifiers interfere with the network by providing nonbridging oxygens. A second class of elements is called *intermediates*. They are capable of infiltrating a preexisting network but are not, by themselves, network formers. Included in this group are Al, Pb, Ti, and Zn. They, too, are introduced in the form of their oxides: Al_2O_3, PbO, TiO_2, and ZnO. A third category are called *colorants*. These are impurity ions, which also reside in the voids and have characteristic absorption spectra that tint the glass. Common colorants and the corresponding colors are Cd (red), Co (blue), Cr (green), Cu (cyan), Fe (brown), and Mn (magenta). To understand how this works, think of white light as being composed of the three primary colors: red, blue, and green. Superpositions of pairs of primary colors generate secondary additive colors (e.g., red + green gives yellow; green + blue gives cyan; and blue + red gives magenta). The ion Cu^{2+} in soda-lime silicate glass absorbs at $\lambda = 790$ nm (near the red). Thus the transmitted light carries the blue and green components, giving a cyan color to the glass. Similarly, Co^{2+} absorbs at 560 and 600 nm, removing the red and green components and leaving blue.

Ions fall into the categories of network formers or network modifiers as a result of the strength of the bond they form with oxygen. The bond strengths for silicate, borate, germanate, and phosphate are typically in the range 4 to 5 eV. The network-modifier bond strengths are only 0.5 to 1.5 eV. Energetically, it is not possible for the modifiers to disrupt the bound network formers once they form in the cooling liquid. The intermediates, Al, Be, and Ti have bond strengths in the range 3 to 4 eV and so can be substituted more readily. One may argue that high bond strengths favor glass formation, since crystallization requires a continuous breaking and reforming of bonds as the atoms arrange themselves into the crystal structure. When a strong bond is formed and none can take its place, however, the geometrical arrangement becomes permanent, making crystal formation unlikely.

Consider the most common glass, silica (SiO_2), a schematic representation of which is given in Fig. 13.11. Each silicon ion is fourfold coordinated to oxygen ions, and each oxygen ion is twofold coordinated to silicon ions. The basic unit is a tetrahedron with the Si^{4+} at the center and the O^{2-} at the vertices. The bond is mixed ionic (41%) and covalent (59%). Silica provides a good example of the applicability of Zachariasen's rules for oxide glasses, as presented in Section 4.3. Rule 1 states that an O atom is bonded to not more than two Si atoms; rule 2 says the number of O atoms surrounding each Si is small (in this case, four); rule 3 requires the oxygen tetrahedra to be corner-shared; and rule 4 shows each oxygen tetrahedron to be linked to at least three (in this case, four) other tetrahedra, guaranteeing the three-dimensional character of the network. It should be noted that Zachariasen's third rule is the same as Pauling's third rule.

Figure 13.11. Structure of silica. The oxygens are at the vertices of the SiO_2 glass tetrahedra, and Si resides at the centers. All vertices are linked to two tetrahedra (not all are shown in the figure.)

The distribution of bond lengths is obtained from the measured radial distribution function. A graph of the measured radial distribution function, RDF, (see Section 3.6) of a-SiO_2 is given in Fig. 13.12. It is related to $g(r)$ of Eq. (3.67) through the relation RDF $= 4\pi r^2 \rho_o g(r)$. The Si–O bond length is 0.163 nm, the O–O bond length is 0.266 nm, and the Si–Si distance peaks at 0.312 nm. The next-to-nearest Si–O distance peaks at 0.415 nm, and the next-to-nearest Si–Si (and O–O) distance peaks at 0.51 nm. The Si–O–Si bond angle takes on a broad range of values between 120 and 180°, with a peak at 139°. The tetrahedra are organized into rings, typically containing between three and seven members.

Silica is characterized by having a low coefficient of thermal expansion. It has high mechanical strength because of the considerable degree of covalency in the bonds. Chemically, it is relatively inert, witnessed by the large number of chemicals stored in glass bottles. The main drawback of pure silica is the high melting temperature (1710°C) needed to form it. This is where network modifiers enter. The addition of Na_2O to silica dramatically lowers the eutectic temperature (to 793°C; Fig. 13.13). The Na^+ ion is effective in terminating portions of the network (Fig. 13.14). The number of bridging oxygens is reduced. This permits liquidization to occur at a lower temperature

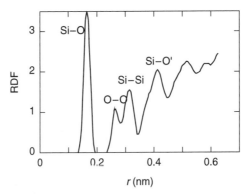

Figure 13.12. Radial distribution function for silica showing NN and next-NN peaks for O and Si atoms. [From R. L. Mozzi et al., *J. Appl. Crystallogr.*, **2**, 164 (1969). Reprinted by permission of the International Union of Crystallography.]

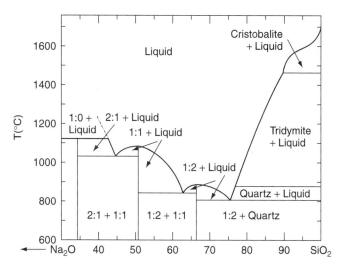

Figure 13.13. Binary phase diagram for sodium silica. The abscissa is in mol %. The abbreviation $m:n$ stands for $m(Na_2O):n(SiO_2)$. (From E. M. Levin et al., *Phase Diagrams for Ceramicists.* Reprinted by permission of the American Ceramic Society. Copyright 1964 by the American Ceramic Society. All rights reserved.)

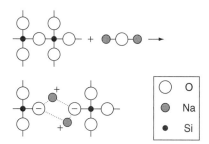

Figure 13.14. Na_2O is effective in breaking the Si–O network.

than would otherwise be possible. It also has the effect of reducing the viscosity. Such an additive which lowers the eutectic temperature and lowers the viscosity is called a *fluxing agent*. Since the potential energy function describing ionic bonds is more anharmonic in the radial direction than for covalent bonds, it also tends to increase the coefficient of thermal expansion. A similar effect can be had by using other alkali oxides as fluxes, such as Li_2O or K_2O, in place of Na_2O. However, Li_2O (because of the high mobility of the Li^+ ion) promotes crystal formation and K^+, being less mobile than Na^+ (because of its size), reduces the amount of flexibility and makes the glass less durable. On the other hand, the presence of K^+ in the surface of a glass introduces a compressive stress which tends to seal surface cracks, thereby improving the strength.

Unfortunately, soda-silica (sodium-silica glass) is not very stable. In the presence of water the sodium ions exchange places with protons, go into solution, and the glass disintegrates. This deficiency may be remedied by adding CaO to the composition to form soda-lime-silica. The Ca^{2+} ion, being more highly charged, forms a tighter bond

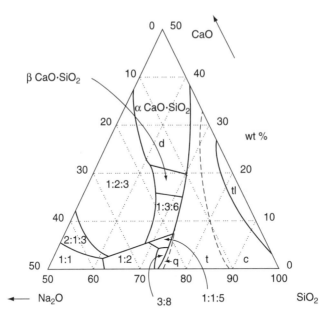

Figure 13.15. Ternary phase diagram for the glass-forming region of $Na_2O \cdot CaO \cdot SiO_2$. [From G. W. Morey et al., *J. Soc. Glass Technol.*, **9**, 232 (1925). Reproduced by permission of the Society of Glass Technology.]

with the O^{2-} ions and is not as mobile as Na^+. Reference to the ternary phase diagram for SiO_2–Na_2O–CaO in Fig. 13.15 shows the various phases and the glass-forming region. Soda-lime-silica is the most common glass in use.

Borate glasses are based on B_2O_3. The crystal forms in two-dimensional sheets, as illustrated in Fig. 4.12a. The interatomic bonds are again mixed ionic and covalent. The individual sheets are bound together by weaker electric-multipole and van der Waals forces. The B^{3+} ions are coordinated to three O^{2-} ions and the O^{2-} ions are twofold coordinated to B^{3+} ions. Instead of the basic building blocks being tetrahedra, they are triangles, which are vertex-shared. The nearest-neighbor B–O bond length is 0.137 nm and the corresponding O–O length is 0.237 nm. Borate glasses are characterized by having a low coefficient of thermal expansion, about an order of magnitude lower than for silica. Since they are essentially two-dimensional, they also have reduced viscosity, as the sheets can slide past each other and so are easier to work with. This bears some similarity to graphite, where the sheet structure and the extremely weak intersheet bonding make it suitable as a lubricant.

When Na_2O is added to borate, the added O converts the triangles to tetrahedra with a B–O distance of 0.148 nm. Thus it acts to build the three-dimensional network rather than to destroy it, as in the case of silica. The result is that the viscosity goes up and the coefficient of thermal expansion falls.

If one tries to alloy silica with borate glass (75% silica, 20% borate, 5% Na_2O) one finds that the two materials are immiscible. As the combination is cooled from the melt, an interlocking mesh of independent glasses is formed. The borate may be leached out by exposing the glass to acid, and what remains is porous silica. This material is valuable in its own right for use in sensors and filters. However, if it is heated, the pores eventually shrink and pure fused silica is formed. This is the *Vycor process*.

Phosphate glass is based on the chemical compound P_2O_5. The bonds are largely covalent. It has a relatively low melting temperature, 580°C. Germanate glasses are based on the compound GeO_2 and are similar in nature to silica but have lower melting points and are hygroscopic.

Glasses for commercial applications generally have composite structures. Additives are combined with the network formers to endow the glass with specific properties or to facilitate processing. For example, PbO increases the index of refraction and mass density of the glass. Materials such as CaO, MgO, and ZnO are found to increase the durability of the glass. Alumina, a network modifier, suppresses the tendency for crystal formation. The cations Al^{3+} create lattice strains which prevent the formation of silica crystallites. To maintain charge balance, when Al^{3+} joins the network there must be a compensating ion, such as Na^+, nearby. Compositions of common glasses are presented in Table 13.5.

Some elementary relations concerning networks in glasses are readily deduced. Let p denote the number of oxygen atoms associated with a polyhedron surrounding a network-forming ion. Of these, b are bridging oxygens (i.e., they are bonded to two network ions) and n are nonbridging. Thus

$$b + n = p. \tag{13.12}$$

If the chemical formula is M_xO_y, there are a total of y/x oxygens per network ion, M. Each nonbridging oxygen contributes a full atom to the tetrahedron, whereas each bridging oxygen contributes only half of an atom, since it is shared between adjacent polyhedra. Thus

$$n + \frac{b}{2} = \frac{y}{x}. \tag{13.13}$$

For example, in SiO_2, $x = 1$, $y = 2$, and $p = 4$. Solving Eqs. (13.12) and (13.13) yields $b = 4$ and $n = 0$. For $Na_2O\cdot2(SiO_2)$, $x = 2$, $y = 5$, and $p = 4$, so $b = 3$ and $n = 1$. In both cases b is large enough to produce a three-dimensional network. In the case of borate glass, B_2O_3, if one were to try to construct tetrahedra, one would run into a difficulty. Thus setting $p = 4$, $x = 2$, and $y = 3$ yields $b = 5$ and $n = -1$, which is an impossibility. However, by assuming triangular coordination of the network former, $p = 3$, one finds $b = 3$ and $n = 0$, which is acceptable. As stated earlier, the structure is given in Fig. 4.12a. Note that in order to form a network, $b \geq 1$, which implies that $y/x \leq p - 1$. For tetrahedra, $y/x \leq 3$, whereas for triangles, $y/x \leq 2$.

TABLE 13.5   Composition (wt %) of Some Common Commercial Glasses

Glass	SiO_2	B_2O_3	Al_2O_3	CaO	K_2O	MgO	Na_2O	PbO
Fused silica	100	—	—	—	—	—	—	—
Soda-lime Window	72	—	—	10	—	3	15	—
Bottle	73	—	1	10	—	—	15	—
Lamp bulb	71	2	—	7	1	3	16	—
Pyrex	81	13	2	—	—	—	4	—
Lead silicate	56	—	—	—	11	—	—	33
Aluminosilicate	55	8	15	17	—	4	1	—

Another family of glasses are the chalcogenides, which are composed of combinations of group IV (Si, Ge, or Sn) and/or column V atoms (P, As, and Sb) in combination with group VI elements (S, Se, and Te). Typical binary compositions are As_2S_3, As_2Se_3, As_2Te_3, GeS_2, and $GeSe_2$. They tend to be good glass formers with strong covalent bonds and a resistance to devitrification. The indices of refraction are relatively high, in the range 2 to 4. Since the ions forming the glass tend to be heavy, the vibrational frequencies are pushed to longer wavelengths and they are transparent in the near infrared. Unlike the oxide glasses the bandgaps tend to be narrow, so they often absorb visible light. The materials are photoconductors since the electrons excited to the conduction band are mobile.

Often, chalcogenide glasses are used in combination with halogen ions, such as Br, Cl, and I. Typical ternary glass systems are of the form (As, Ge, Sb, Si)–(S, Se, Te)–(Br, Cl, I). For example, $As_xTe_yI_z$ can be used as an electro-optical switch. When exposed to light, it switches from an insulator to a conductor. When subjected to a pulsed electrical field, it reverts to an insulator. The mechanism may involve the formation of microcrystallites which are formed when exposed to light and which become unstable when subjected to an electric field.

The rheology of glass-forming liquids is considered along with the rheology of polymers in Chapter 14. It should be stressed that research on glasses is an ongoing endeavor and important discoveries are still being made.

REFERENCES

General

Barsoum, M. W., *Fundamentals of Ceramics*, McGraw-Hill, New York, 1997.

Jaffee, B., W. R. Cook, Jr., and H. Jaffee, *Piezoelectric Ceramics*, Academic Press, San Diego, Calif., 1971.

Klein, C., and C. S. Hurlburt, Jr., *Manual of Mineralogy*, Wiley, New York, 1985.

Kingery, W. D., H. K. Bowen, and D. R. Uhlmann, *Introduction to Ceramics*, 2nd ed., Wiley, New York, 1976.

Reed, J. S., *Introduction to the Principles of Ceramic Processing*, Wiley, New York, 1988.

Schneider, S. J., Jr., *Engineered Materials Handbook*, Vol. 4, *Ceramics and Glasses*, The Materials Information Society, ASM International, Materials Park, Ohio, 1991.

Zeolites

Breck, D. W., *Zeolite Molecular Sieves: Structure, Chemistry and Use*, Wiley, New York, 1973.

Kerr, G. T., Synthetic zeolites, *Sci. Am.*, July 1989, p. 101.

Meier, W. M., and D. H. Olson, *Atlas of Zeolite Structure Types*, 3rd rev. ed., Butterworths, London, 1992.

Rabo, J. A., *Zeolite Chemistry and Catalysis*, American Chemical Society, Washington, D.C., 1976.

Weisz, P. B., A first-hand view of shape selective catalysis in zeolites: new science and technology, *Mater. Res. Soc. Bull.*, Oct. 1989, p. 54.

Refractories

Moore, R. E. ed., Refractories, Mater. Res. Soc. Bull., Nov. 1989.

Shaw, K., *Refractories and Their Uses*, Wiley, New York, 1972.

Glasses

Doremus, R. H., *Glass Science*, 2nd ed., Wiley, New York, 1994.

Holloway, D. G., *The Physical Properties of Glass*, Wykeham, London, 1973.

Paul, A., *Chemistry of Glasses*, Chapman & Hall, London, 1982.

Rawson, H., *Properties and Applications of Glass*, Elsevier, Amsterdam, 1980.

Rawson, H., *Glasses and Their Application*, Institute of Metals, London, 1991.

Weyl, A. W., and E. C. Marboe, *The Constitution of Glasses*, Interscience, New York, 1962.

Wong, J., and A. Angell, *Structure by Spectroscopy*, Marcel Dekker, New York, 1976.

PROBLEMS

13.1 Derive Eq. (13.9).

13.2 For the silicon oxynitride compound Si_2N_2O, assume that Si, N, and O atoms have their usual valences (4, 3, and 2) and that the N and O atoms do not form covalent bonds with each other.

    (a) Given a local bonding unit $Si-N_xO_y$ for Si with $x + y = 4$, determine x (and y) for this crystal structure.

    (b) What are the local bonding units for N and O?

13.3 In the perovskite crystal structure shown in Fig. 13.1, what are the local atomic bonding units for the two types of cations in the cubic unit cell? What are the corresponding coordination polyhedra? (Use $SrTiO_3$ as an example.) Sketch the two bonding units and their coordination polyhedra.

13.4 For the $Si_xN_yO_z$ ternary phase diagram, locate the following compounds: SiO_2, Si_3N_4, Si_2N_2O, and $Si_3N_2O_3$.

13.5 Find the average number of bridging oxygens, b, and nonbridging oxygens, n, for the following glasses:

    (a) $CaO \cdot SiO_2$, and

    (b) soda-lime (i.e., $2CaO \cdot 3Na_2O \cdot 15SiO_2$).

Note: Additional problems are given in Chapter W13.

CHAPTER 14

Polymers

14.1 Introduction

Polymers represent a class of materials currently in use in a broad variety of applications. The first meaningful application occurred in 1935 when Wallace E. Carothers invented nylon. Nylon quickly found application in stockings, fabrics, ropes, combs, tires, and many household items. It is characterized by being very strong, flexible, stretchable, and light in weight. Just as important, it is unaffected by corrosion or bacteria. Furthermore, it is inexpensive to manufacture and is easy to process. Since then, many new polymers have been discovered and they have broadened the range of applications immensely. Along with metals, semiconductors, and ceramics, polymers serve as one of the cornerstones of materials science and engineering.

A recent study by the National Research Council outlined some of the primary applications of polymers. Already, plastics such as Styrofoam (polystyrene) are actively competing with paper in the packaging field. High-strength plastics, such as polycarbonates and acrylonitrile–butadiene–styrene (ABS), serve as alternatives for house walls and roofing materials. Kitchen surfaces employ such plastics as polyethylene terephthalate (PET). Polymers have found application as additives for oils to increase their viscosity. They are also employed in secondary oil-recovery schemes. Tough new materials such as Kevlar are used in bulletproof vests.

Polymers are used in dentistry, where composite fillings undergo polymerization reactions when triggered by ultraviolet light. They are also used in a variety of implantable drug-delivery mechanisms. In electronics the high conductivity of certain classes of polymeric materials make them attractive alternatives to conventional metallic wiring. They are used as photoresists in microlithography processes and are utilized as a packaging material for microelectronics devices. They have been used in optoelectronic devices as nonlinear optical elements and as light-emitting diodes. They serve as sensors for chemical and biochemical agents.

This chapter has two goals. The first is to develop some of the physics and chemistry that are unique to polymers. Then a set of current applications of polymers to some of the areas listed above are studied. An attempt is made to relate the applications to the underlying science developed both in this chapter and in preceding chapters.

The chapter begins with a brief review of the structure, geometry, and topology of polymeric molecules. A more detailed description is given at our Web site[†] in Sections W14.1–W14.3. Then a study is made of some of their mechanical properties.

[†] Supplementary material for this textbook is included on the Web at the resource site (ftp://ftp.wiley.com/public/sci_tech_med/materials). Cross-references to elements of the Web material are prefixed by "W."

This is followed by a discussion of various thermal properties. A discussion of free-volume theory is given in Section W14.4. The applications described in the text are in the fields of structural plastics, ionic conductors, photoresists, piezoelectricity, and liquid crystals. Additional applications such as polymeric foams, porous films, electrical conductors and nonlinear optics are given in Sections W14.5 to W14.8. Although this selection of applications is far from being exhaustive, it serves as a reasonable base from which the reader can gain appreciation for the utility and versatility of polymers.

STRUCTURE OF POLYMERS

Polymers may be classified as being homopolymers or copolymers. Homopolymers have the same monomer repeat unit and may be symbolized by the sequence −A−A−A−A−A−, where A denotes the monomer unit. In copolymers two or more monomers comprise the polymer. Denoting the monomers by A and B, a random copolymer may be constructed in which the sequencing is consistent with "the flips of a coin" (e.g., −A−A−B−A−B−B−A−B−B−), that is, random sequencing. Alternatively, one could create a block copolymer, in which there is a periodic ordering (e.g., −A−A−B−B−B−A−A−B−B−B−). Note that this is identical to considering a homopolymer −C−C−C− with the enlarged monomer unit C = A−A−B−B−B.

The examples above illustrate linear polymers. Graft copolymers have side chains emanating from a linear polymer, such as in the configuration

$$-A-A-A-A-A-A-A-A-.$$
$$BBB$$
$$BBB$$
$$BBB$$

There are also branched polymer chains in which a treelike structure is formed and the monomers form a network.

Polymers are long chainlike molecules of variable length composed of repeating basic structural units called *monomers* which are covalently bonded with each other. Despite this, the polymer molecules are generally quite flexible and assume various contorted shapes in space. These shapes change with time due to thermal fluctuations. The chains may be isolated from each other, interacting only via electrostatic forces, van der Waals forces, or short-range repulsive interactions. Alternatively, they may be cross-linked, in which case the chains interconnect via chemical bonds. The resulting network could be planar or three-dimensional. At low temperatures the polymers often exist in either a crystalline or a glassy state. A large variety of polymers exist, depending on the character of the monomer.

An example of a linear section of the polystyrene polymer is presented in Fig. 14.1. The styrene monomer unit, $CH_2CHC_6H_5$, is outlined by a dotted rectangle. Polystyrene is denoted by the structural formula $-CH_2-CHC_6H_{5-n}$. The crystalline state of polystyrene has a hexagonal crystal structure with $a = 2.19$ nm, $c = 0.665$ nm and has 18 monomers in a unit cell. Figure 14.1 is oversimplified, and the real geometry is more complex. For example, the angle between three successive C atoms along the top row is really 116°.

In Fig. 14.2 some typical monomer units are presented. In the silicones (polysiloxanes) the −Si−O− backbone is attached to side groups, denoted by R, as shown. If

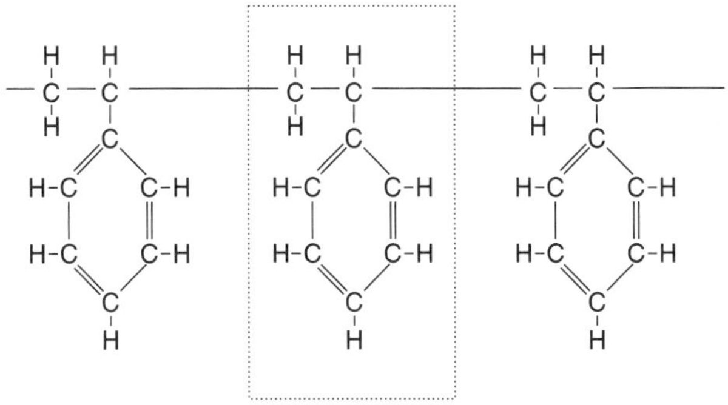

Figure 14.1. Section consisting of three monomers in a polystyrene polymer. In reality the side groups are not all pointing to the same side of the backbone and the atoms do not all lie in the same plane.

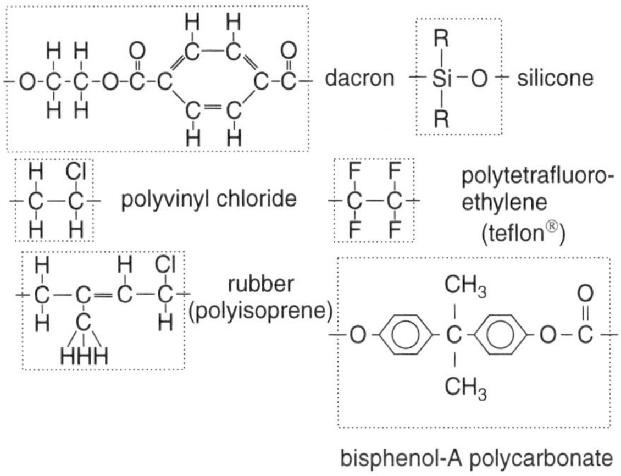

bisphenol-A polycarbonate

Figure 14.2. Typical monomer units. Various side groups may be placed in R, including organic, inorganic, or organometallic units.

the backbone is simply $-Si-$, one has the polysilanes. If the backbone has $-P{=}N-$, one obtains the polyphosphazenes. Note that Teflon (polytetrafluoroethylene) has two C and four F atoms forming the monomer. This is because the monomer is based on the ethylene molecule, which has two carbons. The angle between three successive C atoms is 113.85°. The simplified diagram of Fig. 14.2 does not convey the true geometric structure of the polymer. The F–C–F angle is 108° and the C–C–F angle is 108.7°, slightly distorted from the ideal tetrahedral angle of 109.47°.

Figure 14.3 depicts a node of a branched polymer (branched polyethylene) in which chain segments of sizes j, k, and l monomers meet at a junction. The presence of such junctions are characteristic of cross-linked polymers.

In the process of vulcanization, sulfur is used to cross-link polymer chains lightly, thereby enhancing the resiliency of the material and creating a useful rubber. Short polysulfide bridges are formed, linking together the individual polymer chains. The vulcanization process involves the use of chemical accelerators and activators to control the formation and size of the polysulfide bridges and to reduce the unwanted

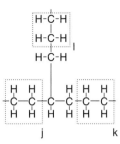

Figure 14.3. Junction in branched polyethylene.

formation of cyclic sulfur structures. The molecular weight of the polymer chain is typically $\approx 2 \times 10^5$, whereas the molecular weight between cross-links may typically be ≈ 5000.

It is possible to create copolymers by linking together two different polymers by means of a polysulfide bridge. Butyl rubber is such an example, where isobutylene is cross-linked to a small amount of isoprene. Styrene–butadiene–rubber (SBR), used for making tires, is another commercially important rubber.

Traditionally, polymers are classified as being of three types: thermoplastics, elastomers, and thermosets. The thermoplastics may be either crystalline or amorphous. Some are composed of single-chain molecules, while others are cross-linked. For temperatures above the melting temperature, $T > T_m$, the molecules retain their identity until a ceiling temperature, T_c, is reached, at which point thermal degradation (such as oxidation or pyrolysis) occurs. For $T_m < T < T_c$ the liquid flows readily and the plastic may be molded into a new shape. The amorphous thermoplastics have a second characteristic temperature, T_g, called the *glass-transition temperature*. For $T_m > T > T_g$ the thermoplastic is elastic, whereas for $T < T_g$ it freezes into a glassy state that is rigid. The thermoplastics are reprocessable and recyclable materials (i.e., they may be remolded at temperatures $T > T_g$). Common examples of thermoplastics are polystyrene (PS), polyethylene (PE), polypropylene (PP), and polyvinyl chloride (PVC). Engineering thermoplastics include polycarbonates (PC), polyacetals, nylons, polyesters, and acrylonitrile–butadiene–styrene (ABS). High-performance engineering thermoplastics include polyamides (PA), polyphenylene sulfide (PPS), polysulfone (PSF), and polyether ketone. Thermoplastics are used for packaging, construction, furniture, electrical equipment, and a variety of consumer products.

Elastomers (from *Elastic* and *Polymer*) are highly elastic polymers that may be stretched to several times their size. They then spring back to their original shape when the external tension is removed. However, they generally lose their elastic properties when heated. They are characterized by a low degree of cross-linking. Examples of elastomers include polybutadiene, polysulfide rubber (brand name Thiokol), polychloroprene (brand name Neoprene), Buna rubbers (composed of butadiene and styrene), polyisoprene, silicone, polynitrile, polyurethane (PUR), and butyl rubber [polyisobutylene (PIB)].

Thermosets have a high degree of cross-linking, which endows the material with a structural rigidity. They form three-dimensional networks when they gel. As in the case of elastomers, the material is degraded at higher temperatures, so they are not recyclable. At temperatures above T_g, the polymers are mobile and begin to link together. But as time goes on and the network continues to grow, the value of T_g increases. When $T_g > T$ the network solidifies and the polymerization reaction is brought to a halt spontaneously. Thermosets find applications in coatings and adhesives.

In the following sections the primary focus is on linear polymers, since these are capable of capturing most of the relevant physical phenomena. Further illustrations of the growth and microstructure of polymers are given in Sections 21.13, 21.14, and W21.21 to W21.25. The discussion here begins with the ideal linear polymer, in which the structure is isomorphic to a random walk path. It then proceeds to self-avoiding random walks. The concept of persistence length is then introduced. Finally, the crystallization of polymers is studied.

14.2   Geometry of Polymers

In this section some of the geometric properties of polymers are introduced. The structure of ideal linear polymers is discussed. The concept of self-avoiding walks is presented as a model for a linear polymer. The mathematical details of deriving the results are presented in Sections W14.1 and W14.2. Additional material relating to persistence lengths is given in Sections W14.3.

An unbranched polymer, called a *linear polymer*, may be parametrized as a single curve in space. An ideal linear polymer is one in which the shape is that of a random walk in three dimensions. The convention used will be that a linear polymer consists of $N + 1$ monomer segments connected by N chemical bonds. These monomer units are often bound to terminating radicals at the ends, such as protons or OH radicals. The terminating bonds are not included in the bond count above. The goal of this section is to explore some of the geometric relations that characterize such structures. It will be assumed that N is sufficiently large so that endpoint corrections are of no importance. In reality N is frequently of order 10^4 or 10^5, so this is a good approximation.

The first quantity characterizing the polymer is the molecular weight. If M_1 is the mass of a monomer unit, then the mass of the polymer molecule is

$$M_{N+1} = (N + 1)M_1. \tag{14.1}$$

Often, there will be a distribution of values of N in a macroscopic sample, so there will be a distribution of masses. We return to this point later.

If one were to travel along the polymer from end to end, one would travel a distance Na, where a is the length of a monomer unit. The end-to-end distance in space, however, would be shorter than this, due to the contorted shape of the polymer. The mean-square end-to-end distance $\langle r_N^2 \rangle$ of a polymer with N intermonomer bonds is given by the formula

$$\langle r_N^2 \rangle = Na^2. \tag{14.2}$$

Due to the isotropy of space the mean-square end-to-end shadow distances of r_N on the yz, xz, and xy planes are

$$\langle x_N^2 \rangle = \langle y_N^2 \rangle = \langle z_N^2 \rangle = \tfrac{1}{3}\langle x_N^2 + y_N^2 + z_N^2 \rangle = \tfrac{1}{3}\langle r_N^2 \rangle = \tfrac{1}{3}Na^2. \tag{14.3}$$

For an ensemble of polymers there will be a distribution of end-to-end distances given by

$$F_N(x_N^2) = \left(\frac{3}{2\pi N a^2}\right)^{1/2} e^{-3x_N^2/2Na^2}, \qquad (14.4)$$

The corresponding distribution of r_N^2 is therefore

$$G_N(r_N^2) = \left(\frac{3}{2\pi N a^2}\right)^{3/2} e^{-3r_N^2/2Na^2}. \qquad (14.5)$$

The center of mass of the polymer is defined (approximately, by neglecting end-group corrections) by

$$\mathbf{R} = \frac{1}{N+1} \sum_{n=0}^{N} \mathbf{r}_n. \qquad (14.6)$$

One may show that

$$\langle R^2 \rangle = \frac{a^2}{6} \frac{N}{N+1} (2N+1). \qquad (14.7)$$

For large N this approaches

$$\langle R^2 \rangle = \frac{Na^2}{3}. \qquad (14.8)$$

It is also possible to obtain a formula for the mean-square distance of a given monomer to the center of mass:

$$\langle s_n^2 \rangle = \langle r_n^2 \rangle - 2\langle \mathbf{R} \cdot \mathbf{r}_n \rangle + \langle R^2 \rangle$$
$$= \frac{N^2 a^2}{N+1} \left[\frac{1}{3}(w^3 + (1-w)^3) + \frac{1}{6N}\right] \longrightarrow N\frac{a^2}{3}[w^3 + (1-w)^3], \quad (14.9)$$

where $w = n/N$.

The probability-distribution function, $P(s_n)$, for the distances s_n is given by

$$P(s_n) = \left(\frac{\gamma}{\pi}\right)^{3/2} \exp(-\gamma s_n^2), \qquad (14.10)$$

where

$$\gamma = \frac{9N^2}{2a^2} \frac{1}{n^3 + (N-n)^3}. \qquad (14.11)$$

The size distribution of the polymers is important in determining their rheological properties (i.e., how they flow). This is of fundamental interest in the processing of polymers. The radius in solution determines the viscosity of the polymer. For high molecular weights the degree of entanglement grows rapidly, and the viscosity becomes very high as the strands intertwine. One defines a critical value of the molecular weight, M_c, for which molecular entanglement occurs. For example, PS has $M_c = 31,200$ and polymethylmethacrylate (PMMA) has $M_c = 27,500$.

The size distribution is also important in determining the rigidity of a plastic. For high molecular weights, when entanglement occurs, a plastic has a high degree of rigidity below the glass-transition temperature, T_g.

One method for measuring the molecular weight distribution is called *gel-permeation chromatography* (GTP), or alternatively, *size-exclusion chromatography* (SEC). A cylindrical column is filled with glass or polystyrene spheres. A solvent is made to flow through the column at a fixed rate. Polymer is injected into the solvent and the solution flows through the column. The smaller polymer molecules are readily trapped in the smaller pore spaces, whereas the larger molecules are less readily trapped there. The trapped molecules eventually leave the traps, flow through the column, and are monitored by an optical detector. Since the order in which the molecules emerge from the column is related to the size of the molecule, with the larger molecules emerging before the smaller ones, the optical detector is able to measure the distribution of sizes. The optical detector could measure the index of refraction of the solution, in which case it is called an *optical diffractometer*, or might examine infrared or UV absorption spectra. The GTP apparatus is generally calibrated using a set of monodisperse polymers (i.e., ones with well-defined molecular sizes).

Other methods for determining the size distribution include measuring the sedimentation rate, the osmotic pressure, or by studying the light scattering from solutions.

There are two constraints that a linear-chain polymer must obey: Each monomer must be attached to the previous monomer in the chain, and no monomer can cross another monomer. The case of a single molecule is considered first, followed by a dense collection of molecules. If only the first constraint is imposed, the result has already been derived: The end-to-end distance grows as $\sqrt{N}$, just as in a random walk. It will be seen that the effect of the second constraint is to transform this to $r_N \sim N^\nu$, where $\nu = 0.588 \pm 0.001$. The fact that the distance grows as a power of N greater than that for the overlapping chain model is expected. After all, since certain backbending configurations are omitted because they lead to self-overlap, it is expected that the chain will form a looser, more-spread-out structure. The precise value of the exponent depends on the results of a more detailed calculation.

Next consider a dense polymer. Each monomer is surrounded by other monomers, some belonging to its own chain and some belonging to others. The no-crossing rule applies to all other monomers. By extending the chain to larger sizes, the chain will avoid itself, but it will probably overlap other chains. Thus there is nothing to gain by having a more extended structure. The net result is that there is a cancellation effect and the chain retains the shape of a random walk. Thus in the dense polymer the mean end-to-end distance grows as $\sqrt{N}$.

14.3 Polymer Crystals

In a liquid polymer the chains fold, convolute, and form dense regions. As the temperature is lowered, these regions may acquire some degree of long-range order as a result of the chain–chain interactions. Such interactions include van der Waals forces, hydrogen bonding (when possible), as well as steric forces. The system tends toward a kinetically attainable state of lowest free energy. A schematic of the resulting structure is presented in Fig. 14.4. It consists of crystalline platelet regions (lamella) and amorphous regions. The thicknesses of the platelets are typically ≈ 10 to 20 nm. Note that the same polymer chain can span both crystalline and amorphous regions.

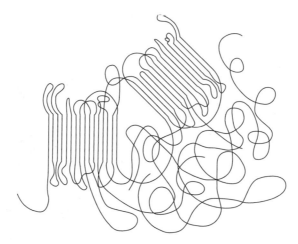

Figure 14.4. Polymer that is partially crystalline and partially amorphous.

The degree of crystallinity, λ, is a parameter giving the volume fraction of material in the crystalline state. The volume of a given mass of the polymer at temperature T may be expressed as $V(T) = \lambda V_c(T) + (1 - \lambda)V_a(T)$, where $V_a(T)$ and $V_c(T)$ are the volumes occupied by same mass of amorphous or crystalline polymer, respectively. The degree of crystallinity may be determined from x-ray diffraction studies, where only crystalline portions contribute to the sharp ring pattern characteristic of a collection of crystallites with long-range order. The amorphous polymer regions contribute to the diffuse x-ray background. When the temperature is elevated above the melting temperature T_m, the platelets melt and the x-ray diffraction pattern is diffuse.

Since there is a density mismatch between the platelets and amorphous regions, a stress field is set up. The energy of the system may be lowered by having several platelets aggregate. This produces spherical regions of aggregated platelets called *spherulites*, which may be observed with an optical microscope using polarized light.

Table 14.1 provides some crystallographic data relevant to polymers. The propensity for high crystallinity is determined by a number of physical factors. If the chain is isotactic (i.e., has a simple, ordered structure of side groups), crystal formation is favored. Thus isotactic polymers will have much higher crystallinity than atactic polymers (i.e., polymers with side groups arranged randomly along the backbone). For example, isotactic linear PE has values of λ in the range 50 to 94%. Commercial isotactic PP, used in pipes and films, has λ in the range 58 to 64%. While isotactic PS has λ near 50%, atactic PS and atactic PVC are amorphous.

The size of the side groups attached to a polymer chain affects crystallinity. Generally, the larger the side group, the more difficult it will be to form crystals.

Figure 14.5 shows a PE crystal with the orthorhombic structure. The polymer chains are aligned along the c axis with a centered, rectangular projection in the ab plane.

In many situations the packing of the chains becomes more efficient if the chains assume a helical shape. A unit cell may contain several helices.

14.4 Defects in Polymers

In crystals the classification of defects is straightforward and has been covered in Chapter 4. In polymers, where there is a high degree of disorder to begin with, the

TABLE 14.1   Crystal Structure Parameters for Some Polymers

| Polymer[a] | Lattice | Cell Dimensions | | n_c^b | Conformation[c] | Density ρ (kg/m³) |
		nm	deg			
PE	Orthorhombic	a = 0.742	—	2	2*1/1	1000
		b = 0.495	—			
		c = 0.255	—			
	Monoclinic	a = 0.809	$\alpha = 90$	2	2*1/1	998
		b = 0.479	$\beta = 90$			
		c = 0.253	$\gamma = 108$			
PS	Hexagonal	a = 2.19	—	18	2*3/1	1111
		b = 2.19	—			
		c = 0.663	—			
PP(α)	Monoclinic	a = 0.665	$\alpha = 90$	4	2*3/1	936
		b = 0.650	$\beta = 90$			
		c = 2.096	$\gamma = 99$			
PP(β)	Hexagonal	a = 1.908	—	9	2*3/1	922
		c = 0.649	—			
PP	Hexagonal	a = 0.638	—	1	2*3/1	939
		c = 0.633	—			

Source: Data from J. Brandrum and E. H. Immergut, eds., *Polymer Handbook*, 3rd ed., Wiley, New York, 1989; and G. L. Anderson, ed., *Physics Vade Mecum*, American Institute of physics, New York, 1981.

[a]PE, PS, and PP are polyethylene, polystyrene, and polypropylene, respectively.

[b]n_c is the number of monomer units per unit cell.

[c]The conformation is given by n^*p/q, where n is the number of atoms in the asymmetric unit of the chain and p is the number of units in q turns of the helix.

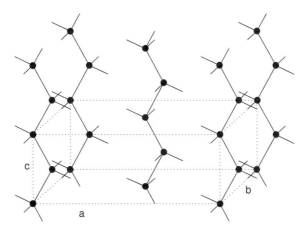

Figure 14.5. Crystal structure of polyethylene. [From M. Mackley, *Mater. Res. Soc. Bull.*, 22(9), 47 (1997).]

identification of defects requires more care. There are zero-, one-, two-, and three-dimensional defects to be found.

One type of three-dimensional defect has already been discussed in Section 14.3, when the concept of crystallinity was introduced. A polymer with a high degree of crystallinity is orientationally disordered and contains significant regions of amorphous polymer.

Surfaces of polymer films provide instances of two-dimensional defects. The melting temperature of a polymer, or any material for that matter, is depressed by the presence of a surface, according to the formula

$$T_m = T_m^0 \left(1 - \frac{2\sigma}{\Delta H_m t} \right), \tag{14.12}$$

where T_m^0 is the melting temperature for the bulk polymer, σ is the surface energy of a large sample, ΔH_m is the heat of fusion, and t is the thickness of the film. For a sphere of radius R the corresponding formula is

$$T_m = T_m^0 \left(1 - \frac{3\sigma}{\Delta H_m R} \right). \tag{14.13}$$

In both cases the fractional correction is given by the ratio of the surface energy divided by the bulk enthalpy change. Typical values for $(\Delta H_m, \sigma)$ for the polymers PE, PS, PP, and PMMA are $(2.8 \times 10^8 \ J/m^3, 0.101 \ J/m^2)$, $(9.1 \times 10^7, 0.035)$, $(2.2 \times 10^8, 0.029)$ and $(6.1 \times 10^7, 0.039)$, respectively. For example, for a 100-nm film of PMMA, the fractional lowering of the melting temperature is approximately 0.1%.

One-dimensional defects include edge and screw dislocations. The Burgers vector for these is typically much larger than those in crystalline solids, being in the range 5 to 50 nm. Zero-dimensional defects include the ends of linear polymer chains and abrupt folds called *hairpin defects*.

Polymers such as PS, PC, and PMMA may develop microscopic cracks in a direction perpendicular to the deformation direction when subject to strain. The phenomenon is called *crazing*. The craze consists of a stretched microfibril of polymer (of diameter ≈ 1 to 10 nm) surrounded by a void space. The critical strain for the production of crazing is typically in the range 0.4 to 2.5%. Brittle cracks can initiate from crazes. For ductile polymers (e.g., PC), instead of crazing occurring, there are shear bands, inclined at 45° to the deformation direction.

Polymer liquid crystals also possess defects that are manifested as abrupt changes in the direction of the director field, $\hat{n}(\mathbf{r})$, which specifies the orientation of the molecule throughout space. There are disclinations, which may be point or line defects. The strength of the disclination, S, is the number of rotations of the vector $\hat{n}$ as one traverses a small circle around the singularity. Examples of defects with $S = \pm \frac{1}{2}$ and ± 1 are presented in Fig. 14.6. For the case $S = \frac{1}{2}$ one notes that upon traveling from point A to B, the direction of the director rotates counterclockwise by 180°. For the case $S = -\frac{1}{2}$, following the path $ABCDEF$ produces a rotation that is clockwise by 180°. Similarly, the $S = 1$ case results in a 360° counterclockwise rotation as the director is followed from A to B and then to A. The case $S = -1$ has a $-360°$ rotation for the sequence $ABCDEFGH$.

Defects interact with each other with long-range interactions. The interaction energy may be obtained from Eq. (14.45) by utilizing a suitable $\hat{n}(\mathbf{r})$ and integrating over all space. It is found that the interaction energy is proportional to the product of the strengths of the defects and varies inversely as the separation between them. The analogy with two-dimensional electrostatics is apparent. Isolated defects have their director fields terminating on a surface of the solid.

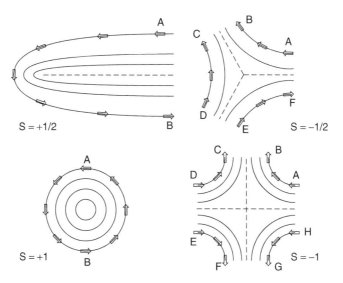

Figure 14.6. Disclinations in two dimensions.

One may have a point defect when the ends of a number of polymer chains diverge from a point in the bulk of the material, much like the electric field lines may be thought of as emanating from a point charge in electrostatics.

MECHANICAL PROPERTIES

The mechanical properties of polymers are different than those of metals or ceramics. For example, the Young's modulus E of a polymer is typically in the range 100 to 4000 MPa, whereas a metal has E in the range 50 to 400 GPa. For polymers, E is a sensitive function of temperature, falling rapidly with increasing T. The maximum tensile strength σ_{TS} of a polymer is typically only 100 MPa, whereas for a metal it may typically go up to 4 GPa. The elongation to break, ϵ_b, for a polymer may be as large is 1000%, whereas for a metal it may typically be less than 100%. Furthermore, polymers are viscoelastic materials, so ε_b depends on the strain rate, $\dot{\varepsilon}$, with faster strain rates producing smaller values of ε_b. In Table 14.2, values are given for some mechanical properties of polymers. Even for a given polymer, these numbers depend on the molecular weight, the polymerization temperature, crystallinity, degree of cross-linking, and other physical and chemical factors which are often not quantified when the data are reported in the literature. This accounts for a wide range of reported values for some of these parameters.

Note that the densities of the polymers are all within a factor of 2 of each other, but the mechanical properties vary widely. The Poisson ratios tend to be considerably larger than for metals, perhaps due to the inherent anisotropy of the polymer molecules themselves. The elongation to break for elastomers is typically larger than for thermoplastics.

The discussion begins by examining polymers under tension, and a formula for the stress as a function of strain for elastomers is derived. We then proceed to consider some of the viscoelastic properties of polymers.

TABLE 14.2 Mechanical Properties of Polymers

Polymer	Density ρ (kg/m^3)	Tensile Strength σ_{TS} (MPa)	Elongation to Break ε_b (%)	Young's Modulus E (GPa)	Poisson Ratio ν	Bulk Modulus B (GPa)
Thermoplastics						
PE						
(Low density)	910–925	15–79	150–600	0.06–0.17	0.49	1–2.8
(Med density)	926–940	12–19	100–150	0.17–0.38	—	—
(High density)	941–965	18–33	12–700	0.41–1.03	—	—
PS (amorphous)	1040–1065	—	1–2.5	3.2–3.4	0.33	3.2
PP (amorphous)	850–854	29–39	11–15	1.03–1.72	—	—
PVC	1390	—	13–210	2.96	0.38	4.1
PMMA	1190	—	2–10	3.3	0.40	5.5
PC	1200	—	60–100	2.4	—	—
Polysulfone						
(B430)	1520	121	2.5	8.96	—	—
(S1000)	1230	66	6.0	2.41	—	—
Elastomers						
Polyisoprene						
Vulcanizate gum	970	17–25	750–850	0.0013	0.5	1.95
Hard rubber	1170	60–80	6	0.003	0.5	4.17
SBR						
Vulcanizate gum	980	1.4–3	400–600	—	—	1.96
Vulcanized	1150	17–28	400–600	—	—	2.5

Source: Data from J. Brandrum and E. H. Immergut, eds., *Polymer Handbook*, 3rd ed., Wiley, New York, 1989; and J. L. Anderson, ed., *Physics Vade Mecum*, American Institute of Physics, New York, 1981; and other sources.

14.5 Polymers Under Tension

An unstretched linear chain polymer is a molecule with a highly contorted shape. The ensemble-averaged end-to-end displacement vector is zero, although any one polymer will have an end-to-end displacement whose length $\propto \sqrt{N}$. There are many configurations that the molecule can have which are consistent with this condition. When a tensile force is applied to it, the molecule becomes elongated. There are fewer configurations consistent with the elongated end-to-end distance. Eventually, when the molecule becomes fully elongated the polymer becomes very stiff. A sufficiently increased tension will ultimately result in the breaking of a covalent bond. One approach to the elastic properties of polymers is the use of thermodynamics. The Helmholtz free energy, $F = U - TS$, is largely entropy dominated. Thus the elongated state, with its lower entropy, has a relatively high free energy, whereas the unstretched state, with its larger entropy, has a lower free energy.

The elastic properties will be obtained by considering the effect of a tensile force, F, applied to the individual polymer molecule. This is illustrated in Fig. 14.7, where a fully stretched molecule and a partially stretched molecule are drawn.

Imagine the individual monomer units subjected to an external force field **F** along the z direction. The potential energy of the polymer, relative to its value when the

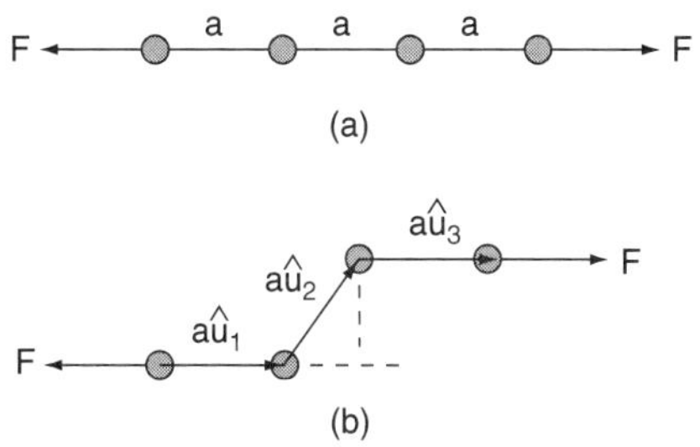

Figure 14.7. Two polymer molecules subjected to a tensile force F: (a) fully stretched; (b) partially stretched.

end-to-end displacement is zero, is

$$U = -\mathbf{F} \cdot \mathbf{r}_N = -F\hat{k} \cdot \sum_{j=1}^{N} a\hat{u}_j, \qquad (14.14)$$

where r_N is the end-to-end distance. The mean value for the z component of the end-to-end displacement is given by averaging over the Boltzmann distribution:

$$\langle z_N \rangle = \langle \hat{k} \cdot \mathbf{r}_N \rangle = \frac{\int d\hat{u}_1 \cdots d\hat{u}_N \hat{k} \cdot \mathbf{r}_N e^{-\beta U}}{\int d\hat{u}_1 \cdots d\hat{u}_N e^{-\beta U}}$$

$$= a \sum_{n=1}^{N} \frac{\int d\hat{u}_1 \cdots d\hat{u}_N \hat{k} \cdot \hat{u}_n e^{\beta F a \sum_j \hat{u}_j \cdot \hat{k}}}{\int d\hat{u}_1 \cdots d\hat{u}_N e^{\beta F a \sum_j \hat{u}_j \cdot \hat{k}}} = Na \frac{\partial}{\partial \theta} \ln Q, \qquad (14.15)$$

where $\theta = \beta a F$, $\beta = 1/k_B T$, and

$$Q = \int \frac{d\hat{u}}{4\pi} e^{\theta \hat{u} \cdot \hat{k}} = \frac{\sinh \theta}{\theta}. \qquad (14.16)$$

Thus the Langevin function $L(\theta)$ is obtained:

$$\langle z_N \rangle = Na \left(\coth \theta - \frac{1}{\theta} \right) = NaL(\theta). \qquad (14.17)$$

A power series expansion in results in

$$\langle z_N \rangle = Na \left(\frac{\theta}{3} - \frac{\theta^3}{45} + \frac{2\theta^5}{945} - \frac{\theta^7}{4725} + \frac{2\theta^9}{93,555} + \cdots \right). \qquad (14.18)$$

A graph of $\langle z_N \rangle / Na$ as a function of θ is given in Fig. 14.8. Note that the maximum value of $\langle z_N \rangle$ is Na. The spring constant for the polymer molecule in the small F, Hooke's law regime is given by

$$K = \frac{F}{\langle z_N \rangle} = \frac{3k_B T}{Na^2}. \qquad (14.19)$$

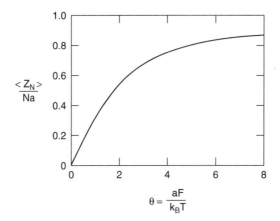

Figure 14.8. Polymer length in units of Na versus the scaled tension, $\theta = aF/k_BT$.

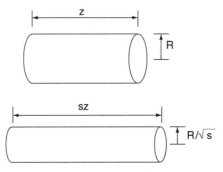

Figure 14.9. Cylindrical sample undergoing an affine, volume-conserving transformation as it is stretched.

Since it is entropy dominated, it varies linearly with absolute temperature and inversely as the size of the polymer.

Next consider a macroscopic sample in the shape of a cylinder and evaluate its Young's modulus, E. Several simplifying assumptions will be made. First, assume that all molecules have the same length, Na. Second, assume that the volume of the sample remains constant (i.e., the sample is incompressible). This corresponds to a Poisson's ratio $\nu = \frac{1}{2}$. The latter assumption is a constraint that is imposed on the system but which is consistent with experiment. It is not a property of the ideal linear polymer molecule. Under the applied force the cylinder undergoes an *affine transformation* (i.e., one that is volume conserving). This is illustrated in Fig. 14.9, in which the length increases by a factor s and the radius decreases by a factor $1/\sqrt{s}$.

The work done in stretching the sample in the linear regime is given by

$$W = \int_z^{sz} F_z\,dz + \int_R^{R/\sqrt{s}} F_R\,dR = \frac{K}{2}\left[z^2(s^2 - 1) + R^2\left(\frac{1}{s} - 1\right)\right]. \tag{14.20}$$

This is to be averaged over an ensemble of configurations. If a single molecule occupied the sample, one could write $\langle z^2 \rangle = \langle r^2 \rangle/3$, $\langle R^2 \rangle = 2\langle r^2 \rangle/3$, and $\langle r^2 \rangle = Na^2$. Inserting the expression for the spring constant would then give

$$\langle W \rangle = \frac{k_B T}{2} \left(s^2 + \frac{2}{s} - 3 \right). \tag{14.21}$$

Let there now be n monomers per unit volume (i.e., n/N polymers per unit volume). Then the work done per unit volume in stretching the cylinder is

$$\frac{\langle W \rangle}{V} = \frac{n k_B T}{2N} \left(s^2 + \frac{2}{s} - 3 \right) \tag{14.22}$$

and is equal to the energy stored in the polymer. The tensile strain is given by the stretch per unit length:

$$\varepsilon = \frac{sz - z}{z} = s - 1. \tag{14.23}$$

The stress is the rate of change of the work per unit volume with respect to the strain:

$$\sigma = \frac{n k_B T}{N} \left(s - \frac{1}{s^2} \right). \tag{14.24}$$

In the elastic limit, where $\varepsilon \ll 1$, the energy stored in the polymer approaches

$$\frac{\langle W \rangle}{V} \longrightarrow \frac{3n}{2\beta N} \varepsilon^2 = \frac{1}{2} E \varepsilon^2, \tag{14.25}$$

so

$$E = \frac{3n k_B T}{N}. \tag{14.26}$$

Note that one may relate the work per unit volume to the change in entropy per unit volume, using $W = -T\Delta S$, so

$$\frac{\Delta S}{V} = -\frac{n k_B}{2N} \left(s^2 + \frac{2}{s} - 3 \right). \tag{14.27}$$

Using the relation for the shear modulus for isotropic materials given in Table 10.4, $G = E/(1 + 2\nu)$, and using $\nu = 1/2$, Eq. (14.24) may be rewritten as

$$\sigma = \frac{2G}{3} \left(s - \frac{1}{s^2} \right) \tag{14.28}$$

which is called the *Staudinger equation*. A phenomenological prefactor is often added to improve the fit to the experimental data. The Mooney–Rivlin formula, involving two empirical constants C_1 and C_2 (see Fried, 1995), is

$$\sigma = 2 \left(C_1 + \frac{C_2}{s} \right) \left(s - \frac{1}{s^2} \right). \tag{14.29}$$

Even this formula has its deficiency. When the polymer approaches a state of full elongation (i.e., when the end-to-end distance approaches Na), the slope of the stress–strain curve should rise very rapidly until the polymer breaks (see Fig. 10.5).

The question arises as to how the stresses applied to the ends of a macroscopic sample of a polymer are communicated between the individual linear chain molecules. If there is cross-linking between the chains due to covalent bonds, and these links percolate along the length of the sample, the stress is transmitted directly through the links. However, even in the absence of cross-linking, stress transmission is possible.

There are two primary interactions between polymer chains. One is the steric interaction that prevents one chain from crossing another. The other is the weak van der Waals interaction between chains. If the polymers become entangled with each other, applying a stress along the sample will lead to situations where one molecule will press against another. Even a bend in one polymer chain butted against a neighboring chain provides a mechanical linkage for stress transmission. In a more entangled situation, a given polymer chain may be wound n times around another chain. As the system is strained, this *winding number* remains an invariant. Thus steric forces provide a means for chain-to-chain stress transmission.

If a single crystal of polymer is formed, such as polyethylene shown in Fig. 14.5, the van der Waals forces bind the chains together. When a shear stress is applied across the crystal, the elasticity is due to the van der Waals forces. In attempting to slide one chain past an adjacent chain, the potential energy will be increased. This provides a restoring force tending to return the crystal to its equilibrium configuration. These van der Waals forces probably also play some role in determining the elastic properties of amorphous polymers as well.

14.6 Viscoelasticity

Polymers exhibit both elastic and viscous behavior when subjected to stresses (above the glass transition temperature), and are therefore called *viscoelastic materials*. To obtain an intuitive feeling for the origin of these effects, a simple mechanical model for a polymer subjected to a shear stress will be constructed. Suppose that a force **F** is applied along the top surface of a polymer, with area A. The shear stress is $\sigma = F/A$, flow will occur parallel to **F**, and there will be a velocity gradient perpendicular to **F**. Partition the polymer into layers of thickness δ, with successive velocities $v_1, v_2, v_3, \ldots$, which are constant at steady state. Assume that the polymers can protrude from their layers somewhat. At a given interface between the layers, two types of behavior can occur. A monomer from one layer may collide with a monomer from a neighboring layer, thereby causing an interchange of momentum between the two layers. It is assumed that the motion of the molecule is along a straight line parallel to the interface. The rubbing is the origin of the viscous force in this model. Alternatively, a molecule may be shared for some time by adjacent layers and become stretched as the layers slip past each other. The stretched polymer gives rise to a type of Hooke's law force which is the origin of the elastic force. These processes are illustrated in Fig. 14.10.

Newton's second law provides the connection between the net force on a layer and its change in momentum. Consider an intermediate layer, n, and let N_s be the number of molecules per unit area shared at an interface. Let K and M be the spring constant between layers and the mass of a molecule, respectively, and let u_n denote

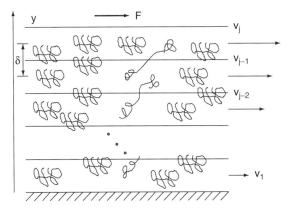

Figure 14.10. Representation of molecules in layers parallel to the surface of a sample of polymer under a shear stress.

the displacement of the layer. For layer n,

$$N_s AK[(u_{n+1} - u_n) - (u_n - u_{n-1})] = \dot{N}_c AM[(v_n - v_{n-1}) + (v_n - v_{n+1})], \quad (14.30)$$

which may be written in continuum form as

$$N_s AK \frac{\partial^2 u}{\partial y^2} = -\dot{N}_c AM \frac{\partial^2 v}{\partial y^2}. \quad (14.31)$$

Here N_c is the effective number of monomers per unit area colliding per unit time. At the top layer the equation is

$$F - N_s AK(u_j - u_{j-1}) = -\dot{N}_c AM(v_{j-1} - v_j), \quad (14.32)$$

or

$$\sigma = G \frac{\partial u}{\partial y} + \eta \frac{\partial v}{\partial y}, \quad (14.33)$$

where $G = N_s K \delta$ is the shear modulus and $\eta = M \delta \dot{N}_c$ is the viscosity. It is convenient to rewrite the equations in terms of the shear strain $\varepsilon = \partial u / \partial y$:

$$\sigma = G\epsilon + \eta \dot{\epsilon}. \quad (14.34)$$

This is a special case of the Zener model for anelasticity described in Chapter W10 and given by Eq. (W10.3). Although this equation is derived here on the basis of a very specific model, one may take this as the starting point of a phenomenological description. This equation was originally proposed by Voigt to describe viscoelastic behavior (see Tobolsky, 1960). Note that in the interior of the polymer, integration of the second-order differential equation results in the same expression:

$$G \frac{\partial^2 u}{\partial y^2} = -\eta \frac{\partial^2 v}{\partial y^2} \longrightarrow G\varepsilon + \eta \dot{\varepsilon} = \text{constant}, \quad (14.35)$$

with the constant being just σ, the shear stress.

The Voigt model contains a characteristic relaxation time

$$\tau = \frac{\eta}{G},$$ (14.36)

so

$$\dot{\varepsilon} + \frac{\varepsilon}{\tau} = \frac{\sigma}{\eta}.$$ (14.37)

Consider the response of a polymer to the sudden imposition of a stress at the surface [i.e. $\sigma(t) = \sigma_0 \Theta(t)$]. Assume that the strain in the polymer was initially zero [i.e., $\varepsilon(0) = 0$]. The following solution to the Voigt equation [see Eq. (W10.4)] is found:

$$\varepsilon(t) = \frac{\sigma_0}{G} \left(1 - e^{-t/\tau}\right).$$ (14.38)

Thus the strain approaches the asymptotic value $\varepsilon(\infty) = \sigma_0/G$ for $t \gg \tau$. In actuality, this behavior is not exactly observed. After an initial stretching phase, obeying the law above, the strain continues to grow rather than saturating. This is an example of creep behavior.

Strain relaxation when a stress is removed is predicted by Voigt to follow an exponential decay law, since [see Eq. (W10.4)]

$$\dot{\varepsilon} + \frac{\varepsilon}{\tau} = 0 \longrightarrow \varepsilon(t) = \varepsilon_0 e^{-t/\tau}.$$ (14.39)

The expressions relating stress and strain (and their rates of change) are called *constitutive equations*. The Voigt equation was derived on the assumption that at each interface, the momentum-transfer and spring-stretching mechanisms act in parallel with each other. Maxwell adopted the alternative viewpoint, that they compete with each other at a given interface. Thus if one interface has stretched molecules, it is unlikely to have colliding molecules, and vice versa. Therefore, the two types of interfaces should be placed in series with each other. The total strain consists of two parts ($\varepsilon = \varepsilon_1 + \varepsilon_2$), one due to elasticity, $\varepsilon_1 = \sigma/G$, and the other due to viscosity, $\varepsilon_2 = \int \sigma/\eta \, dt$. Thus

$$\frac{\dot{\sigma}}{G} + \frac{\sigma}{\eta} = \dot{\varepsilon}.$$ (14.40)

This equation predicts, for example, that if the strain is kept constant for some period of time, the stress will relax with the characteristic time τ:

$$\sigma = \sigma_0 e^{-t/\tau},$$ (14.41)

where again $\tau = \eta/G$. This type of behavior is observed experimentally.

In reality, neither the Voigt model nor the Maxwell model is capable of describing all the possible observed behaviors involving the application of transient stresses to polymers. Empirical models based on both series and parallel combinations of dissipative and elastic interfaces have been developed to help fit the experimental data. This is illustrated in Fig. 14.11, where simple mechanical "circuits" for the Voigt (V) and Maxwell (M) models are presented, along with two other hypothetical composite

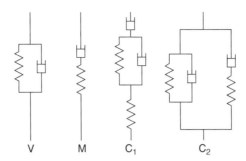

Figure 14.11. Mechanical "circuits" consisting of springs and dash pots for the Voigt (V), Maxwell (M), and two composite (C_1, C_2) models.

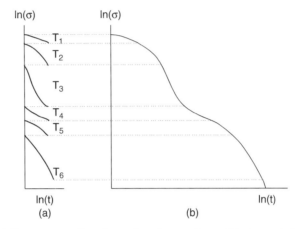

Figure 14.12. (*a*) Stress versus time for various temperatures; (*b*) stress plotted versus rescaled time.

models, labeled C_1 and C_2, respectively. Note that none of these mechanical circuits involve an inertial (mass) element, and hence there is no acceleration term. The materials are just too viscous for inertia to play a role. The dash pot shown represents a viscous friction element such as would be provided by a piston in a leaky cylinder. A practical application of such a dash pot is provided by the shock absorber in an automobile.

A sketch of a typical stress-relaxation curve $\sigma(t)$ for a polymer is presented as a log-log plot in Fig. 14.12*a*, for a set of temperatures $T_1 < T_2 < \cdots < T_6$. These curves extend over the same time range and begin at the same stress. The stresses have been scaled by a multiplicative factor chosen so that for a given value of σ there is only one stress-relaxation curve. The scaling is such that the stress at the top of curve $j + 1$ is set equal to that at the bottom of curve j, for $j = 1, \ldots, 5$. Next, imagine rescaling the time by another multiplicative factor so that the curves line up as shown in Fig. 14.12*b*. The scaling is such that the time at the left of curve $j + 1$ is set equal to the time at the right of curve j, for $j = 1, \ldots, 5$. It is found that they join smoothly to form a single curve, without discontinuities in slope at the joining points. Furthermore, if any other intermediate temperature curve were rescaled with scale factors interpolated from the previous ones, it would overlay the right-hand curve. Alternatively, if a stress curve were measured for a longer time interval at a given temperature and the scale factor for that temperature were used, the same overlay would occur. The conclusion is that the behavior obtained over a long time interval at a low temperature is equivalent to that found for a shorter interval, but at a higher temperature. This relationship is called the *temperature–time superposition principle* (TTSP).

The physics of the TTSP may be understood in terms of a simple model. In order for stress to relax, the polymer strands must move some distance. This motion is compounded out of the segmental motions of the monomers. The bottleneck in this motion is the time required for a monomer (or group of monomers) to take a single step, through some fixed distance, Δa. This is controlled by some free-energy barrier, ΔF. Thus the step time is $\Delta t \sim (1/\nu) \exp(\Delta F / k_B T)$, where ν is some attempt frequency, typically a vibrational frequency of a monomer. The rate at which the stress relaxes is controlled by the temperature. Elevating the temperature is the same as decreasing Δt:

$$\frac{\Delta t_1}{\Delta t_2} = \exp\left[\frac{\Delta F}{k_B}\left(\frac{1}{T_1} - \frac{1}{T_2}\right)\right]. \tag{14.42}$$

Things happen faster at higher temperatures.

The actual motions involved in the relaxation process are complicated. They often involve cooperative motions of a "cage" of nearby monomers opening up their "gate" and letting a monomer pass through. Composite motions of the chain itself, involving coupled rotational and translational degrees of freedom, are believed to be involved as well. In fact, polymers are known to have a large number of different relaxation times which are distributed over many decades of time. For short time intervals or narrow temperature ranges, a small number of values of τ may be sufficient. This is not likely to be true, however, for longer time intervals or wider temperature ranges.

THERMAL PROPERTIES

The mechanisms governing the thermal properties of polymers are much the same as those governing insulators. Typical values for some of these properties are presented in Table 14.3. The high values of the thermal coefficient of volume expansion β are expected, given the flexibility of the molecules and the fact that polymers are very anharmonic.

The melting transition is a first-order phase transition, so the derivatives of the Gibbs free energy are discontinuous at the melting temperature, T_m. Consequently, there is a discontinuity in the entropy, $S = -(\partial G/\partial T)_P$, and the volume, $V = (\partial G/\partial P)_T$, at T_m. This implies the existence of a latent heat of fusion, $\Delta H_m = \Delta G + \Delta(TS) = T_m \Delta S_m$, when the polymer melts, as well as a relative volume change, $\Delta V/V$. The size of these parameters depends on the crystallinity of the polymer.

In the following section the glass transition is studied. This is supplemented on our home page by an introduction to the free-volume model. This model has been historically useful in understanding some aspects of the glass transition.

14.7 Thermal Properties of Polymers

In this section bulk polymer materials are considered, as opposed to polymers in solution. The concern will be with what happens to the properties of the polymer as a

TABLE 14.3   Typical Thermal Properties for Polymers

Polymer	Melting Temperature T_m (°C)	Specific Heat at Room Temperature C_p (J/kg·K)	Thermal Coefficient of Volume Expansion β (10^{-6} K^{-1})	Thermal Conductivity κ (W/m·K)	Latent Heat of Fusion ΔH_m (kJ/kg)	Fractional Volume Change $\Delta V/V$ (%)
PE	137	1.92	287	0.3–0.5	277	—
PS	240	0.95	510–600	0.11	87	—
PP	176	1.8	105	0.12	209	—
PVC	273	0.95	520	0.16	—	—
PMMA	200	1.38	460–500	0.19	—	—
PC			260	0.2	—	—
PTFE	332	—	—	0.38	82	29
Polyisoprene						
Vulcanizate gum	87	1.83	660	0.153	203	—
Vulcanized	—	1.39	190	0.162	—	—
SBR						
Vulcanizate gum	—	1.89	660	0.19–0.25	—	—
Vulcanized		1.50	530	0.30	—	—

function of the temperature, T. The thermal properties are intimately connected with the rheological, or flow, characteristics, and are thus of fundamental interest in the processing of these materials.

At high temperatures the polymer is a liquid with the molecules as a whole at liberty to move around. If shear stress is applied, hydrodynamic flow will take place, with the flow rate limited by the viscosity. As the temperature is lowered, the polymer will eventually solidify into a solid at the melting temperature T_m (see Table 14.3 for typical values). The nature of this solid depends crucially on the rate of cooling, dT/dt. For very slow cooling rates, the polymer will crystallize, with the monomers occupying definite positions in a crystal structure. For rapid rates of cooling the polymer will form an amorphous rubbery material. This rubbery material is a viscoelastic solid (i.e., it has both elastic properties and the ability to flow). For intermediate cooling rates there may be a coexistence of the amorphous and crystalline phases as separate domains. The presence of the crystalline phase may be detected using x-ray diffraction, and the degree of crystallinity (the fractional amount of the crystal phase) may be ascertained by measuring the intensities of the Bragg peaks and normalizing them against a fully crystalline sample. If the crystalline and amorphous phases are compared, it is found that the crystal is denser. For example, the densities of amorphous and crystalline polycarbonate are 1196 and 1316 kg/m^3, respectively. For polyethylene terephthalate the corresponding values are 1335 and 1515 kg/m^3.

As the liquid is cooled, its volume shrinks, and the polymer molecules come into closer proximity to each other. The polymer molecules bind together with weak van der Waals bonds. There is not yet long-range order. Given enough time there will be thermal activation out of these temporary bonds and the system will seek its state of lowest free energy, the crystalline solid. This requires that the molecules remain fairly mobile during this period. However, at rapid cooling rates, the volume shrinks rapidly and the steric hindrance of other molecules frustrates the ability of the molecules to attain that state. Stated bluntly, high potential barriers are thrown up by the presence of other polymer molecules, which prevent thermalization from occurring. The system becomes trapped in a metastable state. Molecular motion is still possible, however, by diffusion.

There are two types of diffusive motion observed for polymers, depending on their degree of polymerization. For short molecules (typically, < 600 monomers) the molecule as a whole is able to diffuse. The distance the center of mass moves in a time t follows the Brownian motion formula $r \sim \sqrt{Dt}$, where D is the diffusion coefficient. The diffusion coefficient and the viscosity η are related by the Stokes–Einstein formula $D = k_B T / 6\pi\eta\xi$, where ξ is a characteristic length scale. The viscosity is found to decrease rapidly with increasing temperature (see Section W14.4), so D will increase rapidly with increasing T.

In addition, there are a host of internal diffusion modes within the molecule itself corresponding to thermal conformational changes. In these polymers it is found that the viscosity varies linearly with N, the number of monomers in the chain. For long polymers, however, the chains become highly entangled with each other, much like spaghetti in a bowl. The possible types of motion become severely restricted. Motions in which monomers move perpendicular to the chain are likely to be forbidden by steric constraints. There is, however, one motion that is relatively easy, and that is displacement along the chain itself. For such a motion to occur, each monomer moves to the position of its NN along the chain, with the leading monomer moving to a vacancy in front of it. This motion, similar to that of a snake, is called *reptation*. The instantaneous shape of a polymer is that of a random walk, so the typical size scales as $\sqrt{N}$. Since the one-dimensional motion of reptation is itself a random walk, the distance the center of mass is displaced in time t varies now as $t^{1/4}$. It is as if there were an inebriated engineer on a train randomly throttling the train backward and forward on a track that was itself laid by drunken track layers, so it meandered in a random-walk pattern. In this realm it is found experimentally that the viscosity varies as $N^{3.4}$. Thus the shorter chains are best characterized as being viscoelastic fluids, while the long chains as rubbery materials.

As the temperature is lowered further, and if $-dT/dt$ is large enough, there exists a temperature T_g at which the amorphous rubbery material has its viscosity diverge and becomes a solid (see Table 14.4). This is the glass-transition temperature. If T_g is above room temperature, one refers to the polymer as a plastic; if below, it is called an elastomer or rubber. Below T_g center-of-mass motion for the molecule is no longer possible. There is a discontinuity in the coefficient of volume expansion at T_g, with its value becoming smaller below T_g. Discontinuities are also observed in the various elastic moduli (where they may change by a factor of a thousandfold over a relatively small ΔT) and the heat capacities. The precise nature of the glass transition is not yet understood. The transition appears to be a thermodynamic phase transition. The clusters of immobile molecules that begin to form in the amorphous phase probably grow in size as T is lowered toward T_g and eventually percolate to form an "infinite" cluster at the glass-transition temperature. There is presumably some small temperature range near T_g over which all molecules join this cluster. The value of T_g is typically 0.5 to 0.7 times the value of T_m.

Below T_g additional phase transitions are often observed, and these are labeled by Greek letters α, β, ... This fraternity of transitions has to do with the freezing out of the motions of various polymer subgroups and side chains, each with its characteristic transition temperature.

TABLE 14.4   Glass-Transition Temperature, T_g, and Melting Temperature, T_m, for Various Common Polymers[a]

Polymer	T_g (°C)	T_m (°C)
Polyethylene (PE)	−21	137
Polypropylene (PP) (isotactic)	−1	176
Polypropylene (syndiotactic)	−8	176
Polycarbonate (PC)	149	220
Polybutadiene (*trans*-1,4)	−90	92
Polyvinyledine dichloride (PVDC)	−18	198
Polyvinyl fluoride (PVF)	−41	200
Polyacrylonitrile (PAN)	97	317
Nylon 6	50	220
Nylon 6, 6	50	255
Polyvinyl acetate (PVAC)	32	—
Polyethylene terephthalate (PET)	69	225
Polymethylmethacrylate (PMMA)	105	200
Polystyrene (PS)	100	240
Polyvinyl chloride (PVC)	81	273

[a]Conflicting values for many of these numbers appear in the literature.

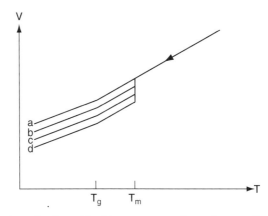

Figure 14.13. Variation of volume V with temperature for various cooling rates. The temperatures T_m and T_g are the melting and glass-transition temperatures, respectively.

The variation of the volume versus temperature is sketched in Fig. 14.13 for various cooling rates: curve d for slow cooling, curve a for rapid cooling, and curves b and c for intermediate rates. In reality the transitions are not as sharp as shown, because it is difficult to attain thermal equilibrium as a sample is being cooled at a finite rate. In curve a the sample remains amorphous below T_m down to T_g, where it solidifies. Curve d corresponds to a case with a high degree of crystallinity. Curves b and c are mixtures of crystallites and amorphous material. If one could achieve a state of perfect crystallinity, there would be no glass-transition temperature for that case and the curve would have a continuous slope below T_m.

The glass-transition temperature depends on both the structural and bonding properties of the polymer. For example, a flexible chain, in which the molecules may make all kinds of gyrations, generally has a low T_g, whereas a stiff one has a high T_g. Massive groups appended to the sides of the backbone impede motion and tend to elevate T_g. Ionic or polar groups, which are likely to form inter- or intramolecular bonds, also serve to elevate T_g.

The glass transition is a second-order phase transition, meaning that the Gibbs free energy $G(T, P)$ suffers a discontinuity in curvature at $T = T_g$. The first derivative of G remains continuous across the transition. There are discontinuities in the second derivatives of G, such as the isobaric specific heat, $C_p = -T(\partial^2 G/\partial T^2)_p$, the isothermal bulk modulus, $B = -V/(\partial^2 G/\partial P^2)_T$, and the volume coefficient of thermal expansion, $\beta = (1/V)(\partial^2 G/\partial P\partial T)$. For example, for PMMA, the values of (C_p, β, B) just below T_g are (1.73 kJ/kg·K, 2.6×10^{-4} K^{-1}, $\approx 2.8 \times 10^9$ Pa), while the values just above T_g are (1.95, 5.8×10^{-4}, $\approx 2.2 \times 10^9$).

APPLICATIONS

Polymers are used for a large number of applications and it is impossible to list all of them. In the following sections representative case studies are presented that illustrate the versatility of polymers. The applications make use of the structural, mechanical, electrical, and optical properties of polymers.

In the present section structural plastics, polymeric ionic conductors, photoresists, piezoelectric polymers, and liquid crystals are discussed. Polymeric foams, porous films, electrically conducting polymers, and nonlinear optical polymers are discussed in Sections W14.5 to W14.8.

14.8   Structural Plastics

Polymers are often used as structural materials for automotive parts, home appliances, telephones, and a myriad of applications from everyday life. In principle, a polymer is strengthened mechanically if it has many cross-links. However, there is a disadvantage in that such a material becomes highly viscous, the flow rate diminishes, and it is difficult to process. A central problem is to try to find the "golden mean" between the two extremes.

The thermoplastic bisphenol-A polycarbonate (PC) (see Fig. 14.2) is commonly used as a structural plastic. The material is tough and fire-resistant. When polycarbonate is combined with acrylonitrile–butadiene–styrene (ABS) its impact strength is improved considerably. The rubbery ABS microparticles serve to cushion the blows that the composite material may suffer.

The viscosity of a polymer grows dramatically as the chain length grows. It is therefore advantageous to limit the growth. Additives that stop the growth of the chains in the polymerization process are used to accomplish this. In practice phenol groups (C_6H_5OH) serve as these "end caps," which terminate the polymerization. The polymerization process itself, and how it is terminated, is considered in Chapter 21.

It is found that polycarbonates can withstand severe impacts. The reason for this is believed to be that there is a microscopic shock absorber built into the polymer. In the molding process the molecule is often prevented from reaching its lowest energy state by the existence of potential barriers. The presence of side groups attached to the polymer chain inhibit crystal formation and also provide cushion space. Thus segments of the molecule may lie in conformations which are in metastable rather than in stable equilibrium states. These potential barriers are set up largely by the steric hindrance presented by other neighboring molecules or even by other portions of the given molecule itself. These metastable states are called *high-energy conformers*. When an impulse is applied to the polymer the barriers may be lowered temporarily and the molecule can assume a new conformation. The new state is called a *low-energy conformer*, although it may also sometimes be pushed to a higher energy state. This rearrangement involves a motion of the segments and this takes time. From elementary mechanics, impulse is the average force multiplied by the duration. By lengthening the duration of impact the average force is diminished and the blow is therefore cushioned.

Another class of structural plastics are the polyimides. These materials are tough and have excellent high-temperature performance characteristics. They may be used as thrust reversers on jets and even in some jet engine parts themselves. The key to their success is the very high glass-transition temperature, T_g, approaching 500°C in some cases. Polyimides are available as both thermosets and thermoplastics. The thermosets have high T_g values; the thermoplastics have lower values but still up to 220°C. Examples of thermosets are phenol resins, urea resins, and epoxies. Examples of thermoplastics are Kapton [$((C_2NO_2)(C_6H_2)(C_2NO_2)(C_6H_4)O(C_6H_4))_n$] and Apical. The thermoplastics are easier to process and are tougher. They may be hardened by exposure to ion beams, a process that presumably promotes cross-linking. Polyimides may be alloyed with polycarbonates to produce tough flame-resistant materials that may be used as fire helmets.

14.9 Polymeric Ionic Conductors

Polymers may act as hosts for ions in much the same way that liquid solvents do. In liquids the ions reside in a cavity formed by the liquid molecules. The ions polarize the liquid and the resulting electrostatic energy is called the *solvation energy*. Ions in polymers can polarize the atoms of the polymers and there will be a corresponding solvation energy. Suppose that a salt is introduced into a polymer. When the solvation energy of the ions in the polymer host exceeds the ionization energy of the salt, the individual ions separate and attach themselves to Lewis base sites on the polymer. Conductivity is possible above the glass-transition temperature, where the polymer molecules are free to move. The free volume also provides room for the ions to move. The ionic conductivity possesses a diffusive liquidlike behavior in addition to the hopping-type behavior characteristic of an electron in a disordered solid. The conductivity is typically somewhat lower than that of a doped semiconductor. For example, polyethylene oxide may form a complex with lithium chlorate salts and this material has a conductivity of $5.6 \times 10^{-4} (\Omega \cdot m)^{-1}$. Higher conductivities are achievable by using "comb" polymers, such as polysiloxane $(SiOR_2)_n$. In these polymers there is a SiO backbone and two floppy side chains, R, extending outward from each Si atom, that can readily transport charge.

One way of thinking about how the conductivity is brought about relics on the dynamic-percolation model. In the static-percolation model one envisages sites on a lattice interconnected by a set of randomly placed conductors, placed with probability p along any bond between NNs. If not enough conductors are present to form an

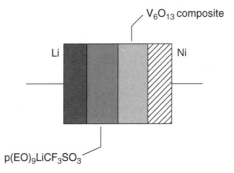

Figure 14.14. Polymer electrolyte battery based on the polyethylene oxide monomer (EO=CH$_2$CH$_2$O). [Reprinted with permission from D. F. Shriver and G. C. Farrington, *Chem. Eng. News*, **63**, 42 (1985). Copyright 1985 by the American Chemical Society.]

infinite connected network, no conductivity occurs. Beyond a critical concentration of resistors, called the *percolation threshold*, the system conducts. In the dynamic percolation model one adopts the same model but allows the distribution of conductors to be changed continuously, with a characteristic renewal time τ_r. Now conductivity is always possible, and one just has to wait long enough. Typically, a time τ_r/p is needed to make a step of size a, the interatomic distance, and a random walk ensues. Thus the mean-square displacement of a charge in time t is $a(pt/\tau_r)^{1/2}$. In the presence of an applied field there is a preference for taking steps along the direction of the electrical field and conductivity results. Now the motion of the charges is diffusive for any concentration of conducting links.

Using polymer electrolytes it is possible to design a solid-state battery. A schematic of such a 3-V battery is given in Fig. 14.14. Lithium ions from the left electrode are able to diffuse through the polymer electrolyte and deposit themselves in the composite electrode. Such batteries may be made paper-thin with good mechanical properties.

14.10 Photoresists

Microlithography is of prime importance in the fabrication of semiconducting integrated circuits. The photochemistry of polymers plays a crucial role in this process. A polymer layer is placed on a substrate material, such as Si, and the image of a circuit section illuminates the polymer. The illuminated portions of the polymer layer have chemical properties that are altered and so may subsequently be removed, leaving behind the desired image. This type of polymer is known as a *positive photoresist*. Subsequent deposition of oxides, metals, additional semiconductor, additional polymer, and so on, is used to build up the three-dimensional circuit. To illustrate how a polymer may be altered by shining light on it, consider the Norrish process for the photodissociation of polymethylmethacrylate. The process is illustrated in Fig. 14.15.

The upper left-hand portion of the figure shows two adjacent monomers being "attacked" by an ultraviolet photon. A C–C bond is broken, leaving behind two radicals, one on the backbone and the other cut off. The latter undergoes a unimolecular reaction to form CO$_2$ and the CH$_3$ radical, or alternatively, CO and the CH$_3$O radical. The backbone proceeds to fission into two products, as shown in the lower left side of the figure. The net result is that the polymer has been reduced in length. Subsequent

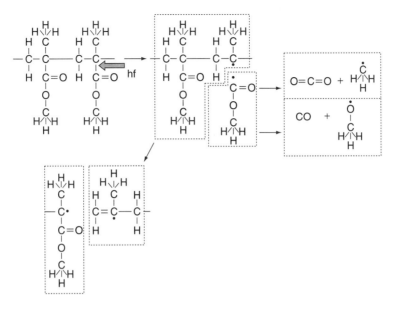

Figure 14.15. Photodissociation of PMMA.

photons can act to shorten the illuminated chains even further. Since short chains are more mobile than long ones, they may be removed selectively. The use of PMMA modified by certain copolymers or modified with particular side chains enhances the cross section for photodissociation.

14.11 Piezoelectric Polymers

Piezoelectricity, the interrelationship between stress and electric fields, is studied in detail in Section 15.6. It is introduced here in the context of polymers. When an electric field is established in a piezoelectric material a stress is produced, and vice versa. The piezoelectric stress tensor g_{ijk} relates the two fields:

$$\sigma_{ij} = \sum_{k=1}^{3} g_{ijk} E_k. \tag{14.43}$$

Alternatively, the strain tensor may be expressed in terms of the electric field:

$$\varepsilon_{ij} = \sum_{k=1}^{3} d_{ijk} E_k, \tag{14.44}$$

with d_{ijk} being the piezoelectric strain tensor.

Certain polymers have proven to be effective piezoelectric materials and the focus here is on one of them, polyvinylidene fluoride (PVDF). This polymer is sensitive to weak mechanical stress, has a wide frequency-response range, a high dielectric strength, and is flexible. There are two phases, α and β (Fig. 14.16). The former is not piezoelectric and so not of interest here. The β phase, however, is piezoelectric. When cooled

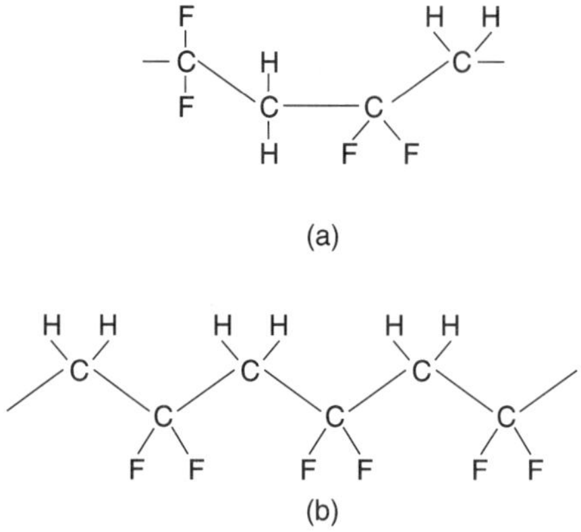

(a)

(b)

Figure 14.16. (*a*) Monomer of the α phase; (*b*) three monomers of the β phase of PVDF.

from the melt the α phase is produced, but when extruded and stretched it is converted from α to β. Alternatively, the material may be poled in an external electric field at a temperature above the glass-transition temperature T_g to align the microscopic dipoles, and then cooled. Fluorine has a high affinity for electrons, so negatively charged fluoride ions are produced, leaving the protons on the other side of the molecule. There is a net dipole moment per unit length across the molecule, so the material forms an electret (i.e., a material with a permanent electric polarization vector in the absence of an applied electric field). The strong potential difference across a sample is obscured, however, by adsorbed layers of charge on the opposing faces. When exposed to heat these charges desorb and the potential difference is uncovered. Thus the material is also a pyroelectric. The charge per unit area divided by the temperature increment is called the *pyroelectric constant*, ρ. The pyroelectric nature makes the material suitable as an infrared sensor. As the material absorbs infrared radiation it warms up and this produces a potential difference across it. In addition, thermal stresses can produce corresponding electric fields. Some properties of β-PVDF are given in Table 14.5.

Typically, films of thickness 25 µm (≈ 1 mil) are manufactured that are stable up to 70°C. The dynamic sensitivity of the film to stress ranges from 10^{-8} to 10^6 N/m^2. The

TABLE 14.5 Material Properties of β-PVDF at $T = 300$ K^a

d_{xyz}	23×10^{-12} m/V
d_{zzz}	-33×10^{-12} m/V
g_{xyz}	0.216 N/V·m
g_{zzz}	-0.339 N/V·m
ρ (pyroelectric constant)	-25×10^6 C/m^2·K
E_{max} (dielectric strength)	8×10^7 V/m
ε_r (dielectric constant)	12
E (Young's modulus)	2 GPa
Resistivity	10^{13} Ω·m
Density	1780 kg/m^3
Color	Transparent

^{a} The electric field is parallel to the backbone direction, z.

frequency response is flat out to 10 MHz. Metallic layers are deposited on opposing faces, or sometimes, sandwich structures with two polymer films between three metallic layers are fabricated (bimorphs).

By combining liquid crystals with polymers it is possible to create easily deformable elastomers with interesting optical and piezoelectric properties. The liquid-crystal molecules are rodlike, with their axes defined by their directors $\{\hat{n}\}$. In the unstressed state the directors are spatially unoriented. The material is random on a length scale comparable to the wavelength of light and appears translucent. When stretched, however, the directors align themselves along an axis. They no longer scatter light in directions other than the forward direction. The material becomes transparent. If cholesteric liquid crystals are employed, they have a chirality (i.e., they are either left- or right-handed). The cholesteric crystals lack inversion symmetry and therefore are piezoelectric. Thus in addition to becoming transparent, a potential difference can develop across the liquid crystal. An example of such a material is polysiloxane (80%) swelled with chiral 4'-(2-methylbutyl)biphenyl-4-carbonitrile (20%) (CB15).

These materials possess a further optical peculiarity. If light is incident along the uniaxial-stress direction, there exists a wavelength where total reflection occurs due to a form of Bragg reflection. This cholesteric crystal has a pitch p, which is analogous to the pitch of a spring. The crystal presents itself as a one-dimensional periodic structure with periodicity p. A ray reflected from a depth Np has an optical path length $2nNp$, where n is the refractive index. When $2np = \lambda$ there is a strong constructive interference, and hence a strong reflection.

14.12 Liquid Crystals

Liquid crystals have properties intermediate between those of liquids and crystalline solids. They are based on *macromolecules*, which are sometimes polymeric in nature but, more commonly, are not. In a true crystal there is complete order of the unit cells in a lattice. This constrains the translational positions of the centers of mass as well as the orientations of the molecules. In the liquid crystal at least one of the translational degrees of freedom melts. Viscosities in the range 10 to 100 centipoise (cP) are common, where 1 poise = 0.1 Pa·s.

There are three basic families of liquid crystals: smectic, nematic, and cholesteric. The smectic family is further broken down into subfamilies: smectic-A, smectic-B,. . . (eight smectic phases, in all). Examples of some of these phases are presented in Fig. 14.17.

The *smectic phases* have translational disorder and orientational order. In the smectic-A phase the liquid crystal becomes stratified into parallel two-dimensional fluids that can flow perpendicular to the c axis. There is a fixed interlayer spacing along the c axis but no correlation of the positions of molecules along the a and b directions in adjacent layers. The molecules are aligned parallel to the c axis. In the smectic-C phase the same stratification exists, but the molecules are aligned along some axis other than the c axis.

In the *nematic phase* there is flow possible in all directions, but the molecules retain their mean orientation in space, although the orientation of any given molecule may fluctuate away from the mean direction.

The *cholesteric phase* has a stratified structure with the molecules in any layer aligned parallel to each other. Flow is possible in the layers. However, the direction of alignment of the molecules varies from layer to layer in a helicoidal manner.

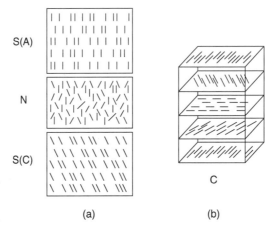

S(A)

N

S(C)

(a)

C

(b)

Figure 14.17. (*a*) Front views of smectic-A, nematic (N), and smectic-C phases; (*b*) skewed view of the cholesteric (C) phase.

One may describe the orientation of a molecule by a unit vector directed along its long axis. This is called the *director*, $\hat{n}$. (In the special case where the molecule has inversion symmetry, it is more properly specified by a dyadic $\hat{n}\hat{n}$, but this fine point will be ignored.) One may regard this director as a hydrodynamic variable and write it as $\hat{n}(\mathbf{r}, t)$. In principle the laws of hydrodynamics need to be supplemented by an additional equation describing the behavior of the director field. For example, one may assign a free energy per unit volume to this field of the form given by Frank and Oseen:

$$f(\mathbf{r}) = \frac{K_1}{2}[\nabla \cdot \hat{n}(\mathbf{r})]^2 + \frac{K_2}{2}\left[\hat{n}(\mathbf{r}) \cdot \nabla \times \hat{n}(\mathbf{r}) + \frac{2\pi}{p}\right]^2 + \frac{K_3}{2}[\hat{n}(\mathbf{r}) \times (\nabla \times \hat{n}(\mathbf{r}))]^2,$$
(14.45)

where K_1, K_2, and K_3 are elastic constants, describing splay, twist, and bend, respectively, and p describes the pitch in a cholesteric liquid crystal. Splay, twist, and bend are depicted in Fig. 14.18.

Liquid crystals are somewhat analogous to magnets. The directors replace the electron spins as the order parameter. One may therefore study phase transitions between the various phases, much as in magnetism. For example, the nematic-to-isotropic phase transition is very similar to the magnetic phase transition at the Curie temperature, where the (domain) magnetization vanishes. In addition, there are dynamical analogs of such objects as spin-wave excitations in magnets. In the liquid crystal, there are low-lying excitations consisting of a wave in the director direction propagating through the crystal, much as the stalks of wheat in a field would wave in a gentle breeze.

The reason that liquid crystals are of such great interest is that they form natural elements for use in displays. This is based on two properties. First, their optical response is very sensitive to the degree of ordering of the director field. The molecules are rodlike and their polarizability is highly anisotropic. Hence there is strong optical birefringence when the molecules are aligned. Typically, the difference in index of refraction between light polarized parallel to the direction of alignment of the molecules compared with perpendicular to the alignment is $\Delta n = n_\parallel - n_\perp = 0.03$ to 0.30. In practice, a glass polarizer and glass analyzer are oriented with their polarization axes perpendicular to

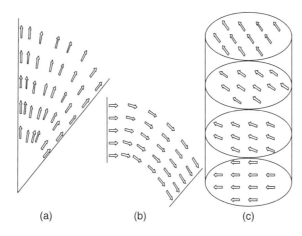

Figure 14.18. Illustration of (a) splay, (b) bend, and (c) twist.

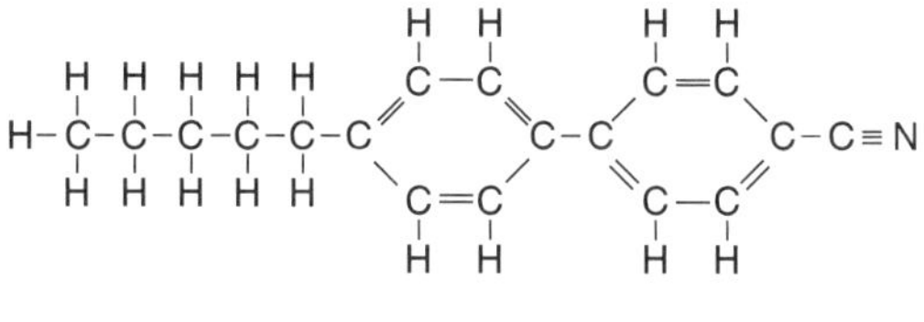

Figure 14.19. Molecule 5CB.

each other. Between them is placed a nematic liquid crystal with its directors parallel to the glass sheets and perpendicular to the direction of propagation of the light. The liquid crystal is gradually twisted so that light which enters the polarizer has its direction of polarization twisted as it propagates through the liquid crystal and is able to pass through the analyzer. Liquid crystals for displays also use dipolar molecules which may be aligned by applying an electric field. The dielectric constant of an aligned sample is highly anisotropic, with $\Delta\varepsilon = \varepsilon_\parallel - \varepsilon_\perp$ in the range -10 to 50. If the molecules are aligned, the polarization vector of light passing through the liquid crystal will not be twisted, and the light will be blocked by the analyzer. Thus one can selectively control the transmission of light through portions of the liquid crystal. Furthermore, the amount of power expended in doing this is very small, typically in the range of 0.01 W/m^2.

An example of a typical liquid crystal used for such purposes is based on the biphenyl 5CB molecule (Fig. 14.19). It has a long, rodlike C_5H_{11} appendage at one end, followed by two benzene rings, followed by the polar CN group. The liquid crystal is in the nematic phase between 22.5 and 35°C. The rodlike character of the molecule is responsible for its propensity to align along a given direction. The delocalized π electrons associated with the benzene rings provide it with a large anisotropy in the polarizability of the molecule. Thus if an ac electric field from a light wave is directed along the plane of the molecule, a large displacement of the electrons may be obtained. If the field were perpendicular to the plane of the molecule, this polarization would be much smaller. This means that there will be a large anisotropy of the index of refraction of the nematic crystal. Finally, the polar group at the end of the molecule couples readily to an applied dc electric field. Recall that the torque is given by $\tau = \mu \times \mathbf{E}$, where μ is the electric dipole moment. It has both permanent- and induced-dipole contributions. Furthermore, the viscosity of the liquid crystal is sufficiently low that it has a rapid response time. If the end chain were made much longer, this would increase the viscosity and slow down the response.

Typical shifts in the index of refraction attainable by potential differences of about a volt across a thin slab of liquid crystal are $\Delta n = 0.1$. Consider the effect on a thin-film interference experiment. The optical phase shift for the round-trip path through the slab at normal incidence is

$$\phi = 2kd + \phi' = 2\omega \frac{n}{c} d + \phi', \qquad (14.46)$$

where k is the propagation vector and ϕ' is the difference in the reflection phase shifts at the two surfaces. A change in phase $\Delta\phi = \pi$ would convert constructive interference into destructive interference, and vice versa, and would thus be a maximal effect. For 500-nm light a slab of thickness 125 nm would do the job.

To create practical displays, the microscopic domains of oriented directors must be aligned parallel to each other. Microscopic electric fields are then imposed by means of a polarizing grid. The alignment may be accomplished, for example, by creating a network of parallel grooves on the electrodes. A number of these technological tricks rely on various physical properties of the liquid crystals themselves.

Ferroelectric liquid crystals (FLCs) are cholesteric crystals in which there is a net electric-dipole moment associated with the molecule. Under normal conditions these dipoles point in spiraling directions when one travels from layer to layer. However, in the presence of an external electric field directed along the layer, the dipoles align and produce a net electric polarization. In addition, the index of refraction ellipsoid changes size and shape. Of crucial importance is the fact that this can be accomplished with little energy expenditure and very rapidly. Thus FLCs have found a wide variety of applications, including fast optical shutters and modulators, and active-matrix liquid-crystal displays (AMLCDs).

REFERENCES

Billmeyer, F. W., *Textbook of Polymer Science*, Wiley, New York, 1966.

Brandrup, J., and E. H. Immergut, eds., *Polymer Handbook*, 3rd ed., Wiley, New York, 1989.

Committee on Polymer Science and Engineering, National Research Council, *Polymer Science and Engineering: The Shifting Research Frontiers*, National Academy Press, Washington, D.C., 1994.

Cowie, J. M. G., *Polymers: Chemistry and Physics of Modern Materials*, 2nd ed., New York, Chapman and Hall, 1991.

de Gennes, P. G., *Scaling Concepts in Polymer Physics*, 2nd ed., Cornell University Press, Ithaca, N.Y., 1984.

de Gennes, P. G., *Introduction to Polymer Physics*, Cambridge University Press, Cambridge, 1990.

Doi, M., and S. F. Edwards, *The Theory of Polymer Dynamics*, Clarendon Press, Oxford, 1986.

Elias, H.-G., *Macromolecules*, Vols 1 and 2, Plenum Press, New York, 1984.

Flory, P. J., *Principles of Polymer Chemistry*, Cornell University Press, Ithaca, N.Y., 1953.

Flory, P. J., *The Statistical Mechanics of Chain Molecules*, Wiley-Interscience, New York, 1969.

Fried, J. R., *Polymer Science and Technology*, Prentice Hall, Upper Saddle River, N.J., 1995.

Martin, D. C., and C. Viney, eds., *Defects in polymers*, Mater. Res. Soc. Bull., Sept. 1995, and articles therein.

Nielsen, L. E., and R. F. Landel, *Mechanical Properties of Polymers and Composites*, 2nd ed., Marcel Dekker, New York, 1994.

Reiser, A., *Photoreactive Polymers*, Wiley, New York, 1989.

Tobolsky, A. V., *Properties and Structure of Polymers*, Wiley, New York, 1960.

Young, R.J., and P. A. Lovell, *Introduction to Polymers*, 2nd ed., Chapman & Hall, London, 1991.

PROBLEMS

14.1 A polymer whose viscoelastic properties are described by Eq. (14.40) (i.e., the Maxwell model) is subjected to a time-dependent stress $\sigma = \sigma_0 \exp(-i\omega t)$. Find the steady-state strain. Compare this result to that of a polymer that obeys the Voigt model, given by Eq. (14.37).

14.2 Consider an elastomer consisting of monomers that are optically anisotropic [i.e., they have a polarizability $\alpha_{\parallel}(\omega)$ for light parallel to the chain axis and $\alpha_{\perp}(\omega)$ for light polarized perpendicular to the chain axis]. Assume that there are N chains per unit volume. Let $\bar{n}(\omega)$ be the mean index of refraction of the material. The elastomer is stretched with a stretching parameter s, as defined in Section 14.5. Show that the elastomer will have a birefringence given by

$$\delta n(\omega) = n_{\parallel}(\omega) - n_{\perp}(\omega) = \frac{2\pi N}{45} \frac{[\bar{n}(\omega) + 2]^2}{\bar{n}(\omega)} \left(s^2 - \frac{1}{s}\right) [\alpha_{\parallel}(\omega) - \alpha_{\perp}(\omega)].$$

Obtain an expression for the stress optical coefficient, $C \equiv \delta n(\omega)/\sigma$, where σ is the applied stress.

14.3 Consider a polymer in which stress relaxation is occurring, as shown schematically in Fig. 14.12. The free-energy barrier for a certain rate-determining process involved in the relaxation of stress is $\Delta F = 2$ eV.

    (a) If the average time Δt_1 for passing over this barrier at $T_1 = 1000°C$ is 10^{-9} s, find the corresponding Δt_2 at $T_2 = 900°C$ and $T_2 = 100°C$.

    (b) Repeat part (a) when $\Delta F = 0.2$ eV.

14.4 Consider a single molecule of the polymer polyethylene, $(C_2H_4)_n$, with $n = 10^4$ monomer units and $a = 0.255$ nm as the length of one monomer unit (see Fig. 14.5).

    (a) Compare the length of the polymer molecule, $\approx Na$, with its root-mean-square size $\langle R^2 \rangle^{1/2}$.

    (b) Calculate the average mass density of this polymer molecule using any assumptions concerning its volume that you feel are reasonable (and can justify!).

    (c) Compare your result in part (b) with the mass density of crystalline polyethylene given in Table 14.2. Explain any difference.

Note: Additional problems are given in Chapter W14.

CHAPTER 15

Dielectric and Ferroelectric Materials

15.1 Introduction

In this chapter we are concerned with dielectric phenomena and materials. The term *dielectric* is used in a broad sense, as relating to various electrical phenomena, primarily in ceramics and polymers. Energy storage is involved, but often energy dissipation and conduction also occur.

The chapter begins by introducing the simple classical-mechanical model for a dielectric proposed by Lorentz. A distinction is made between the field applied to a sample, the average electric field within the sample, and the local electric field that the atoms or ions experience. The Clausius–Mossotti formula is derived. The dielectric properties of ionic crystals are then discussed. This is followed by a discussion of what happens at high electric fields when dielectric breakdown occurs.

A phenomenological discussion of the ferroelectric phase transition is given in terms of the Landau theory of phase transitions. This is followed by a discussion of ferroelectricity and piezoelectricity and their relation to the symmetry of the point groups of the crystal structure. Thermistors with both positive and negative temperature coefficients are then considered. This is followed by a study of varistors, which are materials with voltage-dependent resistivity. The chapter concludes with a discussion of superionic conductors, with particular emphasis on the β-aluminas.

Additional material relating to capacitors (Section W15.1), substrate materials (Section W15.2), nonvolatile random-access memories (Section W15.4), the quartz crystal oscillator (Section W15.5), the lithium battery (Section W15.6), and fuel cells (Section W15.7) appears at our Web site.[†] It also contains some of the more technical points relating to the theory of first-order ferrolectric phase transitions (Section W15.3).

15.2 Lorentz Oscillator Model for the Dielectric Function

The Lorentz model was introduced in Section 8.6. In this model matter is composed of atoms that are represented by classical harmonic oscillators of mass m and charge q, each connected to a spring of stiffness constant $m\omega_0^2$ and subject to a damping rate γ. Although the model is not quantum mechanical, it contains many of the features of the quantum system and is simple to work with. In this section the model is extended by including the effects of the other atoms on the response of a given atom. The

---

[†] Supplementary material for this textbook is included on the Web at the resource site (ftp://ftp.wiley.com/public/sci_tech_med/materials). Cross-references to elements of the Web material are prefixed by "W."

displacement of an oscillator, $\mathbf{r}$, obeys the equation

$$m\left(\frac{d^2\mathbf{r}}{dt^2} + \gamma\frac{d\mathbf{r}}{dt}\right) = -m\omega_0^2\mathbf{r} + q\mathbf{E}_{\text{loc}}, \tag{15.1}$$

where $\mathbf{E}_{\text{loc}}$ is the local electric field at the site of the atom.

There are three electric fields in the problem, and the distinction between them must be clearly understood. One imagines that there is a sample of material placed in a uniform external electric field, $\mathbf{E}_0$, which may be produced, for example, by a parallel-plate capacitor. There is a macroscopic electric field $\mathbf{E}(\mathbf{r})$ inside the sample. The field $\mathbf{E}(\mathbf{r})$ is determined by taking an average of the microscopic electric field over a domain containing a large number of unit cells but whose size is still small compared with the size of the sample. Usually, this electric field varies with location. In some special cases, however, $\mathbf{E}(\mathbf{r})$ will be constant throughout. This will occur in cases where the sample is ellipsoidal in shape, including the special limits of spheroidal, spherical, flat, or needlelike. Three special cases are examined:

1. *Needlelike sample aligned parallel to* $\mathbf{E}_0$. The net electric dipole developed on a needlelike sample will be small, and so will be the field generated by the dipole. The field just outside the needle at its equatorial plane will be approximately equal to $\mathbf{E}_0$. Since the tangential component of the electric field is continuous across the interface and $\mathbf{E}_0$ is tangential to the interface, it follows that the macroscopic field inside the needle is given by $\mathbf{E} = \mathbf{E}_0$.

2. *Semi-infinite sample with its surface oriented perpendicular to* $\mathbf{E}_0$. Continuity of the normal component of the electric displacement vector gives $\mathbf{D} = \epsilon_0\mathbf{E}_0 = \epsilon\mathbf{E} = \epsilon_r\epsilon_0\mathbf{E}$, where ϵ is the permittivity and ϵ_r is the dielectric constant of the sample. Thus $\mathbf{E} = \mathbf{E}_0/\epsilon_r$. The field inside is simply diminished by the layer of polarization charge induced on the interface.

3. *Spherical sample of radius* a. The electrostatic potential outside the sphere may be expressed as a sum of the potential due to the imposed field (assumed to be directed along z) plus a contribution from the electric dipole induced on the sphere, μ:

$$\Phi(\mathbf{r}) = \begin{cases} -E_0 z + \dfrac{\mu z}{4\pi\epsilon_0 r^3} & \text{if } r > a \\ -LE_0 z & \text{if } r < a. \end{cases} \tag{15.2}$$

Here L is called the *depolarization ratio*. Continuity of Φ and the radial component of $\mathbf{D}$ at $r = a$ leads to

$$-E_0 + \frac{\mu}{4\pi\epsilon_0 a^3} = -LE_0, \tag{15.3a}$$

$$-\epsilon_0 E_0 - 2\frac{\mu}{4\pi\epsilon_0 a^3} = -\epsilon LE_0, \tag{15.3b}$$

from which it follows that

$$L = \frac{3}{2 + \epsilon_r}. \tag{15.4}$$

Since the static-field dielectric constant is larger than 1, the depolarization factor is less than 1 and the field is partially shielded by induced polarization charges on the surface of the sphere.

The atoms sit at special sites in the material and not at random sites. The field at the atomic site, $\mathbf{E}_{\text{loc}}$, is therefore not equal to the average field $\mathbf{E}$. Lorentz obtained the relation between the local and average electric fields by dividing the material into two zones centered about a given atomic site. The far zone contains all those atoms outside a sphere of radius b centered about the atomic site. The near zone consists of all the atoms for which $r < b$. Thus

$$\mathbf{E}_{\text{loc}} = \mathbf{E} + \mathbf{E}_{\text{near}} + \mathbf{E}_{\text{far}}. \tag{15.5}$$

The contribution of the field from the far zone, $\mathbf{E}_{\text{far}}$, is obtained by treating it as a macroscopic material with a uniform polarization field: $P_z(\mathbf{r}) = P_0 \Theta(r - b)$. From Gauss's law, $\nabla \cdot \mathbf{D} = 0$ and $\mathbf{D} = \epsilon_0 \mathbf{E} + \mathbf{P}$, it follows that $\nabla \cdot \mathbf{E} = \rho/\epsilon_0$, where ρ is the polarization charge density, given by

$$\rho = -\nabla \cdot \mathbf{P} = -P_0 \frac{\partial}{\partial z} \Theta(r - b) = -P_0 \cos \theta \delta(r - b). \tag{15.6}$$

The presence of the Dirac delta function implies that the polarization charge appears as a charged layer on the spherical surface. It produces a field at the site located at $r = 0$, given by

$$E_{\text{far},z} = -\int d\mathbf{r}' P_0 \cos \theta' \delta(r' - b) \frac{-z'}{4\pi\epsilon_0 r'^3} = \frac{P_0}{4\pi\epsilon_0} \int d\hat{r}' \cos^2 \theta' = \frac{P_0}{3\epsilon_0}. \tag{15.7}$$

The contribution from the near-field atoms was taken by Lorentz as a discrete sum of dipolar fields arising from the atoms in the region $r < b$. By homogeneity, one expects the dipoles on these sites to be equal to each other. The total contribution is therefore

$$E_{\text{near},z} = \sum_n \frac{3\mu z_n^2 - \mu r_n^2}{4\pi\epsilon_0 r_n^5} \Theta(b - r_n), \tag{15.8}$$

where μ is the electric dipole moment per atom. In two cases this sum vanishes. If all atoms sit at sites of cubic symmetry, then $\langle x_n^2 \rangle = \langle y_n^2 \rangle = \langle z_n^2 \rangle = \langle r_n^2 \rangle/3$, and $\mathbf{E}_{\text{near}} = 0$. Similarly, for an amorphous material, the same relations on average can be expected to hold, so again $\mathbf{E}_{\text{near}} = 0$. In the more general case the sum has to be computed and there will be a residual near-field contribution. The final expression for the Lorentz local field for the cubic crystal or amorphous solid is

$$\mathbf{E}_{\text{loc}} = \mathbf{E} + \frac{\mathbf{P}}{3\epsilon_0}. \tag{15.9}$$

Returning to the Lorentz oscillator, assume that it is subjected to a harmonically varying electric field $E \propto \exp(-i\omega t)$. Then the steady-state oscillator amplitude is given by

$$\mathbf{r} = \frac{q}{m} \frac{1}{\omega_0^2 - \omega^2 - i\omega\gamma} \mathbf{E}_{\text{loc}}. \tag{15.10}$$

The instantaneous amplitude of the dipole, $\mu = q\mathbf{r}$, may be written as

$$\mu = \epsilon_0 \alpha(\omega) \mathbf{E}_{\text{loc}}, \qquad (15.11)$$

where the dynamic polarizability is defined by

$$\alpha(\omega) = \alpha_0 \frac{\omega_0^2}{\omega_0^2 - \omega^2 - i\omega\gamma}, \qquad (15.12)$$

and $\alpha_0 = q^2/m\omega_0^2$ may be interpreted as the dc polarizability of the atom. The dynamic polarizability displays a resonance peak at $\omega = \omega_0$. If there are n atoms per unit volume, the polarization is $\mathbf{P} = n\mu = \epsilon_0 \chi(\omega)\mathbf{E}$, where $\chi(\omega)$ is the electric susceptibility. Using the relation $\epsilon(\omega) = \epsilon_0[1 + \chi(\omega)]$, one obtains the Clausius–Mossotti formula

$$\frac{\epsilon_r(\omega) - 1}{\epsilon_r(\omega) + 2} = \frac{n\alpha(\omega)}{3}. \qquad (15.13)$$

If the resonance is sharp (i.e. $\gamma \ll \omega_0$), it is sometimes useful to rewrite the dynamic polarizability in the approximate form

$$\alpha(\omega) = \frac{\alpha_0 \omega_0}{2} \frac{1}{\omega_0 - \omega - i(\gamma/2)} \qquad (15.14)$$

Thus γ is the full-width at half maximum of the function $|\alpha(\omega)|^2$.

In cases where delocalized electronic states are involved, the electron is spread out over several unit cells and the concept of local field loses it meaning. The mean field that the electron experiences is just $\mathbf{E}$, the average field in the solid. In that case

$$\epsilon_r(\omega) = 1 + \frac{ne^2}{m\epsilon_0} \frac{1}{\omega_0^2 - \omega^2 - i\omega\gamma}. \qquad (15.15)$$

This formula will be applied to two limiting cases: free-electron metals and semiconductors.

In the case of a free-electron metal there are no restoring forces and $\omega_0 \to 0$. The parameter γ is the collision rate with phonons and impurities. The dielectric function becomes [see Eq. (8.10)]

$$\epsilon_r(\omega) = 1 - \frac{\omega_p^2}{\omega(\omega + i\gamma)}, \qquad (15.16)$$

where ω_p is the plasma frequency defined by Eq. (8.11). The significance of ω_p for optical reflection is discussed in Section 8.4.

In the case of a semiconductor, one makes the approximate replacement $\omega_0 \to E_g/\hbar$, where E_g is the gap energy, and neglects the collision rate. At zero frequency the result is

$$\epsilon_r(0) = 1 + \left(\frac{\hbar\omega_p}{E_g}\right)^2, \qquad (15.17)$$

which was obtained previously in Eq. (8.32).

15.3 Dielectric Properties of Ionic Crystals

In an ionic crystal there are two sources of polarization. One is due to the electronic polarization of the ions. The other is due to the displacement of the ions themselves from their equilibrium positions. In Section 8.4 a study was made of ionic crystals in which each unit cell contained a two-ion basis. In this section the Lorentz model is extended to ionic crystals and local-field effects are included. For simplicity's sake, attention is restricted to a diatomic crystal and the optical phonon modes are studied in the limit of weak dispersion. This is equivalent to neglecting the forces between atoms of neighboring unit cells.

The classical equation of motion is of the same form as Eq. (15.1), that is, a driven harmonic oscillator with a viscous damping term

$$M\left(\frac{d^2x}{dt^2} + \Gamma\frac{dx}{dt} + \Omega_0^2 x\right) = QE_{\text{loc}}, \tag{15.18}$$

where x is the displacement of the positive ion relative to the negative ion, M the reduced mass of the ion pair $[M = M_1M_2/(M_1 + M_2)]$, Q the ionic charge, Ω_0 the optic phonon frequency, and Γ the damping constant for the phonon mode. Assuming a harmonic driving field, the ionic electric-dipole moment is

$$\mu_r = \frac{Q^2}{M}\frac{1}{\Omega_0^2 - \omega^2 - i\omega\Gamma}E_{\text{loc}}. \tag{15.19}$$

The electronic dipole moment is $\mu_e = \epsilon_0\alpha_e E_{\text{loc}}$. Assume that the electronic and ionic contributions are simply additive. The polarization is then given by $P = n(\mu_e + \mu_r) = (\epsilon - \epsilon_0)E$. Using the Lorentz formula for the local field, one arrives at the formula

$$\frac{\epsilon - \epsilon_0}{\epsilon + 2\epsilon_0} = \frac{n}{3}\left(\alpha_e + \frac{Q^2}{M\epsilon_0}\frac{1}{\Omega_0^2 - \omega^2 - i\omega\Gamma}\right). \tag{15.20}$$

It is convenient to express this in terms of a high-frequency permittivity, $\epsilon(\infty)$, and a low-frequency permittivity, $\epsilon(0)$. Here $\epsilon(\infty)$ is evaluated at a frequency ω such that $\Omega_0 \ll \omega \ll \omega_0$, where ω_0 is a typical interband electronic resonance frequency. The quantity $\epsilon(0)$ is evaluated at $\omega = 0$. Thus

$$\frac{\epsilon(\infty) - \epsilon_0}{\epsilon(\infty) + 2\epsilon_0} = \frac{n}{3}\alpha_e, \quad \frac{\epsilon(0) - \epsilon_0}{\epsilon(0) + 2\epsilon_0} = \frac{\epsilon(\infty) - \epsilon_0}{\epsilon(\infty) + 2\epsilon_0} + \frac{nQ^2}{3\epsilon_0 M\Omega_0^2}, \tag{15.21}$$

$$\frac{\epsilon - \epsilon_0}{\epsilon + 2\epsilon_0} = \frac{\epsilon(\infty) - \epsilon_0}{\epsilon(\infty) + 2\epsilon_0} + \frac{\Omega_0^2}{(\Omega_0^2 - i\omega\Gamma - \omega^2)}\left[\frac{\epsilon(0) - \epsilon_0}{\epsilon(0) + 2\epsilon_0} - \frac{\epsilon(\infty) - \epsilon_0}{\epsilon(\infty) + 2\epsilon_0}\right], \tag{15.22}$$

so

$$\epsilon(\omega) = \epsilon(\infty) + [\epsilon(0) - \epsilon(\infty)]\frac{\Omega_0'^2}{\Omega_0'^2 - i\omega\Gamma - \omega^2}, \tag{15.23}$$

TABLE 15.1   Optical Phonon Energies, $E_L = \hbar\Omega_L$ and $E_T = \hbar\Omega_T$, and Dielectric Constants for Several Materials

Material	E_T (meV)	E_L (meV)	$\epsilon_r(0)$	$\epsilon_r(\infty)$
NaCl	21.1	33.4	5.90	2.35
NaBr	16.8	26.2	6.28	2.59
AlN	82.9	113.7	9.14	4.84
AlSb	23.0	24.4	17.9	15.7
GaP	45.6	50.1	10.2	8.46
CdTe	14.4	17.3	10.2	7.13

Source: Data from S. S. Mitra et al., in W. Paul, ed., *Handbook of Semiconductors*, North-Holland, Amsterdam, 1982.

where the shifted resonance frequency is

$$\Omega_0' = \Omega_0 \left[\frac{\epsilon(\infty) + 2\epsilon_0}{\epsilon(0) + 2\epsilon_0} \right]^{1/2}. \tag{15.24}$$

A plot of the real and imaginary parts of the dielectric function was presented in Fig. 8.6. They are related by the Kramers–Kronig relations discussed in Section 8.7. A list of phonon energies and dielectric constants for representative materials is presented in Table 15.1. Additional phonon energies were presented in Table 8.3. It should be noted that the phonon energies are high for AlN and low for CdTe. Two factors determining the phonon energies are the masses M of the ions (small for AlN, large for CdTe) and the strengths of the restoring forces represented by the spring constant K (i.e., $\Omega \propto \sqrt{K/M}$). These spring constants are determined by both the ionic and covalent nature of the bonds. Covalent bonds generally have stiffer spring constants. The bond in AlN is considerably covalent in nature.

Equation (15.21) shows the high-frequency permittivity to be determined by the electronic contribution. This is to be expected since the high inertia of the ions do not permit them to respond to the high frequency of the electric field. However, at low frequencies both the ions and electrons respond and they both contribute to the permittivity. Note that the ionic frequencies obey the inequality $\Omega_T < \Omega_L$, where Ω_L is the frequency at which the dielectric function is zero and Ω_T is the frequency at which there is a resonance. This is consistent with the Lydanne–Sachs–Teller equation, Eq. (8.25), and the observation that $\epsilon(0) > \epsilon(\infty)$.

Note that in ceramic crystals there are situations involving mixed ionic and covalent bonds. When $\epsilon(0)/\epsilon(\infty)$ is large it indicates that there is a substantial ionic component to the bond. If the ratio were close to 1, it would indicate a largely covalent bond.

15.4   Dielectric Breakdown

The breakdown of a complex system is often determined by the breakdown of a particular component in that system. One example involves a mechanical chain, where the tensile strength is set by the weakest link. In the three-dimensional extension, fracture may begin with the formation of a microcrack, followed by propagation of that crack across the material. Another example is dielectric breakdown, where the onset of conductivity in one part of a system can trigger an electron avalanche that causes the entire material to break down. A convenient way to characterize such systems is by means of Weibul statistics.

Consider a chain consisting of N links. Let $f(T)$ be the probability that a link will fail if the tension is T. The probability of the link not failing is then $1 - f(T)$. The probability for having the chain survive tension T is $p = [1 - f(T)]^N$. If $f(T)$ is small, one may approximate this as $p = \exp[-Nf(T)]$. Thus the longer the chain, the less likely it is to survive.

Weibul empirically modeled the function $f(T)$ by a power law. Thus, letting T_0 denote the minimum tension that every link can withstand, $f(T) = f_0[T - T_0]^m$, where m is called the *Weibul modulus* and determines how wide the distribution of tensions will be. Thus

$$p = \begin{cases} e^{-Nf_0(T-T_0)^m} & \text{if } T > T_0 \\ 1 & \text{if } T < T_0. \end{cases} \qquad (15.25)$$

When a sufficiently strong potential difference is imposed across a material, dielectric breakdown occurs. There are two primary types of breakdown: avalanche and thermal. In the case of *avalanche dielectric breakdown* the process proceeds in two stages: field ionization followed by electron impact ionization. It is possible for an electric field, E, to induce an interband transition by what is called the Zener process: In the presence of an electric field, the valence and conduction bands vary in space with a gradient proportional to the strength of the electric field. Tunneling is possible from a filled valence-band state to the conduction band. If there are impurity states in the gap, these will probably tunnel first. When the electron is in the conduction band it is free to move and be accelerated by the electric field. It may collisionally excite additional electrons. The number of electrons then increases exponentially, and after 40 or so generations, a dense plasma has been produced and avalanche breakdown results.

A simple quantum-mechanical estimate of the tunneling probability P_t may be made (see Appendix WC). Consider a triangular barrier of height E_g and thickness $\Delta x = E_g/eE$ (Fig. 15.1). Then

$$P_t(E) = \exp\left[-2\int_0^{\Delta x} dx \sqrt{\frac{2m^*}{\hbar^2}(E_g - eEX)}\right] = \exp\left(-\frac{4}{3E}\sqrt{\frac{2m^*E_g^3}{e^2\hbar^2}}\right). \qquad (15.26)$$

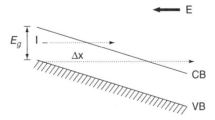

Figure 15.1. Tunneling of an electron from the valence band (or an impurity level, I) to the conduction band in the presence of an electric field.

To induce collisional ionization, the energy that the primary electron gains before an inelastic collision occurs must exceed the gap energy, E_g. Thus $(eE\tau)^2/2m_e^* > E_g$, where m_e^* is the conduction band effective mass. This defines a critical value of the breakdown field. At higher temperatures τ is decreased due to an increased density of thermally activated phonons.

Thermal breakdown occurs when the production of heat per unit volume exceeds the ability of the material to dissipate it. The power produced per unit volume is E^2/ρ. Generally, the heat diffusion equation must be solved for the temperature field $T(\mathbf{r}, t)$

$$-\kappa\nabla^2 T + c_v \frac{\partial T}{\partial t} = \frac{E^2}{\rho}, \tag{15.27}$$

where κ is the thermal conductivity and c_v is the specific heat per unit volume. Equation (15.27) is obtained by combining the continuity equation for the thermal energy flux (including an ohmic heating source term) with Fourier's law for heat flow. The resistivity is usually very sensitive to the temperature, varying as $\rho \sim \rho_0 \exp(\beta E_d)$ for the case of donor ionization (with ionization energy E_d). The actual steady-state temperature profile achieved will depend on the boundary conditions employed in solving Eq. (15.27). One often employs heat sinks with large surface areas or with flowing air or other fluids around them to keep the temperatures low.

Impurities and defects provide local inhomogeneities at which breakdown can be initiated. Weibul statistics again provides a way of characterizing the process. One

TABLE 15.2   Dielectric Strengths for Typical Materials

Material	E_{max} (MV/m)
Porcelain	6.1–13
Spinel	11.9
Mullite	7.8
Zirconia	5.0
Titania	2
Alumina	9.9–15.8
BN	35.6–55.4
Si_3N_4	15.8–19.8
Mica	30.5–79.1
Glass	15–25.0
MgO	8.5–11.0
$BaTiO_3$	4
$PbZrO_3$	7.9
Polystyrene	140
Polyethylene	70
PMMA	30

Source: Data, in part, from R. C. Buchanan, ed., *Ceramics Materials for Electronics*, Marcel Dekker, New York, 1986; and L. L. Hench and J. K. West, *Principles of Electronic Ceramics*, Wiley, New York, 1990.

may imagine a volume V in which there are $N = nV$ separate units, each capable of undergoing a breakdown. Referring to Eq. (15.25), inserting the electric field instead of the tension, and taking the minimum electric field for breakdown as zero yields a formula for the survival probability

$$P = e^{-Vh_0 E^m} \tag{15.28}$$

where h_0 is a constant (or has a mild volume dependence) and m is the Weibul modulus. Values of m in the range 15 to 30 are not unusual.

Values of the breakdown strength E_{max} for typical dielectric materials are given in Table 15.2. A number of factors may influence the actual dielectric strength, including porosity, imperfections, and crystallites. For the materials listed there seems to be a correlation of dielectric strength with bond strength. Covalently bonded materials such as the polymers have higher dielectric strengths than ionically bonded materials such as the oxides.

APPLICATIONS

15.5   Ferroelectric Phase Transitions

In an attempt to provide a phenomenological theory for phase transitions, Landau identified an order parameter and expanded the free energy as a power series in the order parameter. The theory works best for second-order phase transitions where it is known that the order parameter is small near the transition point. However, one may also use it as a basis for discussing the phenomenology of first-order transitions. In the case of ferroelectric phase transitions the order parameter is the electric polarization, P. Above a certain temperature, T_c, the crystal is in the high-symmetry phase and $P = 0$. Below T_c the symmetry is broken and the polarization develops a nonzero value. The materials of primary interest are the perovskites. They have the same crystal structure as the mineral perovskite, $CaTiO_3$. These are ternary oxides of the form ABO_3 with cubic symmetry above T_c and lower symmetry below T_c. Atom A is typically Ba, Ca, Pb, Sr, Na, or K, and atom B is typically Nb, Ti, or Zr. Often, there are a sequence of phase transitions that occur as the temperature is lowered. For example, $BaTiO_3$ undergoes a first-order structural phase transition [cubic (C) to tetragonal (T) at 130°C], followed by additional transitions to the orthorhombic (O) phase at 0°C, and the rhombohedral (trigonal) (R) phase at −90°C, as the temperature is lowered. Other perovskites undergo second-order phase transitions. This is illustrated for $BaTiO_3$ in Fig. 15.2, where the dielectric constant is plotted as a function of temperature (anisotropy and hysteresis effects are neglected).

Before considering the three-dimensional case, it is worthwhile examining the Landau theory of phase transitions in one dimension. The Gibbs free-energy density is expanded as a power series in the order parameter

$$g = g_0 + \frac{a}{2}P^2 + \frac{b}{4}P^4 + \frac{c}{6}P^6 - EP, \tag{15.29}$$

where E is the electric field. The coefficient a is taken to be temperature dependent and to change sign at $T = T_0$. It is convenient to assume a simple linear dependence

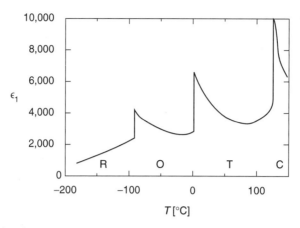

Figure 15.2. Dielectric constant ϵ_r versus temperature for BaTiO$_3$. R, O, T, and C refer to the rhombohedral, orthorhombic, tetragonal, and cubic phases, respectively. [From W. J. Merz, *Phys. Rev.*, **76**, 1221 (1949). Copyright 1949 by the American Physical Society.]

$a = a_0(T - T_0)$. The model is general enough to include both first- and second-order phase transitions.

For the second-order phase transition case, take $c = 0$ and assume that a_0 and b are both positive. Start by assuming that $E = 0$. Minimizing g gives

$$\frac{\partial g}{\partial P} = a_0(T - T_0)P + bP^3 = 0. \tag{15.30}$$

For $T > T_0$ the state with minimum g $(= g_0)$ occurs when $P = 0$ and there is no spontaneous polarization. However, for $T < T_0$ the state with minimum g occurs for

$$P = \pm\sqrt{\frac{a_0}{b}(T_0 - T)}. \tag{15.31}$$

The value of g is given by

$$g = g_0 - \frac{a_0^2}{4b}(T_0 - T)^2. \tag{15.32}$$

The spontaneous polarization grows continuously from zero at high temperatures as the temperature is lowered below the transition temperature, T_0.

The dielectric constant may be derived by looking at the case $E \neq 0$. In place of Eq. (15.30) there is

$$\frac{\partial g}{\partial P} = a_0(T - T_0)P + bP^3 - E = 0. \tag{15.33}$$

For $T > T_0$ the approximate solution is

$$P = \frac{E}{a_0(T - T_0)}, \tag{15.34}$$

from which one may construct the dielectric constant:

$$\epsilon_r = 1 + \frac{P}{\epsilon_0 E} = 1 + \frac{1}{\epsilon_0 a_0 (T - T_0)} = 1 + \frac{C}{T - T_0}. \quad (15.35)$$

where $C = 1/\epsilon_0 a_0$ is called the *Curie constant* for ferroelectricity. The dielectric constant is seen to diverge as T is lowered toward T_0. For $T < T_0$ one may show that the approximate solution is

$$\epsilon_r = 1 + \frac{1}{2\epsilon_0 a_0 (T_0 - T)}, \quad (15.36)$$

which also diverges as T is raised toward T_0.

The behavior of the spontaneous polarization and the susceptibility ($\chi = \epsilon_r - 1$) is depicted in Fig. 15.3, where these parameters are plotted as a function of $\epsilon_0 a_0 (T - T_0)$. One sees the growth of P described by Eq. (15.31) as well as the singularity described by Eqs. (15.35) and (15.36). In Section W15.3, details are given for the analogous case when there is a first-order phase transition.

Table 15.3 presents the polarization properties of some common ferroelectrics. Some of these materials are not single crystals, so there is a range of local environments. Performing an average over the inhomogeneities spreads the transition point over a range of temperatures. The broad maximum of ϵ as a function of T characterizes these "relaxor" materials.

It should be noted that the power series in the order parameter, which is the basis for the Landau theory, is really only meant to be applicable to temperatures near the critical temperature. It is therefore not proper to extend the formulas for the polarization to $T = 0$ K. One expects, on theoretical grounds, that the polarization at absolute zero would be given by $\mathbf{P} = n\boldsymbol{\mu}$ where n is the number of unit cells per unit volume and $\boldsymbol{\mu}$ is the electric dipole moment of the unit cell. Using this notation, the Curie constant is $C = n\mu^2/3k_B\epsilon_0$.

The geometric arrangements of the ions in the various perovskite phases will be studied in more detail in the next section.

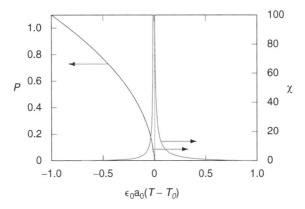

Figure 15.3. Polarization and susceptibility ($\chi = \epsilon_r - 1$) for the Landau model as a function of the scaled temperature. Here $E = 0$.

TABLE 15.3   Polarization Properties of Typical Ferroelectrics

Material	T_C (°C)	C (K)	$P(25°C)$ (C/m²)	$P(T_C)^a$ (C/m²)	
LiTaO₃	618	80,000	0.50	0.36	(450)
LiNbO₃	1210	103,000	0.71	0.67	(450)
BaTiO₃ (BT)	130	76,000	0.26	0.19	
PbTiO₃ (PT)	490	113,000	0.81	—	
PbZr.₉₅Ti.₀₅O₃	220	36,000	0.34	0.23	
PLZT	110	2,300	0.34	—	
KH₂PO₄ (KDP)	−150	9,000,000	0.05	0.03	(−173)
KD₂PO₄	−60	—	0.048	—	
PZN₀.₁BT₀.₂PT₀.₁ ^{b}	35	390,000	—	—	
Triglycine sulfate (TGS)	49	3,500,000	0.03	0.02	

Source: Data from B. A. Tuttle et al., *Mater. Res. Soc. Bull.*, July 1994, p. 20.

^{a}The parentheses on the right indicate the temperatures at which the measurements were made, if not exactly at T_C.

^{b}PZN is lead zinc niobate (see Section 15.6).

15.6   Ferroelectricity and Piezoelectricity

A crystal whose unit cell possesses no spontaneous electric dipole moment is called a *nonpolar crystal*. As in any dielectric, a dipole moment may be induced in response to an applied electric field, but disappears when the field is removed. The net electric dipole moment of the unit cell is determined by the vector sum of the ionic charges multiplied by the basis vectors plus the induced dipoles of each ion (i.e., $\mu = \Sigma z_i e \mathbf{s}_i + \Sigma \mu_i$). If the unit cell has a spontaneous electric dipole moment in the absence of an electric field, it is called a *polar crystal* or *electret*. Polar crystals may be classified as being pyroelectric, ferroelectric, or antiferroelectric. In a *pyroelectric crystal* the dipole moment remains fixed and is largely unaffected by external fields. In a *ferroelectric crystal* the dipoles of neighboring unit cells are aligned in the same direction. When an external electric field is imposed, these dipoles orient themselves parallel to the macroscopic electric field. The direction of the polarization vector may be reversed by applying a field of sufficient strength in the opposite direction. In ferroelectricity there is a competition between the ferroelectric exchange interaction (i.e., chemical forces), which tends to produce aligned dipoles in neighboring cells, and thermal agitation, which tends to destroy this alignment. The Curie temperature, T_C, is determined by which of these tendencies dominates. For $T < T_C$ there is ferroelectricity, whereas for $T > T_C$ there is the paraelectric phase and the dipoles either become randomly oriented (a kind of order–disorder transition) or disappear (e.g., in the case of a structural phase transition, where the high-symmetry phase has no net dipole per unit cell). In an *antiferroelectric crystal* the dipole moments of neighboring cells point in opposite directions, so there is no net polarization.

The existence of a spontaneous polarization implies that there is a preferred spatial direction in the crystal. If the unit cell possesses inversion symmetry there would be no net electric dipole, since there is no reason to prefer one direction over its opposite. Thus a necessary (but not sufficient) condition for a crystal to be ferroelectric is that it lacks inversion symmetry. Of the 32 crystal point groups only 10 can be polar (triclinic: C_1; monoclinic: C_{1h}, C_2; orthorhombic: C_{2v}; tetragonal: C_{4v}, C_4; trigonal: C_3, C_{3v}; and hexagonal: C_6, C_{6v}). In Figs. 15.4 and 15.5 schematic representations of the 32 point-group symmetries are given. Figure 15.4 gives the groups without inversion symmetry, and Fig. 15.5 gives the groups with inversion symmetry. These 10 groups are a subset of the 20 groups presented in Fig. 15.4. Notably absent from the polar group are members of the cubic class of crystals.

Ferroelectricity is closely connected to structural phase transitions and the underlying microscopic mechanism responsible is complex. One may identify several contributing factors that are likely to play a role. They are the Jahn–Teller effect, Earnshaw's theorem, and the long-range interaction between unit cells via their electric dipole–dipole interactions and their strain fields.

The Jahn–Teller effect (see Section W9.1) describes the lowering of the electronic energy of a crystal when it undergoes a structural phase transition from a

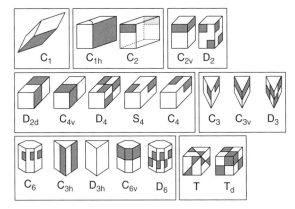

Figure 15.4. Twenty point groups that lack inversion symmetry. The groupings are for triclinic, monoclinic, orthorhombic, tetragonal, trigonal, hexagonal, and cubic crystals. (Adapted from N. W. Ashcroft and N. D. Mermin, *Solid State Physics*, Holt, Rinehart and Winston, New York, 1976.)

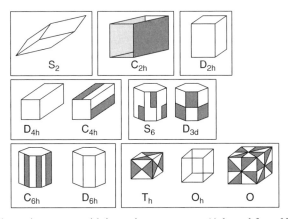

Figure 15.5. Twelve point groups with inversion symmetry. (Adapted from N. W. Ashcroft and N. D. Mermin, *Solid State Physics*, Holt, Rinehart and Winston, New York, 1976.)

high-symmetry state to a state of lower symmetry. It occurs when there is a degeneracy of the electronic states (above and beyond spin degeneracy). As the symmetry is lowered there may be a partial lifting of the degeneracy, with some of the energy levels being raised and others lowered. The "center of gravity" of the multiplet remains unshifted. If the levels are only partially occupied, the crystal can lower its overall electronic energy by undergoing a distortion. If the parameter describing the distortion is d, the lowering of electronic energy will be proportional to it (i.e., $\Delta E_1 = -c_1 d$). But distorting the crystal will also raise its elastic energy by an amount $\Delta E_2 = c_2 d^2$, where c_2 is an appropriate elastic modulus. The overall energy change is $\Delta E = -c_1 d + c_2 d^2$. The state of lowest energy occurs for a finite distortion, given by $d = c_1/2c_2$, where $\Delta E = -c_1^2/4c_2$. In a three-dimensional crystal there may be several equivalent sites to which the atom may move.

Earnshaw's theorem, mentioned in Section 13.2, shows that the equilibrium of an ion in a purely electrostatic field is at best unstable. This is important when one attempts to place a small cation at the center of a large cubic cell, such as occurs with Ti^{4+} in $BaTiO_3$. The central ion could find a state of diminished energy by moving to a site of lower symmetry closer to one or more of the O^{2-} ions. This would result in the formation of an electric dipole moment for the unit cell.

Since bonding is neither purely ionic nor purely covalent in the ferroelectric crystals, some degree of Jahn–Teller transition and some degree of electrostatic instability may determine the structural phase transition.

At elevated temperatures the "central" ion is able to move rapidly from one equilibrium position to another, so the mean position of the ion lies at the center of the unit cell. At lower temperatures, where this motion is hindered, a state of lower symmetry may result.

The long-range interactions between unit cells in a crystal can occur via the electric dipole–dipole interaction, in which case the interaction energy falls off with distance as r^{-3}. It could also occur through the strain fields associated with the distortion of the unit cell shapes.

If the crystal develops an electric polarization in response to an applied stress, it is called a *piezoelectric crystal*. Centrosymmetric groups (having an inversion symmetry point) are always nonpiezoelectric. Thus the only piezoelectric crystals are the ones with the point groups depicted in Fig. 15.4.

There is an interconnection between the three types of response (electric polarization, strain, and entropy) and the three driving "forces" (electric field, stress, and temperature). In the general case, constitutive equations are formulated in which the responses are expressed as linear combinations of the driving forces, or vice versa. In this section the most general case is not studied; rather, the focus is on special but important situations.

Suppose that there are no complicating temperature gradients and no stresses. The application of an electric field may produce a response. The strain, to lowest order in the electric field, is linear in **E** and is given by

$$\epsilon_{ij} = \sum_{k=1}^{3} d_{ijk} E_k \longrightarrow \epsilon_\mu = \sum_{k=1}^{3} d_{k\mu} E_k, \tag{15.37}$$

where $d_{ijk}\{i, j, k = 1, 2, 3\}$ or, in reduced notation, $d_{k\mu}$ $\{k = 1, 2, 3; \mu = 1, \ldots, 6\}$ is the piezoelectric strain tensor. Conversely, one can apply a stress in the absence of an electric field or a temperature gradient and obtain an electric polarization **P**:

$$P_i = \sum_{j=1}^{3} \sum_{k=1}^{3} d_{ijk} \sigma_{jk} \longrightarrow P_i = \sum_{\mu=1}^{6} d_{i\mu} \sigma_{\mu}. \tag{15.38}$$

(It may be shown from thermodynamics that the same $d_{i\mu}$ tensor governs both effects.) Depending on the symmetry of the crystal, some of the d tensors may be zero. Twenty of the 32 point groups (see Fig. 15.4) allow for a nonvanishing piezoelectric tensor (triclinic: C_1; monoclinic: C_{1h}, C_2; orthorhombic: C_{2v}, D_2; tetragonal: D_{2d}, C_{4v}, D_4, S_4, C_4; trigonal: C_3, C_{3v}, D_3; hexagonal: C_6, C_{3h}, D_{3h}, C_{6v}, D_6; and cubic: T, T_d). Symmetry imposes many other constraints on the $d_{i\mu}$ components, as well, including forcing some of them to vanish and others to be the same. In Fig. 15.5 the 12 point groups for which the piezoelectric tensor is zero are shown. Upon inversion of the coordinate system, $\mathbf{r} \to -\mathbf{r}$, $\mathbf{u} \to -\mathbf{u}$ ($\mathbf{u}$ is the ionic displacement), and $\mathbf{P} \to -\mathbf{P}$, but $\overleftrightarrow{\sigma} \to \overleftrightarrow{\sigma}$. This is possible only if the crystal lacks a center of symmetry. Note that the point group O has a vanishing d tensor despite the fact that it has inversion symmetry.

The focus of attention here will be on the technologically important piezoelectric ceramic $BaTiO_3$, for the sake of definiteness. Barium titanate is cubic for $1460°C > T > 130°C$, with $a = 0.4009$ nm. The Ti^{4+} ion sits at the center of a cube with Ba^{2+} ions at the corners and O^{2-} at the centers of the faces. (An equivalent description places Ba^{2+} at the center, Ti^{4+} at the corners, and O^{2-} at the centers of the edges.) An analysis using Pauling's rules gives the lines emanating from Ba^{2+} an ionic strength of $\frac{2}{12}$ and those from Ti^{4+} a strength of $\frac{4}{6}$. The O^{2-} ions act as a sink for four Ba^{2+} lines and two Ti^{4+} lines, so $4(2/12) + 2(2/3) = 2$, the oxygen valence. The ionic radii are $(0.135, 0.068, 0.14)$ nm for $(Ba^{2+}, Ti^{4+}, O^{2-})$. A sketch of the cubic unit cell is given in Fig. 15.6. The crystal has O_h symmetry and is nonpiezoelectric.

As the temperature is lowered toward $T = 130°C$, one of the phonon modes lowers in frequency (goes "soft"), culminating in a phase transition (see Fig. 15.2). The crystal symmetry is reduced to tetragonal with the point group C_{4v}. This is illustrated in Fig. 15.7. The crystal becomes both polar and piezoelectric. The values for the nonvanishing d-tensor components are: $(d_{31} = d_{32} = -35, d_{33} = 86, d_{15} = d_{24} = 390) \times 10^{-12}$ m/V at room temperature. The precise nature of this phase transition is still a matter of debate. It may be displacive or it may be an order–disorder transition or some combination of the two. A further reduction of the temperature to $0°C$ causes a displacive phase transition to orthorhombic with point group C_{2v}. Finally, at $T = -80°C$ another transition occurs to the trigonal C_{3v} group.

The fact that the crystal is always near a phase transition means that it is readily influenced by the presence of an external electric field. The relative dielectric constant

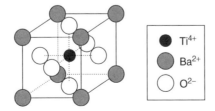

Figure 15.6. Cubic unit cell of $BaTiO_3$, with Ti^{4+} sitting at the body center, Ba^{2+} at the corners, and O^{2-} at the centers of the faces.

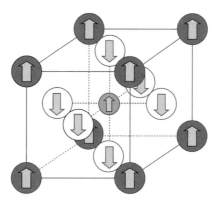

Figure 15.7. The cations move up and the anions move down as the symmetry lowers from cubic to tetragonal.

of $BaTiO_3$ is enormous, being around 10,000 at 120°C. It is very temperature dependent, becoming largest next to the onset of each phase transition. It becomes highly anisotropic below 130°C when the cubic symmetry is destroyed, with the c axis value being typically about 10 times larger than the a axis value. In many polycrystalline piezoelectric materials the dielectric constant displays a broad maximum as a function of temperature and the crystal is termed a *relaxor*. Examples of other types of distortions of the unit cell that occur in perovskites are given in Fig. 15.8. These include various rotations of the oxygen octahedra surrounding each Ti^{4+} ion.

Piezoelectric crystals have found applications in general-purpose actuators and transducers, positioners, microphones, loudspeakers, spark igniters, deformable mirrors, impact printers, and microscopic motors. Numerous crystals display piezoelectricity. For example, α-quartz (α-SiO_2) is a trigonal crystal with the point group D_3 and has d-tensor components $d_{11} = -d_{12} = -d_{26}/2 = 2.3$, $d_{14} = -d_{25} = -0.67$ pm/V. (Application of piezoelectricity to the quartz crystal oscillator is covered in Section W15.5.) Ammonium dihydrogen phosphate, $NH_4H_2PO_4$ (ADP), is a tetragonal

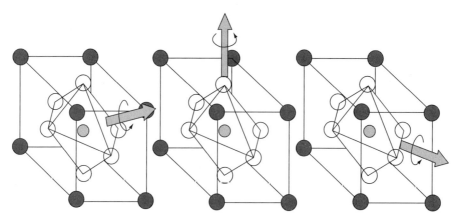

Figure 15.8. Other perovskite distortions involving rotations of the oxygen octahedra. (From P. A. Fleury et al., in K. A. Muller et al., ed., *Structural Phase Transitions*, Vol. I, Springer-Verlag, New York, 1981.)

crystal with symmetry D_{2d} with $d_{36} \approx 23$ pm/V. Rochelle salt ($NaKC_4H_4O_6 \cdot 4H_2O$) is an orthorhombic crystal with $d_{25} = -53$ pm/V and $d_{36} = 11$ pm/V. It has a narrow range of temperature over which it is ferroelectric (-18 to $+24°C$). The value of d_{14} is highly temperature dependent, due to a phase transition, and reaches a maximum value of 511 pm/V at $33°C$. The leading materials currently in mass use are modified PZT: $Pb(Zr, Ti)O_3$ (at the morphotropic phase boundary) and PLZT: $(Pb, La)(Zr, Ti)O_3$. The material $Pb_{1-x}La_x(Zr_yTi_{1-y})_{1-x/4}O_3$ (PLZT) with $(x, y) = (0.07, 0.6)$ has a high value of $d_{33} = 710$ pm/V. These materials are based on the underlying perovskite, $PbTiO_3$ with some Pb being replaced by La and some Ti by Zr. The morphotropic phase boundary is the temperature-independent phase boundary separating the rhombohedral and tetragonal phases. At this boundary the dielectric constant is also very high.

Pyroelectricity was first discovered when certain crystals appeared to develop electric dipole moments when warmed. However, it was found that these crystals had permanent electric dipole moments (i.e., they were electrets) even before they were warmed, but this dipole moment was masked by adsorbed ions on the faces which had an electric dipole moment opposite to that of the crystal and canceled it. When the crystal was heated, the ions desorbed and the underlying moment was revealed. In the pyroelectric effect the electric displacement vector D varies in response to a change in temperature. The pyroelectric coefficient is defined as $p_i = \partial D_i / \partial T = \partial P_i / \partial T$ (holding the electric field and stress constant). Only those crystals for which one can define a preferred direction for the spontaneous electric dipole moment will be pyroelectric. These include 10 groups from Fig. 15.4 (C_1, C_h, C_2, C_{2v}, C_3, C_{3v}, C_4, C_{4v}, C_6, and C_{6v}). The large pyroelectric effect in PLZT ($p = 0.0322$ $C/(m^2 \cdot °C)$ at $T = 145°C$), for example, makes it suitable for infrared sensors and for night-vision devices. Comparable values have been achieved by the material modified-PZN [$PZN_{0.75}ST_{0.15}PT_{0.10}$; with $PZN = Pb(Zn_{0.333}Nb_{0.667})O_3$, $PT = PbTiO_3$, $ST = SrTiO_3$].

15.7 Thermistors

Thermistors are resistors whose resistances are sensitive functions of the temperature. They are classified according to the sign of the temperature coefficient of resistivity, $\alpha = d(\ln \rho)/dT$. For $\alpha > 0$ (<0) they are PTCR (NTCR) thermistors, corresponding to positive (negative) temperature coefficients of resistance. Of course, metals and semiconductors are examples of PTCR and NTCR resistors, respectively. The values of $|\alpha|$ for these common materials are much smaller than for thermistors.

PTCR Thermistors. PTCR thermistors have resistances that suddenly rise by several (three to eight) orders of magnitude when the temperature passes a critical switching temperature T_c. This makes them ideal for current-limiting applications, such as protecting motors. If the current becomes too large, the device gets warmer, the resistance increases, and by Ohm's law, the current must drop.

These devices often consist of polycrystalline n-doped semiconductor ferroelectrics such as $BaTiO_3$, $Ba_{1-x}Pb_xTiO_3$ or $Ba_{1-x}Sr_xTiO_3$. When T_c is exceeded, the thermistor undergoes a semiconductor-to-insulator transition. The Pb and Sr additives serve to move T_c up or down, and thus the switching temperature can be fine-tuned over a several-hundred-degree interval. The switching temperature corresponds to the ferroelectric phase transition temperature. Typical dopant ions include La^{3+} and Y^{3+}, which substitute for Ba^{2+}, or Nb^{5+}, which substitutes for Ti^{4+}. The surfaces of the crystallites are oxidized, which makes them insulating. The oxygen atoms at the surface also serve as electron traps.

According to the Heywang theory, a Schottky barrier is formed at the boundary between two grains. (See Section 20.8, for more details regarding Schottky barriers.) The model for barrier formation is simple. Assume that there is a sheet of trapped electrons of charge per unit area $-\sigma$ at the interface ($z = 0$). On both sides of the interface there is an ionized donor density N_d^+. These positive charges produce an electric field. Within a depletion layer of width $2a$ about the surface, the mobile carriers are swept away by the field. Outside that region, charge neutrality is assumed and the potential is constant. Within the region the charges give rise to an electric field, which may be described by the electrostatic potential through the formula $\mathbf{E} = -\nabla\phi$. The potential obeys Poisson's equation,

$$\epsilon\frac{d^2\phi}{dz^2} = -[N_d^+e - \sigma\delta(z)], \qquad \text{where} \quad -a < z < a. \tag{15.39}$$

It is assumed that the ferroelectric obeys the Curie–Weiss law [see Eq. (15.35)], so that

$$\epsilon_r = \frac{\epsilon}{\epsilon_0} = 1 + \frac{C}{T - T_c} \qquad \text{if} \quad T > T_c. \tag{15.40}$$

The case where $T < T_c$ is discussed later. Integration of the Poisson equation with the boundary conditions $E = 0$ and $\phi = 0$ at $z = \pm a$ results in the formula

$$\phi(z) = \phi_0 + \frac{\sigma}{2\epsilon}|z| - \frac{eN_d^+}{2\epsilon}z^2 \qquad \text{for} \quad -a < z < a, \tag{15.41}$$

where a is determined by the charge densities

$$a = \frac{\sigma}{2eN_d^+}, \tag{15.42}$$

and the barrier potential height is given by

$$\phi_0 = -\frac{\sigma^2}{8\epsilon eN_d^+}. \tag{15.43}$$

Thus the theory predicts that there is potential energy barrier of height $U_0 = -e\phi_0 = \sigma^2/8\epsilon N_d^+$ that the electron must surmount to pass from one grain to the other. At $T = T_c$, $\epsilon \rightarrow \infty$ and the barrier disappears. The electrons simply pass from one grain to the other. For $T > T_c$, however, ϵ will have lower values and U_0 can be substantial. The fraction of electrons with enough energy to surmount the barrier is proportional to $\exp(-\beta U_0)$ and the conductivity is reduced by this factor. Alternatively, the resistivity will vary as $\rho \propto \exp(\beta U_0)$. Thus a sharp rise in the resistivity is expected above T_c as T is lowered. Electron tunneling is neglected here.

Jonker realized that as T is lowered below T_c a structural phase transition occurs and the grains develop spontaneous polarization. For example, BaTiO$_3$ undergoes a cubic-to-tetragonal transition at $T = 130°C$ and a spontaneous polarization develops.

The polarization can orient its direction in each grain in such a manner as to minimize the electrostatic energy of the system. The net result is that the surface charge $-\sigma$ is compensated by discontinuities in the polarization vector. This follows directly from Gauss's law: $\nabla \cdot \mathbf{D} = -\sigma \delta(z) = \nabla \cdot (\epsilon_0 \mathbf{E} + \mathbf{P})$. If $P_z = P_0 - \sigma/2$ for $z > 0$ and $P_z = P_0 + \sigma/2$ for $z < 0$, where P_0 is arbitrary, then $\nabla \cdot E = 0$ and no barrier exists below $T = T_c$. Thus the material remains a semiconductor for $T < T_c$.

NTCR Thermistors. NTCR thermistors are essentially semiconductors with a high negative value of α. They find application in thermometry and as compensators where thermal drift of devices must be stabilized. Suppose that there is n-doped material with electrons that can be thermally activated into the conduction band from states within the gap. The conductivity varies as the number of carriers. Thus $\rho = 1/\sigma = \rho_0 \exp(\beta E_d/2)$, where E_d is the donor ionization energy. Then $\alpha = -E_d/2k_B T^2 = -B/T^2$. The parameter B is essentially the binding energy of a trapped electron. For a typical NTCR thermistor, $E_d \approx 0.8$ eV, so $\alpha \approx 0.05$ K^{-1}. The range of B values thus extends from 2000 K to 20,000 K.

Common materials employed as NTCR thermistors are $NiMn_2O_4$, $Ni_xMn_{3-x}O_4$, and $Ni_xMn_yCo_{3-y-x}O_4$, which have the cubic inverse-spinel crystal structure. The Ni^{2+} and half of the Mn^{3+} ions occupy the tetrahedral interstitial sites in the close-packed array of O^{2-} ions and the other half of the Mn^{3+} ions occupy the octahedral interstitial sites. Note that both Ni and Mn can exist in states with valence 2 or 3. The high conductivity of the spinels has been attributed to the presence of ions that can support multiple valence states. It is not inconceivable that electrons hop from site to site accompanied by valence fluctuations, a kind of impurity band conduction. This may account for the special role that spinels play in NTCR materials. For applications at higher temperatures (up to 1300°C) oxides of rare earths are used, such as La_2O_3, Eu_2O_3, or Y_2O_3.

15.8 Varistors

Varistors are resistors whose resistance is a sensitive function of the voltage difference across them [i.e., $R = R(V)$]. Typical materials employed are ZnO, $SrTiO_3$, and SiC. State-of-the-art varistors are fashioned from a combination of oxides (e.g., $ZnO + Bi_2O_3 + TiO_2 + CoO + MnCO_2 + Al_2O_3$) in various proportions. They make ideal materials for surge protectors because of their nonlinear I–V characteristics. For low voltages the resistance is high and they behave essentially as capacitors. For high voltages, however, the resistance drops to a low value, usually following some empirical power law. The current density also obeys a power law in the electric field, such as $J \sim cE^\alpha$, where α can be as large as 200. The response time is very fast. When shunted across electronic devices they offer protection against a rapid increase in line voltage, such as might occur during a lightning storm or a switching event.

The microstructure of a typical varistor consists of many grains of n-type ZnO surrounded by thin insulating layers (Fig. 15.9). It may be regarded as a complex network of capacitors and voltage-dependent resistors. A typical grain size is ≈ 20 µm, whereas the insulating layer is rich in Bi_2O_3 and may be only ≈ 0.1 µm thick. The bandgap for ZnO is $E_g = 3.2$ eV. If the voltage difference between two adjacent grains exceeds E_g/e, the top of the valence band of one grain will lie above the bottom of the conduction band of the adjacent grain. (This voltage is modified by the contact potential

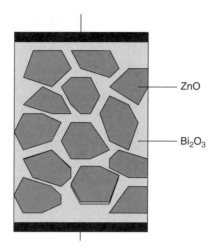

Figure 15.9. Varistor consisting of grains of *n*-type conducting ZnO surrounded by insulating layers of Bi_2O_3.

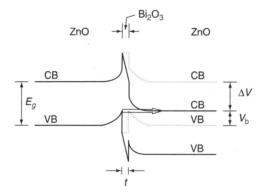

Figure 15.10. Electrons can tunnel from one ZnO grain to another when the dc bias between grains, ΔV, becomes sufficiently large.

difference, as will soon be seen.) Electrons may then tunnel from one grain into the other. Having been injected into the conduction band, they are then free to conduct across the grain. Let the thickness of the varistor be D and the grain size d. There are typically $\approx D/d$ grains across the width of the device. Thus a voltage $V \approx E_g D/de$ will cause a breakdown. In actuality the bands are bent near the interface with the insulating layer. To equalize the chemical potentials of ZnO and Bi_2O_3, some electrons are transferred to the Bi_2O_3. This causes a depletion layer to be formed near the surface of the ZnO. Band bending results and a barrier of height V_b is established. Tunneling will occur through the insulating region of thickness t, as well as the depletion region of the adjacent grain (Fig. 15.10). The tunneling is mostly through the Schottky barrier formed in the depletion layer. It is considerably thicker than the actual Bi_2O_3 layer. Once the difference of potential between the grains reaches the value $(E_g - V_b)/e$, electrons may tunnel from the top of the valence band on the grain surface to the bottom of the conduction band in the interior of the other grain. This rapid onset of tunneling is responsible for the rapid variation of current with voltage change.

ZnO is used primarily to suppress large amplitude pulses of relatively long duration (>1 μs). SrTiO$_3$ is used where the voltage amplitudes are smaller but the duration is shorter (>3 ns). The preference for these materials involves the large heat capacity and high thermal conductivity, which permit Joule heating to be rapidly dissipated.

Some of the materials issues involve the migration of Zn interstitials to the grain boundaries when a current surge heats the varistor and leads to the subsequent degradation of the device. Ions such as Na$^+$ may be introduced to preferentially occupy the interstitial sites and thereby obstruct the Zn mobility. The varistor functions best when the grain sizes are all of comparable size. Certain additives, such as Si, inhibit grain growth, whereas other elements, such as Be and Ti, encourage grain growth.

15.9   β-Aluminas and Ionic Transport in Solids

The rapid transport of ions through solids can be utilized in a variety of applications, including solid-state batteries, ion exchangers, catalysts, fuel cells, water electrolyzers, hydrogen storage reservoirs, filters, and chemical sensors. The mobility of some ions at elevated temperatures can approach those found in ionic solutions. One way of thinking about the phenomenon, in fact, is in terms of the melting of atoms in one sublattice of the crystal while another sublattice retains its long-range order. One class of such *superionic conductors* is the β-aluminas: β-alumina and β″-alumina. Even at room temperature the conductivity of potassium β″-alumina is 10 Ω^{-1}m^{-1}.

The structure of the β-aluminas consists of structural blocks with a modified spinel structure separated by sheets through which ions can move. The chemical formula for the normal spinel block is D$_8$T$_{16}$O$_{32}$, where D is a divalent ion and T is a trivalent ion. The spinel block is formed by assembling eight subcubes of FCC close-packed oxygens in a $2 \times 2 \times 2$ structure. There are 64 possible tetrahedral (t) interstitial sites and 32 possible octahedral (o) interstitial sites. In the normal spinel, D occupies eight of the t sites and T occupies 16 of the o sites. In the inverse spinel, eight D and eight T ions sit on the t sites and eight T sit on the o sites. In β-aluminas there are only T atoms [Al (or Fe or Ga)] instead of both T and D atoms. Sodium β-alumina has the chemical composition Na$_{2+2x}$Al$_{22}$O$_{34+x}$, where $0.15 < x < 0.30$. The β″-alumina has some stabilizing D ions present, with a typical formula being Na$_{2+2x}$Mg$_{2x}$Al$_{22-2x}$O$_{34}$ and $x \approx 0.66 \pm 0.03$. The stabilization facilitates the growing of large single crystals.

The densely occupied spinel blocks are separated by 40-nm-thick sheets with loosely packed bridging oxygens and mobile sodium ions. It is in these sheets that the action takes place. The unit cell of the crystal consists of three spinel blocks interspersed between three sheets. The length of the cell is 339 nm. The sodium ions may be replaced by other monovalent ions, such as K$^+$, Li$^+$, NH$_4^+$, or even hydrated protons, or by divalent ions such as Pb^{2+} or Ba^{2+}. The high conductivity stems from the ease with which the Na$^+$ ions can move around. In the β″-aluminas the ions hop from a site to a vacancy, creating a vacancy in its original site. In the β-aluminas there is a simultaneous jump of two Na$^+$ ions (the interstitialcy mechanism). In Fig. 15.11 a schematic diagram is presented of the atoms in the neighborhood of a conducting sheet in β″-alumina. Due to the imperfect stoichiometry of the materials, they are neither perfect crystals nor are they random structures. If they were perfectly ordered, the ions would tend to be localized at potential minima and it would be difficult to get them to hop from site to site. Large-scale coordinated motions of adjacent ions

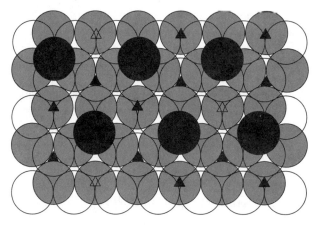

Figure 15.11. β''-alumina: (shaded, dark, open) circles denote (below, in, above)-plane oxygen ions; (dark, light) triangles are (occupied, vacant) Na$^+$ BR sites. Each Na$^+$ ion is covered by another triad of O^{2-} ions. [From B. Dunn et al., *Mater. Res. Soc. Bull.*, **14**(9), 22 (1989).]

TABLE 15.4 Barrier Heights, E_a, and Conductivities, σ, for the β-aluminas

Cation	E_a (eV)	σ ($T = 300$ K) $(\Omega \cdot m)^{-1}$
β-Alumina		
Na$^+$	0.15	3.58
Ag$^+$	0.16	1.30
Li$^+$	0.24	0.30
K$^+$	0.28	0.016
Rb$^+$	0.39	0.0012
β''-Alumina		
K$^+$	0.15	10.0
Na$^+$	0.33	2.50
Li$^+$	0.30	0.50
Ag$^+$	0.19	0.40
H$_3$O$^+$–NH$_4^+$	0.30	0.01
Cd^{2+}	0.61	0.0001

Source: Data from J. B. Bates, G. C. Weng, and N. G. Dudney, Solid electrolytes: the beta-aluminas, *Phys. Today*, July 1982.

assist in effectively lowering the potential barriers for hopping to about 0.15 eV. The coordination numbers of the cation sites are also somewhat peculiar. For example, there exists a sevenfold coordinated site called a *Beevers–Ross* (BR) *site*, in which the cation is surrounded by seven O^{2-} ions.

The ion jump rate, $\Gamma = \nu \exp(-\beta E_a)$, is determined by the barrier height, E_a, and the attempt frequency, ν. Here $\beta = 1/k_B T$. If a is the mean jump distance, $D = a^2 \Gamma$ is the ion diffusivity. The conductivity follows from the Nernst–Einstein relation $\sigma = \beta n q^2 D$. In Table 15.4 the barrier heights and room-temperature conductivities for some of the β-aluminas are tabulated.

A Na/S battery that operates at 300°C can be constructed with β-alumina as the solid electrolyte. A core of liquid sodium is surrounded first by the electrolyte and then by a sheath of sulfur-impregnated carbon felt. The open-circuit voltage is 2.076 V at 350°C. The anode reaction involves creating Na^+ ions which are transported into the β''-alumina. The cathode reaction involves the creation of polysulfide ions S_3^{2-}, S_4^{2-}, and S_5^{2-}. The net product is Na_2S_n, with $n = 3, 4$, or 5.

REFERENCES

Buchanan, R. C., eds., *Ceramic Materials for Electronics*, Marcel Dekker, New York, 1986.

Frohlich, H., *Theory of Dielectrics*, 2nd ed., Clarendon Press, Oxford, 1958.

Hench, L. L., and H. K. West, *Principles of Electronic Ceramics*, Wiley, New York, 1990.

Ikeda, T., *Fundamentals of Piezoelectricity*, Oxford University Press, New York, 1990.

Capacitance

Koteki, D. E., *Semicond. Int.*, Nov. 1996, p. 109.

Beta-Aluminas

Bates, J. B., J. C. Wang, and N. J. Dudney, Solid electrolytes: the beta aluminas, *Phys. Today*, July 1982, p. 46.

PROBLEMS

15.1 Given the Landau free-energy density for a ferroelectric of the form

$$g = g_0 + \frac{a}{2}P^2 + \frac{b}{4}(P_x^4 + P_y^4 + P_z^4) + \frac{c}{2}(P_y^2 P_z^2 + P_z^2 P_x^2 + P_x^2 P_y^2) - EP_z,$$

where $b > c$. Let $a = a_0 (T - T_c)$ and assume that b and c are constant. Find P_z and χ as a function of T for the state of thermal equilibrium.

15.2 Design a piezoelectric actuator that can be used to sweep an STM head over the surface of a solid. What is the area that can practically be covered?

15.3 Adapt Weiss molecular field theory (see Chapter 9) to describe a ferroelectric. Assume that there are just two orientations for the electric-dipole moment of a unit cell and that NN cells interact via an exchange interaction. Obtain the hysteresis curve and values for the coercive field E_c, saturation polarization P_{sat}, and remanent polarization P_{rem}.

15.4 $BaTiO_3$ is a paraelectric for $T > T_C = 130°C$ and has a Curie constant $C = 76,000$ K (see Table 15.3).

  (a) If the lattice constant for the cubic unit cell of $BaTiO_3$ is $a = 0.401$ nm, calculate the electric-dipole moment μ of this unit cell.

  (b) What would the corresponding polarization $P = \mu n$ be at $T = 0$ K?

Note: An additional problem is given in Chapter W15.

![CHAPTER 16]

CHAPTER 16

Superconductors

16.1 Introduction

The discovery of superconductivity was made possible by the liquefaction of helium in 1908 by H. Kamerlingh Onnes at Leiden University in the Netherlands. Following this considerable technological feat, Kamerlingh Onnes then began to use liquid He as a cryogenic fluid to provide the lower temperatures needed for the study of the properties of materials at temperatures below the normal boiling point of liquid He at $T_b = 4.2$ K. During one such investigation in 1911, a student measured the electrical resistance of a rod of solid Hg and noted that the voltage drop along the length of the rod fell rapidly to zero below $T = 4.15$ K, now known to be the *superconducting transition temperature* T_c for Hg. After observing that the voltage reappeared above T_c, it was concluded[†] that Hg had actually lost all of its resistance to the flow of electrical current for $T \leq T_c$. The behavior of the electrical resistance observed in the Hg rod is shown in Fig. 16.1.

Superconductors have been studied actively since their discovery in 1911 and are of great current interest both for the fundamental reason that they exhibit quantum phenomena on a macroscopic scale and also for the important technological applications which they have in superconducting magnets, in sensors based on superconducting quantum interference devices (SQUIDs), and for a wide variety of uses in electromagnetic equipment (e.g., motors, power transmission cables, etc.).

The most important characteristic property of a superconductor is the disappearance of its electrical resistance below the critical temperature T_c. Thus superconductors are first and foremost perfect conductors of electricity. Many other properties also undergo drastic changes when materials, typically metals, enter the superconducting state. For example, superconductors exhibit the Meissner effect in low applied magnetic fields H (i.e., the magnetic flux density B is zero within the bulk of the superconductor). These and other fascinating macroscopic and microscopic properties of bulk superconductors are discussed in more detail in Sections 16.2 and 16.3, where the characteristic intrinsic properties of superconductors are presented. Useful macroscopic and microscopic models for explaining and describing superconductivity are also presented. Examples of the wide variety of known superconducting materials are then introduced and discussed in Sections 16.4 and 16.5. Important present and potential applications of both bulk and thin-film superconductors are described in Sections 16.6 and 16.7.

[†] H. Kamerlingh Onnes, *Leiden Commun.*, **122b**, **124c** (1911).

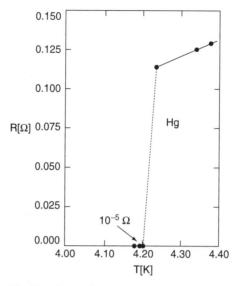

Figure 16.1. Behavior of the electrical resistance for a rod of Hg at its superconducting transition temperature $T_c = 4.15$ K. (From C. Kittel, *Introduction to Solid State Physics*, 7th ed., copyright 1996, John Wiley & Sons, Inc. Reprinted by permission of John Wiley & Sons, Inc.)

CHARACTERISTIC PROPERTIES OF SUPERCONDUCTORS

Since the initial discovery of superconductivity in 1911, an ever-increasing number of superconducting materials has been discovered, with T_c values occurring at ever-higher temperatures. The record for the highest T_c has increased steadily, but slowly, from $T_c = 4.15$ K for Hg in 1911 up to 1986, when the record was held by the intermetallic compound Nb_3Ge, with $T_c = 23.2$ K. The rate of increase of T_c with time then increased dramatically with the discovery by Bednorz and Mueller[†] of the first high-T_c superconductor, lanthanum barium copper oxide ($La_{2-x}Ba_xCuO_4$), with a T_c value close to 36 K. Since then, activity in the field of superconductivity has increased exponentially, and as a result, the number of copper oxide–based or cuprate superconductors has grown steadily, as has the record T_c, which is generally accepted to be 135 K for $HgBa_2Ca_2Cu_3O_8$. Figure 16.2 shows the plot of the record T_c versus time from 1911 to the present. The break in the slope in 1986 stands out quite clearly. The current recordholders, the copper oxide–based superconductors, are discussed in detail in Section 16.5.

Most metals which are not otherwise magnetic, and even some which are, become superconductors when cooled to sufficiently low temperatures. Notable exceptions are the alkali metals (Li, Na, K, etc.), the alkaline earth metals (Mg, Ca, Sr, etc.), and the noble metals (Cu, Ag, and Au). These metals do not become superconducting even down to about $T \approx 10^{-3}$ K and lower. As long as delocalized electrons (i.e., electrons that are not tied up in strong chemical bonds) are present, a material has a good chance of eventually becoming a superconductor as the temperature is lowered.

[†] J. G. Bednorz and K. A. Mueller, *Z. Phys. B*, **64**, 189 (1986).

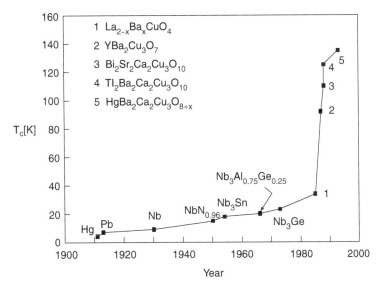

Figure 16.2. Graph of the highest-known superconducting transition temperature T_c versus year of discovery from 1911 to the present.

There exist two distinct classes of superconductors, *types I* and *II*, with type I corresponding to essentially all the elemental superconductors and type II to most alloy and compound superconductors. The distinctions between these two classes will be made clear in the following discussions of the macroscopic and microscopic properties of superconductors.

16.2 Macroscopic Properties and Models

The macroscopic properties of bulk superconductors include their response to applied electric and magnetic fields as well as their thermal and optical properties. These characteristic macroscopic properties are presented next, and some interesting and important connections between the magnetic and thermodynamic properties are discussed. The useful two-fluid, London, and Ginzburg–Landau macroscopic models of superconductivity are also presented.

Electrical Properties. The unique response of superconductors to applied electrical fields is their most characteristic property. The response in the dc or zero-frequency limit corresponds to *perfect conductivity* (i.e., zero resistivity). At higher frequencies in the microwave and infrared regions, however, superconductors can exhibit losses and at sufficiently high frequencies actually behave as normal metals. It should also be mentioned that superconductors in their normal states above T_c are typically good, but not always the best, conductors of electricity.

Perfect Conductivity. It has been well established experimentally that superconductors are perfect conductors with zero electrical resistivity ρ, at least when carrying dc or low-frequency electrical currents and in the presence of magnetic fields that are not too high. There exist, however, critical values of current above which superconductivity is destroyed. To illustrate this, consider the transport current density J flowing through a long type I superconducting wire of radius R. In this case the magnetic field at the surface of the wire is given by $H(r = R) = JR/2$. This type I superconducting wire will make a transition to the normal state when the magnetic field H at its surface is equal to the thermodynamic critical magnetic field H_c, discussed in more detail later. Thus the *critical current density* for the destruction of superconductivity in a long wire of radius R is given by

$$J_c = \frac{2H_c}{R}. \tag{16.1}$$

It will be shown later that the transport current in a superconducting wire can flow only in a layer of thickness λ at its surface where magnetic flux can penetrate. Here λ is the *penetration depth* for magnetic flux. The critical current will be reduced below the value given above if an external field H is also applied to the wire.

In type II superconductors energy dissipation due to the flow of electrical current can occur in the state known as the *superconducting mixed state*. This behavior is quite important for the applications of superconductors and is discussed in more detail in Section 16.6.

Magnetic Properties. Consider the response of a ring of material to a changing magnetic field $\mathbf{H}(t)$ applied along its axis of symmetry. If the material is an electrical conductor, currents will be induced in it whose magnitude $i(t)$ is determined by Faraday's law and by the electrical impedance of the ring. The direction of the induced currents is determined by Lenz's law and is such that their own magnetic field opposes the change in the external field $\mathbf{H}(t)$. In a normal metal these induced currents will decay exponentially when the external magnetic field stops changing, according to $i(t) = i_0 e^{-t/\tau}$. Here the time constant $\tau = L/R$, where R and L are the resistance and the self-inductance of the ring, respectively. If the material is a superconductor below its critical temperature, the currents induced will be superconducting *screening currents*. These screening currents will persist indefinitely even when the external field stops changing, since $\tau = \infty$ when $R = 0$. These induced superconducting currents are responsible for the Meissner effect observed in superconductors and have been found to remain constant, with no observable decay, for several years.

Meissner State. Depending on its external shape and on the direction of an applied magnetic field $\mathbf{H}$, a superconducting material can completely screen or expel magnetic flux from its interior as long as H does not exceed certain critical values which are the *thermodynamic critical field* H_c for type I superconductors and the *lower critical field* $H_{c1} < H_c$ for type II superconductors. This ability of a superconductor to screen magnetic flux from its interior, known as the *Meissner effect*,[†] is as characteristic a property of superconductors as is perfect conductivity. The *Meissner state* is usually considered to correspond to a magnetic flux density $\mathbf{B} = 0$ inside the superconductor, although this is strictly true only if the intrinsic magnetization $\mathbf{M}_i$ of the material in the normal state is zero. The total magnetization $\mathbf{M}$ of a superconductor will in general arise from two contributions: the contribution $\mathbf{M}_s$, due to the superconducting screening

---

[†] W. Meissner and R. Ochsenfeld, *Naturwissenschaften*, **21**, 787 (1933).

currents, and the intrinsic magnetization $\mathbf{M}_i$, present even in the normal state. $\mathbf{M}_i$ can be either paramagnetic or diamagnetic. In some superconductors $\mathbf{M}_i$ cannot be ignored, especially in high magnetic fields. This point is addressed in Section 16.6.

The relationship between $\mathbf{B}$ and $\mathbf{H}$ for a superconductor can be written as follows, assuming that $\mathbf{M}_i = 0$:

$$\mathbf{B} = \mu_0(\mathbf{H} + \mathbf{M}_s) = 0, \tag{16.2}$$

where $\mu_0 = 4\pi \times 10^{-7}$ N/A^2 is the permeability of free space and $\mathbf{M}_s$ is the magnetization resulting from the superconducting screening currents. The observed reversibility of the transition from the superconducting to the normal state in an applied magnetic field is evidence that superconductors are not simply perfect conductors. As the magnetic field is decreased below the critical field, the magnetic flux due to $\mathbf{H}$ would be trapped inside a perfect conductor, whereas in a superconductor the magnetic flux is instead apparently expelled from its interior.

There are two ways of viewing the magnetic response of a superconductor in the Meissner state where $\mathbf{B} = 0$. The conceptually correct view is that there are induced superconducting screening currents of density $\mathbf{J}_s$ flowing just inside the surface whose magnitude and distribution are such as to generate a field $\mathbf{H}_s = -\mathbf{H}$ inside the superconductor. They also produce a magnetic field outside the superconductor which leads to a distorted external field distribution. The magnetization $\mathbf{M}_s$ from this point of view results from the magnetic moment $\mathbf{m}_s$ of the screening current density $\mathbf{J}_s$. The more conventional but technically incorrect view is that the superconductor is a *perfect diamagnet* (i.e., that it has an effective diamagnetic magnetization $\mathbf{M}_s = -\mathbf{H}$ and a corresponding effective diamagnetic susceptibililily $\chi_s = M_s/H = -1$ in SI units). These two distinct ways of viewing the superconducting Meissner state are illustrated in Fig. 16.3. The superconductor does have a net magnetic moment $\mathbf{m}_s$ opposite in direction to $\mathbf{H}$, but this is due to the induced screening currents $\mathbf{J}_s$ and not to any inherent diamagnetic magnetization. With this caution, the more conventional view that $\mathbf{M}_s = -\mathbf{H}$ will be used here[†].

Just as there are limits to the ability of a superconductor to conduct electrical currents without offering resistance to their flow, there are also limits to its ability to screen out magnetic flux completely (i.e., to remain in the Meissner state). When externally applied magnetic fields exceed certain critical values, magnetic flux can enter the bulk of the superconductor in one of three different ways:

1. The entire bulk of the superconductor can make a transition to the normal state. This transition occurs at the critical field $H = H_c$ and is determined by thermodynamics.

2. The superconductor can enter an *intermediate state* at a magnetic field $H < H_c$ in which the bulk is subdivided into alternating superconducting and normal regions. This transition is determined by the shape of the superconductor and by the direction of the applied magnetic field.

3. The superconductor can enter the *mixed state* in which magnetic flux enters the bulk in the form of individual *flux quanta*, that is, in units of $\Phi_0 = h/2e =$

[†] The magnetization $\mathbf{M}_s$ and the screening current density $\mathbf{J}_s$ are related through the Maxwell's equation for $\nabla \times \mathbf{H}$ by $\mathbf{J}_s = \nabla \times \mathbf{M}_s$.

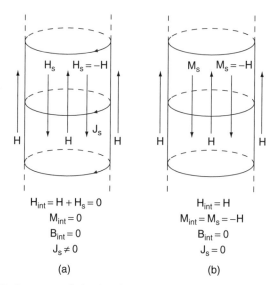

Figure 16.3. Two distinct ways of viewing the Meissner state of a long superconducting cylinder with $B_{int} = 0$ for the case of a magnetic field **H** applied along its axis: (*a*) induced superconducting screening currents J_s flowing just inside the surface of the superconductor generate a field $H_s = -H$ inside the superconductor; (*b*) the superconductor is a perfect diamagnet (i.e., it has a large effective diamagnetic magnetization $M_s = -H$ and a magnetic susceptibility $\chi_s = -1$).

2.068×10^{-15} T · m^2, where the units T · m^2 are equivalent to N · m/A ($\Phi_0 = hc/2e = 2.068 \times 10^{-7}$ G · cm^2 in cgs-emu). Note that the charge $2e$, instead of simply e, appears in Φ_0. The reason for this will become clear when the Bardeen–Cooper–Schrieffer (BCS) theory of superconductivity, in which pairs of electrons play a key role, is presented. This transition to the mixed state occurs when the magnetic field at the surface of the superconductor reaches the critical value $H_{c1} < H_c$.

These three different types of transitions in external magnetic fields serve to distinguish between what are known as type I and II superconductors, with cases 1 and 2 occurring in type I and case 3 occurring only in type II. Other important differences between type I and II superconductors are discussed later.

As mentioned earlier, the phase transition from the Meissner state directly to the normal state is determined by thermodynamics; that is, for H less than the thermodynamic critical field H_c, the Gibbs free energy per unit volume $G_n(H, T)$ of the normal state is higher than the corresponding free energy $G_s(H, T)$ of the superconducting state. The reverse is true for $H > H_c$. This transition in a magnetic field is reversible and is a first-order phase transition. When expressed in terms of the apparent diamagnetic magnetization $M_s = -H$ of the superconducting state, the dependence of G_s on the external field when **H** is applied parallel to the axis of a long superconducting cylinder is given by

$$G_s(\mathbf{H}, T) = G_s(0, T) - \mu_0 \int_0^{\mathbf{H}} \mathbf{M}_s(\mathbf{H}) \, d\mathbf{H} = G_s(0, T) + \frac{\mu_0 H^2}{2}. \tag{16.3}$$

The integral represents the work per unit volume done by the external field **H** in magnetizing the superconductor.

At the thermodynamic critical field H_c the free energies of the two phases are equal, so

$$G_n(H_c, T) = G_s(H_c, T) = G_s(0, T) + \frac{\mu_0 H_c(T)^2}{2}. \tag{16.4}$$

Since the Gibbs free energy of the normal state is ordinarily unaffected by external fields,[†] it follows that $G_n(H_c, T) = G_n(0, T)$. Thus in zero external field, that is, for $H = 0$, Eq. (16.4) can be written in the form

$$G_n(0, T) - G_s(0, T) = \frac{\mu_0 H_c(T)^2}{2}. \tag{16.5}$$

This is the fundamental equation for the thermodynamic treatment of superconductivity as developed by Gorter and Casimir.[‡] The quantity $G_n - G_s$ is often referred to as the *superconducting condensation energy*. Note that for the typical superconductor Pb this condensation energy is quite small, only $\approx 1 \times 10^{-7}$ eV per conduction electron. The dependencies of G_n and G_s on H are shown schematically in Fig. 16.4 for $T < T_c$. Note that $G_s = G_n$ at $T = T_c$ and that $G_s > G_n$ for $H > H_c$.

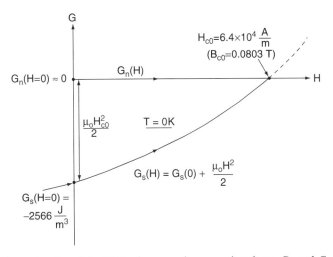

Figure 16.4. Dependencies of the Gibbs free energies per unit volume G_n and G_s of the normal and superconducting states, respectively, on magnetic field H for a long cylinder of Pb at $T = 0$ K.

[†] See Section 16.6 for a description of the effect that the paramagnetic susceptibility of the normal metal has on the stability of the superconducting state in high magnetic fields.
[‡] C. J. Gorter and H. B. G. Casimir, *Physica*, **1**, 306 (1934).

Measurements of the thermodynamic critical field for many superconductors have shown that the parabolic expression known as *Tuyn's law*,

$$H_c(T) = H_{c0}\left(1 - \frac{T^2}{T_c^2}\right),\tag{16.6}$$

provides a useful approximation that is typically accurate to within a few percent. Here H_{c0} is the critical field at $T = 0$ K. Note that in this case the slope of the critical field curve at T_c is given by $dH_c/dT = -2H_{c0}/T_c$. Thus an estimate for H_{c0} can be obtained from measured values of this slope and of T_c. Experimental results for $H_c(T)$ for several type I superconductors are presented in Fig. 16.5. The transition temperatures T_c, critical magnetic fields H_{c0} at $T = 0$ K, and Debye temperatures Θ_D are listed for the elemental superconductors in Table 16.1.

The reversible transition between the Meissner state and the normal state for the case of an external field **H** applied parallel to the axis of a long cylindrical superconductor occurs at the thermodynamic field H_c. This transition is conveniently shown in the form of a magnetization curve (curve a in Fig. 16.6, i.e., as a plot of $-\mathbf{M}_s$ versus **H**).

Intermediate State. In all cases other than that of a magnetic field **H** applied parallel to the axis of a long cylindrical superconductor or parallel to the surface of a superconducting plate or thin film, the superconductor actually enters an intermediate state at a magnetic field $H_c' < H_c$. This is a result of demagnetizing effects which cause the magnetic field H_i inside the superconductor to exceed the applied magnetic field H. If N is the demagnetizing factor of the superconductor for the direction of the applied field **H**, it can be shown that the internal field $H_i = H/(1 - N)$. Thus the superconductor will enter the intermediate state when $H_i = H_c$ [i.e., for an external field $H_c' = (1 - N)H_c$. Some examples of demagnetizing effects are:

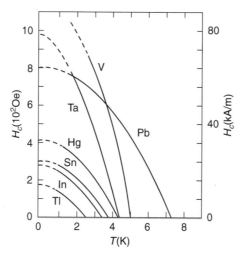

Figure 16.5. Experimental results for the thermodynamic critical field $H_c(T)$ for several type I superconductors. (From G. Burns, *Solid State Physics*, copyright 1985. Reprinted by permission of Academic Press, Inc.)

TABLE 16.1 Superconducting Elements: Transition Temperatures T_c (K), Critical Magnetic Fields H_{c0} at $T = 0$ K (oersteds, with 1 Oe = 79.6 A/m), and Debye Temperatures Θ_D (K)

Each element cell lists (top to bottom): T_c, H_{c0}, Θ_D.

1	2	3	4	5	6	7	8	9	10	11	12	13	14	15	16	17	18
H																	He
Li	Be 0.026 / — / —											B	C	N	O	F	Ne
Na	Mg											Al 1.175 / 105 / 420	Si	P	S	Cl	Ar
K	Ca	Sc	Ti 0.40 / 56 / 415	V 5.40 / 1408 / 383	Cr	Mn	Fe	Co	Ni	Cu	Zn 0.85 / 54 / 310	Ga 1.083 / 59.2 / 325	Ge	As	Se	Br	Kr
Rb	Sr	Y	Zr 0.61 / 47 / 290	Nb 9.25 / 2060 / 276	Mo 0.92 / 96 / 460	Tc 7.8 / 1410 / 411	Ru 0.49 / 69 / 580	Rh	Pd	Ag	Cd 0.517 / 28 / 209	In 3.408 / 282 / 109	Sn 3.72 / 305 / 195	Sb	Te	I	Xe
Cs	Ba	LaFCC 4.88 / 800 / 151	Hf 0.128 / — / —	Ta 4.47 / 829 / 258	W 0.015 / 1.15 / 383	Re 1.70 / 200 / 415	Os 0.66 / 70 / 500	Ir 0.113 / 16 / 425	Pt	Au	Hg 4.154 / 411 / 87	Tl 2.38 / 178 / 78.5	Pb 7.20 / 803 / 96	Bi	Po	At	Rn
Fr	Ra	Ac															

Ce	Pr	Nd	Pm	Sm	Eu	Gd	Tb	Dy	Ho	Er	Tm	Yb	Lu 0.1 / — / 350
Th 1.38 / 160 / 165	Pa 1.4 / — / —	U	Np	Pu	Am	Cm	Bk	Cf	Es	Fm	Md	No	Lr

Source: Data from B. W. Roberts, *J. Phys. Chem. Ref. Data*, **5**, 581 (1976).

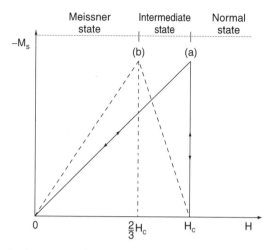

Figure 16.6. Magnetization curves for type I superconductors: curve a, magnetic field **H** is applied parallel to the axis of a long cylindrical superconductor; curve b, magnetic field **H** is applied to a superconducting sphere.

1. $N = 0$ and $H'_c = H_c$ for **H** applied parallel to the axis of a long superconducting cylinder (or parallel to the surface of a superconducting plate or thin film)
2. $N = \frac{1}{3}$ and $H'_c = 2H_c/3$ for **H** applied to a superconducting sphere in any direction
3. $N = \frac{1}{2}$ and $H'_c = H_c/2$ for **H** applied perpendicular to the axis of an (infinitely) long superconducting cylinder
4. $N \approx 1$ and $H'_c \approx 0$ for **H** applied perpendicular to the plane of a superconducting plate or thin film

These values of N are the same as those given in Fig. 17.3; demagnetizing effects in magnetic materials are discussed in Chapter 17.

The dependence of the magnetization $\mathbf{M}_s$ of a superconducting sphere on an external field **H** is illustrated in curve b of Fig. 16.6. For $H > H'_c = 2H_c/3$, the sphere can lower its free energy G_s by entering an inhomogeneous state consisting of alternating superconducting and normal regions, with $\mathbf{B} = 0$ in the former and $\mathbf{B} = \mu_0 \mathbf{H}_c$ in the latter. These alternating regions are laminar in form and have been observed experimentally using techniques similar to those used to view magnetic domains in ferromagnets and antiferromagnets. For $2H_c/3 < H < H_c$ the sphere is in the intermediate state, with the normal regions growing in size as H increases until they occupy the entire volume of the sphere at $H = H_c$.

In type I superconductors there is a net positive energy associated with interfaces between superconducting and normal regions which is in fact required to explain the Meissner effect. The existence of this positive *surface energy* σ_{sn} indicates that energy is required to form nuclei of the normal phase within a superconductor, and vice versa. The metastable existence of the superconducting state for $H > H_c$ and of the normal state for $H < H_c$ is therefore possible. These phenomena are known as *superheating* and *supercooling*, respectively, and can be observed only in type I superconductors.

The surface energy σ_{sn} is discussed in more detail when the Ginzburg–Landau theory of superconductivity is presented later in this section.

Mixed State. For many superconductors the surface energy σ_{sn} is negative, a situation that actually favors the coexistence of normal and superconducting regions. The necessary conditions for this to occur in a given superconducting material are discussed later. Superconductors with $\sigma_{sn} < 0$ are known as type II and have magnetic properties which are quite different from those of type I superconductors. For example, in type II superconductors it is energetically favorable not only for normal regions to be present below the bulk thermodynamic critical field H_c but also for superconducting regions to be present above H_c. There are thus two additional critical fields $H_{c1} < H_c$ and $H_{c2} > H_c$ for type II superconductors, corresponding to the upper limits of the Meissner state and the mixed state, respectively. The field H_{c1} is referred to as the *lower critical field* and H_{c2} as the *upper critical field*.

The reversible magnetization curve of an ideal type II superconducting cylinder is shown in Fig. 16.7a, where it can be seen that $H_{c1} < H_c < H_{c2}$. The dependencies of the magnetic flux density B inside both type I and II superconductors on the applied magnetic field H are shown schematically in Fig. 16.7b, corresponding to the case of **H** applied parallel to the axis of a long superconducting cylinder. The schematic H–T phase diagram for a type II superconductor shown in Fig. 16.7c indicates the regions of existence of the mixed state and the Meissner state. The critical fields H_{c1} and H_{c2} have parabolic temperature dependencies which are similar to that of H_c given in Eq. (16.6).

As mentioned earlier, in the mixed state, magnetic flux enters the bulk of the superconductor, creating normal regions in the form of vortices whose axes are aligned along the direction of the applied field **H**. For this reason the mixed state is also known as the *vortex state*. The superconducting screening currents that circulate around each vortex are the source of the magnetic flux in each vortex. The increase in G_s in the mixed state due to the kinetic energy of the supercurrents is partially compensated by the reduction in the field energy term $\mu_0 H^2/2$.

The density of vortices per unit area of the mixed state is given by B/Φ_0, where B is the average flux density present in the mixed state and Φ_0 is the flux quantum (see Fig. 16.7b). When the density of vortices is high enough for repulsive vortex–vortex interactions to occur, a triangular vortex array is observed. Additional properties of vortices and of the vortex state which have been predicted on the basis of the London model and the Ginzburg–Landau theory are presented in more detail later. The interactions of vortices with each other and with transport currents are described in Section 16.6 in the discussion of critical currents in type II superconductors.

Finally, superconductivity can persist at the surface of a superconductor up to a *surface critical field* $H_{c3} = 1.7 H_{c2}$ (i.e., in magnetic fields for which the bulk of the superconductor has already returned to the normal state). This is known as *surface superconductivity* and can be observed in both type I and II superconductors.

Thermal Properties. The Gibbs free energy of a superconductor relative to its value in the normal state and its relation to applied magnetic fields have been discussed in the preceding section. The thermal properties of a superconductor (i.e., its specific heat and entropy) are discussed next.

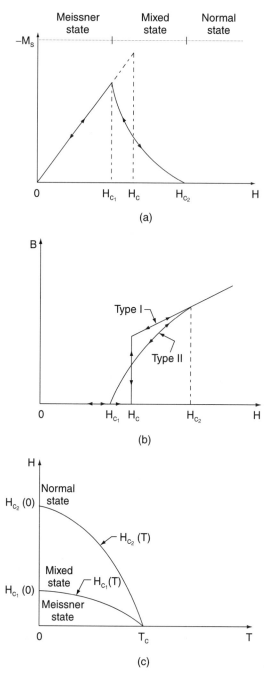

Figure 16.7. (*a*) Reversible magnetization curve for a type II superconducting cylinder for a magnetic field H applied parallel to its axis; (*b*) dependencies of the magnetic flux density B inside type I and II superconductors on applied field H; (*c*) schematic $H-T$ phase diagram for a type II superconductor.

Specific Heat and Entropy. As in the normal state, the specific heat $C_s = C_{es} + C_{ls}$ of a superconductor consists of contributions from the electrons, C_{es}, and from the lattice, C_{ls}. Whereas the lattice or phonon contribution C_{ls} is unchanged from that of the normal state, the electron contribution C_{es} is quite different from the corresponding normal-state value given by $C_{en} = \gamma T$. The measured specific heats C_s in the superconducting state (for $H = 0$) and C_n in the normal state (for $H > H_c$) are shown in Fig. 16.8 for the type I superconductor Al. As T decreases, C_s exhibits a discontinuous jump at T_c, then falls rapidly below C_n and approaches zero exponentially as $T \rightarrow 0$ K according to $\exp(-2\varepsilon(0)/2k_B T)$. At T_c the specific heat jump $\Delta C_s = C_s(T_c) - C_n(T_c) = C_{es}(T_c) - C_{en}(T_c)$ is observed to be approximately equal to $1.4\gamma T_c$ for many type I superconductors and is predicted from thermodynamic arguments to be equal to $4\mu_0 H_{c0}^2/T_c$ when Eqs. (16.5) and (16.6) are used.

The observed exponential dependence of C_{es} on T as $T \rightarrow 0$ K provides strong evidence for the existence of an *energy gap* of magnitude $2\varepsilon(0)$ in the density of electron states in the superconductor, with $C_{es}(T)$ proportional to the concentration $n_n(T)$ of normal electrons or quasiparticles excited across the energy gap. This result has played a key role in the development of microscopic theories of superconductivity.

The entropy $S_{es}(T)$ of the electrons in the superconducting state can be obtained from the measured specific heat $C_{es}(T)$ using the expression

$$S_{es}(T) = \int_0^T \frac{C_{es}(T')\,dT'}{T'}. \qquad (16.7)$$

It is found that $S_{es} < S_{en}$ for $0 < T < T_c$ (i.e., the electrons in the superconducting state are in a more ordered phase than the corresponding normal electrons). According

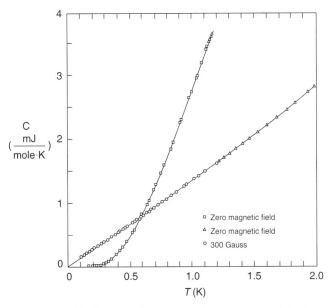

Figure 16.8. Measured specific heats in the superconducting state and in the normal state for the type I superconductor Al. [From N. E. Phillips, *Phys. Rev.*, **114**, 676 (1959). Copyright 1959 by the American Physical Society.]

to the microscopic BCS theory this ordered phase results from a condensation of the superconducting electrons in momentum space.

At $T = T_c$ and at $T = 0$ K it is found that $S_{es} = S_{en}$ [i.e., there is no latent heat $L = T(S_{en} - S_{es})$ associated with the transition at these two temperatures]. The absence of a latent heat at $T = T_c$ shows that the normal-to-superconducting transition in zero applied magnetic field is *second order* (i.e., there is no discontinuous change in the density of superconducting electrons at T_c). The transition is *first order*, however, for $0 < T < T_c$ when a magnetic field is applied.

Thermal Conductivity. As in the normal state, the thermal conductivity κ_s of a super-conductor is given by the sum of contributions of the conduction electrons and the lattice, $\kappa_s = \kappa_{es} + \kappa_{ls}$. Both the electron and lattice contributions are affected in important ways by the transition to the superconducting state because the electrons that are condensed (i.e., superconducting) can no longer carry heat and can no longer participate in electron–phonon scattering processes. It is therefore expected and in fact observed for $T < T_c$ that $\kappa_{es} < \kappa_{en}$ and that $\kappa_{ls} > \kappa_{ln}$ where κ_{en} and κ_{ln} are the electron and lattice contributions in the normal state, respectively. Thus the electrons are able to transport less heat in the superconducting state than in the normal state, while the phonons can transport more heat due to their longer mean free paths in the superconducting state. These effects are illustrated in Chapter W16 at our Web site.[†]

Thermopower. Thermal effects associated with the transport of entropy by the charge carriers, such as the *Seebeck effect* due to a temperature gradient and the *Peltier effect* due to the flow of electric current, are observed in normal metals. These effects are absent in superconductors, however, at least in zero magnetic field. This is due to the fact that the superconducting electrons cannot transport entropy in their condensed state.

Optical Properties. An interesting experimental result is the absence of any obvious change in the visible appearance of a material when it enters the superconducting state (i.e., its optical properties in the visible region are unchanged by the transition). This result is somewhat unexpected since the imaginary part $\varepsilon_2(\omega)$ of the optical dielectric function, which is a measure of the ability of a material to absorb light, is proportional to the electrical conductivity $\sigma(\omega)$ of the material (see Section 8.3). Thus although perfect dc conductivity exists [i.e., $\sigma_s(\omega = 0) = \infty$], the same cannot be said of the high-frequency conductivity $\sigma_s(\omega)$ of the superconductor, which is found to be the same as that of the normal state, $\sigma_n(\omega)$. The absorption of electromagnetic energy by superconductors in the far-infrared and the microwave regions is discussed next.

Far-Infrared Absorption. Direct evidence for an energy gap 2ε in the spectrum of electron energy states of a superconductor can be obtained from measurements in the far-infrared frequency range, where the absorption of electromagnetic energy by a superconducting thin film can fall well below that found in the normal state. This decrease in absorption is observed for frequencies $\omega < \omega_g = 2\varepsilon/\hbar$. Typical frequencies

---

[†] Supplementary material for this textbook is included on the Web at the resource site (ftp://ftp.wiley.com/ public/sci_tech_med/materials). Cross-references to elements of the Web material are prefixed by "W."

corresponding to this absorption edge lie in the microwave or far-infrared range (i.e., $f_g = \omega_g/2\pi \approx 10^{10}$ to 10^{12} Hz). The response of the superconductor is observed to be the same as that of the normal metal at frequencies $\omega > \omega_g$.

The frequency dependencies of the real and imaginary parts of the complex electrical conductivity $\sigma_s(\omega) = \sigma_{1s}(\omega) + i\sigma_{2s}(\omega)$ of the superconductor for $T < T_c$ relative to the normal state conductivity σ_n can be obtained from the measured ratio T_s/T_n of the transmittances for a thin film in the superconducting and normal states, respectively. The real component $\sigma_{1s}(\omega)$ is responsible for the absorption of light and is zero for $\omega < 2\varepsilon/\hbar$, while the measured imaginary component $\sigma_{2s}(\omega)/\sigma_n$ is found to be given approximately by $(4 \text{ to } 5)k_BT_c/\hbar\omega$. This result for $\sigma_{2s}(\omega)/\sigma_n$ is consistent with both the Meissner effect and the existence of the penetration depth λ for the magnetic field. In addition, from a standard Kramers–Kronig analysis, which allows $\sigma_{1s}(\omega)$ to be derived from the measured $\sigma_{2s}(\omega)$, it is found that $\sigma_{1s}(\omega) \to \infty$ as $\omega \to 0$. This is, of course, the infinite zero-frequency conductivity found more directly from standard electrical measurements.

The superconducting energy gap measured from far-infrared absorption studies is observed to be temperature dependent, with $2\varepsilon(T) \to 0$ as $T \to T_c$. As $T \to 0$ K, the gap approaches the value $2\varepsilon(0)$, which is observed to be approximately equal to $3.5k_BT_c$ for a wide range of superconductors (see Table 16.2).

Microwave Surface Resistance. When exposed to high-frequency electromagnetic fields ($f \approx 10^{11}$ Hz), superconductors are no longer perfect conductors and instead exhibit losses. In this microwave region the surface resistance of a superconductor remains finite even at temperatures $T < T_c$, with a value $R_s(T)$ which is proportional to the concentration $n_n(T)$ of electrons that remain "normal." The normal electrons correspond to the conduction electrons which are not condensed into, or have been excited out of, the lower-energy superconducting state. In the direct-current or zero-frequency limit, the normal electrons will be "short-circuited" by the superconducting electrons, and the superconductor will therefore exhibit perfect conductivity. At sufficiently high

TABLE 16.2   Experimental Values[a] of the Superconducting Energy Gap Ratio $2\varepsilon(0)/k_BT_c$

Superconductor (T_c)	$\dfrac{2\varepsilon(0)}{k_BT_c}(C_{es})$	$\dfrac{2\varepsilon(0)}{k_BT_c}$ (IR)	$\dfrac{2\varepsilon(0)}{k_BT_c}(i\text{–}V)$
Al (1.175 K)	2.9	—	3.3–3.43
Sn (3.7 K)	3.6	3.6	3.46–3.65
Pb (7.2 K)	—	4.14	4.18–4.33
	$2\varepsilon(0)/k_BT_c$ (average)		
Nb (9.3 K)	3.8		
Nb$_3$Sn (18 K)	3.0		
Ba$_{0.6}$K$_{0.4}$BiO$_3$ (18.5 K)	3.7		
La$_{1.85}$Sr$_{0.15}$CuO$_4$ (36 K)	4.3		
YBa$_2$Cu$_3$O$_7$ (87 K)	4.0		
Bi$_2$Sr$_2$Ca$_2$Cu$_2$O$_{10}$ (108 K)	5.7		

Source: Data from E. A. Lynton, *Superconductivity*, 2nd ed., Methuen, London, 1964, p. 99; and C. P. Poole, Jr., H. A. Farach, and R. J. Creswick, *Superconductivity*, Academic Press, San Diego, Calif., 1995, p. 167.

[a] Values of the energy gap $2\varepsilon(0)$ at $T = 0$ K are obtained as follows: C_{es}, specific heat; IR, far-infrared absorption; i–V electron tunneling.

frequencies the inertia of the superconducting electrons will prevent them from staying in phase with the applied electromagnetic field, and hence the normal electrons will also be able to participate in the electrical response of the superconductor.

Only at $T = 0$ K when $n_n = 0$ will the surface resistance of the superconductor remain equal to zero for frequencies up to the gap frequency $f_g = 2\varepsilon(0)/h \approx 10^{10}$ to 10^{12} Hz, where $2\varepsilon(0) \approx 0.1$ to 10 meV. The measured ratio R_s/R_n of the microwave surface resistances in the superconducting and normal states, illustrated in Fig. 16.9 for superconducting Al, exhibits a temperature dependence resulting from the following factors:

1. The concentration of normal electrons $n_n(T)$ drops rapidly for $T < T_c$ and decreases to zero as $T \to 0$ K.
2. The superconducting energy gap $2\varepsilon(T)$ increases rapidly from zero at T_c and reaches the value $2\varepsilon(0)$ at $T = 0$ K.
3. The penetration depth $\lambda(T)$ of the microwave field into the superconductor approaches infinity as $T \to T_c$ and decreases rapidly for $T < T_c$.

As a result, the ratio R_s/R_n and the corresponding losses in the superconducting state increase with both increasing frequency and increasing temperature.

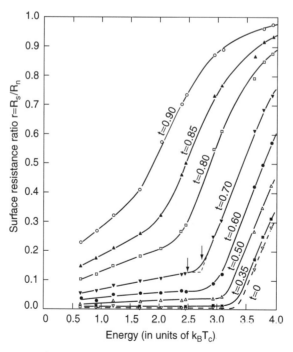

Figure 16.9. Measured ratio R_s/R_n of the surface resistances in the superconducting and normal states for superconducting Al with $T_c = 1.18$ K and $k_BT_c = 0.102$ meV. The symbol $t = T/T_c$ is the reduced temperature. [From M. A. Biondi et al., *Phys. Rev.*, **116**, 853 (1959). Copyright 1959 by the American Physical Society.]

Macroscopic Models. Macroscopic models of superconductivity are usually phenomenological in nature and are capable of providing at best semiquantitative descriptions
of certain properties of superconductors. They have nevertheless played important roles
in the development of our understanding of superconductivity. The macroscopic models
which are described here include the London model for the electromagnetic response of
superconductors and the thermodynamic theory of Ginzburg and Landau. The two-fluid
thermodynamic model of Gorter and Casimir is discussed in Chapter W16.

London Model. F. London and H. London proposed a classical model[†] for the low-
frequency electromagnetic behavior of superconductors in 1935 in order to explain
the Meissner effect, which had been discovered in 1933. The *London equation* for a
superconductor in an external magnetic field can be written as

$$\mathbf{J}(\mathbf{r}) = -\frac{n_s e^2}{m}\mathbf{A}(\mathbf{r}) = -\frac{1}{\mu_0 \lambda_L^2}\mathbf{A}(\mathbf{r}), \tag{16.8}$$

or, equivalently, as

$$\nabla \times \mathbf{J}(\mathbf{r}) = -\frac{n_s e^2}{m}\nabla \times \mathbf{A}(\mathbf{r}) = -\frac{n_s e^2}{m}\mathbf{B}(\mathbf{r}). \tag{16.9}$$

Here n_s is the *concentration of superconducting electrons* and $\mathbf{A}(\mathbf{r})$ is the *vector poten-
tial* of the magnetic field, such that $\mathbf{B}(\mathbf{r}) = \mu_0(\mathbf{H} + \mathbf{M}_s) = \nabla \times \mathbf{A}(\mathbf{r})$. The London
equation is a local expression since $\mathbf{J}(\mathbf{r})$, $\mathbf{B}(\mathbf{r})$, and $\mathbf{A}(\mathbf{r})$ are all evaluated at the same
point $\mathbf{r}$. The *London penetration depth* λ_L is defined by

$$\lambda_L(T) = \sqrt{\frac{m}{\mu_0 n_s(T)e^2}}. \tag{16.10}$$

Using the two-fluid expression $n_s(T)/n_s(0 \text{ K}) = 1 - (T/T_c)^4$, it follows that

$$\lambda_L(T) = \frac{\lambda_L(0)}{\sqrt{1 - (T/T_c)^4}}. \tag{16.11}$$

Thus $\lambda_L(T) \to \infty$ as $T \to T_c$.

For the case of an external field $\mathbf{H}$ applied parallel to the surface of a bulk superconductor (Fig. 16.10), the London equation (16.8) in conjunction with Ampere's law,

$$\nabla \times \mathbf{H} = \mu_0 \nabla \times \mathbf{B} = \mathbf{J}, \tag{16.12}$$

yields for the spatial variation of the resulting flux density $\mathbf{B}_y$ inside the superconductor
and normal to its surface the result

$$B_y(x) = B_y(0)e^{-x/\lambda_L}. \tag{16.13}$$

[†] F. London and H. London, *Proc. R. Soc. A*, **149**, 71 (1935); *Physica*, **2**, 241 (1935).

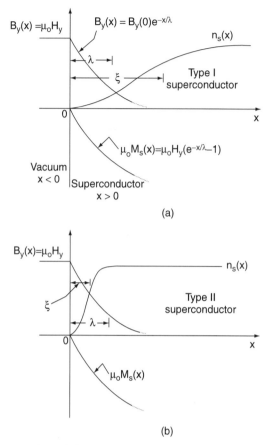

Figure 16.10. The penetration of the magnetic flux density $\mathbf{B}_y(x)$ at the surface of bulk type I and II superconductors: (a) type I superconductor with $\xi > \lambda$; (b) type II superconductor with $\xi < \lambda$. The variations of the density of superconducting electrons n_s and of the effective magnetization $\mathbf{M}_s$ due to the supercurrents are also shown.

The magnetic flux is thus predicted to penetrate into the superconductor only at its surface, with $B_y(x)$ decreasing exponentially over the characteristic distance λ_L. Note that the magnetic field $\mathbf{H}$ inside the superconductor is the same as the applied field (see Fig. 16.3b). The variations normal to the surface of n_s and of the effective magnetization $\mathbf{M}_s$ due to the supercurrents are also shown. Here the superconductor is viewed as a perfect diamagnet.

The supercurrent density $\mathbf{J}_s$ is directed perpendicular to $\mathbf{H}$ according to Ampère's law and also decreases exponentially into the superconductor according to

$$\mathbf{J}_s(x) = \mathbf{J}_s(0)e^{-x/\lambda_L}, \tag{16.14}$$

where $J_s(0) = B_y(0)/\mu_0\lambda_L$. This is the superconducting screening current which prevents the penetration of magnetic flux into the bulk of the superconductor. The Meissner effect thus follows directly from the London model. Transport currents passing through superconductors are also restricted to flowing in surface regions of width λ_L. The predicted penetration of **B** at the surface has been confirmed experimentally and is most easily observable for superconducting samples which have at least one dimension comparable to λ_L (e.g., thin films, thin wires, or small spheres).

The observed penetration depth λ will be temperature dependent through $n_s(T)$ and should approach infinity as $T \rightarrow T_c$, where $n_s \rightarrow 0$. At $T = 0$ K, where $n_s = n$, the prediction for the superconductor Al is that $\lambda_L(0$ K$) = 12.5$ nm. Experimental values for the penetration depth $\lambda(0$ K$)$ are typically about five times greater than the London prediction $\lambda_L(0$ K$)$. Experimental values of $\lambda(0$ K$)$ for a wide range of superconductors are presented in Table 16.4.

When magnetic flux does penetrate into the bulk of a superconductor, the London equation predicts that the corresponding total magnetic flux Φ contained in any normal-state regions will be quantized in units of $\Phi_0 = h/q$ (i.e., $\Phi = m\Phi_0$, where m is an integer and q is the charge of the fundamental conducting units in the superconductor). This prediction for the quantization of flux has been demonstrated experimentally[†] where it has been found that $q = 2e$ (i.e., the conducting units correspond to pairs of superconducting electrons).

The London model has been extended to a nonlocal representation of superconductivity by Pippard[‡]. This nonlocal approach takes into account the finite *coherence length* ξ of the *superconducting order parameter* (i.e., the density n_s of superconducting electrons in bulk superconductors cannot change in space abruptly but only over distances on the order of ξ). Note that $\xi(T)$ also diverges at T_c. In the nonlocal Pippard model the penetration depth in a pure superconductor is given by

$$\lambda(l \gg \xi_0) \approx (\xi_0 \lambda_L^2)^{1/3}, \tag{16.15}$$

where l is the electron mean free path and ξ_0 is the corresponding coherence length in the pure superconductor at $T = 0$ K. When $\xi_0 \gg \lambda_L$, as is often the case in pure type I superconductors, it follows that λ will be much greater than λ_L, in agreement with experiment. Values of ξ_0 are also given in Table 16.4.

Interesting size effects can be observed in superconductors when one or more of the dimensions of a superconducting sample (e.g., a thin film or wire) becomes comparable to or less than either λ or ξ. For example, the critical field of a small superconductor can be increased well above H_c when magnetic flux penetrates into a significant fraction of its volume.

Ginzburg–Landau Theory. In 1950, Ginzburg and Landau (G-L) presented a phenomenological thermodynamic theory§ of superconductivity as a second-order phase transition. The *G-L theory* is expressed in terms of a superconducting order parameter or wavefunction ψ whose physical significance is that $|\psi|^2 = n_s^* = n_s/2$ where n_s is the density of superconducting electrons. As in the approach of Pippard, the G-L theory is nonlocal and in addition includes the surface energy σ_{sn} associated with interfaces between superconducting and normal regions. In the G-L theory both ψ and n_s depend on temperature T and also on position **r**.

[†] R. Doll and M. Naebauer, *Phys. Rev. Lett.*, **7**, 51 (1961); B. S. Deaver, Jr. and W. M. Fairbank, *Phys. Rev. Lett.*, **7**, 43 (1961).

[‡] A. B. Pippard, *Proc. R. Soc. A*, **203**, 210 (1950).

§ V. L. Ginzburg and L. Landau, *Zh. Eksp. Teor. Fiz.*, **20**, 1064 (1950).

The ability of the quantum-mechanical G-L theory to predict the behavior of super-conductors in external magnetic fields $H \approx H_c$ is its main achievement. The Gibbs free energy per unit volume $G_s(H,T)$ of the superconducting state in the G-L theory includes not only the field energy term $\mu_0 H^2/2$ [see Eq. (16.3)], but also a term that accounts for spatial variations of $\psi(\mathbf{r})$ through its gradient $\nabla\psi$. The inclusion of such a term has the effect of requiring that ψ not change too rapidly in space. The G-L expression for G_s, valid just below T_c, is

$$G_s(\mathbf{H}, T) = G_n(0, T) + \alpha|\psi|^2 + \frac{\beta|\psi|^4}{2} + \frac{\mu_0 H^2}{2} + \frac{[-i\hbar\nabla\psi - e^*\mathbf{A}\psi]^2}{2m^*}, \tag{16.16}$$

where $\alpha < 0$ and $\beta > 0$ are parameters of the model that depend only on T. Here e^* and m^* are the charge and mass of the fundamental conducting units in the superconducting state, and $\mathbf{A}$ is the vector potential. The two G-L equations relating ψ and $\mathbf{A}$ are obtained[†] by minimizing $G_s(\mathbf{H}, T)$ with respect to variations in both ψ and $\mathbf{A}$. It has been shown that the central equations of the G-L theory valid near T_c can be derived from the microscopic BCS theory with the results that $e^* = 2e$ and that the real part of ψ is proportional to the superconducting energy gap 2ε. Quantization of magnetic flux Φ in units of $\Phi_0 = h/e^* = h/2e$ occurs naturally within the framework of the G-L theory.

Three important parameters of the G-L theory are the critical magnetic field $H_c(T)$, the penetration depth $\lambda(T)$, and the dimensionless *Ginzburg–Landau parameter* $\kappa(T) \approx \lambda/\xi$, which is related to H_c and λ by

$$\kappa = \frac{2\sqrt{2}\, e\mu_0 H_c \lambda^2}{\hbar} = \frac{2\pi\sqrt{2}\, \mu_0 H_c \lambda^2}{\Phi_0}. \tag{16.17}$$

As $H_c \to 0$ and $\lambda \to \infty$ at $T = T_c$, κ approaches a finite value $\kappa(T_c)$. In general, $\kappa(T) = \lambda(T)/\xi(T)$. For a pure metal in which the coherence length is not limited by mean-free-path effects, $\kappa(T_c)$ can also be expressed in terms of the penetration depth and the coherence length by $\kappa(T_c) = 0.96\lambda(0)/\xi_0$, where $\xi_0 = \xi(0\text{ K})$. Using Eq. (16.17), the critical field H_c can be expressed in terms of λ and ξ by

$$H_c(T) = \frac{\Phi_0}{2\pi\mu_0\sqrt{2}\, \lambda(T)\xi(T)}. \tag{16.18}$$

The surface energy σ_{sn} can be shown to appear naturally within the G-L theory, as follows. Consider the changes in B and n_s that occur at the surface of a superconductor, as shown in Fig. 16.10. Similar changes in B and n_s also occur at the interfaces between normal and superconducting regions. The penetration of B into the superconductor over a length λ lowers the magnetic contribution to G_s of the superconducting state by an amount proportional to λ. In addition, the decrease of n_s over a length ξ within the superconductor at the interface raises G_s due to the loss of condensation energy of the superconducting electrons by an amount proportional to ξ. Both effects contribute to

---

[†] For derivations and discussions of the two G-L equations, see de Gennes (1966) and Lynton (1964).

the net change in energy per unit area σ_{sn} at such an interface, with the result that

$$\sigma_{sn} \approx (\xi - \lambda)\frac{\mu_0 H_c^2}{2}. \tag{16.19}$$

The important result here is that $\sigma_{sn} \propto (\xi - \lambda)$. When $\xi > \lambda$ and $\sigma_{sn} > 0$ (Fig. 16.10a), there is a net energy cost in creating an interface since the loss of condensation energy outweighs the gain due to field penetration. This is the necessary condition for the existence of the Meissner effect and type I superconductivity. The converse is true when $\xi < \lambda$ and $\sigma_{sn} < 0$ (Fig. 16.10b), in which case the formation of superconducting–normal interfaces (i.e., the creation of normal regions containing magnetic flux within the superconductor) is energetically favorable. This is the case in type II superconductors where ξ is small and λ is large due either to low n_s or to mean-free-path effects that destroy the coherence of the superconducting wavefunction ψ.

Ginzburg and Landau showed that $\sigma_{sn} = 0$ for $\kappa = 1/\sqrt{2}$. Therefore, the condition for type II superconductivity can be expressed by $\kappa \geq 1/\sqrt{2}$, while $\kappa < 1/\sqrt{2}$ for type I superconductors. When the electron mean free path l decreases due to impurity scattering or other effects, λ will increase and ξ will decrease according to

$$\lambda(l) = \lambda(\infty)\sqrt{1 + \frac{\xi}{l}} \quad \text{and} \quad \frac{1}{\xi(l)} = \frac{1}{\xi_0} + \frac{1}{l}, \tag{16.20}$$

where $\lambda(\infty)$ and ξ_0 correspond to the clean limit where $l \gg \xi_0$. The parameter κ will therefore also increase as l decreases and for $\xi_0 \gg l$, $\kappa(T_c) = 0.75\lambda_L(0\,\text{K})/l$. This increase in κ can lead to a transition from type I to type II behavior in superconducting alloys (e.g., in superconducting $Sn_{1-x}In_x$ alloys at $x \approx 0.025$).

The upper critical field H_{c2} for type II superconductors is given in terms of κ and H_c in the G-L theory by

$$H_{c2}(T) = \sqrt{2}\,\kappa(T)H_c(T). \tag{16.21}$$

This equation is also valid for type I superconductors where $H_{c2} < H_c$. In this case H_{c2} will be the absolute lower limit for the metastable existence of the normal state in an applied magnetic field below H_c. Using the definition of H_c in terms of κ given in Eq. (16.17), this result can also be expressed as

$$H_{c2}(T) = \frac{\Phi_0}{2\pi\mu_0\xi^2(T)}. \tag{16.22}$$

The ratio of the lower and upper critical fields can be expressed in terms of κ by

$$\frac{H_{c1}}{H_{c2}} \approx \frac{\ln\kappa}{2\kappa^2}, \tag{16.23}$$

which shows that $H_{c1} \ll H_{c2}$ for type II superconductors, which have $\kappa \gg 1$.

The detailed nature of the mixed state in type II superconductors was first predicted by Abrikosov[†] using the G-L theory. As a result of the negative surface energy

---

[†] A. A. Abrikosov, *Sov. Phys. JETP*, **5**, 1174 (1957).

σ_{sn}, interfaces between superconducting and normal regions are energetically favored, leading to the penetration of magnetic flux into the superconductor for $H > H_{c1}$. According to Abrikosov, the flux penetrates in units of the flux quantum Φ_0, with the resulting normal regions or vortices forming a "lattice" that is aligned parallel to the external field. The magnetic flux density $B = \mu_0 H_c$ in the normal core of each vortex. Each vortex has an inner region or normal core of radius ξ where $n_s \approx 0$ and a region of radius λ where magnetic flux is present. Triangular arrays of vortices have, in fact, been observed experimentally, as shown in Fig. 16.11a for the high-T_c superconductor $YBa_2Cu_3O_{7-x}$. The spatial variations of the flux density B and of n_s for an isolated vortex are shown schematically in Fig. 16.11b. Note that superconducting screening currents that are induced by the external magnetic field surround each vortex and are in fact the source of most of the flux in the vortex.

A pair of vortices will interact strongly with each other when they are separated by less than the penetration depth λ. The magnetic flux associated with a given vortex

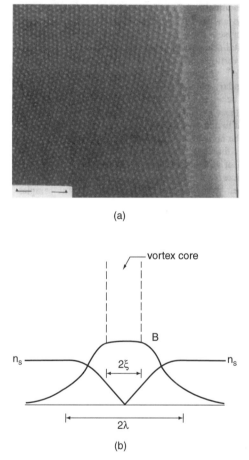

Figure 16.11. (a) Triangular "lattice" of vortices decorated with fine ferromagnetic particles and observed in the mixed state of the type II high-T_c superconductor $YBa_2Cu_3O_{7-x}$. The marker shown is 10 μm long. [From G. J. Dolan et al., *Phys. Rev. Lett.*, **62**, 827 (1989). Copyright 1989 by the American Physical Society.] (b) Variations of the flux density B and of n_s at an isolated vortex.

will exert a Lorentz force on the screening currents that surround neighboring vortices. The resulting vortex–vortex forces are repulsive and decrease exponentially with the distance between the vortices. Additional forces on vortices due to transport currents flowing through superconductors are discussed in Section 16.6. For H just below H_{c2}, the vortices have their maximum density, with a mutual separation $\approx \xi < \lambda$. This result is consistent with Eq. (16.22), which states that the flux density B_{c2} in the superconductor at H_{c2} corresponds to a single quantum of flux Φ_0 within an area $\approx \pi\xi^2$ associated with each vortex.

The predictions of the G-L model for the existence of a negative surface energy σ_{sn}, for type I and II superconductivity, and for the triangular flux lattice have been verified by a wide range of experimental studies. The G-L theory also predicts that superconductivity can persist at the surface of a superconductor up to a field $H_{c3} = 1.7H_{c2}$, where H_{c2} is the critical field for the destruction of superconductivity in the bulk. This superconducting surface sheath has a thickness on the order of ξ and has been observed in both type I and II superconductors using electrical and magnetic measurements. Although successful as a phenomenological theory, the G-L theory is nevertheless unable to explain why electrons enter the superconducting state in the first place.

16.3 Microscopic Properties and Models

While the discussions of the macroscopic properties of superconductors and of some macroscopic theories proposed to explain them provide a useful overall picture of super-conductivity, the microscopic origins of superconductivity still need to be addressed. The goal of this section is to present some of the essential microscopic properties of superconductors, followed by the description of the successful microscopic theory of superconductivity presented by Bardeen, Cooper, and Schrieffer in 1957.

It is interesting to note that while the elements, alloys, and compounds that become superconducting are electrical conductors in their normal states above T_c, they do not have the highest-known normal-state electrical conductivities σ. In fact, the noble metals Cu, Ag, and Au with the highest electrical conductivities do not become superconductors at all, at least not above $T \approx 10^{-3}$ K. The best conductors have room-temperature conductivities $\sigma(\text{RT}) \approx 5 \times 10^7 \ \Omega^{-1} \ \text{m}^{-1}$, while the high-$T_c$ ceramic superconductors have $\sigma(\text{RT}) \approx 10^5$ to $10^6 \ \Omega^{-1} \ \text{m}^{-1}$, factors of 50 to 500 less than Cu. It is clearly a challenge to understand this unexpected behavior on a microscopic basis in terms of the properties of the electrons and their interactions with each other and with the lattice.

Microscopic Properties. The microscopic properties discussed here are the gap of magnitude 2ε in the energy spectrum of electron states in the superconductor, the isotope effect involving the average mass of the ions in the superconductor and its effect on T_c, and the tunneling of electrons into and out of superconductors.

Superconducting Energy Gap. As indicated previously, indirect evidence for the presence of an energy gap 2ε in the spectrum of allowed states for the conduction electrons in a superconductor can be found from measurements of the temperature dependence of the electronic contributions to the specific heat C_{es}. It is found that the temperature dependence is exponential and proportional to $\exp[-2\varepsilon(0)/2k_BT]$ as

$T \to 0$ K. The factor of 2 appears in the denominator of the exponent because two quasiparticles are created when the gap energy 2ε is provided through the excitation process. Thus the energy required per quasiparticle is just ε.

More direct evidence for the existence of an energy gap is obtained from the measured optical response of a superconductor, where, as described earlier, absorption of electromagnetic energy falls well below that of the normal state for frequencies such that $\omega < \omega_g = 2\varepsilon/\hbar$. As can be seen in Table 16.2, the energy gaps determined from thermal, optical, and electron-tunneling measurements on a given superconductor are found to be the same to within experimental uncertainty. Typical measured values for $2\varepsilon(0)$ are as follows: 0.35 meV for Al with $T_c = 1.175$ K, 2.7 meV for Pb with $T_c = 7.2$ K, and 30 meV for $YBa_2Cu_3O_7$ with $T_c = 92$ K. Note that $2\varepsilon(0)$ is proportional to T_c for these superconductors. Also listed in this table are average values of the ratio $2\varepsilon(0)/k_B T_c$ for a variety of superconductors.

The constancy of the ratio $2\varepsilon(0)/k_B T_c$ observed for the elemental metallic superconductors Al, Sn, and Pb with quite different T_c values is clear evidence for the basic similarity of most superconductors when expressed in reduced coordinates. The situation is not as clear with the high-T_c superconductors, where the energy gaps are quite anisotropic, reflecting the strongly anisotropic crystal structures of these materials.

The temperature dependence of the energy gap can be determined from measurements of the microwave surface resistance ratio R_s/R_n as a function of T (see Fig. 16.9). The measured gap $2\varepsilon(T)$ is found to be in good agreement with the predictions of the BCS theory (see Fig. 16.15b).

Anisotropy in the energy gap with respect to the crystalline axes can be inferred from the observed anisotropy in the measured thermal conductivity and also from the dependence of the frequency of the infrared absorption edge on crystal orientation. Typical magnitudes of the energy gap anisotropy in elemental superconductors are $\approx 10\%$. The gap anisotropy can be eliminated when the electron mean free path l is shortened due to the effects of electron scattering from impurities and other defects.

Isotope Effect. For many elemental superconductors such as the metals Hg, Pb, and Sn, the transition temperature T_c is observed to depend on the average isotopic mass M of the atoms in the superconductor according to $T_c \propto M^{-a}$ with $a \approx \frac{1}{2}$. This *isotope effect*, predicted by Froehlich in 1950 and observed experimentally in the same year, and the fact that the Debye temperature Θ_D is also proportional to $M^{-1/2}$ together suggest strongly that interactions between electrons and phonons play an important role in superconductivity. The experimental demonstrations[†] of the isotope effect in Hg in 1950 played a significant role in pointing the way toward the successful microscopic theory of superconductivity proposed by Bardeen, Cooper, and Schrieffer in 1957.

Electron Tunneling. The existence of the superconducting energy gap can have important effects on the tunneling of electrons from a superconductor through an insulating barrier into either a normal metal or another superconductor. As a result, electron-tunneling studies have provided some of the most detailed and direct information about the magnitude and temperature dependence of the superconducting energy gap and also the electron density of states in the superconductor.

[†] E. Maxwell, *Phys. Rev.*, **78**, 477 (1950); C. A. Reynolds, B. Serin, W. H. Wright, and L. B. Nesbitt, *Phys. Rev.*, **78**, 487 (1950).

Consider first the tunneling current of electrons passing from a superconductor S (Pb) through a thin insulating barrier I (aluminum oxide) and into a normal metal N (Al, above $T_c = 1.18$ K), as first measured by Giaever[†]. The experimental configuration, including the densities of electron states in the superconductor and the normal metal, is shown schematically in Fig. 16.12a for $T = 0$ K. A positive voltage V applied to either side of an S–I–N junction will lower the energy levels of the electrons on that side of the junction by an amount equal to eV relative to the electrons on the other side. At $T = 0$ K electrons can tunnel elastically, (i.e., at constant energy), through the junction only when $eV > \varepsilon(0)$. For a positive voltage applied to the normal metal N, electrons will tunnel from just below the energy gap in S to just above E_F in N, while for a positive voltage applied to S, electrons will tunnel from E_F in N to just above the energy gap in S. For $eV \gg \varepsilon(0)$ the tunneling characteristic will approach ohmic behavior with $i = V/R$, as expected for tunneling between two normal metals. Here R is the total resistance of the junction.

The resulting i–V (current–voltage) characteristics shown schematically in Fig. 16.12b demonstrate that the magnitude of the energy gap can be measured with a voltmeter. A tunneling current of electrons can pass through such a S–I–N junction in either direction but only when each tunneling electron can find an unoccupied energy level on the opposite side of the junction at the same or lower energy. It can be seen that for applied voltage $V = 0$, no current can flow at $T = 0$ K in either direction due to the presence of the energy gap of magnitude $2\varepsilon(0)$ centered about the Fermi level E_F in the superconductor.

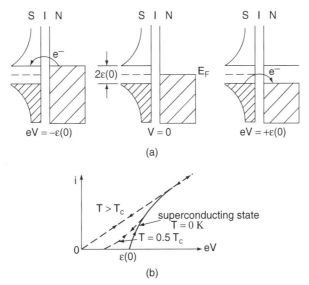

Figure 16.12. Experimental configuration employed for electron-tunneling studies in an S–I–N junction (S, superconductor; I, insulator; N, normal metal): (a) densities of electron states in the superconductor and the normal metal for $T = 0$ K and for $V = 0$, $eV = +\varepsilon(0)$, and $eV = -\varepsilon(0)$; (b) resulting i–V (current–voltage) characteristics for $T = 0$ K, $0 < T < T_c$, and $T > T_c$.

[†] I. Giaever, *Phys. Rev. Lett.*, **5**, 147 (1960).

For $T > 0$ K the excitation of electrons across the energy gap in the superconductor will allow current to flow through the junction in either direction even for $eV < \varepsilon(T)$. A tunneling curve for this case is also shown in Fig. 16.12b. The values of the energy gap obtained from tunneling studies and listed in Table 16.2 are considered to be the most reliable and accurate values available. Measurements of the i–V curves as a function of T allow the temperature dependence of the energy gap to be determined. The measured values of $2\varepsilon(T)$, shown in Fig. 16.15b, have provided an important test of the microscopic BCS theory of superconductivity.

When both sides of a tunnel junction are superconducting (i.e., for an S–I–S junction formed between two identical superconductors separated by a thin insulating layer as shown in Fig. 16.13a), the onset of tunneling at $T = 0$ K occurs at $eV = 2\varepsilon(0)$ instead of at $\varepsilon(0)$ (see Fig. 16.13b). Figure 16.13c shows i–V data obtained for an Al–Al$_2$O$_3$–Al tunnel junction. An S$_1$–I–S$_2$ tunnel junction of two superconductors with different energy gaps $2\varepsilon_1$ and $2\varepsilon_2$ and $2\varepsilon_1 > 2\varepsilon_2$ has an i–V characteristic for $0 < T < T_c$ as shown in Fig. 16.13d. Here the onset of weak tunneling occurs for $eV = 2(\varepsilon_1 - \varepsilon_2)$ while the onset of strong tunneling occurs at the higher voltage $eV = 2(\varepsilon_1 + \varepsilon_2)$. The i–V curves are more complicated in this case and can provide information about both energy gaps when two different superconductors form the junction.

Another contribution to the usual tunneling current of normal electrons can arise from the tunneling of a supercurrent of Cooper pairs across the junction even when the applied voltage $V = 0$ (see Fig. 16.13b). The tunneling of Cooper pairs through an S–I–S junction is at the heart of many important applications of superconductivity involving superconducting quantum interference devices, also known as SQUIDs. The basis for the operation of a SQUID is the Josephson effect, which is discussed in Chapter W16, where applications of superconductivity are presented.

Microscopic Models. The existence of an energy gap and the importance of electron–phonon interactions as indicated by the isotope effect were the two key experimental guideposts which eventually led to the development of a successful microscopic model of superconductivity. An important challenge faced by any successful theory of superconductivity is to identify the mechanism that can lower the Gibbs free energy of the superconducting state relative to that of the normal state by only $\approx 10^{-7}$ eV per conduction electron.

Bardeen–Cooper–Schrieffer Theory. The first successful microscopic theory of superconductivity was presented in 1957 by Bardeen, Cooper, and Schrieffer (*BCS*),[†] who were later awarded the Nobel Prize in Physics for this achievement. It was already known by 1950 that an electron traveling through and interacting with the lattice has a contribution to its energy associated with the emission and reabsorption of *virtual phonons*. These virtual phonons exist for only very short times, $\Delta t \approx 10^{-13}$ s. According to the Heisenberg uncertainty principle, the resulting minimum uncertainty in the energy of the phonon is $\Delta E \approx \hbar/2 \, \Delta t \approx 0.5 \times 10^{-21}$ J. Since a typical phonon energy $E \approx k_B \Theta_D \approx 1.4 \times 10^{-21}$ J with $\Theta_D \approx 100$ K, it follows that $\Delta E/E \approx 1$. Thus energy need not be conserved rigorously in the emission and reabsorption processes involving virtual phonons.

[†] J. Bardeen, L. N. Cooper, and J. R. Schrieffer, *Phys. Rev.*, **108**, 1175 (1957).

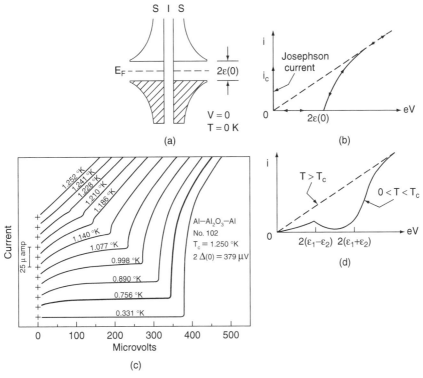

Figure 16.13. S–I–S junction formed between two identical superconductors S separated by a thin insulating layer I: (*a*) Schematic energy level diagram for $T = 0$ K and $V = 0$. (*b*) *i*–*V* characteristic showing the onset of tunneling at eV $= 2\varepsilon(0)$ for $T = 0$ K. The Josephson tunneling current is shown at $V = 0$. (*c*) Tunneling data for an Al–Al$_2$O$_3$–Al S–I–S tunnel junction. [From B. L. Blackford et al., *Can. J. Phys.*, **46**, 141 (1968).] (*d*) *i*–*V* characteristic of an S$_1$–I–S$_2$ tunnel junction of two superconductors with different energy gaps $2\varepsilon_1$ and $2\varepsilon_2$ ($\varepsilon_1 > \varepsilon_2$) for $0 < T < T_c$.

An attractive interaction can occur between a pair of electrons via scattering processes that involve exchanges of virtual phonons. This can be viewed simply as follows. The first electron distorts the lattice instantaneously by pulling in closer to it the neighboring positive ions. The second electron is then attracted to the enhanced region of positive charge created by the first electron. The pairing interaction between the two electrons need not be instantaneous. If the polarization of the lattice lasts for about 10^{-13} s, an electron traveling at $v_F \approx 10^6$ m/s may have traveled about 100 nm before the second electron experiences the lattice polarization. Thus the "paired" electrons can be far apart, which will help to reduce their mutual Coulomb repulsion.

From a more formal point of view, the distortion of the lattice by the first electron with wave vector $\mathbf{k}_1$ corresponds to the emission of a virtual phonon with wave vector $\mathbf{q}$ (i.e., $\mathbf{k}_1 \to \mathbf{k}_1' + \mathbf{q}$ where $\mathbf{k}_1' = \mathbf{k}_1 - \mathbf{q}$). A second electron with wave vector $\mathbf{k}_2$ then absorbs the virtual phonon and is scattered to $\mathbf{k}_2' = \mathbf{k}_2 + \mathbf{q}$. This scattering process is illustrated in Fig. 16.14. The conservation of the total momentum $\hbar\mathbf{k}$ for the overall process is expressed in terms of the wave vectors involved by

$$\mathbf{k}_1 + \mathbf{k}_2 \to \mathbf{k}_1' + \mathbf{k}_2'. \tag{16.24}$$

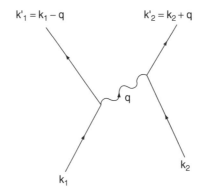

Figure 16.14. Scattering process involving the exchange of a virtual phonon with wave vector **q** by two electrons.

A net attractive interaction and hence an overall lowering of the total energy occurs for those electrons near E_F for which this attractive interaction dominates the usual screened Coulomb repulsion.

The pairs of electrons whose energies are lowered due to the exchange of virtual phonons are the *Cooper pairs*, which play a critical role in the BCS theory. The electrons that can participate in superconductivity are those that lie within an energy $\hbar\omega(q)$ of the Fermi surface, in which case there are empty states available above E_F into which the electrons can be scattered. Here $\hbar\omega(q)$ is an average phonon energy in the material. It has been shown that the lowering of energy for the superconductor is greatest when $\mathbf{k}_1 + \mathbf{k}_2 = \mathbf{k}_1' + \mathbf{k}_2' = 0$ (i.e., when $\mathbf{k}_1 = -\mathbf{k}_2$). It is also energetically favorable for the intrinsic spins **s** of the two electrons to point in opposite directions (i.e., $\mathbf{s}_1 = -\mathbf{s}_2$, so that $\mathbf{s}_1 + \mathbf{s}_2 = 0$). The BCS superconducting state is therefore a singlet ($S = 0$) spin state. Each electron with wave vector **k** and spin **s** ↑ and with an energy within $\hbar\omega(q)$ of E_F can be paired with an electron of wave vector $-\mathbf{k}$ and spin $-\mathbf{s}$ ↓ in a Cooper pair, represented by $(\mathbf{k}\uparrow, -\mathbf{k}\downarrow)$. The transition to the superconducting state can thus be considered to correspond to a condensation of the conduction electrons in momentum space.[†]

When Cooper pairs are present, an energy gap is predicted by BCS to appear at the Fermi surface. This energy gap separates the states of paired electrons below the gap from those of the unpaired normal electrons or *quasiparticles* above the gap. Values of the energy gap $2\varepsilon(0)$ at $T = 0$ K are typically $\approx 10^{-3}$ to 10^{-2} eV, energies that are much greater than the condensation energy $\approx 10^{-7}$ eV of a single electron. Thus exciting a single electron with wave vector **k** across the energy gap effectively disturbs the pairing interaction of $\approx 10^4$ pairs of electrons for which the final state wave vectors $\mathbf{k}_1'$ and $\mathbf{k}_2'$ in principle could have been **k**. This example illustrates clearly that the condensed state is a coherent, highly correlated many-electron state and is not composed simply of distinct, individual Cooper pairs. The BCS prediction for the density of electron states in the superconductor in the region of the energy gap is

---

[†] When transport current flows through a superconductor all of the Cooper pairs will have a net momentum **p**. Cooper pairs should then be represented by $[(\mathbf{k} + \mathbf{p}/\hbar)\uparrow, (-\mathbf{k} + \mathbf{p}/\hbar)\downarrow]$.

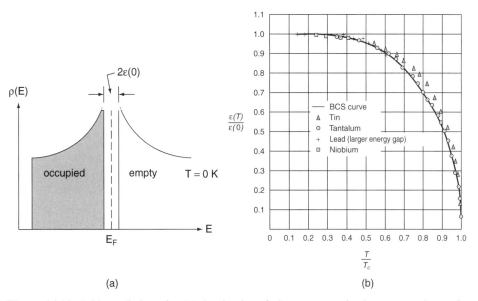

Figure 16.15. BCS predictions for (*a*) the density of electron states in the superconductor for $T = 0$ K, and (*b*) the temperature dependence of the energy gap. Also shown for comparison are data for Sn, Ta, Pb, and Nb. [From P. Townsend et al., *Phys. Rev.*, **128**, 591 (1962). Copyright 1962 by the American Physical Society.]

shown in Fig. 16.15*a* for $T = 0$ K. The states no longer available within the energy gap are shifted to regions just above and below the edges of the gap.

According to BCS, the essential differences between the superconducting and normal states consist only in the net attractive interaction between pairs of electrons due to the exchange of virtual phonons. In the simplest approach the electron-phonon and screened Coulomb interactions are both assumed in the BCS theory to be short-ranged, isotropic in space, and to have a constant net magnitude $U < 0$ within a thin energy shell of width $\approx \hbar\omega(q)$ centered at E_F. The interaction U is assumed to be zero outside this energy shell. Note that the attractive interaction U is to a first approximation independent of the specific electronic properties of any metal [e.g., E_F, $\rho(E_F)$, etc.]. This feature of the BCS theory is in fact consistent with observations made earlier which have stressed that many of the measured properties of superconductors, at least those studied up to 1957 (and with the obvious exception of T_c), show a striking similarity when expressed in reduced coordinates or as functions of $t = T/T_c$.

At finite temperatures an increasing number of electrons are excited across the energy gap. Thermal fluctuations destroy superconductivity for $T > T_c$ just as they destroy ferromagnetism above the Curie temperature T_C. These normal electrons or quasiparticles can be scattered just as in the normal state. There still remains the condensate of superconducting Cooper pairs. A two-fluid point of view is thus consistent with the predictions of the BCS theory. The BCS prediction for the temperature dependence of the energy gap $2\varepsilon(T)$ is shown in Fig. 16.15*b*, where it is compared with the measured $2\varepsilon(T)$ for the superconductors Sn, Ta, Pb, and Nb. It can be seen that $2\varepsilon(T)$ is predicted to decrease with increasing T and to go to zero as $T \to T_c$, in very good agreement with experiment.

The BCS theory correctly predicts the observed thermal and electromagnetic properties of many superconductors, especially those for which $T_c \ll \Theta_D$. This is the *weak-coupling limit* and is appropriate for superconductors in which the electron–phonon interaction is not too strong [i.e., when $\rho(E_F)U \ll 1$]. Since the BCS theory involves complicated many-body quantum-mechanical effects, a detailed discussion is not presented here. Instead, only the important predictions of the standard weak-coupling BCS theory are listed in Table 16.3.

Note that the superconducting energy gap $2\varepsilon(0)$, the condensation energy $G_n - G_s$, and the transition temperature T_c are all predicted to be exponential functions of the product $\rho(E_F)U$ of the electron density of states and the net interaction strength U. Typical values of the net interaction strength in weak-coupling superconductors correspond to $\rho(E_F)U \approx 0.3$, which, from Eq. (16.27), corresponds to $T_c \approx \Theta_D/30$. The weak-coupling limit is not appropriate for superconductors such as Pb and Hg, where $T_c/\Theta_D \approx 0.1$. Pb and Hg are thus said to be *strong-coupling superconductors*. The high-T_c superconductor $YBa_2Cu_3O_7$ with $T_c = 92$ K and $\Theta_D \approx 400$ K is also a strong-coupling superconductor since $T_c/\Theta_D \approx 0.23$.

TABLE 16.3   Important Predictions of the BCS Theory in the Weak-Coupling Limit, $\rho(E_F)U \ll 1$[a]

1. Superconducting energy gap at $T = 0$ K:

$$2\varepsilon(0) = \frac{2\hbar\omega(q)}{\sinh[1/\rho(E_F)U]} \approx 4\hbar\omega(q)\exp\left[-\frac{1}{\rho(E_F)U}\right] = 3.52k_BT_c \qquad (16.25)$$

2. Ground-state condensation energy of a superconductor at $T = 0$ K and $H = 0$:

$$G_n - G_s = \frac{\mu_0 H_{c0}^2}{2} = \frac{2\rho(E_F)[\hbar\omega(q)]^2}{\exp[2/\rho(E_F)U] - 1} \approx 2\rho(E_F)[\hbar\omega(q)]^2\exp\left[-\frac{2}{\rho(E_F)U}\right]$$

$$= \frac{\rho(E_F)\varepsilon(0)^2}{2} \qquad (16.26)$$

3. Superconducting transition temperature:

$$k_BT_c = 1.14\hbar\omega(q)\exp\left[-\frac{1}{\rho(E_F)U}\right] = \frac{2\varepsilon(0)}{3.52} \qquad (16.27)$$

4. Specific heat jump at T_c:

$$C_{es} - C_{en} = 1.43\gamma T_c \qquad (16.28)$$

5. Isotope effect for T_c:

$$T_c \propto \hbar\omega(q) \propto \Theta_D \propto M^{-1/2} \qquad (16.29)$$

[a]H_{c0}, thermodynamic critical magnetic field at $T = 0$ K; $\rho(E_F)$, density of electronic states per unit energy per spin at the Fermi energy E_F in the normal state; $\hbar\omega(q)$, average phonon energy and range of attractive interaction [in the BCS theory $\hbar\omega(q)$ is often approximated by $k_B\Theta_D$, where Θ_D is the Debye temperature]; U, strength of the net attractive interaction between pairs of electrons; γ, coefficient of linear term in electronic specific heat in the normal state, equal to $2\pi^2k_B^2\rho(E_F)/3$ in the free-electron model; M, average ionic mass.

Note that the isotope effect follows naturally from Eq. (16.27) since $\hbar\omega(q) \approx k_B\Theta_D \propto M^{-1/2}$, where M is the average ionic mass. Deviations from an exponent of $-\frac{1}{2}$ can indicate that some of the simplifying assumptions of BCS, such as a cutoff of both the electron–phonon and the screened Coulomb interactions at the same energy $\hbar\omega(q)$, may not be valid for certain superconductors.

One of the most impressive results of BCS is the prediction given in Eq. (16.25) that $2\varepsilon(0) = 3.52k_BT_c$, a result that is in good quantitative agreement with the experimental values for the wide range of superconductors listed in Table 16.2. Strong-coupling superconductors such as Pb and Hg, however, tend to show significant deviations from the quantitative predictions of BCS. For example, Pb and Hg have $2\varepsilon(0) \approx 4k_BT_c$ and exhibit specific heat jumps at T_c which are well above the BCS prediction of $1.43\gamma T_c$. The BCS predictions for $2\varepsilon(T)$, $H_c(T)$, $\lambda(T)$, $S_{es}(T)$, and $C_{es}(T)$ in reduced coordinates have been tabulated[†] as functions of the reduced temperature $t = T/T_c$.

Extensions of the standard BCS theory to account for the observed effects of anisotropy and of strong electron–phonon coupling have included the dependence of the net interaction U on $\mathbf{k}$ and $\mathbf{k}'$ and on energy, respectively. Specific details of the phonon density of states $\rho(\omega)$, defined in Chapter 5, have also been incorporated into the framework of the BCS theory by Eliashberg[‡] who defined a dimensionless *electron–phonon coupling constant* λ_{ep} as

$$\lambda_{ep} = 2\int_0^\infty \frac{\alpha^2(\omega)\rho(\omega)}{\omega}d\omega. \qquad (16.30)$$

Here $\alpha^2(\omega)$ is the strength of the electron–phonon coupling. The phonon density of states $\rho(\omega)$ can be measured by inelastic neutron scattering. The electron–phonon interaction λ_{ep} and T_c are all enhanced in materials such as Pb, where $\rho(\omega)$ is large at low ω. In terms of this coupling constant, the weak- and strong-coupling limits of superconductivity correspond to $\lambda_{ep} \ll 1$ and $\lambda_{ep} \geq 1$, respectively. Typical values of λ_{ep} fall in the range between 0.2 and 2. Since in strong-coupling superconductors the strength of the electron–phonon interaction depends on the details of the phonon density of states, it is not surprising that the universal behavior predicted in the weak-coupling limit of the BCS theory is no longer valid.

When the repulsive screened-Coulomb interaction is also explicitly included in the BCS theory, $\rho(E_F)U$ in Eq. (16.27) should be replaced by the quantity $\lambda_{ep} - \mu^*$, where $\mu^* \approx 0.1$ to 0.2 is a dimensionless parameter characterizing the strength of the Coulomb interaction. The following semiempirical expression[§] for T_c, valid for $\lambda_{ep} < 1.5$ to 2, has often been used,

$$T_c = \frac{\Theta_D}{1.45}\exp\left[\frac{-1.04(1+\lambda_{ep})}{\lambda_{ep} - \mu^*(1+0.62\lambda_{ep})}\right]. \qquad (16.31)$$

[†] B. Muehlschlegel, *Z. Phys.*, **155**, 313 (1959).
[‡] G. M. Eliashberg, *Zh. Eksp. Teor. Fiz.*, **38**, 966 (1960); **39**, 1437 (1960).
[§] W. L. McMillan, *Phys. Rev.*, **167**, 331 (1968).

It can be seen from this equation that a high T_c value is favored by high Θ_D and a high λ_{ep} along with a low μ^*. Typical values[†] for these parameters as illustrated by the type II superconductor Nb are $T_c = 9.22$ K, $\Theta_D = 277$ K, $\lambda_{ep} \approx 0.82$, and $\mu^* \approx 0.13$. Thus the electron–phonon coupling is of intermediate strength in Nb.

The BCS theory has been extended to include the effects of magnetic fields $H \approx H_c$ by Gor'kov. The resulting equations have been shown to be equivalent to the corresponding Ginzburg–Landau expressions near T_c when the penetration depth λ is much greater than the coherence length ξ, as in type II superconductors. In addition, the G-L parameters e^* and m^* for the charge and mass of the charge carriers responsible for superconductivity are identified as $2e$ and $2m$, respectively (i.e., twice the charge and mass of the electron). In addition, the energy gap 2ε has been shown theoretically to be proportional to the magnitude of $\psi(\mathbf{r})$, the G-L order parameter.

It has been shown that the existence of an energy gap in the density of electronic states in a superconductor requires that the relationship between the current density $\mathbf{J}$ and the vector potential $\mathbf{A}$ be nonlocal, as proposed by Pippard. The BCS expression for this relationship is consistent with that of Pippard if the coherence length at $T = 0$ K is given by

$$\xi_0 = \frac{\hbar v_F}{\pi \varepsilon(0)}. \tag{16.32}$$

This expression for ξ_0 is consistent with the range of the superconducting correlations being given by $\xi_0 \approx v_F \tau_C$ when the lifetime τ_C of the Cooper pairs is $\tau_C \approx \hbar / k_B T_c$. This expression for ξ_0 is valid when the electron mean free path $l \gg \xi_0$. Values for ξ_0 calculated from Eq. (16.32) or obtained from measured values of $H_{c2}(0)$ using the G-L expression Eq. (16.22) are presented in Table 16.4 for several superconductors. Although the range of the attractive interaction between pairs of electrons is only ≈ 0.2 nm, the distance ξ_0 over which the electrons are correlated as a result of this interaction can be seen to be much larger, ≈ 10 to 100 nm.

Other Pairing Mechanisms. It has not yet been determined conclusively whether the superconductivity of two new classes of materials, the heavy-fermion and the high-T_c copper oxide–based materials, can be understood in terms of the indirect, attractive electron–phonon interaction mechanism of the usual BCS theory. It does appear, however, that many of the properties of the high-T_c superconductors are consistent with the BCS predictions. This is discussed in more detail in Section 16.5.

If mechanisms other than the usual indirect electron–phonon interaction are responsible for superconductivity, it may still be possible to apply the BCS formalism as long as the appropriate characteristic energy is used instead of the average phonon energy $\hbar \omega_q$. Other suggested mechanisms for superconductivity have included excitons, antiferromagnetic magnons, and optical rather than acoustic phonons.

Coexistence of Superconductivity and Magnetic Moments: Gapless Superconductivity. The strong effects of magnetic moments on superconductivity can be understood, at least qualitatively, on the basis of the BCS theory, in which the pairing mechanism involves electrons with oppositely directed intrinsic spins $\mathbf{s}$. In the spin-flip scattering

---

[†] W. L. McMillan, *Phys. Rev.*, **167**, 331 (1968).

processes described in Chapter 9, the direction of the spin of a conduction electron can be reversed when it is scattered by a localized magnetic moment. Spin-flip scattering is therefore a *pair-breaking* process because a Cooper pair represented by $(\mathbf{k} \uparrow, -\mathbf{k} \downarrow)$ cannot survive the reversal of the spin of only one of the participating electrons.

When impurities such as Mn^{2+} with magnetic moments and spin $S \approx \frac{5}{2}$ are added to a superconductor such as Zn with $T_c = 0.85$ K, the transition temperature of the resulting dilute alloy is rapidly depressed to $T \approx 0$ K by the addition of only 25 parts per million of Mn. The rapid decrease in the electron scattering lifetime τ due to the spin-flip scattering process can also have the interesting effect of producing *gapless superconductivity* in which the superconducting order parameter ψ remains finite while the magnitude of the energy gap 2ε decreases to zero. Thus the existence of an energy gap is not required for the occurrence of superconductivity.

In some superconducting compounds there exist interacting magnetic moments that undergo a cooperative ferromagnetic or antiferromagnetic transition below the superconducting T_c. In some of these materials the superconductivity is destroyed, while in others it can apparently coexist with the magnetic order. In the high-T_c copper oxide–based superconductors magnetic moments within the structure do not have a significant effect on the superconducting properties as long as they are not located in the copper–oxygen planes where the Cooper pairs are formed. The recently discovered family of rare earth–nickel borocarbides (e.g., RNi_2B_2C with R = Y, Dy, Ho, Er, Tm, and Lu) provides examples of superconducting materials with T_c both greater and less than T_N, the Nèel temperature for antiferromagnetic ordering.

EXAMPLES OF SUPERCONDUCTORS

A wide variety of materials including over 20 elements and hundreds of compounds have been found to undergo a superconducting transition as their temperatures are lowered. In this section the superconducting materials discussed include the metallic elements and the oxide-based ceramics. Superconducting alloys of metallic elements, intermetallic compounds, and unusual superconductors are discussed in Chapter W16. These superconductors will be classified according to the nature of their response to applied magnetic fields (i.e., as type I or II superconductors). These two types of magnetic response have already been described in detail in Section 16.2. The great majority of the superconductors which have or are likely to have important technological applications due to their high critical temperatures, critical fields, or critical currents are of the type II variety.

16.4 Metallic Elements

The transition temperatures T_c and the critical magnetic fields H_{c0} at $T = 0$ K of the elements which become superconducting are presented in Table 16.1. All of the superconducting elements with the sole exception of Nb are type I superconductors. Nb is a type II superconductor because of its high T_c value (9.25 K), which leads to a relatively low value of the coherence length ξ (see Table 16.4) and hence to a value of $\kappa \approx \lambda/\xi$, which is greater than $1/\sqrt{2}$.

Note that none of the metallic elements in the first and second groups (i.e., the alkali and alkaline earth metals) or the noble metals Cu, Ag, and Au are superconducting under normal conditions down to at least $\approx 10^{-3}$ K. In these three groups of metals the attractive electron–phonon pairing interaction believed to be responsible for super-conductivity is not sufficiently strong compared to the repulsive screened-Coulomb interaction. This viewpoint is consistent with the observation that the high electrical conductivities which these metals possess also result from weak electron–phonon scattering. In addition, metallic elements (e.g., Fe, Co, Ni, Gd, etc.) that possess permanent magnetic moments (i.e., which are ferro- or antiferromagnetic) also do not exhibit superconductivity. This is due to the destructive spin-flip scattering effect which magnetic moments can have on the superconducting interaction involving Cooper pairs.

Some of the metallic elements become superconducting only under special condi-tions (e.g., as thin films, under pressure, or following irradiation). In these cases the normal structures of the materials are modified so that superconductivity becomes possible. For example, Si and Ge under high pressure undergo transformations from the diamond crystal structure to more compact structures. These structural transforma-tions convert Si and Ge from semiconductors to metals, and as metals, they can then undergo a transition to the superconducting state.

It is interesting to note from Table 16.1 that high T_c values for the metallic elements Pb and Hg tend to be associated with low values of the Debye temperature Θ_D, in apparent contrast to the weak-coupling BCS prediction of Eq. (16.27) or the generalized version Eq. (16.31), which both state that $T_c \propto \Theta_D$. This apparent contradiction can be resolved by noting that high values of the electron–phonon coupling constant λ_{ep} which, according to BCS, favor high T_c values also tend to favor low values of Θ_D. The presence of λ_{ep} in the exponential factor in Eq. (16.31) thus clearly dominates over the effect of a low Θ_D.

Many of the transition metal elements are observed to exhibit superconductivity. Deviations from the value $a = \frac{1}{2}$ for the isotope effect exponent (i.e., $T_c \propto M^{-a}$) are often observed for transition metal superconductors such as Mo and Ru in which d elec-trons participate in the superconductivity along with s (and p) electrons. The metallic elements V ($3d^3$), Nb ($4d^3$), and Ta ($5d^3$) are all hole-type superconductors with open Fermi surfaces, whereas Mo ($4d^4$) and W ($5d^5$) are electron-type superconductors.

In Table 16.4 values of the measured penetration depth $\lambda(0$ K$)$, the calculated coher-ence length $\xi(0$ K$)$, and the Ginzburg–Landau parameter $\kappa(T_c)$ are listed for a selection of elemental and compound superconductors. Although $\kappa(T_c)$ can be determined accu-rately from the ratio of H_{c2}/H_c as $T \to T_c$ using Eq. (16.21), or approximately from λ/ξ, the lengths λ and ξ are in general not as well known.

16.5   Oxide-Based Ceramics

The *high-T_c* oxide-based superconductors differ from other classes of superconductors in that they are ceramics with chemical bonding which is of the mixed ionic–covalent type rather than being primarily metallic. In fact, oxide materials were originally consid-ered to be poor candidates for superconductivity due to their low carrier concentrations. The crystal structures of the oxide-based superconductors are typically tetragonal or orthorhombic rather than cubic. $La_{2-x}Ba_xCuO_4$, the first high-T_c cuprate supercon-ductor, was discovered in 1986 and was followed in 1987 by $YBa_2Cu_3O_7$, the first superconductor developed with $T_c > 77$ K, the boiling point of liquid N_2. An important common feature of the high-T_c cuprates is the presence in these structures of layers of atoms with the composition CuO_2. It is widely believed that the superconductivity

TABLE 16.4 Penetration Depths $\lambda(0\ K)$, Coherence Lengths $\xi(0\ K)$, and Ginzburg–Landau Parameters $\kappa(T_c)$

Superconductor	T_c (K)	$\lambda(0\ K)^a$ (nm)	$\xi(0\ K)^b$ (nm)	$\kappa(T_c)$
Al	1.175	40	1600	0.013^c
Sn	3.72	42	230	0.087^c
Pb	7.2	50	82	0.24^c
Nb	9.25	50	39	1.28
Nb–Ti	9.5	300	4	75
Nb_3Sn	18	65	3	22
Nb_3Ge	23.2	90	3	30
$La_{1.85}Sr_{0.15}CuO_4$	38	200^d	2.0^d	100
$YBa_2Cu_3O_7$	92	170^d	1.8^d	95
$HgBa_2Ca_2Cu_3O_8$	133	130^e	1.3^e	100

Source: Data from C. P. Poole, Jr., H. A. Farach, and R. J. Creswick, *Superconductivity*, Academic Press, San Diego, Calif., 1995, Tables 9.1 and 9.2, p. 271.

[a]The penetration depth $\lambda(0\ K)$ is typically determined experimentally from direct measurements of the penetration of magnetic flux into the superconductor, often in the mixed state.
[b]The coherence length $\xi(0\ K)$ is typically calculated from v_F and $2\varepsilon(0)$ using Eq. (16.32) or from $H_{c2}(0\ K)$ using Eq. (16.22).
[c]These values of $\kappa(T_c)$ have been determined from Eq. (16.21) in the limit $T \to T_c$; see F. W. Smith, A. Baratoff, and M. Cardona, *Phys. Kondens. Mater.*, **12**, 145 (1970).
[d]Averages of polycrystalline data.
[e]Values appropriate to the *ab* plane.

of these materials can be associated with these layers and that the resulting two-dimensional nature of the CuO_2 planes plays an important role in determining the existence of superconductivity at these relatively high temperatures.

These materials are unusual in that they can be transformed from magnetic insulators to metals and from nonsuperconductors to superconductors by varying their compositions over fairly limited ranges via doping or through the control of other processing variables (e.g., oxygen partial pressure or temperature). The structure, bonding, and composition, normal-state properties, and superconducting properties of the high-T_c oxide-based materials are described next.

Structure, Bonding, and Composition. Because of the importance of the *perovskite* structure for the known high-T_c superconductors, the cubic unit cell of the prototypical perovskite material $BaTiO_3$ is shown in Fig. 16.16. There are Ti^{4+} ions at the corners of the unit cell, a Ba^{2+} ion in the body-centered position, and O^{2-} ions at the centers of all 12 edges. The (001) plane containing Ti^{4+} and O^{2-} ions is also shown.

The superconducting materials which are by far of greatest current interest are the high-T_c copper oxide–based superconductors, also known as the cuprates. Band-structure calculations indicate that the superconductivity observed in these materials can be associated with the planes containing copper and oxygen ions and that the oxygen content and doping also play critical roles in determining T_c.

The high-T_c cuprate superconductors have been characterized as metallic, oxygen-deficient, mixed-valence compounds of variable composition. Their structures consist of cations such as Cu^{2+}, Cu^{3+}, Ba^{2+}, La^{3+}, Bi^{3+}, Tl^{3+}, Hg^{2+}, and Y^{3+} surrounded

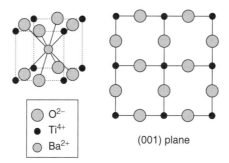

(001) plane

Figure 16.16. Cubic unit cell of the perovskite structure for the compound BaTiO$_3$. The (001) plane containing Ti^{4+} and O^{2-} ions is also shown.

in various coordinations and with various degrees of mixed ionic–covalent bonding with O^{2-} anions. Although the bonding between the Cu and O ions is not purely ionic, it is nevertheless standard practice to refer to them as Cu^{2+} and O^{2-} ions. These materials are chemically and structurally much more complicated than the elemental, alloy, and intermetallic compound superconductors. These complications have made the unraveling of the mechanisms leading to superconductivity in these materials quite difficult.

The important families of cuprate superconductors discovered so far are listed in Table 16.5 along with their T_c values and some of their structural parameters. These materials are usually denoted by the four integers $vwxy$, which specify the number of layers containing a given cation. For example, the record high-T_c compound Hg$_v$Ba$_w$Ca$_x$Cu$_y$O$_z$, also known as Hg-1223, with $vwxy = 1223$, corresponds to Hg$_1$Ba$_2$Ca$_2$Cu$_3$O$_8$. The standard abbreviations for these families of superconductors are also given (e.g., YBCO for the Y$_x$Ba$_w$Cu$_y$O$_z$ family, BSCCO for the Bi$_v$Sr$_w$Ca$_x$Cu$_y$O$_z$ family, etc.).

The most important common structural feature of the high-T_c cuprates listed in Table 16.5 is the layers or planes containing copper and oxygen ions. These copper–oxygen layers are parallel to the ab (xy) planes of the tetragonal and orthorhombic structures and are separated along the c (z) axes in the case of YBa$_2$Cu$_3$O$_7$ by layers containing ions such as the rare earth Y^{3+} or oxides such as Ba^{2+}O^{2-}. The stacking sequences of the layers along the c axis are illustrated schematically in Fig. 16.17 for the prototypical perovskite material BaTiO$_3$, for the first high-T_c host material La$_2$CuO$_4$, for YBa$_2$Cu$_3$O$_7$, for Bi$_2$Sr$_2$Ca$_2$Cu$_3$O$_{10}$, and for HgBa$_2$Ca$_2$Cu$_3$O$_8$, the structure with the highest-known T_c. The unit cell corresponding to the usual chemical formula (e.g., La$_2$CuO$_4$) for the superconductor is shown in brackets. The ionic charges given in this figure are the nominal values. Each high-T_c cuprate superconductor has a unique stacking sequence of layers along the c axis.

The different layers of atoms shown in Fig. 16.17 can be thought of as serving distinct functions with regard to the metallic behavior and superconductivity observed in these materials:

1. The CuO$_2$ layers or planes are the metallic, superconducting layers containing delocalized charge carriers which are essentially always holes. The arrangement of Cu and O atoms in this (001) plane is the same as in the (001) plane of the perovskite structure containing Ti^{2+} and O^{2-} ions (see Fig. 16.16).

TABLE 16.5   Families of Cuprate Superconductors

Compound (Abbreviation)	$vwxy$ (z)	Structure[a]	a^b (nm)	c (nm)	T_c (K)
$La_{2-x}Sr_xCuO_4$	$x \approx 0.16$	BC tetr	0.381	1.32	40
$La_{2-x}Sr_xCaCu_2O_6$	—	—	—	—	60
(LSCO)					
$Y_xBa_wCu_yO_z$	0213 (7)	ortho	0.386	1.17	92
(YBCO)	0214 (8)	ortho	0.384	2.72	80
	0427 (15)	ortho	0.385	5.03	93
$Bi_vSr_wCa_xCu_yO_z$	2212 (8)	BC tetr	0.539	3.06	92
(BSCCO)	2223 (10)	ortho	0.542	3.7	110
$Tl_vBa_wCa_xCu_yO_z$	2201 (6)	BC tetr	0.383	2.32	90
(TBCCO)	2212 (8)	BC tetr	0.385	2.94	110
	2223 (10)	BC tetr	0.385	3.59	125
	2234 (10)	BC tetr	0.385	≈ 4.2	119
	1201 (5)	tetr	0.386	0.95	<17
	1212 (7)	tetr	—	1.27	91
	1223 (9)	tetr	—	1.59	116
	1234 (11)	tetr	—	1.91	122
	1245 (13)	tetr	—	2.23	<120
$Hg_vBa_wCa_xCu_yO_z$	1201 (4)	tetr	0.386	0.95	97
(HBCCO)	1212 (6)	tetr	0.386	1.26	128
	1223 (8)	tetr	0.386	1.77	135

Source: Data from C. P. Poole, Jr., H. A. Farach, and R. J. Creswick, *Superconductiority*, Academic Press, San Diego, Calif., 1995, p. 180; and G. Burns, *High-Temperature Superconductiority*, Academic Press, San Diego, Calif., 1992, p. 57.

[a]ortho, orthorhombic; BC tetr, body-centered tetragonal.

[b]For the orthorhombic materials the third lattice constant b is typically greater than a by about only 1 to 2%.

2. The CuO and BiO layers are believed to donate holes to the superconducting CuO_2 planes.

3. The Y^{3+}, Hg^{2+}, and Ca^{2+} cation layers are insulating and donate electrons to the CuO_2 planes.

4. The BaO, SrO, and LaO layers apparently serve primarily as barriers that isolate groups of CuO_2 planes from each other.

The layers of atoms listed as function 2 and 3 are often referred to as the *charge-reservoir layers* since they control the charge concentrations on the CuO_2 layers.

In the highest-T_c materials (e.g., $YBa_2Cu_3O_7$ and the Bi, Tl, and Hg compound families), the superconducting copper–oxygen planes are separated from each other by single layers containing only metallic ions (e.g., Ca^{2+} or Y^{3+}) but no O^{2-} ions. This is also illustrated in Fig. 16.17. The trend to having higher T_c values with larger unit cells and a greater number of closely spaced CuO_2 planes per unit cell is clear from this figure and also from Table 16.5. For example, T_c increases from less than 17 K up to 122 K in the Tl-based cuprate superconductors as the number of closely spaced CuO_2 planes increases from $n = 1$ for $TlBa_2CuO_5$ up to $n = 4$ for $TlBa_2Ca_3Cu_4O_{11}$. Within the CuO_2 planes the Cu–O distance is $a/2 \approx 0.19$ nm, corresponding to a

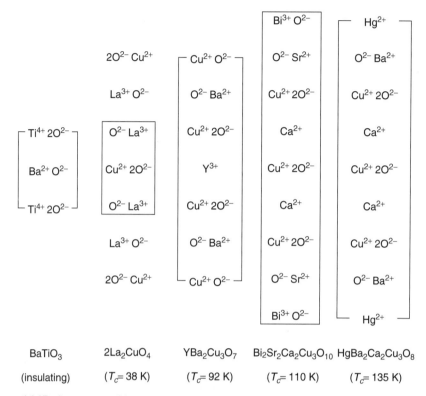

Figure 16.17. Sequences of layers or planes of atoms which are repeated along the c axis for $BaTiO_3$, La_2CuO_4, $YBa_2Cu_3O_7$, $Bi_2Sr_2Ca_2Cu_3O_{10}$, and $HgBa_2Ca_2Cu_3O_8$. The unit cell corresponding to the usual chemical formula (e.g., La_2CuO_4) is shown in brackets.

strong bond of the mixed ionic–covalent type, while for Cu and O ions in the adjacent CuO planes the distance is greater, ≈ 0.24 nm. Pairs of CuO_2 planes are typically ≈ 0.32 nm apart when separated by a single plane (e.g., a plane with Y^{3+} ions as in $YBa_2Cu_3O_7$ or with Ca^{2+} ions as in $HgBa_2Ca_2Cu_3O_8$).

The structures and normal-state properties of the $La_{2-x}Sr_xCuO_4$, $YBa_2Cu_3O_{7-x}$, and $HgBa_2Ca_2Cu_3O_8$ superconductors are discussed in more detail in Chapter W16.

Superconducting Properties. A general result for the cuprate superconductors is that T_c is found to increase with the number n of superconducting CuO_2 layers within the unit cell that are separated from each other by single nonsuperconducting layers containing, for example, Y^{3+} or Ca^{2+} ions. This is illustrated in Table 16.5, where it can be seen that T_c increases from 35 K to 92 K and then to 133 K as n increases from 1 for $La_{2-x}Sr_xCuO_4$, to $n = 2$ for $YBa_2Cu_3O_7$, and then to $n = 3$ for $HgBa_2Ca_2Cu_3O_8$. This behavior is also observed for the BSCCO and TBCCO families. There is, however, an apparent saturation of $T_c \approx 120$ to 130 K for $n = 3$ to 4, which may be due to difficulties in synthesizing materials that have such large unit cells along the c axis or which may be an intrinsic effect.

In addition, T_c also varies with doping or composition, as can be seen in the phase diagrams for $La_{2-x}Sr_xCuO_{4-y}$ and $YBa_2Cu_3O_{7-x}$ shown in Fig. 16.18. In the former

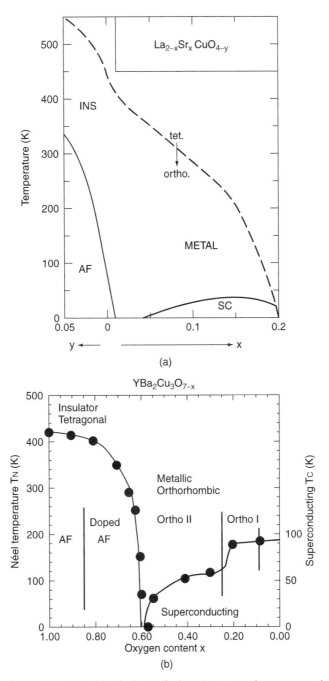

Figure 16.18. The variations with doping of the structure, the superconducting critical temperature T_c, and the Néel temperature T_N are presented here in the phase diagrams for (*a*) $La_{2-x}Sr_xCuO_{4-y}$, and (*b*) $YBa_2Cu_3O_{7-x}$. [(*a*) From K. C. Hass, *Solid State Physics*, Vol. 42, copyright 1989. Reprinted by permission of Academic Press, Inc. (*b*) From P. Burlet et al., *J. Supercond.*, **9**, 357 (1996).]

case, T_c reaches a maximum of ≈ 40 K for $x \approx 0.16$, while in the latter case $T_c = 92$ K from $x = 0$ to $x = 0.2$ before decreasing as x increases further and holes are removed from the CuO_2 layers. For every oxygen removed from $YBa_2Cu_3O_{7-x}$, the two electrons formerly bonded to this O^{2-} ion are redistributed within the structure, with some fraction entering the CuO_2 layer and reducing the hole concentration there. Bond charges associated with the Cu atoms in the CuO_2 layers can be calculated from the Cu–O bond lengths. The average effective valence of the Cu atoms can be determined from the bond charges and shows the same nonlinear dependence on the oxygen vacancy concentration as does T_c. This result confirms the close connection between T_c and the details of the bonding and the hole concentration in the CuO_2 layers in $YBa_2Cu_3O_{7-x}$.

Insulating behavior is observed for $x \leq 0.05$ in $La_{2-x}Sr_xCuO_4$ and for $x \geq 0.6$ in $YBa_2Cu_3O_{7-x}$. Thus the compositions that correspond to superconductivity are not far from the compositions at which these materials become insulating. It should be noted that the concentration of oxygen vacancies is not necessarily simply related to the average number of holes N_{hole} per Cu atom in the CuO_2 layers. In principle $N_{hole} = 1/2 - x$, falling from $\frac{1}{2}$ for stoichiometric $YBa_2Cu_3O_7$ to zero for $YBa_2Cu_3O_{6.5}$. The material $La_{2-x}Sr_xCuO_4$ ceases to be a superconductor for $x \geq 0.26$ even though its metallic character continues to improve as the Sr doping level rises. This behavior, although not understood, may be related to a structural transition from orthorhombic to tetragonal, as is also observed for $YBa_2Cu_3O_{7-x}$ for $x \approx 0.6$.

The high-T_c cuprate superconductors share several important properties in common with the more conventional BCS-like superconductors discussed earlier. For example, the fundamental charge of the superconducting charge carriers has been shown to be $2e$ via measurements of flux quantization. Also, superconducting energy gaps have been found to be present, type II behavior has been observed with vortices present in the mixed state, and Josephson tunneling has been observed. Some important differences are also observed, including the very short coherence lengths ξ and, of course, the high T_c values. As a result, it is still not clear whether the superconductivity of the high-T_c cuprate materials can be explained in terms of the usual, or even a modified, BCS formalism. Instead of the spin singlet ($S = 0$) with angular momentum $l = 0$ (s-state) pairing of BCS, the high-T_c superconductivity in $YBa_2Cu_3O_7$ is apparently primarily the result of $l = 2$ (d state, $d_{x^2-y^2}$) anisotropic pairing. A small s-state component to the electron-pair superconducting wavefunction may also be present due to the orthorhombic symmetry of $YBa_2Cu_3O_7$.

If the high-T_c superconductors are of the BCS type with electron–phonon coupling as the source of the pairing, they will clearly be strong-coupling superconductors. Other possibilities for the pairing interaction within the superconducting CuO_2 layers include those of a magnetic or an electronic origin [e.g., antiferromagnetic magnons (involving residual magnetic moments on the Cu ions), excitons, or optical rather than acoustic phonons]. It is also possible that high-T_c superconductivity can still be explained within the strong-coupling BCS formalism with $\lambda_{ep} \approx 2$.

An important property of the high-T_c cuprate superconductors is the strongly anisotropic nature of the superconductivity that results from the anisotropic tetragonal or orthorhombic structures of these materials. With the obvious exception of the transition temperature T_c, all the superconducting properties (i.e., critical fields and critical currents, superconducting energy gaps, penetration depths, coherence lengths, and so on) have values which are uniform (or nearly so) in the ab plane but which differ considerably from the corresponding values along the c axis. These anisotropies result from the anisotropic structure and effective masses m^* of the charge carriers in the normal state, with $m_c^*/m_a^* \approx 30$ in $YBa_2Cu_3O_7$. Even higher effective-mass anisotropies are observed in the BSCCO and TBCCO families.

The superconducting properties of the high-T_c cuprates are discussed further in Chapter W16.

APPLICATIONS OF SUPERCONDUCTORS

In addition to being interesting from a strictly scientific point of view, it has long been recognized that superconductors can have a wide range of important large-scale technological applications in high-field magnets and electromagnetic equipment as well as small-scale applications in sensors and passive microwave devices. The development and widespread use of many of these applications has been delayed, however, by the intrinsic properties of available superconductors (e.g., their transition temperatures T_c and critical magnetic fields) as well as by extrinsic problems associated with their critical currents and also with their synthesis, processing, and fabrication into useful forms. Some important present and potential applications of superconductors are discussed next, beginning with critical issues related to the development of applications of superconductivity.

16.6   Critical Issues

Superconductors with values of $T_c \leq 20$ K have already found applications in superconducting magnets, in sensors, and in electromagnetic equipment typically operating in liquid He at $T_b = 4.2$ K. These applications have to a large extent been limited for economic reasons (e.g., by the cost of liquid He). It is clearly important to develop superconductors that are suitable for applications at $T = 77$ K and higher, if possible. As a result, the search for higher superconducting T_c values remains an important effort. In addition, for a given T_c, the highest-possible superconducting critical fields H_{c2} and critical currents J_c are also clearly desirable. The materials with these desirable characteristics have been found to be type II superconductors, which are ordinarily used in the mixed state. The main advantage of type II superconductors, in addition to their high T_c values, is the fact that they can remain superconducting even after magnetic flux has entered the bulk.

One useful measure of the "quality" of a superconducting material is the abruptness of its transition to the superconducting state when measured, for example, by the change of resistance R as a function of temperature. Superconductors that are homogeneous and single phase have abrupt transitions, whereas those that are inhomogeneous have broadened transitions. The T_c values are usually quoted as corresponding to the midpoint of the resistive transition (i.e., at the temperature at which the measured resistance has dropped to one-half of its value in the normal state). A more reliable probe of the superconducting/normal transition at T_c, at H_c, or at H_{c2} in the presence of a magnetic field is the measurement of the magnetic susceptibility χ, which actually determines the volume fraction of the sample that has become superconducting. A transition to a state of zero resistance indicates only that a continuous superconducting path exists through the sample. In an inhomogeneous sample this resistive transition does not necessarily establish that the sample is completely superconducting.

Critical Temperatures. The transition temperature T_c of a given superconducting material is determined by its crystal structure and its stoichiometry. Examples of this sensitivity to structure and stoichiometry involving the high-T_c cuprates have been given in Section 16.5. The search for higher T_c values is motivated primarily by two factors: superconductors with higher T_c values will in general have higher critical magnetic fields and critical currents and will require less costly refrigeration to reach their T_c values (which so far have all been below any normally occurring temperatures on earth). There are several readily available cryogenic fluids: liquid He: boiling temperature $T_b = 4.2$ K, liquid H_2: $T_b = 20.3$ K, liquid Ne: $T_b = 27.1$ K, liquid N_2 (LN2): $T_b = 77.4$ K, and liquid Ar: $T_b = 87.3$ K.

The first superconductor developed with $T_c > 77$ K was $YBa_2Cu_3O_7$, discovered in 1987. This discovery represented a major breakthrough in the development of new applications for superconductivity. A list of high-T_c materials, including the current recordholder $HgBa_2Ca_2Cu_3O_8$ with $T_c = 135$ K, has been presented in Table 16.5. Previous estimates that the BCS electron–phonon mechanism for superconductivity could provide T_c values only as high as about 40 K have been revised, with current predictions being that $T_c \approx 200$ K can be achieved for values of the electron–phonon coupling constant $\lambda_{ep} \approx 6$.

A plot of T_c versus γ, the coefficient of the linear term in the electronic specific heat, is shown for a wide variety of superconductors in Fig. 16.19. It can be seen that high T_c values are generally correlated with high values of γ and therefore with high values of $\rho(E_F)$, with the obvious exception of the *heavy-fermion* superconductors, which fall beyond the lower right-hand corner of the plot. The oxide-based superconductors toward the top of the plot have T_c values that are about a factor of 10 higher than those of conventional superconductors with similar values of γ. It is not clear that the superconductivity of the oxide-based materials or of the heavy-fermion materials follows from the standard BCS electron–phonon mechanism. It should be noted that there exists no similar correlation between T_c and the Debye temperature Θ_D, which is characteristic of the average phonon energy in the lattice.

Various approaches for increasing T_c have been attempted based on increasing λ_{ep} and/or $\rho(E_F)$. The successful approach of Bednorz and Mueller was based on the previous observation of superconductivity at $T_c \approx 13$ K in $BaPb_{1-x}Bi_xO_3$ and their conclusion that "within the BCS mechanism, one may find still higher T_cs in perovskite-type or related metallic oxides, if the electron–phonon interactions and the carrier densities at the Fermi level can be enhanced further." The specific material $La_{2-x}Ba_xCuO_{4-y}$ was chosen based on previous work in this material which had demonstrated the presence of oxygen-deficient phases with mixed-valent copper ions and itinerant electronic states. Bednorz and Mueller correctly believed, and also demonstrated, that considerable electron–phonon coupling and metallic conductivity leading to superconductivity with high T_c values could be found in this and related oxides.

The search for superconductors with ever-higher T_c values continues in many laboratories with an ultimate goal of $T_c \approx 300$ K, so that refrigeration of superconductors would in principle no longer be necessary.

Critical Magnetic Fields. The upper critical field H_{c2} of a superconducting material, like its T_c, depends primarily on its crystal structure and stoichiometry but can also depend on the microstructure of the material through the electron mean free path. The search for type II superconductors with higher H_{c2} values has been motivated by the

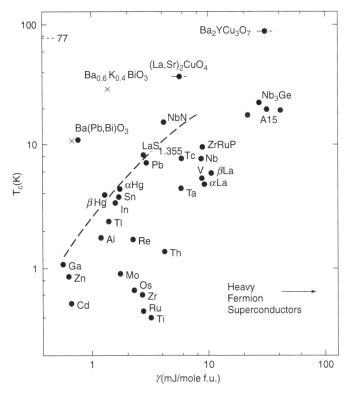

Figure 16.19. Superconducting T_c for a wide variety of superconductors plotted versus γ, the coefficient of the linear term in the electronic specific heat, on a logarithmic plot. [From R. J. Cava et al., *Mater. Res. Soc. Bull.*, **14**(1), 49 (1989).]

desire to produce superconducting magnets for the generation of ever-higher magnetic fields. Magnetic flux densities B of 10 to 20 T, corresponding to field strengths H of about 8 to 16 MA/m, are a typical current goal.

An intrinsic limit to the attainable values of H_{c2} results from the magnetic response of the metal in its normal state. The charge carriers with spins $s = 1/2$ that participate in Cooper pairing in the superconducting state will have their zero-field energy levels in the normal state split by the amount $\Delta E = g\mu_0\mu_B H$ in the presence of an applied magnetic field H (see Fig. 9.8). Here $g = 2$ is the electron g factor and μ_B is the Bohr magneton. This splitting is the source of the Pauli paramagnetic susceptibility χ_P and results in a lowering of the Gibbs free energy of the normal state in a magnetic field by the amount $G_n(H) - G_n(O) = -\mu_0\chi_P H^2/2$. This decrease in G_n is due to the reorientation of the directions of the magnetic moments of electrons with energies near the Fermi energy E_F, as described in Section 9.9, and will compete with the condensation of the electrons into the superconducting state for $T < T_C$. There will be a transition to the normal state when the splitting $\Delta E \approx 2\varepsilon(T)$ (i.e., the superconducting energy gap). The magnetic field H_P for this transition, called the *Pauli limiting field* or *paramagnetic limit*, is given by

$$H_P(T) \approx \frac{\varepsilon(T)}{\mu_0\mu_B}.\tag{16.33}$$

At $T = 0$ K, where, according to BCS, $2\varepsilon(0) = 3.52\,k_B T_C$, this limiting field is equivalent to an upper critical field given in units of tesla by

$$B_{C2}(0\text{ K}) = \mu_o H_P(0\text{ K}) \approx 2.62 T_C \text{ (K)}. \qquad (16.34)$$

For a given material the value of H_P will be determined by the material's actual paramagnetic susceptibility, not just by the Pauli paramagnetic contribution. In addition, penetration of the magnetic field into the mixed state of the superconductor will lessen the effect of the normal-state susceptibility on the upper critical field. The upper critical fields measured for a wide range of superconducting materials agree reasonably well with these expressions. Equation (16.33) can also be used to estimate $H_{c2}(0\text{ K})$ for the high-T_c superconductors when the upper critical field is too high to be measured directly.

Some experimental results for $B_{c2}(= \mu_0 H_{c2})$ versus T are given in Fig. 16.20 both for superconductors with $T_c < 20$ K and also for high-T_c superconductors. In addition, the upper critical fields B_{c2} for a variety of high-field superconductors are listed in Table 16.6. These upper critical fields are essentially never achievable in long lengths of superconducting wire fabricated from these materials, due to critical current limitations, to be described later.

If not limited by the paramagnetic effects outlined earlier, the values of $B_{c2}(0\text{ K})$ predicted using the G-L and BCS theories and Eqs. (16.22), (16.25), and (16.32) are

$$B_{c2}(0\text{ K}) = \frac{\Phi_0}{2\pi\xi^2} = \frac{\pi\Phi_o\varepsilon(0)^2}{2\hbar^2 v_F{}^2} \propto \frac{m^* T_c{}^2}{E_F}. \qquad (16.35)$$

Thus increasing T_c is a very effective way of increasing H_{c2}. Reducing ξ by decreasing the electron mean free path l also has the effect of increasing H_{c2}.

Critical Currents. Higher superconducting critical transport current densities J_c are needed in magnets and also in the superconducting wires and cables to be used for the transmission of electrical energy. The critical transport currents of type II superconductors in the mixed state, while related to the intrinsic properties T_c and H_{c2}, are

TABLE 16.6 Upper Critical Fields $B_{c2} = \mu_0 H_{c2}$ for a Variety of High-Field Supercon- ducting Materials at $T = 4.2$ or 77 K

Superconductor	T_c (K)	B_{c2} $(\mu_0 H_{c2})^a$ (T)	Superconductor	T_c (K)	B_{c2} $(\mu_0 H_{c2})^a$ (T)
Nb wire, RRR^b = 1400	9.3	0.47(4.2 K)	YBa$_2$Cu$_3$O$_7$	92	30^c (77 K)
Nb wire, cold drawn	9.3	0.41(4.2 K)	Bi$_2$Sr$_2$Ca$_2$Cu$_3$O$_{10}$	110	≈30 (77 K)
Nb$_{0.44}$Ti$_{0.56}$	9.0	14 (4.2 K)	Tl$_2$Ca$_2$Ba$_2$Cu$_3$O$_{10+x}$	128	≈80 (77 K)
Nb$_3$Ge	23.2	37 (4.2 K)	HgBa$_2$CuO$_{4+x}$	99	≈20 (77 K)
Nb$_3$Sn	18	23.5 (4.2 K)			

Source: Data from C. P. Poole, Jr., H. A. Farach, and R. J. Creswick, *Superconductivity*, Academic Press, San Diego, Calif., 1995, p. 272; B. W. Roberts, *J. Phys. Chem. Ref. Data*, **5**, 581 (1976); and others.

$^a B_{c2} = \mu_0 H_{c2} = 10$ T is equivalent to $H_{c2} = 7.96$ MA/m.

^{b}RRR, residual resistance ratio ≈ R(333 K)/R(4.2 K).

$^c H$ applied parallel to the c axis.

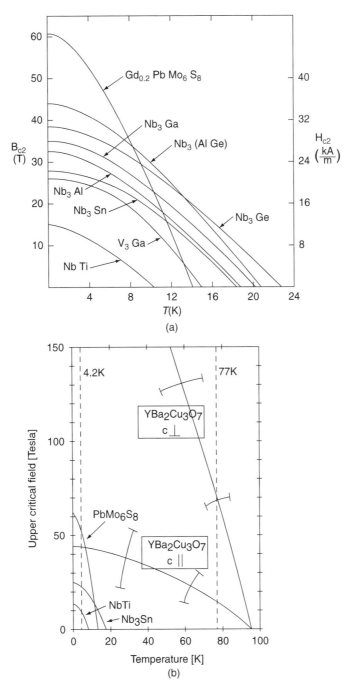

Figure 16.20. Plots of $B_{c2} = \mu_0 H_{c2}$ as a function of temperature for several high-field super-conductors. The lower critical field B_{c1} is too small to be shown here. (*a*) Superconductors with $T_c < 20$ K; (*b*) the same as in part (*a*), but also including high-T_c superconductors. [(*a*) From M. N. Wilson, *Superconducting Magnets*, Clarendon Press, Oxford, 1983. By permission of Oxford University Press. (*b*) From H. J. Scheel, *Mater. Res. Soc. Bull.*, **19**(9), 26 (1994).]

primarily extrinsic properties which are sensitive functions of the microstructure of the material. Much of current high-T_c research is focused on the processing of superconducting materials in order to obtain the most desirable set of properties, especially T_c and J_c.

In a general sense, J_c, the highest transport current density that can flow through a superconductor without loss of energy, is determined by the interaction of the vortices present in the mixed state with the transport current and with the magnetic field in the superconductor. Consider the case of a type II superconductor that is in the mixed state due to the application of a transverse magnetic field H_z. When a transport current i_y flows through the bulk of the superconductor, the current and the vortices with their magnetic flux will exert Lorentz forces on each other which act along the x axis (i.e., at right angles both to the axis of the vortex and to the direction of the current; see Fig. 16.21). When the current i_y flows in a superconducting wire of rectangular cross section and dimensions l_x, l_y, and l_z, the Lorentz force acting on the wire (i.e., on the vortices) per unit length of vortex line is given by

$$\frac{F_x}{L} = \frac{\mu_0 l_y i_y H_z}{L} = |\mathbf{J}_y x \mathbf{\Phi}_0|_x, \tag{16.36}$$

where $\mathbf{J}_y = \mathbf{i}_y/l_x l_z$ and L is the total length of all the vortices in the superconductor. Here $\mathbf{\Phi}_0$ is the vector associated with the quantum of flux Φ_0.

When expressed per unit volume of the superconductor, this force is given by

$$\frac{F_x}{V} = J_y B_z. \tag{16.37}$$

These Lorentz forces tend to push the vortices out of the sample, with new vortices entering the superconductor at the opposite side. If the vortices are pinned and cannot move, the net Lorentz force will act on the wire itself. A useful experimental definition of the average *pinning force density* in a superconductor can be obtained from

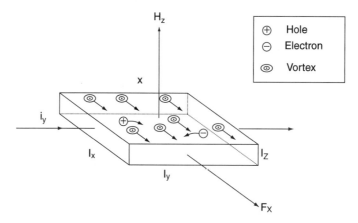

Figure 16.21. Interactions between transport current i_y and vortices are shown in the mixed state of a superconductor in the presence of a magnetic field H_z.

Eq. (16.37) as follows:

$$\frac{F_P}{V} = \mu_0 J_c H, \tag{16.38}$$

where J_c is the measured critical current density in a magnetic field H. In addition, the Lorentz force acting on the charge carriers due to the magnetic flux in the vortices will generate a transverse Hall voltage. As can be seen in Fig. 16.21, the charge carriers, whether electrons or holes, will be pushed to the same side of the sample as the vortices.

The critical transport currents of homogeneous, defect-free type II superconductors are limited by the fact that viscous forces in the material oppose the motion of vortices. The current source must therefore provide the energy needed to keep the vortices moving. The current flow will therefore no longer be lossless (i.e., there will now be an effective resistance to the flow of current in the superconductor). The resulting voltage drop along the length of the sample due to the motion of the vortices will be proportional to the average speed v of the moving vortices and to the magnitude of the magnetic flux density B in the superconductor. The speed v will, in turn, be proportional to the current density J and inversely proportional to the average strength of the viscous forces.

If the superconductor is inhomogeneous, the vortices can be immobilized or *pinned* by pinning centers such as chemical or structural defects (e.g., impurities, precipitates of normal material, grain boundaries, etc.). In this case $v = 0$ and there will no longer be a voltage drop or energy dissipation. With effective pinning of vortices the high currents that are needed for the generation of strong magnetic fields can flow in a lossless manner in a superconducting magnet. While the nature of the pinning centers and of the resulting pinning forces acting on the vortices are not well understood, it is clear that a vortex has lower energy at the site of the pinning center, especially if the center is a normal region. The maximum effect for a given pinning center will occur when its diameter is $\approx \xi$, the size of the vortex core. The optimum area density of defects is equal to the area density of vortices. The vortex spacing is approximately $\sqrt{\Phi_0/B}$, where B is the average flux density.

Further discussions of critical currents, including the phenomena of flux creep and collective vortex effects, are presented in Chapter W16.

16.7 Specific Applications

The important current and potential applications of superconductors involve their use in large-scale applications such as high-field magnets, power transmission cables, and in a variety of electromagnetic equipment. Small-scale applications include sensors based on SQUIDs and also passive microwave applications.

Large-Scale Applications. The development of high-field superconducting magnets and other electromagnetic equipment, such as motors, generators, transformers, solenoids, and power transmission lines, requires superconductors with high critical fields H_{c2} and high critical currents J_c which also have the necessary mechanical properties (e.g., high ductility) to be formed into long lengths of wire or tape of the proper dimensions. For high-T_c superconductors the relevant critical field for large-scale applications is the *irreversibility field* H_{irr} above which losses due to vortex motion in the vortex liquid phase occur. The successful development of superconducting magnets of high-T_c

materials has the potential of allowing replacement of the heavy Fe-based cores now used in motors, generators, and transformers. This would result in large savings both in weight and in decreased electrical losses. So far, the high-T_c superconductors which have the highest-available critical fields and are in principle capable of operating at $T = 77$ K do not possess the high critical currents and mechanical properties which would allow them to replace the standard Nb_3Sn and $Nb-Ti$ materials currently used in magnets operating at $T = 4.2$ K. These Nb-based materials are not granular and can be prepared with strong flux pinning, in contrast to the high-T_c materials, which tend to be granular, very brittle, and in which flux pinning is much weaker.

Plots of $B_{c2}(T)$ for the superconductors in current use as superconducting magnet wire are shown in Fig. 16.20a, with additional plots presented in Fig. 16.20b for some high-T_c superconductors. The useful Nb-based superconductors have $B_{c2} \approx 10$ to 20 T at $T = 4.2$ K, while the high-T_c superconductors can have $B_{c2} \geq 30$ T at $T = 77$ K. The critical fields and currents of some of these materials at $T = 4.2$ K and 77 K are presented in Tables 16.6 and W16.2, respectively.

The superconducting magnet wire needed for current advanced accelerator applications at $T = 4.2$ K must be in the form of 5-μm filaments with $J_c \approx 0.3$ MA/cm^2 at $B = \mu_0 H = 5$ T ($H = 4$ MA/m). The high-T_c superconductors cannot yet reach this value of J_c at $T = 77$ K even though their upper critical fields are quite high. The development of superconducting wire or tape with $J_c \approx 1$ MA/cm^2 at $T = 77$ K will be a major achievement.

A major breakthrough in the development of high-field superconducting magnets was the discovery that the intermetallic-compound superconductor Nb_3Sn has a critical field $H_{c2} \approx 20$ MA/m ($B_{c2} \approx 26$ T). As a result, Nb_3Sn is preferred over $Nb-Ti$ alloys for very high field applications. Nb_3Sn and $Nb-Ti$ are used in the form of multifilament conductors containing up to 6000 μm or sub-micrometer-sized filaments embedded in a Cu matrix. The filaments of superconducting wire must be surrounded by a high-conductivity normal-metal matrix which can carry the current and help to dissipate heat in the event that part of the superconducting wire becomes normal. Examples of multifilament Nb_3Sn wires embedded in a Cu-alloy matrix are shown in Fig. 16.22. Since the transport current flowing along a superconducting wire is restricted to the region just inside its surface where the magnetic field penetrates, it follows that individual filamentary wires with radii $a \approx \lambda$ have the important advantage that more of the cross section of the wire can carry current. If the filaments are well separated, as shown in Fig. 16.22, the effects of their magnetic fields on each other can be minimized.

Further discussions of the potential of high-T_c superconductors for high-field applications and of issues related to the fabrication of long superconducting wires with flux-pinning centers are presented in Chapter W16.

Small-Scale Applications. For the small-scale applications of superconductivity described here, the high critical magnetic fields H_{c2} and critical currents J_c needed for large-scale applications are no longer as necessary. Instead, high T_c values and ease of fabrication of superconducting thin films become critical issues. The small-scale applications of superconductors as sensors are discussed next. Additional small-scale applications, including computer devices, optical detectors, thermal switches, and microwave components and devices, are presented in Chapter W16.

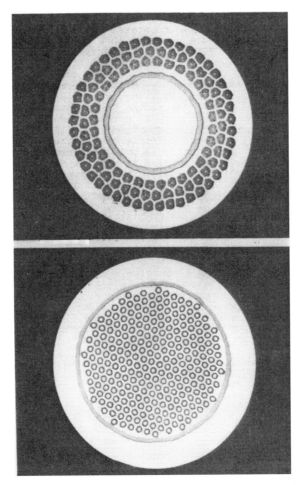

Figure 16.22. Examples of multifilament Nb$_3$Sn wires embedded in a Cu-alloy matrix. [From M. Thoener et al., *IEEE Trans. Magn.*, **27**, 2027 (1991). Copyright 1991 by IEEE.]

Most current small-scale applications of superconductors as sensors involve the use of SQUIDs (*superconducting quantum interference devices*) whose operation involves the *Josephson effect* (i.e., the tunneling of Cooper pairs across an insulating barrier or *weak link* separating two superconductors). A weak link can be an insulating layer but can also be a thin layer of normal metal or a thin constriction separating the two superconductors. Electron tunneling has been discussed in Sec. 16.3. The details of Cooper-pair tunneling and of the Josephson effects that are essential for the operation of SQUIDs are discussed in Chapter W16.

Important applications of the Josephson effects are found in SQUIDs, which can serve as extremely sensitive detectors or sensors of magnetic flux Φ and related electromagnetic quantities. SQUIDs are quite versatile since they can be used to measure any physical quantity that can be converted to a flux (e.g., magnetization, magnetic susceptibility, magnetic field, field gradient, current, and voltage). They have found important applications in fields as dissimilar as studies of biomagnetism, geophysical exploration, and nondestructive testing. In practice, SQUIDS are devices that usually consist of a superconducting loop containing either a single junction or a parallel configuration of two Josephson tunnel junctions, as shown in Fig. 16.23.

The two-junction superconducting loop or dc SQUID consists of a pair of "short" junctions or weak links a and b across which the two superconductors S_1 and S_2 are coupled. The total area A of the loop shown here is much greater than the individual areas A_a and A_b of the two junctions. This device is essentially a magnetometer and is sensitive to a change in magnetic flux $\Delta\Phi$ through its area, which can be a small fraction of one flux quantum Φ_0. For $A = 10^{-5}$ m^2 and a detectable sensitivity of $\Delta\Phi = \Phi_0/10^3$, this corresponds to a change in magnetic flux density $\Delta B = \Delta\Phi/A \approx 2 \times 10^{-13}$ T or $\Delta H = \Delta B/\mu_0 \approx 1.6 \times 10^{-7}$ A/m. This device has a sensitivity greater than that of a single junction by the factor $A/(A_a + A_b)$.

Two practical versions of SQUIDs, dc and radio-frequency SQUIDs, are shown in Fig. 16.23. The dc SQUID corresponds to the two-junction loop and can be used as an extremely sensitive detector of changes of magnetic flux when the loop current

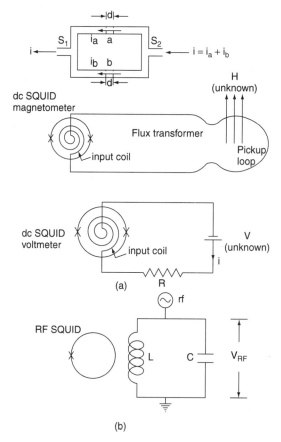

Figure 16.23. Two common configurations of Josephson junctions used in SQUIDS. (*a*) *Dc SQUID.* The superconducting loop consists of two Josephson tunnel junctions or weak links *a* and *b* across which the two superconductors S_1 and S_2 are coupled. A dc SQUID magnetometer and a voltmeter are also shown. (*b*) RF SQUID. The superconducting loop contains a single junction and is coupled to a tuned LC circuit.

or voltage is monitored. Magnetic flux from an external input circuit or pickup loop is usually coupled into the dc SQUID through a flux transformer, as shown. The flux transformer is a multiturn coil inductively coupled to the SQUID and connected to a much larger single-turn pickup loop. Due to the sensitivity of the current to a fraction of a single flux quantum Φ_0, the dc SQUID is a much more sensitive detector than other nonsuperconducting devices, such as those based on the Hall effect. The energy sensitivity of present SQUIDs (i.e., the energy associated with the smallest change of magnetic flux $\Delta\Phi$ detectable in 1 s) is $\Delta E \approx 10^{-32}$ J. If $\Delta\Phi \approx \Phi_0$, then $\Delta E \approx e\Delta\Phi/\Delta t = e\Phi_0/(1\text{s}) = h/2 \approx 3 \times 10^{-34}$ J. Thus the present sensitivity ΔE is approaching the limit set by the Heisenberg uncertainty principle, $\Delta E \Delta t > h/4\pi$.

The dc SQUID can also be employed in a gradiometer configuration in which differences between the magnetic flux in two oppositely wound coils are measured. Modern dc SQUIDs utilize Nb–Al_2O_3-Nb Josephson junctions and are thin-film devices consisting of layers of the superconductor Nb and the insulator Al_2O_3 along with external current connections.

The radio-frequency SQUID consists of a single Josephson junction in a superconducting loop which can be inductively coupled to an *LC* tuned circuit, as shown. Changes in flux in the radio-frequency SQUID loop cause changes in the loading of the tuned circuit which can be detected by measuring the radio-frequency voltage appearing across this circuit. Although radio-frequency SQUIDs are not as sensitive as dc SQUIDs, most SQUIDs in use today are of the radio-frequency variety, due to the fact that they were the first to be developed.

REFERENCES

Burns, G., *High-Temperature Superconductivity*, Academic Press, San Diego, Calif., 1992.

de Gennes, P. G., *Superconductivity of Metals and Alloys*, W.A. Benjamin, New York, 1966.

Lynton, E. A., *Superconductivity*, 2nd ed., Methuen, London, 1964.

Poole, C. P., Jr., H. A. Farach, and R. J. Creswick, *Superconductivity*, Academic Press, San Diego, Calif., 1995.

Roberts, B. W., *J. Phys. Chem. Ref. Data*, **5**, 581 (1976); this reference provides a useful compilation of the properties of the known superconducting elements, alloys, and compounds up to 1976.

Rose-Innes, A. C., and E. H. Roderick, *Introduction to Superconductivity*, 2nd ed., Pergamon Press, Oxford, 1978.

Ruggiero, S.T., and D. A. Rudman, eds., *Superconducting Devices*, Academic Press, San Diego, Calif., 1990.

Vonsovsky, S. V., Yu. A. Izyumov, and E. Z. Kurmaev, *Superconductivity of Transition Metals*, Springer-Verlag, Berlin, 1982.

White, R. M., and T. H. Geballe, Long range order in solids, in H. Ehrenreich et al., eds., *Solid State Physics*, Suppl. 15, Academic Press, San Diego, Calif., 1979.

Wilson, M. N., *Superconducting Magnets*, Clarendon Press, Oxford, 1983.

PROBLEMS

16.1 **(a)** Derive expressions for the difference in entropy $\Delta S(T) = S_n(T) - S_s(T)$ and the difference in specific heat $\Delta C(T) = C_n(T) - C_s(T)$ between the normal and superconducting states in terms of the critical magnetic field $H_c(T)$ and its first derivative dH_c/dT. [*Hint:* Use Eq. (16.3) and standard thermodynamic relationships.]

(b) Evaluate these expressions for $\Delta S(T)$ and $\Delta C(T)$ for the case where $H_c(T)$ can be approximated by $H_{c0}[1 - (T/T_c)^2]$ and show that:

    (i) $\Delta S(T_c) = \Delta S(0) = 0$.

    (ii) $\Delta S(T) > 0$ for $0 < T < T_c$.

    (iii) $\Delta C(T_c) = -4\mu_0 H_{c0}^2/T_c$. Calculate $\Delta C(T_c)$ from this expression using data from Table 16.1 for Al and compare with the measured result $-1.45\gamma T_c$ for Al.

    (iv) $\Delta C = 0$ for $T = T_c/\sqrt{3}$ and $T = 0$ K.

16.2 (a) Using Eq. (16.5), calculate the condensation energy in J/m³ and in eV per electron at $T = 0$ K for the superconductor Pb for which $H_{c0} = 6.39 \times 10^4$ A/m.

(b) Compare your result from part (a) with the expression $\varepsilon(0)(\varepsilon(0)/E_F)$ where the superconducting energy gap $2\varepsilon(0) = 2.6$ meV for Pb. Here $\varepsilon(0)/E_F$ is the fraction of conduction electrons whose energies are actually affected by the condensation.

16.3 Consider the London penetration depth λ_L defined in Eq. (16.10).

(a) Calculate $\lambda_L(0$ K$)$ for the superconductors Al, Pb, and Nb.

(b) If a superconductor has a London penetration depth $\lambda_L(0$ K$) = 200$ nm, what is the concentration n_s of superconducting electrons at $T = 0.5T_c$?

16.4 When transport current i flows through a superconducting wire of radius R, its path is confined to a region of thickness λ, the penetration depth, just inside the surface of the wire.

(a) In this case show that the critical current density $J_c = i_c/A_{\text{eff}}$ is independent of R and can be expressed in terms of the critical field H_c by $J_c = H_c/\lambda$. Here A_{eff} is the effective area through which the current flows, with $A_{\text{eff}} \ll \pi R^2$.

(b) Calculate J_c for superconducting Pb at $T = 0$ K. [*Note:* H_{c0} and $\lambda(0)$ are given in Tables 16.1 and 16.4, respectively.]

(c) Sketch $J_c(T)/J_c(0)$ from $T = 0$ K to T_c using the temperature dependencies of H_c and λ given in Eqs. (16.6) and (16.11), respectively.

16.5 A type II superconductor has $T_c = 125$ K, $\Theta_D = 250$ K, and $\kappa(T_c) = 50$. On the basis of standard theories [free-electron model, Debye model, BCS theory, G-L theory, Pauli limit for H_{c2} given in Eq. (16.33)], estimate the following:

(a) The superconducting energy gap $2\varepsilon(0)$.

(b) The upper critical field $H_{c2}(0) = H_P$.

(c) The coherence length $\xi(0)$ and the penetration depth $\lambda(0)$.

(d) The thermodynamic critical field $H_{c0} - H_c(0)$.

(e) The coefficients γ and A of the electronic and phonon contributions to the specific heat, γT and AT^3, respectively.

16.6 The i–V characteristic of an S_1–I–S_2 tunnel junction of two superconductors with different energy gaps $2\varepsilon_1$ and $2\varepsilon_2$ ($2\varepsilon_1 > 2\varepsilon_2$) is shown in Fig. 16.13d for $0 < T < T_c$. Draw the alignment of the energy gaps of the two superconductors for $V = 0$, $0 < V < V_-$, $V = V_-$, $V_- < V < V_+$, $V = V_+$, and $V > V_+$ where $eV_- = 2(\varepsilon_1 - \varepsilon_2)$ and $eV_+ = 2(\varepsilon_1 + \varepsilon_2)$. Use these alignments to explain the onset of weak tunneling at V_- and the onset of strong tunneling at the higher voltage V_+.

16.7 Use Eq. (16.20) to find the limiting values of $\lambda(l)$ and $\xi(l)$ **(a)** in the clean limit where the electron mean free path $l \gg \xi_0$, and **(b)** in the dirty limit where $l \ll \xi_0$.

16.8 **(a)** Calculate the density of vortices per unit area B/Φ_0 for the following values of B, the average flux density present in the mixed state of a superconductor. Take $H_{c2} = 1.6$ MA/m.

   (i) $B = \mu_0 H_{c2}/2$.

   (ii) $B \approx B_{c2} = \mu_0 H_{c2}$.

   (b) Calculate the average separation d between the vortices from your answers in part (a) and compare your answers with the coherence length ξ. [*Hint:* You can obtain ξ with the help of Eq. (16.22).]

16.9 Prove that the superconducting screening currents which surround each vortex in the mixed state provide most of the flux in the vortex. As a specific example, calculate the flux density B in a single vortex for the case of the type II superconductor Nb_3Sn at $T = 0$ K. Note that $B_{c2}(0 \text{ K}) = \mu_0 H_{c2}(0 \text{ K}) \approx 26$ T for Nb_3Sn. (*Hint:* The coherence length ξ for Nb_3Sn is given in Table 16.4.)

16.10 Calculate the number of holes N_{hole} per Cu ion in the CuO_2 copper–oxygen layers in the superconductor $YBa_2Cu_3O_{7-x}$ for the cases of $x = 0$, 0.25, and 0.5. Assume the following ionic charge states for the ions in this structure: Y^{3+}, Ba^{2+}, Cu^{2+}, and O^{2-}.

16.11 For the compound with the chemical formula $La_{1.7}Sr_{0.3}CuO_{3.9}$:

   (a) What is the total number of electrons per formula unit outside closed shells?

   (b) How many electrons are contributed by each ion?

   (c) What is the average valence of the copper atoms?

   (d) Assuming that all copper ions have a charge of $+2e$, what is the number of holes per formula unit?

16.12 Derive Eq. (16.33) for the Pauli limiting field H_P by setting $G_n(H) = G_s(H)$ at $H = H_P$ and using the Pauli paramagnetic susceptibility $\chi_P = \mu_0 \mu_B^2 \rho(E_F)$ of the conduction electrons in the normal state. [*Hint:* Use $G_n(H) - G_n(0) = -\mu_0 \chi_P H^2/2$, $G_s(H) - G_s(0) \approx 0$ for $H \approx H_{c2}$ when flux penetration into the superconductor is almost complete, and the BCS result $G_n(0) - G_s(0) = \rho(E_F)\varepsilon(0)^2/2$.]

16.13 When a transport current i_y flows through a type II superconducting wire, the current and the vortices present in the mixed state will exert Lorentz forces on each other.

   (a) Using the geometry shown in Fig. 16.21, show that the force per unit length of vortex line due to the current is given by $F_x/L = \mu_0 i_y H_z l y/L = |\mathbf{J}_y \times \Phi_0|_x$, where F_x is the force on the wire and L is the total length of the vortices in the wire. (*Hint:* $L \propto \Phi_z$, the average flux in the superconductor.)

(b) Calculate the total number of vortices for the case of a rectangular supercon-ducting tape of length $l_y = 1$ m, width $l_x = 1$ cm, and thickness $l_z = 1$ mm in a transverse field $H = 10^4$ A/m.

(c) Calculate F_x/L when this tape is carrying a current $i_y = 100$ A.

Note: Additional problems are given in Chapter W16.

CHAPTER 17

Magnetic Materials

17.1 Introduction

Magnetic materials have had a greater financial market than semiconductor materials due to their widespread use in recording media, to the use of magnetic ferrites in electronics and recording heads, and to the use of permanent magnets in a wide variety of applications.[†] The magnetic materials with extensive technological applications discussed in this chapter are those that typically exhibit some form of long-range magnetic order (e.g., ferromagnetism or ferrimagnetism).

Since the fundamental magnetic properties of materials are discussed in Chapter 9, in this chapter we focus instead on some additional characteristic properties of magnetic materials which are important for their applications. These are discussed in Sections 17.2 to 17.6 and include magnetic microstructure and domains, the magnetization process, magnetic anisotropy, and the effects of sample shape and size on magnetic properties. Important magnetic effects corresponding to magnetostriction, magnetoresistance, magneto-optical effects, and dynamic magnetic behavior are presented in Sections 17.7 to 17.10. Finally, some important examples and applications of specific magnetic materials are discussed in Sections 17.11 and 17.12.

CHARACTERISTIC PROPERTIES OF MAGNETIC MATERIALS

The properties and applications of technologically important ferromagnetic and ferrimagnetic materials are controlled by their magnetic microstructure, by the magnetization process, and by effects related to magnetic anisotropy, shape, and size. The focus here is on both the *intrinsic* and *extrinsic* magnetic properties of magnetic materials. Intrinsic magnetic properties are determined primarily by the composition and crystal structure of the material and include the saturation magnetization M_{sat}, the Curie temperature T_C, the magnetocrystalline anisotropy K, and the magnetostriction λ. Important extrinsic magnetic properties which are determined primarily by the microstructure of the material include the coercive field H_c, the magnetic susceptibility χ, the remanent magnetization M_r, and the shape and area of the magnetization hysteresis loop. Magnetic microstructure and its effect on magnetic properties are discussed first, with the emphasis here on crystalline magnetic materials. Examples of the properties and applications of some amorphous magnetic materials are presented in Sections 17.11 and 17.12.

[†] G. A. Prinz, *Mater. Res. Soc. Bull.*, June 1988, p. 28.

17.2 Magnetic Microstructure and Domains

The magnetic microstructure of a ferromagnet or a ferrimagnet (i.e., the distribution and relative orientations of the individual magnetic moments) is almost never uniform throughout the material even below the critical temperature T_C for magnetic ordering. For example, in crystalline ferromagnetic materials the magnetic moments are often ordered with respect to each other only in localized regions known as *magnetic domains*. The magnetic domain microstructure is in general not an intrinsic property but is primarily extrinsic, since it depends on the shape; prior magnetic, thermal, and mechanical treatments; and possible inhomogeneity of the material.

The existence of magnetic domains in ferromagnets under equilibrium conditions was first proposed in 1907 by P. Weiss in his molecular-field model of magnetism.[†] As is shown later, the free energy of a magnetic material is in fact lowered when domains are formed despite the fact that there is a surface energy associated with the interfaces or walls that separate the domains. The most significant effect of domains on the magnetic properties of ferromagnets and ferrimagnets is that the materials can exist in a completely demagnetized state (i.e., with zero net magnetization **M**), even below T_C. This results from the fact that the domains can be oriented so that essentially all of the magnetic flux lies completely within the solid.

Magnetic domains were first directly observed by Bitter in 1931 with a technique that involved the application of a colloid of small ferromagnetic particles to the surface of a ferromagnet below T_C. The resulting spatial arrangement of the ferromagnetic particles revealed the domain structure present at the surface of the ferromagnet. An example of a typical magnetic domain structure can be seen in Fig. 17.1. The material shown is a Fe + 3 wt % Si crystal examined with a scanning electron microscope with polarization analysis (SEMPA), a highly sensitive technique that analyzes the spin polarization of the secondary electrons emitted. The arrows indicate domains with four different magnetization directions. A wide variety of experimental techniques, including the classic Bitter technique, magneto-optical techniques involving polarized light (Kerr effect and Faraday rotation), magnetic force microscopy (MFM, a variation

Figure 17.1. Magnetic domain structure in the ferromagnetic material Fe + 3 wt % Si (single crystal). The arrows indicate domains with four different magnetization directions. The bar in the lower right-hand corner corresponds to 10 µm. (From D. T. Pierce et al., *Mater. Res. Soc. Bull.*, **13**(6), 20 (1988).

[†] P. Weiss, *J. Phys.*, **6**, 661 (1907).

of atomic force microscopy), and SEMPA, are currently in use for the direct or indirect observation of magnetic domain microstructures. Magnetic force microscopy can also be used to modify the magnetic domain structure in thin films when the probe tip is placed in contact with the surface of the film.

As discussed in Chapter 9, at $T = 0$ K the direct exchange interactions between magnetic moments will lead to the alignment of all the moments in the same direction within a given ferromagnetic domain. Each domain thus has a net magnetic moment **m** whose magnitude and direction may be different from those of neighboring domains. To understand how the presence of domains can lower the free energy of a magnetic material, consider the relative energies of the domain structures shown schematically in Fig. 17.2. In Fig. 17.2a a ferromagnetic solid in the shape of a rectangular film of thickness t and area $A = lw$ is magnetized in the $+z$ direction (i.e., perpendicular to the film surface). In this case the z axis is an easy direction of magnetization which is favored by some form of magnetic anisotropy in the material. Here the entire solid consists of a single magnetic domain in which $\mathbf{M} = \mathbf{M}_s$, the spontaneous magnetization appropriate for the temperature of the solid. The measured spontaneous magnetizations of the ferromagnets Fe and Ni have been shown as functions of temperature in Fig. 9.13.

The magnetostatic *self-energy* of the magnetic material in its own magnetic field is given by

$$U_m = -\frac{\mu_0 \int \mathbf{M} \cdot \mathbf{H}_i \, dV}{2}, \tag{17.1}$$

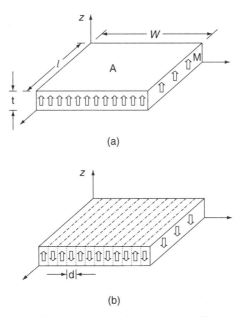

(a)

(b)

Figure 17.2. Ferromagnetic solid in the shape of a rectangular film. (a) The film is magnetized in the $+z$ direction so that the entire film consists of a single magnetic domain. (b) The film is divided instead into a uniform array of parallel magnetic domains. The direction of **M** is reversed in adjacent domains.

where μ_0 is the permeability of free space and $\mathbf{H}_i$ is the internal magnetic field. This internal field is in general given by

$$\mathbf{H}_i = \mathbf{H} + \mathbf{H}_D, \tag{17.2}$$

where $\mathbf{H}$ is the externally applied field and $\mathbf{H}_D$ is any *demagnetizing field* that may be present in the material due to its shape. The demagnetizing field $\mathbf{H}_D$ is proportional to and directed opposite to $\mathbf{M}$ in the material and has the effect of reducing the flux density $\mathbf{B} = \mu_0(\mathbf{H} + \mathbf{M})$ within the material. In symmetric or regular solids the demagnetizing field is given by

$$\mathbf{H}_D = -N\mathbf{M}, \tag{17.3}$$

where N is the dimensionless *demagnetizing factor* determined by the shape of the solid and the direction of $\mathbf{M}$ within the solid. Values for N are presented in Fig. 17.3 for various symmetric solids when uniformly magnetized along the x, y, and z axes. Note that $1 > N > 0$ and that $N_x + N_y + N_z = 1$ for the sum of the demagnetizing factors in three mutually orthogonal directions.

For convenience of calculation $\mathbf{H}_D$ is usually attributed to fictitious positive and negative *magnetic poles* of equal magnitude that are taken to be distributed uniformly over the top and bottom surfaces of the film shown in Fig. 17.2a. These magnetic poles, if they in fact existed, would be the magnetic analog of the electrical charges that generate electric fields. In reality, $\mathbf{M}$ and $\mathbf{H}_D$ arise from the same source (i.e., the individual magnetic dipole moments $\mathbf{m}_i$ present in the material). The sources giving rise to $\mathbf{H}_D$ are then the magnetic fields of each local magnetic moment. These fields are directed opposite to $\mathbf{m}_i$ in its vicinity, as can be seen in Fig. 17.4.

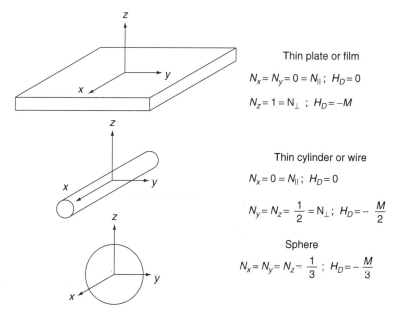

Figure 17.3. Values for the demagnetizing factors N_x, N_y, and N_z for various symmetric solids when uniformly magnetized along the x, y, and z axes, respectively.

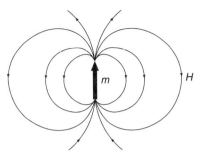

Figure 17.4. The sources of the demagnetizing field $\mathbf{H}_D$ are the magnetic fields of each local magnetic moment $\mathbf{m}_i$. These fields are directed opposite to $\mathbf{m}_i$ in its vicinity.

For the film shown in Fig. 17.2a, the demagnetizing factor is $N \approx 1$ when the magnetization $\mathbf{M}_s$ is directed perpendicular to its surface. Therefore, $\mathbf{H}_D = -\mathbf{M}_s$ (i.e., the demagnetizing field $\mathbf{H}_D$ is equal in magnitude and directed opposite to $\mathbf{M}_s$). It follows from Eq. (17.1) that, in the absence of an external field $\mathbf{H}$,

$$U_m = -\frac{\mu_0 \int \mathbf{M}_s \cdot (-\mathbf{M}_s)\, dV}{2} = \frac{\mu_0 M_s^2 V}{2}, \qquad (17.4)$$

where the volume of the film is $V = At$.

When the film is divided instead into a uniform array of parallel magnetic domains, with the direction of $\mathbf{M}_s$ reversed in adjacent domains as shown in Fig. 17.2b, the magnetostatic energy will be reduced below the value given in Eq. (17.4). In addition, the net magnetization $\mathbf{M}$ of the solid will also be reduced well below $\mathbf{M}_s$. It can be shown in this case that the magnetostatic energy becomes[†]

$$U_m = \frac{0.136\mu_0 M_s^2 V d}{t}, \qquad (17.5)$$

where d is the width of each domain. If only the magnetostatic energy U_m were to be considered, an array of domains would be favored over a uniformly magnetized, single-domain film as long as $(0.136\, d/t) < 0.5$ (i.e., as long as the domain width d is less than about 3.7 times the film thickness t). In fact, on the basis of Eqs. (17.4) and (17.5) it can be seen that $U_m \to 0$ as $d \to 0$. It would thus appear to be energetically favorable for the film to continue subdividing into smaller and smaller ferromagnetic domains.

The division of the magnetic microstructure into increasingly small domains is in fact not observed in ferromagnetic materials due to the finite *surface energy* σ_w associated with the interfaces or walls between domains. Within the domain wall the magnetic moments gradually change their directions from one domain to the next. The surface energy σ_w includes contributions from both the exchange energy $U(exch) = -J\mathbf{S}_1 \cdot \mathbf{S}_2 = -JS_1 S_2 \cos\theta$ and magnetic anisotropy energy density E_a, which is determined by the directions of the spins with respect to either the crystalline axes or some other preferred

[†] C. Kittel, *Rev. Mod. Phys.*, **21**, 541 (1949).

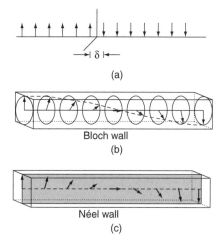

Figure 17.5. Three possible configurations for the directions of spins within a 180° magnetic domain wall of thickness δ: (*a*) the domain wall is as thin as possible; (*b*) a Bloch wall; (*c*) a Néel wall. (From R. M. White and T. H. Geballe, *Long Range Order in Solids*, Suppl. 15 of H. Ehrenreich, F. Seitz, and D. Turnbull, eds., *Solid State Physics*, copyright 1979. Reprinted by permission of Academic Press, Inc.)

direction in the material. With $J > 0$, U(exch) will be increased by the amount $+2JS^2$ for each pair of adjacent oppositely directed spins with $\theta = 180^0$ (Fig. 17.5*a*). This increase in U(exch) will clearly be smaller when the angle θ between the directions of adjacent spins is less than 180° and will tend to 0 as $\theta \to 0$. Exchange effects therefore will tend to lead to domain walls of essentially unlimited width in which the directions of the spins change very gradually from one domain to the next. This was first recognized by F. Bloch[†].

The contribution of the magnetic anisotropy of the material to the domain wall energy opposes the tendency for the formation of wide domain walls. Within the domain wall some of the spins will point away from the directions of easy magnetization (i.e., away from directions in the crystal corresponding to lower magnetic energy). Anisotropy effects will therefore tend to lead to narrower domain walls so that fewer spins will point in energetically unfavorable directions. As a result of the combined effects of exchange and anisotropy, the magnetic moments within the domain wall will in fact prefer to shift gradually from the direction of the magnetization $\mathbf{M}_s$ in one domain to its direction in the adjacent domain, but with as few spins as possible in energetically unfavorable directions. A possible resulting configuration for the directions of the spins in a 180° domain wall is shown in Fig. 17.5*b*. This type of interface between magnetic domains, known as a *Bloch wall*, can be seen to resemble one-half of a period of a type of ferromagnetic spin wave. Another type of domain wall in which the spin configuration differs from that found in Bloch walls is the *Néel wall* shown in Fig. 17.5*c*. Further discussions of the energetics of domains and domain walls are presented in Chapter W17 at our Web site.[‡]

[†] F. Bloch, *Z. Phys.*, **74**, 295 (1932).

[‡] Supplementary material for this textbook is included on the Web at the resource site (ftp://ftp.wiley.com/public/sci_tech_med/materials). Cross-references to elements of the Web material are prefixed by "W."

The domain wall thickness δ and the orientation of the spins in a domain wall in a ferromagnet will in general depend on the relative orientations of the spontaneous magnetizations $\mathbf{M}_s$ in adjacent domains and on the magnitudes of the exchange and anisotropy energies. The domain wall thickness that minimizes the wall energy σ_w can be shown to be given approximately by

$$\delta = \sqrt{\frac{JS^2\pi^2}{Ka}}, \tag{17.6}$$

where K is the appropriate first-order uniaxial or cubic anisotropy coefficient defined in Eq. (9.46) or (9.47), respectively, and a is the lattice constant. Note that δ will be quite wide in magnetic materials with low magnetic anisotropy. Using values of these parameters appropriate for Fe at $T = 300$ K (i.e., $J = 2.7 \times 10^{-21}$ J, $S = 1$, $K = K_1 = 4.2 \times 10^4$ J/m^3, and $a = 0.287$ nm), the value $\delta = 4.7 \times 10^{-8}$ m $= 47$ nm is obtained for the domain wall thickness. This wall thickness corresponds to about 160 lattice constants for Fe.

The domain-wall surface energy σ_w is given approximately by

$$\sigma_w = 2\sqrt{\frac{JS^2\pi^2K}{a}} = 2\delta K, \tag{17.7}$$

which, for Fe at $T = 300$ K, has the value $\sigma_w = 3.9 \times 10^{-3}$ J/m^2. Additional sources of magnetic anisotropy such as strain and particle shape can also contribute to an increase of the surface energy σ_w and a decrease in the wall thickness δ. The surface energy will also depend on the orientation of the wall in the crystal (e.g., whether the wall is parallel to {100}, {110}, or {111} planes). Note that the domain wall width δ depends on σ_w since, from Eq. (17.7), $\delta = \sigma_w/2K$.

It is interesting to compare actual values for the energies per unit volume $u = U/V$ for the two cases discussed here (i.e., for the uniformly magnetized ferromagnetic film with a single domain shown in Fig. 17.2a and for the same film with the array of domains shown in Fig. 17.2b). The material to be used for this example is Fe with $M_s = 1.71 \times 10^3$ kA/m at $T = 300$ K and the values of δ and σ_w calculated earlier. For the single-domain case, Eq. (17.4) yields

$$u_m = \frac{U_m}{V} = \frac{\mu_0 M_s^2}{2} = 1.84 \times 10^3 \text{ kJ/m}^3. \tag{17.8}$$

For the second case, Eq. (W17.4) gives, using $t = 1$ mm,

$$u = u_m + u_w = 0.738 M_s \sqrt{\frac{\mu_0\sigma_w}{t}} = 2.79 \text{ kJ/m}^3. \tag{17.9}$$

This result is three orders of magnitude less than that given above for a single domain. As long as a ferromagnetic film with magnetization directed perpendicular to its surface is not too thin, this result indicates clearly that it can be energetically very advantageous for it to take on a magnetic microstructure consisting of an array of domains. In this case with $t = 1$ mm, Eq. (W17.3) yields $d = 2.8 \times 10^{-6}$ m for the domain width. Thus

$d \approx 60\delta$ for Fe. An even lower energy would result if the ferromagnetic film had its magnetization aligned parallel to its surface, in which case the demagnetizing factor N and the energy U_m would both be essentially zero.

The actual magnetic domain microstructure observed in a ferromagnetic or ferrimagnetic solid will depend on its shape, its crystal structure, and on the presence of inhomogeneities and defects such as voids, inclusions, precipitates, grain boundaries, dislocations, and stacking faults. The surface of a ferromagnetic material can also influence the magnetic microstructure (Fig. 17.6). Here additional domains can be seen to be present at the surface. Such surface domains are called *closure domains* since they "close" the magnetic field lines within the solid at its surface, preventing their leakage outside and thus lowering the magnetostatic energy U_m even further.

Two types of domain walls can be seen in Fig. 17.6: the usual 180° walls within the material and 90° walls between the closure domains and the internal domains. Here the angle mentioned refers to the relative directions of the magnetizations $\mathbf{M}_s$ in two domains. The formation of closure domains will be more difficult when they correspond to a hard direction of magnetization. This would be the case for the HCP ferromagnet Co when the easy direction corresponding to the c axis is perpendicular to the surface of the solid. Even in a cubic ferromagnet such as Fe the formation of closure domains at the surface of the material will cost additional elastic energy since, due to the magnetostrictive effects described in Section 17.7, the internal domains and the closure domains will be elongated along different $\langle 100 \rangle$ axes.

Under certain conditions magnetic domains can take the form of cylindrical domains or bubbles. These localized domains are formed when stripe domains collapse in increasing magnetic fields. Magnetic domains are also found in antiferromagnetic materials, where the justification for their presence is not as obvious as in ferromagnets and ferrimagnets since the macroscopic magnetization M and hence the magnetostatic energy U_m in an antiferromagnet are zero even in the absence of magnetic domains. The presence of magnetic domains is also more difficult to detect in antiferromagnets since their effect on the magnetic behavior is much less noticeable than is the case in a ferromagnet.

In polycrystalline magnetic materials the magnetic domain microstructure will depend on the sizes and relative orientations of the crystallites. When the crystallites are randomly oriented, each can act essentially as a single domain of fixed size that will interact with the domains in neighboring crystallites.

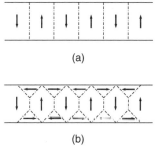

Figure 17.6. Schematic magnetic domain configurations at the surface of a ferromagnetic solid: (*a*) domain structure expected on the basis of minimizing the magnetic free energy of a ferromagnetic film (see Fig. 17.2*b*); (*b*) additional closure domains present at the surface.

The response of magnetic materials to applied magnetic fields involves the development of a nonzero net magnetization **M**. The processes associated with the development of a net magnetization generally involve the motion and pinning of domain walls, which in turn are very dependent on the magnetic microstructure of the material. The discussion of magnetization processes in Section 17.3 is therefore descriptive and general.

17.3   Magnetization Processes and Magnetization Curves

When cooled from above T_C in zero applied magnetic field, ferromagnetic and ferrimagnetic materials can exist in demagnetized states in which the macroscopic magnetization **M** is zero even though individual magnetic domains are magnetized spontaneously with magnetizations $\mathbf{M}_i = \mathbf{M}_s$. The demagnetized state occurs when the vector sum of the magnetic moments of the individual domains in the material is zero. When a magnetic field **H** is applied to such a magnetic solid, a net macroscopic magnetization **M** will appear as the magnetic moments of the individual domains respond to the applied field. Since the internal molecular fields due to the local magnetic moments are so large, $B_{\text{eff}} \approx 10^3$ T ($H_{\text{eff}} = B_{\text{eff}}/\mu_0 \approx 10^6$ kA/m), the principal effect of the much weaker applied magnetic field is to reorient the directions of the individual magnetic moments of the domains and hence their magnetizations. These changes in the domain microstructure can result from the displacement or motion of domain walls and from the rotation of the direction of the magnetization $\mathbf{M}_i$ within a given domain.

The initial *magnetization curve* for a typical ferromagnet at a temperature $T < T_C$ is presented in Fig. 17.7. Here the dependence of the macroscopic magnetization **M** on the applied field **H** is shown schematically. In zero applied field the ferromagnet is completely demagnetized and has no net macroscopic magnetization due to its magnetic domain microstructure. As the applied field is increased, the magnetization in the direction of **H** begins to increase as domains whose magnetizations $\mathbf{M}_i$ are favorably

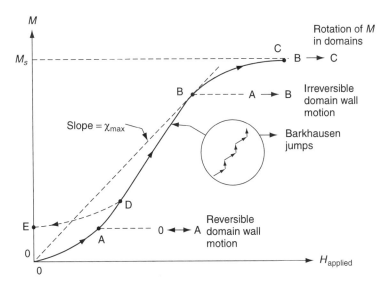

Figure 17.7. Initial magnetization curve for a typical ferromagnet.

oriented with respect to **H** begin to grow in size via domain wall motion, described later. This first portion of the initial magnetization curve between the origin O and point A is nearly reversible, since **M** will return to zero if the field **H** is removed, the demagnetization process again occurring via domain wall motion. The initial magnetic susceptibility χ_i is the initial slope of the M versus H curve, and the initial magnetic permeability $\mu_i = \mu_0 (1 + \chi_i)$ is the initial slope of the corresponding **B** versus **H** curve.

As **H** is increased beyond point A, additional increases in the magnetization can result from irreversible processes in which domain walls move past defects such as grain boundaries which impede their motion. Evidence for the pinning of domain walls by defects can be found from a close examination of the magnetization curve in this region, which often shows discrete jumps or increases in **M** as domain walls become unpinned and move rapidly to new positions. These discrete jumps in M shown schematically in the inset of Fig. 17.7, known as *Barkhausen jumps*, are strong, although indirect, evidence for the reality of the motion of individual domain walls. The more weakly pinned domain walls will be displaced first, followed by the walls that are more effectively pinned.

If the field **H** is decreased to zero from a point D lying on the initial magnetization curve between points A and B, **M** will not reverse itself and return to zero but will instead, retain a finite value at **H** = 0, as indicated at point E. This irreversibility results from defects in the material which prevent the domains from returning to their original configuration corresponding to **M** = 0.

The final approach to saturation (i.e., to the spontaneous magnetization $\mathbf{M}_s$), as the applied field increases from point B to point C typically occurs by rotation of the magnetization within domains into the direction of **H**. Much higher magnetic fields are typically required to reach the completely saturated state where $M = M_{\text{sat}} = ng\mu_B J$. The maximum magnetic susceptibility χ_{max} corresponds to the maximum value of the ratio M/H along the magnetization curve, as shown in Fig. 17.7.

The motion of domain walls and the resulting growth of domains can be understood with the help of Fig. 17.8. Here at the top a 180° wall between two adjacent domains is shown schematically for a linear chain of magnetic moments. The position of the wall is stable in the absence of an applied magnetic field since no external torques are exerted on the individual magnetic moments. When a field **H** is applied in the direction of the magnetic moments $\mathbf{m}_i$ within the left-hand domain, the only spins experiencing torques $\tau = \mu_0 m_i H \sin\theta$ are those within the domain wall for which $\sin\theta$ is not zero (i.e., those for which the angle θ between $\mathbf{m}_i$ and **H** is neither 0° nor 180°). The torques on the magnetic moments in the wall will tend to align them in the direction of **H**. As can be seen from the figure, the result is that the domain wall is displaced to the right and the favorably oriented domain on the left grows in size.

Figure 17.8b shows schematically how the domain structure in a film, shown previously in Fig. 17.6b, changes when a field **H** is applied perpendicular to its surface. Favorably oriented domains grow in size via domain wall motion until the entire film consists of a single magnetic domain.

The work per unit volume $w = W/V$ performed by the applied magnetic field **H** in order to magnetize the ferromagnet to $\mathbf{M} = \mathbf{M}_s$ is given by Eq. (9.16),

$$w = \frac{\mu_0 H^2}{2} + \mu_0 \int_0^{\mathbf{M}_s} \mathbf{H} \cdot d\mathbf{M}. \qquad (17.10)$$

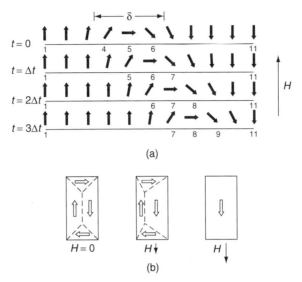

Figure 17.8. Motion of domain walls in the presence of an applied field **H**. (*a*) When a field **H** is applied, the domain wall is displaced to the right. (*b*) The domain configuration changes in the presence of an applied magnetic field **H** due to domain wall motion.

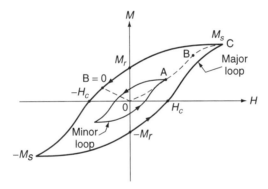

Figure 17.9. Major magnetization or hysteresis loop of a ferromagnet; also, an example of a minor hysteresis loop.

The first term in this expression is just the energy density of the magnetic field H while the second term is equal to the area in the first quadrant between the magnetization curve and the vertical **M** axis. While some of the work done to generate the magnetization M_s is stored as potential energy in the system of magnetic moments, the rest is used in irreversible processes (e.g., to overcome the pinning forces on the domain walls). This work has therefore been dissipated in the material as heat.

When the field **H** is decreased to zero from point C on the initial magnetization curve the magnetization does not return to zero but instead decreases to the value $\mathbf{M}_r$ known as the *remanent magnetization* or *remanence* (Fig. 17.9). When the applied field is removed from a magnetized polycrystalline material, the remaining magnetization $\mathbf{M}_r$ will correspond to the state in which the magnetization of each crystallite will point in the easy direction that is closest to the direction of the applied **H**. The ratio $S = M_r/M_s$ $(0 < S < 1)$ is known as the *squareness* of the magnetization loop.

The macroscopic magnetization of the material can be decreased to zero only by overcoming the effects of the pinning of the domain walls through the application of a magnetic field $\mathbf{H} = \mathbf{H}_c$ in the opposite direction, as shown. The field $\mathbf{H}_c$ is known as the *coercive field* or *coercivity* of the material and is the field necessary to cause the reversal of about one-half of the magnetization present at point C. The portion of the curve in the second quadrant from $\mathbf{M} = \mathbf{M}_r$ to $\mathbf{H} = -\mathbf{H}_c$ is known as the *demagnetizing curve*. It is important to note that the configuration of domains in a ferromagnet will be different at the demagnetized point where $\mathbf{M} = 0$ and $\mathbf{H} = -\mathbf{H}_c$ and at the origin where the material was originally demagnetized.[†] The work per unit volume required to demagnetize the material from $\mathbf{M} = \mathbf{M}_r$ to $\mathbf{M} = 0$ is proportional to the area under the demagnetization curve in the second quadrant, which can be seen to be given approximately by $\mu_0 M_r H_c/2$.

As **H** is increased beyond $\mathbf{H}_c$ in the opposite direction, the magnetization of the ferromagnet will approach $-\mathbf{M}_s$, as shown. The major *hysteresis loop* shown in Fig. 17.9 is obtained by cycling **H** between positive and negative values while **M** cycles between $\pm\mathbf{M}_s$. The net work per unit volume w done by the external field **H** in order to take the ferromagnet around the hysteresis loop in the direction shown in Fig. 17.9 is given by the integral $\mu_0 \oint \mathbf{H} \cdot d\mathbf{M}$ around one complete magnetization cycle. This is the area enclosed by the loop. The net work w done in going around the hysteresis loop in either direction is not stored in the material but instead, is dissipated irreversibly as heat. If the applied field H is changed rapidly, additional dynamic losses can occur as heat is generated in the material due to macroscopic eddy current losses.

A hypothetical square magnetization loop is shown in Fig. 17.10. Note that $M_r = M_s$ for the square loop and that H_c is the field, which switches the magnetization from $+M_s$ to $-M_s$. This square loop is predicted for the case when the external field **H** is applied along the easy axis of magnetization for a single-domain magnetic solid in which changes of the magnetization occur via rotation of $\mathbf{M}_s$. A linear magnetization loop is found when **H** is applied perpendicular to the easy axis. The square loop is the

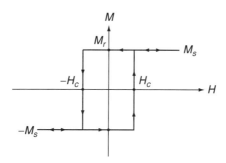

Figure 17.10. Hypothetical square magnetization loop with squareness $S = M_r/M_s = 1$.

[†] This point is discussed in detail in Chikazumi (1964, Chap. 12).

preferred form of the magnetization loop for permanent-magnet materials, as discussed in Section 17.11.

When the magnetic induction B in the magnetic material is plotted versus the applied field H, hysteresis is also observed and the *remanent induction* B_r and the coercive field H'_c at which $B = 0$ can be determined. Note that $B_r = \mu_0 M_r$ but that H'_c and H_c are not simply related, although it can be shown that $H'_c < H_c$. The point on the M versus H hysteresis loop in the second quadrant where $B = 0$ is shown in Fig. 17.9. The difference between H_c and H'_c will be small when the slope dB/dH near H'_c is large.

The actual dependence of the macroscopic magnetization $\mathbf{M}$ on $\mathbf{H}$ can be very complicated in a ferromagnet or ferrimagnet since the magnetization process depends both on the direction of the applied field $\mathbf{H}$ relative to the easy directions of magnetization in the material (see Fig. 9.16) and also on irreversible processes associated with domain wall motion. These irreversible processes depend critically on the shape of the magnetic solid, its demagnetizing factor N, and its microstructure (i.e., its structural uniformity). As an example, see Fig. W17.7, where the effect of heat treatment on the magnetic properties of a permanent magnet Alnico alloy is shown. In addition, there can exist many different magnetic domain configurations in a ferromagnet which are nearly equivalent energetically.

The magnetic domain microstructure present in a ferromagnet will also depend on its past history (i.e., on the applied magnetic fields, temperatures, and even stresses that it has experienced). For example, the macroscopic magnetization $\mathbf{M}$ in zero applied field $\mathbf{H}$ can have any value between $+\mathbf{M}_r$ and $-\mathbf{M}_r$. Also, the magnetic field for which $\mathbf{M} = 0$ can lie anywhere between $-\mathbf{H}_c$ and $+\mathbf{H}_c$. In fact, any $(\mathbf{M},\mathbf{H})$ point within the major hysteresis loop can be reached by applying the proper sequence of magnetic fields. For these reasons it is usually quite difficult to develop simple models that are capable of predicting how the magnetization $\mathbf{M}$ of ferromagnetic or ferrimagnetic materials will depend on the applied magnetic field $\mathbf{H}$, the temperature T, and the state of stress of the material. One property that can usually be predicted accurately is the microstructure-independent intrinsic saturation magnetization $\mathbf{M}_{sat}$ achieved in high applied fields and at low temperatures.

If a magnetic field is applied to a ferromagnetic single crystal such as Fe but not in one of the easy $\langle 100 \rangle$ directions of magnetization, the domains whose magnetizations are closest in direction to $\mathbf{H}$ will grow in size via domain wall motion, essentially as shown earlier in Fig. 17.8. If the applied field is sufficiently strong, the magnetization $\mathbf{M}_s$ within each domain will finally rotate from one the easy $\langle 100 \rangle$ directions to the direction of $\mathbf{H}$, resulting in the magnetic saturation of the solid corresponding to the temperature in question. This behavior has been illustrated in Fig. 9.16, where magnetization curves for ferromagnetic Co are shown. It can be seen there that saturation of the magnetization in the easy [0001] direction corresponding to the c axis in Co can be accomplished in low magnetic fields via domain wall motion while in the hard direction perpendicular to the c axis, much higher fields are required to overcome the effects of magnetocrystalline anisotropy.

In magnetic materials that are not structurally uniform, domain wall motion can be difficult due to the pinning of the walls by defects such as grain boundaries. In these materials the magnetization process requires higher magnetic fields and the initial magnetization curve rises much more gradually than for a uniform material (e.g., the single crystal of Co shown in Fig. 9.16).

17.4 Magnetically Hard and Soft Materials

The suitability of magnetic materials for specific applications is often determined by the values of their coercive fields H_c and the areas of their hysteresis loops. The field H_c depends on both intrinsic and extrinsic properties of ferromagnetic materials and therefore is highly structure sensitive. For example, the easier it is for domain walls to move in a given material (i.e., the less pinning there is), the easier it will be for the material to return to a domain configuration corresponding to $\mathbf{M} = 0$. Thus defect-free materials can be expected to have lower values of H_c and M_r than will structurally imperfect materials.

A logarithmic plot of measured coercive fields H_c versus the initial magnetic susceptibilities $\chi_i = (\mu_i/\mu_0) - 1$ measured for a wide range of magnetic materials is shown in Fig. 17.11; μ_i is the initial magnetic permeability. Both H_c and χ_i are extrinsic properties and can be seen to vary over several orders of magnitude, with high values of χ_i in general corresponding to low values of H_c, and vice versa. This correlation is to be expected since ferromagnets which can easily be magnetized in low applied fields, corresponding to a high susceptibility $\chi = M/H$, can also be demagnetized relatively easily, corresponding to a low H_c. The reverse is expected for materials with low χ and high H_c, which are hard both to magnetize (low χ) and to demagnetize (high H_c). The magnetization and demagnetization processes are both expected to be easier in materials in which domain walls are free to move easily.

Figure 17.11 thus helps to distinguish between what are commonly known as magnetically hard and soft materials. *Soft magnetic materials* have high magnetic susceptibilities χ, typically in the range 10^3 to 10^5, and low coercive fields H_c, typically in the range 0.1 to 100 A/m. They also have narrow hysteresis loops, corresponding to low losses. Since the direction of the magnetization can easily be reversed in soft

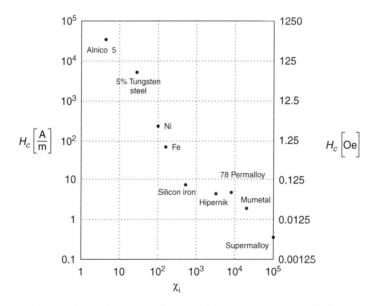

Figure 17.11. Measured coercive fields H_c and initial magnetic susceptibilities χ_i for a wide range of magnetic materials.

magnetic materials, they find use in transformers, motors and generators, microwave components, and so on, where the hysteresis loop is traversed repeatedly and losses must be minimized. It is necessary for domain walls to move easily in soft magnetic materials, so pure, strain-free materials are usually required.

Hard magnetic materials, corresponding to low values of χ, typically in the range 1 to 100, and high values of H_c, typically in the range $10^3 - 10^5$ A/m, can result from the strong pinning of domain walls by defects. The field H_c can also be increased by the presence of a second phase introduced by precipitation or by the use of very small particles. This is referred to as *magnetic hardening*. High values of the remanent magnetization M_r, wide hysteresis loops, and large hysteresis losses are also desirable in hard magnetic materials. These materials are hard to magnetize, but once magnetized their magnetization remains close to M_s. They are used in permanent magnets and as magnetic and magneto-optical recording media, where losses are not an important consideration.

Whether a material is magnetically hard or soft is generally not an intrinsic property of that material. Thermal, mechanical, and magnetic processing treatments often play critical roles in determining H_c and thus the degree of magnetic hardness.

Magnetic anisotropy can also play a significant role in the magnetization process and in increasing H_c. The mechanisms by which this can occur are discussed next.

17.5   Effects of Magnetic Anisotropy

Magnetic anisotropy refers to the fact that for any of a variety of possible reasons, magnetic moments $\mathbf{m}_i$ can find it energetically favorable to point in specific directions in the crystal structure. These are referred to as the *easy directions* or *axes of magnetization*. The contribution of magnetic anisotropy to the magnetic energy of a ferromagnet is expressed by the anisotropy energy density E_a [see Eqs. (9.46) and (9.47)], which can be interpreted as a measure of the work (per unit volume) done by an applied field $\mathbf{H}$ to rotate the spontaneous magnetization $\mathbf{M}_s$ away from an easy direction of magnetization. As mentioned earlier, magnetization curves will have very different shapes according to the direction of $\mathbf{H}$ relative to the easy directions of magnetization.

Magnetic materials with high magnetic anisotropy will typically have high coercive fields H_c and will be magnetically hard. In polycrystalline materials with high magnetic anisotropy, high magnetic fields will be required both to saturate the magnetization and then to demagnetize the material. This is due to the fact that there will be a distribution in space of the easy directions of magnetization due to the random orientations of the crystallites. There will thus always be domains for which the rotation of the magnetization into the direction of the applied field $\mathbf{H}$ will be difficult. This will both delay the approach to saturation of the magnetization and will increase the coercive field H_c.

In materials with high magnetic anisotropy the initial magnetization process occurs primarily by the motion of domain walls (if they are not pinned by defects), so that favorably oriented domains grow at the expense of those that are less favorably oriented. For the case of low magnetic anisotropy, the reversible rotation of the domain magnetization $\mathbf{M}_s$ is often the preferred magnetization process.

The magnetic anisotropy energy density E_a of a uniformly magnetized material can often be expressed simply by

$$E_a = K \sin^2 \theta \qquad (17.11)$$

when the angle θ that the magnetization $\mathbf{M}_s$ makes with respect to a particular direction in the lattice is small. Here K is the effective magnetic anisotropy coefficient, which corresponds to the sum of all relevant contributions to the anisotropy (e.g., magnetocrystalline, magnetoelastic, shape, and possibly surface-related anisotropy). When $K = E_a (90°) - E_a(0°)$ is positive, the easy directions are $\theta = 0°$ and $180°$; when K is negative, the easy directions correspond to $\theta = 90°$. Note that E_a is just the energy needed to overcome the spin–orbit interaction (i.e., to rotate the magnetization M from the easy to the hard direction).

The derivative of E_a with respect to θ is the magnitude of the torque per unit volume τ/V exerted by an effective *internal magnetic anisotropy field* $\mathbf{H}_K$ which tends to keep $\mathbf{M}_s$ pointing in the easy direction (Fig. 17.12). The effective field $\mathbf{H}_K$ points along the easy axis and is in some respects analogous to the molecular field that results from the Heisenberg exchange interaction [see Eq. (9.30)]. It follows that

$$\frac{\partial E_a}{\partial \theta} = K \frac{\partial \sin^2 \theta}{\partial \theta} = 2K \sin \theta \cos \theta = K \sin 2\theta \approx 2K \sin \theta, \tag{17.12}$$

a result that is valid when $\theta \ll 1$. Also, using $\mathbf{M}_s = \mathbf{m}/V$,

$$\frac{|\tau|}{V} = \frac{|\mathbf{m} \times \mathbf{B}_K|}{V} = \frac{|\mathbf{m} \times \mu_0 \mathbf{H}_K|}{V} = |\mathbf{M}_s \times \mu_0 \mathbf{H}_K| = \mu_0 M_s H_K \sin \theta. \tag{17.13}$$

Since $\partial E_a/\partial \theta = |\tau|/V$, it follows that $2K = \mu_0 M_s H_K$ and therefore that

$$H_K = \frac{2K}{\mu_0 M_s} \tag{17.14}$$

is the magnitude of the magnetic anisotropy field. In these expressions $\mathbf{m}$ is the total magnetic moment of the material contained within a volume V (e.g., a magnetic domain). As shown later, applied fields H of the order of H_K are needed to rotate the magnetization $\mathbf{M}_s$ from an easy to a hard direction.

When the magnetic anisotropy in the material is uniaxial, the sources can be either the magnetocrystalline anisotropy as in HCP Co, magnetostrictive anisotropy (discussed in Section 17.7), shape-induced anisotropy, or induced anisotropies due to annealing

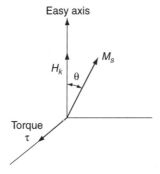

Figure 17.12. Effective internal magnetic anisotropy field $\mathbf{H}_K$. The torque τ which tends to align $\mathbf{M}_s$ with the direction of $\mathbf{H}_K$ along the easy axis of magnetization is also shown.

in an applied field (magnetic annealing) or to cold working which introduces uniaxial texture in the material. The total uniaxial anisotropy present in a ferromagnetic solid can correspond to the sum of more than one of the sources listed above. For example, the effects of magnetocrystalline and shape anisotropies can be additive in long, needle-shaped magnetic particles that are used in magnetic recording media. This is discussed in more detail later when the effects of particle shape are considered explicitly.

Even in the absence of magnetocrystalline anisotropy, the external shape or dimensionality of a magnetic material [e.g., a thin needle (one-dimensional) or a thin film (two-dimensional)], can induce magnetic anisotropy and hence determine the preferred direction for the magnetization **M**. For the case of anisotropy arising from shape effects, the *shape anisotropy constant K_s* is given by

$$K_s = \frac{\mu_0(N_\perp - N_\parallel)M_s^2}{2}. \tag{17.15}$$

Here $N_\perp$ and $N_\parallel$ are the demagnetizing factors for the directions perpendicular and parallel, respectively, to the long or major axis (the axis of symmetry) of a uniformly magnetized ellipsoidal ferromagnetic solid. The magnetic anisotropy field is given by

$$H_s = \frac{2K_s}{\mu_0 M_s} = (N_\perp - N_\parallel)M_s. \tag{17.16}$$

Note that $N_\perp = 1/2$ and $N_\parallel = 0$ for a thin ferromagnetic cylinder or needle-shaped particle (see Fig. 17.3). In this case, when $K_s > 0$ the magnetization **M**$_s$ of such a cylindrical particle will prefer to be aligned along the length of the cylinder. For the case of a thin ferromagnetic film where $K_s < 0$ (since $N_\perp < N_\parallel$), the effects of shape-induced anisotropy will tend to cause the magnetization **M**$_s$ to align itself parallel to the film surface. Finally, there is no shape anisotropy for a sphere since $N_\perp = N_\parallel$.

Magnetic anisotropy can also result from magnetostrictive effects when a stress σ is applied to an isotropic ferromagnetic material. In this case the magnetic anisotropy field can be written as

$$H_\sigma = \frac{2K_\sigma}{\mu_0 M_s}, \tag{17.17}$$

where the *magnetoelastic stress anisotropy constant K_σ* is given by

$$K_\sigma = \frac{3\lambda\sigma}{2}. \tag{17.18}$$

The sign of K_σ is determined by the sign of the magnetostrictive constant λ (see Section 17.7). The sign of K_σ in turn determines the directions of the easy and hard axes of magnetization in the material, with the easy axis coinciding with the direction of the stress when λ and K_σ are both positive.

The dimensionality, shape, and size of a magnetic solid can also have important effects on its magnetic properties, including the details of the magnetization processes, the magnetic domain microstructure, as well as the magnetic anisotropy. The effect of shape on magnetic anisotropy has just been discussed. Some additional effects of shape and size on the magnetic properties are discussed next.

17.6 Effects of Shape and Size

When a magnetic field **H** is applied to a symmetric, regularly shaped magnetic solid so that a net magnetization **M** in the direction of **H** results, the effective internal field is given by

$$\mathbf{H}_i = \mathbf{H} + \mathbf{H}_D = \mathbf{H} - N\mathbf{M}, \tag{17.19}$$

where $\mathbf{H}_D$ is the demagnetizing field and N is the appropriate demagnetizing factor. The applied field **H** will therefore have to overcome $\mathbf{H}_D$ in order to saturate the magnetization of the material. In addition, measured magnetization curves must be corrected for demagnetization effects so that the true magnetic response of the material is obtained. Consider the initial portion of a measured magnetization curve shown in Fig. 17.13. Here **M** is shown plotted against the applied field **H** and also against the internal field $\mathbf{H}_i$ after the appropriate correction for the demagnetizing field has been made. Following Eq. (17.19), the correction is made by subtracting $N\mathbf{M}$ from the applied field **H** at every point along the curve, as shown. The true magnetization curve of **M** versus $\mathbf{H}_i$ whose slope is the magnetic susceptibility χ can thus be obtained from the measured **M** versus **H** curve. The magnetic induction in the material in the case of a regular magnetic solid will be given by

$$\mathbf{B} = \mu_0(\mathbf{H}_i + \mathbf{M}) = \mu_0[\mathbf{H} + (1 - N)\mathbf{M}]. \tag{17.20}$$

Note that $\mathbf{B} = \mu_0(1 - N)\mathbf{M}$ in the absence of an applied field **H**.

As the dimensions of a ferromagnetic or ferrimagnetic solid in the form of a spherical particle, a long rod, or a thin film decrease and become comparable to the domain wall thickness δ, the formation of magnetic domains will eventually become energetically unfavorable, due to the increasing importance of the surface energy σ_w associated with the domain walls. The solid will then always be magnetized in a single domain with the value $\mathbf{M}_s$. This has been demonstrated earlier for the case of a thin ferromagnetic film.

Consider now the case of a ferromagnetic single-crystal sphere of radius r. When the particle consists of a single domain, it follows that the magnetostatic energy will be

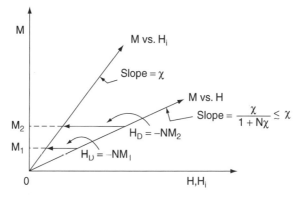

Figure 17.13. Correction of a measured magnetization curve for demagnetization effects. The initial portion of the curve is shown, with **M** plotted against the applied field **H** and also the internal field $\mathbf{H}_i$.

$$U_m = \frac{\mu_0 N M_s^2 V}{2} = \frac{\mu_0 M_s^2 V}{6}, \tag{17.21}$$

where $N = \frac{1}{3}$. For a sphere of a material with a cubic crystal structure and with high magnetic anisotropy there is a *critical radius* r_c below which the particle will consist of a single domain, as shown in Fig. 17.14a. In the case where the spherical particle consists of four domains, as shown in Fig. 17.14b, the magnetostatic energy will be zero while the total wall energy will be

$$U_w = 2\sigma_w \pi r^2. \tag{17.22}$$

An approximate value for the critical radius for obtaining a particle with a single domain instead of one with four domains can be determined by equating the values of U_m and U_w given above, the result being

$$r_c = \frac{9\sigma_w}{\mu_0 M_s^2}. \tag{17.23}$$

The critical radius will tend to decrease with increasing T due to the temperature dependences of σ_w and M_s, both of which also decrease as T increases. The same result for r_c is obtained for a spherical single crystal with uniaxial magnetic anisotropy, for which the domain configuration shown in Fig. 17.14c is appropriate.

The coercive fields H_c of small, single-domain ferromagnetic particles can be quite large, especially when a significant amount of magnetic anisotropy exists in the material. High values of H_c result from the fact that changes in magnetization will no longer proceed via domain wall motion, which can occur in relatively low applied magnetic fields. Changes in magnetization must instead occur via rotation of the magnetization vector **M**, an energetically costly process requiring high magnetic fields when high magnetic anisotropy is present. As discussed later, there are other processes that can lead to the reversal of **M** in much smaller fields. It is observed experimentally that H_c does increase gradually as the particle size decreases, as shown in Fig. 17.15 where the dependence of H_c on particle size is presented for a variety of magnetic materials.

As the particle size becomes very small, the coercive field H_c can start to decrease since very small ferromagnetic particles with correspondingly small magnetic moments can exhibit *superparamagnetic* behavior. In this case it is possible for the thermal energy $k_B T$ to exceed the energy barriers $\approx E_a V$ due to magnetic anisotropy. Here E_a is the magnetic anisotropy energy density and V is the volume of the particle. The same effect can occur in thin magnetic films, in which case V is the volume of a crystallite or grain. The *blocking temperature* T_b, which is the temperature above

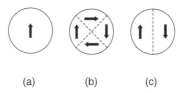

(a)   (b)   (c)

Figure 17.14. Spherical ferromagnetic particles consisting of (a) a single magnetic domain, (b) four magnetic domains, and (c) two magnetic domains.

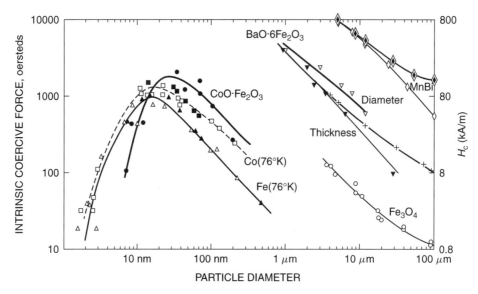

Figure 17.15. Measured dependence of the coercive field H_c on particle diameter for a variety of magnetic materials. The data correspond to room temperature (except as indicated). [Reprinted with permission from F. E. Luborsky, *J. Appl. Phys.*, **32**, 171S (1961). Copyright 1961 by the American Institute of Physics.]

which the magnetic moment of the particle is no longer completely constrained by magnetic anisotropy, can be defined by equating $k_B T_b$ with the anisotropy energy barrier. Ferromagnetic Fe particles behave superparamagnetically at $T = 300$ K when they are smaller than about 3 nm. As can be seen in Fig. 17.15, the measured maxima in the H_c versus particle diameter curves typically occur for values of d between 10 and 20 nm.

Further discussions of size and shape effects in magnetic materials are presented in Chapter W17.

IMPORTANT EFFECTS IN MAGNETIC MATERIALS

In addition to the characteristic magnetic properties described in Sections 17.2 to 17.6, some additional effects that are important for the applications of magnetic materials are discussed next. These include magnetostriction (effects related to stress and strain in magnetic materials), magnetoresistance (the effects of magnetic fields on electrical transport), magneto-optical effects (characteristic optical effects in magnetic materials), and time-dependent or dynamic effects.

Another interesting effect in magnetic materials that will not be described in detail here is the *magnetocaloric effect*. Certain magnetic materials become hotter when placed in a magnetic field and, conversely, cooler when removed from the field. This effect is related to the entropy S of the system of spins and the coupling of the spins to the lattice (i.e., the phonons). The magnetocaloric effect is the basis of magnetic refrigeration or cooling,

17.7 Magnetostriction

Magnetostriction is defined, in general, as the set of magnetoelastic phenomena associated with the interaction of stress and strain with magnetization in a magnetic material. For example, a change in the state of magnetization can induce strain and hence changes in the dimensions (and shape) as well as in the elastic constants of a magnetic material. Conversely, applying stress to a ferromagnet will in general result in a change of its state of magnetization. This is often referred to as the *inverse magnetostrictive effect*. For simplicity, the discussion here focuses primarily on ferromagnetic materials with cubic crystal structures.

The microscopic origin of magnetostriction is the spin–orbit interaction through which the shape and extent of the d (or f) electron orbitals of a magnetic ion depend on the direction of its magnetic moment in the lattice. Thus the distance between neighboring magnetic ions, the strain, and the macroscopic dimensions of a magnetic solid depend on the effects that the applied magnetic field has on the relative directions of the magnetic moment of the magnetic ions through the overlap of their d (or f) orbitals.

The coupling of the magnetization $\mathbf{M}$ to the lattice for a cubic magnetic material can be expressed by the magnetic anisotropy energy density E_a given in Eq. (9.47) as

$$E_a(\text{cubic}) = K_1(\alpha_1^2\alpha_2^2 + \alpha_2^2\alpha_3^2 + \alpha_3^2\alpha_1^2) + K_2\alpha_1^2\alpha_2^2\alpha_3^2 + \cdots.$$

Here K_1 and K_2 are the first- and second-order cubic anisotropy coefficients and $\alpha_1 = \cos\theta_1, \alpha_2 = \cos\theta_2$, and $\alpha_3 = \cos\theta_3$ are the direction cosines of the vector $\mathbf{M}$ relative to the x, y, and z axes of the cubic crystal, respectively. As described in Chapter W9, the magnitude of K_1 is predicted to be proportional to a high power of the spontaneous magnetization M_s for $T \ll T_C$ The coupling of the strain ε to the lattice for a cubic material is expressed by the elastic energy density $u_{el}(\varepsilon)$, given in Eq. (10.32) as

$$u_{el}(\varepsilon) = \frac{C_{11}(\varepsilon_1^2 + \varepsilon_2^2 + \varepsilon_2^2)}{2} + C_{12}(\varepsilon_1\varepsilon_2 + \varepsilon_1\varepsilon_3 + \varepsilon_2\varepsilon_3) + \frac{C_{44}(\varepsilon_4^2 + \varepsilon_5^2 + \varepsilon_6^2)}{2}.$$

Since the magnetization $\mathbf{M}$ and the strain ε are both coupled to the lattice, it follows that they will be indirectly coupled to each other. The coupling between $\mathbf{M}$ and ε is expressed by the magnetoelastic energy density $u(\mathbf{M}, \varepsilon)$ which determines both the effect of the strain ε on the magnetic anisotropy and the effect of the direction of the magnetization $\mathbf{M}$ on the state of strain. This energy density can be defined in terms of a *magnetoelastic tensor* b_{ijkl} by $u(\mathbf{M}, \varepsilon) = \Sigma_{ijkl}b_{ijkl}\alpha_i\alpha_j\varepsilon_{kl}$. For cubic crystals, the magnetoelastic energy density can be expressed by

$$u(\mathbf{M}, \varepsilon) = B_1(\alpha_1^2\varepsilon_1 + \alpha_2^2\varepsilon_2 + \alpha_3^2\varepsilon_3) + B_2(\alpha_1\alpha_2\varepsilon_6 + \alpha_2\alpha_3\varepsilon_4 + \alpha_3\alpha_1\varepsilon_5), \qquad (17.24)$$

where $B_1 = b_{1111}$ and $B_2 = b_{1212}$. *The magnetoelastic coupling constants B_1 and B_2 represent the strain gradient of the anisotropy energy and can in principle be calculated if the spatial dependencies of the interactions between the magnetic moments are known. Note that $u(\mathbf{M}, \varepsilon)$ as defined is zero for an unstrained crystal where all the ε_i are zero.

It is apparent from this expression for $u(\mathbf{M}, \varepsilon)$ that the magnetic anisotropy energy density E_a, and hence the energetically-favored directions for $\mathbf{M}$, will depend on the state of strain in the crystal. This dependence is in fact required for the existence of linear magnetostriction. The magnetic crystal will deform itself in order to reach a state of lower total magnetic and elastic energy. In other words, it will be energetically favorable to increase the elastic energy density $u_{el}(\varepsilon)$ in order to achieve an even greater decrease of the magnetic anisotropy and magnetoelastic energy densities E_a and $u(\mathbf{M}, \varepsilon)$, respectively.

The equilibrium state of strain in a cubic magnetic crystal can be determined by minimizing the sum $E_a + u_{el}(\varepsilon) + u(\mathbf{M}, \varepsilon)$ with respect to the strain components ε_i. The resulting dependence of the fractional change in length $\delta l / l$ (i.e., the *linear magnetostriction* or *magnetostrictive strain*) of the crystal in a direction defined by the direction cosines β_1, β_2, and β_3 is usually expressed by[†]

$$\frac{\delta l(\beta_1, \beta_2, \beta_3)}{l} = \frac{3}{2}\lambda_{100}\left(\alpha_1^2\beta_1^2 + \alpha_2^2\beta_2^2 + \alpha_3^2\beta_3^2 - \frac{1}{3}\right)$$

$$+ 3\lambda_{111}(\alpha_1\alpha_2\beta_1\beta_2 + \alpha_1\alpha_3\beta_1\beta_3 + \alpha_2\alpha_3\beta_2\beta_3). \qquad (17.25)$$

Here λ_{100} and λ_{111} are the saturation values of the magnetostriction in the $\langle 100 \rangle$ and $\langle 111 \rangle$ directions, respectively. They are related to the magnetoelastic coupling constants B_i and the elastic constants C_{ij} of a cubic magnetic material by the following expressions which can be derived by energy minimization:

$$\lambda_{100} = \frac{2B_1}{3(C_{11} - C_{12})} \quad \text{and} \quad \lambda_{111} = \frac{B_2}{3C_{44}}. \qquad (17.26)$$

The signs of λ_{100} and λ_{111} are determined by the signs of B_1 and B_2, respectively, since $(C_{11} - C_{12})$ and C_{44} are always positive.

Note that the fractional change in length in the [100] direction ($\beta_1 = 1, \beta_2 = \beta_3 = 0$) is given by

$$\frac{\delta l(1, 0, 0)}{l} = \frac{\lambda_{100}(3\alpha_1^2 - 1)}{2}. \qquad (17.27)$$

Also, when $\mathbf{M}$ is in a $\langle 100 \rangle$ or a $\langle 111 \rangle$ direction, the magnetostrictive strains in the same directions are

$$\frac{\delta l(1, 0, 0)}{l} = \lambda_{100} \quad \text{and} \quad \frac{\delta l(1, 1, 1)}{l} = \lambda_{111}. \qquad (17.28)$$

Thus it can be seen that the linear magnetostriction is in general anisotropic and corresponds either to an elongation, $\delta l > 0$, or a contraction, $\delta l < 0$, depending on the sign of the λ values involved.

A simplified form for Eq. (17.25) is obtained when the magnetostriction in the material is assumed to be isotropic. This assumption is clearly not always valid but

---

[†] For details, see C. Kittel, *Rev. Mod. Phys.*, **21**, 541 (1949).

may be expected to hold for polycrystalline and amorphous materials. For an isotropic magnetostriction λ, the linear magnetostriction becomes

$$\frac{\delta l(\theta)}{l} = \frac{3\lambda(\cos^2\theta - \frac{1}{3})}{2}, \tag{17.29}$$

where θ is the angle between the direction of **M** and the direction in which δl is measured, independent of any directions in the lattice. For a demagnetized isotropic ferromagnet in a zero magnetic field in which the domain magnetizations $\mathbf{M}_i$ are randomly oriented, it follows that the average of $\langle \cos^2\theta \rangle = \frac{1}{3}$, so $\delta l(\theta) = 0$ in any direction. Magnetostriction in magnetic materials is discussed further in Chapter W17.

When a uniform stress σ is applied to a single crystal of a cubic magnetic material, the resulting strain will make contributions to both the elastic and magnetoelastic energy densities. Consider the case where $\lambda_{100} > 0$. In this case an increase in the magnetization **M** in the [100] direction will cause an expansion [i.e. $\delta l(100) > 0$], in the same direction. Conversely, the magnetization M will be increased by a tensile stress for which $\delta l > 0$ (and decreased by a compressive stress for which $\delta l < 0$), except at $M = 0$ and $M = M_s$. If the material is magnetically isotropic, the contribution of the stress to the magnetoelastic energy density is

$$u(\mathbf{M}, \varepsilon) = \frac{3\lambda\sigma \sin^2\theta}{2}. \tag{17.30}$$

Here θ is the angle between the direction of the applied stress σ and **M**. Terms in $u(\mathbf{M}, \varepsilon)$ that are independent of θ have been omitted from this expression.

Note that $u(\mathbf{M}, \varepsilon)$ in Eq. (17.30) has the same dependence on θ as the magnetic anisotropy energy density $E_a = K_\sigma \sin^2\theta$, given in Eq. (17.11). This form for E_a is valid when the angle θ that **M** makes with respect to an easy axis of magnetization is much less than 1 radian. Thus the directions of an applied stress σ and of the resulting strain ε can determine the direction of an easy axis of magnetization in a magnetic material, thereby affecting the magnetic domain structure and the coercive field H_c in a ferromagnet. It follows that texture resulting from mechanical treatment (e.g., cold working), can create uniaxial magnetic anisotropy. The magnetoelastic anisotropy constant $K_\sigma = 3\lambda\sigma/2$ was defined earlier in our discussion of the effective magnetic anisotropy field $H_\sigma = 2K_\sigma/\mu_0 M_s$ [see Eqs. (17.17) and (17.18)].

Experimental values for λ at saturation in ferromagnets and ferrimagnets are small, typically in the range 10^{-5} to 10^{-6}. Much larger values, $\approx 10^{-3}$ to 10^{-2}, are often observed in rare earth–based ferromagnets. This is referred to as *"giant" magnetostriction* and is discussed in more detail in Chapter W17 where magnetostrictive materials with important applications are discussed. Note that even in the demagnetized state of a ferromagnet below T_C, each domain is strained and hence deformed along the direction of its spontaneous magnetization $\mathbf{M}_s$. Experimental results for λ_{100} and λ_{111} as well derived values for the isotropic magnetostriction λ for some cubic ferromagnets and ferrimagnets are given in Table 17.1.

TABLE 17.1 Magnetostrictions of Ferromagnets and Ferrimagnets

Magnetic Material	$\lambda_{100}(10^{-6})$	$\lambda_{111}(10^{-6})$	$\lambda^a(10^{-6})$
Ferromagnets[b] (Room Temperature)			
Fe (BCC)	+24.1	−22.7	−4.0
Ni (FCC)	−63.7	−28.7	−42.7
Ferrimagnets[c] (T = 293 K)			
Fe_3O_4	−20	+78	+39
$Mn_{0.6}Fe_{2.4}O_4$	−5	+45	+34
$Mn_{1.05}Fe_{1.95}O_4$	−28	+4	−8.8

[a] Values of the isotropic magnetostriction have been calculated from $\lambda = 0.4\lambda_{100} + 0.6\lambda_{111}$; see Eq. (W17.7).
[b] Data from E. P. Wohlfarth, Iron, cobalt, and nickel, in E. P. Wohlfarth, ed., *Ferromagnetic Materials*, Vol. 1, North-Holland, Amsterdam, 1980.
[c] Data from S. Chikazumi, *Physics of Magnetism*, Wiley, New York, 1964, pp. 173–175.

17.8 Magnetoresistance

Magnetoresistance refers to the effects that applied magnetic fields H have on electrical currents flowing in a material. More specifically, magnetoresistance in magnetic materials arises from the effects that an applied field has on the magnetization **M** and on the scattering of conduction electrons by localized magnetic moments and by magnons (i.e., spin waves). These effects are also referred to as *magnetoconductance* or *galvanomagnetic effects*.

There are several possible definitions of magnetoresistance (MR) which are used in the literature. One common definition, also known as the MR *ratio*, is

$$\text{MR} = \frac{\Delta R}{R} = \frac{R(H) - R(0)}{R(0)}, \tag{17.31}$$

where $R(H)$ and $R(0)$ are the electrical resistances measured in a magnetic field H and at $H = 0$, respectively. According to this definition, the magnetoresistance can be either positive or negative. Other definitions include the simple expression $R(H)/R(0)$ and also $[R(0) - R(H)]/R(H)$, which is appropriate when $R(H)$ can be much less than $R(0)$. The magnetoresistance observed in nonmagnetic conductors, as discussed in Chapter 7, is due to the direct influence of the applied field **H** on the motion of the electrons via the Lorentz force $e\mathbf{v} \times \mathbf{B}$. Materials that possess large magnetoresistive effects in low magnetic fields have applications as magnetic sensors, specifically in magnetic recording heads, as is discussed in Section 17.12.

The spin-dependent scattering of conduction electrons in metals by local magnetic moments often plays a critical role in the magnetoresistance observed in magnetic materials. The magnitude of this scattering will depend on whether the spins are in a periodic, ordered state, as at $T = 0$ K, or whether the spins are at least partially disordered, as will be the case for $T > 0$ K. The scattering of electrons and hence the resistance R can both be expected to decrease as the spins become more ordered in a magnetic field **H**. This behavior corresponds to a negative MR.

In any magnetic or nonmagnetic material the conduction electrons can be divided into two groups according to the directions of their spins (i.e., some will have their spins pointing up while the rest will have spins pointing down). In nonmagnetic materials and in zero magnetic field, there will be equal concentrations of up- and down-spin electrons (i.e., $n \uparrow = n \downarrow$). In a ferromagnet below T_C the concentrations of up- and down-spin electrons will be different even at $H = 0$. The up-spin electrons, with spins pointing in the same direction as the local magnetization $\mathbf{M}$, are referred to as *majority-spin electrons*; the *minority-spin electrons* are the down-spin electrons.

The occupation of the energy bands will be different for the majority- and minority-spin electrons in a ferromagnet due to their exchange interactions with the localized magnetic moments. This effect is illustrated in Fig. 9.14 for the 3*d* electrons in Ni. In addition, due to the Pauli exclusion principle, majority- and minority-spin electrons will have different scattering rates, $1/\tau \uparrow$ and $1/\tau \downarrow$, respectively, from localized moments and from magnons. The resistances $R \uparrow$ and $R \downarrow$ associated with majority- and minority-spin electrons in ferromagnets will therefore in general not be equal.

The magnetoresistance is usually anisotropic, with differences observed between the *longitudinal magnetoresistance* $MR(\|) = \Delta R(\|)/R$ measured with $\mathbf{H}$ parallel to the direction of the current, and the *transverse magnetoresistance* $MR(\perp) = \Delta R(\perp)/R$ measured with $\mathbf{H}$ perpendicular to the direction of the current. Typical differences between $MR(\|)$ and $MR(\perp)$ are about 2% in the permalloy $Fe_{0.20}Ni_{0.80}$, a ferromagnetic material commonly used in anisotropic magnetoresistive device applications. This permalloy has the advantage that the MR effects, although not large, occur in low fields, $H \approx 80$ A/m.

Very large MR effects, so-called "giant" and "colossal" MR, are discussed in Chapter W17.

17.9 Magneto-Optical Effects

Magneto-optical (MO) effects in magnetic materials can serve as very sensitive probes of the magnetic order present in magnetic domains and in magnetic thin films. These effects are all related to the anisotropy induced in the optical properties of magnetic solids due to the magnetization $\mathbf{M}$. The microscopic origin of the MO effects is the spin–orbit interaction. The electric and magnetic fields of the incident light affect the orbital motion of the electrons, which in turn is coupled via their spins to the magnetization. A variety of MO effects involving the transmission of polarized light (i.e., *Faraday effects*) and the reflection of polarized light (i.e., *Kerr effects*) can be employed, with typical geometries illustrated in Fig. 17.16. The Faraday and Kerr effects are generally subdivided into *polar, transverse*, and *longitudinal* geometries according to the orientation of the magnetization $\mathbf{M}$ relative to the surface of the material and to the polarization and plane of incidence of the light. The magnitudes of these effects are largest when the direction of propagation of the light is parallel to the direction of the magnetization in the material. It should be noted that magneto-optical effects are the ac analog of the Hall effect discussed in Chapter 7. Both effects involve off-diagonal components of the conductivity tensor σ.

Magneto-optical effects correspond in general to changes in both the amplitude and phase of the reflected or transmitted light. In the case of polar and longitudinal geometries, when the incident light is linearly polarized with its plane of polarization either parallel or perpendicular to the plane of incidence, the transmitted and reflected

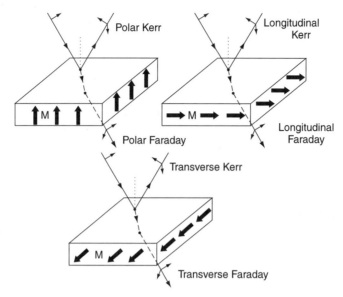

Figure 17.16. Typical geometries for the magneto-optical Faraday and Kerr effects. The directions and relative orientations of the magnetization **M** and the polarization of the incident light utilized are illustrated schematically. The incident light is shown plane polarized while the reflected and transmitted light are in general elliptically polarized.

light will in general be elliptically polarized due to the interaction of the light with the magnetization of the material. In the analysis of MO effects, an incident linearly polarized light wave is often constructed from pairs of circularly polarized waves of equal amplitude, one wave polarized right-handed and the other left-handed. Since both the rotation of the major axis of the ellipse and its eccentricity are proportional to the magnetization M of the material, the polar and longitudinal effects are both useful probes of the magnetic microstructure. When the transverse geometry is used, the observed changes in the intensities of the transmitted and reflected light are proportional to M. Magneto-optical effects are also of current technological interest for the magneto-optical recording of information in thin films, as discussed in Section 17.11.

Since the Faraday and Kerr effects depend not only on the properties of the magnetic material (direction and magnitude of **M**, optical properties, thickness of film, etc.) but also on the properties of the incident light (wavelength, state of polarization, and angle of incidence), only some specific cases of current scientific or technological interest are described in Chapter W17.

17.10 Dynamic Magnetic Effects

The dynamic or time-dependent response of magnetic materials can play an important role in many of their applications. For example, when applied magnetic fields are time dependent, induced electrical currents known as *eddy currents* are a source of energy dissipation. The precession of magnetic moments and of the total magnetization **M** in a magnetic field typically occurs at frequencies in the microwave regime. This follows from the fact that the cyclotron frequency of an electron is given by $\omega_c = eB/m$ and hence $f/B = e/2\pi m = 2.80 \times 10^{10}$ Hz/T. Measurement of the frequency of precession via a resonance technique (e.g., *ferromagnetic resonance*) can provide useful information about the gyromagnetic ratio γ and the local internal magnetic fields in the material.

A phenomenological description of time-dependent processes and the associated energy dissipation in magnetic materials can be developed in terms of a *complex magnetic susceptibility* χ or complex magnetic permeability μ, where

$$\chi = \chi' + i\chi'' \quad \text{and} \quad \mu = \mu' + i\mu''. \tag{17.32}$$

The quantities $\chi'(\mu')$ and $\chi''(\mu'')$ give the components of the magnetization $\mathbf{M}$ (magnetic induction $\mathbf{B}$) which are in phase and out of phase, respectively, with the applied field $\mathbf{H}$. It follows from the definitions $\chi = M/H$ and $\mu = B/H$ that $\mu' = \mu_0(1 + \chi')$ and $\mu'' = \mu_0\chi''$. In weak magnetic fields the magnetic induction B and the magnetization M can lag the time-varying applied magnetic field $H(t) = H_0 e^{-i\omega t}$ according to

$$B(t) = B_0 e^{-i(\omega t + \delta)} \quad \text{and} \quad M(t) = M_0 e^{-i(\omega t + \phi)} \tag{17.33}$$

where $\delta(\omega)$ and $\phi(\omega)$ are frequency-dependent phase angles. Magnetic losses in the material are proportional to the *loss factor* $\tan\delta = \mu''/\mu' = \chi''/(1 + \chi')$. Note that $\tan\phi \approx \tan\delta$ when $\mu_0 H_0 << B_0 \cos\delta$.

Some specific examples of dynamic magnetic effects, including eddy currents, ferromagnetic resonance, magnetic relaxation, and magnetomechanical damping, are described in Chapter W17.

EXAMPLES AND APPLICATIONS OF MAGNETIC MATERIALS

A selection of examples of magnetic materials with important technological applications is presented next. The materials are classified according to whether they are magnetically hard or soft, as defined in Section 17.4. The materials to be discussed include magnetically hard materials with applications as permanent magnets and in magnetic recording media, as well as magnetically soft materials used in electrical equipment, as magnetic recording heads, as magnetostrictive materials, and in microwave applications.

Essentially all technologically important magnetic materials are metals and inter-metallic or ceramic compounds containing two or more elements. They include both crystalline and amorphous materials. The specific magnetic materials discussed here and on our home page are listed in Table W17.2 along with their important applications. They all contain at least one element that is magnetic in its own right, such as the transition metals Fe, Co, and Ni (all elemental ferromagnets), Cr and Mn (both elemental antiferromagnets), and rare earth metals such as Dy, Sm, Nd, and Tb.

The intrinsic magnetic properties (M_{sat}, T_C, K) of these materials as well as their extrinsic magnetic properties (H_c, χ, M_r) determine their suitability for specific applications. The magnetic microstructure (i.e., the sizes, shapes, and orientations of the crystallites as well as the presence of secondary phases), controls their extrinsic properties. The microstructure in turn can be controlled through the use of various thermal, mechanical, and magnetic processing treatments.

Ferromagnetic $3d$ transition metals and alloys comprise an important group of magnetic materials with a wide range of applications as both hard and soft magnetic materials, depending on the alloying elements and on the thermal and mechanical treatments they receive during processing. The dependence of the average atomic magnetic moment $\langle m \rangle$ on alloy composition or, equivalently, on the average total number of $3d$ and $4s$ electrons per atom $z = z_{3d} + z_{4s}$ is presented in the *Slater–Pauling curve* shown in Fig. 17.17. These results are for selected binary alloys of the $3d$ transition metal elements with partially filled $3d$ shells along with the nonmagnetic alloying elements Cu and Zn with nearly filled $3d$ shells. The observed maximum $\langle m \rangle \approx 2.4\mu_B$ occurs for values of z between 8 and 8.5. The spontaneous magnetization M_s shows a similar dependence on alloy composition.

The interpretation of the Slater–Pauling curve is related to the effect on $\langle m \rangle$ and M_s of the occupancy of the $3d$ electron energy levels or bands as determined by Hund's rules. In particular, $\langle m \rangle$ is correlated directly with the densities of unoccupied up-spin and down-spin levels in the partially filled $3d$ energy band, as discussed in Section 9.6 (see Fig. 9.14). As can be seen from the Slater–Pauling curve, alloying elements that are nonmagnetic (V) or antiferromagnetic (Cr, Mn) in their pure, metallic state have the effect of weakening the average atomic magnetic moment in the ferromagnetic $3d$ transition metal alloys. An interesting example of a weak, itinerant ferromagnet in which none of the components are themselves magnetic is the Laves-phase compound $ZrZn_2$. Neutron diffraction studies have shown that the magnetic moments in this material are associated with the Zr–Zr bonds but are not localized.

Magnetic materials are generally classified as being either magnetically hard or soft according to the values of their coercive fields H_c, their magnetic susceptibilities χ

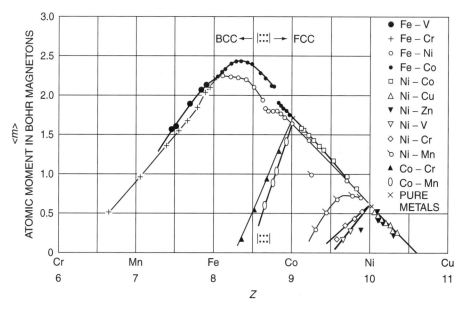

Figure 17.17. Slater–Pauling curve showing the measured dependence of the average atomic magnetic moment $\langle m \rangle$ on alloy composition or, equivalently, on the average total number of $3d$ and $4s$ electrons per atom $z = z_{3d} + z_{4s}$ for binary alloys of the $3d$ transition metal elements. (From R. Bozorth, *Ferromagnetism.* Copyright 1993 IEEE.)

(see Fig. 17.11), and their magnetic losses. Examples of magnetically hard and soft materials with important technological applications are discussed next.

17.11 Hard Magnetic Materials

As discussed in Section 17.4, hard magnetic materials have low values of χ (typically, in the range 1 to 100) and high values of H_c (typically, in the range 10^3 to 10^5 A/m) which result from the strong pinning of domain walls by defects. The field H_c can also be increased by the presence of a second phase, introduced by precipitation or by the use of very small particles. This is referred to as *magnetic hardening*. Hard magnetic materials also often have high values of the remanent magnetization M_r, wide hysteresis loops, and large hysteresis losses. Although these materials are hard to magnetize, once magnetized their magnetization remains close to M_s. They are employed in permanent magnets and as magnetic recording and magneto-optical recording media.

Permanent-Magnet Materials. Although practically any ferromagnetic material has the potential to serve as a permanent magnet (i.e., to produce strong and constant magnetic fields in a given region of space), the materials in actual use are typically those that meet the following criteria:

Intrinsic Properties

1. High T_C (well above the temperature of operation)
2. High M_{sat}
3. High magnetic anisotropy [i.e., high effective magnetic anisotropy field H_a (H_K, H_s, or H_σ)]

Extrinsic Properties

1. High coercive field H_c (usually above 2×10^3 A/m)
2. High remanent magnetization M_r (or remanent induction $B_r = \mu_0 M_r$)
3. High magnetic energy storage as specified by the maximum energy-density product $(BH)_{\text{max}}$.

A high H_c value is required because permanent magnets must be stable (i.e., they should not be easily affected by external magnetic fields). For this reason the nucleation and growth of magnetic domains must be difficult in permanent-magnet materials. In general, a material with high H_c must also have high M_r. High M_r and $(BH)_{\text{max}}$ values are required so that high magnetic fields can be generated by the permanent magnet. Although these extrinsic magnetic properties are determined to some extent by the intrinsic properties of a given material (e.g., its composition, M_{sat}, T_C, and H_a), they are to a large extent controlled by the thermal and mechanical processing treatments that the material receives during fabrication.

The stability of the magnetic field produced by a permanent magnet is greater when the magnetic domain structure in the magnet is stable against external influences such as changes in temperature, external magnetic fields, shock, stress, vibration, and radiation. The domain structure will have greater stability when the magnetic moments of the domains are "locked" in place by strong local magnetic anisotropy fields, which also lead to high coercive fields and magnetic hardness.

The operation of permanent magnets and specific magnetic materials are discussed further in Chapter W17.

The hard magnetic materials that have proven to be useful in permanent magnets typically have microstructures consisting of small magnetic particles embedded in a suitable matrix in order to maximize H_c. They include precipitation-hardened alloys, some steels, rare earth–transition metal intermetallic compounds, and ceramic materials. Most of these magnetically hard materials with excellent magnetic properties have less than ideal mechanical properties, due to their brittleness.

Permanent-magnet materials, discussed in more detail in Chapter W17, are listed in Table 17.2, where typical values of some of the extrinsic magnetic properties that are important for their applications are presented. These are representative values since the properties of a given material will depend on its exact composition and on its microstructure, which, in turn, depends on the specific processing that the material has received during fabrication. As a result, values for these properties can often vary by a factor of 2.

It should be noted that pure Fe is not included because its coercive field H_c is not high enough for most permanent-magnet applications. Magnetically soft Fe with

TABLE 17.2   Properties of Permanent-Magnet Materials

Material	$(BH)_{max}$ $(kJ/m^3)^a$	B_r (T)	H_c' $(kA/m)^b$	T_C (K)
Transition Metal Alloys				
Alnico 5[c] (51Fe, 14Ni, 8Al, 24Co, 3Cu)	35.8	1.25	43.8	1120
Steels[c]				
Cobalt steel (35Co, 0.7C, 4Cr, 5W, bal. Fe)	7.7	0.95	19.1	—
Tungsten steel (5W, 0.3Mn, 0.7C, bal. Fe)	2.5	1.03	5.6	—
Rare Earth–Transition Metal Intermetallic Compounds				
Nd–Fe–B[d]	200–380	1.0–1.4	700–1000	580
SmCo$_5$[e]	130–180	0.8–0.9	600–670	990
Sm(Co,Fe,Cu,Zr)$_7$[e]	200–240	0.95–1.15	600–900	1070
Ceramics				
BaO·6Fe$_2$O$_3$[d]	28	0.4	250	720

[a] Note that $1 kJ/m^3 = 1 kA \cdot T/m$.

[b] The quantity H_c' is the coercive field corresponding to $B = 0$.

[c] Data from D. R. Lide and H. P. R. Frederikse, etc., *CRC Handbook of Chemistry and Physics*, CRC Press, Boca Raton, Fla., 1994. The alloy composition is given in weight percent. See the handbook for methods of fabrication.

[d] Commercial material; from Magnet Sales & Manufacturing catalog.

[e] Data from K. H. J. Buschow, *Rep. Prog. Phys.*, **54**, 1123 (1991). Sm(Co,Fe,Cu,Zr)$_7$ is a two-phase material that can be thought of as a composite of SmCo$_5$ and Sm$_2$Co$_{17}$-type phases.

its high permeability and high saturation magnetization is, however, often used in conjunction with permanent magnets to provide a path for the magnetic flux between the magnet and the region of space where the magnetic field is to be produced. The magnetic properties of pure Fe are discussed in Chapter W17.

The upper limit for $(BH)_{max}$ in hard magnetic materials can be approximated by $B_r H'_c$, which can in principle reach values close to 10^5 kJ/m^3 when target figures of $B_r \approx 1$ to 10 T and $H'_c \approx 1 \times 10^4$ kA/m are used. This stored magnetic energy is "potential" energy in the sense that it is not actually used during operation of the permanent magnet. Permanent magnets are never intentionally cycled around their hysteresis loops, as this would dissipate a considerable fraction of the stored magnetic energy.[†]

The permanent-magnet materials listed in Table 17.2 include low-cost ceramics with relatively poor magnetic properties as well as high-cost rare earth–transition metal intermetallic compounds with superior magnetic properties. In the middle are the transition metal Alnico alloys. These materials are discussed in more detail in Chapter W17.

Materials for Magnetic Recording Media. Magnetic materials play a dominant role in the recording and storage of information, while semiconductors remain the key materials in the primary or active memory [i.e., random-access memory (RAM)] in computers and other electronic devices. The requirements for magnetic recording media are that the materials be magnetically hard, with high H_c, high M_r, a nearly square hysteresis loop, and low noise. The electrical resistivity of the material must be high enough to minimize eddy currents losses but not too high since charging must also be avoided. High resistance to corrosion and wear are also important. The materials currently in use possess magnetic surface layers which are supported either on flexible tapes or on flexible or rigid disks. The transducer or head, which both writes and reads the stored information, has different magnetic requirements from that of the recording media and is discussed in Section 17.12.

The magnetic structure of magnetic recording materials consists of small, individual magnetic domains which ordinarily have only two stable states or directions of magnetization, corresponding to the 0 and 1 bits of binary information storage. These domains can correspond either to small particles in particulate media or to small grains or crystallites in continuous thin-film media. The two states typically correspond to the magnetization **M** of the domain pointing in opposite directions along a preferred or easy axis determined by the local magnetic anisotropy of the material. The easy axis typically lies in the plane of the film. It must be possible to switch the magnetization from one direction to the other by the application via the head of a magnetic field known as the *switching field*. Lower noise operation is possible when the switching occurs by rotation of the domain magnetization rather than by domain wall motion.

A narrow distribution of switching fields in a recording medium is desirable to achieve magnetic transitions in the medium which are both narrow and stable. In general, however, individual domains or particles in the medium will have switching

---

[†] Electromagnets are used to generate magnetic fields, which are meant to be changed repeatedly. They operate on the portion of the B–H loop in the first quadrant.

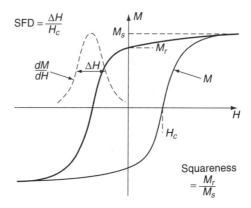

Figure 17.18. Hysteresis loop of a recording medium used to define the switching field distribution parameter (SFD). [From M. P. Sharrock, *Mater. Res. Soc. Bull.*, **15**(3), 53 (1990).]

fields that are not identical, due to their different orientations, shapes and sizes. The parameter known as the *switching field distribution* (SFD) of a typical recording medium is defined using the schematic magnetization hysteresis loop shown in Fig. 17.18. The value of SFD is given by $\Delta H / H_c$, where ΔH is the full width at half maximum of the derivative dM/dH in the second quadrant of the hysteresis loop.

There are important advantages in having the size of the individual magnetic domains be as small as possible without, however, reaching the point at which instabilities due to superparamagnetic effects appear. One advantage is that the maximum-possible density of stored information per unit area is increased, and another is that the magnitude of the noise in the output signal is decreased. Note that the current size of one bit of stored information is $\approx 1\mu m^2$, corresponding to $\approx 10^8$ bits/cm^2. At present each bit of stored information actually consists of ≈ 100 adjacent magnetic domains or distinct magnetizable units in the recording medium. Ideal thin-film recording media should consist of well-separated noninteracting magnetic particles. The particles should also all have sizes corresponding to the maximum H_c.

The high and controllable values of H_c needed in magnetic recording media are determined in large part by the magnetic anisotropy of the magnetic material, as discussed in Section 17.5. Although high coercive fields $H_c \approx 10$ to 100 kA/m are, in general, a prerequisite for magnetic recording materials, the optimum value of H_c depends on several factors. If H_c is too low, the stability of the recorded information will also be low since the unwanted movement of domain walls and the rotation and reversal of domain magnetizations may occur too readily under ambient magnetic field and temperature conditions, thereby causing the loss of stored information. If H_c is too high, the desired reversal of **M** during writing or erasing of information will require high magnetic fields, which are outside the range of currently available heads.

It is important for magnetic media to have high coercivity squareness (i.e., a nearly square or rectangular $M-H$ curve in the second quadrant) to ensure that a majority of the domains will have similar switching fields. This is equivalent to a low value of the SFD parameter. In this case the transitions between bits of stored information will be sharper, thus allowing greater storage densities.

The magnetic materials used in recording media are incorporated into thin films, either as small magnetic particles dispersed in a thin film of an organic binder or as the continuous magnetic thin film itself. Materials representing these two types of media are described in Chapter W17.

Materials for Magneto-Optical Recording. Magneto-optical (MO) recording and storage of information currently involves the use of light both to write information onto a suitable MO thin-film medium and to read the information using the MO Kerr or Faraday effects. In the writing process the MO film medium is heated locally in an area with diameter $\approx 1\mu$m to a temperature near or just below its Curie temperature T_C using a focused diode laser beam. At this elevated temperature the coercive field H_c of the MO medium is reduced either to zero or to a value sufficiently low that an applied magnetic field can orient the magnetization in the heated region or domain in the desired direction, thereby creating a bit of stored data. This method is known as *thermomagnetic recording.*

The information stored in the domain pattern is read by measuring the rotation of the plane of polarization of an incident linearly polarized laser beam as it is either reflected from (Kerr effect) or transmitted by (Faraday effect) the MO medium, as described in Section 17.9. Typical Kerr rotations observed using the MO Kerr effect (MOKE) are of the order of 0.2 to 0.4°.

The magnetic requirements for MO thin-film media are basically the same as for the magnetic recording media described earlier; that is, the materials must be magnetically hard at room temperature, preferably having square hysteresis loops with $M_r \approx M_s$. In addition, the films should possess perpendicular magnetic anisotropy so that the domain patterns can be read using the polar Kerr effect, which requires that $\mathbf{M}$ be perpendicular to the reflecting surface. Their T_C values should be less than about 300°C, so that thermomagnetic recording can be carried out using currently-available laser powers. The temperature T_C should not be too low or the stored information will not be sufficiently stable at ambient operating temperatures. The combination of film reflectivity R and optical rotation angle θ must be large enough to generate a measurable MO effect. The elimination of noise due to inhomogeneities in the MO medium is essential and can be achieved more readily by using either amorphous or fine-grained materials. As for all recording media, long-term stability and chemical stability (i.e., good corrosion and oxidation resistance) are also important attributes. Thermal properties such as specific heat and thermal conductivity control the formation and sizes of domains and are important for the thermomagnetic recording process.

The magnetic materials currently in use in MO recording media which so far have the best combination of magnetic and MO properties are amorphous alloys of rare earths (REs) and transition metals (TMs) (i.e., RE–TM alloy media) in which the RE ions interact antiferromagnetically with the TM ions. See Chapter W17 for further discussions of these materials.

17.12 Soft Magnetic Materials

Soft magnetic materials have high magnetic susceptibilities χ, typically in the range 10^3 to 10^5, and low coercive fields H_c, typically in the range 0.1 to 100 A/m. High resistivity and narrow hysteresis loops are desirable properties in these materials so that both electrical and magnetic losses can be minimized. Since the direction of the magnetization can be reversed easily in soft magnetic materials without excessive losses, they find use in transformers, motors and generators, magnetic recording heads, magnetostrictive devices, microwave components, and so on, where the hysteresis loop is traversed repeatedly. It is necessary for domain walls to move easily in soft magnetic materials, so pure strain- and defect-free materials are usually required. Thermal, mechanical, and magnetic processing treatments play critical roles in determining H_c and the degree of magnetic softness.

Iron Alloys and Electrical Steels. Magnetically soft iron alloys and the "electrical" steels find a wide range of applications in electromagnetic devices, including the cores of transformers, motors, generators, inductors, and electromagnets. The variety of iron-based alloys and steels available for use is extensive, and only the general characteristics of some important examples of these materials are presented here.

The magnetic and electrical properties which are desired in this broad class of materials include very high magnetic permeability $\mu = B/H$ and spontaneous magnetization M_s, very low coercive fields, narrow hysteresis loops, and relatively high electrical resistivity. The last two properties are needed to reduce magnetic and eddy current losses. Low magnetostriction λ is also desired. The specific properties needed for a particular application will also depend on the range of frequencies to which the material will be exposed, from dc up to microwave frequencies, $f \approx 10$ GHz. The ideal material would be a perfect electrical insulator with the high M_s value of Fe. To date no such material has been found.

As for all magnetic materials, the desired magnetic properties listed above depend not only on the composition of the material but also on its microstructure (i.e., the grain sizes and orientations, degree of ordering, defect densities, etc.) as well as its state of mechanical strain.

A wide variety of iron-based alloys and electrical steels is available for these applications in electrical and magnetic equipment. Some important classes of alloys and steels, including Fe–Ni and Fe–Si alloys, are discussed briefly in Chapter W17. The measured maximum relative magnetic permeabilities $\mu_r(\text{max})$, coercive fields H_c, and spontaneous magnetizations M_s of Fe and some important magnetically-soft Fe-based alloys and electrical steels are summarized in Table 17.3. The values presented here are representative since these properties are strongly dependent on the exact compositions and microstructures of the materials. Note that for these materials $\mu_r(\text{max})$ is typically greater than 10^3 and that $M_s \gg H_c$, as expected for materials that are magnetically soft.

Materials for Magnetic Read/Write Heads. Magnetic read/write heads are transducers in which electrical signals are converted by the head in the *write* mode to strong, localized magnetic fields that act to change both the magnitude and direction of the magnetization in the magnetic surface layer on a magnetic tape or disk, thus creating a pattern of stored information. In the *read* mode the reverse operation takes place whereby the pattern of magnetic domains induces electrical signals in the head, which is acting as a magnetic field sensor.

The requirements for the magnetic materials used in inductive recording heads are that they be magnetically soft with high M_s, high permeability over as wide a frequency range as possible, and low coercive fields. They should also have high electrical resistivity, low hysteresis losses, and low magnetostriction. Magnetic head materials must also be mechanically hard, to resist wear when in contact with the tape or disk during operation. High M_s is needed during writing so that magnetic fields which exceed the coercive field H_c of the recording medium can be generated, thus allowing the reversal, if necessary, of the magnetization of the magnetic domains in the medium. The magnitude of the magnetic field required for writing in recording media also depends on the

TABLE 17.3   Magnetic Properties of Pure Fe and Some Magnetically Soft Fe Alloys and Electrical Steels[a]

Alloy (wt %)[b]	$\mu_r(max)$[c]	H_c (A/m)	M_s (10^3 kA/m)
"Pure" α-Fe ($\approx$99%)	$\approx 10^3$	80	1.71
Pure α-Fe ($\approx$99.99%)	2×10^5	0.8	1.71
78 Permalloy (78Ni, 22Fe)	$\approx 10^5$	4	0.86
Supermalloy (79Ni, 16Fe, 5Mo)	$\approx 10^6$	0.16	0.63
Mumetal (77Ni, 18Fe, 5Cu)	2.4×10^5	2	$\approx$0.5
Hipernik (50Ni, 50Fe)	7×10^4	4	1.27
Silicon–iron (97Fe, 3Si) (oriented)	4×10^4	8	1.6
Amorphous $Fe_{80}B_{11}Si_9$	—	2	1.27

Source: Data from S. Chikazumi, *Physics of Magnetism*, Wiley, New York, 1964, except for $Fe_{80}B_{11}Si_9$, which is from N. Cristofaro, *Mater. Res. Soc. Bull.*, May 1998, p. 50.

[a]Data are at room temperature.
[b]The compositions of the alloys are given in weight percent unless otherwise stated.
[c]The maximum relative magnetic permeability $\mu_r(max)$ is expressed here in units of $\mu_0 = 4\pi \times 10^{-7}$N/A^2 and corresponds to the maximum value of B/H on the hysteresis loop in the first quadrant taken in increasing field.

speed of the writing process since rapid magnetization reversals require higher applied fields than are needed for the recording of data for long-term storage.

All writing heads are inductive, while reading heads can be either inductive or magnetoresistive. Inductive reading heads make use of Faraday's law, whereby changing magnetic fields resulting from the magnetic domain pattern of the recording medium induce voltages in the sensing coil of the head. In magnetoresistive heads the resistance of the head changes in response to the local magnetic field of the recording medium. The transitions between the magnetic domains where magnetic flux appears normal to the surface are detected by the read heads.

Magnetic materials currently in use in recording heads are discussed in Chapter W17.

Magnetostrictive Materials. Magnetic materials exhibiting large values of magnetostriction can be used both as sensors and actuators in *smart materials*, materials that are able to respond to their environment. Magnetostrictive sensors can detect very small strains or forces and have sensitivities that can be hundreds of times higher than strain gauges based on semiconductors. Magnetostrictive actuators that convert electrical to mechanical energy can deliver large forces when placed under strain via the application of a magnetic field (e.g., they can provide pressures of greater than 50 MPa at strains of up to 0.6%). Since the applications of magnetostrictive materials involve magnetic cycling, they should be magnetically soft to avoid excessive energy dissipation. It is therefore advantageous for magnetostrictive materials to have low magnetic anisotropy. High values of the spontaneous magnetization M_s are desirable in magnetostrictive materials for two reasons: more magnetic energy can be stored in the material and the magnetostrictive anisotropy field $H_\sigma = 2K_\sigma/\mu_0 M_s$ that must be overcome in order for saturation to be achieved will be lower.

The specific materials with important magnetostrictive applications that are discussed in Chapter W17 typically contain at least one magnetic rare earth element and often

TABLE 17.4   Magnetic Materials with Giant Magnetostrictions[a]

Material	$\dfrac{3\lambda_s}{2}(10^{-6})$
Dy (78 K)	1400
Tb (78 K)	1250
TbFe$_2$	2630
SmFe$_2$	-2340
DyFe$_2$	650
Tb$_{0.3}$Dy$_{0.7}$Fe$_2$ (Terfenol-D)	≈ 2300

Source: Data from K. B. Hathaway and A. E. Clark, *Mater. Res. Soc. Bull.*, Apr. 1993, p. 36.

[a] These data are for polycrystalline materials at room temperature, unless otherwise noted. The saturation magnetostriction $3\lambda_s/2$ is equal to $\lambda_\parallel - \lambda_\perp$. Here $\lambda_\parallel$ is the magnetostriction measured in the same direction as the applied field **H** [i.e., $\delta l(\theta = 0°)/l$ of Eq. (17.29)], and $\lambda_\perp$ is the magnetostriction measured in the same direction in the material but with **H** rotated by 90° [i.e. $\delta l(\theta = 90°)/l$].

a magnetic transition metal element as well. Examples include Tb, Dy, and Tb$_{1-x}$Dy$_x$ alloys; Fe-based intermetallic compounds such as TbFe$_2$, SmFe$_2$, and the pseudobinary compound Tb$_{0.3}$Dy$_{0.7}$Fe$_2$; and Fe-based amorphous metallic glasses. Some values of the "giant" magnetostriction observed in these magnetic materials are presented in Table 17.4. Normal values of the dimensionless magnetostriction λ are in the range 10^{-6} to 10^{-5} for most ferromagnetic and ferrimagnetic materials.

Ferrimagnetic Garnets. The ferrimagnetic garnets, which have significant magnetic and magneto-optical applications, are transparent insulators with the chemical formula RE$_3$Fe$_5$O$_{12}$, where RE is a rare earth ion. They have the same structures as the silicate minerals known as garnets, an example of which is Mg$_3$Al$_2$(SiO$_4$)$_3$. The metal cations present in the ferrimagnetic garnet structure are usually RE ions but can also be yttrium, Y, whose chemical properties are similar to those of the REs. The names of the ferrimagnetic garnets, such as yttrium iron garnet, Y$_3$Fe$_5$O$_{12}$, are usually abbreviated to YIG (yttrium iron garnet), GdIG, and so on.

The applications of the ferrimagnetic garnets are facilitated by the wide range of chemical substitutions that can be made for the metal ions in the structure in order to modify the magnetic, optical, and magneto-optical properties for specific purposes. These properties include the following:

1. *Magnetic*: magnetic anisotropy, M_{sat}, T_C, T_{comp}, and the magnetic field required for saturation

2. *Optical*: optical absorption and optical birefringence

3. *Magneto-optical*: Verdet constant and the corresponding maximum Faraday rotation

In the garnet structure the RE $4f^n$ ions (Sm, $n = 5$, to Lu, $n = 14$) and the Fe ions are all trivalent. The chemical formula for the ferrimagnetic garnets is sometimes written as $RE_3^c Fe_2^a Fe_3^d O_{12}$, which is equivalent to $[(3RE_2O_3)^c \cdot (2Fe_2O_3)^a \cdot (3Fe_2O_3)^d]/2$, where the superscripts c, a, and d refer to the three distinct sites for the trivalent metal cations in the cubic unit cell of 160 atoms which contains eight formula units. The 16 Fe^{3+} ions in the octahedral a sites of the cubic unit cell occupy positions on a BCC lattice and each is also at the center of an octahedron of O^{2-} ions. This local $A-B_6$ bonding unit is therefore $Fe^{3+}-(O^{2-})_6$. The 24 Fe^{3+} ions in the tetrahedral d sites are each surrounded by four O^{2-} ions. Due to their size, the 24 RE^{3+} ions are in the larger dodecahedral c sites surrounded by eight O^{2-} ions. All of these trivalent ion-centered polyhedra are distorted from regular shapes. A portion of the unit cell for the garnet crystal structure which illustrates the local bonding environments of the RE^{3+} and Fe^{3+} ions is shown in Fig. 17.19. Due to their large size, the RE^{3+} ions cannot fit into interstitial positions. As a result, the O^{2-} ions are not able to form a close-packed structure as in other oxides, such as the spinels.

The magnetic behavior of the ferrimagnetic garnets is controlled by antiferromagnetic superexchange interactions between the Fe^{3+} ions on the a and d sites. This indirect interaction, which takes place via the intervening oxygen ions, dominates over the other possible RE–Fe interactions because the $Fe(a)$–$Fe(d)$ distance is shorter and the $Fe(a)$–O–$Fe(d)$ angle of 126° is closer to 180° than is true for RE–Fe pairs. The RE ions on the c sites in turn interact antiferromagnetically with the equal number of Fe ions on the d sites, again via the superexchange interaction. The result is that the RE ions are aligned parallel to the $Fe(a)$ ions, and the net magnetic moment per

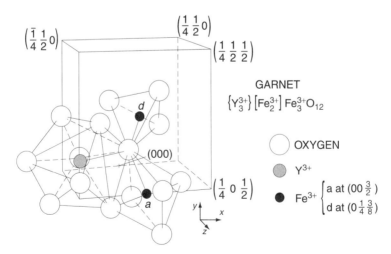

Figure 17.19. Portion of the unit cell for the garnet crystal structure which illustrates the local bonding environments of the RE^{3+} and Fe^{3+} ions. The chemical formula for the ferrimagnetic garnets is often written as $RE_3^c Fe_2^a Fe_3^d O_{12}$, where the superscripts c, a, and d refer to the three distinct sites in the cubic unit cell of 160 atoms containing eight formula units. (From R. M. White and T. H. Geballe, *Long Range Order in Solids*, Suppl. 15 of H. Ehrenreich, F. Seitz, and D. Turnbull, eds., *Solid State Physics*, copyright 1979. Reprinted by permission of Academic Press, Inc.)

$RE_3Fe_5O_{12}$ formula unit is given by

$$m = 3m_{RE(c)} + 2m_{Fe(a)} - 3m_{Fe(d)} = 3m_{RE(c)} - m_{Fe(d)} \qquad (17.34)$$

since the Fe^{3+} ions on the a and d sites have the same magnetic moments. Here $m_{RE} = g_{RE}\mu_B J_{RE}$ and $m_{Fe} = g_{Fe}\mu_B J_{Fe} = 2(\frac{5}{2})\mu_B = 5\mu_B$.

Since in Eq. (17.34) the magnetic moment $3m_{RE(c)}$ of the RE ions and the net Fe magnetic moment $m_{Fe(d)}$ are oppositely directed and since the magnetization of the RE ions decreases more rapidly with increasing temperature than the net magnetization of the Fe ions, several of the ferrimagnetic garnets have compensation temperatures $T_{comp} < T_C$, where the RE and Fe magnetic moments cancel each other and the net magnetization of the material is zero. The temperature dependencies of the magnetizations and the cancellation of the RE and Fe magnetic moments at T_{comp} are illustrated for the ferrimagnetic garnets in Fig. 9.21.

As mentioned earlier, the magnetic properties of the ferrimagnetic garnets can readily be modified by chemical substitutions. Consider YIG, where the Y^{3+} ions are diamagnetic so that the magnetic moment per formula unit is determined completely by the Fe^{3+} ions, with $m = |2m_{Fe(a)} - 3m_{Fe(d)}| = m_{Fe(d)}$. When nonmagnetic Sc^{3+}, Ga^{3+}, or Al^{3+} ions are introduced into YIG, for example, their relative ionic sizes determine whether they substitute for Fe^{3+} ions on a or on d sites. Since Sc^{3+} is larger than Fe^{3+}, it enters the larger octahedral $Fe^{3+}a$ sites preferentially, thereby increasing the magnetic moment of the material. The reverse occurs when the smaller Ga^{3+} or Al^{3+} ions enter the smaller tetrahedral $Fe^{3+}d$ sites, thereby leading to a decrease in the magnetic moment. The ions Sc^{3+}, Ga^{3+}, and Al^{3+} all lead to a reduction in T_C when substituted into YIG, due to the resulting reduction in the $Fe(a)$–O–$Fe(d)$ superexchange interaction.

YIG is an attractive candidate for applications at microwave frequencies, due to its combination of high electrical resistivity and its extremely low magnetic anisotropy, which is evidenced by an exceedingly narrow ferrimagnetic resonance line width of only about 50 A/m, corresponding to $K_1 = 560$ J/m^3 and $H_K = 2K_1/M_s = 6.4$ kA/m. The low magnetic anisotropy is due to the spin-only character of the Fe magnetic moment (Y is diamagnetic). Some of the important magnetic properties of YIG and GdIG are presented in Table 17.5. Many of these properties can be varied for specific applications by chemical substitution, as mentioned earlier.

TABLE 17.5   Magnetic Properties of $RE_3Fe_5O_{12}$ Ferrimagnetic Garnets

RE Ion	Y (YIG)	Gd (GdIG)
Lattice constant (nm)	1.2376	1.2417
$n(4f^n)$	0	7
$n_B(0 \text{ K})^a$	5.01	16.0
M_s (293 K) (kA/m)	139	135
T_{comp} (K)	(none)	286
T_C (K)	553	564

Source: Data from S. Geller, J. P. Remeika, R. C. Sherwood, H. J. Williams, and G. P. Espinosa, *Phys. Rev. A*, **137**, 1034 (1965), and references therein.

$^a n_B = M_{sat}/n\mu_B$.

Note that the Curie temperature T_C is hardly affected by the magnetic character of the RE^{3+} ions on the c sites, changing by only about 2% between the diamagnetic Y^{3+} ion and the strongly magnetic Gd^{3+} ion with $S = 7/2$. This is a clear indication that the interactions between the RE^{3+} ions and the Fe^{3+} ions on the a and d sites are much weaker than the antiferromagnetic $Fe^{3+}(a)$–$Fe^{3+}(d)$ interactions that determine T_C. The Bi^{3+} ion is one of the most useful chemical substitutes for Y^{3+} in YIG or for Gd^{3+} in GdIG for the purpose of enhancing the Faraday rotation. In Bi-doped YIG an unwanted increase of the magnetic anisotropy of the Fe ions also occurs, due apparently to an irregular distribution of the Bi ions in the structure. This perpendicular growth-induced anisotropy is desirable, however, for magneto-optical recording applications.

YIG is indispensable in microwave technology, finding applications in a variety of microwave components, including gyrators, waveguide isolators, switches, tunable filters, phase shifters, amplifiers, and so on.

REFERENCES

Aharoni, A., *Introduction to the Theory of Ferromagnetism*, Clarendon Press, Oxford, 1996.

Bozorth, R. M., *Ferromagnetism*, Van Nostrand, New York, 1951.

Chen, C.-W., *Magnetism and Metallurgy of Soft Magnetic Materials*, Dover, Mineola, N.Y., 1986.

Chikazumi, S., *Physics of Magnetism*, Wiley, New York, 1964.

Craik, D., *Magnetism*, Wiley, Chichester, West Sussex, England, 1995.

Greidanus, F. J. A. M., and W. B. Zeper, *Magneto-optical storage materials, Mater. Res. Soc. Bull.*, Apr. 1990, p. 31.

Hathaway, K. B., and A. E. Clark, *Magnetostrictive materials, Mater. Res. Soc. Bull.*, Apr. 1993, p. 34.

Heck, C., *Magnetic Materials and Their Applications*, Butterworth, London, 1974.

Judy, J. H., *Thin film recording media, Mater. Res. Soc. Bull.*, Mar. 1990, p. 63.

Kittel, C., *Physical theory of magnetic domains, Rev. Mod. Phys.*, **21**, 541 (1949).

Sharrock, M. P., *Particulate recording media, Mater. Res. Soc. Bull.*, Mar. 1990, p. 53.

Tebble, R. S., and D. J. Craik, *Magnetic Materials*, Wiley-Interscience, London, 1969.

White, R. M., and T. H. Geballe, *Long Range Order in Solids*, Suppl. 15 of H. Ehrenreich, F. Seitz, and D. Turnbull, eds., *Solid State Physics*, Academic Press, San Diego, Calif., 1979.

PROBLEMS

17.1 Consider a single-domain uniaxial ferromagnetic particle magnetized along its easy axis with $\mathbf{M} = \mathbf{M}_s$ in zero applied magnetic field. The magnetic anisotropy energy density is given by $E_a = K \sin^2 \theta$ where $K > 0$ and θ is the angle between the magnetization $\mathbf{M}$ and the easy axis. A magnetic field $\mathbf{H}$ is now applied at 90° to the easy axis, as shown in Fig. P17.1.

(a) Show that the sum of the anisotropy and magnetostatic energy densities for this particle is $u(\theta) = K \sin^2 \theta - \mu_0 MH \sin \theta$.

(b) Find the angle θ between $\mathbf{M}$ and the easy axis as a function of the magnitude of the field H by minimizing u with respect to θ. (Note that it will be important to check for the stability of the solution by requiring that $\partial^2 u / \partial \theta^2 > 0$.)

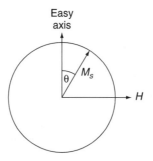

Figure P17.1

(c) Show that the resulting magnetization curve (i.e., the plot of the component of **M** in the direction of **H** versus the applied field H) is a straight line (with slope $\chi = \mu_0 M_s^2/2K$) up to $H = H_K = 2K/\mu_0 M_s$, at which point the magnetization is saturated in the direction of **H**. Here H_K is the effective magnetic anisotropy field defined in Eq. (17.14).

17.2 Show that the slope of the **M** versus applied magnetic field **H** curve shown in Fig. 17.13 is given by $\chi/(1 + N\chi)$, where the magnetic susceptibility is $\chi = M/H_i$. Here H_i is the internal magnetic field and N is the demagnetizing factor of the solid.

17.3 Estimate the work w (in J/m^3), as defined in Eq. (17.10), required to magnetize the ferromagnet Co in a hard direction perpendicular to the c axis from the magnetization curve presented in Fig. 9.16. (It will be necessary to convert magnetization M from the cgs-emu unit of gauss to the SI unit of A/m and magnetic field H from the cgs-emu unit of oersted to the SI unit of A/m.)

17.4 Prove that when the shape anisotropy constant K_s is < 0 (i.e., when $N_\perp < N_\parallel$), the magnetization **M** for a ferromagnetic film will lie in the plane of the film.

17.5 Calculate the radius at which a spherical Fe particle behaves superparamagnetically at $T = 300$ K by setting $K_1 V = k_B T$, where $K_1 \approx 4.2 \times 10^4$ J/m^3 is the first-order magnetocrystalline anisotropy coefficient for Fe and V is the volume of the sphere

17.6 Calculate the increase in temperature ΔT of a magnetic material with a square magnetization loop, with $M_s = 1370$ kA/m and $H_c = 1100$ kA/m, when the loop is traversed once. Assume that the material is thermally-isolated from its surroundings and that its specific heat is 4×10^6 J/m^{3}K.

17.7 For a magnetically isotropic material with magnetostriction λ, prove that $B_2(C_{11} - C_{12}) = 2B_1 C_{44}$. Show, in fact, that if the material is also elastically isotropic, then $B_1 = B_2$. (*Hint:* See Section 10.8.)

17.8 Show that $B(t)$ and $M(t)$ both lag the applied magnetic field $H(t) = H_0 e^{-i\omega t}$ by the same phase angle δ when $\mu_0 H_0 \ll B_0 \cos\delta$. [*Hint:* Start by substituting the expressions for $B(t)$ and $M(t)$ from Eq. (17.33) into the expression $B = \mu_0(H + M)$.]

17.9 For the Slater–Pauling curve shown in Fig. 17.17, assume that the average atomic magnetic moment per atom is $\langle m \rangle = (z\uparrow - z\downarrow)\mu_B$ and that the average total number of 3d and 4s electrons per atom is $z = z\uparrow + z\downarrow$. Here $z\uparrow = z_{3d}\uparrow + z_{4s}\uparrow$ and $z\downarrow = z_{3d}\downarrow + z_{4s}\downarrow$.

(a) For the $Fe_{0.7}Co_{0.3}$ alloy with $\langle m \rangle = 2.5$ μ_B, find the values of $z\uparrow$ and $z\downarrow$.

(b) If the $3d\uparrow$ energy band is filled, so that $z_{3d}\uparrow = 5.0$ for this alloy, calculate $z_{3d}\downarrow$, $z_{4s}\uparrow$, and $z_{4s}\downarrow$. You may assume that $z_{4s}\uparrow = z_{4s}\downarrow$.

(c) Consider the case of pure Co. If $z_{3d}\downarrow$, $z_{4s}\uparrow$, and $z_{4s}\downarrow$ have the same values as in the $Fe_{0.7}Co_{0.3}$ alloy, calculate $z_{3d}\uparrow$ and $\langle m \rangle$.

(d) Compare your predicted value for $\langle m \rangle$ with the observed value for Co on the Slater–Pauling curve.

(e) Explain the shape of the Slater–Pauling curve on the basis of the filling of the $3d$ energy band. (*Hint*: See the discussion of the itinerant model of magnetism as applied to Ni in Section 9.6.)

Note: Additional problems are given in Chapter W17.

CHAPTER 18

Optical Materials

18.1 Introduction

All materials interact with light in some manner and thus theoretically qualify as optical materials. However, certain materials are technologically more important than others, so the purpose of this chapter is to use several case studies to illustrate concepts and applications. Optical properties of materials are discussed in Chapter 8, and this chapter builds on that discussion by applying them to practical situations. The focus is on materials used for the propagation, generation, and detection of photons.

The discussion begins with the simple propagation of light through a passive medium. Optical fibers provide convenient channels for telecommunication, so this is the first topic of discussion. The goal in designing a good optical fiber is to make it as passive as possible (i.e., to maintain the fidelity of the output signal to the input signal). As will be seen, this requires considerable engineering of the optical fiber.

Light is characterized by various physical parameters, including frequency, polarization, and intensity. To the extent that one is able to control these parameters, one is able to make maximum use of light. Section W18.1 at our Web site[†] is devoted to optical polarizers, which provide the technological basis for precise manipulation of the polarization of light. Related to this is a description of the phenomenon of Faraday rotation, which concerns the rotation of the polarization vector of the light in the presence of a magnetic field in a material. This is discussed in Section W18.2.

The discussion then turns to controlling the frequency and intensity of light. Light-emitting diodes are introduced. Lasers provide a means for producing highly monochromatic light beams which are also very well collimated and often have high intensities. After discussing lasers in general, we look at particular lasers, including semiconductor lasers and ceramic-based lasers. A section is assigned to the evolving field of bandgap engineering of optical materials. Descriptions of the cascade laser and the quantum-dot laser are given.

The ability to fabricate materials with precise compositional control allows engineers to create modulated optical structures. The topic of optical band structure is studied in Section W18.3 and shows the potential applicability of such devices.

The interaction of intense radiation fields with matter brings into play various nonlinear optical effects. A section is devoted to nonlinear optical materials, which utilize such effects to create useful devices. High intensities also produce damage in optical materials. The laser-induced damage mechanism is discussed in Section W18.4.

[†] Supplementary material for this textbook is included on the Web at the resource site (ftp://ftp.wiley.com/public/sci_tech_med/materials). Cross-references to elements of the Web material are prefixed by "W."

Light is often used to convey information for permanent or temporary storage. Three aspects of information storage are considered in this chapter. The first is conventional photography. The second is xerography, which is introduced in the context of the study of photoconductors. The third concerns photorefractive materials. The electro-optic effect provides the physical mechanism for understanding such materials.

PROPAGATION OF LIGHT

18.2 Optical Fibers

The telecommunications industry took a great stride forward with the advent of optical fibers. Infrared carrier waves with wavelengths on the order of 1 μm offer a potential four order-of-magnitude increase in bandwidth over microwaves ($\lambda \approx 0.01$ m) and a million-fold increase over ultrahigh frequency (UHF) signals. In analogy with the human nervous system, engineers now have the ability to use a bundle of individual fibers to provide a massive parallel flow of information in the form of optical (instead of electro-chemical) signals from one system to another. Although the basic idea of using total internal reflection to confine a light beam is an old one, the modern breakthrough came about by improving fiber materials to the point where signal losses due to absorption and scattering became tolerable, and in some cases, negligible. By making the fibers comparable in diameter to the wavelength of light, the fiber must be regarded as a waveguide. Geometric optics must be replaced by physical optics to obtain a precise understanding of the propagation.

There are several ways that optical fibers are constructed and utilized. In a stepped fiber there is an inner transparent core surrounded by an outer transparent cladding. Additional layers of coating material surround these to provide mechanical support, chemical insulation, and to protect the fiber against the elements (including gnawing rodents). The inner core, typically of radius $r_1 = 5$ μm, has the higher index of refraction, n_1. The cladding, whose radius is typically $r_2 = 60$ μm, has an index n_2, where $n_2 < n_1$. In the geometric-optical description, total internal reflection occurs for beams traveling with an angle of incidence (measured relative to the normal to the surface) $\theta > \theta_c = \sin^{-1}(n_2/n_1)$, where θ_c is the critical angle. The fiber may be utilized in either single-mode or multimode operation. In the simplest case the optical signal passes through the fiber as a set of bright (and dark) bits (i.e., binary ones and zeros). This method is commonly used in intercomputer communications, via optical networks. It is also possible to send analog signals via modulated carrier waves. For small enough r_1 the higher-frequency modes are cut off. To get a proper description of the mode structure, one must solve Maxwell's equations. One finds that if the number $V = 2\pi(r_1/\lambda)(n_1^2 - n_2^2)^{1/2} < 2.4$, only a single mode propagates. For larger V the number of modes grows rapidly, as $V^2/2$. A geometric-optical schematic of a higher mode is given in Fig. 18.1, where the ray inside the fiber (of wavelength λ/n_1) reflects off the wall a number of times (at an angle of incidence greater than the critical angle) as it "spirals" its way forward. The corresponding wavelength outside the fiber is λ/n_2.

Another type of optical fiber is the graded-index (GRIN) fiber. Here the index of refraction $n(r)$ varies as a function of the radial distance from the axis, r, with the index being higher toward the center. The need to clad one fiber onto the other is thereby eliminated, but is replaced by the technological problem of creating the $n(r)$ variation.

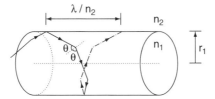

Figure 18.1. Ray undergoing total internal reflection as it "spirals" along a polygonal helix inside an optical fiber.

One of the limiting factors in the use of fibers is their dispersion [i.e., the indices of refraction depend on the frequency of the light: $n = n(\omega)$]. The phase velocity is given by $v_{ph} = c/n(\omega)$. The group velocity (the velocity of a pulse) is given by

$$v_g = \frac{\partial \omega}{\partial k} = \frac{c}{n + \omega(\partial n/\partial \omega)} = \frac{c}{\partial(n\omega)/\partial \omega} \tag{18.1}$$

and is also dispersive (i.e., depends on the frequency ω). A pulse of duration Δt has spatial extent $\Delta s = v_g \, \Delta t$. After passing through a fiber of length L, its extent will be broadened by amount Δx, where

$$\Delta x = \frac{Lc}{nv_g^2 \, \Delta t} \frac{\partial v_g}{\partial \omega} = -\frac{L}{n \, \Delta t} \frac{\partial^2}{\partial \omega^2}(n\omega). \tag{18.2}$$

If the time between successive pulses is $\Delta t'$, the condition for the pulses not to overlap is

$$\Delta t' > -\frac{L}{2nc \, \Delta t} \frac{\partial}{\partial \omega} \left[\frac{\partial}{\partial \omega}(n\omega) \right]^2. \tag{18.3}$$

Thus the longer the fiber, the slower the pulse rate must be. This limitation may be greatly reduced by operating in the vicinity of the point of inflection, where $\partial^2(n\omega)/\partial \omega^2 = 0$. For silica this point is at $\lambda = 1.3 \ \mu m$. However, this does not coincide with the wavelength at which losses are a minimum, which is closer to 1.55 μm. A dispersion-shifting additive could be employed to make these points coincide.

Most systems in current use employ the fiber as a linear transmission medium. Some thought has been given to the use of fibers with nonlinear refractive indices. For example, suppose that there is a medium where n grows with increasing intensity. The peak of the pulse has a higher intensity, experiences larger n values, and hence moves more slowly than the leading and trailing edges of the pulse. The trailing edge will catch up with the central part of the pulse. (The leading edge of the pulse could be reduced by the use of rapid switching on of the pulse.) The effect is called *self-steepening* and the nonlinearity tends to compress the pulse. This is opposite to the tendency of dispersion, which will lengthen the pulse. The resulting steady-state pulse shape is called a *soliton*. The soliton represents a stable pulse structure in which these two effects cancel each other and the pulse propagates without distortion. In such a fiber a train of solitons will retain its integrity independent of the distance L and will therefore preserve a digital signal with high fidelity.

The key limitations to fiber utility come from absorption and scattering. A pulse of initial intensity I_0 will find its intensity diminished to $I = I_0 \exp(-\alpha L)$ after propagating a distance L, according to Beer's law. Here α is the absorption coefficient of the material. There are four obstacles that must be overcome. Fibers absorb in the infrared and ultraviolet. As the photon energies approach the bandgap, typically several electron volts, phonon-assisted absorption processes come into play. The photon may combine with one or more thermally excited phonons to bridge the bandgap (of size E_g) and elevate an electron from the valence to the conduction band. This gives rise to what is called the *Urbach edge*. Note that the Urbach edge can also result from static bond disorder in a glass, as described in Chapter W11. Since the probability of borrowing energy from thermal excitations falls off exponentially with energy, there is a contribution to the absorption coefficient of the form

$$\alpha_U(\omega) = \alpha_0 \exp\left(\frac{\hbar\omega}{E_0}\right),$$

as given in Eq. (W11.25).

In the infrared region of the spectrum, fibers absorb strongly, due to the ionic vibrational modes. Since there is a large ionic component to chemical bonds in ceramics, the phonons are strongly infrared active. For example, in silica there is a 400 cm^{-1} mode corresponding to the rocking back and forth of the O ion in Si–O–Si. An 800 cm^{-1} mode is associated with the bending of this complex. The stretch of the Si–O bond produces an absorption band of 1100 to 1200 cm^{-1}. Other bands at 500 and 600 cm^{-1} have been traced to defects, such as threefold coordinated O ions or planar triangular rings. Low-frequency modes at 200 cm^{-1} are believed to be caused by the concerted motion of several Si ions. Multiple harmonics of all the modes and intercombination bands extend out to higher frequencies and produce some absorption in the near infrared. This absorption is a highly nonlinear process that is describable by the formula $\alpha_{\mathrm{IR}}(\omega) = B \exp(-b\omega)$. To understand this formula note that on the order of $N = \omega/\Omega$ phonons will be excited, where Ω is an average phonon frequency. One may imagine an electron being virtually excited by the photon and emitting N phonons before returning to its initial state. If the electron–phonon coupling constant is denoted by g, the effect can be expected to go as g^{2N}. This may be rewritten as $\exp(-b\omega)$, where $b = (2/\Omega)\ln(1/g)$. For SiO$_2$ it is found that $b = 2.9 \times 10^{-14}$ s.

A third limitation arises from static fluctuations in the density or composition of the fiber. A small region may have a higher or lower density than the mean. This will produce a locally greater or lesser dielectric constant. When subjected to the optical field, an electric-dipole moment will be induced and this will radiate in all directions, including at angles outside the critical angle. This is called *Rayleigh scattering*, and the attenuation constant due to scattering grows as $\alpha_R(\omega) = C\omega^4$. It can be shown that the constant C is given by $C = (n^2/3\pi c^4)\langle \delta n^2 \rangle \xi^3$, where n is the mean index of refraction, $\langle \delta n^2 \rangle$ its mean-square fluctuation, and ξ its spatial correlation length. It is found for SiO$_2$ that $\langle \delta n^2 \rangle \xi^3 = 3 \times 10^{-4}$ nm^3.

The combined extinction coefficient from the foregoing three effects is of the form

$$\alpha(\omega) - Ae^{a\omega} + Be^{-b\omega} + C\omega^4. \tag{18.4}$$

This rises rapidly at both high and low frequencies, has a minimum in between, and defines what is called the *V-curve*. The deeper the minimum, the more suitable the material is for an optical fiber. An example of such a curve is given in Fig. 18.2 for

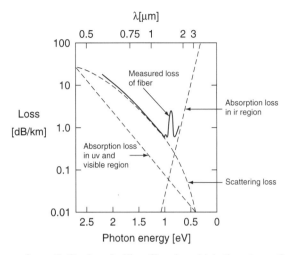

Figure 18.2. V-curve for a GeO_2-doped silica fiber in which the attenuation (proportional to the absorption coefficient) is plotted as a function of photon energy and wavelength. [Reprinted with permission from H. Osanai et al., *Electron. Lett.*, **12**, 549 (1976). Copyright 1976 by IEE Publishing.]

germania-doped silica fiber. The peak at about 1400 nm is due to the second harmonic of the fundamental vibration of OH radicals at 2720 nm. They are believed to diffuse into the glass during the process of forming the fiber. If this loss peak were eliminated, it would be possible to get the minimum loss down to ≈ 0.2 dB/km.

A fourth limitation is posed by the presence of impurities and defects. Great care must be used to eliminate absorbers such as OH^-, hydrogen, and such impurities as Fe^{2+}, Fe^{3+}, Co^{2+}, and Cr^{3+}. Typically, one reduces these impurities to the few parts per billion level. Typical defects that can cause problems are O vacancies and O interstitials. One must also make sure that the fiber is free from cracks, crystallites, and bubbles, as these will scatter light.

One type of optical fiber is based on high-purity silica glass ($n = 1.456$), which is then doped with impurities. Many dopants may be employed, such as Al, B, F, Ge, N, P, Pb, and Ti. A popular fiber is $SiO_2/GeO_2/F$. The core is doped with Ge since this tends to elevate the refractive index (typically by 1%). The cladding is doped with F, as this lowers the refractive index (typically, by 2 to 3%). Early fibers achieved attenuation factors of 20 dB/km. Now they are ≈ 0.15 dB/km, and efforts are being made to reduce them still further.

Another type of glass used in fibers is based on fluorides and other halides. It is possible to push the infrared phonon modes to lower frequencies by incorporating heavy metals in the structure. Thus one has such glasses as ZrF_4, ZB (ZrF_4-BaF_2), ZBL [$ZrF_4(62\%)-BaF_2(33\%)-LaF_3(5\%)$], and ZBLAN [$ZrF_4(53\%)-BaF_2(20\%)-LaF_3(4\%)-AlF_3(3\%)-NaF(20\%)$]. At lower frequencies Rayleigh scattering becomes less important (falling as ω^4). In forming a glass one must cool it sufficiently rapidly from the melt to avoid crystallization. There exists a critical cooling rate below which the glass will crystallize. It is found that the stability of a glass is enhanced by increasing the number of components in the glass. The more components there are, the longer it takes for a crystal to form properly, and the longer the amorphous structure of the liquid persists. This is sometimes called the *confusion principle*.

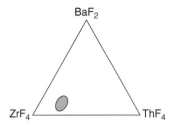

Figure 18.3. Ternary phase diagram for a fluoride glass. The glass-forming region is shaded. (Reprinted from M. Poulain et al. *Mater. Res. Bull.*, **12**, 151, copyright 1977, with permission from Elsevier Science.)

The actual combinations of the multicomponent glasses is determined by the phase diagram. Consider, for example, the ternary compound ZrF_4–ThF_4–BaF_2. The glass tends to form near the eutectic point in a region called the *vitreous area*. This is illustrated in Fig. 18.3, where the glass-forming region is shaded.

Interest has also centered on the chalcogenide glasses, such as As_2S_3, As_2Se_3, GeS_2, Ge_2P_2S, $GeSe_2$, and Ge_2As_2Te.

GENERATION OF LIGHT

Standard methods for generating light include flames and other thermal sources, incandescent and fluorescent bulbs, vapor lamps, arc lamps, light-emitting diodes, and chemiluminescent and bioluminescent sources. From the standpoint of modern technology, however, few sources offer the versatility and control that the laser does. The laser (light amplification by stimulated emission of radiation) provides the scientist or engineer with a well-collimated, monochromatic, high-intensity light source for a host of applications. Perhaps the most important of these is communication, where, combined with optical fibers, it permits a high volume of information to be reliably transported great distances at close to the speed of light.

In the following sections the primary focus of attention is on the laser, although the light-emitting diode will also be introduced. The discussion begins with a description of stimulated emission. This is followed by a study of light-emitting diodes and semiconductor lasers. Then ceramic-based lasers are discussed. Examples of bandgap engineering of optical materials are given, including the cascade laser and the quantum-dot laser.

18.3 Lasers

Any laser has three essential components: (1) an active medium in which a population inversion may be established; (2) an energy source, which establishes (pumps) the population inversion; and (3) a resonant optical cavity, which determines the condition for and the frequency of the lasing. Consider a set of two-level systems ("atoms") in a material that is in thermal equilibrium with electromagnetic radiation. Let N_1 be the number of atoms in the lower energy level E_1 and N_2 be the number in the upper energy level E_2. The blackbody-radiation energy density per frequency interval is given by the Planck spectrum

$$\rho(\omega, T) = \frac{\omega^2 n^3}{\pi^2 c^3} \frac{\hbar\omega}{e^{\beta\hbar\omega} - 1}, \tag{18.5}$$

where $\beta = 1/k_B T$ and n is the index of refraction of the material. The interest will be in the resonance case, where $\hbar\omega = E_2 - E_1$. Transitions occur between the levels of the atoms and lead to the following kinetic equations:

$$\frac{dN_1}{dt} = -BN_1\rho(\omega) + B'N_2\rho(\omega) + AN_2, \tag{18.6a}$$

$$\frac{dN_2}{dt} = BN_1\rho(\omega) - B'N_2\rho(\omega) - AN_2. \tag{18.6b}$$

The first term on the right-hand side of these equations is due to absorption, which drives atoms from state 1 to state 2 at a rate proportional to both the population N_1 and the radiation density, as would be expected for a linear process. The second term, introduced by Einstein, is called *stimulated emission* and is needed to preserve thermo-dynamic equilibrium. It causes downward transitions at a rate also proportional to the population N_2 and the radiation density. The final term represents spontaneous emission, with the lifetime of the excited state being given by $\tau = 1/A$. At thermal equilibrium the population ratio is given by the Boltzmann distribution $N_2/N_1 = \exp{(-\beta\hbar\omega)}$ and steady-state conditions must apply, so $dN_i/dt = 0$. Solving the resulting equations leads to the Einstein A–B relations

$$B = \frac{\pi^2 c^3}{\hbar\omega^3 n^3}A, \qquad B' = B. \tag{18.7}$$

These equations are basic to the operation of a laser. Although these equations have been derived using thermal equilibrium arguments they are general relations governing the radiation processes.

Next consider a nonequilibrium situation in which the laser-active medium, described by the two-level atoms, and the radiation are confined to an optical cavity. The rate of increase in photon energy per unit volume and unit frequency range is given by

$$\frac{d\rho(\omega)}{dt} = \frac{dN_1}{dt}\frac{\hbar\omega}{V}g(\omega), \tag{18.8}$$

where V is the volume and $g(\omega)$ is a line-shape function which is usually adequately described by a Lorentzian

$$g(\omega) = \frac{\gamma/2\pi}{(\omega - \omega_0)^2 + (\gamma/2)^2}. \tag{18.9}$$

The radiation is assumed to be peaked around frequency $\omega_0 = (E_2 - E_1)/\hbar$ with line width γ. The quantity γ is referred to as the *homogeneous line width* and is associated with the natural lifetime of the excited state. Assuming that stimulated emission and absorption dominate over spontaneous emission one finds that

$$\frac{d\rho(\omega)}{dt} = B(N_2 - N_1)\frac{\hbar\omega}{V}g(\omega)\rho(\omega). \tag{18.10}$$

For a fixed population inversion, $N_2 - N_1 > 0$, this leads to exponential growth of the cavity radiation. Alternatively, for radiation of frequency ω_0 passing through the cavity there is exponential growth with propagation distance, since $dz = (c/n)dt$. Thus $\rho(\omega) = \rho_0 \exp(\alpha z)$, where the gain constant is given by

$$\alpha = \frac{N_2 - N_1}{V}\frac{\pi^2 c^2}{\omega^2 n^2 \tau}g(\omega). \tag{18.11}$$

Normally, radiation confined to a cavity will decay exponentially in time due to various absorption processes, including leakage through the mirrors at the ends of the cavity. Let the time constant for such processes be denoted by τ_c. When there is a laser-active medium present in the cavity, the growth rate of the intensity is $\alpha c/n$. The threshold for laser action occurs when these two rates balance each other (i.e., $\alpha c/n = 1/\tau_c$). The threshold population inversion density is

$$\frac{N_2 - N_1}{V} = \frac{\omega^2 n^3 \tau}{g(\omega)\pi^2 c^3 \tau_c}. \tag{18.12}$$

A two-level system turns out to be impractical for a laser. The same levels that give rise to stimulated emission, when the upper state is initially populated, also gives rise to absorption when the lower state is initially populated. It would be necessary to maintain a population greater than 50% in the excited state, and this would be energetically very costly. With a four-level system, however, one may operate the laser transition between the intermediate two levels and still have the majority of atoms in the ground state. All that is required is to have a population inversion between the intermediate levels (i.e., $N_3 > N_2$). The upper level (or collection of levels), 4, is populated by some external pump, as in Fig. 18.4b. The atoms fall to level 3 by very fast nonradiative transitions. Then they make transitions to level 2 by emitting stimulated-emission photons. Finally, the system relaxes to the ground state (usually, by a nonradiative process) so as to help preserve the inversion. In the four-level laser $E_2 - E_1 \gg k_B T$, so the absorption process $2 \rightarrow 3$ is unlikely, due to a lack of thermal population in level 2.

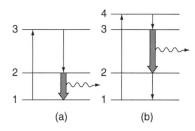

Figure 18.4. (a) Three- and (b) four-level laser. In both cases the upward-pointing arrow denotes the pump. The thick arrows denote the respective laser transitions.

Sometimes one speaks of a three-level system, in which the laser transition is directly from the intermediate state to the ground state (or a level within $k_B T$ of the ground state). This is illustrated in Fig. 18.4a. It is assumed that there is an strong pumping mechanism operative which populates level 3. Using a four-level system, however, allows one to pump the population inversion with less energy.

18.4   Light-Emitting Diodes and Semiconductor Lasers

The light-emitting diode (LED) is constructed from a $p - n$ junction. The n-type semiconductor is usually heavily doped so that the chemical potential of the electrons lies inside the conduction band. Similarly, the p-type material is also heavily doped so that the chemical potential of the holes lies within the valence band. This is illustrated in Fig. 18.5a for the case in which the two materials are far from each other. When the materials are brought into contact, a depletion region is established and the chemical potentials equilibrate (Fig. 18.5b). If the junction is now forward-biased by a potential V, a current flows and the chemical potentials for the electrons and holes no longer coincide. They are displaced by an energy eV (Fig. 18.5c). Electrons from the conduction band of the n-type material are injected into the depletion region. Similarly, holes from the p-type semiconductor are injected into the depletion region, and radiative recombination of electrons and holes occurs. In reality this is a form of electroluminescence. Photons of energy approximately equal to the gap energy E_g are emitted. Recently, semiconductor LEDs have been made which extend the emission from the infrared (GaAs, 868 nm; InP, 806 nm) and red [GaAs$_x$P$_{1-x}$ (with $x < 0.55$), 627 nm] into the green (GaP, 554 nm), blue (ZnSe, 490 nm) and even the violet (GaN, 417 nm).

In attempting to make a laser from an LED, one runs into the difficulty that the semiconductors are heavily doped. The dopants change the lattice constants of the semiconductors, and when the junction is formed, many misfit dislocations are produced. In addition, the high concentration of impurities introduces traps. These open nonradiative decay channels which broaden the laser transition and make it difficult to select a single cavity mode, as the line width encompasses many modes.

The semiconductor laser is made by sandwiching one type of semiconductor between two pieces of doped semiconductor, of n- and p-type, respectively (Fig. 18.6). It is called a *double heterostructure laser* because of the two interfaces involved. Typically, one may place a 0.1-μm layer of Ga$_{1-x}$Al$_x$As between n- and p-type GaAs or

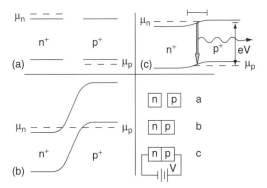

Figure 18.5. Homojunction light-emitting diode. In (c) the width of the depletion region is indicated.

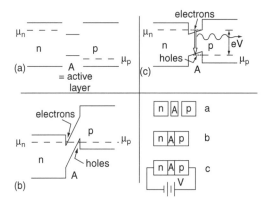

Figure 18.6. Double heterostructure semiconductor laser.

$In_{1-x}Ga_xAs_{1-y}P_y$ between n- and p-type InP. The central layer, called the *active layer* (denoted by A), is where the lasing takes place. The index of refraction of the central layer is larger than that of the bounding media, so light can be trapped by total internal reflection within it. The ends of the device are usually flat, so as to function as mirrors that define the laser cavity. In addition, there are often blocking layers of semiconductor on the sides of the device. The active region is typically a rectangular parallelepiped of width 1 μm, length 300 μm, and height 0.1 μm. Typical threshold currents for lasing are in the range 20 to 50 mA.

The primary advantage of the double heterostructure laser is that it is possible to achieve a matching of the lattice constants to better than 1%, thereby minimizing the number of misfit dislocations produced. A more detailed description of how this is done is given below. Since neither semiconductor need be heavily doped, the lattice constants are not altered much by the dopants, and lattice matching can by achieved. Typical methods of fabrication include chemical vapor deposition, molecular beam epitaxy, or liquid-phase epitaxy.

In the device, the energy gap in the active layer is less than in the surrounding semiconductors, as illustrated in Fig. 18.6a. The concentrations of dopants in the surrounding semiconductors are much less than in the case of the LED, so the chemical potentials for the electrons and holes lie within the bandgap. When the semiconductors are brought in contact with the active layer (Fig. 18.6b), the chemical potentials of the electrons and holes equilibrate. A strong electric field is set up across the active layer. Electrons from the conduction band of the n-type material can enter the layer, fall to the bottom of the conduction band, and have difficulty surmounting the barrier to get out. The same is true for the holes. They can enter the valence band of the active layer and are trapped there.

The primary means of operation of the device is the forward-bias mode (Fig. 18.6c). The p-type semiconductor is connected to a high potential and the n-type semiconductor is grounded. Electrons from the n-type material are injected into the conduction band of the active layer and holes from the p-type material also flow there. Radiative recombination occurs, with the process stimulated by photons already present in the cavity modes. If the material is of high purity, with few defects, the stimulated radiative lifetime can be much shorter than the nonradiative lifetime caused by phonons. The laser efficiency is therefore very high.

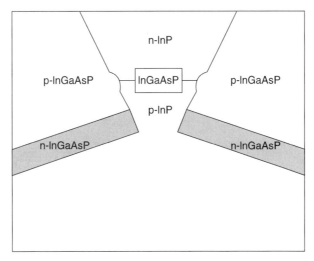

Figure 18.7. Cross section of buried heterostructure laser. [Reprinted with permission from Y. Suematsu, *Phys. Today*, **38**, 32 (May 1985). Copyright 1985 by the American Institute of Physics.]

The wavelength of the light that is emitted depends primarily on the bandgap energy in the active layer, although it is influenced by the chemical potential of the electrons as well. In the case of the GaAs/Ga$_{1-x}$Al$_x$As structure it lies in the range 0.77 to 0.88 μm. For the In$_{1-x}$Ga$_x$As$_{1-y}$P$_y$/InP laser it is in the range 1.2 to 1.65 μm. The latter range is highly desirable since it includes the window of transparency for optical fibers, making it suitable for telecommunications purposes.

A sketch of the cross section of a buried heterostructure laser is given in Fig. 18.7. The active region, GaInAsP, is epitaxially sandwiched between n- and p-type InP. Blocking layers of n- and p-type GaInAsP confine the currents to the channel containing the active layer.

Returning to the problem of lattice-constant matching, it is desirable to grow a layer of one type of material on another in such a way that their atoms align perfectly at the interface. Consider, for example, the deposition of the quaternary group III–V semiconductor In$_{1-x}$Ga$_x$As$_y$P$_{1-y}$ on an InP substrate. Both materials are of the same crystal structure, so all that is needed is for the atoms to have the same interatomic spacings (assuming that there is no surface reconstruction). The lattice constants for some group III–V semiconductors are given in Table 18.1a. One makes use of a bilinear interpolation scheme, called *Vegard's law*, to find the suitable compositional parameters x and y. Vegard's law for pseudobinary alloys such as Al$_{1-x}$Ga$_x$As is described in Section 11.11. Thus the lattice constant is predicted to be

$$a(x, y) = xya_{GaAs} + x(1 - y)a_{GaP} + (1 - x)ya_{InAs} + (1 - x)(1 - y)a_{InP}. \quad (18.13)$$

Lattice matching [i.e., $a(x, y) = a_{InP}$] is accomplished along the curve

$$x = \frac{0.452y}{1 - 0.024y}. \quad (18.14)$$

TABLE 18.1   Parameters for III–V Semiconductors at $T = 300\ K$

	P	As	Sb		P	As	Sb
	a. Lattice Constant a (nm)				*g. High-Frequency Relative Permittivity $\epsilon_r(\infty)$*		
Al	0.545	0.566	0.614	Al	—	8.16	10
Ga	0.545	0.565	0.610	Ga	9.11	10.9	14
In	0.587	0.606	0.648	In	9.6	12.3	15.7
	b. Bandgap Energy E_g (eV)				*h. Thermal Conductivity κ (W/m·K)*		
Al	2.45	2.15	1.612	Al	92	84	60
Ga	2.272	1.424	0.725	Ga	75	56	27
In	1.350	0.355	0.170	In	80	29	16
	c. Conduction Band Mass of Electrons m_e^*/m				*i. Coefficient of Linear Thermal Expansion $\alpha(10^{-6}K^{-1})$*		
Al				Al	—	3.5	4.2
$\quad m_{Le}^*/m$	—	1.1	—	Ga	5.3	5.4	6.1
$\quad m_{Te}^*/m$	—	0.19	—	In	4.6	4.7	4.7
Ga	0.82	0.67	—				
In	0.077	0.022	0.145				
	d. Light-Hole Band Mass of Electrons m_{lh}^*/m				*j. TO Phonon Frequency Ω_{TO} (cm^{-1})*		
Al	—	0.153	—	Al	—	364	319
Ga	0.14	0.12	0.06	Ga	366	268	230
In	0.12	0.025	0.015	In	307	217	179
	e. Heavy-Hole Band Mass of Electrons m_{hh}^*/m				*k. LO Phonon Frequency Ω_{LO} (cm^{-1})*		
Al	—	0.41	0.4	Al	—	402	340
Ga	0.79	0.8	0.23	Ga	402	292	240
In	0.45	0.41	0.4	In	348	239	190
	f. Static Relative Permittivity ϵ_r (0)				*l. Band-Offset Parameter ϕ (eV)*		
Al	—	10.1	11	Al	1.27	1.002	0.47
Ga	11.1	13.1	15.7	Ga	0.797	0.562	0.07
In	12.6	15.2	16.8	In	0.857	0.58	0.04

Source: Data from G. P. Agrawal and N. K. Dutta, *Semiconductor Lasers*, 2nd ed., Van Nostrand, New York, 1993; and M. Bass, ed., *Handbook of Optics*, 2nd ed., McGraw-Hill, New York, 1995; as well as other literature. Data for part (*l*) from S. Teveri and D. J. Frank, *Appl. Phys. Lett.*, **60**, 630 (1992).

For such a matched system the bandgap energy $E_g(x, y)$ is observed to vary with energy approximately as

$$E_g = 1.35 - 0.72y + 0.12y^2 \quad \text{eV.} \tag{18.15}$$

Here $E_g(\max) = 1.35$ eV and $E_g(\min) = 0.75$ eV. This allows one to engineer quantum-well heterostructural devices by creating layers with customized bandgaps.

The interpolation formula may be generalized to include more complicated alloys. Consider, for example, a group III–V compound with as many as six elements of the form $A_\alpha B_\beta C_\gamma X_\lambda Y_\mu Z_\nu$. Here A, B, and C each have valence III and X, Y, and Z have valence V. To satisfy the valence requirements

$$\alpha + \beta + \gamma = 1, \qquad \lambda + \mu + \nu = 1. \tag{18.16}$$

Denote any physical property of the binary compounds by [AX], [BX], ..., [CZ]. The trilinear interpolation formula is then

$$[A_\alpha B_\beta C_\gamma X_\lambda Y_\mu Z_\nu] = \alpha\lambda[AX] + \alpha\mu[AY] + \alpha\nu[AZ] + \beta\lambda[BX] + \beta\mu[BY]$$
$$+ \beta\nu[BZ] + \gamma\lambda[CX] + \gamma\mu[CY] + \gamma\nu[CZ], \tag{18.17}$$

subject to the foregoing constraints on the stoichiometric coefficients. (For the six-element case there are four independent coefficients.) Using this formula it is possible to obtain an estimate for a large number of properties of compounds. Formula (18.17) includes the quaternary compound as a special case ($\gamma = \nu = 0$). Table 18.1 provides data for a number of physical properties for the situation where A, B, and C represent Al, Ga, and In and X, Y, and Z represent P, As, and Sb. The independent parameters may be set, for example, to design heterostructures of material with specific energy gaps, lattice constants, and indices of refraction.

In Table 18.1 data are presented that permit one to compute the discontinuity in valence-band energy across an interface between two semiconductors. Suppose that semiconductors A and B are in contact. In general, the gap energies will be different and the conduction and valence bands will not line up with one another. The valence-band discontinuity is $\Delta E_v = \phi_A - \phi_B$, where ϕ is tabulated in Table 18.1l. The discontinuity in conduction-band energies follows from the formula $\Delta E_c = \Delta E_g - \Delta E_v$, where $\Delta E_g = E_g(A) - E_g(B)$. Semiconductor heterostructure superlattices are studied further in Section 20.9.

18.5   Ceramics for Lasers

Solid-state lasers are often constructed from single-crystal ceramic materials or glasses. The reason for not using polycrystalline ceramics is that there would be too much light scattering from the grains. Examples of ceramic lasers include the ruby laser, the Nd-YAG laser, the Nd-glass laser, and the Er-doped silica fiber. The electronic levels of an impurity ion provide the quantum system which is to be pumped (excited) and from which the laser transition occurs. Pumping of the laser is usually achieved by means of an external flashlamp (e.g., a xenon or krypton lamp) or another laser.

The first solid-state laser was the ruby laser. Ruby is alumina (Al_2O_3) doped with chromia (Cr_2O_3). The index of refraction is $n = 1.76$. For lasers the Cr doping is at the level of 5×10^{-4} part by weight. In pure alumina the crystal structure is hexagonal. The Al^{3+} ions reside at the centers of octahedra of oxygen ions. For every two octahedra containing an Al^{3+} ion, there is one that is empty. The Cr^{3+} dopant ion substitutes for the Al^{3+} ion.

Obviously, the energy levels of the impurity ion play a crucial role in determining both the frequencies of the laser transitions and the spontaneous lifetimes. Since the ions often sit at substitutional sites in the crystal, the crystal field of the host crystal causes a Stark splitting of the ionic energy levels. The systematics of this splitting depend on both the angular momentum states of the excited levels of the ion as well as the point-group symmetry of the crystal field. For a free atom, where there is full rotational symmetry, the wavefunctions are described in terms of the spherical harmonics, and atomic-term notation is used to label the states. For a crystal, where there are a finite set of symmetry operations, one introduces linear combinations of the spherical harmonics called *crystal harmonics*. These crystal harmonics transform among themselves when the symmetry operations are applied to them. For a first-order perturbation to be caused by a crystal field V in a state with wavefunction ψ, the product $\psi^*\psi$, when expanded in the crystal harmonics, should contain the representation to which V belongs. For glasses, in which there is no point-group symmetry, one simply uses atomic-term notation to describe the ionic states.

Referring to Figs. 18.8a and W9.1, the ground state of the Cr^{3+} ion is labeled $^4A_2$ (the crystal field notation is appropriate to the octahedral group). Photons are absorbed from a flashlamp near 0.55 μm and 0.4 μm, pumping the ion to the $^4T_2$ and $^4T_1$ states, respectively. There is an intersystem crossing (i.e., a change in net electronic spin) in which excitation is passed from the $^4T_2$ state to the ^2E state (which is actually two closely spaced lines). The laser transition occurs when a photon is emitted and the ion passes from the ^2E state back to the $^4A_2$ state. The wavelength of the photon emitted is 694.3 nm, and the spontaneous lifetime is 3 ms. The laser involves primarily the states $^4A_2$, $^4T_2$, and ^2E, so the system is a three-level laser. Other valence states of chromium are available for laser transitions. For example, Cr^{4+} ions in forsterite or in Ca_2GeO_4 have been found to lase in the near infrared (1.3 to 1.6 μm).

The Nd–YAG laser consists of Nd^{3+} dopants in an yttrium aluminum garnet (YAG) host. The index of refraction is $n = 1.8$. The usual chemical structure for a garnet is $A_3B_2(SiO_4)_3$, where A has valence 2 and B has valence 3. In YAG the composition is $Y_3Al_2(AlO_4)_3$. The cubic crystal structure is complex, with tetrahedra, octahedra, and polyhedra with eightfold coordination present. The garnet crystal structure is given in Fig. 17.19. The Nd^{3+} substitutes for an Al^{3+} ion.

Referring to Fig. 18.8b, the ground state of the Nd^{3+} ion is the $^4I_{9/2}$ state. The laser material absorbs over a broad range of wave numbers, extending from 13×10^3

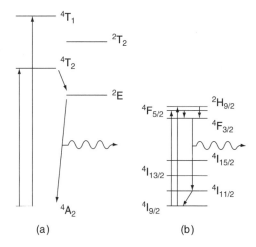

(a)   (b)

Figure 18.8. Level schemes for (*a*) the ruby laser (Cr^{3+}) and (*b*) the Nd–YAG laser (Nd^{3+}).

to 25×10^3 cm^{-1}, including the $^2H_{9/2}$ and $^4F_{5/2}$ states (using atomic-term notation), which relax by nonradiative processes to the $^4F_{3/2}$ state. The actual laser transition occurs between the $^4F_{3/2}$ and the $^4I_{11/2}$ state and has a wavelength of 1.0641 μm with a spontaneous lifetime of 0.55 ms. In free space this would be a highly forbidden transition, since $\Delta L = 3$ and $\Delta J = 4$, but the crystal field invalidates the selection rules ($\Delta J = 0, \pm 1; \Delta L = \pm 1$). The $^4I_{11/2}$ state lies 2.111×10^3 cm^{-1} above the ground state. This is sufficiently far above the ground state that the probability of thermal excitation is very small. To this end it should be noted that a thermal photon has a wave number $k_B T/hc = 210$ cm^{-1} at $T = 300$ K. The Nd–YAG system comprises a four-level laser system.

For substitutional impurity ions it is best to have the valence of the ion match the valence of the host cation. For example, Nd^{3+} could be placed in hosts such as $LaCl_3$, $LaBr_3$, $Y_3Al_5O_{12}$ (YAG), or Y_2O_3. The laser wavelengths for the $^4F_{3/2}-^4I_{11/2}$ transition in these materials are 1.0641, 1.068, 1.0646, and 1.073 μm, respectively. The differences are due to the differences in local crystal electric fields. Other popular host materials include MgO, fluorides (e.g., MgF_2), tungstenates (e.g., $CaWO_4$), niobates, germanates, silicates, phosphates, and molybdenates. Common substitutional ions include Cr^{3+}, Cr^{4+}, V^{2+}, Co^{2+}, Ti^{3+}, Pr^{3+}, Ho^{3+}, and Eu^{3+}.

The Nd-glass laser utilizes a glass ($n = 1.5$) rather than a crystalline host material. Typically, the glass is composed of silica, alumina, GeO_2, and/or P_2O_5. The laser frequency is relatively insensitive to the nature of the host material. High quantum efficiencies, on the order of 40 to 80%, are found. Unlike the case of the Nd–YAG laser, the Nd ion finds itself in a continuous range of local environments, so there is considerable inhomogeneous broadening. Nd-glass lasers are usually operated in a pulsed mode rather than in a continuous-wave (CW) fashion.

Lasers at different frequencies may be obtained by replacing Nd by some other rare earth dopant. Typically, Ce, Er, or Yb is used in place of Nd. The Er-doped silica fiber has proven to be particularly useful as a laser amplifier for fiber optical communications. The fiber may be pumped by high-power semiconductor lasers at 0.81, 0.98, or 1.48 μm, some nonradiative relaxation occurs, and radiation is produced at 1.535 μm. The system operates as a three-level laser. If an input signal at 1.535 μm enters one end of the fiber, it stimulates emission and is amplified before emerging from the other end. This serves to regenerate the optical signal and makes long-distance communication feasible.

Obviously, when using crystals as hosts, the crystal must be transparent in the region of the laser transition. In the case of high-power lasers, the vulnerability of the crystal to laser damage is of primary concern. For this reason, among others, glasses are often used as the host material. Since glass is already disordered, local damage simply transforms one disordered state into another. In glass lasers there are a variety of possible local environments at which the ion can sit. This leads to a superposition of possible laser transitions, a phenomenon known as *inhomogeneous broadening of the laser line*.

18.6   Bandgap Engineering of Optical Materials

The ability to create semiconductor heterostructures in a highly controlled manner has spawned the field of bandgap engineering. In this section several examples are presented of how this can be used to design optical materials with interesting properties.

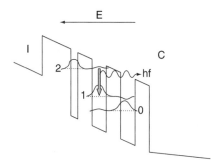

Figure 18.9. Cascade laser, showing the conduction band of the semiconducting heterostructure. The direction of the electric field is denoted. [Reprinted with permission from J. Faist et al., *Science*, **264**, 553 (1994). Copyright 1994 by the American Association for the Advancement of Science.]

Quantum Cascade Laser. Rather than relying on the energy levels provided by nature for the elemental ions, it is now possible to do *engineering* of energy levels by fabricating quantum-well devices. A particular example of this is the quantum cascade laser, in which a three-level laser is constructed by alternating layers of lattice-matched $Ga_{0.47}In_{0.53}As$ and $Al_{0.48}In_{0.52}As$ in a group III–V semiconductor heterostructure sandwich (Fig. 18.9). The bandgap in $Al_{0.48}In_{0.52}As$ is larger than in $Ga_{0.47}In_{0.53}As$ by 0.706 eV, so the AlInAs forms the barriers and the GaInAs forms the wells. The actual barrier height in the conduction band is 0.50 eV (and the corresponding barrier for holes in the valence band is 0.206 eV). Electrons are introduced into level 2 from the injection zone, I. A photon of energy $\hbar\omega$ is emitted in the laser transition when the system makes a quantum transition from state 2 to state 1. Final relaxation to state 0 is followed by the electron tunneling into the collector region, C. The values and relative positions of the energy levels are determined largely by the width of each allowed region and by the bias voltage. The figure shows schematically the wavefunctions associated with the three quantum-well states. The 1–0 transition is assisted by phonons, giving state 1 a short nonradiative lifetime. On the other hand, the 2–1 transition is higher in energy than phonon frequencies, so that there is not a strong inhomogeneous broadening of the optical line. The laser was first constructed to operate at a wavelength of 4.2 μm. A later modification extended this to 5.2 μm, with a peak pulsed power output of 200 mW.

An approximate formula for the energy levels in each well is obtained by using the relation that applies to a particle in a well with infinite walls. There the energy levels are given by $E_n = \hbar^2\pi^2 n^2 / 2m^* w^2$, where w is the width of the well, m^* is the carrier effective mass, and $n = 1, 2, 3, \ldots$. Thus the narrower the well, the higher the energy levels will be. Recent versions of this laser can emit at two or more wavelengths simultaneously.

Quantum-Dot Laser. Research is now being performed to try to create *quantum-dot* lasers. In such devices isolated islands of InGaAs are embedded in layers of GaAs. The energy levels of such a structure are determined by the diameter of the dot and the effective-mass tensor of the electron. The challenge is to maintain the uniformity of size and shape of the islands so as to give a well-defined laser frequency for a macroscopic piece of material.

Experiments on colloidal spherical quantum dots of CdSe (a direct-gap semiconductor) show anomalously long lifetimes of photogenerated electron–hole pairs (≈ 1 μs as opposed to the bulk lifetime of ≈ 1 ns). The valence band is a degenerate p-like band and the conduction band is a nondegenerate s-like band. An electron–hole pair may therefore be characterized by the atomic quantum numbers ($s_{1/2}$, $p_{3/2}$). There is an eightfold degeneracy of these states, which is lifted by the exchange interaction between the electron and the hole. According to the rules for the addition of angular momenta, states with total angular momentum $J = 1$ and 2 will be formed. The lowest state has $J = 2$. Such a state does not have an allowed radiative transition to the ground state in which the electron and hole recombine, so the recombination involves two photons. By the same token, the pair cannot be resonantly excited in a photoluminescence experiment. One refers to the electron–hole pair as a dark exciton.[†] The quantum-dot laser is discussed further in Section 20.10.

Nonlinear Optical Materials. The nonlinear optical properties of homogeneous materials are considered in Section 8.9. There has also been recent interest in creating nonlinear optical materials using the technology of semiconductor heterostructure quantum wells. Suppose, for example, that one wished to construct a material with a high second-order nonlinear optical coefficient $d^{(2)}$. A simple symmetric well in the conduction band, created by depositing a monolayer of GaAs between two layers of $Ga_{1-x}Al_xAs$, will not suffice because it presents a potential to the electrons with reflection symmetry, and it was seen that $d^{(2)}$ is zero for such a potential. However, if one were to construct an asymmetric well, this restriction would no longer apply. In Fig. 18.10 several such well configurations are depicted.

By adjusting the thicknesses of the various portions of the well, it is possible to tune the resonance frequencies corresponding to the $E_2 \to E_1$ transitions between wavelengths of 5 and 20 μm. The two parameters that are readily varied are the thicknesses of each part of the well and the concentration of the Al. In Fig. 18.10 a sketch of the wavefunctions for the lowest two quantum states is shown. The dipole-transition moment is the matrix element of the electric-dipole operator taken between these

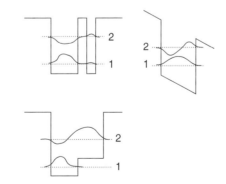

Figure 18.10. Several asymmetric semiconductor quantum wells. [Reprinted with permission from E. Rosencher et al., *Science*, **271**, 168 (1996). Copyright 1996 by the American Association for the Advancement of Science.]

[†] See D. J. Norris et al., *Phys. Rev. Lett.*, **75**, 3728 (1995).

two states: $\mu_{21} = \langle 2|ez|1 \rangle$, and enters in the formulas for the nonlinear susceptibility, Eqs. (8.47) and (8.48), as a cubic dependence. Since the wavefunction lobes in adjacent wells may be typically a few tens of nanometers apart, very large transition moments may be engineered.

RECORDING OF LIGHT

When light interacts with matter, it often leaves a permanent imprint. This could either be undesirable, as in the case of damage, or desirable, as in the cases of photography or xerography. In the following sections various aspects of these topics are considered.

18.7   Photography

Photographic film consists of a thin emulsion of silver halide crystallites (usually AgBr, AgCl, or more complex halides such as $AgBr_{1-x}I_x$ and $AgBr_{1-x}Cl_x$) in a gelatin host. The grains have typical dimensions $0.1 \times 10 \times 10$ μm. Small amounts of sensitizing material, such as Au and S, are also present on the surfaces of the grains. Those parts of the film that are exposed to light undergo a physical and chemical change, and a latent image consisting of a few clustered Ag atoms is formed. This is transformed to a real image when the emulsion is reduced by a developer and upward of a billion silver atoms coalesce around the latent image nucleus. The unexposed regions, with native silver halide, just dissolve away. The latent image serves as a catalyst for the formation of a microscopic silver particle. Effectively the development process amplifies the presence of the latent image by a factor on the order of a billion. In this section these processes are discussed in some detail.

In black-and-white photography there is only one emulsion layer, and the gray tone is determined by the concentration of developed grains. Color photography involves the creation of a multilayer emulsion structure, around 20 μm thick, with each layer sensitized by an array of dye molecules that absorb in a different region of the spectrum. There are typically 14 layers in the structure. The top layers absorbs in the blue, the middle layers in the green, and the bottom layers in the red. Separating these are filter layers. The color-separated latent images are formed in the separate layers. The final image is formed by using a mixture of dye molecules of different colors. A given colored molecule reacts with its matched latent image layer.

The first step is the absorption of a single photon by a grain of silver halide. An electron–hole pair is produced with a high quantum efficiency. The transition is indirect, with a zone-boundary (L-point) valence-band electron being promoted to the center (Γ-point) of the conduction band, assisted by a zone-boundary phonon. The band structure for AgBr is depicted in Fig. 18.11. The direct transitions at the zone center occur well into the ultraviolet, at 4.3 and 5.1 eV for AgBr and AgCl, respectively. The indirect transition for AgBr occurs at 2.7 eV. The band-edge phonons for AgBr have energies of 8 and 12 meV for the TO and LA phonon, respectively. With judicious use of adsorbed dye molecules, the threshold absorption energy may be lowered. A molecule is chosen such that the highest-occupied molecular orbital (HOMO) lies above the top of the valence band and the lowest-unoccupied molecular orbital (LUMO) lies above the bottom of the conduction band. Thus an electron from the dye molecule may be photoexcited from the HOMO and tunnel directly into the conduction band of

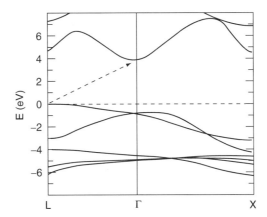

Figure 18.11. Electronic band structure of AgBr. The indirect absorption transition is depicted. [From A. B. Kunz, *Phys. Rev. B*, **26**, 2070 (1982). Copyright 1982 by the American Physical Society.]

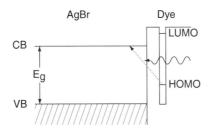

Figure 18.12. Photoexcited charge transfer from a dye molecule to the conduction band of AgBr.

the adjacent crystal. Such a situation is depicted in Fig. 18.12. Instead of a hole in the valence band, one is left with a positively charged dye molecular ion.

The electron is mobile and travels through the lattice of the AgBr grain until it is trapped at a defect site. The hole is relatively immobile. In AgBr the electron mobility is $\mu_e = 6 \times 10^{-3}$ m^2/V·s, whereas the hole mobility is only $\mu_h = 1.1 \times 10^{-4}$ m^2/V·s. This permits the electron in the conduction band to wander away from the hole in the valence band and avoid radiative recombination. The low mobility of the hole is due to its strong interaction with the lattice, resulting in the formation of a self-trapped *polaron*, a composite structure of a hole with a local distortion of the ionic lattice. The distance that the electron travels, L, can be computed by assuming that it takes a random walk: $L = (D\tau)^{1/2}$, where D is the diffusion coefficient and τ is the recombination lifetime (3 µs, which is quite long). The diffusion constant is given by the Einstein relation, $D = \mu_e k_B T/e = 1.6 \times 10^{-4}$ m^2/s (at $T = 300$ K). Thus $L = 22$ µm, which is much larger than the width of the grain. Multiple bounces off the surface of the grain are possible.

In principle, exciton formation could bind the electron and hole together, but the binding energy of the $1s$ exciton in AgBr is sufficiently low that it is readily ionized. This low binding energy is due largely to the high static relative permittivity, $\epsilon_r(0) = 10.6$, and low band effective electron mass, 0.22 m. (If polaronic corrections are included, the effective mass changes to $m^* = 0.29$ m.) Thus $E_{1s} = (m^*/m)/\epsilon_r^2(0)\text{Ry} = 35$ meV, where 1 Ry $= 13.6$ eV.

The defect that captures the mobile electron is a positively charged interstitial silver ion, Ag^+, near the surface of the grain. A neutral Ag atom is formed which has a high-enough electron affinity, EA $= 1.30$ eV, that it can trap a second electron according to the reaction $Ag + e^- \rightarrow Ag^-$. As will soon be seen, the interstitial Ag^+ ions themselves have a high mobility in the crystal. Therefore the reaction $Ag^+ + Ag^- \rightarrow Ag_2$ is possible. This dimer can grow to a larger size via the reactions $Ag_{n-1} + e^- \rightarrow (Ag_{n-1})^-$ and $Ag^+ + (Ag_{n-1})^- \rightarrow Ag_n$. Typically, $n \leq 5$. These clusters of silver atoms form the latent image. The density of clusters that comprises the latent image in a local region of the film is directly proportional to the light intensity that originally fell on that portion of the film. In this way the light intensity is translated into cluster density.

The AgBr crystal has the NaCl crystal structure with a lattice constant $a = 0.576$ nm. In the Frenkel-defect mechanism a Ag^+ cation resides at an interstitial site rather than at its proper lattice site. The interstitial silver ion, Ag_i^+ (whose radius is only 0.126 nm) resides at the tetrahedral site coordinated to four Br^- ions (Fig. 18.13). The energy for forming a Frenkel defect in AgBr is 1.16 eV. Once formed, two processes may be imagined to be responsible for Ag_i^+ migration: direct hopping (or vacancy migration) or the interstitialcy mechanism. In direct hopping the ion simply jumps across a potential barrier to a neighboring tetrahedral interstitial site. The barrier height is approximately 0.3 eV, so this process is unlikely. In the interstitialcy mechanism, two Ag^+ ions make a simultaneous jump (aided by a favorable quadrupolar lattice distortion). The lattice Ag^+ ion jumps to the new interstitial site, while the original Ag_i^+ jumps to the vacated lattice site. The activation energy for the latter transition is anomalously low, being approximately $E_a = 30$ meV. The diffusion constant for Ag_i^+ may be estimated using the formula

$$D_i = \left(\frac{a\sqrt{3}}{4}\right)^2 f e^{-\beta E_a}, \tag{18.18}$$

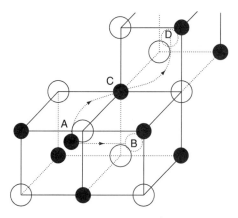

Figure 18.13. Diffusion processes of an interstitial Ag^+ cation (dark sphere). Direct hopping: $A \rightarrow B$. Interstitialcy mechanism: $A \rightarrow C$ simultaneously as $C \rightarrow D$. [From C. R. A. Catlow, *Mater. Res. Soc. Bul.*, **14**(5), 23 (1989).]

where the attempt rate, f, may be approximated by using the frequency of the TO band-edge phonon. The hopping distance is $a\sqrt{3}/4$. One finds that $D_i = 1.5 \times 10^{-7}$ m^2/s. The Ag$_i^+$ ion is sufficiently mobile to be attracted to an Ag$^-$ ion before the electron recombines with a hole.

When a Frenkel pair is formed at the grain surface, some Ag$_i^+$ ions can migrate to the interior, whereas the Br$^-$ anions are immobile. The state of lowest free energy has an excess of Ag$_i^+$ in the interior and a negatively charged surface. A barrier of 150 meV is set up which repels photoexcited electrons from the surface. However, there are still some Ag$_i^+$ ions at the surface. In their neighborhood the barrier is lowered. Thus photoelectrons that bounce around in the interior of the grain behave like billiard balls on a pool table. They bounce off the surface but can fall into the "holes," in this case the surface Ag$_i^+$ ions. If there is a sensitizing Au or S atom near this interstitial, the electron is able to bind more tightly to the site and it is trapped there, forming AgAuS. The "speed" of the film is related directly to the concentration of sensitizing agents. It is defined in terms of the mass of silver per unit area in the developed film, D, and the light energy per unit area, E, also called the *exposure*. It is found that

$$D \propto \ln(E/i), \tag{18.19}$$

where the constant i is called the *inertia* of the film. The speed is proportional to $1/i$. If i is large, the speed is slow, and a large exposure is needed to create a given silver cluster density in the film.

Development is a chemical amplification process in which the Ag cluster associated with the latent image, consisting of approximately four Ag atoms, is replaced by a cluster of several billion silver atoms. Those grains that have been exposed to light contain such clusters; those that have not been exposed do not contain them. The developer is a reducing agent that replaces the Ag$^+$ ions in AgBr by neutral Ag atoms. The latent image cluster serves as a catalyst for this reaction, so the exposed regions form the large silver clusters much more rapidly than do the unexposed regions. The clusters that form in the unexposed regions, referred to as *fog*, diminish the contrast.

Typical chemicals used as developers are illustrated in the upper row of Fig. 18.14. They include quinanone (1,4,-dihydroxybenzene), labeled H$_2$Q; p-phenylenediame, labeled H$_2$D; and methyl-p-aminophenol (metol), labeled H$_2$A. In the second row the corresponding oxidized forms of these molecules are illustrated and are labeled Q, D, and A, respectively. The reduction reactions are

$$H_2Q + 2AgBr \longrightarrow Q + 2H^+ + 2Ag + 2Br^-, \tag{18.20a}$$

$$H_2D + 2AgBr \longrightarrow D + 2H^+ + 2Ag + 2Br^-, \tag{18.20b}$$

$$H_2A + 2AgBr \longrightarrow A + 2H^+ + 2Ag + 2Br^-. \tag{18.20c}$$

In the course of the reaction the clusters grow in size as more Ag atoms are adsorbed onto them. The development process is stopped by adding a stop-bath solution, such as acetic acid. The increased proton concentration in the acid blocks the forward reactions in Eqs. (18.20). The excess AgBr may be removed by adding a fixer solution such as sodium thiosulpfate, to form a water-soluble salt which may be washed away.

The photographic process is rather remarkable in that, with the absorption of only a handful of photons, a process is begun in which a billion silver atoms can be made to coalesce into a mesoscopic particle.

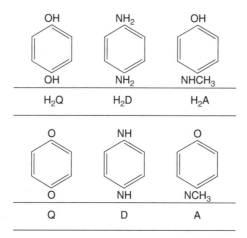

Figure 18.14. Typical chemicals used as developers (top row) and their oxidized forms (bottom row).

18.8   Photoconductors and Xerography

Photoconductivity was discovered in 1873 by Willoughby Smith. It reached its first major commercial application in 1938 when Chester Carlson invented the xerography (electrophotography) process. A photoconductor is a material, such as a semiconductor or an organic material, which is insulating when in the dark, but becomes conducting when light of sufficiently high frequency illuminates it. If the photon energy exceeds the bandgap energy, electron–hole pairs will be produced. These carriers contribute to the material's conductivity, according to the formula $\sigma = e(n\mu_e + p\mu_h)$, where (n, p) and (μ_e, μ_h) are the carrier concentrations and mobilities for the electrons and holes, respectively. It is important that the electron and hole are able to diffuse away from each other before they recombine.

In the xerography process an unilluminated thin-film photoconductor is uniformly charged by means of a corona discharge. The image of the object to be copied is then cast on the thin film. Charge is conducted away from the illuminated area through the photoconductor and is neutralized. The dark areas retain the charge. A charged latent image remains. The photoconductor is then brought into contact with charged pigment particles, which are attracted to the latent image. The thin film is then brought in contact with a sheet of paper and an electric field causes the particles to transfer to the paper. The pigment particles are then heated until they melt and fuse together. Finally, the thin film is cleaned and is ready to make the next copy.

Imagine the photoconductor to be a capacitor with an initial charge per unit area q_0/A. At time $t = 0$ the conductivity is suddenly turned on by light and the capacitor discharges with the time dependence

$$\frac{q(t)}{A} = \frac{q_0}{A} \exp\left(-\frac{t}{\tau_n}\right), \tag{18.21}$$

with the discharge time being given by

$$\tau_n = \frac{\epsilon}{\sigma}, \tag{18.22}$$

where ϵ is the electric permittivity and σ is the conductivity of the medium. The initial charge per unit area is determined by the voltage of the corona discharge, V_0, and the capacitance of the film, $C = \epsilon A/d$, according to the formula $q_0 = CV_0$:

$$\frac{q_0}{A} = \frac{\epsilon V_0}{d} \tag{18.23}$$

where d is the film thickness. For a short burst of light the exposure E (the energy/area, as in Section 18.7) determines the number of carriers produced, and hence the conductivity. The number of electrons and holes generated per unit volume are the same and are given by $n = p = \eta E \alpha / \hbar \omega$, where ω is the photon frequency, α the photon absorption coefficient, and η the quantum efficiency for creating an electron–hole pair upon photon absorption. The conductivity is $\sigma = ne(\mu_e + \mu_h) \equiv ne\mu$, so the discharge time may be expressed as

$$\tau_n = \frac{\epsilon \hbar \omega}{e\mu\eta\alpha E}. \tag{18.24}$$

The electrification of the pigment particles (toner), whose size is typically 10 μm, is brought about by mixing them with carrier beads of size around $R = 200$ μm. The carrier beads are magnetic and serve as a "brush" to deposit the pigment on the photoconductor. Magnetic forces keep the beads from being attracted to the thin film. When dissimilar materials (beads and pigment) come in contact with each other, there is often a transfer of electric charge that occurs in order to equalize the chemical potentials. For metal–metal systems this gives rise to a contact potential, given by the difference in Fermi levels between the metals. For insulating systems there may be charge hopping from one material to the other, accompanied by band bending. One may set up an empirical triboelectric series which determines qualitatively in what direction charge will be transferred. This series is purely qualitative and can be determined by rubbing dissimilar materials together and seeing which becomes positively charged and which becomes negatively charged. The triboelectric series is tabulated for representative materials in Table 18.2. Due to the extremely high resistance of these materials, it is difficult to correlate this series experimentally with the ionization energies and electron affinities of the respective materials. The chemical potential depends on the radius of the particle, so corrections to this series can occur for small particles.

The carrier mobilities of photoconductors used in xerography are fairly low, being typically around 10^{-9} m^2/V·s. Since the materials are amorphous, the carriers are localized. Conduction proceeds by thermally activated hopping from atom to atom. There are other materials, such as anthracene, which have much higher mobilities, about 10^{-4} m^2/V·s. These materials are not used, however, because of the difficulty in making thin films. In the early days of xerography, amorphous selenium was used. It had several drawbacks, however, including its tendency to crystallize and its poor sensitivity to red light. The materials commonly used now include chalcogenide glasses, a-Si:H, and polymers containing molecular dopants. One generally desires a low dark conductivity, a high efficiency for photogenerating carriers, and a high carrier mobility. Practical considerations argue for ease in manufacturing the thin film. It is also desirable to avoid having deep traps in the bandgap, which could inhibit the cleaning phase of the reproduction cycle. Polymers are chemically stable, flexible, and are very good insulators. The dopant molecules serve as the sites between which the carriers can hop. Typical photoconductors are the polyvinylcarbazole polymer doped

TABLE 18.2 Tribolelectric Series

Positive[a]

    Silicone elastomer with silica filler
    Borosilicate glass, fire polished
    Window glass
    Polyformaldehyde
    Polymethylmethacrylate (PMMA)
    Polyamide 11
    NaCl
    Wool, knitted
    Silica, fire polished
    Silk, woven
    Cotton, woven
    Polyurethane elastomer
    Polystyrene
    Polyethylene glycol terephthalate
    Polychlorobutadiene
    Natural rubber
    Sulfur
    Chlorinated polyether
    Polytrifluorochloroethylene
    Polytetrafluoroethylene (Teflon)

Source: Data from J. Henniker, *Nature*, **196**, 474 (1962).

[a]When two materials are rubbed together, the material closer to "positive" becomes positively charged.

with trinitrofluorenone molecules (PVK–TNF), the polycarbonate polymer doped with triphenylamine (Polycarbonate–TPA), or polycarbonate doped with N, N'-diphenyl-N, N'-bis(3-methylphenyl)-[1,1'-biphenyl]-4,4'-diamine (polycarbonate–TPD). These molecules are illustrated in Fig. 18.15.

The molecular-doped polymers are generally used in conjunction with a second layer which serves as a photosensitizer. The sensitizer absorbs lower-frequency photons and produces carriers that are injected into the polymer. Molecules such as phthalocyanine, which is even sensitive to the infrared, are used for this purpose.

Recent attention has also been directed at hydrogenated amorphous silicon for xerography purposes. Another candidate is a photoconducting liquid crystal, 2,3,6,7,10,11-hexahexylthiotriphenylene (HHTT), with a high mobility (around 10^{-5} m^2/V·s), which can readily be formed into a thin film. Liquid crystals have properties intermediate between solids and liquids and are discussed in Chapter 14. The HHTT material consists of self-organized stacks of disklike molecules arranged in a hexagonal array. The molecules have a strong overlap of π-orbitals, which act as the conduction conduits. A somewhat simplified representation of the HHTT liquid crystal is given in Fig. 18.16.

18.9 Electro-optic Effect and Photorefractive Materials

In Sections 18.7 and 18.8 methods were described for preserving images either as photographs or in inked form. In this section a method is described for creating an

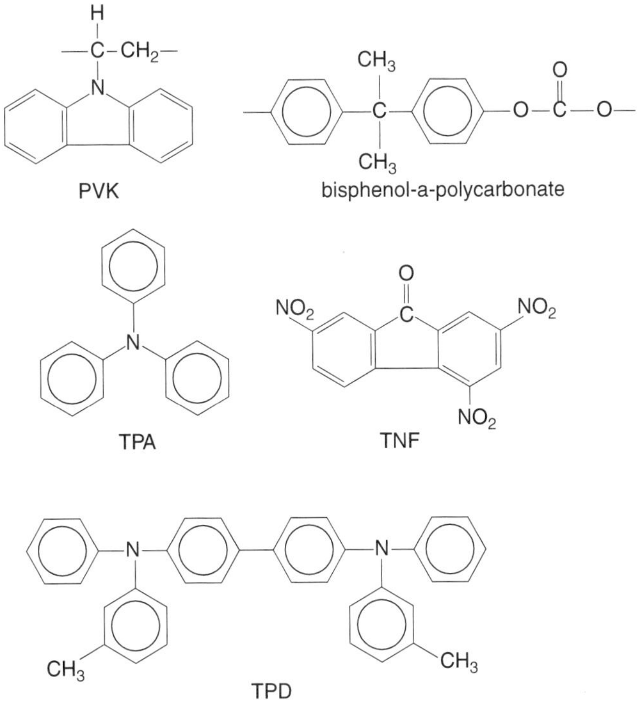

PVK

bisphenol-a-polycarbonate

TPA

TNF

TPD

Figure 18.15. Two polymers and three molecular dopant molecules used as photoconducting transport channels.

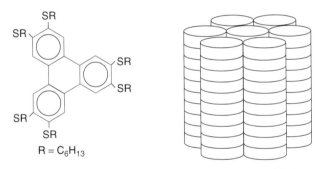

R = C$_6$H$_{13}$

Figure 18.16. Liquid-crystal HHTT. The platelike molecules are assembled in an array of stacks. [Reprinted with permission from D. Adam et al., *Nature*, **371**, 141 (1994). Copyright 1994 by Macmillan Magazines Ltd.]

image in a crystal which can then be referenced by means of a laser. Before the method is described, however, the underlying physics needs to be understood.

The *electro-optic effect*, also called the *Pockels effect*, involves the modification of the index of refraction tensor by an electric field. It permits electro-optical modulation of the phase of a light beam using a time-varying electric field. It also permits the deflection of a light beam in response to an electric field, which allows for photonic switching between optical channels or fibers. It thus provides a valuable link between optics and electronics. The effect is linear in the electric field and may be defined through the equation

$$\left(\Delta \frac{1}{n^2}\right)_{\alpha\beta} = \sum_{\gamma=1}^{3} r_{\alpha\beta\gamma}E_\gamma, \qquad \alpha = 1, 2, 3, \quad \beta = 1, 2, 3, \tag{18.25}$$

where $r_{\alpha\beta\gamma}$ are the electro-optic tensor coefficients. Since the index of refraction tensor is symmetric, one may also use the compact notation

$$\left(\Delta \frac{1}{n^2}\right)_i = \sum_{\gamma=1}^{3} r_{i\gamma}E_\gamma, \qquad i = 1, \ldots, 6. \tag{18.26}$$

The effect vanishes for centrosymmetric crystals, such as $BaTiO_3$ in the cubic phase (i.e., at high temperatures). For noncentrosymmetric crystals, the pattern of nonvanishing coefficients is related to the symmetry of the crystal. For example, in the crystal $LiNbO_3$, which has a threefold axis of symmetry (C_{3v} symmetry), the nonvanishing components of $\mathbf{r}$ are

$$\mathbf{r} = \begin{bmatrix} 0 & -r_{22} & r_{13} \\ 0 & r_{22} & r_{13} \\ 0 & 0 & r_{33} \\ 0 & r_{51} & 0 \\ r_{51} & 0 & 0 \\ -r_{22} & 0 & 0 \end{bmatrix}, \tag{18.27}$$

where the four independent components are $(r_{13}, r_{22}, r_{33}, r_{51}) = (9.6, 6.8, 30.9, 32.6) \times 10^{-12}$ m/V at $\lambda = 633$ nm. The index of refraction tensor itself is diagonal with elements $n_{11} = n_{22} = 2.286$ and $n_{33} = 2.200$. The photorefractive shifts in the $1/n^2$ tensor elements are thus given by the six equations

$$\Delta(1/n^2)_{xx} = -r_{22}E_y + r_{13}E_z, \tag{18.28a}$$

$$\Delta(1/n^2)_{yy} = r_{22}E_y + r_{13}E_z, \tag{18.28b}$$

$$\Delta(1/n^2)_{zz} = r_{33}E_z, \tag{18.28c}$$

$$\Delta(1/n^2)_{yz} = \Delta(1/n^2)_{zy} = r_{51}E_y, \tag{18.28d}$$

$$\Delta(1/n^2)_{xz} = \Delta(1/n^2)_{zx} = r_{51}E_x, \tag{18.28e}$$

$$\Delta(1/n^2)_{xy} = \Delta(1/n^2)_{yx} = -r_{22}E_z. \tag{18.28f}$$

The coefficients r_{22} and r_{33} determine the shifts of the diagonal elements of $1/n^2$ along the direction of the applied electric field, and r_{13} determines the shift of these elements perpendicular to the field. Similarly, r_{22} and r_{51} govern the shifts of the off-diagonal elements. A schematic drawing of $LiNbO_3$ is presented in Fig. 18.17. The crystal essentially has a perovskite crystal structure with face-shared oxygen octahedra. There is a actually a small staggered rotation of the oxygen octahedra about the c axis (not shown) so that the unit cell contains three formula units.

A table of the electro-optic coefficients for some other common crystals is given in Table 18.3. Since $\Delta(1/n^2) \sim 1/n^3 \sim rE$, it is sometimes convenient simply to report

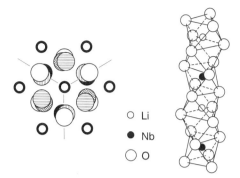

Figure 18.17. View of LiNbO$_3$ from the *ab* plane using an oblique perspective. In the left-hand diagram the lithium ion at the center lies directly above a niobium ion. [From L. Arizmendi et al., *Mater. Res. Soc. Bull.*, **19**(3), 32 (1994).]

TABLE 18.3 Nonvanishing Electro-optic Tensor Components for Typical Electro-optic Materials

Crystal	Symmetry	Component (10^{-12} m/V)									$\lambda\,(\mu m)$
		r_{41}	r_{51}	r_{42}	r_{52}	r_{13}	r_{23}	r_{22}	r_{33}	r_{63}	
GaAs	I$_d$	1.43	0	0	1.43	0	0	—	0	1.43	1.15
BaTiO$_3$	C$_{4v}$	0	1640	1640	0	8	8	—	28	0	0.546
KH$_2$PO$_4$(KDP)	D$_{2d}$	8.77	0	0	8.77	0	0	—	0	−10.5	0.546
KNbO$_3$	C$_{2v}$	0	105	380	0	28	1.3	—	64	0	0.633
Bi$_{12}$SiO$_{20}$	I	5.0	0	0	5.0	0	0	—	0	5.0	0.633
Ba$_{1/4}$Sr$_{3/4}$Nb$_2$O$_6$	C$_{4v}$	0	42	42	0	67	67	—	1340	0	0.633
LiNbO$_3$	C$_{3v}$	0	32.6	0	0	9.6	0	6.8	30.9	0	0.633

Source: Data from A. Yariv, *Optical Electronics*, 3rd Ed., Holt, Rinehart and Winston, New York, 1985.

the figure of merit $n^3 r$, where r is an effective electro-optic coefficient for some preferred orientation of the crystal relative to the electric field. One finds $n^3 r = 11,300, 2460, 320, 82$, and 68×10^{-12} m/V for BaTiO$_3$, Ba$_{1/4}$Sr$_{3/4}$TiO$_3$, LiNbO$_3$, Bi$_{12}$SiO$_{20}$, and GaAs, respectively. For the organic compounds 2-methyl 4-nitroaniline and *m*-nitroaniline, these numbers are 530 and 97×10^{-12} m/V, respectively. The value of r for PLZT materials is a sensitive function of composition. For example, (PbLa$_{0.085}$)(Zr$_{0.65}$Ti$_{0.35}$)$_{0.915}$O$_3$, PbLa$_{0.08}$(Zr$_{0.4}$Ti$_{0.6}$)$_{0.92}$O$_3$, and PbLa$_{0.12}$(Zr$_{0.4}$Ti$_{0.6}$)$_{0.88}$O$_3$ have r values of 38.6, 100, and 120×10^{-12} m/V, respectively.

Despite its modest figure of merit, LiNbO$_3$ is often the material of choice for photonic modulation and switching applications because of the ability to grow large single crystals. In addition, it has an acousto-optic effect and is piezoelectric and pyroelectric. The acousto-optic effect means that the index of refraction can be changed by straining the crystal. Such a strain, for example, could be provided by acoustic phonons generated by a transducer. The phenomena of piezoelectricity and pyroelectricity are discussed in Section 15.6. By varying the index of refraction periodically along the direction of propagation of the light, it is possible to modulate the frequency of light. By varying the off-diagonal matrix elements it is possible to deflect a light beam passing through the crystal in response to the electric field applied.

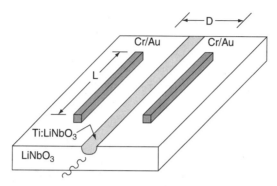

Figure 18.18. Electro-optic modulator consisting of a Ti:LiNbO$_3$ optical waveguide embedded in a LiNbO$_3$ substrate and Cr/Au electrodes. [From R. A. Becker, *Mater. Res. Soc. Bull.*, **13**(8), 20 (1988).]

Lithium niobate is an excellent insulator, so field-induced migration of defects is not a problem. Also for wavelengths longer than 1 μm, photo-induced charge migration and photoconductivity are not important effects. A schematic view of an electro-optical modulator is given in Fig. 18.18. A LiNbO$_3$ substrate has a semicylindrical region of Ti-doped LiNbO$_3$ running along the surface. The Ti:LiNbO$_3$ channel has a higher index of refraction than the substrate, so it acts as a light pipe and confines the light. It is flanked on both sides by deposited electrodes of length L separated by a distance D, whose definitions are given in Fig. 18.18. These electrodes may consist of an alloy of Cr and Au. The optical phase shift, $\delta\phi(t)$, developed for light of wave vector k (in vacuum) as it progresses along the channel may be expressed as

$$\delta\phi(t) \approx n^3 r \frac{kL}{2D} V(t), \tag{18.29}$$

where $V(t)$ is the time-dependent voltage. The electric field is proportional to $V(t)/D$, neglecting terms that depend logarithmically on D.

An example of a device that exploits this phase shift is the Mach–Zehnder interferometer. A schematic diagram of the Mach–Zehnder interferometer is given in Fig. 18.19. Light is split into two paths which travel along adjacent optical channels. One channel is subjected to an electric field and the other is not. The light beams are then recombined into a single channel. The interference will be destructive if the phase shift is $\delta\phi = \pi$. In this way phase modulation is readily converted to amplitude

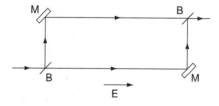

Figure 18.19. Mach–Zehnder interferometer. M, mirror; B, beam splitter (i.e., a half-silvered mirror.)

modulation. Typically, modulation rates of several gigahertz may be imposed on light with voltages on the order of 10 V.

In the photorefractive effect, one uses light to create a transient latent image in a macroscopic crystal. The latent image takes the form of a spatially modulated index of refraction. Those regions undergoing illumination have their index of refraction changed by an amount $\Delta n(\mathbf{r}, t)$. Once formed, $\Delta n(\mathbf{r}, t)$ may be "read" using a probe laser. The image is usually a hologram rather than a direct optical image.

The process begins with the absorption of light by a donor atom to produce a mobile electron and a positive ion (Fig. 18.20). The donor level D lies somewhere within the bandgap of the host crystal. The electrons are mobile and diffuse until they encounter an acceptor atom and are trapped. The energy level A of the electron on the resulting anion also lies somewhere in the gap region. If there is a gradient in the illumination, microscopic electric fields will be set up as a consequence of the spatial inhomogeneity of electrical charge. The electric fields cause changes in index of refraction due to the electro-optic effect. Unless steps are taken to fix the image permanently, the inverse process can also occur and the electron can be thermally activated from the acceptor site back to the donor ion and the image will disappear. In one of the common photorefractive crystals, $LiNbO_3$, the recombination time is typically on the order of 10^6 s. In $BaTiO_3$, however, it is only 10^3 s. Nevertheless, for image-processing applications, it may not be necessary to retain the image for very long times, or alternatively, it may be refreshed periodically.

Image fixing (i.e., making the image permanent and not subject to erasure) may be accomplished by using a crystal doped with ions, such as Fe in $LiNbO_3$. If the crystal is hot while it is illuminated, the ions will drift in response to the photorefractively produced electric field. The actual drift mechanism involves thermal activation of the ions over barriers that are lowered in the direction of the electric field. The latent image is impressed on the ionic concentration field. By cooling back to room temperature, the ions are no longer free to diffuse and the image becomes permanent.

Impurity ions in $LiNbO_3$ usually reside on the sites provided by the Li ions. Iron, however, could sit on either site. If it is on the Nb^{5+} site, it assumes the form Fe^{2+} and is an acceptor. If it sits on the Li^+ site, the ion becomes Fe^{3+} and serves as a donor.

Although single crystals such as $LiNbO_3$ have desirable photorefractive properties, they are expensive to grow. An effort has been made to try to find organic systems

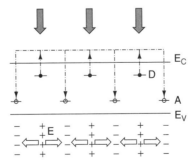

Figure 18.20. Photorefractive effect. The conduction- and valence-band edges are denoted by E_c and E_v and the photogenerated electric field by E. The thick downward-pointing arrows depict regions of illumination. [From L. Arizmendi et al., *Mater. Res. Soc. Bull.*, **19**(3), 32 (1994).]

exhibiting the same effect.[†] These would be easy to grow and be flexible as well. In addition, the static dielectric constants are inherently lower in covalently bonded materials, so stronger electric fields can exist, enhancing the photorefractive effect. One such system is DMNPAA:PVK:ECZ:TNF (in the weight percent ratios 50:33:16:1). The DMNPAA molecules [2,5-dimethyl-4-(p-nitrophenylazo)anisole] act as the chromophores (i.e., the material with a large photorefractive effect — hyperpolarizability). The PVK [poly(N-vinylcarbazole)] is a photoconducting polymer and the TNF (2,4,7-trinitro-9-fluorenone) is a photosensitizer, pushing the absorption band into the visible. Together, the TNF and PVK form a charge-transfer complex. The ECZ (N-ethylcarbazole) serves to lower the glass-transition temperature, T_g, of the polymer. This permits the DMNPAA molecules to reorient in response to the internally developed dc electric field at ambient temperatures. Since DMNPAA has a large anisotropy in its polarizability, it also gives rise to photo-induced birefringence of the material.

Photorefractive materials have also been found to exhibit four-wave mixing (FWM). Light from two primary beams directed at an angle with respect to each other establishes an interference pattern. The photorefractive grating that results is shifted slightly in space, due to the photocarrier charge migration. A secondary beam directed at the grating can diffract from this grating and, at the same time, extract energy from the primary beam. Growth of the secondary beam results and efficient energy transfer can occur, provided that the region of overlap of the beams is sufficiently large.

REFERENCES

General References

Glass, A. M., ed., *Photonic Materials*, Mater. Res. Soc. Bull., Aug. 1988, p. 14.

Weber, M. J., ed., *CRC Handbook of Laser Science and Technology*, 5 vols. plus supplements, CRC Press, Boca Raton, Fla., 1982.

Photography

James, T. H. ed., *The Theory of the Photographic Process*, 4th ed., Macmillan, New York, 1977.

Tan, Y. T. ed., *Silver Halides in Photography*, Mater. Res. Soc. Bull., May 1989, p. 13.

Tani, T. Physics of the photographic latent image, *Phys. Today*, Sept. 1989, p. 36.

Xerography

Burland, D. M., and L. B. Schein, Physics of electrophotography, *Phys. Today*, May 1986, p. 46.

Mort, J., Xerography, *Phys. Today*, Apr. 1994, p. 32.

Shaffert, R. M., *Electrophotography*, Focal Press, London, 1980.

Williams, E. M. *The Physics and Technology of Xerographic Processes*, Wiley, New York, 1984.

Photorefractive Effect

Agullo-Lopez, F., ed., *Photorefractive materials*, Mater. Res. Soc. Bull., Mar. 1994, p. 29.

Gambino, R. J., *Optical storage disk technology*, Mater. Res. Soc. Bull., Apr. 1990, p. 20.

[†] See, for example, K. Meerholz et al., *Nature*, **371**, 497 (1994).

Gunter, P., and J. P. Huignard, eds., *Photorefractive Materials and Their Applications*, Vol. 1, Springer-Verlag, Berlin, 1988.

Semiconductor and Solid-State Lasers

Agrawal, G. P., and N. K. Dutta, *Semiconductor Lasers*, 2nd ed., Van Nostrand Reinhold, New York, 1993.

Gan, F., *Laser Materials*, World Scientific, Singapore, 1995.

Pressley, R. J., ed., *CRC Handbook of Lasers*, CRC Press, Boca Raton, Fla., 1971.

Suematsu, Y., Advances in semiconductor lasers, *Phys. Today*, May 1985, p. 32.

Suematsu, Y., and A. R. Adams, eds., *Handbook of Semiconductor Lasers and Photonic Integrated Circuits*, Chapman & Hall, London, 1994.

Nonlinear Optics

Butcher, P. N., and D. Cotter, *The Elements of Nonlinear Optics*, Cambridge University Press, New York, 1990.

Shen, Y. R., *The Principles of Nonlinear Optics*, Wiley, New York, 1984.

PROBLEMS

18.1 Suppose that a quantum dot has the shape of a two-dimensional circular disk. A model that is often used to describe the potential of an electron confined in such a dot is $V(r) = m^*\omega_0^2 r^2/2$. Suppose a magnetic induction **B** is imposed perpendicular to the plane of the dot. Show that the electron energy levels are given by the formula

$$E_{n,l} = (2n + |l| + 1)\sqrt{(\hbar\omega_0)^2 + \left(\frac{\hbar eB}{2m^*}\right)^2} - \frac{l\hbar eB}{2m^*},$$

where $n = 0, 1, 2, \ldots$ and $l = \ldots, -2, -1, 0, 1, 2, \ldots$.

18.2 Consider a Lorentz oscillator model for an electron moving in a one-dimensional anharmonic potential described by the Toda potential $V(x) = Ae^{-ax} + Bx$, where A, a, and B are constants. The equation of motion is

$$m\left(\frac{d^2x}{dt^2} + \gamma\frac{dx}{dt}\right) = aA(e^{-ax} - e^{-ax_0}) + qE\cos\omega t.$$

Derive expressions for the linear polarization P, at frequency ω and the nonlinear polarization at frequency 2ω, $P(2\omega)$.

18.3 Consider a particle of mass m moving in the anharmonic symmetric potential

$$V(x) = A\cosh[a(x - x_0)]$$

subject to a damping force $-\gamma v$ and a driving force $qE\cos\omega t$. Find the Fourier coefficients for the dipole moment at frequencies ω and 3ω.

18.4 Consider the lattice matching of a layer of $In_{1-x}Ga_xAs_yP_{1-y}$ on an InP substrate.

    (a) Derive Eq. (18.14) giving the composition for lattice-matching. Use the lattice constants given in Table 18.1a.

    (b) Find the Ga atomic fractions x_{min} and x_{max} for lattice matching. What are the compositions of the resulting "alloys"?

18.5 Using Vegard's law, derive an expression for the bandgap energy $E_g(x, y)$ of a layer of $In_{1-x}Ga_xAs_yP_{1-y}$ which is lattice matched to an InP substrate. Compare your expression with the experimental result given in Eq. (18.15) and comment on any differences.

Note: An additional problem is given in Chapter W18.

SURFACES, THIN FILMS, INTERFACES, AND MULTILAYERS

CHAPTER 19

Surfaces

19.1 Introduction

A macroscopic sample of a solid may be regarded as a giant molecule containing N atoms, where N is typically on the order of Avogadro's number. The atoms fall into four categories: bulk atoms, atoms at interstitial or defect sites, impurity atoms, and surface atoms. The majority are bulk atoms. The impurity atoms and atoms at defect sites are distributed throughout the solid and depend on the quality of the crystal. The surface atoms include not only the outer layer of the solid but also those atoms sufficiently close to the surface to feel its presence. The number of surface atoms is typically on the order of $N^{2/3}$. In this chapter special attention is paid to the role played by the surface in describing the physics and chemistry of materials.

There are two common surface preparation techniques. One involves cleaving preexisting bulk material to expose two new surfaces. Obviously, substantial energy input is required to make this cut; otherwise, the solid would already have spontaneously cleaved. Chemical bonds have to be broken, and this requires energy input. This process is the solid-state analog of molecular dissociation. In the simplest (overidealized) case the atoms of the separated solids would simply retain their previous relative positions in space. However, this rarely happens. The lattice planes near the surface are likely to undergo spatial displacements perpendicular and parallel to the surface called *relaxations*. Often, there is a readjustment of individual atomic positions and a local change of the symmetry of the surface, a process called *reconstruction*. These processes may be regarded as solid-state generalizations of the bond-length adjustments and conformational changes that occur when large molecules dissociate.

A second way of forming the surface involves condensation or deposition of individual atoms from a plasma, vapor, or liquid onto a preexisting solid or microscopic cluster. The solid thereby grows in extent. The interatomic distances and symmetry of the resulting solid depend on the interplay of the various chemical forces involved.

To describe the physics of surfaces a simplifying model must be employed, and several are available. Each has its benefits, in that it allows one to obtain a simplified description for a limited set of physical phenomena. Each also has its deficiencies, in that there will be phenomena that could not be understood in terms of that model alone. Hopefully, by having a range of models and some understanding as to their applicability, one could try to get a thorough understanding of the physics and chemistry of surfaces.

At the most fundamental level, where the solid is treated as a giant molecule, the temptation is to apply the ab initio techniques of quantum chemistry directly. Consider, for example, a cluster in the shape of a simple cube with L atoms on a side, so the total number of atoms is $N = L^3$. The atoms may be classified into categories: bulk, surface, edge, and corner. There are $(L - 2)^3$ bulk atoms, $6(L - 2)^2$ true surface atoms, $12(L - 2)$ true edge atoms, and eight corner atoms on the cube. [Note that $N = [(L - 2) + 2]^3$]. For the bulk atoms to constitute a majority, as in a macroscopic sample, it would require that $L \geq 10$. This would involve a computational chemistry problem involving at least 1000 atoms, well beyond the reach of present-day computers. On the other hand, for a laboratory sample, several centimeters on a side, $L \approx 10^8$, so the fraction of surface atoms is $\approx 6 \times 10^{-8}$. Thus considerably more than 1000 atoms would be needed in a simulation to ensure that surface atoms represent an insignificant minority.

Some models often used include the following: the dielectric half-space, the Sommerfeld free-electron model with a step potential, the jellium model in conjunction with the density-functional formalism, the tight-binding model, and the molecular cluster. Some of these are employed to analyze surface phenomena.

The chapter begins by considering the ideal surface. We then proceed to study real surfaces, in which relaxation, reconstruction, or defects occur. This is followed by a study of the electronic properties of surfaces. After that, various surface modification procedures are discussed. A section is included on surface phonons. Finally, some elementary aspects of adhesion and friction and their relation to elementary excitations are discussed.

Material relating to surface states appears at our Web site[†] in Section W19.1. This is followed by a discussion of surfactants in Section W19.2. Dynamical processes on surfaces, including adsorption, desorption, diffusion, and catalysis, are covered in Sections W19.3 to W19.6. Additional material concerning friction may be found in Section W19.7.

19.2   Ideal Surfaces

Imagine slicing an infinite crystal along a lattice plane, thereby splitting the crystal into two parts. Discard all the atoms on one side of the plane, so that a semi-infinite crystal remains. Keep these remaining atoms rigidly fixed at their original positions. The exposed plane will be called an *ideal surface*. Within this solid there are an infinite number of equally spaced Bravais lattice planes parallel to the surface. Similarly, the atomic locations of the surface-plane atoms are similar to those of corresponding atoms lying on the inner parallel planes. The major difference is that the surface atoms have missing neighbors, whereas the "bulk" atoms have their full complement of neighbors. The ideal surface is a convenient point of reference for the analysis of realistic surfaces that include relaxations and/or reconstructions.

It is also possible to slice the crystal along planes that are not lattice planes. The resulting surfaces will consist of sets of terraces and steps. These are nonideal surfaces and are not considered in this section.

In the case of a crystal structure consisting of a lattice with a basis, the nature of the surface is also determined by the choice of the atoms through which the slice is taken. For example, consider the cubic CsCl crystal. A slice along a (100) plane containing Cs^+ ions will have the Cs^+ ions lying on the surface and the Cl^- ions lying half a lattice constant below the surface. The surface would be positively charged. On the

---

[†] Supplementary material for this textbook is included on the Web at the resource site (ftp://ftp.wiley.com/public/sci_tech_med/materials). Cross-references to elements of the Web material are prefixed by "W."

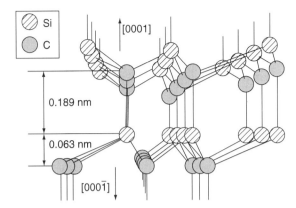

Figure 19.1. Ideal SiC surfaces terminated along the [0001] and [000$\bar{1}$] directions. [Reprinted with permission from L. Muehlhoff et al., *J. Appl. Phys.*, **60**, 2842 (1986). Copyright 1986 by the American Institute of Physics.]

other hand, a slice along a (100) plane containing Cl^- ions will produce a surface containing Cl^- ions with the Cs^+ ions lying below the surface. In such a case the surface would be negatively charged. In both cases, ideal surfaces are created, but they are different surfaces.

Another example is the family of hexagonal SiC polytypes described in Section 11.10. Here a crystal of hexagonal SiC oriented with its surface perpendicular to the [0001] direction has one Si-terminated (0001) surface with the opposite (000$\bar{1}$) surface being C-terminated (Fig. 19.1).

It is simplest to begin with a pure Bravais lattice in two dimensions, which we call a *surface net*. In three dimensions it was seen that there are 14 independent types of Bravais lattices. In two dimensions there are only five: the square lattice, the rectangular lattice, the centered-rectangular lattice, the hexagonal lattice, and the oblique lattice (Fig. 19.2).

The surface net positions are specified by two primitive lattice vectors, v_1 and v_2. Thus the surface net is defined by the lattice vectors

$$\mathbf{R}' = n_1\mathbf{v}_1 + n_2\mathbf{v}_2, \qquad (19.1)$$

where n_1 and n_2 are integers. Table 19.1 lists the five surface nets along with the definitions of the primitive lattice vectors.

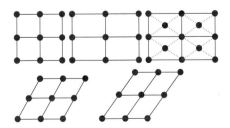

Figure 19.2. Five Bravais lattices in two dimensions.

TABLE 19.1 Surface Nets and Primitive Lattice Vectors

Surface Net	$\mathbf{v}_1$	$\mathbf{v}_2$
Square	$a\hat{i}$	$a\hat{j}$
Rectangular	$a\hat{i}$	$b\hat{j}$
Centered-rectangular	$a\hat{i}$	$(a\hat{i} + b\hat{j})/2$
Hexagonal	$a\hat{i}$	$a(\hat{i} + \hat{j}\sqrt{3})/2$
Oblique	$a\hat{i}$	$b(\hat{i}\cos\theta + \hat{j}\sin\theta)$

Technical details regarding how to construct the surface net for a given crystal structure are presented in Appendix W19A.

In the case of a crystal structure consisting of a lattice with a basis, the positions of the atoms are given by

$$\mathbf{R}'_j = n_1\mathbf{v}_1 + n_2\mathbf{v}_2 + \mathbf{s}_j, \tag{19.2}$$

where the three-dimensional vectors $\mathbf{s}_j$ specify the positions of the various atoms in the basis. These atoms may lie in the surface plane or may lie below it.

REAL SURFACES

The ideal surface is not the state of lowest free energy for the solid. If a solid were suddenly to be cleaved in half, atomic motions will result as the material searches for the state of minimum free energy. In the case of metals this will often lead to the relaxation of the distances between successive lattice planes. For semiconductors it could lead to surface reconstructions in which groups of atoms undergo some form of rearrangement. At finite temperatures surface defects will also be introduced. This section is concerned with these three aspects of realistic surfaces: relaxation, reconstruction, and defects.

19.3 Relaxation

In many metals it is observed that the distance between the surface layer of ions and its adjacent layer is smaller than in the ideal crystal. This disturbance of periodicity persists to several layers beneath the surface, often with alternating compression and rarefaction of the crystal, until the lattice spacing eventually converges to the bulk value. The phenomenon is referred to as *surface relaxation*. It is believed to be associated with the quantum-mechanical penetration of the tails of the electron wavefunctions into vacuum and the tendency for electrons to delocalize over the surface. In this section some issues related to this problem are addressed.

First, one needs to know the electron-density profile near the surface. An approximate description is given by the Sommerfeld model for a metal in conjunction with the simple one-dimensional step potential of depth V_0:

$$V(z) = \begin{cases} 0 & \text{if } z > 0, \\ -V_0 & \text{if } z < 0. \end{cases} \tag{19.3}$$

This step represents, in a crude way, the confining effect of the ions of the solid. The Schrödinger equation,

$$-\frac{\hbar^2}{2m}\nabla^2\psi + V(z)\psi = E\psi, \tag{19.4}$$

simplifies to a one-dimensional form with the substitution

$$\psi(\mathbf{r}_\parallel, z) = \frac{\phi(z)}{\sqrt{A}} \exp(i\mathbf{k}_\parallel \cdot \mathbf{r}_\parallel), \tag{19.5}$$

where A is the surface area. Thus

$$\left[\frac{p_z^2}{2m} + V(z) - \epsilon\right]\phi(z) = 0, \tag{19.6}$$

where

$$\epsilon = E - \frac{\hbar^2 k_\parallel}{2m}. \tag{19.7}$$

The solution is

$$\phi(z) = \begin{cases} B\exp(-\kappa z) & \text{if } z > 0, \\ \dfrac{B\sin(qz + \delta)}{\sin\delta} & \text{if } z < 0, \end{cases} \tag{19.8}$$

where

$$\kappa = \frac{1}{\hbar}\sqrt{-2m\epsilon}, \tag{19.9a}$$

$$q = \frac{1}{\hbar}\sqrt{2m(V_0 + \epsilon)}. \tag{19.9b}$$

The wavefunction is continuous at $z = 0$. Continuity of the first derivative requires that

$$\cot\delta = -\frac{\kappa}{q}. \tag{19.10}$$

The wavefunction will be normalized to unity provided that

$$B = \sqrt{\frac{2\sin^2\delta}{L}}, \tag{19.11}$$

where L is the thickness of the crystal sample.

In the Sommerfeld model the electron density is obtained by summing the absolute squares of the wavefunctions for the various states, weighted by the Fermi–Dirac distribution function

$$n(z) = \sum_s \sum_{k\parallel, q} |\psi(\mathbf{r})|^2 f(E, T), \tag{19.12}$$

where s is the spin projection. As noted in Section 7.4, in the limit of zero temperature the Fermi factor reduces to

$$f(E, T) \longrightarrow \Theta(-W - E), \tag{19.13}$$

where W is the work function [i.e., the energy needed to raise an electron from the Fermi level to the vacuum level ($E = 0$)]. In Eq. (19.13), $\Theta(z)$ is the unit step function (1 for $z > 0$ and 0 for $z < 0$). Those states below the Fermi level are filled ($f = 1$) and those above it are empty ($f = 0$). Thus

$$n(z) = 2 \int \frac{d^2 k_\| A}{(2\pi)^2} \int_0^\infty \frac{dkL}{\pi} f(E, O)$$

$$\times \left[\frac{2}{AL} \sin^2(k_2 z + \delta)\Theta(-z) + \frac{2}{AL} \sin^2 \delta \exp(-2\kappa z)\Theta(z) \right]. \tag{19.14}$$

The transverse wave-vector integral is readily performed:

$$\int d^2 k_\| \, \Theta \left[-W + V_0 - \frac{\hbar^2}{2m}(k_\|^2 + k_z^2) \right] = \frac{2\pi m}{\hbar^2} \left(E_F - \frac{\hbar^2 k_z^2}{2m} \right) \Theta \left(E_F - \frac{\hbar^2 k_z^2}{2m} \right), \tag{19.15}$$

where $E_F = V_0 - W$. Letting

$$k_F = \frac{1}{\hbar} \sqrt{2m E_F}, \tag{19.16}$$

the density finally becomes

$$n(z) = \frac{1}{\pi^2} \int_0^{k_F} dk_z (k_F^2 - k_z^2)[\sin^2(k_z z + \delta)\Theta(-z) + \sin^2 \delta \exp(-2\kappa z)\Theta(z)]. \tag{19.17}$$

This integral may be evaluated numerically. Results for aluminum ($W = 4.25$ eV, $V_0 = 15.9$ eV) are given in Fig. 19.3, in which the electron density, n (in nm^{-3}), is plotted as a function of z (in nm). The value of W may be obtained from photoemission data.

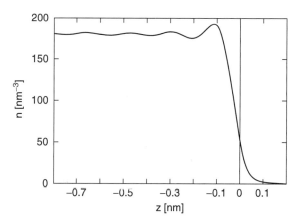

Figure 19.3. Electron density versus position for surface of Al. Theoretical calculation using Eq. (19.17).

There are three features worthy of note. First, the electron density approaches that of the uniform-background positive-charge density deep in the solid. Second, it decays exponentially into the vacuum region ($z > 0$), consistent with the quantum-mechanical behavior of barrier penetration. Finally, there are characteristic oscillations, called *Friedel oscillations*, near the surface. Their wavelength is π/k_F, where k_F is the Fermi wave vector.

It is an oversimplification to assume that the background potential depends only on z and not on the transverse (x,y) coordinates. In reality, the potential is produced by ions and their associated core electrons and has strong x and y variations along the surface. Barrier penetration therefore occurs not only in the z direction but also in the x and y directions, with electrons spilling into the void spaces between surface ions. The net result is that the electron cloud around the ions is spread out and is not fully effective in screening the ions. Strong electric fields develop in the surface region, with a corresponding increase in electrostatic energy. This energy would be minimized if the ions were to withdraw into the solid (i.e., if surface relaxation were to occur). Of course, there are competing effects that limit the extent of this withdrawal, including the repulsion of other ions.

It is found that among the low-index planes, the low-density faces of a free-electron metal experience the greatest relaxations. For example, the number of atoms per unit area, N, is related to the lattice constant by the formula $N = f/a^2$, where f is a face-dependent number. For the FCC crystal structure $f = 1.41$, 2, and 2.31 for the (110), (100), and (111) faces, respectively. For Al, an FCC metal, the corresponding relative relaxations, $\Delta a/a$ are -8%, 0%, and 1%, respectively. For Al(111), $\Delta a/a > 0$ indicates that some relaxations occur outward rather than inward. For Cu they are -9%, -1%, and -1%. For the BCC metals the numbers are $f = 1.0$, 1.0, and 3.08 for the (100), (110), and (111) faces. For example, for Mo (hardly a free-electron metal), $\Delta a/a = -10\%$, -2%, and ?. This behavior is probably fortuitous. Things are more complicated when the highly directional d-band electrons become involved. For example, in Fe the ratios are $\Delta a/a = -2\%$, -1%, and -17%, opposite to the previous trends.

19.4 Reconstruction

The ideal surface formed by slicing a solid is often not the surface found in nature. The reason is that such a slice may leave the solid in a configuration with an elevated free energy. If the system is allowed to relax to its minimum free energy state, conformational changes often ensue and a new structure emerges. These changes could involve the migration of surface atoms to new positions or even desorption of some atoms. It is analogous to molecular chemistry, where the fragments often undergo conformational changes to seek the lowest-energy state when a molecule is split.

Examples of reconstruction are found most readily when working with periodic solids. A transition from one ordered state to another is easy to identify, especially when the surface symmetry is changed. However, it is also possible to have reconstruction of amorphous materials, polymers, glasses, and so on. If the average geometric arrangement on the real surface differs from what would have been obtained by taking a slice through the material, reconstruction has occurred.

In the absence of a detailed solution to the Schrödinger equation, it is difficult to point precisely to the mechanism responsible for reconstruction. Three contributions to the surface free energy can be identified, however. One involves the energy associated with dangling bonds remaining after slicing the solid. The second involves the electrostatic energy stored in the material. The third is the elastic energy of the solid. It is the competition between these energies and the search for the state of lowest free energy that determines the eventual structure of the surface. (Metastable states may also be reached by starting with specific initial conditions.)

Consider, for example, the (001) face of Si. The ideal surface is illustrated in Fig. 19.4a and the 2 × 1 reconstructed surface is given in Fig. 19.4b. The large circles depict Si atoms in the surface plane, and the small circles correspond to Si atoms in the closest plane below the surface. Bulk Si has the diamond structure, with each Si atom coordinated to four nearest neighbors. In slicing through a (001) plane the surface atom loses two of its bonding partners, and two dangling bonds per Si atom result. In the reconstruction, neighboring surface atoms approach each other (dimerize) and the size of the unit cell is doubled. The surface Si–Si bond distance is lowered to the point where the incremental gain in binding energy is balanced by the incremental gain in elastic strain energy. After dimerization there is only one dangling bond per surface atom. The dangling bond energy is thereby lowered. There is evidence that under certain surface preparation conditions buckling occurs and the altitudes of the pair of dimerized Si atoms are not equal.

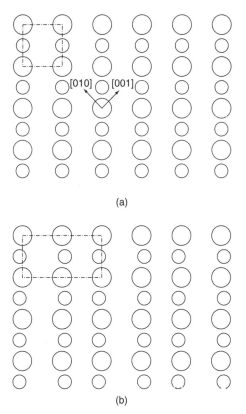

Figure 19.4. (a) Ideal Si (001) surface with a 1 × 1 unit cell; (b) reconstructed Si (001) surface with a 2 × 1 unit cell.

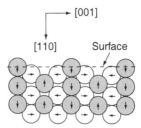

Figure 19.5. Missing row reconstruction for the Pt(110)-(1 × 2) surface. [From P. Fery et al., *Phys. Rev. B*, **38**, 7275 (1988). Copyright 1988 by the American Physical Society.]

The reconstructions are not always so simple. In Fig. 19.5 the profile of a reconstructed Pt (110)-(1 × 2) surface is given. Bulk Pt has an FCC crystal structure. The surface is at the top of the figure. The shaded circles lie in the viewing plane, the unshaded circles beneath them. Every other row of Pt atoms is missing on the surface, making the surface consist of (111) microfacets. The region near the surface also shows distortions of the interatomic distances from what they would be in the bulk of the crystal. The top-layer spacing is decreased by 20% from its bulk value. There is a buckling of 0.017 nm in the third layer and a 0.004-nm lateral shift in the second layer. The arrows show the directions of the atomic displacements. Based on density-functional calculations it has been argued that transfer of electron density from the *d* bands to the *sp* bands near the surface of Pt can account for the surface reconstruction.

As another complex example, consider the case of the (100) face of InSb. The bulk crystal has the zincblende structure [two interpenetrating FCC atomic lattices (one In, the other Sb) displaced from each other along the main diagonal of the cube by one-fourth of the main diagonal distance]. The structure one obtains by taking a (100) slice through the crystal is shown in Fig. 19.6a. The indium atoms are depicted by open circles. They lie behind the plane of the antimony atoms, represented by the dark circles.

Scanning tunneling microscopy has revealed a centered 4 × 4 structure for the surface above in which there is an additional overlayer of Sb (darkened circles), as in Fig. 19.6b. The Sb atoms in the overlayer dimerize and every fourth dimer is missing. Obviously, the In atoms from the uppermost layer had to have been expelled in order to have this Sb-rich overlayer.

19.5 Surface Defects

An ideal surface is formed by slicing a solid through a lattice plane. If no reconstruction takes place, the surface can be expected to be atomically smooth at $T = 0$ K. At finite temperatures, however, atoms may be thermally excited and moved from their equilibrium configuration to occupy metastable positions. This is illustrated in Fig. 19.7, in which there appear an adsorbed atom, *a*, a diffusing adsorbed atom, *b*, and a vacancy, *v*, on the surface.

A stepped surface may be generated either as a form of reconstruction of a sliced surface or as a result of the crystal growth process. On such a surface there are terraces, *T* (which are essentially atomically smooth), separated from each other by ledges (steps), *L*, as shown in the figure. A defect in the ledge appears as a kink, *K*. Figure 19.7 illustrates what is called the *TLK model of a surface*. In addition,

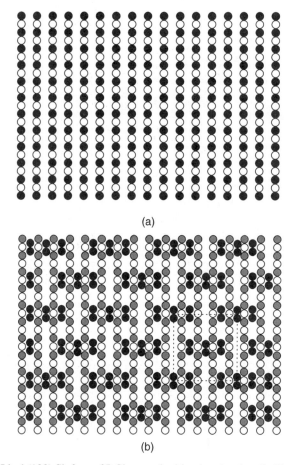

(a)

(b)

Figure 19.6. (*a*) Ideal (100) Sb face of InSb crystal with a 1×1 unit cell; (*b*) reconstructed InSb (100) surface. The dotted lines surround a 4×4 unit cell. [Reprinted from C. F. McConville et al., *Surf. Sci.*, **303**, copyright 1994, with permission from Elsevier Science.]

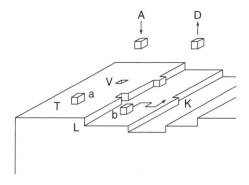

Figure 19.7. Features of the terrace–ledge–kink model of a real surface.

the figure illustrates the possibility of adsorption (A) and desorption (D) processes occurring.

Other types of defects are possible. There can be line defects in which a crack runs partially through the solid. The crack ends on a line penetrating the solid. The crack itself appears on the surface as a line defect that runs along and terminates abruptly. On either side of the crack the atoms are displaced sideways. The region around the end of the crack is marked by atoms with appreciable displacements. These become smaller as one moves farther away from the crack into the unbroken solid.

Another type of defect is a screw dislocation that has emerged through the surface. As one encircles the axis of the screw dislocation, one climbs to a higher or lower terrace, depending on the helicity of the dislocation. One then climbs down a ledge to return to the original elevation when the circuit is completed. Edge dislocations can terminate at the surface, as can grain boundaries in polycrystals.

How is a surface with defects described? At low temperatures there is a mean elevation and orientation to the surface even though there may be local fluctuations caused by the defects above. The mean elevation and orientation remain the same for small or large sample areas. This is illustrated in Fig. 19.8 by the sketch labeled S, for "smooth."

At high temperatures, on the other hand, the fluctuations in elevation are such that there is no defined mean elevation or orientation. As the area size is increased, the mean elevation and orientation no longer converge on fixed values. Such a surface is said to be rough. It is labeled R in Fig. 19.8. The temperature at which the surface becomes rough is denoted by T_R and the change is called the *roughening transition temperature*.

The roughening transition arises due to the competition between the energy and entropy contributions to the free energy of the surface. Creating steps costs energy. Figure 19.9 illustrates a top view of a stepped surface, denoting the elevations of the terraces by positive or negative integers. If one assigns an energy per unit length, u, to the formation of a step, a step of length L will have energy $U = uL$. The fact that the figure shows disjoint steps is really irrelevant. One could as well have been talking about one giant terrace of length L.

How many different ways can a terrace of perimeter L be constructed? Let w denote the number of ways one can turn after taking an elementary step of size l along the surface. Then at each of the $N = L/l$ steps around the loop there are w choices for the direction to turn. There are $W = w^{L/l}$ possible paths to draw. Although it is true that most of these will not be closed, they can be expected to cross each other many times, and therefore many terraces will be created. The entropy associated with these

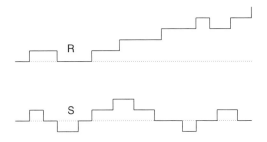

Figure 19.8. Side view of a smooth (S) and a rough (R) surface.

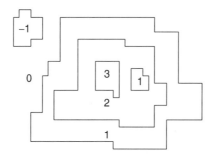

Figure 19.9. Top view of a terraced surface.

choices is $S = k_B \ln(W) = k_B(L/l)\ln(w)$. Thus the Helmholtz free energy is given approximately by

$$F = U - TS = L\left[u - \frac{k_BT}{l}\ln(w)\right] = a(T_c - T), \qquad (19.18)$$

where the critical temperature is given by $T_c = ul/k_B \ln(w)$ and the constant of proportionality is $a = k_BL \ln(w)/l$. For $T < T_c$, F will grow with increasing length and $L = 0$ will be favored. The thermodynamic equilibrium state favors no steps and the lowest-energy surface will be smooth. Only individual adatoms and vacancies can be expected. On the other hand, for $T > T_c$, the entropy term in the free energy will dominate and F will decrease with increasing L. This will favor infinite L. In other words, the surface will become rough.

At still higher temperatures the surface atoms become mobile and surface melting can occur. Since surface atoms are less tightly bound to the solid than are bulk atoms, this temperature is considerably less than the temperature for bulk melting. Of course, as the temperature of the solid increases, so does the probability of desorption. At sufficiently high temperatures the solid will eventually vaporize, as described in Chapter 6. (Recall that not all solids pass through the molten stage before vaporizing. *Sublimation* is the name given to the direct transition from the solid to the vapor state.)

ELECTRONIC PROPERTIES OF SURFACES

Having discussed the geometric structure of surfaces in previous sections, attention is now directed toward their electronic properties. In the following sections the work function is defined and two methods for determining it are discussed: thermionic emission and photoemission. The closely related phenomenon of field emission is studied. The properties of surface states are also described.

19.6 Work Function

In atomic physics, *ionization energy* has a well-defined meaning — it is the minimum energy that must be supplied to an atom to remove an electron to infinity. As the electron is drawn away from the ion there is work that must be done to overcome the Coulomb force. However, the arrangement of the electrons in the resulting ion is not the same as it was originally. The atoms will rearrange themselves and additional work (positive or negative) may have to be supplied by the external force to effect this rearrangement.

Contrast this with the corresponding process in the case of a solid. In place of the ionization energy, there is the work function, W. Often, the work function potential, $\Phi = W/e$ is used. The *work function* could be defined as the energy needed to remove the electron to infinity. However, a solid is often surrounded by electrostatic fields. The work function would then depend on the effect of these fields and would not be characteristic of the solid. A more useful definition is to use the energy needed to remove an electron from the solid to a small distance away from the surface. By *small* one means large compared to a Bohr radius, yet small compared with the macroscopic size of the sample. This definition makes the work function characteristic of the particular solid as well as the particular face of the solid.

As in the atomic case, the energy of an electron is determined by the kinetic energy and the Coulomb interaction with the ions and other electrons. Suppose that an electron is removed slowly by means of some external force. One may identify some specific physical interactions. First, there is the penetration of the electron wavefunctions into the vacuum. Tails of the wavefunctions describing the occupied levels penetrate some distance outside the solid. The net effect of this vacuum penetration is to create an electric dipole layer at the surface. An electron that is removed from the interior of the solid must cross this layer. The dipole is oriented so that its direction is from the vacuum toward the solid (i.e., the vacuum is at a negative potential relative to the solid). A surface barrier is set up by this potential. This is an important contribution to the work function.

Having crossed the surface barrier the electron must still contend with the image force. For the case of a perfect conducting metal the interaction with the image charge is given by the formula

$$U(z) = -\frac{e^2}{16\pi\epsilon_0 z}, \tag{19.19}$$

where z is the distance to an effective plane defining the location of the surface. This formula is obtained by computing the work needed to move a charge $-e$ from z to infinity and noting that the distance appearing in the Coulomb force law is $2z$. For the case of a semiconductor or insulator, formula (19.19) may readily be modified to include the electric permittivity of the material. Note, however, something that is different here from the case of ionization of an atom. As the electron is pulled away from the solid, the positive image charge recedes deeper into the solid. This is equivalent to saying that another electron is being taken from the depths of the solid to replace the one that was removed.

Another part of the work function has to do with the readjustment of the other electrons of the solid. It is often referred to as the exchange and correlation energy, introduced in Chapter 12. A detailed calculation of the work function for various solids has been one of the goals of surface physics.

Different faces of a solid often have different work functions. The different atomic packing and bond orientations on the various surfaces permit the electron tails to extend to a greater or lesser extent and also influence the exchange and correlation contributions. How does one reconcile this with the fact that the potential an infinite

TABLE 19.2 Work Function and First Ionization Energy for Various Metals and Different Crystal Faces

Material	Face	W (eV)	IE(1) (eV)
Li	Polycrystal	2.90	5.39
Na	Polycrystal	2.75	5.14
K	Polycrystal	2.30	4.34
Rb	Polycrystal	2.16	4.18
Cs	Polycrystal	2.14	3.89
Al	(100)	4.41	5.99
	(110)	4.06	5.99
	(111)	4.24	5.99
Ag	(100)	4.64	7.58
	(110)	4.52	7.58
	(111)	4.74	7.58
Cu	(100)	4.59	7.48
	(110)	4.48	7.48
	(111)	4.98	7.48
Au	(100)	5.47	9.23
	(110)	5.37	9.23
	(111)	5.31	9.23
W	(100)	4.63	7.98
	(110)	5.25	7.98
	(111)	4.47	7.98
Ni	(100)	5.22	7.64
	(110)	5.04	7.64
	(111)	5.35	7.64
Mo	(100)	4.53	7.10
	(110)	4.95	7.10
	(111)	4.55	7.10

Source: Data from H. L. Anderson, ed., *Physics Vade Mecum*, American Institute of Physics, New York, 1981.

distance removed from the sample should be constant (e.g., zero)? The answer is that there are electrostatic fields surrounding the solid connecting different faces.

In Table 19.2 some representative work functions for various metals are presented. The work functions for semiconductors and insulators are much harder to measure. One reason is that these solids are easy to charge up and difficult to discharge. A second reason is that their work function (or, more generally, chemical potential) is a sensitive function of dopant impurity-atom concentrations. The first ionization energies IE(1) of the atoms are also listed in Table 19.2. Comparing their values with the corresponding work functions show consistently that IE(1) > W. This is due to the screening effect of the electrons from neighboring atoms in metals which cut off the Coulomb potential as one proceeds away from the metallic ion.

There are three principal ways of measuring the work function. One could heat the material to boil the electrons off, a process called *thermionic emission*. One could shine light on the material and monitor the ejected electrons. This is *photoemission*. One could apply an external field in such a way as to lower the potential energy in vacuum below the Fermi energy and have the electron tunnel through the resulting barrier. This process is called *field emission*. These processes are described briefly in the following sections.

19.7 Thermionic Emission

In this section the Richardson–Dushman formula describing the thermally generated current emitted from a metal at finite temperature is derived. At absolute zero there would be no thermionic emission. The electron sea would be filled to the Fermi level which lies below the vacuum level by the work function, W. At finite temperatures, however, there is a thermal distribution of electron energies characterized by the Fermi–Dirac distribution function (see Appendix WB), $f(E_k, T) = (\exp[\beta(E_k - \mu)] + 1)^{-1}$.

When electrons emerge from the surface their energies are usually measured relative to the vacuum level, so the zero of E_k will be taken here as lying at the vacuum level. The inverse temperature, β, is defined through the formula $\beta = 1/k_B T$. The energy level corresponding to the chemical potential, μ, lies approximately at $E = -W$, to order $(k_B T/E_F)^2$. Since there is a finite thermal occupancy for positive energy states, some electrons will leave the metal. These electrons constitute the thermionically emitted current. One may make the approximation, valid for $\beta(E_k - \mu) \gg 1$, that

$$f(E_\mathbf{k}, T) = \exp(-\beta W)\exp(-\beta E_\mathbf{k}). \tag{19.20}$$

The electric current density normal to the surface is given by

$$J_z = -e \sum_s \sum_\mathbf{k} f(E_\mathbf{k}, T)v_z|\mathbf{T}|^2\Theta(v_z), \tag{19.21}$$

where e is the magnitude of the electron charge, v_z is the component of the velocity normal to the surface, and $\mathbf{T}$ denotes the transmission amplitude through the surface barrier. The sum is over spin states s and wave vectors $\mathbf{k}$ describing motion within the solid. For nonmagnetic metals the spin sum just leads to a factor of 2. The Θ function ensures that there is only an outward current. In Fig. 19.10 a sketch is made of the surface barrier and the Fermi function. Note that the width of the step region for the

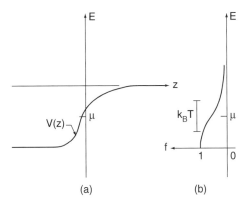

Figure 19.10. (*a*) Surface barrier as a function of distance; (*b*) Fermi function as a function of energy.

Fermi function is determined by the temperature. For low temperatures the step is very sharp. For high temperatures it is gradual. Naturally, at high temperatures more of the electron sea has sufficient energy to leave the solid. The assumption will be made here that the step potential is sufficiently gradual to warrant the replacement of $|T|$ by 1.

The current density may then be written as

$$J_z(T) \approx -e \exp(-\beta W) \int \rho(E) \exp(-\beta E) \langle v_z \Theta(v_z) \rangle \, dE. \tag{19.22}$$

The density of states $\rho(E)$ has been introduced according to Eq. (7.67). An expression for the free-electron density of states is given in Eq. (7.70):

$$\rho(E) = \frac{1}{2\pi^2} \left(\frac{2m}{\hbar^2} \right)^{3/2} \sqrt{E}. \tag{19.23}$$

One may write $v_z = v \cos\theta$ and average over all angles using

$$\left\langle \cos\theta \cdot \Theta\left(\frac{\pi}{2} - \theta\right) \right\rangle = \frac{\int_0^{\pi/2} d\theta \sin\theta \cos\theta}{\int_0^\pi d\theta \sin\theta} = \frac{1}{4}. \tag{19.24}$$

This leads to the expression

$$J_z(T) = -\frac{em}{2\pi^2\hbar^3} \exp(-\beta W) \int_0^\infty E \exp(-\beta E) \, dE, \tag{19.25}$$

and finally to the *Richardson–Dushman formula*:

$$J_z(T) = -\frac{emk_B^2}{2\pi^2\hbar^3} T^2 \exp\left(-\frac{W}{k_B T}\right). \tag{19.26}$$

If one takes into account the reflection coefficient, r, for electrons impinging on the surface in the absence of an applied electric field, one may write this formula in the slightly more general form

$$J_z(T) = -R(1 - r)T^2 \exp\left(-\frac{W}{K_B T}\right), \tag{19.26a}$$

where $R = 1.2 \times 10^6 \text{A m}^{-2} K^{-2}$. The formula shows that the thermionic current depends essentially on two quantities, temperature and work function. For there to be an appreciable current the work function should be small and the temperature should be high. The work function of a pure material is fixed by the chemical composition and the surface plane that is exposed, as shown in Table 19.2.

By putting overlayers of low-W alkali metals on the surface of a metal with a high work function value it is possible to lower the work function appreciably. The lowering comes about due largely to a transfer of charge between the alkali adsorbate and the underlying substrate. From a macroscopic point of view this charge transfer is needed to equilibrate the Fermi levels of the two metals. An internal electric dipole is developed due to the charge transfer and the potential across the charge-transfer region accounts for the lowering of the work function. For example, depositing a monolayer of Li on W(111) reduces the work function by 1.73 eV, from 4.47 eV to 2.74 eV. The value for W predicted from Table 19.2 based on the work function of Li is 2.90 eV.

Cesium is often used as the alkali metal of choice to lower work functions. For example, by depositing an areal density of 4×10^{18} m^{-2} Cs atoms on W(111) it is possible to reduce the work function to about 1.5 eV. This corresponds to less than a monolayer coverage and is actually lower than the value expected for bulk coverage (2.14 eV). At this coverage the dipoles formed by the transfer of electrons from the individual Cs atoms to the W are strong. As the coverage is increased, the surface dipoles experience the electric fields due to neighboring dipoles. The direction of these electric fields is such as to reduce the dipole moments.

Cesium adsorption on semiconductor surfaces can have even more dramatic effects. By depositing Cs on GaAs(110) the work function can be reduced almost to zero. By using Cs_2O rather than Cs, the work function can actually be made negative (i.e., to lie below the top of the valence band).

It is also possible to increase the work function of a metal by using an oxidizing adsorbate. An example is when a monolayer of Cl adsorbs on the (111) surface of Cu and increases W by more than 1 eV.

19.8 Field Emission

Field emission results when a static electric field, E_0, is established outside a metal surface and results in a large enough potential drop that electrons are able to tunnel from the metal into the vacuum. The geometry is illustrated in Fig. 19.11. The work function is denoted by W, the Fermi energy by E_F and the depth of the step-potential well by $V_0 = E_F + W$. The current per unit area is given by the *Fowler–Nordheim formula*:

$$J = 6.15 \times 10^{-6} \frac{E_0^2}{E_F + W} \sqrt{\frac{E_F}{W}} \exp\left(-\frac{6.83 \times 10^9 W^{3/2}}{E_0}\right), \qquad (19.27)$$

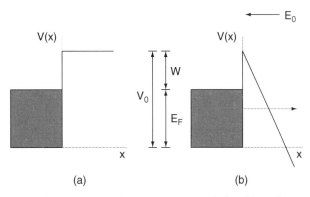

Figure 19.11. The potential step at a surface (*a*) without and (*b*) with an imposed electric field. The symbols are defined in the text.

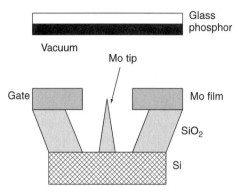

Figure 19.12. Thin-film field-emission cathode. [Reprinted from C. A. Spindt, *J. Appl. Phys.*, **39**, 3504 (1968). Copyright 1968 by the American Institute of Physics.]

where J is in A/m^2, E_F and W are in eV, and E_0 is in V/m. The derivation of this formula is presented in Appendix W19B. Note that this current is independent of the temperature, T.

Flat-panel displays have been designed based on field emission. Electrons tunnel from an array of cathodes past controlling gates toward an anode and impinge on a phosphor-coated glass screen. To obtain high electric fields the cathodes are fashioned into identical conical shapes. The geometry of a single cathode is illustrated in Fig. 19.12. The local field in the neighborhood of a cone tip is enhanced by the sharpness of the tip, in much the same way that a lightning rod is able to create a strong electric field near its point. In actuality the tips of the cones have a radius of curvature of several tens of nanometers. The net result is that the electric field E_0 in the vicinity of the tips is typically two orders of magnitude larger than the imposed electric field (which is in the range 10^7 to 10^8 V/m). In this case the field-enhancement factor $\beta = E_0/E_{\text{applied}}$ is used in the Fowler–Nordheim formula. The glass faceplate often has a transparent conducting layer of indium tin oxide [In$_x$Sn$_y$O$_2$ (ITO)] so it can function as the anode.

The emitting tip is commonly made of Mo, Si, or W, grown on a Si substrate. Typical phosphors include ZnS (blue), Cd-doped ZnS (red), ZnO (green), Y$_2$SiO$_5$, and Gd$_2$O$_2$S. The vacuum must be such that the accelerated electrons are able to strike the phosphor without colliding with intervening gas molecules. Typically, a vacuum of 10^{-7} torr is established between the cathode and the anode.

It has been suggested in the literature that diamond be used as the emitting tip since some of its faces [e.g., the (111) face] possess a negative electron affinity (NEA), meaning that the bottom of the conduction band actually lies above the vacuum level. By doping the diamond n-type one is able to put some electrons into the conduction band. Due to the NEA these electrons may readily leave the diamond in response to an applied electric field without the need for tunneling. Nitrogen doping (with $N_d \approx 2 \times 10^{25}$ m^{-3}) or hydrogenation is often used to create the NEA material. The existence of NEA materials means that a cold cathode can be used because thermionic emission is not necessary.

More recently it was found that the polymer regioregular poly(3-octylthiophene) has an extremely low threshold field of only 2×10^6 V/m for producing field emission. It has been suggested that surface states may play an important role in the field-emission process in this material.

19.9 Photoemission

In the photoelectric effect, incident light of frequency ω liberates an electron when the photon energy $\hbar\omega$ exceeds the work function of the material. Electrons with a broad distribution of energies can be emitted, but the maximum kinetic energy of the energy distribution curve (EDC) is given by the simple formula

$$\left[\tfrac{1}{2}mv^2\right]_{\max} = \hbar\omega - W. \tag{19.28}$$

A graph of the maximum kinetic energy versus the photon energy allows one to determine the work function from the intercept.

A much more difficult problem is to predict the photoelectric yield and the energy distribution curve of the electrons emitted. The yield is the ratio of the number of electrons emitted to the number of incident photons. The problem is complicated by the fact that there are both a surface and a volume contribution to the effect. In its simplest description, the volume effect is due to the photoionization of the individual bulk ions, appropriately corrected for electron overlap, and the surface effect is due to the shape of the confining potential of the solid. In this section the primary focus is on the surface photoelectric effect and we show how an understanding of it may be developed based on elementary quantum mechanics and electrodynamics. An additional complication stems from the fact that the amplitude of the electromagnetic wave in the solid is not constant but is attenuated as it penetrates the metal. This is not considered here.

A first step in this direction would be to invoke the Fermi golden rule (see Appendix WC) for the rate of absorption of light:

$$\Gamma(\omega) = \frac{2\pi}{\hbar} \sum_f \sum_i |\langle \psi_f | \boldsymbol{\mu} \cdot \mathbf{E} | \psi_i \rangle|^2 \delta(E_f - E_i - \hbar\omega). \tag{19.29}$$

Here $\boldsymbol{\mu}$ is the electric-dipole operator of the solid and $\mathbf{E}$ is the incident electric field. The final-state energy E_f is elevated above the incident energy E_i by the photon energy. This expression clearly is inadequate. First, not every excited electron has enough energy to leave the solid. Only if the electron can leave with a positive component of velocity in vacuum normal to the surface is photoemission possible. The factor $\Theta(v_z)$ can be added to enforce this condition. Second, when the electron is excited it must be within an inelastic collision mean free path, λ_e, of the surface or its energy will be seriously degraded and it may not emerge. This could be introduced empirically by including a survival factor $\exp(z/\lambda_e)$ in the evaluation of the matrix element, it being assumed that the solid occupies the half-space $z < 0$. Finally, it must be noted that the electric field itself has a normal component that is strongly spatially dependent near the surface. Details of deriving an expression for the photoelectric yield are given in Appendix W19C.

Experimental results are given in Fig. 19.13 for the case of Al. One notes a rise in the photo yield with increasing energy followed by a precipitous drop at high energy, corresponding to electrons emerging from the Fermi surface. There is evidence for band-structure features in the experimental data.

The EDCs give detailed information concerning the density of states of the electrons near the surface of a metal. The density of states is sensitive to the bonding

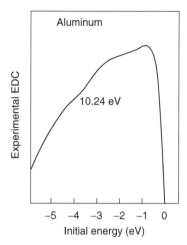

Figure 19.13. Experimental differential yield of emitted electrons for Al. [From R. H. Koyama et al., *Phys. Rev. B*, **2**, 3049 (1970). Copyright 1970 by the American Physical Society.]

state of the surface atoms. For example, large shifts in the density of states occur if atoms are chemisorbed onto the surface, and these can be observed in the EDCs using photoemission. Even more detailed information may be extracted from angular-resolved photoemission spectroscopy (ARPES), where the angular distribution of the emitted electrons is monitored. In this case there is directionality associated with the matrix elements appearing in Eq. (19.29). If x-rays are employed, the state ψ_f is essentially that of a plane wave, since the electrons will be photoemitted with high energies. The matrix elements reduce to Fourier transforms of the solid-state wavefunctions (multiplied by the dipole operator, which is not highly directional). Each orbital has its own distinct energy and distinct directionality properties, and these may be observed using ARPES.

19.10 Surface States

In a bulk crystal the electron energy eigenvalues lie in allowed bands, which are often separated by forbidden gaps. The eigenvalues are labeled by the crystal momentum, **k**, and a discrete band index, n. As **k** varies through the Brillouin zone the energy eigenvalues $E_n(\mathbf{k})$ map out a surface in the four-dimensional space defined by the three components of **k** and energy. In a crystal bounded by a surface the spectrum is modified. In addition to bulk bands there are surface states. They also form bands but in the two-dimensional space spanned by the components of the wave vector parallel to the surface, $\mathbf{k}_\parallel$.

A bulk state is delocalized over the volume of the solid. The probability of finding an electron within a distance dz of the surface would be on the order dz/L, where L is the thickness of the sample. Surface states, on the other hand, are localized at the surface. They usually decay exponentially into the bulk.

To see how surface states come about, it is useful to review the derivation of Bloch's theorem given in Section 7.6. From Eq. (7.40) it was found that for a periodic solid $\psi(\mathbf{r} + \mathbf{R}) = \tau_\mathbf{R} \psi(\mathbf{r})$, and it was argued that in the bulk of the solid the factor $\tau_\mathbf{R}$ must have magnitude 1, for otherwise the wavefunction would not be normalizable. This led to the concept of allowed bands and forbidden bands. At the surface, however, this argument breaks down. Periodicity perpendicular to the surface is broken and waves that decay exponentially into the solid will still be normalizable since the wavefunction will be attenuated in vacuum. Such solutions will occur in the bandgaps. In the bandgaps only states with complex k exist.

Since the electrons are confined to the interior of the solid, they may be regarded as being in bound states. These states decay exponentially into the vacuum region surrounding the solid, as for example, in Eq. (19.8). Suppose that an energy for an electron in the solid is selected to lie somewhere in a forbidden gap. Furthermore, assume that this particular state decays in magnitude as one goes deeper into the crystal. If this wavefunction can be smoothly matched onto the function describing the decay into vacuum, one has identified a surface state. Such a state is called a *Shockley state*.

Another possibility for a surface state occurs when the potential well near the surface is more attractive than elsewhere in the bulk. This can come about, for example, when there is a dangling bond at the surface. The affinity that this bond has for electrons can be modeled as if there were some added potential well present at the surface. If this brings about an additional bound state, it is sometimes referred to as a *Tamm state*. For example, a Tamm state could be produced by a broken bond at the surface of a semiconductor. Additional details may be found in Section W19.1.

Occupied surface states may be detected in angular-resolved photoelectron spectroscopy (ARPES) experiments. The direction of the emitted electron is determined by the wave vector of the electron in its initial state. The energy of the electron shows up as a peak in the energy distribution curve at some energy below the Fermi surface edge. It is also possible to perform inverse photoemission experiments to probe the unoccupied surface states. In these experiments electrons are directed at the surface and the light emitted is monitored.

Since surface states are localized at the surface they play an important role in bonding to adsorbed molecules. The orbitals of the adsorbed species can hybridize with these states to form bonding and antibonding states, just as in the case of two atoms coming together to form a molecule. Because of the surface states, the solid therefore behaves as a chemically active species!

SURFACE MODIFICATION

Having been exposed both to ideal and real surfaces in previous sections, the discussion now proceeds to some common methods for modifying surfaces. Two methods are considered here: anodization and passivation. In Section W19.2 a third method is considered: modification through the use of surface-active agents (surfactants). These techniques add a dimension to the usefulness of surfaces by allowing one to customize their properties. The discussion involves electrochemistry and is related to that given in Section W12.4.

19.11 Anodization

Anodization is the process of modifying the surface of a conducting solid, S, by means of an electrolytic reaction. Typically, the solid will be in contact with a liquid (e.g., water) with ions A^{n-} (of charge $-ne$) and B^{n+} (of charge $+ne$) in solution. As an electrical current is passed through the electrolytic cell, the oxidation half-reaction

$$S + A^{n-} \longrightarrow SA + ne^- \tag{19.30}$$

occurs at the anode, resulting in the buildup of an electrodeposited film at the anode. Electrons flow from the electrolyte into the anode, so the conventional current flows in the opposite direction. At the cathode, S, the complementary reduction half-reaction

$$S' + B^{n+} + ne^- \longrightarrow S'B \tag{19.31}$$

occurs with electrons flowing from the cathode into the electrolyte. It is also possible for the solid to react directly with the water at the anode according to the formula

$$S_n + mH_2O \longrightarrow S_nO_m + 2mH^+ + 2me^-. \tag{19.32}$$

At the cathode the corresponding reaction is

$$S' + 2me^- + 2mH_2O \longrightarrow S' + 2mOH^- + mH_2. \tag{19.33}$$

The net reaction in this case is

$$S_n + mH_2O \longrightarrow S_nO_m + mH_2. \tag{19.34}$$

If the deposited solids are insulating, the situation becomes complex, due to the need for the ions to diffuse through the insulating layers to reach the solution. This tends to limit the growth of the electrodeposited layers to thicknesses of about 1 μm. An example of anodization is provided by the deposition of a protective oxide layer on the surface of aluminum in water, in which case $2Al + 3H_2O \rightarrow Al_2O_3 + 3H_2$.

Depending on the relative standard-electrode potentials, there are situations in which the solid is eroded and goes into solution rather than a film being deposited on the surface.

The morphology of the surfaces produced by anodization is often quite rough. This can be traced to the fact that the deposition or desorption rates are site dependent and are sensitive to surface defects.

19.12 Passivation

Coating a surface of a semiconductor such as Si with a thin oxide layer serves to stabilize the surface. It tends to isolate the surface mechanically and chemically and to make it immune to contamination by adsorbed molecules. In addition, it arrests the tendency for the surface to undergo diffusion processes and/or reconstructions. The encapsulated surface is then said to be *passivated*.

Exposed surfaces often have surface states that are unfilled or low-lying surface states above the Fermi level. The gradual filling of these states by accumulating adsorbates or by spontaneous reconstructions causes the electrical properties (such as carrier velocities or surface recombination velocities) to vary with time. This leads to degradation of the electrical properties of the surface and could prove to be fatal to a large-scale integrated device.

When oxygen adsorbs on the surface the molecular orbitals of oxygen hybridize with the surface states of the semiconductor. Bonding and antibonding molecular orbitals

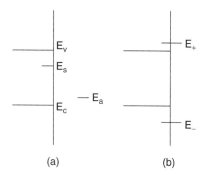

Figure 19.14. Energy levels for (a) unadsorbed and (b) adsorbed situations.

are produced that are separated in energy. The splittings tend to be large enough to push the bonding orbitals below the top of the valence band. Similarly, the antibonding states are pushed well above the Fermi level into the conduction band. The bonding orbitals become occupied and the antibonding orbitals are vacant. The net result is that surface states are removed from the bandgap region and the surface becomes passivated. This is illustrated in Fig. 19.14, where the surface state, with energy E_s, lies inside the bandgap. Also shown is the oxygen orbital, with energy E_a. When the oxygen is brought close to the surface, hybridization takes place and the "molecular" orbitals become E_+ and E_-.

A simple description of this can be obtained by writing the hybridized state as a linear combination of a surface state $|s\rangle$ and an atomic state $|a\rangle$:

$$|\Psi\rangle = \alpha|s\rangle + \beta|a\rangle, \tag{19.35}$$

$$H = \begin{vmatrix} E_s & t \\ t & E_a \end{vmatrix}. \tag{19.36}$$

The Hamiltonian matrix for the two-state manifold is H where t is the strength of the tunneling interaction. (In reality, t is an intersite matrix element of the Coulomb interaction.) The eigenvalues are obtained from the secular equation

$$\begin{vmatrix} E_s - E & t \\ t & E_a - E \end{vmatrix} = 0, \tag{19.37}$$

so

$$E_\pm = \frac{E_s + E_a}{2} \pm \sqrt{\left(\frac{E_s - E_a}{2}\right)^2 + t^2}. \tag{19.38}$$

This formula shows that the effect of the tunneling interaction is to cause level repulsion between the surface state and the adsorbate level. In the limit of weak tunneling, $t \to 0$ and $E_+ - E_- \to E_s - E_a$, as expected. In the limit of strong tunneling one obtains $E_+ - E_- \to 2|t|$.

While in many cases passivation is regarded as a desirable feature, there are some cases in which it is undesirable. This occurs, for example, in the homoepitaxial growth of silicon, which will not occur if the surface is passivated by an adlayer of SiO_2.

It is found experimentally that if the surface temperature is sufficiently high and the oxygen partial pressure is sufficiently low, passivation does not occur. On the other hand, for low temperatures or high partial pressures of oxygen, passivation does occur. The mechanism for this is related to how the oxygen reacts with the silicon surface. When oxygen adsorbs on silicon at sufficiently high surface temperatures, the oxygen molecule dissociates. The oxygen atoms diffuse on the surface and eventually form SiO_2 complexes. This formation occurs preferentially at a preexisting SiO_2 island and leads to island growth. Competing with this is the reaction $Si + SiO_2 \rightarrow 2SiO(g)$, in which the product evaporates from the surface. At high surface temperatures this process leads to destruction of the island. At low oxygen pressures the rate of growth of the island is less than the rate of its evaporation and the surface becomes clean.

Passivation of silicon can also be accomplished with hydrogen. If the oxide layer is etched off with hydrofluoric acid (HF), what is left is a layer of silicon hydride on the surface. This surface remains stable even if exposed to the atmosphere, indicating that the SiH bond is very strong.

The role of surfactants as surface-modifying agents is discussed in Section W19.2.

19.13 Surface Phonons

Just as there are surface states corresponding to bulk electronic states, there are surface phonons corresponding to bulk phonons. The surface phonons propagate along the surface plane. Unlike the bulk phonons, which extend over the entire solid, the amplitudes of the atomic displacements of the surface phonons are attenuated as one progresses deeper into the solid. The dispersion curves for the surface phonons extend over the two-dimensional Brillouin zone corresponding to the surface net. As in the case of the bulk phonons, there are both acoustic and optic branches. However, since the Brillouin zone is two-dimensional, there are only two acoustic branches. In the limit of long wavelengths these are referred to as *Rayleigh waves* or *surface acoustic waves* (SAWs). The optic phonon branches are called *Fuchs–Kliewer waves*. In this section these two limiting cases are considered.

Surface-Acoustic Phonons. Attention here is restricted to the important class of cubic crystals. Suppose that one has a cubic solid occupying the half-space $z < 0$ with the (100) face along the plane $z = 0$. One looks first for a solution with the atomic displacements of the form

$$u_x = A \exp[i(k_x x - \omega t) + \kappa z], \tag{19.39a}$$

$$u_y = 0, \tag{19.39b}$$

$$u_z = B \exp[i(k_x x - \omega t) + \kappa z]. \tag{19.39c}$$

where κ is an attenuation constant, to be determined. From Eq. (10.36) the equation of motion for u_x is

$$\rho \frac{\partial^2 u_x}{\partial t^2} = C_{11} \frac{\partial^2 u_x}{\partial x^2} + C_{44} \left(\frac{\partial^2 u_x}{\partial y^2} + \frac{\partial^2 u_x}{\partial z^2} \right) + (C_{12} + C_{44}) \left(\frac{\partial^2 u_y}{\partial x \, \partial y} + \frac{\partial^2 u_z}{\partial x \, \partial z} \right), \tag{19.40a}$$

and the corresponding equation for u_z is

$$\rho \frac{\partial^2 u_z}{\partial t^2} = C_{11} \frac{\partial^2 u_z}{\partial z^2} + C_{44} \left(\frac{\partial^2 u_z}{\partial y^2} + \frac{\partial^2 u_z}{\partial x^2} \right) + (C_{12} + C_{44}) \left(\frac{\partial^2 u_y}{\partial z \, \partial y} + \frac{\partial^2 u_x}{\partial z \, \partial z} \right).$$

$$(19.40b)$$

Here ρ is the density of the crystal and C_{ij} are the elastic constants. Inserting Eqs. (19.39a–c) into Eqs. (19.40a–b) gives

$$-\rho \omega^2 A = -C_{11} k_x^2 A + \kappa^2 C_{44} A + i k_x \kappa (C_{12} + C_{44}) B, \qquad (19.41a)$$

$$-\rho \omega^2 B = C_{11} \kappa^2 B - k_x^2 C_{44} B + i k_x \kappa (C_{12} + C_{44}) A. \qquad (19.41b)$$

The condition for there to be a solution is the vanishing of the determinant

$$\begin{vmatrix} \rho \omega^2 - C_{11} k_x^2 + C_{44} \kappa^2 & i k_x \kappa (C_{12} + C_{44}) \\ i k_x \kappa (C_{12} + C_{44}) & \rho \omega^2 + C_{11} \kappa^2 - C_{44} k_x^2 \end{vmatrix} = 0. \qquad (19.42)$$

There will be two independent positive solutions for κ. For the case of the isotropic solid, where $C_{44} = (C_{11} - C_{12})/2$ [see Eq. (10.20)], the solutions are simple to state:

$$\kappa_1 = \sqrt{k_x^2 - \frac{\rho \omega^2}{C_{44}}}, \qquad (19.43a)$$

$$\kappa_2 = \sqrt{k_x^2 - \frac{\rho \omega^2}{C_{11}}}. \qquad (19.43b)$$

For $\kappa_1 = \kappa_2 = 0$, (i.e., no attenuation in the z direction into the solid), these correspond to the transverse and longitudinal elastic waves described in Table 10.5. In the anisotropic cubic case the solutions are somewhat more complicated and so will not be given explicitly.

To obtain the Rayleigh waves one must satisfy the boundary condition that the stress at the surface ($z = 0$) vanishes. A linear combination of the two solutions is formed; that is,

$$u_x = A_1 \exp[i(k_x x - \omega t) + \kappa_1 z] + A_2 \exp[i(k_x x - \omega t) + \kappa_2 z], \qquad (19.44a)$$

$$u_z = B_1 \exp[i(k_x x - \omega t) + \kappa_1 z] + B_2 \exp[i(k_x x - \omega t) + \kappa_2 z], \qquad (19.44b)$$

with $u_y = 0$. The coefficients A_1 and B_1 are related through Eq. (19.41a) (using κ_1) and similarly A_2 and B_2 are also related through Eq. (19.41a) (using κ_2) [see Eqs. (19.46c) and (19.46d)]. Using Eqs. (10.15) and (10.18) gives

$$\sigma_3 = C_{12} \varepsilon_1 + C_{12} \varepsilon_2 + C_{11} \varepsilon_3 = 0, \qquad (19.45a)$$

$$\sigma_4 = C_{44} \varepsilon_4 = 0, \qquad (19.45b)$$

$$\sigma_5 = C_{44} \varepsilon_5 = 0. \qquad (19.45c)$$

Thus

$$i k_x C_{12}(A_1 + A_2) + C_{11}(\kappa_1 B_1 + \kappa_2 B_2) = 0, \qquad (19.46a)$$

$$C_{44}(\kappa_1 A_1 + \kappa_2 A_2 + i k_x B_1 + i k_x B_2) = 0. \qquad (19.46b)$$

These, together with

$$-\rho\omega^2 A_1 = -C_{11}k_x^2 A_1 + \kappa_1^2 C_{44}A_1 + ik_x\kappa_1(C_{12} + C_{44})B_1, \qquad (19.46c)$$

$$-\rho\omega^2 A_2 = -C_{11}k_x^2 A_2 + \kappa_2^2 C_{44}A_2 + ik_x\kappa_2(C_{12} + C_{44})B_2, \qquad (19.46d)$$

allow one to determine the frequencies of the two Rayleigh waves.

Surface-acoustic waves on piezoelectric crystals are employed in the operation of electromagnetic filters. Typical values of the speed of Rayleigh waves for the technologically important ceramics $LiTaO_3$, $LiNbO_3$, ZnO, and $Pb(Mn_{1/3}Nb_{2/3})_{1/3}Zr_{1/3}Ti_{1/3}O_3$ are 3300, 4000, 3200, and 2400 m/s, respectively.

Surface-Optic Phonons. Since the optic phonons have an associated electric field, one may find their frequencies by searching for an electric potential wave that propagates parallel to the surface but is attenuated inside the solid and also in vacuum. For long wavelengths the potential wave must satisfy Laplace's equation everywhere but on the surface, so

$$\nabla^2 \phi(\mathbf{r}, t) = 0. \qquad (19.47)$$

It will be assumed that the dielectric function $\epsilon_r(\omega)$ depends only on frequency. The solution is of the form

$$\phi(\mathbf{r}, t) = \phi_0 \exp[i(k_x x - \omega t) - |k_x z|]. \qquad (19.48)$$

The potential is continuous at $z = 0$. The normal component of the electric displacement vector $\mathbf{D}$ must also be continuous, so

$$[\epsilon_r(\omega) + 1]\phi_0 = 0. \qquad (19.49)$$

From Eq. (W8.15),

$$\epsilon_r(\omega) = \epsilon_r(0) + \frac{[\epsilon_r(\infty) - \epsilon_r(0)]\omega^2}{\omega^2 - \omega_T^2 + i\gamma\omega}, \qquad (19.50)$$

where ω_T is the frequency of the transverse optic phonon in the bulk of the solid and γ is the phonon damping constant. When damping is neglected the surface-optic phonon frequency is given by

$$\omega_s = \omega_T \sqrt{\frac{\epsilon_r(\infty) + 1}{\epsilon_r(0) + 1}}. \qquad (19.51)$$

Since $\epsilon_r(\infty) < \epsilon_r(0)$, the frequency is less than that of the bulk optic phonon.

The ion amplitudes may be obtained by relating them to the electric dipole moment of a unit cell, which is given by

$$\mathbf{u}(\mathbf{r}, t) = -\frac{\epsilon(\omega) - \epsilon_0}{n}\nabla\phi(\mathbf{r}, t), \qquad (19.52)$$

with n being the number density of unit cells.

For GaAs the predicted value for the energy of the Fuchs–Kliewer mode is 30.9 meV. Experiments on GaAs(110) show a mildly dispersive surface phonon mode extending from 24 to 27 meV, which may correspond to the Fuchs–Kliewer mode.

Although the discussion above has centered around two special types of surface modes, surface phonons are a general excitation and exist for all types of crystals. The thermal excitation of surface phonons is important in that they govern the displacement of adsorbed atoms. As such, they promote surface diffusion, reaction, and desorption processes. The amplitudes of the surface atoms at a given temperature are substantially larger than the amplitude of the bulk atoms at the same temperature. One may define a surface Debye temperature, Θ_{Ds}, in much the same way as one defines a bulk Debye temperature, Θ_D. It may be shown that the root-mean-square vibrational amplitudes of surface atoms vary as $T^{1/2}$ and inversely with Θ_{Ds}. Typical values for Θ_D and Θ_{Ds} are for Ni(110), 390 and 220 K; and for Pt(100), 234 and 110 K. This confirms that the surface "melts" before the bulk.

19.14 Surface Processes

A variety of processes involving atoms can occur on a surface (e.g., adsorption, desorption, and surface diffusion). Adsorption is the term used to describe the deposition of particles onto a surface. Consider a surface exposed to a fluid environment. In general, the fluid may be a gas, liquid, or plasma. Atoms or molecules impinge on the solid and either stick to it or bounce off, elastically or inelastically. The probability of the particle remaining on the surface is called the *sticking probability*, *s*. In some cases the adsorbate forms a single layer and then the sticking probability falls to zero. In other cases the adsorbate can develop into multilayered structures. In some cases there may be island formation.

The inverse process, in which an adsorbed atom departs from the surface, is termed *desorption*. In thermal equilibrium, both adsorption and desorption occur, with the corresponding fluxes being equal in magnitude. The desorption rate depends on the surface temperature.

The atoms adsorbed on a surface are often not statically bound to a particular site but can hop from site to site. Such a process is termed *surface diffusion*. The degree of surface diffusion is also related to the surface temperature.

The three processes — adsorption, desorption, and surface diffusion — are quite similar. The similarity has to do with the form of the potential energy well that an atom or molecule finds itself in and the mechanism by which it is able to be trapped by or freed from it. The processes are similar to the molecular processes of association, dissociation, and conformational change, familiar from chemistry.

Consider the potential energy, V, of interaction of an atom with a solid. Figure 19.15 illustrates two possible potential energy curves for an atom approaching the solid. Matters are simplified by ignoring the variation of the potential energy with the coordinates along the surface plane and treating the potential as a one-dimensional function, $V(z)$, the distance of the atom from the surface being denoted by z. Figure 19.15a depicts a long-range attraction and a short-range repulsion. At large distances $V(z)$ is governed by van der Waals forces. Closer to the surface, more specific chemical forces come into play. The short-range repulsion is due to Coulomb repulsion of the nuclei and the repulsion of the overlapping electrons. Part of this repulsion is due to the Pauli exclusion principle. Figure 19.15b shows two potential minima. The weaker well,

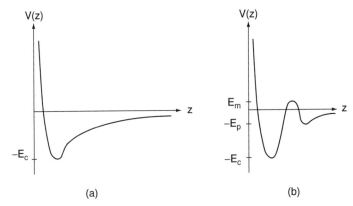

Figure 19.15. (*a*) Potential energy, $V(z)$, as a function of atom-surface distance, z: one-well case; (*b*) same as (*a*), but for the case of a (deep) chemisorption and (shallow) physisorption well.

occurring at larger distances, is often called the *physical adsorption* or *physisorption well*. Such a well commonly occurs when molecules adsorb on surfaces. The second well, closer to the surface, is much deeper and is called the *chemical adsorption* or *chemisorption well*. The depth of the chemisorption well is $-E_c$, the depth of the physisorption well is $-E_p$, and the height of the potential maximum between the two is denoted by E_m, which can have either sign. Adsorption involves trapping an impinging atom in either a physisorption or a chemisorption well.

In Fig. 19.16 a plot is shown of a hypothetical potential energy curve along a direction parallel to the surface, x, for fixed y and z. The zero of energy has been chosen to correspond to the minima. The height of the barrier separating these minima is denoted by E_b. Surface diffusion involves the thermally activated jumping of an atom from one potential minimum to another.

For illustrative purposes it is convenient to simplify the problem by describing it in terms of two potentials, $V(z)$ and $V(x)$ (which are independent functions). In reality, the atom interacts with the solid with a full three-dimensional potential function $V(x, y, z)$. Adsorption sites correspond to the local minima of this function. The processes occurring in the general case will still be the same, but the analysis will be more complex.

Adsorption, desorption, and surface diffusion processes are considered in more detail in Sections W19.3 to W19.5. The surface process(es) known as heterogeneous catalysis are also described in Section W19.6.

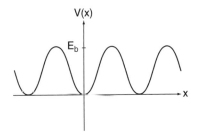

Figure 19.16. Variation of potential along a direction parallel to the surface, x.

ADHESION AND FRICTION

In the following sections some topics related to the adhesion of two solids to each other are discussed. The friction force between two surfaces will also be studied. The interactions between solids may be classified as being either chemical or physical in nature. The chemical bonding of solids is naturally discussed in terms of the hybridization of the surface states of the two solids with each other. The theory is a natural extension of the concepts introduced earlier in the discussion of passivation of surfaces. Obviously, the smoothness of the surface will play an important role in the formation of such chemical bonds. So will the registry of the surfaces (i.e., how well the lattices line up with each other).

The main thrust of the following sections is not on chemical interactions, however, but on the physical attraction of two solids for each other. Particular surface excitations called *surface plasmons* are introduced, and it is shown that they are responsible for the physical forces of attraction between solids. The role of surface phonons, localized vibrations, and electronic excitations play an important role in the study of friction.

19.15   Surface Plasmons

A plasmon is a collective excitation of the solid in which energy oscillates between two forms: electrostatic energy and kinetic energy of the electrons. The theoretical description follows directly from a knowledge of the dielectric function, ϵ_r, of the solid. The wave representing the collective motion of the valence electrons may be described in terms of the electrostatic potential produced as it propagates. Thus consider the complex potential wave

$$\Phi(\mathbf{r}, t) = \Phi_0 \exp[i(\mathbf{k} \cdot \mathbf{r} - \omega t)] \tag{19.53}$$

and insert it into Laplace's equation:

$$\nabla \cdot [\epsilon_r(\mathbf{k}, \omega)\nabla\Phi] = 0. \tag{19.54}$$

A solution can occur only when

$$\epsilon_r(\mathbf{k}, \omega) = 0. \tag{19.55}$$

The root to this equation, $\omega(\mathbf{k})$, defines the dispersion relation for the plasmon, giving the frequency, ω, in terms of the propagation vector, $\mathbf{k}$.

Surface plasmons correspond to plasmon modes which propagate along the surface and are localized near it. They are similar to the Fuchs–Kliewer phonon modes discussed in Section 19.13, but they represent electronic excitations rather than phonon modes. The spatial variation is given by the potential wave

$$\Phi(\mathbf{R}, z, t) = \Phi_0 \exp[-k_\| |z| + i(\mathbf{k}_\| \cdot \mathbf{R} - \omega t)]. \tag{19.56}$$

Here $\mathbf{k}_\|$ is the two-dimensional propagation vector along the surface, $\mathbf{R} = x\hat{i} + y\hat{j}$ locates a point on the surface, and z is the coordinate perpendicular to the surface. The wave is exponentially attenuated in both directions away from the surface. Inserting this into Laplace's equation yields

$$\Phi_0 \delta(z)[\epsilon_r(\mathbf{k}_\|, \omega) + 1] = 0, \tag{19.57}$$

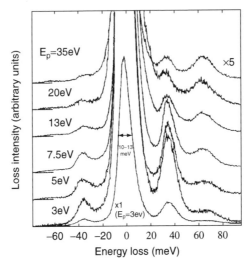

Figure 19.17. HREELS data for n-type GaAs(100). The peaks closer to $E = 0$ are due to surface phonons, while the peaks farther from $E = 0$ are due to surface plasmons. (Reprinted from M. Noguchi et al., *Surf. Sci.*, **271**, 260, copyright 1992, with permission from Elsevier Science.)

where $\epsilon_r(\mathbf{k}_\parallel, \omega)$ is an effective dielectric function for the surface region. The roots of the equation

$$\epsilon_r(\mathbf{k}_\parallel, \omega) = -1 \qquad (19.58)$$

define the dispersion relation for the surface plasmon.

For example, suppose that the dielectric function appropriate to a plasma is employed:

$$\epsilon_r(\omega) = 1 - \frac{\omega_p^2}{\omega^2}. \qquad (19.59)$$

The plasmon and surface-plasmon frequencies are then given by ω_p and $\omega_p/\sqrt{2}$, respectively.

The plasmon energies show up as loss peaks when electrons are inelastically scattered from surfaces. In addition to the elastically scattered electrons, there are electrons scattered with lower energies. Prominent among these are electrons whose energy is diminished by one or more surface plasmons of energy $\hbar\omega_p/\sqrt{2}$. Figure 19.17 illustrates the results of high-resolution electron energy-loss spectroscopy (HREELS) measurements on n-type GaAs. Evidence is seen for both the surface plasmon and the Fuchs–Kliewer phonon modes. Note that the modes appear both as losses and as gains, corresponding to plasmon (or phonon) emission and absorption processes, respectively.

19.16 Dispersion Forces

Dispersion forces are important in physisorption. When two solids are brought into proximity, an attractive force is established between them. It is analogous to the case of the attraction caused by the van der Waals force between atoms familiar from chemistry. In the case of solids it is called *dispersion force*. Unlike short-range chemical bonding forces, it is of purely physical origin and may be understood in terms of surface plasmons.

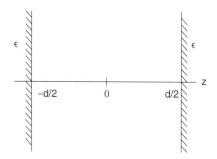

Figure 19.18. Two flat solid surfaces separated by distance d.

Consider two identical flat surfaces separated by distance d (Fig. 19.18). (Identical solids are chosen here to simplify the discussion.) The solid is assumed to be characterized by a dielectric function $\epsilon_r(\omega)$. Because of the reflection symmetry of the problem, the even- and odd-symmetric solutions (under $z \to -z$) are analyzed independently. One may talk in terms of coupled interfacial surface plasmons belonging simultaneously to both solids.

The even-symmetric plasma wave is described (for $z > 0$) by

$$\Phi = \begin{cases} \Phi_0 \cosh(k_\parallel z) \exp\left[i\mathbf{k}_\parallel \cdot \mathbf{R}\right] & \text{if} \quad 0 < z < d/2 \\ \Phi_0 \cosh(k_\parallel d/2) \exp[i\mathbf{k}_\parallel \cdot \mathbf{R} - k_\parallel(z - d/2)] & \text{if} \quad d/2 < z < \infty. \end{cases} \tag{19.60}$$

The corresponding odd-symmetric wave is given (for $z > 0$) by

$$\Phi = \begin{cases} \Phi_0 \sinh(k_\parallel z) \exp\left[i\mathbf{k}_\parallel \cdot \mathbf{R}\right] & \text{if} \quad 0 < z < d/2 \\ \Phi_0 \sinh(k_\parallel d/2) \exp[i\mathbf{k}_\parallel \cdot \mathbf{R} - k_\parallel(z - d/2)] & \text{if} \quad d/2 < z < \infty. \end{cases} \tag{19.61}$$

The condition that the normal component of the electric displacement vector D_z at the surface must be continuous provides the relations

$$\epsilon_r(\omega_e) = -\tanh\frac{k_\parallel d}{2} \tag{19.62}$$

for the even-symmetric wave (whose solution is denoted by ω_e) and

$$\epsilon_r(\omega_o) = -\coth\frac{k_\parallel d}{2} \tag{19.63}$$

for the odd-symmetric wave (whose solution is denoted by ω_o).

These plasmons are bosons, so each mode is analogous to an independent quantum-mechanical oscillator. Even at absolute zero there is energy associated with the zero-point motion of these plasmons (corresponding to the ground state of the oscillators). The expression for this energy is

$$U = \sum_{k_\parallel, \lambda} \frac{\hbar \omega_{k_\parallel, \lambda}}{2} = \int \frac{A d^2 k_\parallel}{(2\pi)^2} \frac{\hbar}{2}(\omega_e + \omega_o), \tag{19.64}$$

where A is the area of one of the faces. For the simple plasma model the frequencies are given by

$$\omega_e = \frac{\omega_p}{\sqrt{1 + \tanh(k_\| d/2)}}, \tag{19.65}$$

$$\omega_o = \frac{\omega_p}{\sqrt{1 + \coth(k_\| d/2)}}, \tag{19.66}$$

where ω_p is the plasma frequency. The force per unit area is obtained by differentiating the energy with respect to distance and dividing by A:

$$\frac{F}{A} = -\frac{\partial}{\partial d}\frac{U}{A}. \tag{19.67}$$

The resulting expression is

$$\frac{F}{A} = \frac{\hbar \omega_p}{2\pi d^3} \int_0^\infty dx\, x^2 \left[\frac{\operatorname{sech}^2 x}{(1 + \tanh x)^{3/2}} - \frac{\operatorname{csch}^2 x}{(1 + \coth x)^{3/2}} \right], \tag{19.68}$$

which yields

$$\frac{F}{A} = -\frac{3}{124}\frac{\hbar \omega_p}{d^3}. \tag{19.69}$$

The force is attractive and grows stronger with decreasing d. For very small d the atomistic nature of the adhesive bond predominates and the dielectric treatment used is inadequate. Then chemical bonding forces prevail.

A simple explanation for the attractive interaction is that it is due to the electrostatic interaction of fluctuating charge waves along the surfaces of the solids, where the fluctuations themselves are quantum mechanical in origin. If the phasing is such that a positive charge fluctuation on one surface is located opposite a negative charge fluctuation on the other surface, there will be a net attraction of the two surfaces toward each other.

19.17   Friction

Tribology is the study of solid-state friction, wear, and lubrication. The macroscopic laws of friction were first stated by Leonardo da Vinci and were later rediscovered by Guillaume Amontons, with added contributions by Charles Augustin de Coulomb. They may be summarized as follows:

1. The static friction force exerted on a body by a flat surface, $\mathbf{F}_s$, is always of such a magnitude and direction as to maintain equilibrium. It is only able to do this if $F_s < \mu_s N$, where N is the normal force exerted by the surface on the body and μ_s is the coefficient of static friction. The force $F_{s,\max} = \mu_s N$ is called the *static limit*.

2. The kinetic friction force is directed opposite to the velocity, $\mathbf{v}$, and is given by $\mathbf{F}_k = -\mu_k N \hat{v}$, where μ_k is the coefficient of kinetic friction. The quantity μ_k is independent of $\mathbf{v}$. (It is now known that this independence is valid only for small velocities. For larger values of v, μ_k decreases as a power law in velocity.)

3. The friction force is independent of the contact area between the object and the surface.

The numbers μ_s and μ_k are characteristic of the physical and chemical states of both the object and the surface. Typical values for μ_s are presented in Table 19.3. The value of μ_k is always smaller than μ_s.

These laws are not exact. For example, one knows that when two identical atomically flat surfaces are brought together and left in contact for some time, the surfaces will cold-weld together. In that case μ_s becomes, for all intents and purposes, infinite. More generally, it is found that for many contacts the value of μ_s increases quasilogarithmically with the time of contact $[d\mu_s/d(\ln t) \approx 0.01]$.

There is particular interest in determining the behavior of friction at microscopic scales. Nanotribology is the study of friction at the atomic scale. It is known from experiment that macroscopic surface roughness has little to do with friction. Hence a distinction must be drawn between the macroscopic, or apparent contact area A_a, and the microscopic, or true contact area A_t, determined by how many interatomic bonds are actually formed at the interface. When two rough surfaces are in apparent contact, only a small fraction of the surface atoms of each solid bond together, so the true contact area is much smaller than the apparent contact area. The static limit is proportional to the true contact area (i.e., $F_{s,\max} = GA_t$). Here G is a proportionality constant with the units of shear stress. As the normal force is increased, elastic deformation (or plastic deformation) of the contacts occurs, more atoms form bonds at the interface, and the true contact area is increased proportionately (i.e., $A_t = N/G'$). Combining these relations gives $F_{s,\max} = \mu_s N$, where $\mu_s = G/G'$. Thus the friction force grows in proportion to N, but without change in the apparent area. Note that changing the apparent contact area, for fixed N, does not change the true contact area. It is observed that changing the degree of surface roughness does not change the true contact area. Thus friction is independent of both the apparent contact area and the degree of macroscopic roughness.

Creep of surface asperities along the interface under load may lead to a gradual increase of A_t with time. This may account for the observed growth of μ_s with time.

From Table 19.3 it is seen that metals generally have higher friction coefficients than ceramics. This may be related to the fact that the metallic bond is not saturated. Ceramics, with the electrostatic nature of their bonding, have weaker surface bonds, so the friction forces are weaker. Polymers, with saturated covalent bonds, have only weak van der Waals forces, so the friction forces are still weaker.

Friction is very sensitive to the presence of oxide layers or surface contaminants. For example, the metal–metal values of μ_s for Cu, Au, Ag, Fe, carbon and alloy steels, and Ni in Table 19.3 are for an air environment. For outgassed materials in vacuum, μ_s is increased dramatically and there is seizure of the surfaces. This is the basis for the process known as *friction welding*. Other materials capable of being friction-welded include Al, Ti, brass, bronze, and polyethylene (but not Teflon). Another example is provided by passivating Ni(100) with S and then adsorbing some ethanol on the surface. It is found that μ_k drops dramatically, from a value of 4.2 for no coverage to 0.2 for a coverage of one monolayer of ethanol. Furthermore, as the coverage is increased, the nature of the friction changes from stick-slip motion to smooth gliding. Thus the study of friction is often masked by the effects of oxide layers or adsorbed organic materials.

Kinetic friction differs from static friction in that there is dissipation of mechanical energy. As one rough surface is moved past another, asperities from the harder surface plow into the softer surface asperities and are sheared away. Eventually a situation is reached in which old interfacial chemical bonds are destroyed in one part of the surface

TABLE 19.3 **Values of μ_s for Typical Solid/Solid Contacts**

Solid/Solid	Remarks	T (°C)	μ_s
Al/Al	In air	—	1.9
Cu/Cu	In air	—	1.6
Au/Au	In air	—	2.8
Ag/Ag	In air	—	1.5
Fe/Fe	In air	—	1.2
Ni/Ni	In air	—	3.0
Steel/steel	Clean	20	0.58
Glass/glass	Clean	—	0.9–1.0
Diamond/diamond	Clean	—	0.1
Sapphire/sapphire	Clean	—	0.2
WC/WC	Outgassed	25	0.58
	Clean	25	0.17
	Clean	820	0.35
	Clean	970	0.40
	Clean	1010	0.45
	Clean	1160	0.5
	Clean	1220	0.7
	Clean	1440	1.2
	Clean	1600	1.8
WC/C(graphite)	Outgassed	25	0.62
	Clean	25	0.15
WC/C	Clean	800	0.32
	Clean	910	0.30
	Clean	1000	0.25
	Clean	1120	0.29
	Clean	1220	0.26
	Clean	1300	0.25
	Clean	1410	0.25
	Clean	1800	0.24
	Clean	2030	0.25
WC/steel	Clean	—	0.4–0.6
PMMA/PMMA	Clean	—	0.8
PS/PS	Clean	—	0.3–0.35

Source: Data from D. R. Lide, ed., *CRC Handbook of Chemistry and Physics*, 71st ed., CRC Press, Boca Raton, Fla., 1990.

and new bonds are created simultaneously in another part of the surface. There is no efficient way for the energy released in the formation of a new bond at one contact location to be transferred for the destruction of a bond at another remote location. The breaking of an old bond requires that work be done, and this comes from the mechanical energy of the solids. The creation of a new bond releases energy to the surface excitations of the solids. Ultimately, the surface excitations, which include localized vibrations, surface phonons, and electron–hole pairs, transfer their energy to bulk phonons. The net result is that the thermal energy of the solids is increased. Thus friction is associated with the irreversible transfer of mechanical energy to thermal energy.

Additional aspects of friction are presented in Section W19.7.

REFERENCES

Duke, C. B., ed., *Surface Science: The First Thirty Years*, North-Holland, Amsterdam, 1994.

Hudson, J. B., *Surface Science: An Introduction*, Butterworth-Heinemann, Boston, 1992.

Lannoo, M., and P. Friedel, *Atomic and Electronic Structure of Surfaces: Theoretical Foundations*, Vol. 16 of Springer-Series in Surface Sciences, Springer-Verlag, Berlin, 1993.

Luth, H., *Surfaces and Interfaces of Solids*, Vol. 15 of Springer Series in Surface Sciences, Springer-Verlag, Berlin, 1991.

Prutton, M., *Surface Physics, 2nd ed.*, Clarendon Press, Oxford, 1983.

Shah, I., Field-emission displays, *Phy. World*, June 1997, p. 45.

Somorjai, G. A., *Introduction to Surface Chemistry and Catalysis*, Wiley, New York, 1994.

Zangwill, A., *Physics at Surfaces*, Cambridge University Press, New York, 1988.

Friction is discussed in *Materials Research Society Bulletin* issues devoted to friction, Oct. 1991 (F. A. Nichols, ed.), and June 1998 (J. Krim, ed.).

PROBLEMS

19.1 Consider a thin slab of metal occupying the space $0 < z < D$. Let the slab be infinite in extent in the x and y directions. The metal is described by the dielectric function of Eq. (19.59).

  (a) Find the frequencies of the two surface plasmons as a function of the wave vector parallel to the surface, $k_{\parallel}$. (*Hint*: Look for waves traveling in the xy plane which are attenuated in space away from the metal.)

  (b) Show that in the limit of small d one of the modes above reduces to the two-dimensional plasmon, with frequency proportional to the square root of the wave vector.

19.2 (a) Draw a picture of the surface layer of atoms on the (100) face of a BCC crystal with lattice constant a.

  (b) Draw a diffraction pattern for electrons that are incident perpendicular to the surface. (Refer to Chapter W22 for more details.) How are the wave vectors of the diffracted beams related to the lattice constant?

  (c) Draw a centered 2×2 overlayer of adsorbed atoms occupying the centered sites.

19.3 A surface is exposed to N_2 at pressure $P = 10^{-6}$ torr and temperature $T = 300$ K. Assume that there are N_s adsorption sites per unit area available on the surface.

  (a) Estimate the mean free path of a molecule in the gas.

  (b) Assume a sticking probability $s = 1$ (independent of coverage). How much time is needed to form a monolayer of adsorbate on the surface?

  (c) Now assume that the sticking probability is $s = 1 - \theta$, where θ is the coverage (fraction of adsorption sites occupied). How much time is required to build up half a monolayer?

19.4 Consider a two-dimensional electron gas in a uniform magnetic field.

  (a) Write the Hamiltonian describing the system. Work in the $\mathbf{A} = \frac{1}{2}\mathbf{B} \times \mathbf{r}$ symmetric gauge.

  (b) Find the energies of the electrons.

(c) Find the wavefunction for the electrons.

(d) What is the effect of adding an impurity described by the potential

$$V(\mathbf{r}) = -V_0 \delta(x)\delta(y)$$

(i.e., what is the effect on the spectrum of the energy levels).

19.5 Suppose that a metal has work function W. Atoms are adsorbed on the surface of the metal. Each adsorbed atom produces a dipole moment p directed perpendicular to the surface. There are N_s adsorption sites per unit area on the surface. Assume that N atoms are adsorbed per unit area. Find an expression for the work function as a function of the coverage, $\theta = N_s/N$.

19.6 Derive an expression for the density of states for the step potential of height V_0.

19.7 **Computer Problem** A free-electron gas is confined to the half-space $z < 0$ by the potential

$$V(z) = \begin{cases} 0 & \text{if } z < 0 \\ \infty & \text{if } z > 0. \end{cases}$$

Assume that deep in the crystal there are n electrons per unit volume.

(a) Find the properly normalized form of the wavefunction of an electron.

(b) Find a formula for the electron density by summing the individual electron contributions over the Fermi sea. The undulations that occur in the vicinity of $z = 0$ are called *Friedel oscillations*.

19.8 Find the dispersion formula for the two-dimensional plasmon for an electron gas with N electrons per unit area confined to move along the planar interface between two dielectrics, with dielectric constants ϵ_{r_1} and ϵ_{r_2}, respectively.

19.9 Use the field-emission data listed in Table P19.9 which were obtained for Al(100) to determine the work function W for this surface. [*Hint:* Plot the data using a Fowler–Nordheim plot in which $\ln(J/E_0^n)$ is plotted versus E_0^{-m}. The values of n and m can be deduced from the Fowler–Nordheim formula, Eq. (19.27).]

TABLE P19.9 Data for Problem 19.9

E_0 (V/m)	J (A/m^2)
1.0×10^9	2.0×10^{-16}
1.5×10^9	6.6×10^{-7}
2.0×10^9	4.5×10^{-2}
3.0×10^9	3.9×10^3
4.0×10^9	1.34×10^6

CHAPTER 20

Thin Films, Interfaces, and Multilayers

20.1 Introduction

Much of the technological utility of materials stems from their use in configurations with finite geometries. In many cases their special properties depend in a critical way on the details of the configuration. In this chapter the focus is on some of the key concepts and applications of thin films, interfaces, and multilayers. A primary focus is on issues relating to how the properties of materials in finite configurations differ from the bulk properties considered previously. The areas of thin films, interfaces, and multilayers are very broad and cannot be covered completely here. A case study approach is presented of topics of current interest in the field.

The chapter begins with a discussion of the surface energies of thin films on substrates and the wetting of surfaces. The method of nanoindentation to measure the mechanical properties of thin films is discussed briefly. Morphology maps, describing the topography of films prepared by sputtering deposition under varying conditions, are then studied. This is followed by an analysis of dislocations in thin films. The formation of thin films by the Langmuir–Blodgett technique is discussed. Grain boundary effects are then studied. Grain boundaries are of interest because of both their prevalence in thin films and their importance in understanding interfacial phenomena.

Interfaces refer to the boundary regions between two materials. They may be formed between semiconductors, metals, ceramics, polymers, or electrolytes. The charge transfer and band bending occurring in the region near the interfacial boundary of a semiconductor are discussed first. The semiconductor–metal interface, known as the *Schottky barrier*, is then analyzed. This is followed by a study of the semiconductor–heterostructure superlattice, in which semiconductor–semiconductor interfaces are involved. The quantum dot, which represents a configuration of semiconductors resulting in the three-dimensional confinement of electrons, is studied next. Although this is not, strictly speaking, an interfacial device, the physics of its operation is quite similar to that of the semiconductor heterostructure, so it is included at this point.

The discussion of multilayers begins with a description of how multilayers may be used to tailor physical properties of materials to specific needs. Applications of multilayers to x-ray optical devices are considered. The hardness of multilayer materials is then briefly discussed.

The chapter concludes with a discussion of the stoichiometric optimization of physical properties by means of creating thin film "libraries" on substrates. This represents a promising technique for probing an entire class of materials with different compositions simultaneously.

The topics discussed in the text are supplemented by additional material at our Web site.[†] This includes sections on strength and toughness (Section W20.1), critical thickness (Section W20.2), ionic solutions (Section W20.3), the solid–electrolyte interface (Section W20.4), multilayer materials (Section W20.5), second-harmonic generation in phase-matched multilayers (Section W20.6), organic light-emitting diodes (Section W20.7), quasiperiodic optical crystals (Section W20.8), and graphite-intercalated compounds (Section W20.9).

THIN FILMS

Thin films form the basis of much of present advanced technology. For example, the magnetic film on a hard disk is typically a 40-nm layer of CoCr alloy sandwiched between a 15-μm amorphous NiP base and a 25-nm amorphous C protective coating. The films are deposited on an Al disk. They must be able to withstand contact with the magnetic read/write head traveling at a relative speed of 80 km/h. The carbon layer with a polymeric overcoating provides the low-friction surface for such encounters. Other examples of thin-film technology are the CMOS transistor and typical VLSI devices, which are composed of layers of n- and p-type semiconductors, polycrystalline silicon, metals, oxides, glass, and other dielectric materials. In this section various issues relating to thin films are addressed.

20.2   Surface Tension

The phenomenon of surface tension is generally first encountered in the study of liquids. It is used, for example, to explain the physical properties of droplets and bubbles. For liquids no distinction is made between the surface tension and the surface energy. An increase in surface area results in an increase in surface energy. The surface acts as if it is in a state of tension, always trying to minimize the area and thereby minimize the surface energy. In solid-state physics the situation is more complicated. An increase in the area of the surface generally entails creating strains in the solid, so a more careful definition of surface tension and its relation to surface energy, surface strain, and surface stress is needed.

In Appendix WA the laws of thermodynamics of bulk materials are reviewed. When a surface or interface is present, the laws must be amended to include them. Thus the Euler relation, Eq. (WA.8), is replaced by

$$U_s = TS + \gamma A, \tag{20.1}$$

where S is the surface (or interface) entropy, A is the surface (or interface) area, and γ is the surface (or interfacial) tension. The terms PV and $N\mu$ are omitted since they are bulk terms. The units of γ may be expressed either in N/m or J/m^2. Some values of surface tensions of solids and liquids are given in Table 20.1.

For the sake of convenience, introduce a two-dimensional Cartesian coordinate system on the surface defined by s_i, where $i = 1, 2$. A strain carries a coordinate

---

[†] Supplementary material for this textbook is included on the Web at the resource site (ftp://ftp.wiley.com/public/sci_tech_med/materials). Cross-references to elements of the Web material are prefixed by "W."

THIN FILMS 717

TABLE 20.1 Surface Tensions of Some Materials

Material	T (°C)	γ_{LV} (or γ_{SV}) (mN/m)
Li	180	398
Na	98	191
K	64	101
Rb	39	76
Cs	29	60
Al	660	860
Ag	961	785
Ag (solid)	750	1140
Cu	1083	1270
Cu (solid)	1080	1430
Au	1063	754
Fe	1535	1650
Ni	1455	1756
Pt	1773	1740
C	—	3700
Si	1410	730
Ge	959	600
In	156	559
Ga	30.5	735
Sb	635	383
Pb	327	470
Ti	1725	1650
W	3419	2319
Polyethylene	150	28
Polystyrene	150	31
H_2O	20	72.8

Source: Data from R. C. Weast, ed., *CRC Handbook of Chemistry and Physics*, 47th ed., CRC Press, Boca Raton, Fla., 1966; and W. D. Kingery et al., *Introduction to Ceramics*, 2nd ed., Wiley, New York, 1976.

s_i to a new coordinate $s_i + u_i$, where u_i is a component of the displacement vector. The components of the surface strain tensor, ε_{ij}, are defined by

$$\varepsilon_{11} = \frac{\partial u_1}{\partial s_1}, \qquad \varepsilon_{22} = \frac{\partial u_2}{\partial s_2}, \qquad \varepsilon_{12} = \varepsilon_{21} = \frac{\partial u_1}{\partial s_2} + \frac{\partial u_2}{\partial s_1}. \tag{20.2}$$

The 2 × 2 strain matrix may be diagonalized by choosing an appropriate coordinate system. The directions in this coordinate system are called the *principal axes*. Consider the special case of a rectangular area element whose corners lie at the principal-axis coordinates $(0,0)$, $(s_1,0)$, (s_1,s_2), and $(0,s_2)$, so $A = s_1 s_2$. Imagine stretching the sides of the rectangle by a small amount, letting $\delta u_1 = s_1 \delta a$ and $\delta u_2 = s_2 \delta b$. The new area will be $A + \delta A$, where $\delta A = A(\delta a + \delta b)$. Using Eq. (20.2), this leads to

$$\delta A = A(\varepsilon_{11} + \varepsilon_{22}). \tag{20.3}$$

Thus the sum of the diagonal elements of the differential strain tensor gives the fractional change in the area. This sum, called the *trace* of the strain tensor, is an invariant (i.e., has the same value in any coordinate system that it has in the principal-axis coordinate system).

The internal energy of the surface may be expressed as a function of the surface entropy and the surface strains [i.e., $U_s = U_s(S, \{\varepsilon_{ij}\})$]. In analogy with Eq. (WA.1), the differential of this function gives the first law of thermodynamics:

$$dU_s = \left(\frac{\partial U_s}{\partial S}\right)_{\varepsilon_{ij}} dS + \sum_{ij} \left(\frac{\partial U_s}{\partial \varepsilon_{ij}}\right)_S d\varepsilon_{ij} \equiv T dS + A \sum_{ij} \sigma_{ij} d\varepsilon_{ij}, \qquad (20.4)$$

where the surface stresses σ_{ij} have been introduced. The second term in this equation represents the change in the surface energy at constant entropy. Combining Eqs. (20.1) and (20.4) and making use of Eq. (20.3) yields the analog of the Gibbs–Duhem formula, Eq. (WA.10),

$$SdT + A\left[d\gamma - \sum_{ij} \sigma_{ij} d\varepsilon_{ij} + \gamma(d\varepsilon_{11} + d\varepsilon_{22})\right] = 0. \qquad (20.5)$$

From this it follows that the surface stress at constant T is given by

$$\sigma_{ij} = \gamma\delta_{ij} + \left(\frac{\partial \gamma}{\partial \varepsilon_{ij}}\right)_T, \qquad (20.6)$$

and the surface entropy by

$$S = -A\left(\frac{\partial \gamma}{\partial T}\right)_{\varepsilon_{ij}}. \qquad (20.7)$$

Thus, unlike the case of liquids, surface tension and surface stress are not identical. If the surface tension varies as a function of strain, its effect must be included in determining the stress.

20.3 Thin-Film Fabrication

The choice of substrate upon which a film is to be deposited is obviously crucial. Issues such as adhesion, chemical compatibility, mechanical stability, cracking and buckling, processing, microstructure, electrical properties, thermal expansion, defects, and epitaxy must be confronted. For example, the deposition of a passivation layer on a film can lead to cracking of the film. Grain-boundary slippage may produce local protuberances called *hillocks* which can interfere with the functionality of devices. To have a framework in which to control such effects, the physics and chemistry of thin films must be thoroughly understood. The discussion begins by asking: What is it that makes a thin film stick to the substrate?

Wetting. One of the fundamental concepts involved in the deposition of a film is wetting. If a liquid is placed on a solid surface, the wetting angle θ is determined by the surface tensions. Figure 20.1 displays a wetting and nonwetting case, corresponding

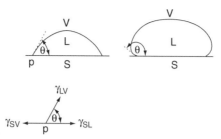

Figure 20.1. A liquid drop on a solid surface. The left figure is for wetting and the right is for nonwetting.

to $\theta < \pi/2$ and $\theta > \pi/2$, respectively. For $\theta = 0$ the film spreads to cover the entire surface.

In Fig. 20.1, γ_{SL}, γ_{LV}, and γ_{SV} denote the surface tensions for the solid–liquid, liquid–vapor, and solid–vapor interfaces, respectively. The horizontal equilibrium condition for the three-phase junction is Young's relation:

$$\gamma_{SL} + \gamma_{LV} \cos \theta = \gamma_{SV}. \tag{20.8}$$

Wetting occurs when $\gamma_{SV} > \gamma_{SL}$ and nonwetting occurs when $\gamma_{SV} < \gamma_{SL}$. Note that if $|\gamma_{SV} - \gamma_{SL}| > \gamma_{LV}$, no solution exists and hence no droplet forms and the surface becomes completely wet. Not shown are the normal force and the gravity force acting on the droplet. If the surface is inclined at an angle relative to the horizontal or is vertical, the droplet can move downward along the plane. Such effects are not considered here.

By studying grain boundaries and their intersection with vapor or liquid phases one may gain additional information. If the dihedral angle α formed by two adjacent grains in equilibrium with a vapor is measured, one is able to obtain a relation between the surface tension for the grain boundary, γ_{GB}, and γ_{SV}. Similarly, the angle β formed by a sessile droplet overlapping a grain boundary on a solid is related to γ_{SL} and γ_{GB}. These situations are depicted in Fig. 20.2.

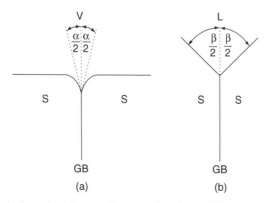

Figure 20.2. (*a*) Grain boundary intersecting a surface in equilibrium with a vapor. (*b*) Grain boundary edge in contact with a liquid.

The corresponding static equilibrium formulas are

$$2\gamma_{SV} \cos \frac{\alpha}{2} = \gamma_{GB}, \tag{20.9}$$

$$2\gamma_{SL} \cos \frac{\beta}{2} = \gamma_{GB}. \tag{20.10}$$

From a knowledge of γ_{SL} and a measurement of the angles α, β, and θ, the surface tensions γ_{GB} and γ_{SV} may be deduced. Note that the angles α and β in Fig. 20.2 depend on the nature of the liquid as well as the solid. Empirically, it is found that $\gamma_{SV}/\gamma_{LV} \approx 1.13$ for all pure metals.

Values of surface tensions for some materials are presented in Table 20.1. For most of the elements the values are those at the melting points. One notes a correlation in the numerical value of γ with the cohesive energy or the bond strength of the materials. To create a surface, one must sever some bonds with neighboring atoms, and this costs energy. The strong covalent bond in C leads to a very high value of γ. The transition metals are also seen to have high values. The ionically bonded materials have intermediate values. The free-electron metals have smaller values. The polymers and hydrogen-bonded water have the lowest values.

Some typical values for $\gamma_{SS'}$ are 2300 mN/m for Al_2O_3–Fe at 1570°C, 1600 mN/m for MgO(s)–Fe(l) at 1725°C, and 5.7 mN/m for PE–PS at 150°C.

Nanoindentation. A technique often used to measure the mechanical properties of thin films, such as the hardness and the elastic modulus, is called *nanoindentation*. A very sharp diamond tip is pressed against the film, of thickness t, and the force on the tip, F, and its displacement, h, are measured as a function of time. Because of the sharpness of the tip, both elastic deformation and plastic flow occur. One must be careful not to penetrate too deeply (more than 10%) into the film, or the substrate will affect the result. Resolutions for the force and displacement can be as small as $\delta F = 10^{-8}$ N and $\delta h = 0.1$ nm. A sketch of typical F versus h curves for both loading and unloading is given in Fig. 20.3.

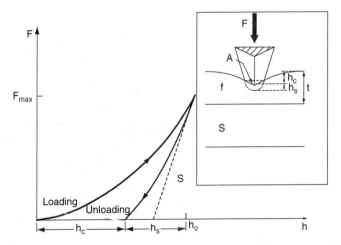

Figure 20.3. Load versus displacement for nanoindenter insertion into a thin film. The inset shows the geometry. [From G. M. Pharr et al., *Mater. Res. Soc. Bull.*, 17(7), 28 (1992).]

Upon insertion of the nanoindenter to a depth h_0 and subsequent removal, one finds that plastic deformation produces a permanent impression in the film to a depth h_c. The elastic deformation of amount $h_s = h_0 - h_c$ is recovered, as shown in Fig. 20.3. From a knowledge of the area A of the nanoindenter as a function of distance from the tip, and a measurement of F versus h, one may compute the stress F/A as a function of displacement. For a symmetric tip the formula for the initial slope of the unloading curve is

$$S = \frac{dF}{dh} = 2E_r\sqrt{\frac{A}{\pi}}, \tag{20.11}$$

where the reduced Young's modulus E_r is expressible in terms of the elastic moduli and Poisson's ratios of the film and indenter, respectively, by

$$\frac{1}{E_r} = \frac{1 - v_f^2}{E_f} + \frac{1 - v_i^2}{E_i}. \tag{20.12}$$

For diamond, $E_i = 1100$ GPa and $v_i = 0.07$. For a triangular pyramidal tip $A(d) = 24.5d^2$, where d is the distance from the tip of the indenter. In reality a detailed experimental study, including the use of electron microscopy, must be used to establish $A(d)$. The hardness is given by $H = F_{\max}/A(h_0)$.

20.4 Morphology Maps

The morphology (i.e., microstructure) of thin films depends on the method of production and a number of physical parameters. To illustrate this, consider a common process, the deposition via sputtering of metal and semiconductor films. The relevant physical parameters include the substrate temperature, T, the pressure of the Ar gas, P, the energy of the incident ions, E, and the direction of the ion beam relative to the surface normal. There is little dependence on the deposition rate, at least for metals, and the nature of the substrate.

The morphology map displays the grain structure of sputtered metal films as a function of the physical variables. It is convenient to introduce a reduced temperature $t = T/T_m$, where T_m is the melting temperature of the film. Figure 20.4 gives a schematic of the morphology as a function of t and P. The map was assembled from scanning electron microscope studies of various metal films.

At small values of t ($t < 0.25$) it is found that the surface consists of thin tapered columns separated by void spaces. The tops of the columns are covered with rounded domes. At intermediate temperatures ($0.25 < t < 0.45$) the columns thicken and meet to form grain boundaries, and the tops of the columns become smooth and eventually become faceted. At still higher temperatures ($t > 0.45$) the columns grow wider, fewer in number, and are equiaxed (i.e., when three grains meet the angle between any two of them is $120°$). For $t \approx 0.75$ the surface becomes bright and consists of smooth grains joined together. These three regions are referred to as zones 1, 2, and 3, respectively, and the overall picture is referred to as the *structure zone model* (SZM). A fourth zone, labeled zone T, consists of a dense array of fibrous grains and is likely to be a transition region between zones 1 and 2.

Much of the morphology may be understood in terms of atomic mobility. At low coverage, islands begin to nucleate and then grow by adsorption. The preferred direction

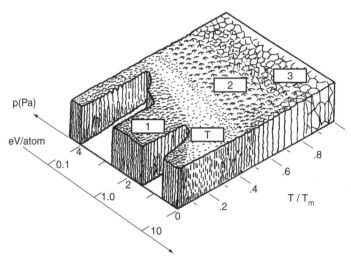

Figure 20.4. Morphology as a function of reduced temperature, $t = T/T_m$, (where T_m is the melting temperature) and argon pressure, P. Also shown is the energy per deposited atom. [From W. D. Westwood, *Mater. Res. Soc. Bull.*, 13(12), 46 (1988).]

for growth is normal to the surface. At low temperatures the islands remain fixed in lateral size and shape. Atomic diffusion is inhibited by the high pressure of the Ar gas in the environment, since the adsorbed gas atoms tend to occupy sites to which the film atoms could jump. In zone 1 the shadowing effect of one column on another brings the directionality of the beam into the story. A column will not grow in the shadow of another column, due to the lack of ample incident flux, and this tends to generate columns that are far apart from each other.

At higher temperatures one enters zone 2 and the surface mobility of atoms is enhanced. Atoms can migrate on the surface and join existing columns which grow laterally until they touch each other. Much of the pore space is eliminated in this densification process. Since atoms may readily move over the columns, the system tends to approach thermal equilibrium. It is found that faceting will occur, with the surfaces of lowest free energy enjoying the most rapid growth. This is explained further in Chapter W21.

In zone 3, as the melting point is approached, bulk diffusion becomes important as well as surface diffusion. The mobility is sufficiently high that Ostwald ripening and recrystallization occur. Ostwald ripening is the phenomenon where smaller grains shrink and larger grains grow at the former's expense. Recrystallization involves the elimination of disorder and the growth of single crystals. More pore space is eliminated and the grains become equiaxed.

A more detailed study[†] of the morphology for low t shows the development of a self-similar, fractal-like structure of small columns grouping together to form larger columns, and these grouping together to form still larger columns (Fig. 20.5). At least five levels of scaling have been recorded for some materials. The structure can be classified as being in zone 1 of the SZM.

[†] R. Messier et al., *J. Vac. Sci. Technol.*, **A2**, 500 (1984).

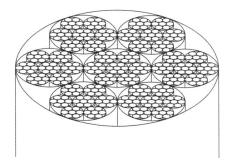

Figure 20.5. Nested columnar structure in a film prepared under conditions of low surface mobility.

20.5 Langmuir–Blodgett Films

Thin films of fatty acids on the surface of water may be transferred to substrates, one molecular layer at a time, by an immersion technique developed by Langmuir and Blodgett. Typically, the adsorbed molecules have long hydrophobic alkyl chains terminated by a hydrophilic ionic group. The terms *hydrophobic* and *hydrophilic* refer to the tendency for molecules to find their state of lowest free energy outside or inside a water environment. Examples of such molecules are illustrated in Fig. 20.6, where behenic acid (B), arachidic acid (A), stearic acid (S), and palmitic acid (P) are shown. These molecules are of the form $CH_3(CH_2)_n-COOH$. Other molecules involve the insertion of organic side groups, such as in ω-tricosenoic acid $[CH_2CH(CH_2)_{20}COOH]$, or modifying the ion group by the incorporation of metal ions. Henceforth the molecules will be symbolized simply by drawing a circle for the hydrophilic group and a line for the hydrophobic tail.

If a droplet of the fatty acid is placed on the surface of water, a film will form that may spread to the thickness of a single monolayer. Suppose that this layer is confined by walls to a finite area and the walls are brought closer together. Figure 20.7 depicts various stages of compression. In part (*a*) the molecules are separated from each other and may have the chains oriented in random directions. Part (*b*) shows that in some situations, the molecules may develop orientational ordering. Upon further compression the chains are pressed against each other and align themselves perpendicular to the

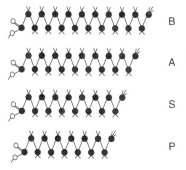

Figure 20.6. Typical fatty acids forming Langmuir-Blodgett films include behenic, arachidic, stearic and palmitic acids.

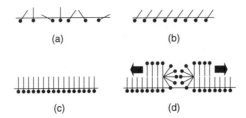

Figure 20.7. Orientation and alignment of fatty acid molecules during several stages of compression of the Langmuir-Blodgett film.

liquid surface, as in part (*c*). A further increase in the lateral surface pressure could result in the surface buckling and having a second molecular layer form, as in part (*d*). The forces responsible for these interactions are the long-range dipole–dipole, image, and van der Waals forces, and the short-range steric interaction between the molecules.

One may define the surface pressure, Π, as the difference in the surface tension of the water without and with the adlayer, $\Pi = \gamma - \gamma_0$. (Note that Π is measured in units of N/m and not Pascal.) If this is plotted (at constant temperature) as a function of the areal density of the adlayer, N, one obtains the *adsorption isotherm*. This isotherm displays breaks in slope when the transition between the various phases in Fig. 20.7 occur, as would be expected for phase transitions. These may be monitored by more direct probes, such as x-ray and neutron diffraction.

The preparation of a Langmuir–Blodgett film is illustrated by the sequence of operations in Fig. 20.8. A hydrophilic substrate (such as glass) is inserted into water upon which floats a monolayer of fatty acid molecules (diagram 1). The water partially wets the surface and forms a concave meniscus. As the substrate is withdrawn, it adsorbs a monolayer of molecules with the hydrophilic ends sticking to the surface (diagram 2). The areal density of molecules on the substrate may be different from that on the water. The substrate now has the hydrophobic ends sticking out from it. A further insertion of the coated surface will now form a convex meniscus. The van der Waals attraction between the hydrophobic chains cause an epitaxial bilayer to be formed as shown in diagram 3. The withdrawal of the substrate from the liquid allows the dipoles of the second layer to attract a third layer of molecules. Repeated application of this procedure results in a buildup of what is termed the *Y-structure*. The thickness of the adlayer is controlled by the number of immersion and withdrawals.

Other structures may be built and some are illustrated in Fig. 20.9. In the X-structure the molecules are all aligned parallel to each other with the hydrophobic ends of the

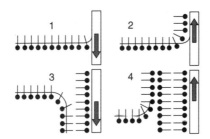

Figure 20.8. Repeated immersion and withdrawal of a hydrophilic substrate results in the formation of the Y-structure.

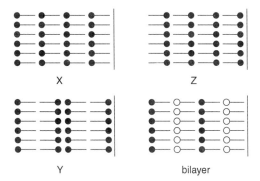

Figure 20.9. Various structures for multilayer Langmuir-Blodgett films: X-, Y-, Z-, and bilayer structures.

first layer coating the substrate. In the Z-structure the molecules are flipped in the direction relative to the X-structure and the hydrophilic ends stick to the surface. In the diagram labeled "bilayer" an X-structure is formed by alternately attaching two different molecular adlayers. The various layers arrange themselves to form a crystalline lattice (usually triclinic or herringbone (twinned-triclinic)). The thickness of the film can be as large as several thousand layers. It is readily monitored by thin-film optical interference techniques.

One way to characterize the formation of the structures above is to introduce a parameter ϕ, defined as the ratio of the number of molecules deposited on a withdrawal of the substrate to the number of molecules deposited on an insertion. For the Y-structure the two numbers are equal and $\phi = 1$. For the X-structure deposition occurs only upon withdrawal, so $\phi = 0$. For the Z-structure the reverse is true and $\phi = \infty$. Hydrophilic surfaces, such as glass or GaAs, tend to adsorb upon the first withdrawal, whereas hydrophobic surfaces, such as Si, tend to adsorb upon the first insertion. Thereafter, there is adsorption each time the substrate crosses the air–water interface. It should be noted that the Si surface is usually terminated with an oxide layer to form SiO_2 or else by a hydride layer to form Si–H bonds.

Langmuir–Blodgett films tend to be very fragile. However, by choosing molecules that can polymerize, the layers may be endowed with a fair degree of mechanical rigidity. Typical polymerizable molecules include ω-tricosenoic and diacetylinic acids, which polymerize when irradiated by electron beams and ultraviolet radiation, respectively. Layers of ω-tricosenoic acid are used to grow films on both Si and GaAs substrates, where they serve as a negative photoresist for lithographic purposes.

Other applications of Langmuir–Blodgett films include coatings on photovoltaic cells, molecular electronics, transducers, electro-optical devices, sensors, lubrication, magnetic layers, mechanical filters, image reproduction, electroluminescent displays, nonlinear-optical materials, biological membranes, field-effect transistors, charge-injection devices, and catalysis.

INTERFACES

The boundaries between different materials, or even different orientations of the same material, are called *interfaces*. Studies of interfaces may be directed at short- or long-range phenomena. The former involve the anomalous chemical bonding associated with the mismatch of chemical composition, reconstruction, and localized defects that occur. The latter are due to the electrostatic and elastic fields that accompany charge transfer and lattice strains. Often, the localized phenomena may be explained by introducing interfacial electronic states, similar to the surface states at the solid–vacuum boundary.

The study of interfaces begins with a study of grain boundaries in Section 20.6. These are formed by the contact of finite crystals of the same composition and structure but different orientation. Grain boundaries are common features of most engineering materials, including metals, ceramics, and composite materials.

After this, the focus is on semiconductors in Section 20.7. Metals are more readily described because the Thomas–Fermi screening length is small and electrostatic fields tend to be rapidly healed. In semiconductors, however, the electric fields often extend over fairly large distances. In Section W20.3 we also study electrolytes.

First, a semiconductor in isolation is considered and the band bending accompanying charge injection or extraction is discussed. The metal–semiconductor interface, or Schottky barrier, may readily be understood in terms of this concept (see Section 20.8). Applications to the semiconductor–semiconductor interface are made in the discussion of the heterostructure superlattice (Section 20.9) and the quantum dot (Section 20.10). Finally, the technologically important Si/a–SiO_2 interface is described (Section 20.11).

In Section W20.4 the electrolytic solution will be joined together with the semiconductor to obtain the solid–electrolyte interface.

20.6 Grain Boundaries

When crystals precipitate from the melt, or nucleate from solution or the vapor phase, a number of nucleation sites develop and crystal growth proceeds around each nucleus. The orientations of the crystallites are often randomly distributed. When they join together, grain boundaries (GBs) are formed. Many of the physical properties of the resulting material are determined by the properties of these GBs. The GBs are often imaged directly using high-resolution electron microscopy (HREM), which can show the positions of the atoms. The GBs can be categorized as low angle or high angle. For low-angle GBs the difference in orientations is small and the interface is readily described as an ordered array of dislocations. For large-angle GBs there is an abrupt and large change of orientation at the interface.

A physical description of the GB involves specifying eight *degrees of freedom* (DOF). These arise as follows. Consider two crystal lattices, each filling all of space, one of which will be held fixed and used as a reference system. The second crystal may be translated or rotated relative to the first. Each of these operations involves three DOFs. The translation is described by a displacement vector **T** and corresponds to the three microscopic DOFs. The rotation is described by a unit vector $\hat{n}$ and an angle of rotation θ about that axis. Next introduce the grain-boundary plane, specified by a unit normal $\hat{n}'$. This introduces another two DOFs. On one side of the GB plane, discard all the atoms of one crystal lattice. On the other side of the plane, discard all the atoms of the other crystal lattice. The parameters $\hat{n}$, θ, and $\hat{n}'$ describe the macroscopic DOFs.

An illustration of the formation of a grain boundary is given in Fig. 20.10. The atoms of one lattice are denoted by open circles, and those of the other lattice are denoted by shaded circles. The reference triangle in this example is a 5–12–13 Pythagorean triangle. In the upper right-hand drawing the second crystal is rotated relative to the first through the angle $\tan^{-1}(\frac{5}{12})$. The crystals are brought together and overlapping atoms are discarded. Reconstruction and relaxation of the atoms at the interface may

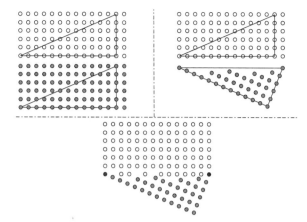

Figure 20.10. Formation of a grain boundary.

be expected. The resulting grain boundary displays a coincidence site lattice (CSL) (depicted by black circles) with a lattice constant ratio (relative to the original crystal) $\Sigma = 13$. The CSL includes atoms that are common to both lattices. The existence of a CSL leads to GBs with lower free energies and stabilizes the structure.

Note that removal of overlapping atoms (which is sterically necessary) leads to microscopic roughness at the grain boundary. This means that in covalent materials there are likely to be dangling bonds present at the interface, which give it chemical characteristics different from that of either crystal. The coordination number of ions in metals and ionic crystals is often anomalous at the interface. Impurities from within the crystal can diffuse to the GB and be bound there. This leads to the phenomenon of *segregation*, where the impurity concentration may be much higher at the GB than within the bulk. Common impurities in steel, for example, include B, N, O, P, S, and Sn. At low, temperatures the diffusivity of these elements is low so the impurities remain dispersed throughout the crystals. At high temperatures the entropic contribution to the free energy also favors dispersion through the bulk. However, in the range 400 to 600°C, segregation to the interface is important. The presence of the impurities at the grain boundaries weakens the chemical bonds between adjacent grains and weakens the steel considerably.

The local environment at the grain boundary often involves structural units not found in bulk crystals. These include the tetrahedron, pentagonal bipyramid, capped trigonal prism, and the Archimedian antiprism (Fig. 20.11). For example, the Archimedian antiprism is found at the $\Sigma = 13$ (001) twist boundary in Ag.

An alternative way to describe the microscopic DOFs is through the use of two unit vectors, both describing the normal to the grain boundary. The first vector, $\hat{n}_1$, is described relative to the lattice-based coordinate system of one of the crystals. The vector $\hat{n}_2$ is described relative to the lattice-based coordinate system of the other crystal. The fifth coordinate is the twist angle θ, which describes the rotation of one of the coordinate systems about the grain boundary normal. It is possible to introduce the tilt unit vector, $\hat{n}_t$, as a vector perpendicular to the plane containing $\hat{n}_1$ and $\hat{n}_2$, and defined by the relation

$$\hat{n}_t = \frac{\hat{n}_1 \times \hat{n}_2}{\sin \psi}, \tag{20.13}$$

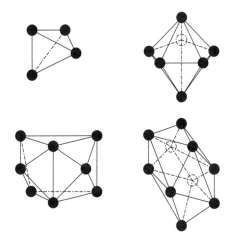

Figure 20.11. Common structural units at grain boundaries. [From C. L. Briant, *Mater. Res. Soc. Bull.*, 15(10), 26 (1990).]

where the tilt angle, ψ, is defined by $\sin \psi = |\hat{n}_1 \times \hat{n}_2|$. For the case $\theta = 0$, the unit cell formed on the grain boundary has the smallest area. It is called an *asymmetric-tilt grain boundary* (ATGB). If there is an *m*-fold rotational axis of symmetry in the plane of the grain boundary, the same ATGB will occur if $\theta = \pi/m$, since this will simply invert the stacking order of the lattice planes.

Symmetrical grain boundaries occur when $\hat{n}_1$ and $\hat{n}_2$ are parallel or antiparallel. In that case there are only three microscopic DOFs. For $\theta = 0$ one obtains a perfect crystal; for other angles one obtains a pure twist boundary. The special case $\theta = \pi/m$ results in twinning of the crystals at the grain boundary. The resulting structure is called a *symmetric-tilt grain boundary* (STGB).

As mentioned, low-angle GBs form ordered arrays of dislocations separated by a large number of lattice constants. Each dislocation is characterized by a Burgers vector. The grain boundary energies per unit area are usually given in terms of the Read–Shockley formula:

$$E(\theta) = \frac{1}{b}[E_c - E_s \ln(\sin \theta)] \sin \theta, \qquad (20.14)$$

where b is the magnitude of the Burgers vector and E_c and E_s represent empirical core and strain field energies per unit length. There are cusp minima at $\theta = 0$ and $\theta = \pi$. This stability is due to the relative smallness of the unit cells in the grain-boundary plane and the closer matching of atomic positions than for other values of θ. Typical values of $E(\theta)$ are 0.2 to 0.5 J/m^2 for Cu and ≈ 0.9 J/m^2 for Au.

The surface tension γ_{GB} can be measured as described in Section 20.3. The energetics of grain boundaries are complex. There is a tendency for atomic alignment to take place, albeit at the expense of strain energy. Sometimes there will be regions of good alignment separated by disordered patches, which account for the overall mismatch of the lattices. All the complexities of surface reconstruction can occur, although now two surfaces are involved. As in the former case, the atomic positions may be altered in such a way as to minimize the areal density of dangling bonds. Lattice relaxation, with an expansion of the lattice volume normal to the interface, often occurs at grain boundaries.

20.7   Band Bending in Semiconductors

When a semiconductor is placed in contact with a metal, insulator, ionic solution, or another semiconductor, a charge transfer between the two materials will take place across the interface in order to equalize the chemical potentials. A space-charge region will form beneath the interface, which leads to band bending in the semiconductor. Various cases are illustrated in Fig. 20.12. The top three diagrams refer to n-type semiconductors and the lower three to p-type semiconductors. The dashed line is the chemical potential (μ). The conduction-band edge (E_c) lies below the vacuum level by an energy equal to the electron affinity, χ.

The various types and degrees of band bending that can occur at semiconductor interfaces are described here. Applications to specific structures (e.g., semiconductor–metal interfaces and semiconductor–electrolyte interfaces) are described in a later section and at our Web site.

In the diagram labeled n_1, electrons are repelled away from the semiconductor interface, leaving a *depletion layer* near the interface. In that layer the majority-carrier electrons are reduced in number and the positive charge of the donor ions is left uncompensated. The resulting electric field leads to band bending and the formation of a potential barrier of height V_B.

In diagram n_2 electrons are attracted into the conduction band toward the interface, giving rise to what is called an *accumulation layer*. Here the bands bend downward due to the excess negative space charge near the interface, and V_B will be negative.

In diagram n_3 the depletion layer is so wide that the valence band is bent above the chemical potential level. In that case there are minority carriers (holes) present in the space-charge region as well as uncompensated donor ions. This is known as *inversion*. Diagrams p_1, p_2, and p_3 give the corresponding pictures for p-type material.

In nondegenerate n-type semiconductors, where the chemical potential is removed from the band edge of the conduction or valence band by at least several $k_B T$, the carrier concentrations are given by

$$n(z) = N_c(T) \exp[-\beta(E_c - e(\phi(z) - \phi(-\infty)) - \mu)], \qquad (20.15a)$$

$$p(z) = N_v(T) \exp[-\beta(\mu - E_v + e(\phi(z) - \phi(-\infty)))], \qquad (20.15b)$$

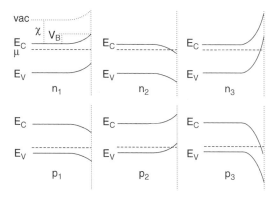

Figure 20.12. Energy-band bending in the space-charge region of semiconductors.

where $\phi(z)$ is the electric potential and $\beta = 1/k_B T$. The product of the carrier concentrations may be written as

$$n(z)p(z) = n_i^2 \equiv N_c(T)N_v(T)\exp[-\beta(E_c - E_v)], \tag{20.16}$$

where $E_c - E_v = E_g$ [see Eq. (11.28)]. The net charge density is given by

$$\rho(z) = e[N_d^+(z) - N_a^-(z) + p(z) - n(z)], \tag{20.17}$$

where $N_d^+(z)$ and $N_a^-(z)$ are the concentrations of ionized donors and acceptors, respectively. In the depletion-layer case (n_1) it is assumed that the space-charge region exists in the zone $-d_n < z < 0$ and that for $z < -d_n$ charge neutrality [i.e., $\rho(z) = 0$], is imposed. The value of d_n depends on how much charge removal (i.e., repulsion) actually takes place between the two materials, and will be determined from boundary-value matching conditions (i.e., from the continuity of the z-component of the electric displacement vector across the interface). In the space-charge region, $N_d^+(z) = N_d^+$ (constant), $N_a^-(z) = 0$, $\beta[E_c - e\phi(z) + e\phi(-\infty) - \mu] \gg 1$, so $n(z) \approx 0$, and $\beta[\mu - E_v + e\phi(z) - e\phi(-\infty)] \gg 1$, so $p(z) \approx 0$. Poisson's equation becomes

$$\frac{d^2\phi}{dz^2} = -\frac{N_d^+ e}{\epsilon}. \tag{20.18}$$

Noting that $\phi(-d_n) = \phi(-\infty)$, one obtains

$$\phi(z) = \phi(-\infty) - \frac{N_d^+ e}{2\epsilon}(z + d_n)^2. \tag{20.19}$$

The energy barrier that is formed by the band bending has the height

$$w_B = eV_B = -e[\phi(0) - \phi(-\infty)] = \frac{N_d^+ e^2 d_n^2}{2\epsilon}. \tag{20.20}$$

For the accumulation-layer case (n_2) electrons are injected into the conduction band, so the charge density becomes

$$\begin{aligned}\rho(z) &= e[N_d^+ - n(z)] \\ &= e\{N_d^+ - N_c(T)\exp[-\beta(E_c - e(\phi(z) - \phi(-\infty)) - \mu)]\}\end{aligned} \tag{20.21}$$

As $z \to -\infty$, $\phi(z) \to \phi(-\infty)$ and $n \to N_d^+$ (for neutrality), so this may be rewritten as

$$\rho(z) = eN_d^+\{1 - \exp[-\beta e(\phi(-\infty) - \phi(z))]\}. \tag{20.22}$$

Poisson's equation becomes

$$\frac{d^2\phi}{dz^2} = -\frac{eN_d^+}{\epsilon}\{1 - \exp[\beta e(\phi(z) - \phi(-\infty))]\}. \tag{20.23}$$

A first integral of this leads to the expression for the electric field:

$$(E_z)^2 = \left(\frac{d\phi}{dz}\right)^2$$

$$= -\frac{2eN_d^+}{\epsilon}\left[\phi(z) - \phi(-\infty) + \frac{1}{\beta e}\{1 - \exp[\beta e(\phi(z) - \phi(-\infty))]\}\right]. \quad (20.24)$$

For the inversion layer case (n_3) the depletion layer has holes present in addition to the donor ions. As $z \to -\infty$, $\phi(z) \to \phi(-\infty)$, $n \to N_d^+$, and $p \to n_i^2/N_d^+$, so

$$\rho(z) = e[N_d^+ + p(z)] = e\left\{N_d^+ + \frac{n_i^2}{N_d^+}\exp[-\beta e(\phi(z) - \phi(-\infty))]\right\}. \quad (20.25)$$

Poisson's equation becomes

$$\frac{d^2\phi}{dz^2} = -\frac{e}{\epsilon}\left\{N_D + \frac{n_i^2}{N_D}\exp[-\beta e(\phi(z) - \phi(-\infty))]\right\}. \quad (20.26)$$

A first integral of this equation leads to the formula

$$(E_z)^2 = -\frac{2e}{\epsilon}\left[N_d^+(\phi(z) - \phi(-\infty)) - \frac{n_i^2}{N_d^+\beta e}\{\exp[-\beta e(\phi(z) - \phi(-\infty))] - 1\}\right].$$
$$(20.27)$$

In all cases the chemical potential is related to the asymptotic potential through the formula

$$\mu = E_c + k_BT \ln \frac{n(T)}{N_c(T)}, \quad (20.28)$$

as expressed in Eq. (11.39). Similar formulas can be derived for band bending in the p-type semiconductor illustrated in Fig. 20.12.

A closer look at the accumulation layer (case n_2 in Fig. 20.12) will now be made. The bending of the bands provides a local potential in which the conduction electrons may move. This one-dimensional potential well may support one or more quantized states bound in the z direction. If the bottom of the well dips sufficiently below the chemical potential, a quantum-mechanical analysis of the charge distribution is called for. Although the motion in the z direction is bound, the motion in the x and y directions is that of free electrons. Therefore it is appropriate to talk about two-dimensional bands.

One approximate approach is through use of the variational principle.[†] One writes a trial ground-state wavefunction of the form

$$\psi(x, y, z) = Bze^{\alpha z + i\mathbf{K}\cdot\mathbf{R}}\Theta(-z). \quad (20.29)$$

Here it is assumed that the electron's wavefunction does not extend out of the solid. The parameter B is a normalization constant, α is an attenuation constant, $\mathbf{K}$ is a

[†] F. Stern and W. E. Howard, *Phys. Rev.*, **163**, 816 (1967).

two-dimensional propagation vector in the xy plane, and $\mathbf{R} = x\hat{i} + y\hat{j}$. One finds that $B = 2\alpha^{3/2}/\sqrt{A}$, where A is the area of the interface. Poisson's equation is

$$\frac{d^2\phi}{dz^2} = \frac{Ne}{\epsilon}|\psi|^2 = \frac{4Ne}{\epsilon}\alpha^3 z^2 e^{2\alpha z}, \tag{20.30}$$

where N is the number of conduction electrons per unit area. Integration leads to

$$\phi(z) = \frac{eN}{4\alpha\epsilon}[-6 + (6 - 8\alpha z + 4\alpha^2 z^2)e^{2\alpha z}]. \tag{20.31}$$

The total Coulomb energy per unit area is

$$\frac{U}{A} = \frac{1}{2A}\int \rho\phi d\mathbf{r} = \frac{33N^2 e^2}{64\epsilon\alpha}. \tag{20.32}$$

The total kinetic energy per unit area is

$$\frac{T}{A} = \frac{\hbar^2}{2m^*}(\pi N^2 + N\alpha^2), \tag{20.33}$$

where m^* is the carrier effective mass. Variation of the total energy per unit area, $E/A = (T + U)/A$, with respect to α gives

$$\alpha = \left(\frac{33Ne^2 m^*}{64\epsilon\hbar^2}\right)^{1/3}. \tag{20.34}$$

The attenuation constant therefore depends on the areal concentration of carriers.

20.8 Schottky Barrier

Assume that an interface is formed between a metal and an n-type semiconductor as shown in Fig. 20.13. The work function of the isolated metal is $W_m = e\phi_m$, and the electron affinity of the semiconductor is χ_s (Fig. 20.13a). It will be assumed that the separation between the metal and semiconductor is made sufficiently small that any potential drop across this separation may be neglected. For the sake of definiteness, assume that the chemical potential in the semiconductor is higher than the Fermi level of the metal. As the two solids are brought together, electrons are transferred from the semiconductor to the metal. These electrons may come either from the interior of the semiconductor or from occupied surface states. In the former case, band bending occurs due to the formation of a depletion space-charge region in the semiconductor and the Schottky barrier is formed (Fig. 20.13b). As the Schottky barrier is formed, the flow of electrons is slowed down and eventually stops. In the latter case a microscopic double layer is formed between the semiconductor and the metal and no band bending occurs (Fig. 20.13c). In this case a layer of positive ions will be on the surface of the semiconductor and excess electrons will reside at the surface of the metal. It is also possible that both effects occur together, so that a reduced barrier is formed (Fig. 20.13d). If an insulating layer were interposed between the metal and the semiconductor, the

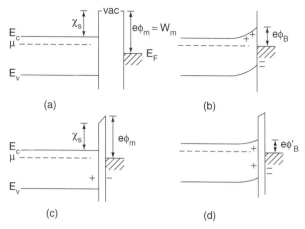

Figure 20.13. Schottky barrier formation: (*a*) isolated semiconductor and metal; (*b*) the semiconductor and metal are in contact and the Schottky barrier is formed; (*c*) elimination of the barrier by Fermi-level pinning; (*d*) partial elimination of the barrier by surface states.

trapezoidal barrier shown in Fig. 20.13*c* and 20.13*d* would be present. If there is no insulating layer, the barrier is absent. Its presence in the diagrams merely signifies that there is a microscopic electric-dipole layer present at the interface with a corresponding electric field present in the insulating layer.

Consider first the case where surface states are not present. The top of the bent conduction band in Fig. 20.13*b* at the interface is an energy χ_s below the vacuum level. The vacuum level is an energy $e\phi_m$ above the Fermi sea of the metal. Thus the height of the barrier from the metal side is

$$e\phi_B = e\phi_m - \chi_s, \qquad (20.35)$$

as shown in Fig. 20.13*b*. The actual amount of band bending, and hence the height of the Schottky barrier from the semiconductor side, is given by the formula

$$-e[\phi(0) - \phi(-\infty)] = e\phi_m - \chi_s - (E_c - \mu). \qquad (20.36)$$

Note that in this formula E_c is a constant. The band bending is described by the electrical potential. The width of the depletion layer for case n_1 of Fig. 20.12 may be determined by combining Eqs. (20.20) and (20.36), giving

$$d_n = \sqrt{\frac{2\epsilon}{N_d^+ e^2}(e\phi_m - \chi_s + \mu - E_c)}. \qquad (20.37)$$

The presence of surface states complicates matters. In the semiconductor these states have energies that lie in the gap region. At finite temperatures those states with energies below the chemical potential are largely occupied and those above are largely vacant. If electrons are removed from the surface states and transferred to the metal, a small double layer will be formed and a potential drop will develop across the intervening space. The bands will not be bent since a depletion region will not be formed. The

Fermi level is said to be *pinned*. This case is illustrated in Fig. 20.13c. In that case $\phi(0) = \phi(\infty)$ and $e\phi_m - \chi_s = E_c - \mu$. If some surface states are emptied, a very small movement of μ can occur. The Fermi level is pinned, but not by 100%.

In reality, surface states are always present and the situation is intermediate between the two extremes. This is illustrated in Fig. 20.13d. Let S represent the fraction of electrons transferred from the depletion region and $1 - S$ be the fraction transferred from surface states. The height of the Schottky barrier from the metal side becomes $e\phi'_B = S(e\phi_m - \chi_s) + (1 - S)(E_c - \mu)$. The amount of band bending is given by

$$-e[\phi(0) - \phi(-\infty)] = e\phi'_B - (E_c - \mu) = Se(e\phi_m - \chi_s) - S(E_c - \mu). \qquad (20.38)$$

Typical values of S for common semiconductors (Ge, Si, GaAs, GaP, ZnO, SiO_2, $SrTiO_3$) are 0.09, 0.09, 0.15, 0.29, 0.94, 0.14, and 0.46, showing that surface states have a very significant effect. Typical Schottky-barrier heights for semiconductor-metal systems at $T = 300$ K are 0.72 V for n-Si/Al, 0.58 V for p-Si/Al, 0.80 V for n-GaAs/Al, and 0.82 V for n-GaAs/Cu.

Schottky barriers are formed when metallic leads are attached to semiconductor elements in electronic devices. They function as rectifiers. The rectification behavior may be understood by examining Eq. (20.36) for the case where an additional bias voltage V is supplied across the metal–semiconductor junction. The width of the depletion region changes to

$$d_n = \sqrt{\frac{2\epsilon}{N_d^+ e^2}(e\phi_m - \chi_s - eV + \mu - E_c)}, \qquad (20.37')$$

For forward biasing, $V > 0$, the width of the depletion region is reduced. Since the thickness of the barrier becomes smaller, the tunneling current is larger. The opposite is true for reverse bias when $V < 0$.

Schottky barriers have found application as optoelectronic devices. For example, there are Si Schottky-barrier solar cells and infrared detectors. An examination of Fig. 20.13b–d shows that electrons may be excited from the Fermi surface of the metal to the conduction band of the semiconductor by a photo-induced tunneling process. The threshold for the process is $\hbar\omega = E_c - \mu$.

Schottky barriers are used as the basis for MESFETs, as described in Chapter W11. Arrays of Schottky barriers may be used for infrared imaging.

A tunable diode was recently introduced[†] based on the interface between n-type InP and the polymer polypyrrole. A value $S = 0.44$ was measured. By connecting the polymer to an electolyte solution held at various potentials, the chemical potential could be changed by as much as 0.6 V, allowing the diode threshold to be tuned similarly.

20.9 Semiconductor–Heterostructure Superlattices

Alternate layers of two types of semiconductors may be grown epitaxially upon each other if the lattice constants and symmetry are not too different. In this way one constructs a solid with a one-dimensional periodicity in the growth direction. Each

---

[†] M. C. Longgran, *Science*, **278**, 2103 (1997).

individual layer of the solid consists of many atomic layers, so local band structures may be assigned to them. For the sake of definiteness the focus will be on a superlattice composed of $Ga_{1-x}Al_xAs/GaAs$, although P or Sb may be substituted for As and In for Al. The semiconductors GaAs and AlAs are very close in lattice constant (i.e., 0.565 nm versus 0.566 nm; see Table 11.9 and Fig. 11.23). Note that the bandgaps are 1.42 and 2.16 eV, respectively. The periodicity is denoted by a. The thickness of the $Ga_{1-x}Al_xAs$ layer is b, and the thickness of the GaAs layer is $a - b$. The bottoms of the conduction bands are E_c and $E_c + \Delta E_c$, respectively, where ΔE_c is called the *offset*. The corresponding values for the tops of the valence bands are E_v and $E_v - \Delta E_v$. This is illustrated (for an idealized case) in Fig. 20.14. In reality, some band bending occurs near the interfaces and the interfaces are not perfectly sharp.

If the distance $a - b$ is sufficiently large so that tunneling between the wells is not important, one refers to the structure as a *multiple quantum well* (MQW). Quantum confinement in low-dimensional systems is discussed in Section 11.6. The structure depicted in Fig. 20.14 is called a *type I superlattice*. It occurs in $Ga_{1-x}Al_xAs/GaAs$ when $a - b > 2.0$ nm and $x < 0.3$. If the band structure of the valence band is inverted, so that the bands both rise or fall simultaneously, it is called a type II superlattice. Such a structure occurs, for example, for $a - b < 2.0$ nm and $x = 1$.

The depressed regions in the conduction band of GaAs are a set of square potential wells for the electrons. Similarly, the raised regions of the valence band act as potential wells for holes. When isolated, the square wells may each support one or more bound states. When the wells are in proximity, tunneling links the wells in a given band and the bound states are broadened into subbands. Although this description is adequate for the conduction band, things are more complicated near the top of the valence band, where fine structure due to spin–orbit coupling must be taken into account and a multiplicity of bands are involved. The discussion is limited to the conduction band.

Expressions for determining the quasibound states in the conduction band follow from the form of the wavefunctions and as consequences of periodicity. They are obtained by extending the Kronig–Penney model introduced in Section W7.3. Each of these bound states actually corresponds to a subband, with crystal wave vector

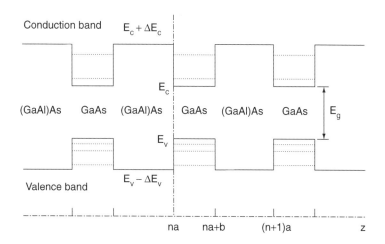

Figure 20.14. Idealized band structure for a $Ga_{1-x}Al_xAs/GaAs$ superlattice.

components k_x and k_y, since the electrons are still free to move in the directions perpendicular to the superlattice. In the region $na < z < na + b$, the wavefunction is

$$\psi(z) = A_n e^{ip(z-na)} + B_n e^{-ip(z-na)}, \tag{20.39a}$$

and when $na + b < z < (n+1)a$,

$$\psi(z) = C_n e^{-q(z-na)} + D_n e^{q(z-na)}. \tag{20.39b}$$

Here the wave vector is

$$p = \sqrt{\frac{2m_1^*}{\hbar^2}(E - E_c) - k_x^2 - k_y^2}, \tag{20.40a}$$

where $E < E_c + \Delta E_c$, and the attenuation constant is

$$q = \sqrt{\frac{2m_2^*}{\hbar^2}(E_c + \Delta E_c - E) + k_x^2 + k_y^2}. \tag{20.40b}$$

A difference in effective masses in the two semiconductors is allowed for, but the bands are assumed to be parabolic. The Bloch periodicity condition requires that

$$A_{n+1}e^{-ipa} = A_n e^{ika}, \tag{20.41a}$$

$$B_{n+1}e^{ipa} = B_n e^{ika}. \tag{20.41b}$$

where $\hbar k$ is the crystal momentum along the z axis. The continuity conditions on wavefunction and velocity, $(-i\hbar/m^*)\partial\psi/\partial z$, at $z = na + b$ yield

$$R\begin{bmatrix} A_n \\ B_n \end{bmatrix} = S\begin{bmatrix} C_n \\ D_n \end{bmatrix}, \tag{20.42a}$$

and these conditions at $z = (n+1)a$ give

$$U\begin{bmatrix} A_n \\ B_n \end{bmatrix} = T\begin{bmatrix} C_n \\ D_n \end{bmatrix}, \tag{20.42b}$$

where the matrices R, S, T, and U are

$$R = \begin{bmatrix} e^{ipb} & e^{-ipb} \\ \dfrac{ip}{m_1^*}e^{ipb} & -\dfrac{ip}{m_1^*}e^{-ipb} \end{bmatrix}, \qquad S = \begin{bmatrix} e^{-qb} & e^{qb} \\ -\dfrac{q}{m_2^*}e^{-qb} & \dfrac{q}{m_2^*}e^{qb} \end{bmatrix} \tag{20.43a}$$

$$T = \begin{bmatrix} e^{-qa} & e^{qa} \\ -\dfrac{q}{m_2^*}e^{-qa} & \dfrac{q}{m_2^*}e^{qa} \end{bmatrix}, \qquad U = e^{ika}\begin{bmatrix} e^{ipa} & e^{-ipa} \\ \dfrac{ip}{m_1^*}e^{ipa} & -\dfrac{ip}{m_1^*}e^{-ipa} \end{bmatrix}. \tag{20.43b}$$

Introducing the two-component vectors

$$\xi_n = \begin{bmatrix} A_n \\ B_n \end{bmatrix}, \qquad \eta_n = \begin{bmatrix} C_n \\ D_n \end{bmatrix}, \tag{20.44}$$

one has

$$\xi_n = R^{-1}S\eta_n, \qquad \eta_n = T^{-1}U\xi_n. \tag{20.45}$$

The solvability condition is

$$\det(R^{-1}ST^{-1}U - I) = 0, \tag{20.46}$$

where I is the unit matrix. After some algebra one arrives at the dispersion equation

$$\cos(ka) = \cosh[(a-b)q]\cos[(a-b)p] - \left(\frac{m_1^* q}{2m_2^* p} - \frac{m_2^* p}{2m_1^* q}\right) \times$$
$$\sinh[(a-b)q]\sin[(a-b)p], \tag{20.47}$$

which is an implicit relation for the subbands $E_j(k)$. These are illustrated by the dashed lines in Fig. 20.14 and are usually rather flat bands. There will always be at least one subband in the well.

Multiple quantum-well structures find applicability in various electronic devices, including quantum-cascade lasers and solar cells.

20.10 Quantum Dot

A quantum dot is a structure composed of several layers of semiconductors that is capable of confining one or more electrons in the conduction band in all three dimensions. Its spectrum will therefore consist of a series of discrete energy levels. A typical geometry is illustrated in Fig. 20.15. Sandwiched between the n-doped GaAs are layers of insulating AlGaAs and a central layer of InGaAs. Electrons in the conduction band of InGaAs are trapped by the insulating layers in the vertical direction. An insulating sleeve around the structure prevents electrons from leaking out in the radial direction. Thus the electrons are confined to a thin disk of InGaAs. Typical dimensions for this disk might be 10 nm in the vertical direction and 100 nm for the radius. The n-GaAs jackets are connected to metallic layers to which external electrodes are attached. The base, which is usually part of the substrate, is called the *source* and the top is called the *drain*.

A conducting ring, called the *gate*, surrounds the structure and may be negatively charged by connecting it to an external voltage V_g. The purpose is to provide

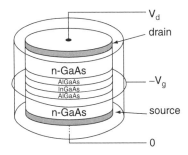

Figure 20.15. Schematic depiction of a typical quantum dot.

a confining potential well which may be considerably smaller in radius than that of the disk. To see how this works, consider the electrostatic potential along the axis of a ring of radius a embedded in a medium of permittivity ϵ. It is given by $\phi(0, z) = Q/4\pi\epsilon\sqrt{a^2 + z^2}$, where Q is the charge on the ring. Near $z = 0$, $\phi(0, z) \approx Q[1 - z^2/2a^2]/4\pi\epsilon a$. The potential satisfies the Laplace equation $\nabla^2\phi = 0$. Near the center of the disk the off-axis potential is therefore given by $\phi(R, z) \approx \phi(0, z) + QR^2/16\pi a^3\epsilon$, which is a harmonic potential in the radial direction. The *spring constant* of this potential is proportional to the charge on the ring. This, in turn, is equal to the product of voltage V_g and the capacitance of the ring. (Note that the ring must have a finite thickness in order to possess a finite capacitance.) The natural frequency of the harmonic oscillator is given by $\omega = (eQ/8\pi m^*\epsilon a^3)^{1/2}$.

Due to the cylindrical symmetry, the wavefunction may be expressed as the product of a function of z and a function of R. If the cylinder is thin enough, the z wavefunction will correspond to a particle in the lowest energy state of a three-dimensional quantum well. The radial wavefunction corresponds to a state of the two-dimensional harmonic oscillator. The radial extent of the wavefunction is determined by the gate voltage V_g. Electrons from the n-GaAs layer can tunnel through the AlGaAs layers and populate the discrete quantized states of the InGaAs quantum dot. The separation between the energy levels will be determined by the spring constant, and hence by V_g. All states lying below the chemical potential of n-GaAs will be populated. The number of occupied states is therefore also controlled by V_g.

The discrete energy spectrum of the dot will be given by the formula

$$E_{n_x, n_y, n_z} = \hbar\omega\left(n_x + \frac{1}{2}\right) + \hbar\omega\left(n_y + \frac{1}{2}\right) + \frac{\hbar^2}{2m^*}\left(\frac{n_z\pi}{b}\right)^2. \tag{20.48}$$

where n_x and n_y are the quantum numbers of the two-dimensional harmonic oscillator (each taking on the values $0, 1, 2, \ldots$), b is the thickness of the InGaAs layer, and n_z is the quantum number of the one-dimensional well along the z direction ($n_z = 1, 2, 3, \ldots$). For simplicity's sake the well is taken to have infinite walls. The tunneling of electrons into and out of the quantum dot may be accounted for by adding explicit tunneling interactions into the Hamiltonian describing the system.

The quantum dot may be regarded as an artificial atom. Like the atom, it exhibits shell structure. Assume that b is sufficiently small so that only the $n_z = 1$ state is populated. The first shell has the quantum numbers (0,0,1) and can accommodate two electrons, one with spin up and the other with spin down. The next shell can hold four electrons in the states (1,0,1) and (0,1,1), each with two spin states. The next shell has six electrons, corresponding to the states (2,0,1), (1,1,1) and (0,2,1). As a function of the total number of conduction electrons on the dot, N, there will be "magic" numbers at $N = 2, 6, 12, 20, \ldots$ at which shells are completed.

Suppose that one starts with V_g so large that no quantized states lie below the chemical potential of the n-GaAs. Then suppose that one lowers V_g and allows first one electron, then another electron, and so on, to enter the well. On the basis of a one-electron picture it would be expected that the ground state will be occupied by two electrons, one with spin up and the other with spin down. One would then expect the next excited state to be filled pairwise in the same way, and so on. But this is not what happens. It is often found that only one electron goes into one spatial state, the next electron goes into the next spatial state with its spin parallel to the first, and the pattern continues until a "shell" is half-filled. After that, additional electrons enter with opposing spins until the entire shell is filled. This is reminiscent of atomic physics, where Hund's rule says that the state of lowest energy corresponds to one in which the electron spins are all parallel to each other until the shell is half-filled (see Section 9.2). The reason for this has to do with the Coulomb interaction between the electrons. There is a charging energy associated with placing the electrons on the disk. A state that has all spins parallel is symmetric under the interchange of any two spins. Since the overall wavefunction must be antisymmetric under interchange of two electrons, the spatial part of the wavefunction must be antisymmetric. This favors configurations where the charges stay far apart from each other. In such configurations the Coulomb repulsion energy is less than it would be if the electrons were to congregate in a symmetric state. This could also be seen by examining the charge density of a charged conducting disk, which is $\sigma = \text{const} \times (a^2 - R^2)^{-1/2}$, where a is the radius of the disk. In this case the disk edges are preferred over the interior for charge accumulation.

By changing the voltage V_d across the upper and lower electrodes one may arrive at situations in which varying numbers of electrons are made to reside in the well (Fig. 20.16). The charging energy ΔU may be obtained by comparing the energy with $N + 1$ electrons residing on the central disk with the energy for having N electrons residing there. The energy for $N + 1$ electrons is given by $U(N + 1) = (N + 1)^2 e^2/2C$, where C is the total capacitance of the disk. The corresponding energy for having N electrons on the disk and one on the electrode is $U(N) = N^2 e^2/2C - eV_d$. Here V_d is the potential on the electrode and is given by $V_d = -eN/C$. Hence $\Delta U = e^2/2C$. A similar, inverted picture would hold for p-GaAs layers and holes as for n-GaAs layers and electrons.

If the potential of the electrode is raised, no current will flow until there is enough potential energy difference to provide the charging energy. When no current flows, the situation is referred to as a *Coulomb blockade*. When current does flow, it proceeds by having a single electron hopping onto the disk. The electron can subsequently hop onto the other electrode. If the potential is raised enough so that two hopping electrons can be accommodated on the disk at once, the current-carrying capacity of the system is increased. In fact, the quantum dot exhibits a Coulomb staircase, in which the current increases in a sequence of steps as the potential difference is increased. The gate potential also controls the conductivity of the quantum dot. Sometimes one refers to the device as a *single-electron transistor*. A single-electron transistor has been used as a scanning probe to map work-function variations along the surface.

Examples of compositions of quantum dots that have been manufactured are presented in Table 20.2. Also given are the stoichiometric ranges and the energies of some of the interband-transition photons.

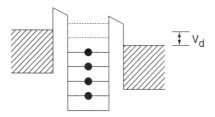

Figure 20.16. Control of the population of electrons in the quantum well by means of an external bias.

TABLE 20.2   Typical Quantum-Dot Systems

System	Stoichiometric Coefficient	Interband Transition Energy (eV)
$In_xGa_{1-x}As/GaAs$	$0.2 < x < 0.1$	1.3–1.14
$Al_xIn_{1-x}As/Ga_yAl_{1-y}As$	$0 < x < 0.7$	1.89–1.42
	$0 < y < 1$	
Ge_xSi_{1-x}/Si	$x = 0.5$	0.82, 0.92
$InP/In_xGa_{1-x}P$	$x = 0.61$	1.655
$GaSb/GaAs$	—	1.07

Source: Data from P. M. Petroff and G. Medeiros-Ribeiro, *Mater. Res. Soc. Bull.*, Apr. 1996, p. 50.

20.11   Si/a-SiO$_2$ interface

The interface between crystalline Si and its insulating amorphous oxide a-SiO$_2$ is one of the most intensively studied systems in materials science and in condensed-matter physics and chemistry, due to the critical role that it plays in passivating the Si surface in modern Si-based electronic devices, especially in the metal–oxide–semiconductor field-effect transistor (MOSFET). Indeed, one of the most important reasons for the dominance of Si in microelectronics is the excellent set of properties that this interface can have. These properties include low defect densities and high physical and chemical stability under most conditions.

As dimensions of semiconductor devices continue to shrink into the nanometer region, oxide thicknesses of ≈ 1 to 5 nm will eventually be required. For such thin oxides the interface and near-interface regions can be expected to play even more important roles in the operation of devices. Although it is clear that Si/a-SiO$_2$ interfaces can be fabricated which are essentially defect-free, the few remaining defects and the actual atomic-scale structure of the interface are controversial topics and remain areas of active research. The focus here will be on the physical and chemical structure of the interface[†] on the atomic level and its effect on the electrical properties of the underlying Si.

The formation of the a-SiO$_2$ layer on Si is described in Section 21.7. The usual procedure involves the reaction of the Si surface with O$_2$, H$_2$O, or moist O$_2$ at temperatures from 800 to 1200°C, the net reaction being $Si(s) + O_2(g) \rightarrow$ a-SiO$_2$(s). The oxidizing species diffuses through the growing oxide layer and attacks the Si–Si bonds of the substrate, resulting in the growth of additional oxide at the existing Si/a-SiO$_2$ interface.

The Si/a-SiO$_2$ interface is the region in which the bonding of Si atoms undergoes a transition from tetrahedral Si–Si$_4$ bonding units in bulk Si to Si–O$_4$ units in the bulk oxide. Although often taken to be atomically abrupt, there is clear evidence from Si 2p core-level photoelectron spectroscopy for the presence of intermediate oxidation states of Si (i.e., Si–Si$_3$O, Si–Si$_2$O$_2$, and Si–SiO$_3$) in the interface region. The interface can therefore be thought of as a region corresponding to a suboxide of Si (i.e., a-SiO$_{2-x}$). The interface is typically observed to have a finite thickness and does not correspond

---

[†] A useful recent review of the Si/a-SiO$_2$ system has been presented by C. R. Helm and E. H. Poindexter, *Rep. Prog. Phys.*, **57**, 791 (1994).

to a single atomic plane across which Si–O bonds are formed. The minimum-possible thickness of the interface is the thickness of the region in which Si with its intermediate oxidation states is found. The effective width of the regions near the interface in which the properties of the substrate Si and oxide deviate from their bulk values can, however, be wider than this.

A very simplified schematic model for the cross section of a Si/a-SiO$_2$ interface is presented in Fig. 20.17. The interface or transition region is proposed to be quite narrow, ≈ 0.2 to 0.3 nm wide, containing $\approx 10^{19}$ m^{-2} Si atoms bonded to one, two, three, or four oxygens. The near-interface region on the oxide side is about 2 nm wide, has the SiO$_2$ stoichiometry, but has properties that deviate from those of bulk SiO$_2$. The near-interface region on the Si side consists of a few monolayers of Si, containing $\approx 10^{19}$ m^{-2} Si atoms which deviate appreciably from their bulk positions. A cross section of a Si/a-SiO$_2$ interface obtained via high-resolution transmission electron microscopy (HRTEM) is presented in Fig. 4.1e.

Various structural models of the Si/a-SiO$_2$ interface have been proposed. One proposal for Si(100)/a-SiO$_2$ consists essentially of a continuous random network (CRN) model in which all Si and O atoms at the interface are fully bonded. The interface is ideally flat and contains 10^{19} m^{-2} Si atoms, all in Si–Si$_2$O$_2$ bonding units. [The density of Si atoms on the ideal Si(100) surface is 6.78×10^{18} m^{-2}.] Real interfaces with Si having other intermediate oxidation states are clearly more disordered than this "ideal" CRN interface.

The concentration of Si atoms decreases by a factor of 2 in going from bulk Si to bulk a-SiO$_2$. As a result, there is a large volume expansion of the Si lattice as Si is converted to a-SiO$_2$. Although large stresses can be associated with this expansion, it is observed that oxide films grow stress-free above about $T = 950°C$ via "viscous" flow. Due to this oxidation-induced lattice expansion, the oxide adjacent to the interface should be under compressive stress, while the Si substrate adjacent to the interface should be under tensile stress. As a result of this stress, Si interstitials are emitted from the interface region into the Si bulk during oxidation.

Less than one defect per 10^4 Si atoms is typically present at device-quality Si/a-SiO$_2$ interfaces. This corresponds to surface-defect densities of $\approx 10^{14}$ m^{-2}. The most common defects associated with the interface consist of interface states with energies in the Si bandgap, fixed-charge states in the oxide, trap states associated with recombination centers which can capture electrons or holes, and mobile ions. Only electrical and electron-spin resonance (ESR) characterization techniques are capable of detecting

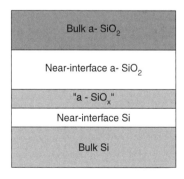

Figure 20.17. Simplified schematic cross-section of the Si/a-SiO$_2$ interface region.

defects at these low levels, the latter technique being appropriate only when the defects are paramagnetic. The dominant defect center identified by ESR spectroscopy and associated with interface traps is an unpaired electron localized in an sp^3-like dangling bond on a Si atom, apparently in an Si–Si$_3$db unit (where "db" stands for the dangling bond). These dangling bonds, known as P_b centers, relieve the strain present at the interface in their vicinity.

Only the P_b center has been identified by ESR at the Si(111)/a-SiO$_2$ interface. For the Si(100)/a-SiO$_2$ interface, the dominant system used in technological applications, two centers labeled P_{b0} and P_{b1} have been identified. The P_b and P_{b0} centers are electrically active, but apparently the P_{b1} center is not. The P_b and P_{b0} centers have multiple charge states and hence multiple levels in the bandgap, depending on whether or not they have trapped carriers. A suggestion is that the P_{b0} centers at the Si(100)/a-SiO$_2$ interface are actually P_b centers which are associated with Si(111) microfacets. Since the P_{b1} centers are apparently not electrically active, this suggestion may explain why the Si(100)/a-SiO$_2$ surface dominates semiconductor device fabrication. This is because appropriate processing conditions can minimize (111) facets and therefore the electrically active and hence undesirable P_{b0} centers associated with this interface.

Passivation of these P_b centers can occur when atomic or molecular hydrogen reacts with the dangling bonds, forming Si–Si$_3$H bonding units. Hydrogen is usually present at or near the interface and passivates the P_b centers via the reaction (a) $P_b + H \rightarrow P_bH$ or (b) $P_b + H_2 \rightarrow P_bH + H$. It is extremely difficult to detect directly the presence of such low concentrations of interface H. Annealing in hydrogen is observed to regenerate or depassivate previously hydrogen-passivated dangling bonds via the reverse of reaction (b).

MULTILAYERS

Techniques such as molecular-beam epitaxy (MBE) and sputtering deposition allow for the deposition of layers of material of controlled thicknesses upon substrates or existing films. Multilayer films are often periodic structures built up by alternating the chemical composition of the constituent layers. Their physical properties may be quite different from those of the bulk materials, especially if the layers are thin, since a substantial fraction of the atoms reside at or near the interfaces. Multilayer deposition is a process that is an alternative to alloying together a mixture of materials to try to fashion a composite material.

Some applications of multilayer materials have already been encountered in earlier chapters, such as the multilayer ceramic capacitor (MLCC) in Section W15.1, magnetic multilayers for sensors in read heads in Chapter W17, and multiwell semiconductor heterostructures and photonic-bandgap materials in Section W18.3. Others studied here include the fabrication of mirrors for use in the soft x-ray and vacuum ultraviolet regions of the spectrum and the creation of mechanically hard films.

20.12   X-ray Mirrors

Reflectors for the soft x-ray (0.4 to 12 nm) and vacuum ultraviolet (12 to 80 nm) portions of the spectrum may be constructed by alternating low- and high-Z layers of atoms in a multilayer structure. If the high-Z layers are spaced $\lambda/2$ apart, the reflected rays at normal incidence will all be in phase and constructively interfere. By varying the angle of incidence one can tune through different wavelengths for constructive interference. Obviously, the degree of roughness at the outer surfaces and the interfaces will influence the reflectivity. In well-controlled cases the reflectivity can be made to be high ($\approx 50\%$). While surface and interface roughness do diminish the reflectivity, one finds that it is still possible to maintain some degree of specular reflection as long as one can define a mean-interface location independent of the size of the mirror.

Consider a structure created from a stack of N_L equally spaced parallel slabs of high-Z material. The center-to-center distance between the slabs is D. Each slab consists of n layers of atoms separated by distance a, and each layer has N atoms. The scattering amplitude for the x-rays is given by

$$F(\mathbf{q}) = \sum_{l=1}^{N_L} \sum_{\mathbf{R} \in V_l} e^{i\mathbf{q} \cdot (\mathbf{R} + lD\hat{k})} f(\mathbf{q}), \tag{20.49}$$

where V_l is the volume of slab l, $f(\mathbf{q})$ is the atomic form factor and $q = 2k \sin(\theta/2)$, θ being the scattering angle and k the propagation vector. The contribution from the low-Z layers is neglected, for simplicity. Let $\mathbf{R} = \mathbf{R}_\parallel + va\hat{k}$ and do the $\mathbf{R}_\parallel$ sum and the l sum, to obtain

$$F(\mathbf{q}) = \sum_{l=1}^{N_L} \sum_{\mathbf{G}_\parallel} N \delta_{\mathbf{q}_\parallel, \mathbf{G}_\parallel} \sum_{v=1}^{n} f(\mathbf{q}) e^{iq_\perp(va+lD)}. \tag{20.50}$$

The most important transverse wave vector $\mathbf{q}$ will generally be the one that produces specular reflection (i.e., $\mathbf{G}_\parallel = 0$). Restrict attention to backscattering, $\theta = \pi$ and $q = 2k$. For the case where $kD = \pi$, this reduces to

$$F = NN_L \sum_{v=1}^{n} e^{-i2kva} f(-2k). \tag{20.51}$$

This gives the scattering intensity:

$$|F|^2 = (NN_L)^2 |f(-2k)|^2 \left[\frac{\sin(n\pi a/D)}{\sin(\pi a/D)} \right]^2. \tag{20.52}$$

Even if $f(-2k)$ is small, the coherent factor $(NN_L)^2$ can lead to appreciable backscattering.

Typical multilayers used for x-ray mirrors are W/C, Ni/C, Co/C, Rh/C, Mo/Si, W/Si, W/B$_4$C, NiTi, AgAl, and W/Mg$_2$Si. In addition to fabricating multilayer mirrors, it is also possible to create other optical elements for x-rays, such as polarizers, zone plates, gratings, and filters. By choosing to look at wavelengths close to x-ray resonances, the atomic structure factor f can be made large and scattering is further enhanced.

20.13 Hardness of Multilayers

Substantial improvement of the hardness and yield strength of materials may be achieved by the use of a multilayer geometry. The focus here is on TiN/NbN multilayers with superlattice periodicities of size $\Lambda \approx 10$ nm, where a Vickers hardness

(VHN) (whose scale is based on deformation by a symmetric diamond tip) of 50 GPa has been reported. This hardness is equal to that of BN and is in a league with diamond, whose Vickers hardness is ≈ 100 GPa. The value exceeds by more than a factor of 2 that which would be expected based on averaging the separate hardnesses for TiN (VHN $= 20$ GPa) and NbN (VHN $= 14$ GPa).

The reason for the symbiotic increase may be understood in terms of dislocation motion. The harder a material is, the more difficult it is for the dislocations to move. Motion of dislocations results in plastic flow and permanent deformation. In a dislocation the atomic positions differ from what they would be in an ordered crystal. The energy of the associated strain field is proportional to the shear modulus G. If a dislocation were to move from a material with a lower G to a material with a higher G, it would have to obtain the additional energy required from the applied external stress. The threshold for the dislocation stress in the multilayer would therefore be proportional to the difference in the G values between the two materials. In the multilayer, the abrupt change in elastic modulus in going from one material to another therefore produces a barrier for dislocation motion. The stress needed to surmount this barrier is proportional to the discontinuity parameter $Q = (G_A - G_B)/(G_A + G_B)$, which is substantial for TiN/TiB ($Q = 0.15$). For NbN/TiN, $Q = 0.098$.

As the parameter Λ decreases, the distance over which dislocations can move also decreases and the hardness increases. However, the interfaces are not atomically flat and, in reality, have a finite extent. In addition, most of the strain energy of the dislocation is localized within a finite volume. If Λ is decreased below these ranges, the material becomes essentially homogeneous and there are no longer boundaries to act as barriers to dislocation motion. Then the hardness will drop with decreasing size. For TiN/NbN the optimum Λ is roughly 5 nm.

The hardness enhancement found in multilayers is reminiscent of the Hall–Petch expression, Eq. (10.43), in which the strength of polycrystalline materials is found to increase with decreasing grain size. There, too, the effect is due to the inhibition of dislocation motion by the grain boundaries. However, for small-enough grains the strength becomes diminished by the sliding motion of the grains. The optimum strength typically occurs at grain sizes of around 10 nm. The material is plastically deformed by a series of slide-and-stick events. Whether such slippage is operative between layers in multilayer materials remains to be seen.

20.14 Stoichiometric Optimization of Physical Parameters

Often, one is interested in optimizing a specific physical parameter by varying the stoichiometric composition of a material. Vapor deposition combined with masking techniques provide a method for synthesizing a continuous range of compositions (a "library") on a single substrate. An appropriate microprobe scanning over the film determines the geometric location of optimum physical response and the stoichiometric composition may be read directly. Such a method is called *combinatorial screening*. In principle it could be employed to optimize piezoelectric materials, nonlinear-optical materials, high-temperature superconductors, dielectrics, and so on.

The method is illustrated here for a ternary compound of the generic form $A_u B_v C_w$ with $u + v + w = 1$. The ternary phase diagram is illustrated in Fig. 20.18. Three materials, labeled A, B, and C, are to be deposited as a film on a surface in such a way that the stoichiometric composition varies continuously over the area of the film. The

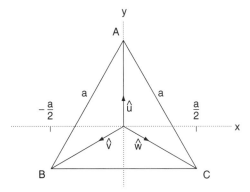

Figure 20.18. Ternary triangle describing the compound $A_u B_v C_w$.

film occupies an equilateral triangular region of the surface, of side a. Let $\hat{u}$, $\hat{v}$, and $\hat{w}$ denote three unit vectors, 120° apart, directed toward the vertices, as shown. A point in the triangle may be expressed as $\mathbf{r} = (a/\sqrt{3})(u\hat{u} + v\hat{v} + w\hat{w})$, with u, v, and w in the range 0 to 1.

Moving masks have their velocities programmed so as to allow the material densities deposited on the surface to vary spatially in a linear fashion. The number of moles per unit area of materials A, B, and C are made to follow the relations

$$\sigma_A = \frac{\sigma}{3}\left(1 + 2\sqrt{3}\frac{\hat{u} \cdot \mathbf{r}}{a}\right), \tag{20.53a}$$

$$\sigma_B = \frac{\sigma}{3}\left(1 + 2\sqrt{3}\frac{\hat{v} \cdot \mathbf{r}}{a}\right), \tag{20.53b}$$

$$\sigma_C = \frac{\sigma}{3}\left(1 + 2\sqrt{3}\frac{\hat{w} \cdot \mathbf{r}}{a}\right). \tag{20.53c}$$

In this case the concentration of A grows in the $\hat{u}$ direction. Similarly, the B material is deposited so as to increase in the $\hat{v}$ direction and material C to increase in the $\hat{w}$ direction. Note that this guarantees a uniform net number of moles per unit area; that is, the net number of moles per unit area is constant:

$$\sigma_A + \sigma_B + \sigma_C = \sigma. \tag{20.54}$$

The area of the film is $a^2\sqrt{3}/4$.

In Fig. 20.19 a sketch is presented of a discovery library that was deposited on a substrate in order to find useful photoluminescent materials yielding blue-white light when illuminated with ultraviolet radiation. Adjacent films of SnO_2, V, Al_2O_3:V, and Al_2O_3 were deposited, followed by graded layers of La_2O_3, Y_2O_3, and $SrCO_3$. On top, and across these, graded layers of Eu_2O_3, Tb_4O_7, Tm_2O_3, and CeO_2 were placed. Of the approximately 25,000 compounds screened, it was found that Sr_2CeO_4 was an excellent photoluminescent material with a quantum efficiency of 0.48. This is illustrated by the shaded area in Fig. 20.19.

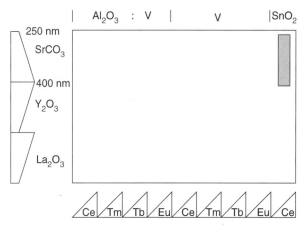

Figure 20.19. Discovery library for photoluminescent materials. The trapezoids and triangles represent graded deposition profiles. [Reprinted with permission from E. Danielson et al., *Science*, **279**, 838 (1998). Copyright 1998 by the American Association for the Advancement of Science.]

Combinatorial screening has been used in a number of problems, including catalyst design, the construction of fuel cells, and the generation of new dielectric materials.

REFERENCES

Thin Films

Gupta, V., An evaluation of the interface tensile strength–toughness relationship, Mater. Res. Soc. Bull., Apr. 1991, p. 39.

Morphology Maps

Messier, R., et al., J. Vac. Sci. Technol., **A2**, 500 (1984).

Thornton, J. A., J. Vac. Sci. Technol., **11**, 666 (1974); **A4**, 3059 (1986).

Langmuir–Blodgett Films

Petty, M. C., *Langmuir–Blodgett Films: An Introduction*, Cambridge University Press, New York, 1996.

Roberts, G., eds., *Langmuir-Blodgett Films*, Plenum Press, New York, 1990.

Grain Boundaries

Bollman, W., *Crystal Defects and Crystalline Interfaces*, Springer-Verlag, Berlin, 1970.

Briant, C. L., Grain boundary chemistry and reactions in metals, Mater. Res. Soc. Bull., Oct. 1990, p. 26.

Merkle, K. L., and D. Wolf, Structure and energy of grain boundaries in metals, Mater. Res. Soc. Bull., Sept. 1990, p. 43.

Interfaces

Luth, H., *Surfaces and Interfaces of Solid Materials*, 3rd ed., Springer-Verlag, Berlin, 1995.

Schottky Barriers

Rhoderick, E. H., and R. H. Williams, in P. Hammond and R. L. Grimsdale, eds., *Metal–Semiconductor Contacts*, Vol. 19, Oxford University Press, Oxford, 1988.

Sze, S. M., *Physics of Semiconductor Devices*, Wiley, New York, 1981.

Semiconductor Heterostructure Superlattices

Singh, J., *Semiconductor Devices: An Introduction*, McGraw-Hill, New York, 1994.

Weisbuch, C., *Quantum Semiconductor Structures: Fundamentals and Applications*, Academic Press, San Diego, Calif., 1991.

Yu, P. Y., and M. Cardona, *Fundamentals of Semiconductors: Physics and Material Properties*, Springer-Verlag, Berlin, 1996.

Quantum Dot

Kouwenhoven, L. P., et al., Electron transport in quantum dots, *Proceedings of the Summer School on Mesoscopic Electron Transport*, Kluwer Academic, Norwell, Mass., 1997.

Multilayer Materials

Clemens, B. M., and R. Sinclair, Mater. Res. Soc. Bull., Feb. 1990, p. 17.

Tu, K. N., et al., *Electronic Thin Film Science: For Electrical Engineers and Materials Scientists*, Macmillan, New York, 1992.

X-ray Mirrors

Barbee, T. W., Jr., Mater. Res. Soc. Bull., Feb. 1990, p. 37.

Hardness of Multilayers

Barnett, S., and A. Madan, Superhard superlattices, Phys. World, Jan. 1988, p. 45.

Stoichiometric Optimization of Physical Parameters

Danielson, E., et al., *Nature*, **389**, 945 (1997).

Xiang, X. D., et al., *Science*, **268**, 1738 (1995).

PROBLEMS

20.1 A cylindrical quantum dot whose conducting region has radius a and height b is placed in an axial magnetic field. Compute the absorption spectrum of the dot for the cases where there are one, two, three, and four electrons.

20.2 Find the capacitance of a Schottky barrier as a function of voltage.

20.3 Find the shape of a drop of liquid on a planar interface.

20.4 Find an expression for the difference of pressure between the inside and outside of a fluid droplet of radius r and surface tension γ.

20.5 A double quantum well for the conduction electrons in a semiconductor hetero-structure is constructed as shown below. Find the electronic energy levels for the bound states.

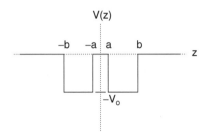

20.6 A crystal has tetragonal symmetry. A cubic sample of the crystal of side a is allowed to approach thermal equilibrium. The surface energy of the (100) and $(\overline{1}00)$ faces is γ_1. The surface energy of the (010), (001), $(0\overline{1}0)$ and the $(00\overline{1})$ faces is γ_2. It is found later that the crystal is tetragonal in shape with dimensions a', b', and b'. Find their values.

20.7 A liquid is forced into the pore space between the grains of a solid. Let S be the surface area per unit volume of the grains. Show that the pressure needed to accomplish this must exceed

$$P = (\gamma_{sl} - \gamma_{sv})S,$$

where γ_{sl} and γ_{sv} are the interfacial energies per unit area for the solid–liquid and solid–vapor, respectively.

20.8 An idealized model of the β-cristobalite form of crystalline SiO_2 has a diamond like cubic unit cell with lattice constant $a = 0.7543$ nm. In this idealized structure Si atoms are placed at the lattice sites of the diamond crystal structure with oxygen atoms midway between pairs of Si atoms, the Si–O–Si bond angle taken to be 180°.

   (a) If the (100) face of the idealized β-cristobalite lattice is rotated by 45° with respect to the (100) face of the Si surface, show that the resulting mismatch between the two lattices is only $\approx 2\%$.

   (b) What fraction of the Si atoms on the Si(100) surface are bonded to oxygen atoms in the β-cristobalite structure? This model of the Si/a-SiO_2 interface was proposed by F. Herman and R. V. Kasowski, *J. Vac. Sci. Technol.*, **19**, 395 (1981).

20.9 Show that the surface energy per unit area is given by

$$\frac{u_s}{A} = \gamma - T \left(\frac{\partial \gamma}{\partial T} \right)_{\{\epsilon_{ij}\}},$$

where γ is the surface tension.

Note: An additional problem is given in Chapter W20.

SYNTHESIS AND PROCESSING OF MATERIALS

Synthesis and Processing of Materials

21.1 Introduction

The terms *materials synthesis* and *processing* refer to the development and use of procedures that result in arrangements of atoms, molecules, and molecular aggregates in appropriate configurations which are desired for specific applications. These procedures involve control of the structure of materials at all levels, from the atomic to the macroscopic. Whereas the term *synthesis* often refers to the chemical and physical procedures by which atoms and molecules are assembled, *processing* procedures such as those employed for electronic materials can also involve the assembly of atoms and molecules. Processing also refers to the manipulation of materials on larger scales and includes procedures involved in the manufacture of materials. The distinctions between materials synthesis and processing procedures are often not clear and the terms are sometimes used interchangeably. It seems clear that the synthesis and processing of materials consist of a continuous range of activities.

The synthesis and processing of materials have been important activities ever since humans first began making and using tools and are currently very active areas of research. New procedures such as rapid thermal processing (RTP), the processing of microelectromechanical systems (MEMS), and the synthesis of new materials such as high-T_c superconductors, carbon-based fullerenes and nanotubes, and rare earth–transition metal ferromagnets such as $Nd_2Fe_{14}B$ have been developed recently. These new materials are often complex in both composition and structure and are typically prepared using procedures involving several steps under carefully controlled conditions. Control of the types and concentrations of desired physical and chemical defects can also play an important role in synthesis and processing.

Discussion of the synthesis and processing of materials to be presented in this chapter focuses on procedures used to prepare materials with essentially their final microstructure, but not necessarily with their final external shape or form. Thus topics related to forming, welding, cutting, joining, and so on, are not stressed here. Since the subject area of synthesis and processing is too large to be presented completely, a case study approach to procedures and materials of current interest is used here. The chapter is organized as follows: general thermodynamic, chemical, and kinetic effects related to synthesis and processing and a general discussion of crystal growth and annealing are presented first, followed by specific examples of the synthesis and processing of semiconductors, metals, ceramics, and polymers in the remaining sections. Further discussions of general issues related to heteroepitaxial growth and processing using ion beams as well as additional examples of the synthesis and processing of a wide range of materials are presented in Chapter W21 at our Web site.[†]

ISSUES IN SYNTHESIS AND PROCESSING

The synthesis and processing of materials refer to the deliberate control of the microstructures and configurations of materials which can be brought about through a wide variety of procedures. The thermodynamic and chemical variables which can be controlled during synthesis and processing include temperature, pressure, and the chemical species involved. Electric and magnetic fields can also be used to influence the motions of charged particles (i.e. electrons and ions), which are involved in procedures such as plasma deposition and etching or in ion implantation and sputtering. *Thermodynamic, chemical*, and *kinetic* effects can all play important roles in the achievement of the desired final microstructure.

A wide variety of final microstructures and configurations are possible in useful materials, including the following:

1. Single crystals
2. Polycrystalline materials with controlled grain size, including microcrystalline and nanocrystalline materials
3. Amorphous materials
4. Composite materials
5. Structured materials (e.g., superlattices, multilayers, etc.)
6. Combinations of the above (e.g., Si-based electronic devices consisting of single-crystal Si, amorphous SiO_2, and polycrystalline metals such as Al, Cu, and W)

These types of microstructures have been discussed in Chapter 4. Examples of synthesis and processing procedures that yield materials with these microstructures are discussed in various sections of this chapter which focus on specific classes of materials.

Several general topics that are relevant to synthesis and processing are introduced in other chapters, including the following:

1. Cohesive energies of solids (Chapter 2)
2. Thermodynamics of defect formation and the law of mass action (Chapter 4)
3. Thermally activated processes such as diffusion and vaporization, binary phase diagrams, the Gibbs phase rule, structural phase transitions, melting, and order–disorder transition (Chapter 6)
4. Ternary phase diagrams (Chapter W13)
5. Temperature–time superposition principle (Chapter 14)
6. Surface adsorption and desorption (Chapter W19)

The general issues that are discussed next include thermodynamic, chemical, and kinetic effects and descriptions of crystal growth and annealing.

[†] Supplementary material for this textbook is included on the Web at the resource site (ftp://ftp.wiley.com/public/sci_tech_med/materials). Cross-references to elements of the Web material are prefixed by "W."

21.2 Thermodynamic and Chemical Effects

Equilibrium thermodynamics is a very useful and powerful tool for analyzing and predicting how materials will behave under the conditions involved in their synthesis and processing. The usefulness of thermodynamic predictions is sometimes underestimated by those who believe, correctly, that kinetic factors more often dominate the evolution of materials systems. A study of thermodynamics is, nevertheless, always a useful starting point for understanding the phenomena involved in the synthesis and processing of materials. This is true even if the results provide only general guidelines for what to expect or how to proceed and even if thermodynamic equilibrium is not achieved in the system under consideration. Kinetic effects can then be added to the framework of thermodynamics as needed to predict reaction rates and pathways. It should, of course, be recognized that kinetic effects can determine not only the final microstructure of the product but also its crystal structure. An example is the successful chemical vapor deposition (CVD) of diamond under conditions of temperature and pressure where graphite is the thermodynamically stable form of solid carbon. The CVD of diamond is described in Section W21.14.

The thermodynamic properties of materials which are of interest here include the *enthalpy H*, *entropy S*, *internal energy U*, and *Gibbs free energy G*. These properties are related by

$$G = H - TS = U + PV - TS, \tag{21.1}$$

where P is the absolute pressure, V the volume, and T the absolute temperature of the material. Although H, S, U, and G can depend either explicitly or implicitly on pressure P, this dependence is usually negligible for solids and liquids under ordinary circumstances and can even be neglected for gases to the extent that they are "ideal." For a system with more than one component, G can also be expressed in terms of the numbers of particles N_i and the *chemical potentials* $\mu_i = (\partial G/\partial N_i)_{T,P,N_j}$ of each component by

$$G(N_i, T) = \sum_i N_i(T)\mu_i(T). \tag{21.2}$$

The chemical potential of a given component is just its Gibbs free energy per particle. A review of thermodynamics is presented in Appendix WA.

It is primarily changes in H, S, U, and G that are important in calculations of thermodynamic or chemical equilibria. For example, the enthalpy change ΔH is the heat absorbed or released in a certain process while ΔS is the entropy change for the process. For any isothermal, isobaric process the free-energy change is given by either

$$\Delta G = \Delta H - T\Delta S = \Delta U + P\Delta V - T\Delta S \tag{21.3}$$

or

$$\Delta G = \sum G(\text{products}) - \sum G(\text{reactants}). \tag{21.4}$$

The *standard Gibbs free energy of formation* of a material, defined as $\Delta_f G^\circ(T)$ in the JANAF tables,[†] is the free-energy change corresponding to the formation of

---

[†] M. W. Chase et al., eds., *JANAF Thermochemical Tables*, 3rd ed., American Chemical Society, Washington, D.C., 1985; *J. Phys. Chem. Ref. Data*, **14**, Suppl. 1 (1985).

the material (e.g., quartz) from the elemental reactants [e.g., Si(s) and O$_2$(g)] in their *standard states*. The standard state of a material is usually chosen to be its equilibrium state at atmospheric pressure and at the temperature in question. For example, the standard states of the elements silicon and oxygen at $P = 1$ atm and $T = 300$ K are crystalline Si(s) and molecular O$_2$(g), respectively. Also, $\Delta_f G^\circ$[SiO$_2$(s), 300 K] $=$ -853.315 kJ/mol, while $\Delta_f G^\circ$[Si(s), 300 K] $\equiv 0$ and $\Delta_f G^\circ$[O$_2$(g), 300 K] $\equiv 0$. The values of $\Delta_f G^\circ$ for the elements in their standard states are thus defined to be equal to zero at all temperatures. The standard state of an element at $P = 1$ atm can, of course, change from solid to liquid to vapor as the temperature is increased.

The overall direction in which thermodynamic processes such as chemical reactions, for example,

$$\text{Si}(s) + \text{O}_2(g) \longleftrightarrow \text{SiO}_2(s), \tag{21.5}$$

or phase changes, for example, solid Si(s) $\leftrightarrow$ liquid Si(l) or solid Fe(s) $\leftrightarrow$ vapor Fe(g), will proceed depends on the sign of ΔG for the process (i.e., on whether the products or the reactants have the higher total free energy). The condition $\Delta G < 0$ ($\Delta G > 0$) indicates that a process is *exothermic (endothermic)* and will proceed faster in the forward (reverse) direction. When $\Delta G = 0$, [i.e., when $\Delta H = T \Delta S$ or ΣG(products) $= \Sigma G$(reactants)], the process will be in equilibrium and will proceed at equal rates in the forward and reverse directions. Thus a given thermodynamic process will always tend to proceed in the direction that lowers the total Gibbs free energy of the system.[†] The thermodynamic driving force for a given process can be a decrease in enthalpy H, an increase in entropy S, or both. The cohesive energies ΔH_c which are the heats of formation of a solid from the constituent atoms are described in Chapter 2.

It is useful to introduce and define the *equilibrium constant K(T)* for a thermodynamic process in terms of the *standard Gibbs free-energy change* $\Delta_r G^\circ(T)$ for the process by

$$\Delta_r G^\circ(T) = \Delta_r H^\circ(T) - T\Delta_r S^\circ(T) = -RT \ln K(T). \tag{21.6}$$

This can also be written as

$$K(T) = \exp\left(\frac{-\Delta_r G^\circ}{RT}\right) = \exp\left(\frac{\Delta_r S^\circ}{R}\right) \exp\left(-\frac{\Delta_r H^\circ}{RT}\right), \tag{21.7}$$

where $R = 8.3145$ J/mol $\cdot$ K is the gas constant. Here $\Delta_r G^\circ$ is the change in Gibbs free energy, expressed in kJ/mol, needed to transform the reactants and products from their equilibrium states to their standard states, where their activities a are equal to unity. Note that $\Delta_r G^\circ = 0$ and $K = 1$ when the reactants and products are in their standard states. The activities of the reactants and products are not in general equal to unity when the process is in equilibrium.

The *activity* of a substance is a dimensionless quantity that relates its thermodynamic reactivity to its reactivity in its standard state. For ideal gases, standard states and unit activities correspond to the gas at 1 atm of pressure. When not in the standard state,

---

[†] [†]Even though a process such as the reaction A + B $\leftrightarrow$ C may proceed faster in the forward direction, with the net result of producing more of species C, the reverse reaction will also be occurring at any given time (i.e., some A and B will be formed from C).

the activity of an ideal gas X is equal to the magnitude of its pressure, expressed in atmospheres, by

$$a(\text{ideal gas } X) = \frac{P(X)}{1 \text{ atm}}. \tag{21.8}$$

The activity of a solid or a liquid under a given set of conditions is usually defined as the ratio of its actual vapor pressure to the equilibrium vapor pressure that it would have in its standard state under the same set of conditions, that is, by

$$a(\text{solid or liquid } X) = \frac{P(X, T)}{P_{eq}(X, T)}. \tag{21.9}$$

Since the vapor pressures of solids are usually not far from their equilibrium values, the activities of solids are often taken to be close to unity. Deviations of the activity of a solid substance from unity can occur when it is not in its standard state. Examples include situations when a solid material is in an amorphous state rather than in a crystalline state (e.g., amorphous Si) or when it is in a crystalline state which is not the state of lowest free energy (e.g., solid carbon in the form of diamond rather than graphite).

The actual free-energy change ΔG for a given process can be expressed in terms of $\Delta_r G^{\circ}$ for the process by

$$\Delta G = \Delta_r G^{\circ} + RT \ln Q = -RT \ln K + RT \ln Q = RT \ln \frac{Q}{K}, \tag{21.10}$$

where Q is the *activity quotient* for the process. As an example, consider a hypothetical process (e.g., a chemical reaction) involving the four substances B, C, D, and E (each of which can be a solid, liquid, or a gas in molecular or atomic form):

$$bB + cC \longleftrightarrow dD + eE. \tag{21.11}$$

This process has an activity quotient given by

$$Q = \frac{a_D{}^d a_E{}^e}{a_B{}^b a_C{}^c}, \tag{21.12}$$

where b, c, d, and e are the set of smallest integers consistent with the process. When $Q = K$ it follows that $\Delta G = 0$ and the process is in equilibrium. Thus K is the equilibrium value of the activity quotient Q.

For solid $\leftrightarrow$ vapor or liquid $\leftrightarrow$ vapor processes, $K(T)$ can be expressed in terms of the equilibrium vapor pressures $P_{eq}(T)$ (expressed in atmospheres) of the reactants and the products by

$$K(T) = \frac{\Pi_i P_{eq}^{m_i}(\text{product } i, T)}{\Pi_i P_{eq}^{m_i}(\text{reactant } i, T)}. \tag{21.13}$$

The exponents m_i correspond to the numerical coefficients of the reactants and products in the equation for the chemical reaction representing the process [e.g., the integers b, c, d, and e in Eq. (21.11)].

As an example, consider the vaporization of an equimolar mixture of $Si(s)$ and $SiO_2(s)$, represented by the reaction

$$Si(s) + SiO_2(s) \longleftrightarrow 2SiO(g). \tag{21.14}$$

The equilibrium constant for this heterogeneous reaction involving both solid- and vapor-phase species is given by

$$K(T) = \frac{P_{eq}^2(SiO(g), T)}{a[Si(s), T]a[SiO_2(s), T]}, \tag{21.15}$$

where $a(Si)$ and $a(SiO_2)$ are the activities of $Si(s)$ and $SiO_2(s)$, respectively, at temperature T. This reaction will be in equilibrium when the activity quotient $Q = K$, that is, when

$$-RT \ln Q(T) = -RT \ln K(T) = \Delta_r G^\circ(T)$$
$$= 2\Delta_f G^\circ[SiO(g), T] - \Delta_f G^\circ[Si(s), T] - \Delta_f G^\circ[SiO_2(s), T]. \tag{21.16}$$

The condition for the equilibrium of reaction (21.14) is expressed in terms of chemical potentials μ by

$$\mu[Si(s), T] + \mu[SiO_2(s), T] = 2\mu[SiO(g), T]. \tag{21.17}$$

The equilibrium constant for the case of a solid material in equilibrium with its own vapor is discussed in Section 6.3.

Equilibrium phase diagrams are important components of the thermodynamic analysis of any synthesis or processing procedure. Their usefulness includes indicating the limits of stability of the various possible phases (solid, liquid, or vapor) as functions of temperature, pressure, and chemical composition. Information on the structural phase transitions between different solid phases is also contained in these diagrams. The phase transitions corresponding to melting, vaporization, and order–disorder transitions in alloys are discussed in Chapter 6. When equilibrium is achieved in a system of reactants for a given set of conditions, the different product phases present and their compositions can be determined from the equilibrium phase diagram. Discussions and examples of the use of binary phase diagrams are given in Section 6.5, and ternary phase diagrams are discussed in Chapter W13. Even when thermodynamic equilibrium is not achieved in a system, as in the continuous deposition of a solid from the vapor phase, as long as steady-state conditions have been achieved it will still be possible to determine the phases present.

The *Gibbs phase rule* provides an important condition for determining how many different thermodynamic phases can be present at equilibrium and which processes are allowed in a given system. Its application to equilibrium phase diagrams is discussed in Section 6.5. The relationship $F = C - P + 2$ between the number of components C, the number of phases P, and the resulting number of degrees of freedom F involving compositions as well as temperature and pressure has been emphasized there. Equilibrium thermodynamic systems and processes must be consistent with both phase equilibria and the phase rule.

Thermochemical databases such as the JANAF and CODATA tables help to provide the necessary input for thermodynamic calculations which determine which species will be present in a system under given conditions of temperature, pressure, and composition. Standard computer programs also exist which can be used to perform the calculations needed to minimize the Gibbs free energy for such systems and to determine the equilibrium phase diagrams and phases.

21.3   Kinetic Effects

Thermodynamic, chemical, and kinetic effects are often interrelated in procedures used for the synthesis and processing of materials. For example, the attainment of thermal equilibrium and of the thermodynamically preferred products having the lowest total Gibbs free energy can usually be more easily realized at higher temperatures, where thermally activated kinetic processes such as diffusion proceed at much higher rates. At lower temperatures the relevant rates of processes may be so slow that equilibrium cannot be achieved within a reasonable length of time. What corresponds to a "higher" or "lower" temperature clearly depends on the processes being considered. Often, one or more kinetic processes can be identified as being rate limiting in procedures such as crystal growth or annealing. Conversely, the thermodynamic properties of materials can have important effects on the kinetics of physical and chemical processes because, as is discussed later, the magnitudes and signs of the relevant enthalpy changes ΔH and entropy changes ΔS of processes often determine their directions and rates.

Kinetic effects determine the rates and the paths of processes that occur as systems approach equilibrium or steady-state conditions, such as those occurring during the deposition of a solid film from the vapor phase at a constant growth rate. The rates of processes (i.e., reaction rates) can be related to thermodynamic quantities through the use of equilibrium constants. Examples of solids in equilibrium with their own vapors have already been presented in Section 6.3. As an additional example, consider an elementary bimolecular reaction between two reacting species B and C which leads to products D and E, as expressed by Eq. (21.11). Such a reaction could occur between atoms or molecules in the gas or liquid phases or even in the solid state, as, for example, the reaction between a vacancy and an interstitial. At equilibrium this reaction will proceed with equal rates in the forward and reverse directions. The condition for equilibrium can therefore be expressed by

$$k_f(T)[B(T)]_{\text{eq}}^b[C(T)]_{\text{eq}}^c = k_r(T)[D(T)]_{\text{eq}}^d[E(T)]_{\text{eq}}^e, \qquad (21.18)$$

where $k_f(T)$ and $k_r(T)$ are the *forward* and *reverse reaction rates* or *rate constants*, respectively, with units of s^{-1}, and $[X(T)]_{\text{eq}}$ is the equilibrium concentration of substance X at temperature T.

For the case of the growth of a thin film from the vapor phase, the net growth rate R_g can often be expressed as the difference between the rate of film deposition and the rate of etching of the film (i.e., by $R_g = R_d - R_e$). In the more general case when a thermodynamic system undergoes an isothermal, solid-state transformation to a new phase (e.g., the sintering of powders to form a ceramic material or the heat treatment of a steel to form the pearlite phase), the formation of the new phase often follows a

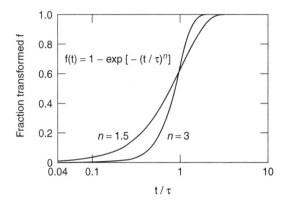

Figure 21.1. The isothermal transformation of a thermodynamic system to a new phase often follows a sigmoidal curve, examples of which are shown here. The fraction f of the system converted to the new phase, calculated using Eq. (21.19), is plotted versus the logarithm of time.

sigmoidal curve, examples of which are shown in Fig. 21.1. The following expression, known as the *Avrami equation*, is often used to describe the kinetics of the nucleation and growth of the new phase:

$$f(t) = 1 - e^{-(t/\tau)^n}. \tag{21.19}$$

Here $f(t)$ is the fraction of the system converted to the new phase at time t. The parameters τ and n are independent of time but can be temperature dependent. Also, τ is a characteristic time determined by microscopic properties of the transformation (e.g., the density of nucleation sites). The value of the exponent n in the Avrami equation is determined by the nature of the transformation (i.e., by the rate of nucleation of discrete regions of the new phase and also by their shape and subsequent growth rate). The case $n = \frac{3}{2}$ is applicable when the growth of particles of the new phase in a matrix is controlled by diffusion, with nucleation occurring only at the beginning of the transformation. Other values of n between $\frac{1}{2}$ and 4 are also possible. The rate of transformation to the new phase is defined as $R = 1/t_{0.5}$, where $t_{0.5}$ is the time at which $f = 0.5$ (i.e., the time for the transformation to be 50% complete). As discussed later, this transformation rate R is usually thermally activated with an exponential dependence on temperature [i.e., $R(T) = A\exp(-E_a/k_B T)$].

By analogy with Eqs. (21.12) and (21.13), the equilibrium constant for reaction (21.11) is given by

$$K_c(T) = \frac{[D(T)]_{eq}^d [E(T)]_{eq}^e}{[B(T)]_{eq}^b [C(T)]_{eq}^c}, \tag{21.20}$$

where the subscript c indicates that concentrations have been used in place of activities. This expression can also be regarded as a statement of the law of mass action. A specific application of this law that is presented in Section 4.6 involves the determination of the equilibrium concentration of Frenkel defects. Its applications to Schottky defects and to interstitials are described in Chapter W4.

The relationship between kinetic rates and thermodynamics can be found by combining Eqs. (21.18) and (21.20), which yields

$$K_c(T) = \frac{k_f(T)}{k_r(T)}.$$ (21.21)

If reaction (21.11) is the only possible process that can occur in the system consisting of species B, C, D, and E, the system will be in thermodynamic equilibrium when $k_f(T) = K_c(T)k_r(T)$. This statement is consistent with the *principle of detailed balancing*, which states that for a system in thermodynamic equilibrium, the probability of occurrence of any process must be equal to the probability of occurrence of the reverse process. If the two probabilities are not equal to each other but are constant in time, the process is said to be occurring under *steady-state conditions*. How far the steady-state conditions are from equilibrium depends on the magnitude of $\ln(Q/K) = \Delta G/RT$ for the process.

When the reaction rates k_f and k_r correspond to processes such as diffusion or vaporization which are thermally activated, they can usually be written in the general Arrhenius form as

$$k_f(T) = A_f \exp\left(-\frac{E_{af}}{k_B T}\right),$$

$$k_r(T) = A_r \exp\left(-\frac{E_{ar}}{k_B T}\right).$$ (21.22)

Here E_{af} and E_{ar} are the *activation energies* for the rate-limiting steps involved in the forward and reverse processes, respectively. The energies E_{af} and E_{ar} as well as the exponential prefactors A_f and A_r can be weakly temperature dependent. The states of a system corresponding to the reactants and to the products are shown schematically in Fig. 21.2, where the energy of the system is plotted versus an appropriate

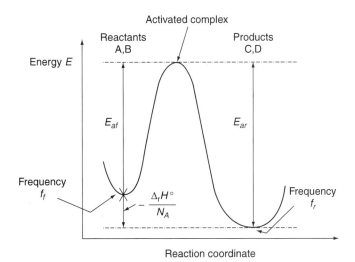

Figure 21.2. Two states of a reacting system corresponding to the reactants and the products, with the energy of the system plotted versus an appropriate reaction coordinate. The enthalpy change $\Delta_r H^O$ for the process connecting these two states is indicated here as being proportional to the difference between the activation energies E_{af} and E_{ar} defined in Eq. (21.22).

reaction coordinate. The state of the system corresponding to the top of the energy barrier between the products and reactants is often referred to as the *transition state* or *excited state* and also as the *activated complex*. A process is nonactivated when an external source of energy is not required for the process to occur or when energy is actually released during the process (i.e., when the process is exothermic). Examples of nonactivated processes are the recombination of interstitials and vacancies in materials or of electrons and holes in semiconductors.

When the forward and reverse processes are both activated, as expressed in Eq. (21.22), then Eq. (21.21) becomes

$$K_c(T) = \frac{A_f}{A_r} \exp\left(-\frac{E_{af} - E_{ar}}{k_B T}\right). \tag{21.23}$$

The connection between this expression and thermodynamics can be made by comparison with Eq. (21.7). This comparison is valid only for simple elementary processes in which the total number of species is left unchanged and where the forward and reverse processes are each dominated by a single activation energy. If this is the case, it follows that

$$\begin{aligned} \frac{\Delta_r H^\circ(T)}{R} &= \frac{E_{af} - E_{ar}}{k_B} \longrightarrow \Delta_r H^\circ(T) = N_A(E_{af} - E_{ar}) \\ \frac{\Delta_r S^\circ(T)}{R} &= \ln\frac{A_f}{A_r} \end{aligned} \tag{21.24}$$

where $N_A = R/k_B = 6.022 \times 10^{23}$ is Avogadro's number. The enthalpy change $\Delta_r H^\circ$ is indicated in Fig. 21.2 as being proportional to the difference between the two activation energies. Even for the exothermic reaction shown with $\Delta_r H^\circ < 0$, extra energy is still required to form the excited state (i.e. the activated complex). In most cases, however, processes involved in the synthesis and processing of materials consist of a large number of competing elementary steps involving diffusion, adsorption, desorption, and so on.

As shown in Chapter 6, the exponential prefactors A_f and A_r for the thermally activated processes of diffusion and vaporization typically contain frequency factors f_f and f_r, respectively, which correspond to the vibrational frequencies of the product and reactant species in their respective potential wells, as shown schematically in Fig. 21.2. These frequencies correspond to the attempt frequencies for reaching the excited state separating the two states.

Kinetic processes such as the bulk diffusion of impurities and point defects in solids and the vaporization of atoms from surfaces are quite important in the synthesis and processing of materials. These processes are discussed in Chapter 6, where it is shown how they are affected by the associated changes in enthalpy and entropy of the species involved. The flux of species incident on a surface from the vapor phase is derived in Chapter W19. Adsorption and desorption rates for surfaces are also discussed there, along with the Langmuir adsorption isotherm for surface coverage. The quasiequilibrium model for the interaction of vapor species with a solid surface that can be used for the description of growth and etching phenomena is summarized in Section W21.14, where it is applied to the chemical vapor deposition of diamond.

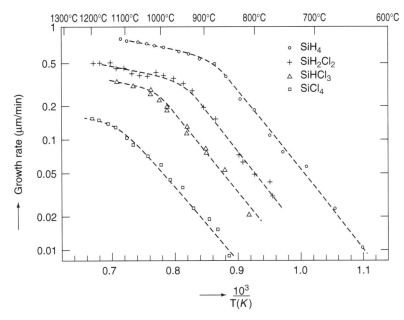

Figure 21.3. Dependence of growth rate $R(T)$ on inverse temperature $1/T$ for the chemical vapor deposition of Si for four different precursor gases. The rate $R(T)$ can be controlled by different mechanisms in different temperature ranges: thermodynamics at higher T (lower $1/T$), mass transport or hydrodynamics at intermediate T, and kinetics at lower T (higher $1/T$). [From F. C. Eversteyn, *Philips Res. Rep.*, **29**, 45 (1974)].

A wide range of materials are prepared in the form of thin films, typically via deposition from the vapor phase. The dependence of the film growth rate on temperature is in general controlled by three different types of mechanisms: thermodynamic, kinetic, and *mass transport* (e.g., vapor-phase diffusion or hydrodynamics). The resulting film growth rate often shows a characteristic dependence on the inverse temperature $1/T$ (Fig. 21.3). Often one or more kinetic processes can be identified as being rate limiting in procedures such as thin-film growth. At low T (high $1/T$) kinetic effects usually limit the growth rate, which therefore is often observed to increase exponentially with increasing T, as shown. As an example, in the CVD of Si(s) either the dissociation of a reactant molecule such as SiH_4 or the desorption from the Si surface of chemically bonded H atoms can be the rate-limiting kinetic process at low T. At higher T, all of the reactant molecules will eventually be effectively dissociated at the surface. The growth rate will then be limited by their rate of arrival at the surface of the growing film (i.e., by mass transport effects, which are often hydrodynamic in nature at higher pressures). In this temperature range the film growth rate is observed to be essentially independent of T.

At even higher T the growth rate is often observed to decrease. This decrease can result from a variety of causes. One possibility is that the desired product [e.g., Si(s)], may become thermodynamically unstable relative to some other product [e.g., vapor Si(g)] and another is that growth precursors such as SiH_2 may desorb more rapidly from the surface at higher T. Yet another possibility is that depletion of the precursor gas via reaction on the reactor walls may be occurring more rapidly. The details of the growth rate curve shown in Fig. 21.3 will clearly depend on the nature of the reactants and of the desired product and also on the details of the system in which the film is deposited. Similar behavior can be expected for synthesis reactions occurring in the solid state where diffusion of reacting species would be the rate-limiting process at low T. A general statement that can be made concerning this temperature-dependent behavior is that kinetic effects associated with enthalpy changes will tend to dominate at lower T, whereas thermodynamic effects associated with entropy changes will tend to dominate at higher T.

If the deposition rate is too high or if T is too low, it will usually be difficult to achieve crystalline growth. Amorphous films will instead be deposited due to the lack of time for defects to be eliminated from the growing film.

In Section 14.6 the effects of annealing on polymer viscoelasticity are discussed in terms of the temperature–time superposition principle, which states that behavior obtained over a long time interval at a lower temperature is equivalent to that obtained for a shorter interval but at a higher temperature. A useful formula relating times and temperatures is given there in which the free-energy barrier for the relaxation process plays a critical role. The rate of cooling of an alloy is an important parameter for determining the composition and microstructure of the resulting solid material. As pointed out in Section 6.5, the faster the rate of cooling through a given phase domain, the less time there is available for the nucleation and growth of that phase to occur.

The phenomena associated with the transport of heat and mass in a reacting system often play critical roles in the achievement of the desired products. Due to their complexity and also due to lack of space, these phenomema are not discussed in detail here. Discussions of heat and mass transport phenomena can be found in Eckert and Drake (1972) and in Kou (1996).

The synthesis and processing of materials at high temperatures opens a wide range of possibilities and procedures not available at room temperature. As Spear (1976) has pointed out, at high temperatures vaporization processes and vapor species become more important, unusual compounds may form, and the composition ranges of most phases increase. It is clear from the form of the term $-T\Delta S$ in ΔG that the effects of entropy become increasingly important at higher temperatures.

21.4 Crystal Growth

Consider a system consisting of N atoms (or molecules) of the same type in a container of volume V with concentration $n = N/V$. The atoms could be in a vapor, about to form a droplet of condensate. Alternatively, the atoms could be in a melt, about to nucleate a solid-state particle; or they may be atoms in solution, which nucleate to form a small particle. For the sake of definiteness, the focus of attention will be on the vapor–liquid transition. If the pressure of the gas exceeds the vapor pressure (for that temperature), the system is thermodynamically unstable and droplets will begin to nucleate.

Let μ_V be the chemical potential of a vapor atom and μ_D be the chemical potential of an atom in the droplet. The number of atoms in vapor is N_V and the number in the droplet is N_D. The droplet is assumed to be a sphere with radius r and surface energy (assumed to be uniform) σ. In forming a droplet, one begins with a Gibbs free energy $G_i = N_V \mu_V$ and changes it to $G_f = (N_V - N_D)\mu_V + N_D\mu_D + 4\pi\sigma r^2$.

The change in G is

$$\Delta G = N_D(\mu_D - \mu_V) + 4\pi\sigma r^2. \tag{21.25}$$

If Ω denotes the volume of an atom in the droplet,

$$N_D = \frac{4\pi r^3}{3\Omega}. \tag{21.26}$$

Therefore,

$$\Delta G(N_D) = -N_D\delta\mu + 4\pi\sigma\left(\frac{3\Omega N_D}{4\pi}\right)^{2/3} \tag{21.27}$$

where $\delta\mu = \mu_V - \mu_D$. Assuming that $\delta\mu > 0$, the function rises and then falls as N_D is increased from zero. The function $\Delta G(N_D)$ will peak at

$$N_D^* = \frac{32\pi}{3}\frac{\Omega^2\sigma^3}{(\delta\mu)^3} \tag{21.28}$$

and have the maximum value

$$\Delta G^* = \frac{16\pi}{3}\frac{\Omega^2\sigma^3}{(\delta\mu)^2} = \frac{1}{2}N_D^*\delta\mu. \tag{21.29}$$

The number N_D^* represents the critical size that a cluster must reach before it becomes established. When this is large enough to have physical significance, the corresponding critical radius is given by $r_c = 2\sigma\Omega/\delta\mu$. If $N_D < N_D^*$ (or $r < r_c$), the cluster will shrink, whereas if $N_D > N_D^*$, it will continue to grow in size. The probability that a given thermal fluctuation will nucleate a particle of a given size is proportional to a factor $P = \exp[-\beta\Delta G(N_D)]$ multiplied by the probability of having N_D particles found together within some critical distance of each other.

A volume V contains nV atoms, and any of these atoms could serve as the site at which nucleation could begin. The number of nuclei of a given size that will form will be given by nVP. The number of nuclei per unit volume of size N_D is therefore

$$n(N_D) = ne^{-\beta\Delta G(N_D)}. \tag{21.30}$$

The same analysis given for droplet formation in a vapor applies to crystals nucleating in the melt. Crystals, however, possess a lattice and there is an anisotropy associated with the solid. After nucleation the crystals will continue to grow as atoms adsorb on the nucleated crystallite. As the crystals grow in size, their shapes will also change. Facets will form as the crystals attempt to find the state of thermodynamic equilibrium (i.e., their state of lowest Gibbs free energy). The stable crystals assume polyhedral shapes and adjust themselves so as to minimize their total surface energies, subject to the constraint that the volume is fixed. Consider one such crystallite. Let h_i denote the distance from a face of area A_i to the center of mass of the polyhedron and σ_i denote the corresponding surface energy. The higher the surface energy is, the less stable a given surface is. The net surface energy is $E_s = \sum \sigma_i A_i$

and the volume is the sum of a collection of pyramidal volumes $V = \sum A_i h_i / 3$. Minimizing the net surface energy subject to the constraint of fixed volume yields the equation $\delta(E_s - \lambda V) = 0$, with λ being a Lagrange multiplier. Variation with respect to A_i gives

$$h_i = \frac{3}{\lambda} \sigma_i. \tag{21.31}$$

One finds that $\lambda = E_s / V$. Those surfaces with the higher surface energies will have higher growth velocities v_i, will be further away from the center of mass, and will generally have smaller areas. As one constructs the resulting polyhedral shape, it may turn out that the geometry is such that some surface faces will be completely absent. In describing the approach to equilibrium, one may assign growth velocities to the various faces to describe the kinetics.

In the early phases of crystal growth the crystals draw their material from the melt (or solution or vapor) and the crystallites grow independently. Later, as the melt becomes depleted, the crystallites compete with each other for growth. In the phenomenon of Ostwald ripening, small crystallites shrink and supply atoms to larger crystallites, which grow. To see how this comes about, consider two spheres with numbers of atoms N_1 and N_2, respectively. The total number of atoms is $N = N_1 + N_2$ and is fixed. The Gibbs free energy for the pair is

$$G(N_1, N - N_1) = N\mu_D + 4\pi\sigma \left(\frac{3\Omega}{4\pi}\right)^{2/3} [N_1^{2/3} + (N - N_1)^{2/3}]. \tag{21.32}$$

This function has a maximum at $N_1 = N_2 = N/2$. If $N_1 > N_2$ crystal 1 will grow at the expense of crystal 2 until crystal 2 disappears, and vice versa. The final energy drops to

$$G(N, 0) = N\mu_D + 4\pi\sigma \left(\frac{3\Omega}{4\pi}\right)^{2/3} N^{2/3}. \tag{21.33}$$

For example, the difference in energy between having one combined sphere and two separate spheres is

$$\Delta G = -4\pi\sigma \left(\frac{3\Omega}{4\pi}\right)^{2/3} (2^{1/3} - 1)N^{2/3}. \tag{21.34}$$

The degree of crystal growth depends to a large extent on the time spent at high temperatures when the atoms are free to move around. If the material is cooled prematurely, the growth of the crystals is stunted and the microstructure existing at that time is frozen into the solid. The number of grains, grain sizes, and grain boundary area depend to a large extent on how far one permits this crystallization to proceed before quenching the system.

21.5 Annealing

Heating a material close to the melting point, holding it at that temperature for an extended period of time, and slowly cooling it back to room temperature is a process called *annealing*. The high mobility of the atoms, reflected in the high diffusion rates, permits the material to search for its state of lowest free energy. The material may be returned to its most stable crystalline state by annealing. Two things occur in the annealing process. There is the recovery stage, in which the dislocation density decreases. In some materials there may be a recrystallization, consisting of crystal nucleation, in which microcrystallites of a new phase are formed within the material, followed by growth of these nuclei into larger grains. In amorphous metals the crystal formation and growth may proceed homogeneously, without the need for prior nucleation. In the early stages of recrystallization the driving force is provided by the residual elastic energy built up in the material by its previous mechanical history. In the latter stages the driving force is provided by the need to minimize the free energy and return to thermodynamic equilibrium.

When a new phase is formed during annealing the kinetics of recrystallization at the early stages is often described by the Johnson–Mehl equation,

$$f(t) = 1 - \exp\left[-\frac{\pi}{3}\frac{dn}{dt}\left(\frac{dR}{dt}\right)^3 t^4\right]. \tag{21.35}$$

Here $f(t)$ is the volume fraction of the material that has recrystallized by time t, dn/dt is the number density of new crystallites nucleating per unit time, and dR/dt is the rate of change of radius of a crystallite with time. These rates are assumed to be constant at a given T. A derivation of this equation proceeds as follows. Idealize the crystallites as a set of spheres of radii $\{R_j\}$ and imagine that they grow by adding a layer of thickness δ at a time, so $R_j = j\delta$, where j is an integer equal to the number of layers. Let Γ be the rate of increase in the number of crystallites in the sample, Δ be a time step, and Ω be the sample volume. Note that Γ is proportional to Ω. For the moment assume that the spheres are isolated from each other. At time $t = 0$ there are no crystallites present, whereas at time $t = j\Delta$ there are $N = j\Gamma\Delta$ spheres present, with radii equally distributed between R_1 and R_j. The total volume occupied by the spheres is

$$V(t) = \frac{4\pi}{3}\Gamma\Delta\left[\delta^3 + (2\delta)^3 + \cdots + (j\delta)^3\right] \approx \frac{\pi}{3}\Gamma\Delta\delta^3 j^4 \tag{21.36}$$

for $j \ll 1$. Using $\delta = \Delta dR/dt$ this gives

$$V = \frac{\pi}{3}\Gamma\left(\frac{dR}{dt}\right)^3 t^4, \tag{21.37}$$

so

$$\frac{dV}{dt} = \frac{4\pi}{3}\Gamma\left(\frac{dR}{dt}t\right)^3. \tag{21.38}$$

Next, include, in a mean sense, the effect of the other spheres. Since f is the fraction of space occupied by spheres, $1 - f$ is the fraction of space not occupied by them. Spheres can only grow into unoccupied regions. Thus

$$\frac{df}{dt} = \frac{1}{\Omega}\frac{dV}{dt}(1 - f), \tag{21.39}$$

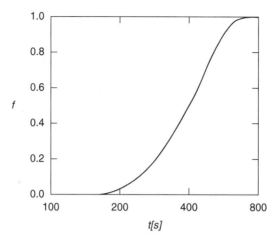

Figure 21.4. Nucleation of the pearlite phase in austenite, [Adapted from R. F. Mehl and W. C. Hagel, *Prog. Met. Phys.*, **6**, 84 (1956).]

where the relation $dn/dt = \Gamma/\Omega$ has been used. Assuming that $f(0) = 0$, this may be integrated to give the Johnson–Mehl equation.

In Fig. 21.4 data are presented for growth of the pearlite phase in the austenite phase in steel as a function of time. The data are in qualitative agreement with the Johnson–Mehl equation, but there are significant quantitative differences. In particular, the assumptions that dn/dt and dR/dt are independent of time need not be valid. Furthermore, the shapes of the crystallites are not spherical and there may exist correlations between nucleation sites associated with strain fields that invalidate the assumption of randomness of nucleation events. Nucleation on grain boundaries and their associated irregularities, such as steps and interface defects, are likely to be important. Including such corrections modifies the form of the Johnson–Mehl formula. Usually, the data may be fit to a formula such as the Avrami equation, given in Eq. (21.19)

$$f = 1 - e^{-(t/\tau)^n}, \tag{21.40}$$

where τ and n are constants, with τ being a characteristic time.

As the temperature T is increased, the growth of the recrystallization fraction $f(t)$ slows down for two reasons. First, the nucleation rate decreases since it varies exponentially with the amount of supercooling. Second, the growth rate will also fall since it, too, is driven by the amount of supercooling. The amount of supercooling is measured relative to the temperature at which the phase transition occurs. For example, in the case of the growth of pearlite in austenite (see Fig. 21.4) the supercooling is measured relative to the eutectic temperature.

At a later stage of grain growth the surface free energy tends to be minimized. Suppose that there are N grains, represented by spheres, of radius R. They occupy a volume $V = 4\pi NR^3/3$ and have a surface energy $U = 4\pi NR^2\gamma$, where γ is the energy per unit area associated with the interface between the crystallite and parent material. Minimizing the surface energy U might involve grains coalescing or, equivalently, having large grains swallowing small grains. One may determine how the energy varies with changing number of grains for fixed grain volume V. Thus

$$\left(\frac{\partial U}{\partial N}\right)_V = \frac{\partial}{\partial N}\left[4\pi N\gamma\left(\frac{3V}{4\pi N}\right)^{2/3}\right]_V = \frac{4\pi\gamma}{3}R^2, \qquad (21.41)$$

$$\frac{\partial U}{\partial R} = \frac{\partial N}{\partial R}\frac{\partial U}{\partial N} = -\frac{3\gamma V}{R^2}. \qquad (21.42)$$

Imagine R as a generalized coordinate and the system relaxing to the state of minimum surface energy by varying R. The relaxation takes place by a series of irreversible processes, so introduce a viscous friction term and write the force balance equation as

$$-K\frac{\partial R}{\partial t} - \frac{\partial U}{\partial R} = 0, \qquad (21.43)$$

with K being the viscous coefficient. This results in the equation

$$\frac{\partial R}{\partial t} = \frac{3\gamma V}{KR^2}. \qquad (21.44)$$

This integrates to

$$R(t) = \left(R_0^3 + 9\gamma Vt/K\right)^{1/3}, \qquad (21.45)$$

where R_0 is the radius at $t = 0$. Equation (21.45) shows that a $t^{1/3}$ growth rate can be expected for long times. As $K \to \infty$, $R \to R_0$, as expected.

The growth of grains of a β-spodumene (LiAlSi$_2$O$_6$) solid solution in a alumina–silica melt was studied as a function of time and the functional form given by Eq. (21.45) was confirmed experimentally.[†]

SYNTHESIS AND PROCESSING OF SEMICONDUCTORS

The synthesis and processing of semiconductors rival those of metals in terms of the variety of procedures and materials involved and also in terms of technological and economic importance. Whereas the synthesis of semiconductors typically involves the growth of single crystalline materials in the form of either bulk crystals or epitaxial layers on appropriate substrates, the processing of semiconductors involves a wide range of procedures, from doping and oxidation to pattern formation and the multiple steps involved in modern semiconductor device fabrication. The growth of amorphous films of semiconductors is also an important area of synthesis. Only a few of the technologically most important synthesis and processing procedures are addressed here. These include the growth of single-crystal Si via the Czochralski method, the thermal oxidation of Si to form surface layers of a-SiO$_2$, and an introduction to Si-based integrated-circuit fabrication. In addition, the following procedures are described in Chapter W21: the float-zone purification of Si, the growth of Si epitaxial layers via chemical vapor deposition (CVD), the growth of the compound

---

[†] C. K. Chyung, *J. Am. Ceram. Soc.*, **52**, 242 (1969).

semiconductor GaAs via molecular-beam epitaxy (MBE), the deposition of films of amorphous semiconductors via plasma-enhanced CVD, an outline of some of the most important steps involved in Si-based integrated-circuit fabrication, and the processing of microelectromechanical systems (MEMS) based on Si.

21.6   Czochralski Growth of Single-Crystal Silicon

The *Czochralski (CZ) growth* of single-crystal Si consists of pulling high-purity, carefully doped crystals in the form of cylindrically shaped *boules* from a bath of molten Si. This is the standard growth procedure for producing the material used for the large-diameter Si wafers which serve as substrates for the fabrication of semiconductor devices. These Si wafers, sliced from the boules using diamond cutting wheels, have increased in diameter from less than 1 inch up to the current standard of 12 inches (300 mm). A typical CZ growth system and a boule of single-crystal CZ Si being withdrawn from the molten Si bath are shown in Fig. 21.5. The molten Si is contained in a fused-silica crucible fitted into a graphite susceptor and is heated either by a radio-frequency induction coil or by a resistance heater surrounding the crucible. In some cases strong axial magnetic fields are applied to the electrically conductive molten Si bath to damp out unwanted convective heat currents in the liquid, thereby helping to minimize temperature gradients and fluctuations. The atmosphere above the molten Si usually consists of an inert gas such as Ar.

The growth of the Si boule is initiated by the use of a seed (i.e., a small, oriented single crystal of Si), which is inserted into the molten bath and then rotated to promote cylindrical uniformity of the growing boule. The crucible containing the molten Si can also be rotated to achieve uniformity of growth. When the dislocation-free Si seed, typically ≈ 12 mm in diameter, is placed into the slightly supercooled molten bath, dislocations are generated due to thermal shock. It is important for these dislocations to be eliminated before growth of the full-diameter crystalline boule begins. The process of dislocation–elimination, known as *necking*, involves reduction of the diameter of the growing crystal to ≈ 3 to 4 mm at a high pull rate of 4 to 6 mm/min. A dislocation-free crystal is usually achieved after a few centimeters of growth. At temperatures above the "plastic" temperature of Si, $T \approx 900°C$, thermal stresses can cause the dislocations to move through the growing crystal by the mechanisms of glide, cross slip, and climb, and thus to grow out of the crystal when they reach its surface. It is also possible at high pull rates and lower thermal stresses for the crystal to have a growth velocity greater than the velocity of dislocation motion. In this case the dislocations are simply left behind.

Following the necking process, the seed is pulled from the molten bath at a rate of 0.1 to 1 mm/min. The time required to grow a 1-m-long boule of CZ Si is thus at least 10^3 to 10^4 min. By both properly adjusting the pull rate and reducing the temperature of the bath, the diameter of the growing Si crystal increases from the size of the seed to the value desired. The growth of a Si boule having the same orientation as the seed crystal occurs in this case under conditions that are close to equilibrium.

Carefully controlled temperature gradients are needed to produce the stable solidification front required for the growth of large single crystals. The shape of this solidification front (i.e., of the solid–liquid interface) is also an important characteristic of the growth since it has a controlling influence on the shape of the growing crystal. The solidification of Si at its melting point, $T_m = 1414°C$, releases the *heat*

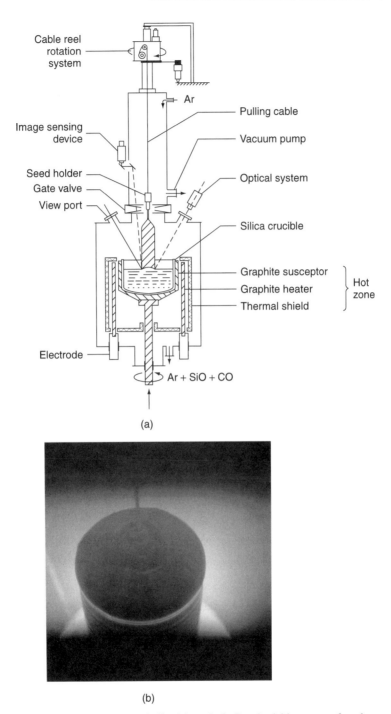

(a)

(b)

Figure 21.5. Czochralski growth of Si: (*a*) typical Czochralski system for the growth of single-crystal Si; (*b*) boule of single-crystal CZ Si being drawn out of the molten Si bath. [(*a*) From W. Lin, *Semiconductors and Semimetals*, Vol. 42, copyright 1994. Reprinted by permission of Academic Press, Inc.; (*b*) from Mitsubishi Materials Silicon Corporation Web site: *www.msil.mmc.co.jp/english/index.html.*]

of fusion $\Delta H_m = 50.21$ kJ per mole of Si. For the growth process to continue, this energy must be effectively removed from the vicinity of the solid–liquid interface. Both heat and mass transport in the molten Si play critical roles in the CZ growth method, with thermal convection being especially important. Axial and radial temperature gradients in the growing Si boule, if large enough, can lead to high thermal stresses and the formation of defects such as dislocations. Under ideal growth conditions the dislocation densities in CZ Si can be close to zero.

The starting material used in the bath is very pure polycrystalline Si obtained by the hydrogen reduction of liquids such as $SiHCl_3$ or $SiCl_4$ which have been purified by repeated distillation. The typical dopant impurities P and B can be added directly to the molten Si. Since the distribution coefficients K of dopant atoms in Si are generally less than unity, as discussed in the description of float-zone purification in Chapter W21, the dopant concentrations in the Si bath, and hence in the growing crystal, will change with time. These effects can be minimized by adjusting the dopant concentration in the bath as growth proceeds and also by increasing the seed rotation rate in order to decrease the buildup of dopant in the liquid boundary layer adjacent to the growth interface.

The dominant impurities in the molten Si bath are usually oxygen and carbon. The oxygen results from the interaction of the molten Si with the SiO_2 crucible; the carbon comes from the graphite present in the hot zone of the growth system. Typical levels of incorporation of O and C in the Si crystal are 10^{24} and 10^{22} atoms/m^3, respectively. Since the solubility of interstitial oxygen in solid Si at $T = 300$ K is $\approx 10^{23}$ m^{-3}, CZ Si tends to be supersaturated with respect to oxygen. The oxygen impurities and associated defects that are generated in the Si lattice are potentially a major problem of the CZ method, although they can also have the beneficial effects of *gettering* or suppressing other defects and also of increasing the mechanical strength and the resistance to thermal strain of CZ wafers. It has been found to be very important to control the concentration and uniformity of the oxygen incorporated into CZ Si during growth when this material is to be used ultimately for device fabrication. Crucible rotation has a significant effect on oxygen incorporation and is used to control its distribution. Although it has not been demonstrated that incorporated C affects Si device performance, it has been observed that C enhances the precipitation of oxygen in Si. Metallic impurities have a very negative effect on the performance of Si devices, due to the fact that they act as traps (i.e., recombination centers for charge carriers). Since their distribution coefficients in Si are typically very low ($\approx 10^{-6}$ or less) they are not effectively incorporated into CZ Si even if they are present in the melt.

The CZ method is also used to grow large single crystals of other materials which can be pulled from a molten bath of the same composition as the desired crystal, for example, GaAs, Li_2GeO_4, Al_2O_3, and solid-state laser host materials such as Ca_2GeO_4 and the garnets $Gd_3Ga_5O_{12}$ (GGG) and $Y_3Al_5O_{12}$ (YAG). In the case of YAG, the starting materials for the molten bath are Al_2O_3 and Y_2O_3. It is also important that the crystal being grown not undergo a structural phase transition between its melting point and room temperature. In the case of the growth of GaAs, the technique is known as *liquid-encapsulated Czochralski* (LEC) since the GaAs melt is covered by an inert layer of molten B_2O_3 to prevent decomposition of the GaAs (i.e., arsenic evaporation). Thermal stresses present during crystal growth are more of a problem for GaAs, which has a lower yield strength and a lower critical resolved shear stress than those of Si, which is mechanically stronger.

21.7 Thermal Oxidation of Silicon

One of the basic steps involved in the fabrication of Si integrated circuits is the growth of thin layers of amorphous SiO_2 (a-SiO_2) on the surfaces of Si wafers via thermal oxidation. The a-SiO_2 layer serves as an excellent protective and passivating coating and is one of the main reasons that Si is the dominant semiconductor in use in electronic devices, even though other semiconductors, such as GaAs and Ge, have superior electrical properties. This oxidation process has been analyzed in detail, and essentially all of the physical and chemical mechanisms, as well as the basic thermodynamic and kinetic aspects of the growth, are now at least qualitatively understood. As a result, a-SiO_2 layers on Si with controlled thicknesses and properties can be prepared reproducibly. In addition to providing chemical passivation for the Si surface, the a-SiO_2 layers help to control the electrical characteristics of the device. The structure and properties of the Si/a-SiO_2 interface are critically important in Si-based devices and will become even more critical as device dimensions continue to shrink and oxide thicknesses approach 1 nm. The Si/a-SiO_2 interface is discussed in more detail in Section 20.11. There remain several important unanswered questions concerning the details of the oxidation process on the atomic level, as indicated later.

The focus here is on a description of the dry (H_2O-free) thermal oxidation of Si in O_2 at atmospheric pressure. Wet oxidation using H_2O and high-pressure oxidation are not discussed in detail. All three processes — dry, wet, and high-pressure oxidations — have been studied very widely. Oxidation typically takes place inside a fused-quartz tube placed within a furnace, with dry O_2 flowing over the surface of the Si wafers at temperatures T_s between 800 and 1200°C. The higher values of T_s lead to increases in the oxide growth rate but can also cause undesired redistribution of dopants in the substrate. Typical thicknesses are $\approx$ 15 to 100 nm for gate oxides and $\approx$ 0.3 to 1 µm for field oxides.

The net chemical reaction for the formation of a-$SiO_2(s)$ is very simple:

$$Si(s) + O_2(g) \longleftrightarrow a\text{-}SiO_2(s). \qquad (21.46)$$

As described in Chapter W2, the standard enthalpy of formation of $SiO_2(s)$ according to this reaction is $\Delta_f H° \approx -906$ kJ/mol. This reaction is therefore strongly exothermic. It has been shown by radioactive-tracer experiments that oxygen diffuses inward through the a-SiO_2 layer to the Si/a-SiO_2 interface instead of Si diffusing outward to the a-SiO_2/vapor interface. The growth of a-SiO_2 consists of the following steps:

1. Transport of O_2 through the vapor phase to the surface of the growing a-SiO_2 film

2. Diffusion of O_2 through the oxide film to the Si/a-SiO_2 interface

3. Dissociation of O_2 and reaction with Si to form the Si–O–Si bonding units present in a-SiO_2.

According to this reaction mechanism, when an a-SiO_2 film of thickness d is grown thermally, a layer of Si at the surface of thickness $0.45d$ is consumed.

The schematic diagram presented in Fig. 21.6, which shows the spatial distribution of O_2 in the system, is the appropriate starting point for analysis of the kinetics of

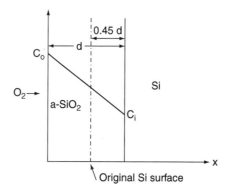

Figure 21.6. Spatial distribution of O_2 in the a-SiO$_2$ layer during the thermal oxidation of Si.

a-SiO$_2$ growth, according to the one-dimensional *Deal–Grove model.*[†] The approach used here will be to recognize from the outset that the oxidation process under normal conditions is controlled not by vapor-phase mass transport but by the diffusion of oxygen through the growing film and by the dissociation and reaction of oxygen at the Si/a-SiO$_2$ interface. This approach is therefore simpler than that used in the original model of Deal and Grove. The flux of O_2 molecules diffusing through the a-SiO$_2$ film is given by

$$F_1 = \frac{D(c_o - c_i)}{x}, \tag{21.47}$$

and the flux of O_2 molecules reacting with Si at the Si/a-SiO$_2$ interface is given by

$$F_2 = k_S c_i. \tag{21.48}$$

Here D is the diffusivity of O_2 through the a-SiO$_2$ layer, c_o and c_i are the concentrations of O_2 molecules in the a-SiO$_2$ layer at the a-SiO$_2$/vapor and Si/a-SiO$_2$ interfaces, respectively, x is the film thickness, and k_S is the combined rate constant for the dissociation and reaction of O_2 molecules at the Si/a-SiO$_2$ interface. Note that k_S has units of m/s. In this simple model c_o is expected to be proportional to the partial pressure of the oxidizing species in the vapor phase.

The following expression for c_i is obtained under steady-state growth conditions (i.e., when $F_1 = F_2$):

$$c_i(x) = \frac{c_o}{1 + k_S x/D}. \tag{21.49}$$

The growth rate of the a-SiO$_2$ layer is given in terms of the flux F_2 by

$$\frac{dx}{dt} = \frac{F_2}{2N(O)} = \frac{k_S c_i}{2N(O)}, \tag{21.50}$$

where $N(O)$ is the concentration of O atoms in the growing oxide. When c_i from Eq. (21.49) is substituted into Eq. (21.50), the following expression for $x(t)$ can be

[†] B. E. Deal and A. S. Grove, *J. Appl. Phys.*, **36**, 3770 (1965).

obtained:[†]

$$x^2(t) + A'x(t) = B't, \tag{21.51}$$

where

$$A' = \frac{2D}{k_S} \quad \text{and} \quad B' = \frac{Dc_0}{N(O)}. \tag{21.52}$$

The thickness of the a-SiO_2 layer as a function of time can be obtained from Eq. (21.51):

$$x(t) = \frac{D}{k_S}\left[\sqrt{1 + \frac{k_S^2 c_0 t}{DN(O)}} - 1\right]. \tag{21.53}$$

For short times, $t \ll t_o = DN(O)/k_S^2 c_0 = A'^2/4B'$; that is, for thin layers, the oxide thickness increases linearly with time according to

$$x(t) = \frac{k_S c_0 t}{2N(O)} = \frac{B't}{A'}. \tag{21.54}$$

For long times $t \gg t_o$ and thicker layers, the following parabolic relationship is obtained:

$$x^2(t) = \frac{Dc_0 t}{N(O)} = B't. \tag{21.55}$$

The quantities B'/A' and B' are known as the *linear* and *parabolic rate constants*, respectively. Note that the characteristic time $t_o = DN(O)/k_S^2 c_0$ is temperature dependent through the kinetic factors $D(T)$ and $k_S(T)$. In the linear-growth region for $t \ll t_o$, when the film is thin, the growth rate is controlled by the reaction at the Si/a-SiO_2 interface and is independent of the diffusion of O_2 through the growing layer. In the opposite, parabolic limit when the layer is thicker, the growth rate is controlled by diffusion and so is independent of the interface reaction. There is still some controversy concerning whether the interface reaction could be slow enough to be the rate-limiting step in the initial oxidation of Si. Other suggestions for the mechanism giving rise to the linear-growth kinetics include the effects of strain at the interface and in the oxide layer. Strain is present due to the fact that the concentration of Si atoms in the substrate is twice that in the oxide layer. There must therefore be a large volume expansion of the Si lattice at the interface as a result of the oxidation.

The results of various experimental investigations of the thermal oxidation of the (111) surface of Si using both O_2 and H_2O and for a wide range of temperatures and partial pressures are presented in Fig. 21.7. The data presented are quite consistent with the predictions of the Deal–Grove model for the linear and parabolic growth regions given in Eqs. (21.54) and (21.55), thus confirming the general validity of this model. The diffusivity $D(T)$ and the reaction rate constant $k_S(T)$ are found to be thermally

---

[†] The more general expression given in the original Deal–Grove model for $x(t)$ is $x^2 + Ax = B(t + \tau)$, where the constants A, B, and τ include the effects of vapor-phase transport and also the possibility of the presence of an initial oxide with thickness $x_i = B\tau/A$ (see Fig. 21.7).

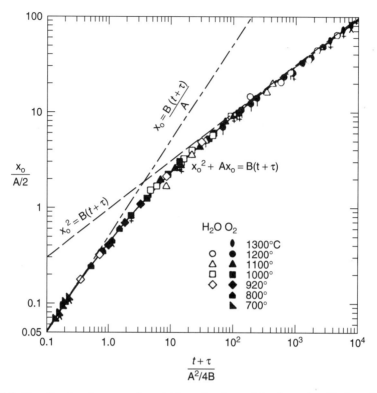

Figure 21.7. Results of various experimental investigations of the thermal oxidation of the (111) surface of Si using both O_2 and H_2O and for a wide range of temperatures and partial pressures. [Reprinted with permission from B. E. Deal et al, *J. Appl. Phys.*, **36**, 3770 (1965). Copyright 1965 by the American Institute of Physics.]

activated, as expected. For the dry oxidation of lightly doped Si (100) substrates at 1 atm of O_2 the rate constants defined in Eqs. (21.54) and (21.55) can be represented by

$$B' = C_1 \exp\left(-\frac{E_1}{k_B T}\right) \quad \text{and} \quad \frac{B'}{A'} = C_2 \exp\left(-\frac{E_2}{k_B T}\right). \qquad (21.56)$$

Here $C_1 = 7.72 \times 10^2 \ \mu m^2/h$, $E_1 = 1.23$ eV, $C_2 = 6.25 \times 10^6 \ \mu m/h$, and $E_2 = 2.0$ eV are obtained experimentally. Note that E_1 is the activation energy for the diffusion of the oxidizing species in the a-SiO_2 layer, while E_2 is the activation energy for the interface reaction. The linear rate constant B'/A' is proportional to the interface reaction rate constant k_S and is observed to depend on the orientation of the Si substrate.

Oxidation with H_2O is up to 100 times faster than with O_2, due to the weakening of the oxide network as a result of the formation of network-terminating Si–OH bonds. Oxidizing species are therefore able to diffuse more rapidly through the oxide. Wet oxidation is used to grow thicker oxide films. The rate of thermal oxidation of heavily doped Si has been found to be much higher than the rate measured for lightly and moderately doped Si. This enhancement can be attributed to excess vacancies present in the Si near the interface, due to the shift of the Fermi level from midgap due to the high doping level.

The data for the oxidation of Si using O_2 presented in Fig. 21.7 also provide evidence for an initial period of rapid growth preceding the region of linear growth for thin layers. It appears that the first $\approx$ 20 nm of oxide grow at a much faster rate than the subsequent oxide. This rapid initial growth period may be related to electric fields resulting from space-charge effects in the oxide, which can accelerate the motion of negatively charged O_2^- ions toward the Si/a-SiO$_2$ interface. It has not been established, however, whether the oxygen-containing species diffusing through the oxide layer to the interface with Si are charged or neutral.

Redistribution of the dopant impurities present in the Si substrate occurs during thermal oxidation, with the acceptors Ga, B, and In being depleted from the interface region and the donors P, Sb, and As piling up in the Si. The extent of the dopant redistribution will depend on the relative solubilities of the dopant in Si and in a-SiO$_2$ and also on the diffusivity of the dopant through the growing layer into the gas phase. This redistribution can clearly have important effects in Si-based devices.

In addition to thermal oxidation, other methods of preparing thin layers of a-SiO$_2$ include CVD or PECVD from mixtures of SiH_4 and O_2 and electrochemical oxidation or anodization. The layers deposited using PECVD at lower temperatures contain hydrogen and so have the chemical composition $SiO_{2-x}H_{2x}$ when all atoms are fully bonded.

21.8 Fabrication of Silicon Devices

The fabrication of Si-based electronic devices [e.g., integrated circuits (ICs)] consists of a very large number of synthesis and processing steps involving procedures such as oxidation, lithography, diffusion, ion implantation, deposition of thin layers of semiconductors and insulators, metallization, etching, and annealing. The *integration* of circuit elements such as transistors, *p-n* junctions, dielectric layers, and so on, into planar structures began in 1960. The era of *VLSI* (very large scale integration) began in the late 1970s, when the number of devices reached 10^5 per chip [where a *chip* is a complete IC fabricated on Si and designed for a specific purpose (microprocessors, memory or data storage, etc.)]. As device densities have continued to increase into the range 10^6 to 10^7 per square centimeter, the term *ULSI* (ultra large scale integration) has become appropriate.

The Semiconductor Industry Association is an industrial consortium that prepares and updates a technology "road map" for the development of IC fabrication. Some of the details of the 1999 road map are presented in Table 21.1. Over the period 1999–2008 it is predicted that a typical *feature size*, as measured by the dynamic random-access memory (DRAM) $\frac{1}{2}$ pitch, will continue to decrease, from 180 nm to 70 nm. This feature size also defines the technology generation of Si-based devices. As transistors continue to decrease in size, their density on the chip will increase correspondingly, as will their switching speeds. This road map predicts an exponential decrease in feature size and exponential increases in the number of transistors per chip and their speed. Thus *Moore's law*, first stated in 1965, which predicts that microprocessor performance will double every 18 to 24 months, is predicted to remain valid over this time span.

Extremely high-purity, high-resistivity single-crystal Si is the "blank slate" upon which the microstructures and multilayer configurations of ICs can be "written." The Si wafers in use today, which provide the mechanical and thermal support for

TABLE 21.1   International Technology Road Map for Semiconductors

	1999[a]	Predicted		
		2002	2005	2008
DRAM $\frac{1}{2}$ pitch (nm) (minimum metal half-pitch)	180	130	100	70
Gbits/cm^2 for memory (at introduction)	0.27	—	1.63	4.03
DRAM cell area (μm^2)	0.26	0.10	0.044	0.017
Maximum substrate diam. (mm)	200	300	300	300
Maximum number of wiring levels[b]	7	8	9	9
Minimum power supply (V) (logic, for maximum performance)	1.8	1.5	1.2	0.9
Chip frequency (MHz) (on-chip local clock)	1250	2100	3500	6000

[a]Data from Semiconductor Industry Association, *International Technology Roadmap for Semiconductors: 1999*, 1999 ed., SEMATECH, 1999, Austin, Texas.

[b]*Wiring level* refers to the vertical electrical connections or vias between different metal interconnect planes in the device.

the device, must be flat, clean, smooth, and free of impurities and crystallographic defects to incredible standards and must meet standards that are expected to be even stricter in the future. Typical wafer thicknesses are 0.3 to 0.5 mm and may increase in the future to provide the mechanical stability needed, for example, in 300-mm wafers.

The materials used in Si-based devices include the following: Si substrates and epilayers; dielectrics such as silicon oxide, nitride, and oxynitride, phosphorus-doped SiO$_2$ [phosphosilicate glass (PSG)], and polyimide; electrical interconnecting conductors such as Al and Al + 0.5 wt % Cu, W, Ti, TiN, transition metal silicides, and low-resistivity polycrystalline Si. New materials will be needed as device sizes continue to shrink. An important example is the need to increase on-chip speeds and to decrease interconnect time delays due to the time constant $\tau = RC = R\epsilon A/d$. Here R is the metallic interconnect line resistance and C is the capacitance of the dielectric between the lines, both of which will increase as line and dielectric layer thicknesses continue to decrease. New low-ϵ dielectric-constant insulating materials, with $\epsilon_r = \epsilon(0)/\epsilon_0 < 3$, are needed to replace a-SiO$_2$ with $\epsilon_r \approx 4$. Possible candidates for the low-ε dielectric are fluorinated a-SiO$_2$ (i.e. SiO$_{2-x}$F$_{2x}$), and fluorinated amorphous carbon (i.e., a-C:F), which is a highly cross-linked amorphous polymer. Another requirement is that these new dielectrics must also be able to withstand processing temperatures up to $T = 400°$C. In addition, because of its lower electrical resistivity and higher resistance to electromigration, Cu may replace Al as the metal of choice for metallic interconnecting lines. High-dielectric-constant insulating materials such as Ta$_2$O$_5$, (Ba$_x$Sr$_{1-x}$)TiO$_3$ (BST), or Pb(Zr$_x$Ti$_{1-x}$)O$_3$ (PZT) will be needed as memory storage capacitors continue to decrease in size.

Electrical characterization of Si-based electronic devices is the preferred method for detecting processing-related defects such as interface and bulk Si traps, oxide

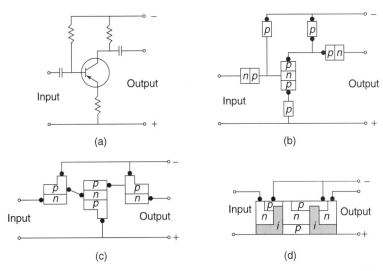

Figure 21.8. Concept of integration of the circuit elements in a Si-based amplifier using a schematic *pnp* transistor circuit: (*a*) circuit consisting of the *pnp* transistor, three resistors, and two capacitors; (*b*) representation of all the resistors and capacitors as Si elements; (*c*) the connecting wires between the elements are eliminated; (*d*) the circuit elements are arranged in a planar configuration. (From James G. Brophy, *Semiconductor Devices*, copyright 1964, by McGraw-Hill. Reprinted with permission of the McGraw-Hill Companies.)

fixed charge, dislocations, and stacking faults. These defects can be probed through measurements of carrier lifetimes, reverse-bias leakage currents, diode ideality factors, oxide breakdown voltages, and ac and dc transistor characteristics. Some testing can be done in situ (i.e. on the wafer), while the devices are being fabricated.

To aid in understanding the concepts of *planar technology* and the integration of circuit elements in a Si-based electronic device, consider the *pnp* transistor circuit shown schematically in Fig. 21.8. This amplifier circuit consists of several discrete elements (i.e., a *pnp* transistor, three resistors, and two capacitors). Since the electrical resistivity of high-purity Si can be controlled via doping, the resistors can be formed from the same Si material as the transistor. In addition, *p-n* junctions based on Si have the ability to act as capacitors when reverse biased, due to the depletion of carriers from the region of the *p-n* junction, as discussed in Chapter 11. All of the resistors and capacitors can therefore be formed from Si, as shown. The elimination of the connecting wires between similar elements indicates that the amplifier circuit consists of a *pnp* transistor and two reverse-biased *p-n* junctions. The final conceptual step consists of arranging the circuit elements in a planar configuration. All of the discrete circuit elements required have thus been integrated into the same structure based on regions of Si with appropriate doping levels, which, when necessary, are isolated from each other by layers of insulating (i.e., undoped) Si. Note that in this planar technology, the electrical contacts are made to the top, planar surface of the device.

A brief overview of the important steps involved in the fabrication of Si-based electronic devices from Si wafers of sufficiently high resistivity is presented in Chapter W21. To illustrate the complexity of the process, consider the fabrication of a 256-Mbit dynamic random-access memory (DRAM). A wafer yields 16 chips, each 25 mm square and consisting of $\approx 3 \times 10^8$ devices with features as small as 0.25 μm. Due to the large number, ≈ 300, of synthesis and processing steps involved in IC fabrication, it is not possible here to describe these procedures in detail. The books by Wolf and Tauber (1990) and by Maly (1987) provide useful descriptions of the steps involved in IC fabrication. Some of the important steps are also described in Chapter W21 (e.g., the CVD of epitaxial Si films and the PECVD of silicon nitride dielectric films). The thermal oxidation of Si to form passivating and protecting a-SiO_2 layers is described in Section 21.7. Other steps, such as diffusion (Chapter 6) and ion implantation (Section W21.3), are also discussed elsewhere.

SYNTHESIS AND PROCESSING OF METALS

The focus in this section is primarily on the metallurgical procedures that are used in the synthesis and processing of metals to achieve the desired final microstructures, which are typically polycrystalline. Processing rather than synthesis is usually the critical issue for these materials. The important processing variables include the temperatures, times, and cooling rates used in heat treatments and the nature of any mechanical treatments (e.g., plastic deformations). An important component of the synthesis and processing of metals is the design of a material for a particular purpose, which includes the choice of composition and alloying elements.

With regard to synthesis, the solidification of even pure metals from the liquid phase is not a simple process. Solidification is made even more complicated when the metal is an alloy. In addition to the formation of dendrites and other cooling-related columnar and equiaxed macrostructural features, chemical segregation of the different compositional phases of an alloy can be expected to occur even under equilibrium conditions. This is illustrated in detail for the case of steels. Since the useful properties of metals are often enhanced through the introduction of defects such as dislocations and precipitates, the synthesis of pure, defect-free single crystals is usually not required or even desirable. This is in distinct contrast to the synthesis of single crystals of highly pure semiconductors such as Si and GaAs. Important exceptions to this are the turbine blades and vanes made of Ti or NiAl alloys, which are now grown as single crystals in order to achieve desired mechanical properties such as high-temperature creep strength.

A wide range of procedures involved in the synthesis and processing of specific metals are discussed in other chapters:

Chapter 12

1. A brief discussion of steels, the Fe–C phase diagram, and the different homogeneous (i.e., single-component) and heterogeneous (i.e., multicomponent) phases of Fe–C alloys (Section 12.7)

2. Intermetallic compounds and the principles of superalloy design and processing (Section 12.8)

3. Coatings for the modification of the surface properties of metals and their preparation via electroplating, chemical reactions, vapor deposition, ion implantation, and thermal reactions (Chapter W12)

4. Synthesis of metallic glasses via the extremely rapid quenching of a liquid metallic alloy (Chapter W12)

5. Synthesis of porous metals (Chapter W12)

Chapter 16

1. Synthesis of multifilament Nb_3Sn wire for high-field superconducting magnets (Chapter W16)

Chapter 17

Note: Processing is very important for magnetic materials since Fe and Fe-based alloys can behave as either hard or soft magnetic materials, depending on the alloying elements present and on the thermal and mechanical processing which they receive.)

1. Synthesis and processing of a wide range of hard magnetic materials (precipitation-hardened alloys, steels, and rare earth–transition metal compounds) for permanent magnets via the formation of carbides of W, Cr, and Co and via martensitic lattice transformations, powder metallurgy, rapid solidification, and sintering (Chapter W17)

2. Synthesis and processing of soft magnetic materials (Fe alloys and electrical steels, Fe–Ni alloys, and Fe–Si alloys) via annealing, rapid cooling, and hot rolling (Chapter W17)

Chapter 19

1. Electrolytic anodization of A1, resulting in the formation of a protective oxide layer (Section 19.11)

As a result of these prior discussions, only two additional important topics are presented in this section: the synthesis and processing of steels and of stainless steels. The synthesis and processing of aluminum alloys, the synthesis of metals via rapid solidification, and surface treatments for metals are described in Chapter W21.

21.9 Synthesis and Processing of Steels

Steels are the technologically most important and complex group of metal alloys. They are extremely versatile since they can be processed to possess a wider range and combination of desirable mechanical properties than any other material, usually at relatively low cost. For example, steels with moderate yield stresses ($\sigma_y \approx 200$ to 300 MPa) can be prepared with both excellent ductility and toughness. High-strength steels, with yield stresses $\sigma_y \approx 5500$ MPa, can also be prepared with sufficient ductility for many applications. The goal in the processing of structural steels is to obtain materials with high strength, good ductility, and adequate corrosion resistance at a reasonable cost.

As mentioned in Chapter 12, the generic term *steel* refers to a wide range of microstructurally heterogeneous alloys of iron and carbon [i.e., $Fe_{1-x}C_x$ alloys with

$0.002 < x < 0.06$ (0.2 to 6 at % C)].[†] Additional elements such as Si, Ni, and Cr are usually added to steels for a variety of purposes. It has been known for over 2500 years that the addition of only about 0.5 at % C to Fe can greatly increase its strength. In addition to *plain carbon steels*, which contain only Fe and C, and *alloy steels*, which also contain up to a few percent of other alloying elements, other important ferrous-based structural materials include *cast irons*, which contain ≈ 8 to 16 at % C, and *stainless steels*, which contain up to ≈ 20 at % of additional alloying elements, such as Ni, Cr, Si, Mn, Mo, and N. Steels are usually specified by giving the allowed ranges of composition of the important alloying elements, together with upper limits for impurity elements such as oxygen and sulfur.

While the simplest steels are just Fe–C alloys, steels in general can be very complex materials in both composition and microstructure. This complexity makes the design of a steel with a given set of properties quite challenging. It is useful to review first how the complex phases that may be present in steels are related to the simpler phases of pure Fe and Fe–C compounds and alloys. Table 21.2 summarizes the properties of these important phases and also of their multicomponent mixtures, which are found in the steels commonly used today. Most of these phases are described in more detail in Chapter W21.

The equilibrium phases listed in Table 21.2 can be seen in the Fe-rich side of the Fe-C phase diagram presented in Fig. 21.9. This phase diagram shows the metastable equilibrium between Fe and *cementite*, Fe_3C. The true equilibrium Fe–C phase diagram would correspond to the stable equilibrium between Fe and graphite. Although graphite is often present at the higher C contents found in cast iron, it is not readily obtained in steels, due to the slow rates of nucleation and growth of graphite. As a result, Fe_3C is formed instead of graphite except at very low cooling rates or when alloying elements such as Si, Al, Ni, and Cu, which tend to destabilize Fe_3C, are present. Cast irons are inhomogeneous alloys, due to the presence of graphite inclusions and are not discussed here.

For pure Fe the terms *ferrite* and *austenite* correspond to BCC α-Fe and FCC γ-Fe, respectively. These terms are also used more generally for BCC and FCC Fe-based alloys and steels. Note that the maximum solubility of interstitial C is ≈ 9 at % at $T = 1150°C$ in austenite, 100 times greater than its solubility in the low-temperature ferrite phase at $T ≈ 727°C$. This much higher solubility of C in austenite is due to the larger octahedral interstitial sites present in the FCC crystal structure. Other interstitial impurities, such as N and B, also have much higher solubilities in austenite than in ferrite. The C atoms in ferrite also occupy the octahedral interstitial sites even though these sites are smaller than the tetrahedral sites in this BCC crystal structure. This occurs because the smaller octahedral sites can be deformed more easily when occupied by C atoms than can the tetrahedral sites. Larger atoms, such as Ni and Cr, which cannot readily occupy the interstitial sites tend to enter the Fe structure substitutionally.

When ferrite with more than ≈ 0.01 at % C is cooled from high T, the precipitation of C in the form of iron carbides with the stoichiometry $Fe_{3-x}C$ occurs in the temperature range $T = 300°C$ down to 20°C. This is due to the lower solubility and higher diffusivity of C in the more open BCC crystal structure of ferrite, compared to the FCC crystal structure of austenite. When aged (i.e., annealed at $T = 200°C$ or above)

[†] Note that the atomic % of C in dilute $Fe_{1-x}C_x$ alloys is equal to ≈ 4.5 times the weight percent.

TABLE 21.2 Important Phases of Fe, Fe–C Compounds and Alloys, and Their Multi-component Mixtures Found in Steels

Phase	Structure and Description[a]	How Phase Is Obtained
	Equilibrium Phases of Pure Fe	
α-Fe (ferrite)	BCC, $a = 0.286$ nm at $T = 20°C$; stable up to $T = 912°C$; $T_{Curie} = 769°C$	Stable phase at STP
γ-Fe (austenite)	FCC, $a = 0.364$ nm at $T = 912°C$	Stable phase for $912 < T < 1394°C$
δ-Fe (δ-ferrite)	BCC, $a = 0.293$ nm at $T = 1394°C$; $T_m = 1538°C$	Stable phase for $T > 1394°C$
	Equilibrium Fe–C Compound	
Fe$_3$C (cementite)	Orthorhombic, $a = 0.509$, $b = 0.674$, $c = 0.452$ nm; a complex interstitial compound	Present in Fe–C alloys under conditions of metastable equilibrium; see Fig. 21.9
	Equilibrium $Fe_{1-x}C_x$ Alloys	
α-Fe–C (ferrite)	Solubility limit of C in α-Fe at $T = 27°C : x = 1.2 \times 10^{-6}$ (0.00012 at % or 1.2 ppm)	Present in Fe–C alloys under equilibrium conditions; see Fig. 21.9
γ-Fe–C (austenite)	Solubility of C in γ-Fe at $T = 1150°C : x \approx 0.09$ (9 at %)	Present in Fe–C alloys under equilibrium conditions; see Fig. 21.9
	Nonequilibrium Multicomponent Phases	
Pearlite	A coarse, lamellar form of cementite in ferrite; a eutectoid structure	Formed between $T = 720$ and $550°C$ during cooling of austenite
Bainite	An intermediate structure composed of fine aggregates of ferrite plates (laths) and cementite particles	Formed between $T = 550$ and $\approx 250°C$ during cooling of austenite
Martensite	BC tetragonal, $c/a = 1 + 0.045$ wt % C; a supersaturated solid solution of interstitial C in ferrite, having a lath or lenticular microstructure	Rapid quenching of austenite to keep C in solution; formed between $T \approx 250°C$ and room temperature or below
Acicular ferrite	A disorganized structure of randomly oriented ferritic plates in a matrix such as martensite	Nucleation of ferrite at small, nonmetallic inclusions during cooling of austenite

[a]The range of thermal stability is given at $P = 1$ atm.

cementite or Fe$_3$C is formed. The dendritic appearance of this compound in ferrite is shown in Fig. 21.10.

It should be noted that the smaller interstitial impurities C and N have diffusivities in both ferrite and austenite which are several orders of magnitude greater than those of larger substitutional impurities, such as Ni and Cr and also of Fe itself. As a result, steels alloyed with metallic elements with low diffusivities must usually be given homogenizing heat treatments at $T = 1200$ to $1300°C$ for long periods of time, in

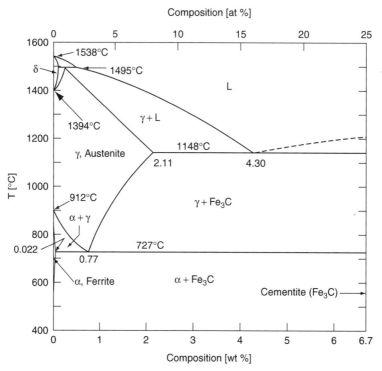

Figure 21.9. The metastable equilibium between Fe and Fe₃C is shown here on the Fe-rich side of the Fe–C phase diagram. [From *Metals Handbook*, 8th ed., Vol. 8, *Structures and Phase Diagrams*, ASM International, Materials Park, Ohio, 1973, p. 275, C-Fe.]

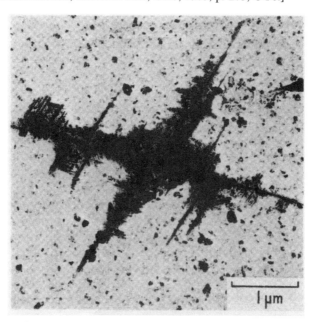

Figure 21.10. Dendrite of cementite, Fe₃C, in quenched and aged α-Fe (26 h at $T = 240°C$). Platelets of cementite can be formed on the {110} planes of ferrite along the eight possible ⟨111⟩ directions. (From R. W. K. Honeycombe et al, *Steels: Microstructure and Properties*, 2nd ed., copyright 1995. Reprinted by permission of Arnold Publishers.)

order to achieve uniform spatial distributions of these elements. In addition, at a given temperature, interstitial and substitutional impurities both diffuse much faster in the relatively open BCC ferrite crystal structure than in the more closely packed FCC austenite crystal structure.

The important mechanical properties of steels, such as strength, ductility, toughness, and hardness, are determined by their microstructures, which, in turn, are to a large degree controlled by the *austenite–ferrite phase transition*, which occurs upon cooling of the steel from high temperatures. At this transition, which occurs at $T = 912°C$ for pure Fe, the atomic volume increases by $\approx 1\%$, an expansion that can generate large stresses within the low-temperature ferrite phase.

The lowest equilibrium temperature for the existence of austenite in the Fe–C system is the *eutectoid temperature* T_e at which austenite transforms to ferrite and cementite. It can be seen from Fig. 21.9 that $T_e \approx 727°C$ and that the eutectoid composition is ≈ 0.77 wt % (3.5 at % C). At this temperature and composition the following eutectoid reaction occurs:

$$\text{austenite } (\approx 3.5 \text{ at } \% \text{ C}) \longrightarrow \text{ferrite } (\approx 0.10 \text{ at } \% \text{ C})$$

$$+ \text{cementite (Fe}_3\text{C}, 25 \text{ at } \% \text{ C}). \quad (21.57)$$

This mixture of reaction products is known as *pearlite*, a lamellar composite consisting of alternating layers of cementite and ferrite (Fig. 21.11). When austenite containing less than ≈ 3.5 at % C is cooled slowly, hypo-eutectoid ferrite starts forming at a temperature between $T = 912$ and $727°C$. At $T = 727°C$ the remaining austenite is transformed to pearlite. The slow cooling of austenite with a C content between 3.5 and ≈ 9 at % C, on the other hand, leads first to the transformation of austenite to Fe$_3$C above $T = 727°C$. The remaining austenite then transforms to pearlite at $T = 727°C$.

Figure 21.11. The slow cooling of austenite in a 0.8 wt % C steel (the eutectoid composition) leads to the formation of pearlite, a lamellar mixture of alternating layers of cementite (Fe$_3$C), and ferrite. The line at the lower left is 10 μm long. (From R. W. K. Honeycombe et al, *Steels: Microstructure and Properties*, 2nd ed., copyright 1995. Reprinted by permission of Arnold Publishers.)

Thus the three primary phases present in slowly cooled plain carbon steels are ferrite, cementite, and pearlite. The relative fractions of these intimately mixed phases and their morphologies or distributions within the steel depend on the C content and on the details of the thermal treatment that the steel receives.

Alloying elements such as Ni, Mn, Cu, and N are often added to steels to stabilize the austenite phase both to higher C concentrations and to lower temperatures, thus facilitating the transformation of austenite to martensite, as described later. Other alloying elements, such as Cr, Ti, Mo, W, Si, and Al, have the opposite effect of destabilizing austenite. Stainless steels that are austenitic at room temperature are obtained when large amounts of Ni are added to the steel in addition to Cr.

The kinetics of the transformation of austenite to ferrite, cementite, and pearlite can be represented in an isothermal *time–temperature-transformation* (TTT) diagram, the data for which are obtained by monitoring these isothermal transformations at a series of temperatures. The characteristic "C" shape of the resulting TTT curves is illustrated in Fig. 21.12, where the two solid curves indicate the times on the horizontal axis

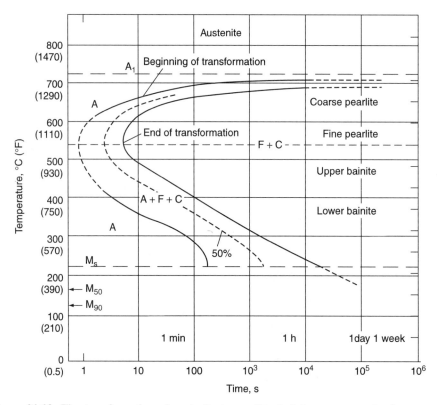

Figure 21.12. The transformation of austenite to pearlite, bainite, or martensite is presented here in an isothermal time–temperature-transformation (TTT) diagram. For the Fe–0.76Mn–0.79C wt % steel shown the regions of formation of pearlite at higher T, bainite at intermediate T, and martensite at lower T are illustrated. Note that $A_1 = T_e$ (the eutectoid temperature), $M_s = T_{Ms}$ (the martensitic start temperature; see Chapter W21); A, austenite; F, ferrite; C, cementite. (From *ASM Handbook*, Vol. 1, *Properties and Selection: Iron, Steels, and High-Performance Alloys*, ASM International, Materials Park, Ohio, 1990, p. 128, Fig. 4.)

corresponding to the beginning and end of the transformation from austenite to the final products at a given temperature on the vertical axis. The "nose" of the curve at the left-hand side indicates the temperature at which the transformation occurs most rapidly. For the Fe–0.76Mn–0.79C wt % steel shown here, the transformations of austenite to pearlite at higher T, to *bainite* at intermediate T, and to *martensite* at lower T are illustrated.

In general, the transformations to pearlite and bainite are slower (i.e., occur at longer times) at higher T, due to the weaker driving force resulting from the small enthalpy change for the process. The transformations are also slower at lower T, due to the exponential slowing down of the thermally activated diffusion of C in the alloy, which plays an essential role in the transformations. Pearlite forms at temperatures down to the "nose" of the curve, $T \approx 550°C$, while bainite and martensite form at lower T. Carbon, with its much higher solubility in austenite than in ferrite, tends to stabilize austenite and to slow down the kinetics of its transformation to other phases, such as ferrite, pearlite, bainite, and martensite.

21.10   Synthesis and Processing of Stainless Steels

Stainless steels are steels that have more than 11 to 12 wt % Cr in solid solution and good corrosion resistance, due to the formation at their surfaces of a stable and protective Cr-based oxide film. As a result, stainless steels are widely used in corrosive environments. Stainless steels are usually austenitic, as a result of the stabilization of the FCC crystal structure down to room temperature due to the incorporation of the alloying elements Ni or Mn. *Austenitic stainless steels* containing 18 to 30 wt % Cr, 8 to 20 wt % Ni, and 0.03 to 0.1 wt % C are readily fabricated and do not undergo the ductile-to-brittle transition observed in BCC ferritic steels as the temperature is lowered. In addition to the austenitic variety, *ferritic* and *martensitic stainless steels* also exist. Whereas martensitic stainless steels can be strengthened via heat treatment, (i.e., by tempering), austenitic and ferritic stainless steels are not heat treatable, so must be strengthened via mechanical work hardening.

The most common austenitic stainless steels contain 18 wt % Cr and 8 wt % Ni and are known as 18–8 stainless steels. The addition of Ni leads to the desired retention of austenite to room temperature for steels containing 18 wt % Cr. Higher Ni contents are required for the stabilization of austenite as the Cr content is either increased or decreased, as illustrated in Table 21.3, where the compositions of several typical austenitic stainless steels are presented, and in Fig. 21.13, where the possible structures of stainless steels based on Fe, Cr, and Ni are conveniently shown in the *Schaeffler diagram*. Here the single-phase regions corresponding to austenite, martensite, and δ-*ferrite* can be seen along with the expected binary and ternary multiphase regions. The axes are labeled in terms of Cr and Ni equivalents, since other elements, such as Si and V, are effective in aiding Cr in the stabilization of δ-ferrite, while elements such as Co, Mn, and C are effective in aiding Ni in the stabilization of austenite. For example, 1 wt % V is equivalent to 5 wt % Cr, and 1 wt % Mn is equivalent to 0.5 wt % Ni in this regard. It can be seen from Fig. 21.13 that Cr actually helps to stabilize austenite when it is added to a steel containing Ni.

In the design of austenitic stainless steels for specific applications, each element plays an important role (or roles) in determining the properties of the steel:

TABLE 21.3 Compositions of Typical Austenitic Stainless Steels[a]

Element	Composition[b] (wt %)						
	301	302	304	310	316	321	347
C	<0.15	<0.08	<0.08	<0.25	<0.08	<0.08	<0.08
N	0.03	0.03	0.03	0.03	0.03	0.03	0.03
Cr	16–18	17–19	18–20	24–26	16–18	17–19	17–19
Ni	6–8	8–10	8–12	19–22	10–12	9–12	9–13
Mo	—	—	—	—	2–4	—	—
Ti	—	—	—	—	—	≈5 times wt % C	—
Nb	—	—	—	—	—	≈10 times wt % C	—
Mn	1.5	1.5	1.5	1.5	1.5	1.5	1.5

[a]By American Iron and Steel Institute (AISI) designation number.
[b]The balance of the composition is Fe.

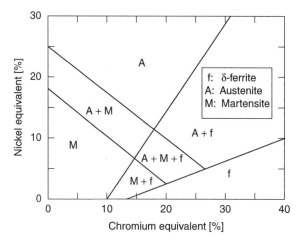

Figure 21.13. The possible structures of stainless steels based on Fe, Cr, and Ni are conveniently presented in a Schaeffler diagram. The single-phase regions corresponding to austenite, martensite, and δ-ferrite can be seen along with the expected multiphase regions. The axes are labeled in terms of Cr and Ni equivalents. [From *ASM Handbook*, Vol. 1, *Properties and Selection: Iron, Steels, and High-Performance Alloys*, ASM International, Materials Park, Ohio, 1990, p. 899 (Fig. 42.)]

1. *C and N*: expansion of the austenite stability region and also solid-solution strengtheners; low levels of C (N) are desirable to minimize the formation of $Cr_{23}C_6(Cr_2N)$, which ties up Cr in an undesirable form.

2. *Cr*: corrosion protection via the formation of a stable oxide surface coating; also, in conjunction with Ni, tends to increase the stability of austenite.

3. *Ni*: expansion of austenite stability region down to room temperature; added resistance to corrosion, especially stress-corrosion cracking.

4. *Mo*: added resistance to corrosion, especially pitting corrosion.

5. *Ti and Nb*: formation of TiC and NbC, thus helping to minimize the formation of $Cr_{23}C_6$; enough Ti and Nb (e.g., ≈ 5 times the wt % of C for Ti and ≈ 10 times

the wt % of C for Nb) is usually added to react with all of the C and to allow for some solid-solution and dispersion strengthening.

6. *Mn*: removal or deactivation of sulfur via the formation of MnS; expansion of the austenite stability region.

Type 304 stainless steel is the material of choice for the construction of ultrahigh-vacuum (UHV) systems since it is very strong, highly corrosion resistant, and has a very low vapor pressure. A shiny surface, compatible with the UHV environment, can be obtained by annealing in a dry H_2 atmosphere. In addition, type 304 austenitic stainless steel has very low magnetic permeability, due to its FCC crystal structure. BCC α-Fe (ferrite) is ferromagnetic with a magnetic moment of $2.2\mu_B$ per atom, while FCC γ-Fe (austenite) is antiferromagnetic with a magnetic moment of only about $0.6\mu_B$ per atom.

The annealing of stainless steels in the range $T = 1050$ to $1150°C$ puts all the C into solution, and is followed by rapid cooling to prevent the formation of undesirable Cr-rich carbides such as $Cr_{23}C_6$. The temperature range $T = 500$ to $850°C$ should be avoided because in this range the nucleation and growth of $Cr_{23}C_6$ occur readily in the stainless steel. As mentioned earlier, the addition of Ti and Nb can also retard the formation of $Cr_{23}C_6$ by leading to the formation of TiC and NbC instead.

Thermal and thermomechanical processing are very important for *duplex stainless steels*, which are mixtures of ferritic and austenitic phases. With the correct composition of alloying elements, a region in the phase diagram can be found in a range of temperatures (typically, $T = 1000$ to $1300°C$) where both the ferrite and austenite phases are stable (see Fig. 21.13).

An interesting method of synthesizing stainless steels involves the mechanical alloying of the constituent elemental metals or alloys in powder form in a ball mill. The atomic mixing apparently occurs without the melting of the constituents through the high pressures generated when the powders are trapped between the balls during collisions. Other materials, such as refractory oxides (e.g., yttria), which are used for dispersion strengthening can also be introduced very effectively in this way. Following extrusion of the resulting completely mixed powder, the bulk alloys obtained can have an extremely small grain size (<1 μm), which is impossible to achieve by any other method.

SYNTHESIS AND PROCESSING OF CERAMICS AND GLASSES

Having discussed the synthesis and processing of semiconductors and metals, attention now turns to ceramics and glasses. Methods of synthesis from solids, liquids, and gases are considered. The following sections begin with a study of traditional preparation techniques, starting with solid-state powders. Synthesis from solutions by means of sol-gel synthesis is then discussed. Then gas-phase synthesis is introduced and the chemical vapor deposition of diamond is studied in Section W21.14. Applications to several representative materials of current interest are also made at our Web site, including the high-temperature superconductor YBCO (Section W21.15), the mechanically hard materials Si_3N_4 (Section W21.16) and SiC (Section W21.17), the zeolite catalyst ZSM-5 (Section W21.18), and the ferroelectric crystal PLZT (Section W21.19).

The discussion concludes with a study of the Pilkington process for the manufacture of glass (Section W21.20).

21.11   Powder Synthesis

The processing of ceramics into functional forms involves a series of transformations, starting with the preparation of a powder by communition (grinding). This is facilitated by the natural brittleness of ceramics. Other methods, such as thermal-plasma synthesis, are used to prepare ceramic powders. The reactants injected into the plasma can be gases, liquids, or solids. The high temperatures in the plasma and the high supersaturation of the reactants provide the driving force for condensation and lead to the synthesis of fine particles via homogeneous nucleation.

The powder is characterized by its size and shape distribution. These distributions may be analyzed through the use of appropriate sieves or by using various microscopies. A mix of particles with different sizes may be useful in creating a ceramic with many binding sites and little pore space. A uniform distribution is desirable when a homogeneous material is desired. The grain size is often critical in determining the mechanical and electrical properties of the ceramic.

One may then wish to suspend the powder in a liquid. The use of wetting agents often enables this to occur. Surfactants are generally employed. If the solvent is water, one adds molecules to the solution with a hydrophobic group (usually a chain hydrocarbon) at one end and a hydrophilic group (usually polar) at the other end. These molecules dramatically decrease the surface tension between the liquid and the powder particles and permit them to become wet. In the case of solvents other than water, one talks instead of lyophilic and lyophobic groups.

If the powder agglomerates, one employs a deflocculant to achieve homogeneity. The van der Waals forces between the powder particles cause them to attract each other. This attraction is offset by Coulomb repulsion. Typical deflocculant molecules used are sodium polyacrylate and sodium pyrophosphate. They coat the particles and give them a net charge (which is balanced by counterions in solution). The Debye screening length is the characteristic distance over which electric fields exist in an ionic solution before they are screened by the counterions. This length is defined by Eq. (W20.28), $\lambda_D = (\epsilon k_B T / \sum z_j^2 e^2 n_j)^{1/2}$, where ϵ is the electric permittivity of the nonionic solution, and $z_j e$ and n_j are the charge and concentration of the jth ion, respectively. If λ_D is sufficiently large, full screening is not achieved until the particles are far apart. At those distances, the thermal energy is larger than the van der Waals interaction, so the particles stay in suspension. At closer distances the Coulomb interactions of the charges enforce mutual repulsion.

In other circumstances, coagulation may be desirable: for example, when one wishes to eliminate liquid or create a dense structure. One could introduce polymers that attach themselves to the particles and bridge them together. Another possibility is to adsorb molecules onto the powder with hydrophobic groups projecting into the solution. When the particles agglomerate, the liquid is squeezed out of the interparticle zones. It is also possible to modify the ionic composition of the solution to the point where the double layer surrounding a particle becomes sufficiently thin that the Coulomb repulsion between particles is weakened. Coagulation agents are called *binders* or *flocculants*. Binders include clays such as kaolin, polymers such as polymethylmethacrylate (PMMA), ethylamine, and xanthan gum. An example where binding is of great importance is in the process of tape casting, which is used in the fabrication of substrates for electronic devices.

Plasticizers are used to make binders flexible. Suppose that the processing temperature, T, is lower than the glass-transition temperature of the binder, T_g. The plasticizer lowers T_g until $T_g < T$. Typical plasticizers include water, ethylene glycol, and glycerol. The molecules reside between the polymer molecules, separate them, and reduce both the steric hindrances posed by their side groups and the van der Waals interactions between them.

Foaming agents serve to allow the solution to retain gas in bubbles. They are employed to decrease the overall density of the ceramic. Polypropylene glycol ether is an example of a foaming agent. Antifoaming agents function in the opposite way. By eliminating bubbles, the ceramic is made to be more homogeneous and less likely to have defects. Surfactants with low surface tension, such as aluminum stearate, are used.

Lubricants are often added to assist in the processing. For example, one may wish to reduce the erosion of an extrusion die.

Thorough mixing of the various components is essential to obtain a homogeneous ceramic. One criterion for mixing is that any two molecules or particles that were initially adjacent to each other are equally likely to be found anywhere in the mixture afterward. It is best to try to achieve a turbulent flow in the process, although this is very difficult for high-viscosity slurries.

Sometimes the powder particles are spray dried and form granules. These are basically agglomerated powders where the grain size is controlled. They have the advantage of being safer to process than powders. The size of the granule also provides a natural limit to the size of the grain that will ultimately grow.

The ceramic may be cast into a particular shape by a pressing process. *Hot isostatic pressing* (HIP) is a method that is often employed. The imposition of a high pressure lowers the sintering temperature, where the particles partially melt and merge together, and also results in a more densified product. Typically, temperatures in the range $T = 800$ to $2700°C$ and pressures up to 350 MPa are employed. Additional benefits of HIP include the preservation of isotropy and elimination of residual stresses which would weaken the ceramic. One could describe the change in density with increasing pressure by assuming that the change is proportional to the porosity,

$$\frac{d\rho}{dP} = A\left(1 - \frac{\rho}{\rho_{max}}\right), \tag{21.58}$$

where ρ_{max} is the density of the fully compactified powder and A is a constant. Integration gives an empirical formula for describing the saturation of density as the pore space is removed:

$$1 - \frac{\rho}{\rho_{max}} = \left(1 - \frac{\rho_0}{\rho_{max}}\right)\exp\left(-\frac{AP}{\rho_{max}}\right), \tag{21.59}$$

where ρ_0 is the density of the powder at $P = 0$. Other pressing techniques include isostatic pressing, hot pressing, and dry pressing.

Before firing, the mixture must be dried. Typical methods include the use of heat convection, vacuum pumps, infrared or microwave radiation, freeze drying, or spray drying. Since shrinkage occurs during drying, an inhomogeneous rapid-drying procedure can lead to unwanted strains, which could result in cracks.

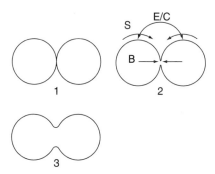

Figure 21.14. Growth of a neck during sintering. The letters B, S, and E/C label the bulk diffusion, surface diffusion, and evaporation/condensation transport mechanisms.

Sintering is a process in which the powder particles or granules start merging together to form a unified solid. The temperature of the mixture is raised sufficiently so that various transport mechanisms begin to operate. Diffusion of atoms and molecules can take place along the surfaces or through the bulk. Evaporation and condensation of molecules can also occur. Actual melting of the powder can be used, a process called *vitrification*. The net result is that molecular bridges between the particles are established. This is illustrated in Fig. 21.14 where necks are formed between particles and begin to grow in size. Note that the surface area of the merged object is less than that of the two starting particles, so the surface energy is decreased. At the same time the various organic components of the mixture are either boiled or burned away.

In liquid-phase sintering, powder particles partially dissolve in solution and the solute molecules redeposit themselves on the particles.

As the ceramic is sintered, the pores, which were initially percolated to form a connected network, become smaller and fewer in number, and eventually become isolated from each other. The ceramic undergoes densification. Grain growth occurs at elevated temperatures but is blocked by the presence of the pores.

The final process involves cooling. Due to the poor thermal conductivity of ceramics, the cooling rate must not be so rapid that thermal stresses nucleate cracks or propagate existing defects.

21.12 Sol–Gel Synthesis

In sol–gel synthesis one starts with a mixture of metal alkoxide precursors dissolved in an alcohol. Ideally, one would like the alcohol to be based on the same alkyl group as the alkoxide. The alkyl groups typically used are C_2H_5, C_4H_9, and C_6H_{13}, although larger groups are also employed. When water is added to the solution, hydrolysis takes place. The microparticles that result from condensation reactions between the hydrated metal ions constitute the sol. The sol is a stable suspension of small colloidal particles. The particles are so small that their weight is negligible compared with the fluctuating forces provided by the liquid, so sedimentation does not occur. In the gel phase the hydrated alkoxides condense into a polymer in which the backbone consists of alternating metal and oxide ions. The resulting porous network percolates throughout the volume of the liquid and results in rigidity.

Consider the case of a monometallic alkoxide, such as TEOS (tetraethylorthosilicate), which is a precursor for the formation of silica by the sol–gel process. The initial set of hydrolysis reactions is

$$M(OR)_n + jH_2O \longrightarrow M(OH)_j(OR)_{n-j} + jROH, \qquad (21.60)$$

where $M = Si$, $n = 4$ (the valence of Si), $R = C_2H_5$, and $j = 1, \ldots, n$. The initial set of condensation reactions is

$$M(OH)_i(OR)_{n-i} + M(OH)_k(OR)_{n-k} \longrightarrow M\text{--}O\text{--}M(OH)_{i+k-1}(OR)_{2n-i-k-1} + ROH. \qquad (21.61)$$

Further condensation steps result in the formation of polymer chains of the form

$$(M\text{--}O\text{--})_m M(OH)_j(OR)_{n(2m+1)-2m-j}. \qquad (21.62)$$

These chains tend to be largely linear, with few cross-links.

The polymerization can be accelerated by the use of an acid catalyst, which promotes the formation of cross-links between the chains. When the network spans the volume of the solution, it becomes rigid. After polymerization terminates, the material is dried and calcined and the organic components are pyrolyzed. Calcination is typically an endothermic reaction taking place at high temperatures in which an oxide is produced and volatile reaction products are released. [An example of a typical calcination reaction is $CaCO_3(s) \rightarrow CaO(s) + CO_2(g)$.] One is left with the metal oxide. Annealing at a high enough temperature permits crystallization to occur.

The method may be generalized to include several metal alkoxides, so as to build up a crystal with a number of different metal ions in the unit cell. In those cases where all metal oxides are not soluble in the alcohol, one may substitute some metallic salts for the precursors.

Sol–gel synthesis may be used to deposit optical coatings on glass surfaces such as lenses and mirrors. These coatings include silica, alumina, zirconia, titania, V_2O_5, and In_2O_3, as well as various transition metal oxides. One may also deposit films composed of mixtures of two or more oxides. Ceramic crystals for use as ferroelectrics, piezoelectrics, and nonlinear optical materials may be synthesized by the sol–gel route. These include barium titanate (BT), lead zirconate titanate (PZT), and lead lanthanate zirconate titanate (PLZT). A number of binary glasses may be produced, including SiO_2–Na_2O, SiO_2–Al_2O_3, SiO_2–GeO_2, and SiO_2–P_2O_5. Similarly, ternary glasses and glasses with many components may be prepared using sol–gel synthesis. Alkoxides such as $Si(OCH_3)_4$, $Si(OC_2H_5)_4$, $Ge(OC_2H_5)_4$, $Li(OC_2H_5)$, $Ti(OC_4H_9)_4$, $Ti(OC_2H_5)_4$, $Ti(OC_5H_7)_4$, $Al(OC_3H_7)_3$, $Al(OC_4H_9)_3$, $Zr(OC_3H_7)_4$, $Zr(OC_4H_9)_4$, and others are used.

Zeolites have been produced from sol–gels, as have refractory materials and electronic materials. Recently, it has been shown that mesoporous films may be deposited on substrates using sol–gel dipcoating.[†] These could be used for chemical separation, catalysis, and sensors. More applications are given by Jones (1989).

[†] A. I. Cooper et al., *Nature*, **389**, 368 (1997).

SYNTHESIS AND PROCESSING OF POLYMERS AND CARBON MOLECULES

Earlier sections of this chapter were concerned primarily with inorganic materials. In the following group of sections, attention turns to organic chemistry. Some aspects of the synthesis and processing of polymers are studied. Since this is a very broad area, it is not possible to provide a complete survey of the field. Rather, some key concepts and applications are chosen as being representative of the methodology being employed in polymer science. The study of polymers is followed by a brief discussion of carbon molecules, with particular emphasis placed on carbon nanotubes.

A brief discussion of polymerization is followed by a section dealing with catalysts used in polymer synthesis. This, in turn, is followed by a section concerned with the synthesis of nanotubes.

On our home page case studies of the synthesis of polycarbonate (Section W21.21) and polystyrene (Section W21.22) are made, followed by a section concerning the synthesis of electroactive polymers (Section W21.23). Some techniques that are used to process polymers are then discussed, beginning with the technique of spin coating (Section W21.24). This is followed by a discussion of microwave and plasma processing of polymers (Section W21.25).

21.13 Polymerization

Polymerization is a process by which the monomers self-assemble into a polymer. There are two main routes for this to occur, step polymerization and chain polymerization. In *step polymerization* two polymers of length m and n may combine to form a polymer of length $m + n$, for any integers m and n. In *chain polymerization* the chain can grow only by one monomer at a time (i.e., $m \rightarrow m + 1$). The net result in either case is the buildup of long polymer chains. For example, to produce a polymer of length 10 by step polymerization the preceding step may have involved the combination of molecules of lengths $9 + 1$, $8 + 2$, $7 + 3$, $6 + 4$, or $5 + 5$. For the chain process the preceding step involves lengths $9 + 1$. Since growth is determined by collisions with monomers or other chains, growth is a stochastic process. A statistical distribution of lengths will be present at any time. If the reactions are terminated by lowering the temperature or by the use of chemical terminators, a distribution of polymer lengths will result. A major challenge to polymer science has been to devise methods to control the size distribution of the chains.

Addition polymerization is a fusion reaction in which smaller polymer molecules grow simply by joining together, in either a step- or chain-growth fashion. In *condensation polymerization* the combination of the two molecules involves a rearrangement in which a small molecule, such as H_2O, is ejected from the growing polymer.

Examples of some addition polymers include polyethylene (PE), polypropylene (PP), polyvinylchloride (PVC), polytetrafluoroethylene (PTFE or Teflon), and polymethylmethacrylate (PMMA). Examples of polymers prepared by condensation reactions include polycarbonate (PC), polyimides, and polyethersulfone (PES). An example of a condensation reaction is the formation of polyester involving monomers of the form HO–R–COOH, where R is a divalent group. The basic condensation reaction,

$$HO-R-COOH + HO-R-COOH \longrightarrow HO-R-COO-R-COOH + H_2O,$$

when iterated through either a step- or chain-growth process, will ultimately lead to polymers of the form $HO-(RCOO)_n-H$. Another example involves the formation of the polyamide nylon-6,6, whose reaction is

$$n\,(H_2N-(CH_2)_6-NH_2) + n\,(HOOC-(CH_2)_4-COOH) \longrightarrow$$

$$H-(NH-(CH_2)_6-NHOC-(CH_2)_4-CO-)_n OH + (2n-1)\,H_2O. \tag{21.63}$$

Consider the chain-polymerization process. Imagine that there is a collection of three types of molecules: initiators (I), monomers (M), and terminators (T). These molecules combine according to certain rules. For example, there may be the following set of rules for the chain-growth process:

1. Polymers always start forming with I.
2. I may be joined either to M or to T.
3. IM...M may be joined to either M or T.
4. IT, IMT, IM...MT are inert and stop growing.

In this way, chains of the type IMMM...MMT may build up. Let N be the total number of polymer chains and N_n be the number of chains of n bonds, where $N = \sum N_n$. Let p be the probability of an unterminated chain IM...M eventually meeting another M and $(1-p)$ be the probability of meeting a T and terminating. Then

$$N_1 = N(1-p), \quad N_2 = Np(1-p), \ldots, N_n = Np^{n-1}(1-p). \tag{21.64}$$

The probability of finding a chain of length n, given by $P_n \equiv N_n/N = (1-p)p^{n-1}$, is called the *Schultz–Flory distribution*. Note that this is properly normalized since

$$N_1 + N_2 + N_3 + \cdots = N(1-p)(1+p+p^2+\cdots) = N\frac{1-p}{1-p} = N, \tag{21.65}$$

so $\sum P_n = 1$. Using the distribution N_n, the mean value of n may be calculated:

$$\langle n \rangle = \frac{1}{N}\sum_{n=1}^{\infty} N_n n = (1-p)\sum_{n=1}^{\infty} n\,p^{n-1} = \frac{1}{1-p}. \tag{21.66}$$

The quantity $1-p$ is the probability of a growing polymer meeting a terminator and is proportional to the terminator concentration. Thus an increased concentration of terminators results in a decrease in the mean polymer size. By varying this concentration, any mean size may be obtained.

Typical initiators are symmetrical molecules which will dissociate into two identical radicals. An example is benzoyl peroxide $(C_6H_5COO)_2$. It dissociates readily in benzene to form two radicals, $C_6H_5COO\cdot$, and these act as the initiators, $I\cdot$. Their role is explored further in Section W21.22, concerning the synthesis of polystyrene.

Polymers of much more complex design may be constructed if more than two functional groups on a monomer are capable of joining to other monomers. For example, *star polymers*, also called *dendrimer polymers*, may be produced if there are three groups. The topology is of the form of the Bethe lattice, pictured in Fig. 7.16. One way to produce star polymers is to begin with an ammonia molecule, NH_3, and add methyl acrylate, $C_4H_6O_2$, and methanol, CH_3OH. The methanol catalyzes the substitution of three methyl acrylate molecules for the three hydrogens in ammonia. One then adds ethylene diamine, $C_2H_8N_2$, which attaches itself to the methyl acrylate. The net result is that each of the three branches is now terminated in a NH_2 group. Originally, in ammonia, there were three active H atoms. Now there are six. This process may be iterated n times. The number of active hydrogens doubles each time according to the sequence $3, 6, 12, 24, \ldots, 3 \times 2^n$. The resulting structure is a macromolecule with a huge number of branches. It is possible to control the growth of the polymer by controlling the number of generations of doubling that takes place. It is also possible to substitute various chemical groups into any given generation, thereby controlling the size and shape of the dendrimer.

21.14   Catalysts in Polymer Synthesis

When polymer synthesis takes place, the reactions are often slow, and several products, some more useful than others, are produced. Catalysts are used to improve both the rapidity and selectivity of the reactions. The catalysts can be heterogeneous (e.g., consist of fine particles) or homogeneous (e.g., consist of molecules or ions in solution). In this section the synthesis of two polymers of commercial importance, polyethylene (PE) and polypropylene (PP), is examined. PE consists of a chain of carbon atoms connected by single bonds. Each C atom is coordinated to two hydrogen atoms. In polypropylene every other C atom is coordinated to a hydrogen atom and a methyl group instead of two H atoms. Adjacent methyl groups can all be on the same side of the backbone (isotatic: IPP), on opposite sides (syndiotactic: SPP), or randomly fluctuating between the two sides (atactic: APP). Both IPP and SPP molecules are one-dimensional crystals; APP lacks this long-range order. IPP and SPP are more valuable than APP since the chains fit together better, allowing for stronger interchain van der Waals bonding and resulting in a harder polymer with a higher melting temperature. SPP has the advantage that it is more transparent and has higher impact strength. The polymers are illustrated in Fig. 21.15.

The free-radical polymerization of PP results in an admixture of APP, IPP, and SPP, which is not very useful. A catalytic reaction was discovered by Ziegler and Natta which would favor either IPP or SPP. It also promotes the polymerization of PE. The typical catalyst involved was a mixture of $TiCl_4$ and $Al(CH_2CH_3)_3$. The metal atoms react with the carbon atoms of both the monomer and the polymer and convert the $C=C$ double bond of the monomer to a single bond

$$
\begin{array}{ccc}
\underset{\underset{H}{|}}{\overset{\overset{H}{|}}{C}}=\underset{\underset{H}{|}}{\overset{\overset{H}{|}}{C}} & \xrightarrow[\text{atom}]{\text{metal}} & \cdot\underset{\underset{H}{|}}{\overset{\overset{H}{|}}{C}}-\underset{\underset{H}{|}}{\overset{\overset{H}{|}}{C}}\cdot
\end{array}
\qquad (21.67)
$$

The metal atom permits a transfer of some charge between the polymer and monomer C atoms, and they electrically attract each other. A new $C-C$ bond is produced, the metal atom is released, and the polymer has grown by one unit.

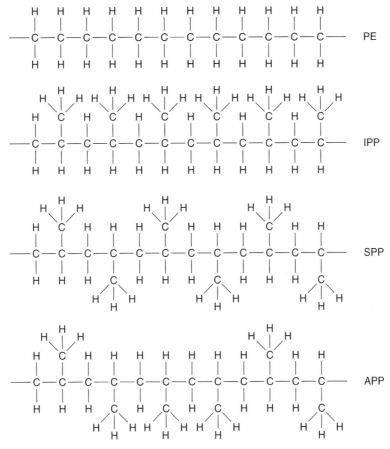

Figure 21.15. Polymers polyethylene (PE) and polypropylene: isotactic (IPP), syndiotactic (SPP), and atactic (APP). The atoms do not all lie in the same plane.

Metallocenes are homogeneous catalysts that have superseded the Ziegler–Natta catalysts. They are single metal atoms (e.g., Ti, Zr, or Si) coordinated to two five-member carbon rings through two carbon atoms. The carbon rings enclose the metal atom as a clamshell would enclose a pearl. The shell can be opened up by adding organic side groups to the carbon rings. Depending on the symmetry of the resulting molecule, one may selectively promote the production of IPP or SPP. For example, if the molecule has a twofold rotational symmetry but no mirror symmetry (e.g., the king of diamonds playing card), it is found to favor the production of IPP. If it has mirror symmetry (e.g., the diagonal of a chess board), it is found to favor the production of SPP. The key to this selectivity is found in the steric hindrance presented by the five-membered ring of the metallocene, which permits the approach of the monomer to the polymer (in the "clam shell") only if it has the proper spatial orientation. Repulsive interactions simply prevent the two carbon atoms from reacting with each other if they are too close to the ring.

An example of a metallocene catalyst is ansa-zirconocene, with two isomers (Fig. 21.16). The metallic atom is Zr. The C and H atoms are omitted and only the

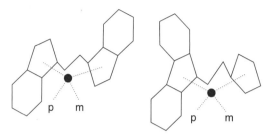

Figure 21.16. Two isomers of the metallocene catalyst ansa-zirconocene.

bonding pattern is shown. The molecule on the left favors the production of IPP and that on the right (which is viewed from an oblique perspective but actually possesses a plane of reflection symmetry) favors SPP. The metal atom forms temporary bonds simultaneously with both the monomer (m) and the polymer (p).

21.15   Synthesis of Carbon Nanotubes

The original method for the synthesis of carbon nanotubes involved the use of a carbon arc discharge in a He atmosphere. Typically, a potential difference of 20 V is maintained across a gap of 1 mm between carbon electrodes, and the arc current is 100 A. The pressure of the He gas is 500 torr. A bundle of carbon nanotubes about 5 mm in diameter is found to be grown on the cathode at the expense of the anode. The bundles consist of smaller bundles, and this pattern is repeated hierarchically until the size of individual nanotubes is reached. The growth rate is substantial, on the order of 1 mm/min. Mixed in with the nanotubes is a distribution of nanoparticles consisting of fullerenes. About 25% of the volume of the material is nanotubes and 75% is fullerenes. If the He pressure is reduced, the overall yield of fullerenes and nanotubes is reduced and the relative yield of fullerene molecules to nanotubes is increased. The individual nanotubes consist of n concentric cylinders, with each cylinder consisting of an atomic layer of carbon with the interlayer distance being 0.34 nm, similar to that of graphite. There is a statistical distribution of values of n. By introducing catalytic particles made of Co, Fe, or Ni, the case $n = 1$ can be made to predominate and one obtains a high yield of single-walled carbon nanontubes. The carbonaceous deposit on the cathode is ultimately dissolved in ethanol and sonicated to disperse the tubes. The nanoparticles can be burned away preferentially, leaving behind pure nanotubes. The individual tubes are up to several micrometers in length.

In planar graphite the C atoms are bonded to each other by sp^2 bonding. Comparing the energy of a rectangular sheet of graphite of a given width to a cylinder of the same circumference, one sees that the dangling bonds on the edge are removed and the energy is lowered when a cylinder is formed. However, to bend the sheet requires increasing its elastic energy. For small-radius tubes the elastic energy is too large. This puts a lower limit on the diameter of a nanotube of ≈ 0.7 nm. At the high temperatures of the cathode (typically, 3000 K) thermal fluctuations can cause the sheet to bend into a cylinder temporarily, and the chemical bonds formed lock in this configuration.

The growth mechanism for the nanotube has not yet been identified but may be the following. Graphite forms a honeycomb lattice with two atoms per unit cell. If a

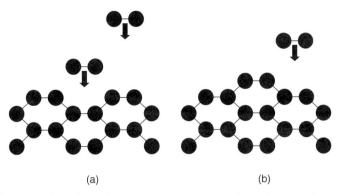

(a)   (b)

Figure 21.17. Adsorption of a carbon dimer on the leading circumference of a nanotube (*a*) to produce growth (*b*).

carbon dimer C_2 were to adsorb on the growing edge of the nanotube, it could add one additional hexagon to the structure and the lattice would grow. Competing with this process is one in which there are adsorption and reconstruction of the leading circumference of the nanotube to form a pentagonal cap. When this cap is formed, the ability of the nanotube to grow is greatly reduced. It is believed that the role of the He gas is to cool the nanotubes sufficiently to prevent this from happening, perhaps by removing thermal energy needed to overcome activation barriers for the rearrangement. The hypothetical growth mechanism is illustrated in Fig. 21.17. Multiwalled nanotubes presumably involve epitaxial growth of additional cylinders on an existing cylinder.

The role of the catalysts in producing single-walled nanotubes has yet to be clarified. It has been suggested that the catalysts are in the form of small nanoparticles and that they form templates for the nucleation and growth of the nanotubes. One should also note, however, that Fe, Ni, and Co are dehydrogenation catalysts and may serve to keep the frontier edges of a growing tube free of impurity hydrogen atoms. This would assure a rapid growth for the tube and would effectively compete with the presumably slower growth of additional concentric carbon tubes.

Additional methods exist for fabricating nanotubes, including pyrolysis of hydrocarbons with catalysts, electrolysis, ion bombardment, laser evaporation of carbon, and chemical vapor deposition. For example, it was found that laser-induced pyrolysis of 2-amino-4,6-dichloro-*s*-triazine on a patterned cobalt catalyst (on a silica substrate) results in aligned nanotubes with lengths and widths determined by the patterning geometry. Nanotube lengths up to 50 μm were generated.[†]

The molecule buckminsterfullerene, C_{60}, and the closely related molecules C_{70}, C_{76}, C_{78}, and C_{84} are produced practically identically to the way that carbon nanotubes are produced. The He gas pressure is lower, being around 100 torr, but the current in the carbon electrodes is the same. The molecules appear as a carbonaceous soot which is readily dissolved in toluene or benzene. The fullerene molecules are extracted by spinning in a centrifuge or may be filtered out of the solution. Solid C_{60} may be crystallized. Some physical properties of the solid are listed in Table 21.4. One of the interesting features of the solid is the continual thermal rotation of the C_{60} molecules.

[†] M. Terrones et al, *Nature*, **388**, 52 (1997).

TABLE 21.4   Physical Properties of Solid C_{60}

Density	1700 kg/m^3
Crystal structure	FCC
NN distance	1.004 nm
Cage diameter	0.71 nm
Lattice constant	1.4198 nm
Index of refraction	2.2 at 630 nm
IR-active modes	1429, 1183, 577, 528 cm^{-1}
Bulk modulus	18 GPa
Ionization energy	7.6 eV
Cohesive energy	
per molecule	1.5 eV
per atom	7.4 eV
Electrical conductivity	Insulator
Electronic bandgap	1.5 eV
Effective mass m_e	1.3 m
Superconducting T_c	
K_3C_{60}	19 K

Source: Date from D. R. Huffman, *Phys. Today*, Nov. 1991, p 22.

REFERENCES

General Issues in Synthesis and Processing

Chase, M. W., et al., eds., *JANAF Thermochemical Tables*, 3rd ed., American Chemical Society, Washington, D. C., 1985; *J. Phys. Chem. Ref. Data* 14, Suppl. 1, (1985).

Chaudhari P., and Flemings. M., *Materials Science and Engineering for the* 1990s, National Academy of Sciences Press, Washington, D. C., 1989.

Cox, J. D., D. D. Wagman, and V. A. Medvedev, *CODATA Key Values for Thermodynamics*, Hemisphere, New York, 1989.

Eckert, E. R. G., and R. M. Drake, Jr., *Analysis of Heat and Mass Transfer*, McGraw-Hill, New York, 1972.

Kou, S., *Transport Phenomena and Materials Processing*, Wiley, New York, 1996.

Reif, F., *Fundamentals of Statistical and Thermal Physics*, McGraw-Hill, New York, 1965.

Spear, K. E., High-temperature reactivity, Chap. 3 in Hannay, N. B., ed., *Treatise on Solid State Chemistry*, Vol. 4, Plenum Press, New York, 1976.

Wagman, D. D., et al., *The NBS Tables of Chemical Thermodynamic Properties*, *J. Phys. Chem. Ref. Data* 11, Suppl. 2, (1982).

Synthesis and Processing of Semiconductors

Brice, J. C., *The Growth of Crystals from Liquids*, North-Holland, Amsterdam, 1973.

Evans, J. W., and L. C. DeJonghe, *The Production of Inorganic Materials*, Macmillan, New York, 1991.

Grove, A. S., *Physics and Technology of Semiconductor Devices*, Wiley, New York, 1967.

Maly, W., *Atlas of IC Technologies: An Introduction to VLSI Processes*, Benjamin-Cummings, Menlo Park, Calif., 1987.

Mathews, J. W., ed., *Epitaxial Growth*, Part A, Academic Press, San Diego, Calif., 1975.

Sapoval, B., and C. Hermann, *Physics of Semiconductors*, Springer-Verlag, New York, 1993.

Tiller, W. A., *The Science of Crystallization; Microscopic Interfacial Phenomena*, Cambridge University Press, Cambridge, 1991.

Vere, A. W., *Crystal Growth: Principles and Progress*, Plenum Press, New York, 1987.

Wolf, S., and R. N. Tauber, *Silicon Processing for the VLSI Era*: Vol. 1, *Process Technology*; S. Wolf, *ibid.*, Vol. 2, *Process Integration*, Lattice Press, Sunset Beach, Calif., 1986 (Vol. 1), 1990 (Vol. 2).

Synthesis and Processing of Metals

Callister, W. D., Jr., *Materials Science and Engineering: An Introduction*, 2nd ed., Wiley, New York, 1990.

Evans, J. W., and L. C. De Jonghe, *The Production of Inorganic Materials*, Macmillan, New York, 1991.

Guy, A. G., *Introduction to Materials Science*, McGraw-Hill, New York, 1972.

Honeycombe, R. W. K., and H. K. D. H. Bhadeshia, *Steel: Microstructure and Properties*, 2nd ed., Edward Arnold, London, 1996.

Lakhtin, Yu. M., *Engineering Physical Metallurgy and Heat-Treatment*, Mir Publishers, Moscow, 1983 (English translation).

Pearson, W. B., *The Crystal Chemistry and Physics of Metals and Alloys*, Wiley-Interscience, New York, 1972.

Catalysis

Britzinger, H. H., et al., Stereospecific olefin polymerization with chiral metallocene catalysts, Angew. Chem. Int. Ed1. Engl., **34**, 1143 (1995).

Ewen, J. A., New chemical tools to create plastics, Sci. Am., May 1997, p. 86.

Polymer Synthesis

Cheremisinoff, N. P., ed., *Handbook of Polymer Science and Technology*, Marcel Dekker, New York, 1989.

Flory, P. J., *Principles of Polymer Chemistry*, Cornell University Press, Ithaca, N. Y., 1953.

Sol-Gel Synthesis

Jones, R. W., *Fundamentals of Sol–Gel Processing*, Institute of Metals, London, 1989.

Carbon Nanotubes

Dresselhaus, M. S., G. Dresselhaus, and P. C. Eklund, *Science of Fullerenes and Carbon Nanotubes*, Academic Press, San Diego, Calif, 1996.

PROBLEMS

21.1 Consider the relationship between the kinetic and thermodynamic aspects of an activated process.

(a) Derive the expressions for $\Delta_r H^\circ$ and $\Delta_r S^\circ$ given in Eq. (21.24).

(b) Determine $\Delta_r H^\circ$ and $\Delta_r S^\circ$ for the vaporization process $Si(s) \leftrightarrow Si(g)$ at $T = 1300$ K using data from the JANAF thermochemical tables (Chase, 1985).

(c) Assuming that the activation energy $E_{ar} = 0$ for the adsorption of $Si(g)$ on $Si(s)$, determine E_{af} from $\Delta_r H^\circ$. Also, determine A_f/A_r from $\Delta_r S^\circ$.

21.2 Calculate the equilibrium constant K for the reaction $Si(s) + SiO_2(s) \leftrightarrow 2SiO(g)$ at $T = 1300$ K and also the $SiO(g)$ equilibrium vapor pressure $P_{eq}[SiO(g)$, 1300 K] using Eqs. (21.15) and (21.16). Use $\Delta_f G^\circ[SiO(g)$, 1300 K] = -211.744 kJ/mol and $\Delta_f G^\circ[SiO_2(s)$, 1300 K] = -678.535 kJ/mol for high cristobalite, the stable form of $SiO_2(s)$ at $T = 1300$ K. You may assume that the activities of the solid phases are equal to unity. This decomposition reaction is important in the annealing of a-$SiO_2(s)$ films on $Si(s)$ substrates. (*Note*: The values of $\Delta_f G^\circ$ given here are not actually accurate to six significant figures, but are presented with this precision in the internally self-consistent JANAF tables.)

21.3 Calculate the equilibrium constant K for the vaporization reaction $Si(s) \leftrightarrow Si(g)$ at $T = 1300$ K and also the $Si(g)$ equilibrium vapor pressure $P_{eq}[Si(g)$, 1300 K]. Use $\Delta_f G^\circ[Si(g)$, 1300 K] = $+257.766$ kJ/mol.

21.4 Calculate the critical size for the nucleation of a drop on the surface of a solid (heterogeneous nucleation).

21.5 According to the reaction $Si(s) + O_2(g) \leftrightarrow SiO_2(s)$ for the thermal oxidation of $Si(s)$, a layer of Si at the surface of the wafer is converted to a-SiO_2. Show that when an a-SiO_2 film of thickness d is grown on Si, a layer of Si of thickness $0.45d$ is consumed. Use the atomic weights of Si and O and the densities $\rho(Si) = 2330$ and $\rho(SiO_2) = 2270$ kg/m^3.

21.6 Find the a-SiO_2 layer thickness $x(t_o)$ that is the crossover between the linear and parabolic limits. [*Hint*: Set the expressions for $x(t)$ given in Eqs. (21.54) and (21.55) equal to each other.] How would you expect this thickness to vary with temperature?

21.7 For binary alloys such as $Fe_{1-x}C_x$ and $Al_{1-x}Cu_x$:

(a) Calculate the at %s of C in $Fe_{1-x}C_x$ alloys with 0.1, 1.0, and 9 wt % C.

(b) Calculate the at %s of Cu in $Al_{1-x}Cu_x$ alloys with 0.1, 1.0, and 9 wt % Cu. (*Hint*: See Problem 6.4.)

21.8 Consider the FCC γ-Fe and the BCC α-Fe crystal structures.

(a) Assuming a constant density of atoms in these two crystal structures, express the lattice constant a(BCC) in terms of a(FCC).

(b) Show that the octahedral interstitial sites located at the centers of the cube edges and in the body-centered positions in the more closely-packed FCC γ-Fe crystal structure are actually much larger in volume than the corresponding octahedral interstitial sites located at the centers of the cube faces and at the centers of the cube edges in the more open BCC α-Fe crystal structure. These interstitial sites are shown in Figs. 1.19 and 1.21. (*Hint*: Calculate the radii r of the largest hard-sphere atoms which can occupy these sites in terms of the hard-sphere radius R of the Fe atom.)

(c) Using the metallic radii given in Table 2.12, decide if B or C atoms can enter the FCC octahedral sites without distorting the Fe lattice.

21.9 Consider the slow cooling of a $Fe_{1-x}C_x$ alloy from $T \approx 1000°C$ to room temperature.

(a) At what carbon composition $x = x_e$ will the austenite present at $T = 1000°C$ transform completely to pearlite at $T_e = 727°C$? (*Hint*: See the Fe–C phase diagram shown in Fig. 21.9.)

(b) What will be the ratio of the volume fractions of the resulting ferrite and cementite components of the pearlite at $T = 20°C$ and $x = x_e$? [*Hint*: See Eq. (21.57).]

(c) What phases will be present in the slowly cooled alloy when the carbon composition x is greater than x_e and less than x_e?

(d) If the carbon composition is 0.5 wt %, what will be the total weight fractions of ferrite and of pearlite in the steel?

21.10 In a $Fe_{1-x}C_x$ alloy, assume that there are 10^8 dislocation lines per square centimeter present. Carbon atoms migrate to the dislocation cores where they find lower energy sites. How much C is required to saturate all the dislocation cores (i.e., to place one C atom per atomic plane along each dislocation line)? Express your answer as a concentration of C atoms per cubic meter, as an atomic fraction x, and as a weight percent.

Note: Additional problems are given in Chapter W21.

Binary compounds and alloys

Ternary compounds and alloys